T0145238

Smart Innovation, Systems and Technologies

Volume 101

Series editors

Robert James Howlett, Bournemouth University and KES International,
Shoreham-by-sea, UK
e-mail: rjhowlett@kesinternational.org

Lakhmi C. Jain, University of Technology Sydney, Broadway, Australia;
University of Canberra, Canberra, Australia; KES International, UK
e-mail: jainlakhmi@gmail.com; jainlc2002@yahoo.co.uk

The Smart Innovation, Systems and Technologies book series encompasses the topics of knowledge, intelligence, innovation and sustainability. The aim of the series is to make available a platform for the publication of books on all aspects of single and multi-disciplinary research on these themes in order to make the latest results available in a readily-accessible form. Volumes on interdisciplinary research combining two or more of these areas is particularly sought.

The series covers systems and paradigms that employ knowledge and intelligence in a broad sense. Its scope is systems having embedded knowledge and intelligence, which may be applied to the solution of world problems in industry, the environment and the community. It also focusses on the knowledge-transfer methodologies and innovation strategies employed to make this happen effectively. The combination of intelligent systems tools and a broad range of applications introduces a need for a synergy of disciplines from science, technology, business and the humanities. The series will include conference proceedings, edited collections, monographs, handbooks, reference books, and other relevant types of book in areas of science and technology where smart systems and technologies can offer innovative solutions.

High quality content is an essential feature for all book proposals accepted for the series. It is expected that editors of all accepted volumes will ensure that contributions are subjected to an appropriate level of reviewing process and adhere to KES quality principles.

More information about this series at http://www.springer.com/series/8767

Francesco Calabrò · Lucia Della Spina
Carmelina Bevilacqua
Editors

New Metropolitan
Perspectives

Local Knowledge and Innovation Dynamics Towards
Territory Attractiveness Through the Implementation
of Horizon/E2020/Agenda2030 – Volume 2

 Springer

Editors
Francesco Calabrò
Mediterranea University of Reggio Calabria
Reggio Calabria
Italy

Carmelina Bevilacqua
University of Reggio Calabria
Reggio Calabria
Italy

Lucia Della Spina
Mediterranea University of Reggio Calabria
Reggio Calabria
Italy

This Volume is part of a project that has received funding from the European Union's Horizon 2020 research and innovation programme under grant agreement N°645651

ISSN 2190-3018 ISSN 2190-3026 (electronic)
Smart Innovation, Systems and Technologies
ISBN 978-3-030-06363-4 ISBN 978-3-319-92102-0 (eBook)
https://doi.org/10.1007/978-3-319-92102-0

© Springer International Publishing AG, part of Springer Nature 2019, corrected publication 2019
Softcover re-print of the Hardcover 1st edition 2019
This work is subject to copyright. All rights are reserved by the Publisher, whether the whole or part of the material is concerned, specifically the rights of translation, reprinting, reuse of illustrations, recitation, broadcasting, reproduction on microfilms or in any other physical way, and transmission or information storage and retrieval, electronic adaptation, computer software, or by similar or dissimilar methodology now known or hereafter developed.
The use of general descriptive names, registered names, trademarks, service marks, etc. in this publication does not imply, even in the absence of a specific statement, that such names are exempt from the relevant protective laws and regulations and therefore free for general use.
The publisher, the authors and the editors are safe to assume that the advice and information in this book are believed to be true and accurate at the date of publication. Neither the publisher nor the authors or the editors give a warranty, express or implied, with respect to the material contained herein or for any errors or omissions that may have been made. The publisher remains neutral with regard to jurisdictional claims in published maps and institutional affiliations.

Printed on acid-free paper

This Springer imprint is published by the registered company Springer Nature Switzerland AG
The registered company address is: Gewerbestrasse 11, 6330 Cham, Switzerland

Preface

This volume contains the proceedings for the third International *"NEW METROPOLITAN PERSPECTIVES. Local Knowledge and Innovation dynamics towards territory attractiveness through the implementation of Horizon/Europe2020/Agenda2030"*, which took place on 22–25 May 2018 in Reggio Calabria, Italy.

The Symposium is jointly promoted by LaborEst (Evaluation and Economic Appraisal Lab) and CLUDs (Commercial Local Urban Districts Lab), Laboratories of the PAU Department, *Mediterranea* University of Reggio Calabria, Italy, in partnership with a qualified international network of academic institution and scientific societies.

The third edition of *"NEW METROPOLITAN PERSPECTIVES"* aims to deepen those factors which contribute to increase cities and territories attractiveness, both with theoretical studies and tangible applications.

It represents the conclusive event of the Multidisciplinary Approach to Plan Smart Specialisation Strategies for Local Economic Development (MAPS-LED) Research Project funded by the European Union's Horizon 2020 Research and Innovation Programme under the Marie Skłodowska Curie Actions—RISE 2014.

This edition of the Symposium is going to give a specific attention to those linkages between innovation dynamics and territories attractiveness, as will be better explained by our colleague Carmelina Bevilacqua.

In the last decades, metropolitan cities have been studied from different perspectives, according to diverse academic and scientific points of view, but under the common attitude towards their spatial dynamics.

Recent economic and political developments press the scientific community addressing two issues of current relevance:

- The spatial implications of the economic and demographic decline of large areas in Europe and Western Countries;
- The impact of ICT dissemination on urban/rural environment and, broadly, on the idea of society.

For decades, technical tools, especially in the field of urban planning, have been developed to allow urban and territorial transformations in a context characterized by expansive dynamics. Looking at the productive system and, within it, the job's organization as one of the discriminating elements of territorial transformation, the following question arises: what is the destiny of the industrial and post-industrial city, known as a place of concentration of workforce and market?

These considerations focus the attention of the academic community on the dimension of distant future. The spread of new communication technologies and new production systems is increasingly pushing everywhere towards the progressive "liquefaction" of the social structures, organizational models and systems that have been known so far as Bauman's intuition. Such a long horizon necessarily requires the renewal of a visionary, utopian vision that imagines society of the future through a dreamlike dimension of avant-garde. It becomes crucial to debate the running direction; the profound changes are going on in contemporary society and its impact on urban/rural environment of the future.

We might suggest the anthropic desertification is a phenomenon shared among the lagging Regions: increasingly, many people move from their hometowns to reach better places, such as the metropolitan areas, to improve their conditions of life. Such process inevitably contributes to the general poorness of those Regions, already weakened, by increasing such declining status even more.

One of the most important topics to be considered, which more than others characterizes all metropolitan regions, is surely their capacity to attract people, and consequently capitals. Indeed, territorial policies aim mostly to catch investments in order to enhance job creation and to positively influence socio-economic indicators. Nevertheless, attractiveness, as explained, is also about people: that is the indicator which can really synthesize a concept which includes both competitiveness and receiving capability.

If we go deeper, competitiveness means: research and innovation, public administration efficiency, skilled workforce, facilities, accessibility, credit access, international perspectives, energy cost consumption. As far as the receiving capability, it could be explained as carefulness for urban quality, housing policies, mobility, welfare, health care, security and, of course, job opportunities.

Particularly, the papers accepted, about 150, allowed us to develop six macro-topics, about *"Local Knowledge and Innovation dynamics towards territory attractiveness"* as follows:

1   Innovation dynamics, smart cities, ICT;
2   Urban regeneration, community-led practices and PPP;
3   Local development, inland and urban areas in territorial cohesion strategies;
4   Mobility, accessibility, infrastructures;
5   Heritage, landscape and identity; and
6   Risk management, environment, energy.

We are pleased that the International Symposium NMP, thanks to its interdisciplinary character, stimulates growing interests and approvals from the scientific community, at national and international levels.

We would like to take this opportunity to thank all who have contributed to the success of the third International Symposium "NEW METROPOLITAN PERSPECTIVES. *Local Knowledge and Innovation dynamics towards territory attractiveness through the implementation of Horizon/Europe2020*": authors, keynote speakers, session chairs, referees, the scientific committee and the scientific partners, the "Associazione ASTRI" for technical and organizational support activities, participants, student volunteers and those ones that with different roles have contributed to the dissemination and the success of the Symposium; particularly, the academic representatives of the University of Reggio Calabria: the Rector Prof. Pasquale Catanoso, the Vice Rector Prof. Marcello Zimbone, the responsible of internationalization Prof. Francesco Morabito, the Chief of PAU Department Prof. Francesca Martorano.

Thank you very much for your support.

Last but not least, we would like to thank Springer for the support in the conference proceedings publication.

Francesco Calabrò
Lucia Della Spina

Local Knowledge and Innovation Dynamics: The MAPS-LED Perspective

The third edition of the International Symposium "New Metropolitan Perspectives" aims at facing the challenges of Local Knowledge and Innovation dynamics towards territory attractiveness through the implementation of the Horizon/EU2020 Agenda. The Symposium is jointly promoted by the LaborEst and the CLUDsLab Laboratories of the PAU Department, Università Mediterranea of Reggio Calabria (IT), in partnership with a qualified international network of prestigious academic institutions and scientific associations. It represents the conclusive event of the Multidisciplinary Approach to Plan Smart Specialisation Strategies for Local Economic Development (MAPS-LED) Research Project funded by the European Union's Horizon 2020 Research and Innovation Programme, under the Marie Skłodowska Curie Actions—RISE 2014. The main aim of RISE Action is to favour the mobility of experienced and early-stage researchers between Europe, associated and third countries. The project empowers the strong international research network built up with the CLUDs Project (7FP) through the exchange of researchers, ideas and practices between EU and USA. To date, about 40 experienced and early-stage researchers benefited by the project mobility towards USA, at the Northeastern University of Boston and the San Diego State University. The researchers, coming from the Higher Education Institutions (HEIs) belonging to the MAPS-LED network, had the opportunity to increase their research, training and networking skills thanks to the high exposure to the international scientific community. The majority of the early-stage researchers belong to the International Doctorate Program in Urban Regeneration and Economic Development (URED), active since 2012 at the Università Mediterranea of Reggio Calabria (Project Coordinator). The Program is funded by the Calabria Region European Social Fund (ESF), making effective the operative linkage between Horizon 2020 (Research) and Cohesion Policy (ESIF).

The MAPS-LED Symposium represents an important event for disseminating research findings and for stimulating a fruitful debate among scientific and policy-makers' community.

The core of the research activities has earmarked for exploring how Smart Specialisation Strategies (S3) can be implemented by incorporating the place-based approach towards regenerating local economies.

The S3 has been designed in order to capture knowledge and innovation dynamics strictly connected with characteristics of context. According to the Maps-led perspective, the key concepts of S3 lie in the mutual correlation among entrepreneur, innovation and economic development. The entrepreneur is pushed by a local entrepreneurial culture activated by enhancing local knowledge. This process is called "entrepreneurial discovery" towards knowledge convergence and informational spillover for clustering phase, as precondition of competitive advantages.

Among the theoretical standpoints that explained how cluster policy and S3 share many similarities in their rationale, the research activities led to focus on the place-based approach as nexus in spurring the innovation process towards emphasizing the role of the city.

Thanks to the exchange scheme of the RISE programme, the MAPS-LED project has delivered a methodology to spatialize economic clusters in Boston and San Diego, as expression of how innovation is experimented in the modern economy and how the "place" works.

The "spatialization cluster methodology" has brought about a proxy for innovation concentration, by turning clusters in physical configurations at city level. This interpretation comes from the rationale grounded into cluster definition, validated by Porter with the model in which innovation, specialization and job creation are connected among those productive sectors related to shaping a cluster. The preliminary research findings pushed towards the explanation of how cluster performance factors can be combined with the context characteristics, by highlighting the spatial implications of knowledge dynamics. The case studies have been grouped into two frameworks of cluster rationale—Traded, to enhance competitive advantages, and Local, to reinforce comparative advantages. In synthesis, the first framework considers innovation as the main drive to define the relativeness of productive sectors to shape traded cluster, and the second ones bring into specialization the main impulse in forming local cluster.

The spatially oriented methodology adopted for Traded clusters in the Boston area analysed the occurrence of "innovation spaces" in the places characterized by the presence of cluster, in order to identify specific urban areas (target areas) in which investigating the interaction of cluster (demand of innovation) with the urban fabric, its sociability and sustainability. The findings from "target areas" analysis allowed, on one hand, at identifying the link between city and S3 by introducing the innovation-driven urban policy as an important phase of the Entrepreneurial Discovery Process (EDP). On the other hand, gentrification and inequality issues resulted as the main negative effects in both cities, Boston and Cambridge, due to the evident increase, more than proportional, of the rent and property values.

The link between city and S3 is mainly stemmed by the emerging business environment or the atmosphere for innovation that acquires an important role in what Foray calls *structuring entrepreneurial knowledge*. Inside the "target areas", anchors institutions, public and private research centres, the entrepreneurs' community and citizens concentrate their efforts supported by public policies (economic development and urban planning). The occurrence of such dynamic forces, able to trigger socio-economic and physical transformation, has brought to investigate how innovation policy can be harnessed in driving growth in specific localities. This aspect called for a better understanding and the exploration of innovation as a source for socio-economic and urban transformation, highlighting urban regeneration initiatives driven by the increasing demand for innovation[1].

The analysis of surrounding conditions has been considered important to give a practical explanation of how the EDP could be structured as policy action. The role of the city has emerged in spurring the innovation process and, in particular, how it can be the start point of the EDP, in terms of public policy action. The possible result of these research activities lies in finding a new concept of *urban dimension* within S3. The urban dimension inside the S3 implementation could be part of the EDP as engine of the quadruple helix model for knowledge dynamics. It is possible to group under the innovation-oriented urban policy's concept the increasing phenomena of innovation districts (in a broadly sense) to refine a different perspective of the role of the city in the creation of an innovation ecosystem. Another aspect emerged from the research activities in Boston is connected to how innovation has become a source of urban form and its transformation, pushing urban regeneration initiatives driven by the demand for innovation.

The spatially oriented methodology adopted for Traded clusters in the Boston area has been implemented also for the spatialization of Local clusters in San Diego. Here, the focus shifted from mapping innovation concentration towards mapping specialization in the innovative milieu perspective. Clusters and knowledge networking reveal how territorial milieu can influence the knowledge dynamics and how knowledge can be shared along the territorial milieu. The aim was to find a connection between urban and inland areas through the territorial milieu as an explanation of innovative milieu. Local Clusters have been examined through Dynamic Analysis, Innovation Ecosystems and their relationship with Community Plans and Zoning providing interesting insights into the activation of social innovation thanks to the interaction of three driving elements: knowledge, innovation and place. The different socio-economic and spatial configuration allowed to identify different development dynamics for local innovation ecosystems. In San Diego, harnessing innovation ecosystem is not limited only to local actors, even regulatory agencies and municipal or regional governments that create a dynamic, innovation-driven economy can be involved in the orchestration process.

In both cases (Boston and San Diego), innovation-oriented public policies pivot around the entrepreneurial spirit, in line with the desired entrepreneurial knowledge convergence of the S3 approach. The MAPS-LED project proposes the Entrepreneurial Discovery Process as a trigger for the coordination of the efforts at local level—public administrations, research institutions, entrepreneurs,

---

[1]The MAPS-LED has been appointed as "success story" in European Commission: New thinking to drive regional economic development. EU Cordis Research and Innovation success stories available at: http://ec.europa.eu/research/infocentre/article_en.cfm?id=/research/headlines/news/article_17_11_15-2_en.html?infocentre&item=Infocentre&artid=46436.

communities—in boosting the local knowledge convergence and generating the expected change.

The MAPS-LED project emphasized how the linkage between planning and innovation policy empowers EDP through bottom-up approaches. In other words, local communities and organizations are in the best position to know what can drive a city's regeneration and deliver economic change reinforcing the urban dimension of S3. The research activities highlighted how EDP could be the mean to design tailor-made policy acting on the fruitful relationship among knowledge, innovation and place. This process should be managed at local level and embedded in the urban development agenda due to its ability to activate urban regeneration mechanisms and expand innovation in distressed areas through public–private partnership and innovative financial instruments. In this sense, the MAPS-LED approach works as cross-cutting element in the understanding of knowledge dynamics, which are complex and difficult to trigger in specific places. The interaction of knowledge, innovation and place—and the related potential **output** indicators provided by the MAPS-LED project—attributes the local asset to the entrepreneurial discovery process activated by urban policy aiming at regenerating urban areas through innovation-led processes. In synthesis, the analysis of the local context shed the light on EDP as evidence-based and horizontal policy for S3 by considering two drivers: the urban regeneration mechanism joint with Knowledge-Based Urban Development to guide the identification of output indicators of EDP; the cluster life cycle analysis to guide the result indicators of the EDP.

Furthermore, the cluster spatialization methodology could help in finding out the regional areas of innovation towards focusing on public and private financial resources. The methodology developed could help in the understanding "where" entrepreneurial knowledge and forces are active and concentrated, lighting up the potential for the discovery phase. This is a cross-sectorial approach because the identification of potentials with respect to the local context allows to discover concentration of knowledge and feed innovation at local level. The identification of local potential areas of innovation, coherently with the principle of Smart Specialisation, can favour the discovery of new domains through an evidence-based territorial perspective rather than a mere analysis of regional economies. Further insights from these findings reveal the potential transformation of these urban areas of innovation in Economic Special Zones.

The results coming from the MAPS-LED project research activities stimulated the scientific debate around the key elements able to trigger the desired change through S3 as well as the understanding of its (current and potential) limits. The participation of international experts involved in the S3 design and the RIS3 implementation, as well as the academic contributions coming from different disciplines, highlighted the potentials of the multidisciplinary approach proposed by the project, allowing to boost up knowledge convergence in an a sectorial rationale. The Symposium represents the opportunity to stimulate the development of innovation-oriented models for the exploitation and valorization of local assets involving different disciplines and in a multilevel governance perspective.

The contribution offered through the Symposium, by either enriching the academic debate or providing evidence-based solutions for the implementation of economic development strategies, is attributed by wide scope and marked cross-cutting dimension. The multidisciplinary nature of the MAPS-LED project is reflected by the structure of the Symposium itself. Each session presents topics and arguments which are, to some extent, ingrained within overall framework of the MAPS-LED project, while they are expected to open up windows of opportunity for further studies and research. Consistently, the Symposium focuses on analysing, at different scales and under numerous perspectives, the strategies, objectives and impacts of local economic development and innovation processes, to achieve a smart sustainable and inclusive growth. In a sentence, the Symposium, and the contributions to its sessions, manifests the effort to re-proposing the multidisciplinary approach implemented within the MAPS-LED research project in a conference-based-dimension.

While the Symposium encompasses a number of sessions dealing with specific topics, it is reasonable framing them within the streamlined "smart, inclusive and sustainable" growth paradigm enacted by the EU 2020 strategy (EU, 2010). By following this logic, the Symposium kicks off by trying to overcome current limits and gaps in the implementation of plans and models (session TS01, TS13 and TS16), while it further develops by bringing to light the importance of the place-based approach to deliver successful urban regeneration processes (session TS02 and TS23). In this regard, the prominent role played by territorial peculiarities in affecting decision-making processes is taken into high consideration (session TS04). Drawing on the belief that the "place" matters, the Symposium devoted different sessions to the study of the territorial-specific developmental mechanisms aiming at identifying logical-operational tools that can interpret urban phenomena (i.e. urban safety with the session TS11), but also evolutionary and community involvement processes in coastal areas (session TS19). The dichotomy urban–rural is treated in more than one session, as it is further scrutinized under the lens of geospatial analysis and modelling tools in the way of identifying how landscapes transition from rural-to-urban (session TS25) as well as under the lens of inclusive knowledge and innovation networks (session TS08). Still on the territorial dimension, the sustainable-led and ecological approach is analysed (thanks to the contribution in session TS14, TS15 and TS17) as well as the cultural heritage territorial network valorization perspective (session TS18), as a mean to favour the discovery of territorial-specific developmental opportunities. The discovery of opportunities in general, and hidden economic potentials in particular, is also the central point of discussion in the session on urban and regional development (TS20). This session refers to innovation spaces as catalyst for the disclosure of latent economic strengths of territories, at different geographic scales. The importance of innovation spaces seems especially relevant in the context of the knowledge economy, where and when the re-combination of "pieces of knowledge" can drive towards unveiling novel products and processes novel products and processes. The challenges and potentials posed by innovative activities are also investigated under an economic perspective by catching up with the complexity of knowledge dynamics. Following their cross-cutting rationale, innovation and knowledge are also the focal point of the session TS05 focused on ICT and heritage for a sustainable development as well as for the territorial innovative networks for public services (session TS26).

This synthetic description of the sessions gives a clear idea of the complexity of the themes treated as well as their alignment with respect to the MAPS-LED project. Moreover, the participation to the Symposium of international experts as well as academics from different disciplines provides interesting insights for the RIS3 evaluation and monitoring processes for the post-2020 programming period. The multidisciplinary approach to plan Smart Specialisation Strategies proposed with the MAPS-LED project emerged as crucial to properly pursue the local economic development in the S3 perspective. Hence, the MAPS-LED project appears at forefront into this research domain.

Carmelina Bevilacqua

Organization

Programme Chairs

Carmelina Bevilacqua	Mediterranea University of Reggio Calabria
Francesco Calabrò	Mediterranea University of Reggio Calabria
Lucia Della Spina	Mediterranea University of Reggio Calabria

Scientific Committee

Stefano Aragona	INU—Istituto Nazionale di Urbanistica
Angela Barbanente	Politecnico di Bari
Filippo Bencardino	SGI—Società Geografica Italiana
Jozsef Benedek	RSA—Babes-Bolyai University, Romania
Christer Bengs	SLU/Uppsala Sweden and Aalto/Helsinki, Finland
Andrea Billi	Università La Sapienza di Roma
Adriano Bisello	EURAC Research
Nicola Boccella	Università La Sapienza di Roma, Presidente SISTur
Mario Bolognari	Università degli Studi di Messina
Kamila Borsekova	Matej Bel University, Slovakia
Nico Calavita	San Diego State University, USA
Roberto Camagni	Politecnico di Milano, Presidente Gremi
Farida Cherbi	Institut d'Architecture de Tizi Ouzou, Algeria
Maurizio Di Stefano	Icomos Italia
Yakup Egercioglu	Izmir Katip Celebi University, Turkey
Khalid El Harrouni	Ecole Nationale d'Architecture, Rabat, Morocco
Gabriella Esposito De Vita	CNR/IRISS Istituto di Ricerca su Innovazione e Servizi per lo Sviluppo

Rosa Anna Genovese	Università degli Studi di Napoli "Federico II"
Giuseppe Giordano	Università degli Studi di Messina
Olivia Kyriakidou	Athens University of Economics and Business, Greece
Ibrahim Maarouf	Alexandria University, Faculty of Engineering, Egypt
Lívia M. C. Madureira	Centro de Estudos Transdisciplinares para o Desenvolvimento—CETRAD, Portugal
Tomasz Malec	Istanbul Kemerburgaz University, Turkey
Ezio Micelli	IUAV Istituto Universitario di Architettura di Venezia
Nabil Mohäreb	Beirut Arab University, Tripoli, Lebanon
Mariangela Monaca	Università di Messina
Bruno Monardo	Università degli Studi di Roma "La Sapienza"
Pierluigi Morano	Politecnico di Bari
Fabio Naselli	IEREK International Experts for Research Enrichment and Knowledge Exchange
Peter Nijkamp	Vrije Universiteit Amsterdam
Davy Norris	Louisiana Tech University, USA
Leila Oubouzar	Institut d'Architecture de Tizi Ouzou, Algeria
Sokol Pacukaj	Aleksander Moisiu University, Albania
Aurelio Pérez Jiménez	University of Malaga, Spain
Keith Pezzoli	University of California, San Diego, USA
María José Piñera Mantiñán	University of Santiago de Compostela, Spain
Fabio Pollice	Università del Salento
Vincenzo Provenzano	Università di Palermo
Ahmed Y. Rashed	Founding Director Farouk ElBaz Centre for Sustainability and Future Studies
Riccardo Roscelli	Politecnico di Torino
Michelangelo Russo	SIU—Società Italiana degli Urbanisti
Alessandro Saggioro	Università La Sapienza di Roma
Helen Salavou	Athens University of Economics and Business, Greece
Paolo Salonia	CNR—Istituto per le tecnologie applicate ai beni culturali, Rome, Italy
Stefano Stanghellini	IUAV, Presidente SIEV
Luisa Sturiale	Università di Catania
Chiara O. Tommasi Moreschini	Università di Pisa
Claudia Trillo	University of Salford, UK

Internal Scientific Board

Giuseppe Barbaro	Mediterranea University of Reggio Calabria
Concetta Fallanca	Mediterranea University of Reggio Calabria
Giuseppe Fera	Mediterranea University of Reggio Calabria
Massimiliano Ferrara	Mediterranea University of Reggio Calabria
Giovanni Leonardi	Mediterranea University of Reggio Calabria
Francesca Martorano	Mediterranea University of Reggio Calabria
Domenico E. Massimo	Mediterranea University of Reggio Calabria
Carlo Morabito	Mediterranea University of Reggio Calabria
Gianfranco Neri	Mediterranea University of Reggio Calabria
Francesco S. Nesci	Mediterranea University of Reggio Calabria
Simonetta Valtieri	Mediterranea University of Reggio Calabria
Santo Marcello Zimbone	Mediterranea University of Reggio Calabria

Scientific Partnership

Regional Studies Association, Seaford, East Sussex, UK
A.I.S.Re—Associazione Italiana di Scienze Regionali, Milan, Italy
Eurac Research, Bozen, Italy
Icomos Italia, Rome, Italy
INU—Istituto Nazionale di Urbanistica, Rome, Italy
Società Italiana degli Urbanisti, Milan, Italy
Società Geografica Italiana, Rome, Italy
SIEV—Società Italiana di Estimo e Valutazione, Rome, Italy
Urban@it, Bologna, Italy
Federculture—Federazione Servizi Pubblici Cultura Turismo Sport Tempo Libero, Rome, Italy
FICLU—Federazione Italiana dei Club e Centri per l'Unesco, Turin, Italy
Rete per la Parità—APS per la Parità uomo-donna secondo la Costituzione Italiana, Rome, Italy

Organizing Committee

Associazione ASTRI—Associazione Scientifica Territorio e Ricerca Interdisciplinare
URBAN LAB S.r.l.

Contents

Contents

Contents

Local Development, Inland and Urban Areas in Territorial Cohesion Strategies

Evaluation Approach to the Integrated Valorization of Territorial Resources: The Case Study of the Tyrrhenian Area of the Metropolitan City of Reggio Calabria

Giuseppina Cassalia$^{(\boxtimes)}$ ⓘ, Carmela Tramontana ⓘ,
and Francesco Calabrò ⓘ

Mediterranea University of Reggio Calabria, 89100 Reggio Calabria, Italy
`giuseppina.cassalia@unirc.it`

Abstract. The paper deals with the topic of integrated territorial valorization strategies, presenting the first step of a research project on the case study of the Tyrrhenian area of the Metropolitan City of Reggio Calabria. In a theoretical view that leads to rethinking European development models in areas characterized by considerable marginality, the paper copes with the weakness/opportunities of lagging areas and the vision of the territory as a system of social, cultural, economic, and natural features. The methodology is based on the definition of evaluative tools enabling the identification of effective and coherent concepts for understanding, and planning to enhance, territorial attractiveness. This preliminary stage of the project aims to verify the preconditions for the establishment of a local, attractive system, combining socioeconomic analysis, identification of thematic attractors, and the study of financial territorial allocation. The paper's results contribute to the scientific debate on the role of the "knowledge phase" in the dynamics of sustainable development, highlighting critical elements and opportunities for implementing an integrated valorization strategy of the case-study area.

Keywords: Territorial resources · Evaluation approach
Valorization strategies · Attractiveness · Inner areas

1 Introduction

The institutional debate of recent years has highlighted the centrality of the territory in the process of economic development [1]. The remarkable differentiation of development processes and the inadequacy of a unique development model at any time, place and scale have highlighted a roadmap in the interpretive schemes: the territory becomes a crucial variable to explain the opportunities that are inherent in some regions and the

---

The paper is the result of the joint work of the authors. Although scientific responsibility is equally attributable, the abstract and Sects. 1, 2, 3.3, 4 were written by G. Cassalia; Sects. 3.2, 3.3 were written by C. Tramontana; Sects. 3, 3.1, were written by F. Calabrò.

© Springer International Publishing AG, part of Springer Nature 2019
F. Calabrò et al. (Eds.): ISHT 2018, SIST 101, pp. 3–12, 2019.
https://doi.org/10.1007/978-3-319-92102-0_1

roadblocks slowing down the development process of others [2]. At the European and national levels, integrated local development policies are now considered indispensable to achieve smart, sustainable and inclusive growth[1].

Nowadays, in this context, the integrated valorization of resources for territorial-scale identity represents a challenging topic of analysis both at the institutional and academic levels [3]. This is a topic strongly linked with the growing need for consistent and systematic evaluation processes, capable of supporting the public decision makers facing the need to choose between alternative programs and projects, under the general-and current-conditions of scarce resources [4]. According to the orientation of the European Commission and the Organization for Economic Co-operation and Development, strategic evaluation is perceived as an extremely important tool for planning and designing policies and interventions and for the effectiveness and efficiency of their implementation, as well as for creating consensus among the actors involved in the decision-making process. The challenge is to develop an effective methodological practice that enables public administrations to undertake multi-objective decisions, while at the same time ensuring active participation in the decision-making process of all stakeholders.

The paper contributes to this current issue, presenting the first stage of a research study applied to the territory of Tyrrhenian area of the Metropolitan City of Reggio Calabria (southern Italy). The study aims at the implementation of local development models based on the endogenous resources of a marginal area. Presenting the results of the context analysis related to tangible and intangible attractors' investigation, socioeconomic and financial dynamics, the assessment defines the baseline needed to draw upon policy considerations inherent in the design of appropriate territorial enhancement policies, highlighting the importance of the evaluation approach to the decision-making process. In conclusion, this preliminary study's results provide a contribution to the scientific debate on the role of the "knowledge phase" in the dynamics of sustainable development, stressing the critical elements and opportunities for building an integrated valorization strategy.

2   The Integrated Valorization of Territorial Resources

The local dimension of development is increasingly important both globally, in the relationship between industrialized countries and underdeveloped countries, and locally, in the relationship between lagging or declining regions [5].

These concepts and principles can be found in the guidelines of the European Commission on regional policy for 2014–2020, which has led to the adoption of "EU Territorial Agenda 2020" [6]. The Territorial Agenda 2020, building on the ESDP, identifies cohesion policy as a 'key framework' through which the EU can address

---

[1] See: Niestroy, I.: Sustainable Development Governance Structures in the European Union, in Institutionalising Sustainable Development, 67–88, OECD Publishing, Paris (2007); European Commission: Europe 2020: A strategy for smart, sustainable and inclusive growth. COM (2010) 2020 final; MUVAL: A Strategy for Inner Areas in Italy: Definition, Objectives, Tools and Governance, Issue 31, Year 2014.

territorial development challenges and help unleash territorial potential at local, regional, national and transnational levels. The debate on models and approaches to territorial development over the last few decades, despite the emphasis placed on the local area as a new level of strategic planning, has encountered considerable difficulties. At the local level and, in particular, lagging areas such as south Italy, the question of weakness/opportunities of "marginality" is still open. For instance, Leone sees the lack of development in the southern regions as the historical result of a wider process of unbalanced development based on tangible and intangible ingredients [7]. Mollica underlined the fragile socio-demographic dimension, the neglect of historic towns, and natural and human-induced risks as a consequence of insufficient maintenance, treating the problem of neglected inner areas in southern Italy [8]. While Alcaro (1999) hypothesized that some features of southern Italy could prove successful in meeting the challenge of development and competitiveness, and Cassano, with his "pensiero meridiano" [Southern thought], an advocate of Southern identity, defend a civilization that does not need to pay any tribute to external contributions [9, 10].

In this dialectical context of weakness vs. opportunities of lagging areas, the vision of the territory as a system of social, cultural, economic, and natural features has become a common thought [11]. This is the case even if, in practice, the management of this system has always contained an intrinsic conflict, especially for the difficulty of combining the economic convenience with a dynamic and harmonious environmental and socio-cultural development [12]. The risk is that local development is still determined only by financial (mostly public) resources, and not by real local dynamics, aimed at enhancing the specificities of the territories.

The following part of the paper seeks to address this challenging issue through the exemplification of the case study of the Tirreno Reggino, southern Italy, part of the Metropolitan City of Reggio Calabria. The first results of a research project aimed at implementing a local development model based on the integrated enhancement of endogenous resources are presented. It is the case of a less-developed territory still facing economic, social, and environmental problems that result in unemployment, depopulation, marginalization, disengagement, and abandonment of historic and architectural capital [13]. Regions like Calabria have been the target during the last three European programming cycles of substantial public investments, but interventions have not yet produced significant socioeconomic effects [14]. The complexity of the choices that governmental bodies at sub-regional levels are facing requires an ability to implement complex decision-making processes. Funding pressures reinforce the need for a better understanding of how local governments can invest in the mitigation of territorial disparities and what the communities get in return.

As underlined in the scientific literature, fragmentation and fragility of marginal areas is due to a variety of factors that can be variously sectoral, structural, transactional, technological, behavioral, and related to resources, capabilities, and cultural perceptions [15]. The preliminary results of the study presented below emphasize how the challenge of smart and sustainable territorial development is today more linked than ever with the local capital and the territory's real needs. The culture of evaluation-in this sense-is an essential step towards achieving the overriding key objectives of efficiency, transparency, and accountability, with the aim of determining both the costs and the benefits of publicly financed projects, and to improve the efficiency and effectiveness of that investment [16, 17]. Under the general-and current-conditions of scarce resources, the evaluation approach presented subsequently support the public decision makers facing the choices between alternative programs and projects [18].

3 Evaluation Models Towards the Integrated Valorization of the Tyrrhenian Area of the Metropolitan City of Reggio Calabria

As stated above, this study is part of a research project for the enhancement of the Tyrrhenian Area of the Metropolitan City of Reggio Calabria aimed at the definition of a tools capable of triggering effective and sustainable development processes based on endogenous resources [19]. This approach would decode the territory according to sociocultural, economic, and environmental declinations, defining the attractiveness of this specific territory, and would focus on the identification of appropriate knowledge, planning, and feasibility studies aimed at the sustainable development of the area. Territorial valorization is considered to be a process composed of various phases ranging from the protection and mobilization of the local resources, taking into account the qualification of the product, the marketing of the same, and its integration with the territory. Sustainability is proposed as the guiding principle for drawing up valorization strategies. From a methodological point of view, the model is segmented into three phases:

- Step A: Knowledge;
- Phase B: Programming;
- Phase C: Feasibility-Sustainability.

This contribution forms an integral part of the knowledge phase undertaken, a phase seen by authors as the indispensable basis for integrating all the possible processes aimed at enhancing the endogenous territorial resources [20]. Furthermore, it's seen as the priority phase for the implementation of valuation actions, enabling the identification of effective and coherent concepts for understanding, and planning to enhance, territorial attractiveness.

3.1   Identification of the Case-Study Area

The case-study area includes 43 municipalities on the Tyrrhenian side of the Metropolitan City of Reggio Calabria. It extends over a total area of 1,111.76 km^2 with an average density of 210 inhabitant/km^2. According to ISTAT census of 2011, the case-study area includes a 199,632 inhabitants. It deserves to be noted that the municipality of Reggio Calabria is not part of the study.

The case-study area has been divided into seven areas according to territorial boundaries already defined by the Integrated Planning Unit 2007/2013 of the Calabria Region, in order to best differentiate the essence of the municipalities involved in the case study. With a view to future enhancement actions, this identification of sub-areas offers established cooperation structures, given the previous experience of Local Development Integrated Projects (LDIP).

3.2 The Socioeconomic Context

The results of large-scale analysis confirm that the substantial population migration occurring in peripheral municipalities is closely linked with employment opportunities: Most municipalities with negative population variations are also those in which unemployment rates are among the highest of the sectors [21]. Substantial depopulation has affected the 16 municipalities classified as "peripheral"[2], which, with a 16% negative population variations, have lost approximately 7,000 inhabitants in recent decades. The 2015 census of the region's population confirms the data of the 1981–2011 period, pointing out that the actions arising from the previous EU Cohesion programming did not appear to have had a major impact on this problem.

Another issue emerging from the analysis is that, with a youth unemployment rate of 51.70%, the Tyrrhenian area thus has one youth out of every two unemployed while the Calabria region registered a rate of 40.5%, the national average 29.1%, and, Europe wide, the rate is 22.62%. As far as the inactivity rate is concerned, none of the seven sub-areas has a youth unemployment rate below 50%. As testimony to how migratory flows are indissolubly linked to employment, light inbound migration has been recorded in those municipalities classified as "outlying", such as Villa San Giovanni, Campo Calabro, and Gioia Tauro, which have acceptable rail-link infrastructure and so offer some job opportunities. In this sense, the employment domain that remains dominant (excluding the public-administration sector) is the agricultural one. This sector registering a good number of PGI-certified product manufacturers proves to be one of the key assets of the entire territorial economy. The research pointed out how products, such as bergamots, clementines, and various PGI wines, would surely benefit from the enhancement and networking of the existing cultural heritage in the context of an integrated resource-enhancement identity. Despite covering only a small area (139 ha) within the various municipalities, vine cultivation comprises 20 IGP and IGP grape producers, particularly in Area 9.

As far as the receptive system is concerned, some places, such as those of Santo Stefano in Aspromonte (ski resort) Area 7 and Villa San Giovanni (main connection port with Sicily) Area 1, register a rate of accommodation capacity close to 50%, and so are areas where the tourist component plays a crucial role in the general economy. Nevertheless, Area 9, incorporating municipalities with a strong seaside tourist vocation such as Palmi, Scilla and Bagnara, is the primary area in terms of available beds. On the contrary, areas in the hinterland such as Area 10 and 14 show almost nonexistent indicators, as well as the lack of structures and receptive systems, making the exploitation of their territories very problematic. Finally, the employment fields of computerization, logistics and information technology represent the lowest employment rates in the whole case-study area.

---

[2] The classification of Municipalities showed in Fig. 1, has been analyzed according to the Strategy for Inner Areas in Italy (MUVAL 2014).

3.3   Identification and Analysis of Cultural Attractors

The second stage of analysis was addressed to the cultural attractors'[3] identification and investigation, considering these elements as key factors for designing integrated valorization programs and for implementing directions and guidelines towards territorial renaissance within a sustainable, local development strategy. The data collection (starting in June 2016) is considered in accordance to two UNESCO conventions [22], and is structured in the following thematic areas:

- Immaterial Attractors-includes data collection and analysis aimed at describing the consistency and attractiveness of intangible cultural resources.
- Landscape Attractors-includes collection, analysis, and identification of elements of particular natural and environmental value, with a view to describing the real potential for the development of the landscape sector.
- Archaeological Attractors-include the collection, analysis, and identification of archaeological sites and/or archaeological finds.
- Historical-Architectural Attractors-includes the collection, analysis, and timely identification of immovable properties of particular artistic, cultural, and historical value.

In greater details, the most significant concentration of thematic attractors per area is the endowment of historical and architectural heritage assets (50%), followed by intangible assets (29%), landscaping (19%), and archaeological assets (2%). These data take into account—beyond the known concentrations in the municipalities located along the coast with tourist vocation (A.9–32% of the total)-above all the widespread presence of the so-called "minor" (rural and industrial architecture) heritage, which is one of the distinctive features of the territory's identity, as well as an asset of great value in economic competition. The concentration of landscape attractors is significant in Area 9 (32%), and in Area 8 (25%). The sites of significant historical and artistic interest in the study area are identified as Areas 8 and 9; those are centers located predominantly along the coastal strip, accompanied by a rather limited number of municipalities that affect the more inner areas. As showed in Table 1, inner areas are lacking in attractors, with percentages below 20% (for instance, Area 8). However, by extending the geography of archaeological attractors, Area 1 owns 33% of archaeological sites and/or archeological finds; the percentage falls to 17% for Area 15-however, that is characterized by the significant presence of the Tauriani Archaeological Park.

3.4   Public Investments' Appraisal

The third step of analysis dealt with the public investments in the case-study area during the 2007–2013 EU programming period. The Calabria Regional Operational

---

[3] The term "cultural attractor" in this study does not refer to the anthropologic–philosophical field (see Buskell, A. 2017, What are cultural attractors? In Biology & Philosophy, Vol. 32, Issue 3, pp. 377–394, Springer Netherlands), but instead to the capacity of cultural resources to act as a driving force for the economic, cultural, and social growth of a territory.

Table 1. Report on the percentage of thematic attractors per area (Author: Arch. I. Lorè)

Thematic attractors/Areas	A.1	A.7	A.8	A.9	A.10	A.14	A.15
Intagible attractors	13%	10%	8%	**29%**	12%	20%	8%
Natural attractors	9%	11%	25%	**32%**	7%	15%	*1%*
Archeological attractors	**33%**	17%	*0%*	17%	8%	8%	17%
Cultural attractors	6%	6%	29%	**32%**	11%	12%	4%

Note: **higher values**; *lower values*

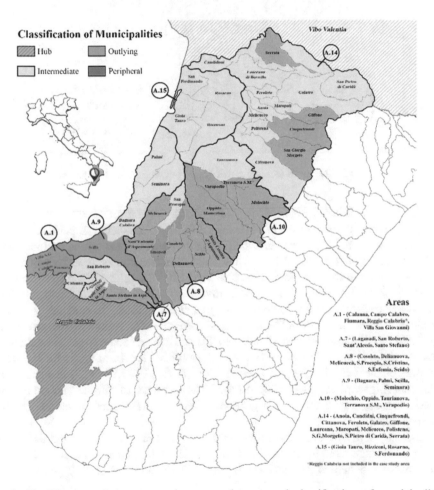

Fig. 1. Identification of the case-study area, sub-areas and classification of municipalities (Author: Arch. C. Zavaglia)

Programme for 2007–2013 comes under the Convergence Objective, and around 58% of the program funds have been earmarked for investments directly linked to sustainable growth and jobs in line with the Lisbon and Gothenburg agendas.

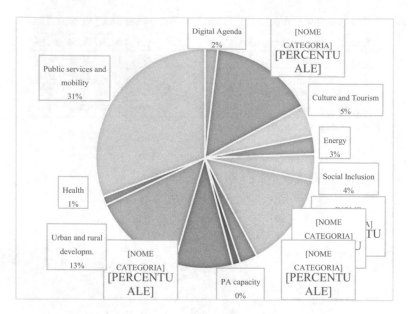

Fig. 2. Investments per categories in the case-study area

The term "public investments" in this research context means the contribution of European, national, and regional funding. Details of the allocation of public funding in the case-study area, have been calculated using the source of the OpenCoesione dataset [23]. In greater detail, Fig. 2 shows the investments made in the 2007–2013 programming cycle in the case-study area per "categories". Looking at the case study as a whole, 31% of the public funding was invested in "public services and mobility", followed in decreasing order by "environment and risk" (15%), "education" (14%), and "urban and rural development" (13%). The categories "investments for growth and jobs" and "SME competitiveness" received 2% and 1% respectively of the total budget. Most of the budget of "public services and mobility" has been spent for the infrastructures connected with the port area of Gioia Tauro (€185,771,863) in Area 15 and Villa San Giovanni (€36,444,100) Area 1. The category of "urban and rural development" mostly refers to urban regeneration and restoration/new functions of abandoned buildings.

Looking at the funding allocation per sub-area, it deserves to be highlighted that Area 15 benefited from the most of the budget (65%), followed by the Area 14 (11%). The areas that benefited the least are the no. 7 (2%) and no. 1 and 10 (4%).

4   Concluding Remarks

The first limiting factor, as far as the case study is concerned, relates to the reliability and comprehensiveness of the data used. In particular, socioeconomic data was obtained mainly from the national official statistic census of 2011, so there is no parameter to investigate the impact of 2007–2013 public investments in the area.

Analysis may be improved by a multi-perspective and systematic view that consultation with other stakeholders (internal, territorial stakeholders) could offer.

Despite these limitations, the data collected was sufficient to implement the conceptual framework, substantially raising the significance of the "knowledge stage" in order to identify the weaknesses and opportunities to deal with designing efficient and sustainable local development strategies.

Nonetheless, the territorial system identifiable in the case-study area is an intervention area suitable for the field testing of new approaches to planning and governance for the achievement of the European Territorial Agenda objectives. The combined analysis of economic, social, cultural, and environmental studies mapped on the territory has therefore led to the definition of a cultural geography within the area and to the understanding of the critical points:

- the gap between the coastal areas in terms of population, employment, financial investments, and cultural attractors;
- the need to re-calibrate the 2014–2020 public investment according to the real socioeconomic needs of the territory;
- the significance of an evaluation approach, when handling territorial-scale valorization programs.

The further objective is to understand and to evaluate whether territorial cultural attractors are able to interact with the whole service system and with the productive chains that could spin off. Besides that, there is the need to evaluate the capability to activate local projects, to start network policies, and to identify local systems or districts, making explicit the active element of the territorial resources valorization.

The elaboration of integrated valorization strategies, particularly in marginal areas, will have to evolve towards a general vision of projects in the territory that should take into account the socioeconomic and cultural resources that contribute to characterize and to identify an area. This is even more significant and strategic for the lagging regions because, to transcend physical marginality, economic immobility, and lack of innovation and technology, one can be more flexible in the planning of the protection and preservation of their own territorial resources.

References

1. United Nations General Assembly: Transforming our world: the 2030 Agenda for Sustainable Development (2015)
2. Todling, F., Tripl, M.: One size fits all? Towards a differentiated regional innovation policy approach. Res. Policy **34**, 1023–1209 (2005)
3. Fusco Girard, L., Nijkamp, L. (eds.): Cultural Tourism and Sustainable Local Development. Ashgate, Aldershot (2009)
4. Della Spina, L.: Integrated evaluation and multi-methodological approaches for the enhancement of the cultural landscape. In: Gervasi, O., et al. (eds.) Computational Science and Its Applications, ICCSA 2017. LNCS, vol. 10404, pp. 478–493. Springer, Cham (2017). https://doi.org/10.1007/978-3-319-62392-4_35

5. OECD: Sustainable Development Governance Structures in the European Union. In: Institutionalising Sustainable Development, pp. 67–88. OECD Publishing, Paris (2007)
6. European Commission: Europe 2020: a strategy for smart, sustainable and inclusive growth. COM (2010) 2020 final (2010)
7. Leone, U. (ed.): Scenari del XXI secolo. Giappichelli, Torino (1999)
8. Mollica, E.: Le aree interne della Calabria: una strategia e un piano quadro per la valorizzazione delle loro risorse endogene. Rubbettino, Soveria Mannelli (1995)
9. Alcaro, M.: Sull'identità meridionale. Forme di una cultura mediterranea. Bollati Boringhieri, Torino (1999)
10. Cassano, F.: Il Pensiero meridiano. Laterza, Roma-Bari (1996)
11. Evans, A.: EU Regional Policy. Richmond Law and Tax Ltd., Richmond (2005)
12. Barca, F.: Towards a place-based social agenda for the EU. Report Working Paper (2009)
13. MUVAL: A Strategy for Inner Areas in Italy: Definition, Objectives, Tools and Governance, Issue 31 (2014)
14. Rodríguez-Pose, A.: Is R&D investment in lagging areas of Europe worthwhile? Theory and empirical evidence. Reg. Sci. **80**, 275–295 (2001)
15. McCann, P., Ortega-Argilés, R.: Smart specialization, regional growth and applications to European Union cohesion policy. Reg. Stud. **49**(8), 1291–1302 (2015)
16. Giuffrida, S., Ventura, V., Trovato, M.R., Napoli, G.: Axiology of the historical city and the cap rate. The case of the old town of Ragusa Superiore, Valori e Valutazioni **18**, 41–55 (2017)
17. Nesticò, A., De Mare, G.: Government tools for urban regeneration: the cities plan in Italy. A critical analysis of the results and the proposed alternative. In: Murgante, B., Misra, S., Rocha, A.M.A.C., Torre, C., Rocha, J.G., Falcão, M.I., Taniar, D., Apduhan, B.O., Gervasi, O. (eds.) ICCSA 2014, Part II, LNCS, vol. 8580, pp. 547-562, Springer, Heidelberg (2014). http://doi.org/10.1007/978-3-319-09129-7_40
18. Della Spina, L., Lorè, I., Scrivo, R., Viglianisi, A.: An integrated assessment approach as a decision support system for urban planning and urban regeneration policies. Buildings **7**, 85 (2017). https://doi.org/10.3390/buildings7040085
19. Calabrò, F., Cassalia, G., Tramontana, C.: The mediterranean diet as cultural landscape value: proposing a model towards the inner areas development process. Procedia-SBS **223**, 568–575 (2016)
20. Cheshire, P., Magrini, S.: Endogenous processes in European regional growth: convergence and policy. Growth and Change **3**, 455–479 (2000)
21. Cassalia, G., Lorè, I., Tramontana, C., Zavaglia, C.: L'analisi socio-economica a supporto dei processi decisionali: il caso dell'Area Tirrenica della Citta' Metropolitana di Reggio Calabria. LaborEst **14**, 26–33 (2017)
22. Cassalia, G., Tramontana, C., Lorè, I., Zavaglia, C.: Statistiche culturali - il censimento del patrimonio culturale nell'Area Tirrenica della provincia di Reggio Calabria. LaborEst **13**, 12–18 (2016)
23. Opencoesione. http://www.opencoesione.gov.it/. Accessed 27 Sept 2017

Disaster Management in the Inner Areas: A Window of Opportunity for National Strategy (SNAI)

Annalisa Rizzo[✉]

Mediterranea University of Reggio Calabria, 89100 Reggio Calabria, Italy
annalisa.rizzo@unirc.it

Abstract. The acceleration of negative demographic and economic trends brought on by cyclical recurrence of catastrophe in inner areas makes National Strategy developmental goals more challenging. The triggering role of disasters can instead be used to catalyze strategies and to introduce radical innovations. This paper presents a research design regarding the integration of risk reduction strategies within the development of National Strategy for Inner Areas (SNAI). An overview of the emerging question of natural hazards and disaster impacts on the inner areas introduces the main topic: the role of disasters within development path trajectory. Drawing from Transition Management and disaster and post-disaster management literature, the research design uses "window of opportunity" concept both as a bridge between theories and as an entry point for developing an analytical structure based on the concepts of development trajectory, transition, trajectory break and trajectory reshape. Finally the research aims and objectives are discussed in light of the ultimate goal of contributing to the resilience-building goal of the National Strategy.

Keywords: Inner areas · Natural hazards · Risk · Transition management

1 Introduction

Crisis brings out built-in systemic contradictions and weaknesses. The reconstruction process following the Central Italy earthquake (2016) revealed the conflict and mismatch between the National Strategy and the Emergency commissioner government, despite the fact that both are expressions of the central government. Lock-in mechanisms and pressure from different lobbies, at regional as well as local levels, prevailed on the long term vision defined with the SNAI and, through the emergency special laws, are re-shaping the regional development trajectory. This crisis highlighted the weaknesses of the system but also of the Strategy. As far as we could observe until the present, it represents a missed "window of opportunity". What are the factors and the features of the Strategy that lead to this output? What kind of tools are needed in order to coordinate the long term vision and the tactical aspect of emergency management?

In the recent past a growing interest in transition and process theories both in academic and policy literature has been shown. One reason is that the uncertainty and complexity of long term development processes lead to questioning of the deterministic

© Springer International Publishing AG, part of Springer Nature 2019
F. Calabrò et al. (Eds.): ISHT 2018, SIST 101, pp. 13–21, 2019.
https://doi.org/10.1007/978-3-319-92102-0_2

approach. Transition Management represents the approach that better serves the purpose of investigating the development path-break represented by disaster.

Drawing from the disaster literature and 'post-disaster window of opportunity' research stream, inscribed into the wider frame of transition theories which combine Actor-Network and evolutionary economics, a new theoretical framework will be traced for investigating the issue.

Both Transition Management and Disaster Management are rooted in pure theoretical frameworks, economic studies for the former and sociology for the latter, but have then been developed as management tools. Hence, both serve either as an analytical framework or as policy support. Window of opportunity is the key concept where these two streams converge, representing the entry point for the research and for the application to the case study.

With this new theoretical approach, which will be tested on the case study of Central Italy earthquake (2016), the research seeks to answer the following questions:

- Can The National Strategy for Inner Areas be ascribed to Transition Management framework?
- How can the Strategy be improved and implemented through the Transition Management scheme?
- How can the cyclic post-disaster window of opportunity intentionally catalyze the Strategy?

This paper aims to question the role of disasters within development strategies. Through a literature review covering topics from disaster management to transition studies, we define the boundaries of the knowledge gap that future research aims to fill.

2 Reasons for Addressing the Issue of Risk Management in the Context of Inner Areas

The cyclic occurrence of earthquakes and other disasters related to natural hazards in Italy regularly bolsters the political and academic attention on the issue of risk reduction and disaster management; unfortunately the momentum is fated to decrease faster than the effects of the disaster. The establishment of Protezione Civile in 1992 and the plans for prevention represent an attempt at adopting a strategic approach at the national level. As the seismic hazard classification (P.C.M. 20/03/2003 n. 3274) shows, Italian regions are not affected the same way by natural hazards, but what's more important they do not have the same vulnerability or the capacity for reaction. These data are not represented on maps or reported.

The Strategy for Inner Areas (2014) has drawn the perimeter of the more fragile regions from a socio-economic point of view, through accessibility and remoteness criteria, linking geomorphological aspects with development trends. If we overlay a hazard map with this classification the strong correlation between the "structural liabilities" and the geomorphological and the socio-economic ones becomes clear.

European development strategies for more inclusive growth based on innovation proved to be less effective in the lagging regions, which needed the most in order to catch up with the leading ones [1].

Given the catalyst role of crises, the widening development gap can even be increased if a disaster occurs. The window of opportunity [2–4] brought up by the crisis can instead be used to trigger radical change towards sustainability. Disasters can accelerate [5] ongoing negative trends as well as catalyze development strategies.

Developing tools for capitalizing on the cyclic post-disaster window of opportunity is a great opportunity and a necessity for the Strategy for Inner Areas. The research presented in this paper aims to investigate how management of natural hazards can be integrated into sustainable development strategies.

3 Development Strategies and Natural Hazards in Inner Areas

The specific polycentric urban pattern spread all over the Italian territory, combined with an unstable, hilly geomorphology, creates the conditions for different challenges, like inclusive growth, risk reduction and sustainable development, to overlap. Even though at the EU level strategies are often conceived and developed following separate paths, at the local level a synthesis that can lead to an integrated strategy is required. The SNAI, launched by the central Italian government aiming for sustainable development of those areas characterized by negative demographic and economic trends, can represent a step in this direction. With the aim of pointing out the need to better integrate the vulnerability reduction challenge into development strategies, the following paragraphs will discuss the correlation between inner areas and risk related to natural hazard, investigated by the author with previous studies.

3.1 Place-Based Strategies for Inclusive Growth

Ecological, social and economic sustainability represents one of the main challenges for EU [6]. The reduction of the development gap among frontier and lagging regions has been identified as one of the core factors of this challenge. In spite of the actions undertaken with the Cohesion Policy, the gap is widening: it has grown by 56 percent between 1995 and 2014, hence the strategy needs to be improved and reshaped. As pointed out by Bachtler (2017) "the adaptation to the specific shocks on regional economies generated by globalization and market integration require differentiated (or place-based) strategies".

The National Strategy for Inner Areas has been developed within this framework, overcoming the compensatory approach of the previous programs. The SNAI redraws the perimeter of this development gap adopting innovative criteria: it uses the peripherality and remoteness as a proxy for phenomena such as "population decline below the critical threshold; job cuts and falling land use; decline in local provision of public and private services; social costs for the nation as a whole, such as hydro-geological instability, and degradation of the cultural and landscape heritage" [7]. The Strategy aims to reverse the demographic trends of Inner Areas, both in terms of resident numbers and in terms of age and birth group. This ultimate objective is pursued through 5 intermediate objectives: (a) Increased per capita wellbeing of residents; (b) Increased employment; (c) Reuse of territorial capital; (d) Cutting the social costs of de-anthropisation; and (e) Bolstering local development factors. The first action undertaken to achieve the intermediate objectives serves the purpose of creating the pre-conditions for territorial development through the upgrading of essential services. Improving living conditions and services should attract new residents. The second action, the launch of local development projects, should provide job opportunities and foster the re-activation of the local "territorial capital".

3.2   Natural Hazards and Inner Areas: An Emerging Issue

The Italian polycentric system is mainly composed by a close-knit network of minor municipalities and rural areas connected to the towns-major centre of services. The territorial pattern often reflects the geo-morphological structure, with the major settlements in the most stable-plane areas and the fine network that covers the hilly and mountainous parts. Hence, the depopulation and abandonment trend which affects inner areas contributes to the widening of the development gap and to reducing the inclusiveness of economic growth, moreover it bears several risk and social costs in terms of hydro-geological instability, decay and soil consumption [7]. A cyclic occurrence of natural hazards exacerbates the phenomena, adding complexity to the issue. Natural hazards catalyze tendencies and accelerate trends, for instance giving input to the ongoing emigration flows. Previous studies [8] carried on by the author discuss the correlation between risk and inner areas.

Based on the definition of risk, given by UNISDR [9], as the product of hazard, vulnerability and exposure, the study focused on seismic risk in the inner areas. While methods for assessing the vulnerability and exposure of buildings and structures are widely discussed both among scientists and practitioners, we can't state the same for the urban or territorial scale. Nevertheless some considerations on the relation between natural hazards and inner areas have been drawn through analysis of the database of the national seismic classification of municipalities and the database of the inner areas. Almost 50% of the municipalities classified as inner areas present a medium-high seismic risk (class 1–2). Compared to major towns, the share of the inner areas in class 1 is three times higher. In other terms, 63% of the territorial surface exposed to the highest risk belongs to inner areas [8].

One of the characteristics of the hazard concept, which has largely been investigated in academia but also in the industrial and management fields [10], is the slow-run of evolvement (e.g. the regional hazard maps are updated every 7 years). On the contrary vulnerability depends on fast-changing variables, therefore the measurement of socio-demographic phenomena and the abandonment trend cannot be based on the decennial national survey. The difficulty of combining demographic surveys with real estate analysis and economic profile represents another constraint for measuring urban vulnerability. Literature on disasters gives more emphasis to short-cycle extreme events and to expected effects of climate change [11–13], while studies on seismic risk often follow the long-cycle of the phenomenon, with a high scientific and political momentum right after the shock followed by decreasing interest. This pattern contradicts the necessity and the will to develop a preparedness approach.

The results of the study underline the necessity to develop a project of vulnerability reduction, tailored to the specific characteristics and needs of inner areas.

4 Reshaping Development Trajectory

The common goal of long term strategies for development is to steer regional economic trajectory. Different approaches have been developed to achieve this through incremental changes rather than introducing radical innovative policies. The first paragraph outlines these different perspectives introducing the heuristic approach of Transition Management. The concept of "window of opportunity" as the critical moment for path breaking that allows radical changes is then discussed, utilizing it as the theoretical bridge between transition studies and disaster literature.

4.1 From Development Strategies to Transition Management

The link between innovation and development, based on Schumpeter's theories and used as theoretical basis for evolutionary models of economic growth, constitutes one of the building blocks of European development strategy [14]. This approach should not only provide more sustainable development for UE regions but also guarantee more inclusive growth by stimulating the lagging regions. Within this framework most development strategies usually try to steer the development process step-by-step with small incremental changes, starting from the existing context. This approach may work for competitive regions which present a development trajectory coherent with the UE strategy, but it becomes more challenging for those regions affected by negative socio-economic trends. We refer here to the 'innovation paradox' as defined by Oughton [15] as "the apparent contradiction between the comparatively greater need to spend on innovation in lagging regions and their relatively lower capacity to absorb public funds earmarked for the promotion of innovation and to invest in innovation related activities, compared to more advanced regions".

Without questioning from a theoretical point of view the development waves and cycles, Transition Management (TM) has been developed as an analytical and policy framework to elaborate development strategies for those goals, in contrast to the existing development paradigm, which implies a path-reshaping or trajectory-breaking. The transition governance approach positions itself within the broader field of complexity science [16, 17], governance [18, 19], sustainability science [20] and social innovation [21]. Outlining a development strategy that aims to change the ongoing trends means managing a transition, where transition is "defined as the result of co-evolving processes in the economy, society, ecology, and technology that progressively build up toward a revolutionary systemic change on the very long term" [22]. Transitions are complex processes which are impossible to predict or control directly, only some recurrent patterns can be anticipated. Managing transitions does not mean planning the process but rather trying to guide and adapt it in order to accelerate and reorient the strategy during the windows of opportunity created by inevitable nonlinear shifts and associated crises. As underlined by Loorbach (2015), adopting Transition Management governance institutions "need to shift from seeking generic solutions to offering more generic frameworks that create space for and help to enable bottom-up social innovation".

Without going into detail about the structure, development and experiences of Transition Management, on which exists a rich literature [2, 3, 21, 23], the aim of this study is to investigate the National Strategy for Inner Areas from the perspective of transition management. As mentioned before SNAI attempts to "correct" and reverse the ongoing economic and demographic tendencies. The approach adopted has common features with TM as well as with more traditional incremental development strategies. Respecting one of the three main characteristics of "transitions towards sustainability" the Strategy is goal oriented, where the goal is related to a collective good and it requires changes in economic frame conditions and in policies in order to allow radical innovation in conflict with vested interests, both in the public and private sectors [2]. Structural change deals with different lock-in mechanisms which stabilize the existing system together with "institutional commitments, shared beliefs and discourses, power relations, and political lobbying by incumbents" [24].

4.2 Window of Opportunity: The Role of Disasters on Development Trajectory

Beyond the direct impact, disasters set the ground for changes that affect the structural aspects of the system [25, 26]. These changes can be seen as an acceleration of ongoing trends as well as an opportunity to reshape a path that would otherwise develop in a "spontaneous" way. The qualitative aspect of the change, the output or the coherence with the existing programs, has to be determined by the observer. In general, crises trigger transformations and unveil latent processes but can also catalyze strategies and new policies. In "Reconstruction Kobe" Edgington (2010) cites a zen proverb to introduce the broad span of opportunities carried by all crises. Destruction and spatial transformations imply the possibility of urban regeneration, especially for those areas neglected by official planning. Scientific literature defines this juncture as a "policy window" or more generally "window of opportunity".

Kingdon [27] claims that when "separate streams of problems, policies, and politics come together at certain critical times, then solutions become joined to problems, and both of them are joined to favorable political forces" [28]. Abrupt events like catastrophes or the emergence of new policy discourse opens up a "policy window", in this critical moment the coupling of problem-solution with coherent contextual factors are more likely to occur.

Heading from Kingdon's conceptualization, several authors, like Oliver-Smith (1996), Van Eijndhoven (2001) and Olson and Gawronski (2003), investigated and explored the "window of opportunity" concept, opening a sub-stream of literature in the field of development/change/transition dynamics. Such a concept proves to be meaningful only when framed in a wider socio-economic and spatial analysis with a dynamic approach. Moreover a long term perspective is required in order to understand the role of the window of opportunity in the development trajectory.

The potentiality of reshaping the development path and introducing innovative practices (for instance more sustainable or inclusive) within the window of opportunity [2, 3, 29] is often underlined from the policy point of view. On the contrary, disasters open up windows of opportunity for speculative actions and for accelerating the previous trends as well [28, 29], spontaneously like in the case of out migration flows due to the direct impact and destructions, but also forwarded by special policies like in the case of Belice earthquake (1968). In the specific context of inner areas and lagging regions, boosting existing negative trends can lead to the "accidental" loss of settlements as an outcome. The triggering role played by disasters increases the already alarming vulnerability of inner areas.

5 Next Steps

The acceleration of negative demographic and economic trends brought on by the cyclic recurrence of catastrophes in inner areas makes National Strategy development goals even more challenging.

The link between long-term development strategy and risk/disaster management should be investigated on a systematic and continuing basis overcoming the "occasional" approach. A change in the time perspective is required in order to grasp the effect that the cyclic nature of disaster has on the development trajectory.

With a new theoretical approach, drawn by transition literature and disaster management, tested on the case study of Central Italy earthquake (2016), the research aims to investigate the link between: (a) long term strategies and risk/disaster management; (b) SNAI as expression of Cohesion Policy and Transition Management.

The ultimate goal is to contribute to the resilience-building purpose of the Strategy. The research design illustrated by this paper proposes three contributions:

(1) theoretical contribution: a new framework for investigating the link between development strategies and risk related to natural hazards;
(2) a case study on the Central Italy earthquake with the comparative analysis of the regional governance among the four regions affected;
(3) the definition of a set of implementation tools developed through the analysis of SNAI through the Transition Management perspective.

The two presented themes, the relationship between inner areas, natural hazards and resilience on one side and between the development strategy for lagging regions and transition management on the other, are in the early stage of investigation but have shown to have strong foundations. The research finds its larger framework in the topical issue of Cohesion Policy and resilience debate. The aim is to explore the feasibility of transferring a management tool designed for ordinary times to the extraordinary context of post-disaster management.

References

1. Bachtler, J., Martins, J.O., Wostner, P., Zuber, P.: Towards Cohesion Policy 4.0: Structural Transformation and Inclusive Growth. Regional Studies Association, Brussels (2017)
2. Geels, F.W.: The multi-level perspective on sustainability transitions: responses to seven criticisms. Environ. Innov. Soc. Transit. **1**(1), 24–40 (2011)
3. Kemp, R., Loorbach, D.: Governance for sustainability through transition management. In: Open Meeting of Human Dimensions of Global Environmental Change Research Community, Montreal, Canada (2003)

4. Solecki, W.D., Michaels, S.: Looking through the post-disaster policy window. Environ. Manag. **18**, 587 (1994)
5. Edgington, D.W.: Reconstructing Kobe: The Geography of Crisis and Opportunity. University of British Columbia Press, Vancouver (2010)
6. Fagerberg, J., Laestadius, S., Martin, B.R.: The triple challenge for Europe: the economy. Clim. Change Gov. Chall. **59**(3), 178–204 (2016)
7. National Strategy for Inner Areas. http://www.dps.gov.it/it/pubblicazioni_dps/materiali_uval/. Accessed 05 Sep 2017
8. Annesi, N., Rizzo, A.: Dalla ricostruzione alla transizione. Guidare i territori in una strategia di lungo periodo. In: X Giornata Studio INU, Napoli (2017)
9. UNISDR: Terminology on Disaster Risk Reduction. United Nations Office for Disaster Risk Reduction, Geneva (2009)
10. Irwin, A., Simmons, P., Walker, G.: Faulty environments and risk reasoning: the local understanding of industrial hazards. Environ. Plan. A **31**(7), 1311–1326 (1999)
11. Davies, M., et al.: Climate change adaptation, disaster risk reduction and social protection: complementary roles in agriculture and rural growth?, IDS Working Papers 2009 (320), pp. 01–37 (2009)
12. Mercer, J.: Disaster risk reduction or climate change adaptation: are we reinventing the wheel? J. Int. Dev. **22**(2), 247–264 (2010)
13. O'Brien, G., et al.: Climate change and disaster management. Disasters **30**(1), 64–80 (2006)
14. Nelson, R.R., Winter, S.G.: An evolutionary theory of economic change. Harvard University Press, Cambridge (1982)
15. Oughton, C., Landabaso, M., Morgan, K.: The regional innovation paradox: innovation policy and industrial policy. J. Technol. Transf. **27**(1), 97–110 (2002)
16. Gell-Mann, M.: Complex Adaptive System. Addison-Wesley, Reading (1994)
17. Midgley, G.: System thinking: an introduction and overview. In: Midgley, G. (ed.) Systems Thinking, p. Xvii. Sage, Thousand Oaks (2003)
18. Kooiman, J. (ed.): Modern Governance: New Government-Society Interactions. Sage, London (1993)
19. Pierre, J.: Debating Governance: Authority, Steering and Democracy. Oxford University Press, Oxford (2000)
20. O'Riordan, T., Lenton, T.: Addressing Tipping Points for a Precarious Future. Oxford University Press, Oxford (2013)
21. Westley, F., Olsson, P., Folke, C., Homer-Dixon, T., Vredenburg, H., Loorbach, D., Thompson, J., Nilsson, M., Lambin, E., Sendzimir, J., Banerjee, B., Galaz, V., van der Leeuw, S.: Tipping toward sustainability: emerging pathways of transformation. Ambio **40** (7), 762–780 (2011)
22. Loorbach, D., Frantzeskaki, N., Huffenreuter, R.L.: Transition management: taking stock from governance experimentation. J. Corp. Citizensh. **2015**(58), 48–67 (2015)
23. Geels, F.W., Schot, J.: Typology of sociotechnical transition pathways. Res. Policy **36**(3), 399–417 (2007)
24. Unruh, G.C.: Escaping carbon lock-in. Energy Policy **30**(4), 317–325 (2002)
25. UN/ISDR, UN/OCHA: Disaster Preparedness for Effective Response Guidance and Indicator Package for Implementing Priority Five of the Hyogo Framework. United Nations secretariat of the International Strategy for Disaster Reduction (UN/ISDR) and the United Nations Office for Coordination of Humanitarian Affairs (UN/OCHA), Geneva, Switzerland (2008)
26. Wisner, B., Blaikie, P., Cannon, T., Davis, I.: At risk: natural hazards, people' vulnerability and disasters. Routledge, London (2004)

27. Kingdon, J.W.: Agendas, Alternatives, and Public Policies. Harper Collins College Publishers, New York (1995)
28. Birkmann, J., Buckle, P., Jaeger, J., et al.: Extreme events and disasters: a window of opportunity for change? analysis of organizational, institutional and political changes, formal and informal responses after mega-disasters. Nat. Hazards **55**, 637 (2010)
29. Scamporrino, M.: Governare la ricostruzione. Università degli Studi di Firenze, Firenze (2013)

The Integrated Coast to Coast Development of Basilicata, Southern Italy

Maria Assunta D'Oronzio$^{(\boxtimes)}$, Mariacarmela Suanno,
and Domenica Ricciardi

CREA - Consiglio per la ricerca in agricoltura e l'analisi dell'economia
agraria - Centro Politiche e Bioeconomia - Postazione regionale Basilicata,
85100 Potenza, Italy
{massunta.doronzio, mariacarmela.suanno,
domenica.ricciardi}@crea.gov.it

Abstract. This study aims to highlight how the Basilicata "Coast to Coast" FLAG has an important role in local sustainable development actions using a bottom-up approach. The FLAG represents one of many significant innovative elements in the Lucanian coastal territory. New development activities stemming from European Sea Policies, favor the fishing and aquaculture sector and represents a driving force in economic development employment, tourism and social-culture growth along the Lucanian coasts with strong rural traditions.

Integrated territorial planning approach is required using an intersectoral and multidisciplinary approach, and a variety of programming tools can be integrated in order to achieve common objectives. Informal and formal collaboration between the FLAG and the Lucanian LAGs is necessary in order to carry out joint projects which are able to connect the blue economy with other productive sectors.

Keywords: FLAG · Rural and fisheries local development · Blue economy

1 Introduction

Over the last decade, the use of various territorial development tools have increased in rural and coastal areas to strengthen the relations between the actors who work at local and supra-local level. Community policies, thanks to the LEADER method, have supported the territorial development strategies designed to be managed by local partnerships [1]. These areas have play an important part in local development and a strategic role in stimulating "the characteristics of social life - networks, norms, trust - enabling participants to act more effectively in the pursuit of shared objectives" [2].

A "strategic" approach has been adopted using the various combination of components which have roots in the territory [3]. A theoretical framework was also developed for the rural, coastal and agri-food fields [4]. Pressures from people and associations regarding innovation, modernization and change affect rural and coastal areas and require the development of actions in order to create a "shared vision for a desirable future and achievable through material changes and not by various means (programs, plans, projects in the strict sense)" [5].

© Springer International Publishing AG, part of Springer Nature 2019
F. Calabrò et al. (Eds.): ISHT 2018, SIST 101, pp. 22–28, 2019.
https://doi.org/10.1007/978-3-319-92102-0_3

The FLAG "Coast to Coast" study of the Basilicata rural and coastal areas, extends over a total area of 1694,4 square kmq and has 113,331 inhabitants [6]. There are nineteen Municipalities, eleven belong to the province of Matera and eight to Potenza. Six municipalities are located on the Ionian sea (Bernalda, Pisticci, Scanzano, Policoro, Rotondella and Nova Siri) and Maratea on the Tyrrhenian sea. Grottole and Miglionico are located near the artificial basin of S. Giuliano and Colobraro, Valsinni and Tursi in the inner rural area. Castelluccio Inferiore, Castelluccio Superiore, Latronico, Lagonegro, Trecchina, Rivello e Lauria, near Maratea, are rural and inner areas (Fig. 1).

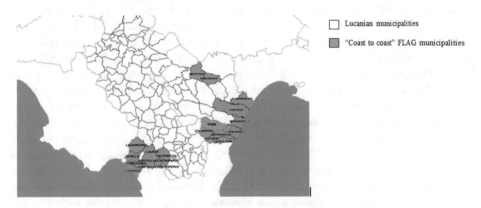

☐ Lucanian municipalities

■ "Coast to coast" FLAG municipalities

Fig. 1. "Coast to Coast" FLAG

The last eight municipalities were included in the 2007–2013 development strategies through the Local Action Group (LAG) "La Cittadella del Sapere". Bernalda, Pisticci, Scanzano, Policoro, Rotondella, Nova Siri, Colobraro, Valsinni and Tursi joined the COSVEL LAG, now known as START 2020.

New development prospects offered by European sea policies aim to enhance endogenous resources at an international level, recognizing the quality of "strategic global inputs", generating wealth, employment and innovation according to a participatory, collaborative and sustainable model [7].

A variety of national and community planning initiatives have begun in the area, however, with a lack of integration local actors have had to make a great deal of effort in their management as they remain tied to the specifics of the Community funds on which they are based. Furthermore, new strategic guidelines aimed at strengthening the artisan fishing systems, aquaculture, production chain and diversification of the sector have been introduced for the first time.

This study illustrates the integrated and sustainable local development "Coast to Coast" FLAG territories, and the results from the effects of the previous LEADER local policies and the new local development objectives to be pursued through the CLLD (Community-Led Local Development). In addition, a new opportunity, in the form of the "River Contract", has been offered to the FLAG partnership qualifying it as one of the main actors on the waters at a regional level.

2 Local Development Integrated into 2007–2013 Programming

The area under examination in the 2007–2013 programming period did not directly test the specific local development experience for coastal territories, and as a result have been the focus of numerous development actions achieved by both the Lucanian LAGs in the area and from other partnerships or subjects involved in the territorial development routes [8].

On the Tyrrhenian sea, the "La Cittadella del Sapere" LAG has initiated interventions to support companies for the development of multi-compartment micro-supply chains and contributed to an increase in the quality of life and social well-being. It has also focused on the enhancement and use of environmental resources and natural spaces through the implementation of thematic interventions and educational-informative and informational paths able to bring rural communities closer to the environment and its natural and landscape resources. Initiatives for the activation of the "River Contract" have already been launched in the south-western area characterized by the Sinni and Noce rivers and by the suggestive habitat. In addition, "La Cittadella del Sapere" co-founded the project "Maratea film festival", cinema week in Maratea, which promoted the territory through cinema in the presence of personalities from the entertainment world, bringing Maratea to the center of Italian culture. The LAG, along with five other Basilicata LAGs also co-financed the film "Basilicata coast to coast" by Rocco Papaleo, set in Maratea. The "La Cittadella del Sapere" LAG is also the sole beneficiary of the Leader Inspired Network Community (LINC) 2015, a transnational co-operation project. After Austria, Germany, Estonia and Finland, the city of Maratea was chosen to host the 2015 edition of the project and LAGs from seventeen European countries participated, making it the "European Capital" of co-operation. In addition, Maratea hosted the Mediterranean Co-Innovation "Heroes" meeting to discuss developments regarding local start-ups with innovation as the focus.

The Cosvel LAG on the Ionian sea focused mainly on using rural tourism as a tool for diversifying the local economy, through the creation of a hospitality system in the historical centers to safeguard and enhance the local heritage from abandonment and degradation. It also worked to support local micro-enterprises. The LAG also carried out territorial animation and supported small coastal fishing sector in the Ionian Sea, which led to the establishment of the first local fishing sector consortium. The consultation activity carried out by the Cosvel LAG was consolidated through a partnership between Nova Siri, Pisticci, Policoro, Scanzano Jonico and Rotondella municipalities to design the integrated planning of coastal interventions.

The Integrated Tourism Packages (ITP) which represented the new regional tourism policy strategic projects contributed to territorial development. As a result, the public and private actors from the business world have been joined together with the aim of sharing lines of development and creating different cultural, natural and landscape resource networks. The "Metapontino" and "Maratea Terramare" ITP along with the involvement of the fishermen activated tools to enhance the cultural and natural heritage in Basilicata to structure a better more competitive tourism in domestic and foreign markets.

The local development of the two areas was achieved using sensitization and involvement activities implemented by local actors and managed by Lucanian environmental Education Centers for Sustainability and by the Environmental Observatories for Sustainability which operate in the educational-training field, territorial planning and information and communication.

The European Fisheries Fund (EFF) Axis IV dedicated to the development of the local coastal territory was not activated by in Basilicata [9]. However, strategic lines were implemented by the region which were aimed at strengthening the artisan fishing system, developing aquaculture activities and promoting investments to encourage diversification and supply chain. In 2014, the "Lucanian Tyrrhenian Shoreline Fishing Companies Co-ordination Association" and the "Metapontino Fishermen's Association" were formed. The first included entrepreneurs and fishermen operating from the navy port in Maratea, the second were, groups of fishermen from the Ionian area.

In support of the fishermen and their associations, the region and LAG actions were aimed at constructing four landing points along the Ionian coast forming a network of small ports and fishing shelters. The coastal territorial network economic improvement of local fisheries represents the recognition of the small coastal fishing sector and the identification of a specific area dedicated to them which can be jointly exploited by both fishermen and by other interested operators.

The main results achieved by the Operational Program for the EEF 2007/2013 include the return of boats registered to maritime areas in other regions, the access of young fishermen into the sector and the purchase of new boats used for fishing tourism activities.

3 "Coast to Coast" FLAG Local Development Strategy

The starting point of the "Coast to Coast" FLAG strategy is to develop the economic potential of the two Lucanian coasts in a sustainable and integrated way, enhancing fish and aquaculture products through short supply chains for consumers and local catering operators, as well as promoting marketing activities aimed at a wider market.

The "Blue Strategy" aims to enhance the coastal and sea culture through specific lines of intervention which will improve quality of life, promote economic prosperity and social development and the implementation of diversified activities. Actions are also closely linked to the promotion of professional skills and national and transnational co-operation between groups and fishing areas. The thematic areas identified are:

- Development and innovation of local supply chains (spinnerets) and production systems
- Sustainable tourism
- Development and management of environmental and natural resources.

The objectives pursued by this strategy are summarized as follows:

(a) Valuing, attracting and creating jobs for young people and promoting innovation at all stages of the fishing and aquaculture product supply chain;

(b) Support diversification, inside or outside commercial fishing, lifelong learning and job creation in fisheries and aquaculture areas;
(c) Strengthen the role of fishing communities through local development and governance of fisheries resources and maritime activities.

The FLAG financial resources are equal to 1 Meuro; 25% is allocated to management and operating costs. The Coast to Coast FLAG will allocate 220,000 euros to develop co-operation activities which will focus on the exchange of experiences in a supra-territorial dimension. These co-operation activities will reinforce strategies and add value by comparing, at national and international level, those subjects who share a common interest, finding solutions to local problems and promoting common resources etc (Table 1).

Table 1. "Coast to Coast" FLAG financial plan

Actions	Public expenditure	EMFF	National resources
A – Supply chain	€ 270.000,00	€ 135.000,00	€ 135.000,00
B - Diversification	€ 318.000,00	€ 159.000,00	€ 159.000,00
C - Governance	€ 170.000,00	€ 85.000.00	€ 85.000,00
Running cost	€ 242.000,00	€ 121.000,00	€ 121.000,00
Total	€ 1.000.000,00	€ 500.000,00	€ 500.000,00
Co-operation	€ 220.000,00	€ 110.000,00	€ 110.000,00

Source: "Coast to Coast" FLAG

The Local Action Plan includes nine actions which will be implemented over a three-year period and mainly concern. support for diversification, new forms of income and the creation of blue partnerships and networks to enhance natural habitats. These actions will receive the largest investment of financial resources.

Support for income diversification will be aimed at fishermen and intends to integrate the fisheries sector with other territorial production segments, using marine resources as examples of new complementary activities in other economic sectors, including on-board investments, tourism linked to sport fishing, catering, environmental services related to fishing and learning activities. The other actions will promote the establishment of partnerships, which obligatorily include the presence of fishermen and/or operators in the aquaculture sector and will be aimed at promoting and enhancing marine biodiversity, river and lagoon ecosystems, knowledge and use of habitats for educational, recreational and educational purposes.

No less important will be the actions aimed at supporting new youth fishing enterprises and the development of new innovative products and processes. Other, actions will include support associations and port networks, co-operation between fisheries, sustainability of blue economy competences and knowledge systems, certification of fish products and, finally, the "fishing markets". The latter will be aimed at strengthening competitiveness in the fishing areas, encouraging diversification of fishermen's economic activities, consolidating interaction with the natural tourism sector and enhancement of the environment. The action also aims to offer viable tourism linked to the "Blue Resource" in order to integrate activities and services (extra accommodation, catering, fishing tourism, aquaculture tourism) to effectively promote and market actions for specific targets and markets.

The FLAG execute the "Blue Strategy" via the River Contract, improving waters in inner areas. Financial resources, equal to 2 Meuro, will be funded by the Cohesion Fund, to activate the River Contract. This will represent an agreement between subjects who have responsibilities in the management and use of water, spatial planning and environmental protection. It is a negotiated strategic planning tool which pursues conservation including the correct management of water resources, the enhancement of fluvial territories and protection from hydraulic risk. Rivers play an important role in the development of ecosystem functions (ecological, cultural and environmental), it is at the hub of historical and landscape identities (green infrastructure) and is a new eco-compatible form of economy.

The "Coast to Coast" FLAG, via the Community-Led Local Development (CLLD) EMFF and the River Contract, is today one of the main actors on Basilicata waters, creating constructive dialogue aimed at tackling the issue of climate change recently threatening the region by flooding, drought and the destruction of ecosystems.

4 Conclusion

The role of the FLAG in the strategy enhances the local dimension: the territory acts as an interrelationship system between economic, socio-cultural and political factors influencing development [10]. The interaction networks between subjects at different local and global levels transform some of the potential activities offered by the territory and puts the levers in place to support specific paths of development [11]. The territory is characterized by a network of economic, social, cultural and institutional relationships which local actors have established and cemented over time, rooting practices and knowledge which are difficult to transfer elsewhere.

The development of the territory cannot be based on the adoption of a single solution model that is considered universally valid and therefore applicable in the most diverse territorial contexts. It can be based on using a medium and long term integrated, cross-sectoral and multidisciplinary approach. The new regulations regarding the European Union structural development policies show a clear desire to relaunch integrated territorial planning (local participatory development), which in the Lucanian territory has led to support for participatory planning and governance practices with the LAGs on local a scale and with the FLAG in the new programming phase. In a rural context the experience of the LAGs represents an element of continuity and in the coastal area the FLAG represents an element of innovation for the Lucanian territory.

The "Coast to Coast" FLAG has embarked on a well-defined development path, despite not having the opportunity to integrate development opportunities offered by the other European Structural and Investment Funds (ESIF) into its strategy. It has nevertheless secured the Cohesion Fund's resources via the River Contract without precluding the possibilities that arise from the marine strategy, whose objectives are to preserve ecological diversity and vitality of the seas and oceans so that they are clean, healthy and productive while keeping the use of the marine environment at a sustainable level and safeguarding the potential for uses and activities for current and future generations [12]. Local Development Strategies (LDS) were approved by two "bottom-up" public-private partnerships: the "Coast to Coast" FLAG and the recently approved LAG "La Cittadella del sapere". The LAG START 2020 strategy is currently being approved.

The hope is that both the LAGs and the FLAG will be able to deal in a synergic and integrated way with the challenges that the different territories pose using inter-territorial cooperation to manage the so-called "blue" themes. Added value for local fisheries and aquaculture products, improvement in the integrated management of the coastal zones and support to the fishing, aquaculture and processing communities can all benefit from and drive the blue economy.

References

1. Di Napoli, R., D'Oronzio, M.A., Verrascina, M.: Il ruolo di leader nella formazione di capitale sociale a livello territoriale: alcune esperienze. In: XXXII Conferenza scientifica annuale AISRe, Torino, 15–17 settembre 2011
2. Putnam, R.D.: Bowling Alone: The Collapse and Revival of American Community. Simon and Schuster, New York (2000)
3. Commissione Europea: PO FEAMP ITALIA 2014–2020, Bruxelles (2015)
4. Becattini, G., Bellandi, M., Dei Ottati,G., Sforzi, F.: (a cura di): Il coleidoscopio dello sviluppo locale. Rosenberg & Sellier, Torino (2001)
5. Dematteis, G.: Reti globali, identità territoriali e cibersbazio, Baskerville, Bologna (2001)
6. Piano di Azione Locale: FLAG Coast to Coast (2017)
7. D'Oronzio, M.A., Licciardo, F.: La Blue Economy e lo Sviluppo Sostenibile in Basilicata, CREA (2016)
8. Regione Basilicata: Programma di Sviluppo Rurale della Regione Basilicata per il periodo 2007–2013, Potenza (2008)
9. Regione Basilicata: Documento regionale per l'attuazione del Fondo Europeo Pesca (FEP) per il periodo 2007/20013, Potenza (2008)
10. Trigilia, C.: Sviluppo locale. Un progetto per l'Italia, Laterza, Bari (2005)
11. Dematteis, G.: Il tessuto delle cento città. In: Coppola, P. (ed.) (a cura di): Geografia politica delle Regioni Italiane, Einaudi, Milano, pp. 192–229 (1997)
12. D'Oronzio, M.A., Licciardo, F.: La Blue economy in Basilicata. Risorse locali per lo sviluppo regionale, Rivista di Economia Agraria, Anno LXXII 1, pp. 45–61 (2017)

Action Research and Participatory Decision-Aid Models in Rural Development: The Experience of "Terre Locridee" Local Action Group in Southern Italy

Claudio Marcianò[✉] and Giuseppa Romeo

Mediterranea University of Reggio Calabria, Reggio Calabria 89100, Italy
claudio.marciano@unirc.it

Abstract. The increasing involvement of local stakeholders in planning processes implies the decentralization of decisional power through participatory activities which make it possible to define development strategies that take into account a wide variety of perspectives. This paper, in the context of Action Research, focuses on the planning process of a Local Action Plan (LAP) within the Leader Axis that involves stakeholders from 36 municipalities belonging to the "Locride" area, in Southern Italy. At the basis of the LAP is the principle of consultation and cooperation between public and private actors, in order to identify common paths of development, thus activating the latent potential of the territories and leveraging the real knowledge of local actors on the issues of the areas designated for intervention. In such a context, a decision-aid model based on a multicriteria method and a convergence process was used in order to control the escalation of conflicts that arose among different stakeholders in the identification of shared thematic areas to be included in the LAP. The results show that the conflicts were effectively managed, and that the main drivers commonly chosen for the development of the territory combine interventions related to the fields of agro- food and artisan products, the social context, and the historical and environmental landscape.

Keywords: Action research · Decision-aid model · AHP
Integrated rural development · Local governance

1 Introduction

As has been recently highlighted by Trencher et al. [1–3], there is a growing debate about the new mission of the academic world to achieve what is called "sustainability co-creation" between Universities and Society in a place-based perspective. Partnerships and collaboration "between academia, industry, government and civil society are now seen as a prerequisite for the knowledge flow and knowledge exchange" [1, 4].

This depicts a role where the university "collaborates with various social actors to create societal transformations into the goal of materializing sustainable development in a specific location, region or societal sub-sector" [3]. In this context the university, through Action Research, is called to craft useful knowledge and bind it to societal

© Springer International Publishing AG, part of Springer Nature 2019
F. Calabrò et al. (Eds.): ISHT 2018, SIST 101, pp. 29–41, 2019.
https://doi.org/10.1007/978-3-319-92102-0_4

action. Moving in this new paradigm of "sustainability co-creation, greater emphasis is given to knowledge production as a vehicle for creating solutions to societal problems and triggering societal transformations towards greater sustainability" [3]. Since in this situation knowledge is a key resource, the role of the university becomes fundamental to foster the creation of networks between research, local government and stakeholders in a bottom-up perspective through collaborative research [4].

The rural development programmes that have been implemented in Europe since the 1990s have consolidated the vision of local-territorial development promoted through participatory and bottom-up processes, by drawing up development policies that are calibrated to the actual needs and characteristics of the territories. This change has led to the abandonment of the traditional top-down approach to move towards development processes which involve more stakeholders in the planning stages. This implies the decentralization of decisional power through participatory activities that make it possible to define development strategies which take into account a wide variety of perspectives.

In this context, participation assumes the role of functional utility, because it enables the learning ability of stakeholders to be used in planning the development of interventions that correspond to the actual needs of the territory [5–7]. In this perspective it is important to integrate three key elements: area, partnership and strategy, which distinguish the local development approach [8].

In this context, the Leader programme, in the form of Community Initiative in the first three programming cycles, took on over time the characteristics of Approach, in a logic of mainstreaming. This transition was aimed to strengthen and to consolidate the typical peculiarities of the participation and the capacity to plan bottom-up strategies for the development of the territory. Within the Leader, Local Action Groups (LAGs) represent the "territorial governance of rural development" [9], which give concrete form to local development policies by planning and subsequently implementing Local Action Plans (LAPs). LAGs play a key role in organizing and carrying out concertation activities between the members of the Socioeconomic Partnerships (PSE), the promoters of the LAPs, in order to identify the strategic priorities to be pursued for the development of the areas concerned.

The overall objective of the LAP is to promote the integrated development of the territory, through the definition of a set of coordinated interventions, designed to increase the economic competitiveness of the territory, and its social cohesion and employment, consistent with the needs of environmental sustainability. It follows that, in a participatory multi-stakeholder process, in order to plan a shared development model, it is essential to integrate and harmonize the potential divergent interests of the groups of actors involved. The correct interpretation of the different opinions and the integration of the different needs, in a development strategy shared by the partnership as a whole, is the key factor for the success of the intervention [10]. However, this process is rather complex, and divergences may arise between different actors, leading to conflict escalation that may slow down or also interrupt the planning process. In such cases, decision aid models may help in controlling conflict escalation by pointing out compromise solutions reached with the contribution of the different decision makers.

This study focuses on a decision-aid model used in the planning process of a Local Action Plan within the Leader Measure that involves stakeholders of 36 municipalities belonging to the "Terre Locride" LAG, a marginal area in Calabria where, in the previous EU programming period, the implementation of the development plan run by the LAG "Locride" was interrupted by administrative problems, and the LAG was dismissed. Therefore, with the new EU programming period 2014-2020, it was very important to make a joint effort to start a new process of integrated development in the area, supported also by GASTROCERT, a Project on Gastronomy and Creative Entrepreneurship in Rural Tourism, funded by JPI Cultural Heritage and Global Change - Heritage Plus.

2 Methodology

The following is a methodological approach designed to identify an analytical-deliberative type of participation, the result of the combination of the functionalist and the deliberative approach. The first of these approaches has as its objective the integration of knowledge and values through the inclusion of heterogeneous actors, who reinforce the legitimacy and effectiveness of the decision-making process. The second objective is to identify a possible mutual agreement between all the parties potentially involved. The latter is a more discursive approach that adds more legitimacy to the entire process [11]. This design aims to simplify the consultation process for the identification of the best strategy adapted to the area, "built" on the basis of local knowledge and shared by all the members of the partnership.

In this decisional context, multicriteria analysis becomes a useful tool in supporting decision making, facilitating the convergence of the actors of the partnership toward a common development strategy [12]. In particular, in this case, a decision-aid model based on the Analytic Hierarchy Process (AHP) and a convergence process was used to quantify individual and collective preferences in relation to the prioritization in the choice of thematic areas on which to focus the strategy of the LAP of the LAG "Terre Locride". In the planning of the LAP, therefore, a modified methodology already used in other experiences of integrated development was followed [10, 13, 14]. This methodological approach can be summarized in three phases: territorial analysis; involvement of the local community; and concertation. The methodological steps are shown schematically in Fig. 1, where the actors appear in the first column, and in the next two columns, respectively, the activities and the results related to each phase. As shown in the figure, the result of an intermediate phase becomes an input for the next phase.

The analysis involved a general study of the environmental, historical, and cultural aspects related to the territory and a more detailed socioeconomic analysis. The results of the analysis were synthesized in a SWOT analysis. Along with the activities of territorial analysis, there were substantial efforts to engage the local population by inviting them to several public meetings in order to communicate the willingness to build a rural development plan, and to motivate and involve the participation of local stakeholders by discovering their main requirements and needs. A conflict escalation occurred in the animation phase, during a territorial assembly in which most of the private and public actors of the area participated. The decision-making problem was related to the choice of three main thematic areas of development among those

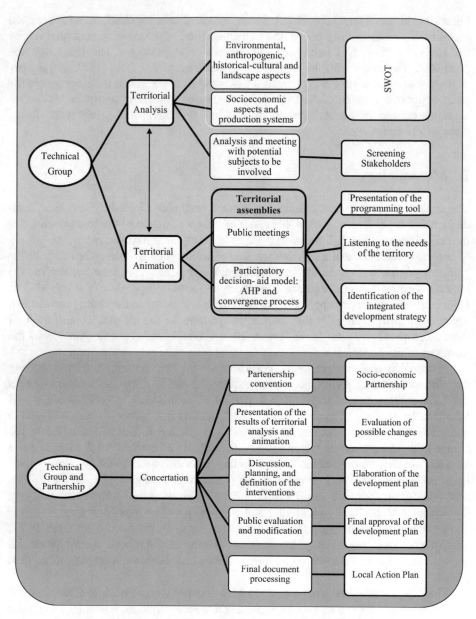

Fig. 1. Integrated planning methodology for the Local Action Plan.

proposed by the regional administration. During this conflict, it was proposed to the assembly to deal with the decisional problem by the adoption and adaptation of a decision-aid model used successfully in other integrated plans (Fig. 2).

The model is based on a convergence process that uses the AHP method [15] in order to elicitate the preferences of:

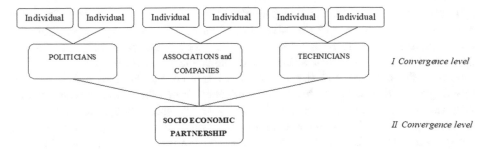

Fig. 2. The convergence process.

- the different stakeholders on *an individual basis*, by gathering the preferences of each interviewed member;
- *the three categories of politicians, technicians, and associations* of the Partnership, through the average of the individual data (first convergence level);
- *the partnership as a whole* (second level of convergence). This level of convergence is realized by compiling a further questionnaire, in which each interviewed actor is asked to express his/her own opinion on the weight that each category should assume in the final group decision. This last questionnaire provides the weights of each category whereby the preferences of each category, resulting from the first level of convergence, are gathered together to provide one shared solution to be presented and discussed in the assembly for final modifications.

There follows a concertation phase that involved the institutional, territorial and thematic meetings which led to the definition, according to the main strategies, of the Actions and Interventions of the LAP to be presented to the Partnership for final approval.

3 Results

The planning methodology was applied for the definition of the Local Action Plan of the "Terre Locridee" LAG in Calabria, South Italy, an area situated on the Ionic side of the Province of Reggio Calabria. The area includes 36 municipalities, covering an area of 1164 km² with a resident population of 120,859 inhabitants in 2015. The average population of the area is equal to 3,357 inhabitants per municipality, but 11 municipalities register less than 1,000 habitants. Emerging from the analysis is a high old-age dependency ratio, and the unemployment rate of young people is almost 40%. The labor market highlights the importance of the primary and tertiary sector which employs, respectively, 29.7% and 29.4% of the workforce. The SWOT analysis highlights the phenomenon of depopulation and territorial abandonment, especially in mountainous areas subject to soil erosion and hydrogeological instability. The area is rich in environmental, historical and archaeological resources which still have to be properly valorized in touristic terms. The territorial animation phase was based on 20 public meetings held in several municipalities of the LAG area between November

2015 and September 2016. The phase involving the local population started with a first screening of the subjects to be involved in the partnership, initially by identifying the representatives of the different categories of these local actors potentially interested in the plan. Through a series of bilateral and public meetings, it was possible to communicate the contents and the opportunities offered by the new LAG. Several low-degree conflicts occurred between the members during the public meetings, until a very serious one occurred when it was time to decide which strategy to choose from those proposed by the administrative Region. Only three thematic areas could be considered in the LAP, and the conflict arose on the issue of how the development strategy of the LAP should be decided, and by whom. The conflict escalation was managed by proposing to the assembly a period of time in which to gather information about the strategies to be adopted directly from the stakeholders in the following public meetings, by means of a decision-aid model to be applied in order to facilitate the definition of a common strategy of development in a participative way. During the next territorial meetings, a specifically targeted questionnaire was administered to 41 actors, and the data gathered were processed through the AHP. The results of three interviews were excluded because their low degree of consistency. Table 2 shows the results of the process that has enabled a first identification of the thematic areas considered to be more strategic for the construction of the plan. The first three columns show the aggregated results of the three categories considered and represent the first level of the process of convergence. There then followed the second phase of the process of convergence that, through the weights assigned to each category by the stakeholders interviewed, represented the choice of the partnership as a whole. In detail, the results highlight that "Legality and social promotion", with 13.39%, is the most important one. This was followed by "Care and protection of landscape, land use and biodiversity" and "Valorization of cultural heritage and artistic heritage" with 11.72% and 10.61%, respectively. It should be noted that the gap between the third thematic area and the next two is quite small. However, for "Valorization and management of environmental and natural resources" this gap is only 0.17 percentage points (p.p.) while for "Access to essential public services" it is 0.27 p.p. From the same table, it can be seen that the highest weight (38.88%) was given to the category of technicians. These results were communicated to the assembly, with advice on some possible integrations that could be important for the local development plan. In particular, during the assembly, the members of the University who led the decision-aid model proposed to integrate the agro-food and artisanal production (production chains) as one of the three main thematic areas of the development strategy. The final results show that the main drivers chosen for the development of the territory combine interventions related to the fields of agro-food and artisan products, the social context, and the historical and environmental landscape (Table 1).

The engagement with the local population continued, leading to the constitution of a wide socioeconomic partnership of 104 members, among whom were the municipalities of the area, the "Mediterranea" University of Reggio Calabria, and the Water Users' Associations of the Province of Reggio Calabria. The Partnership is characterized by a wide variety of members: agricultural, agro-food and artisanal entrepreneurs and their associations; professional organizations; trade unions; regional and local organizations; and the voluntary and non-profit sector. The activity of concertation went on with

Table 1. Sample of local stakeholders interviewed

Categories surveyed	Total respondents	Total sample
Politicians	16	15
Associations & Companies	19	18
Technicians	6	5
Total	**41**	**38**

Source: Our elaboration

institutional, territorial and thematic meetings attended by the political, economic, and social representatives of the area involved. The final stage involved designing, according to the main strategies, the actions and interventions that are part of the LAP, which are outlined in Table 3, where the three main thematic areas are represented in the first three strategic goals. In the final stage the draft plan was discussed and approved by the assembly.

4 Discussion and Conclusion

The gap between academicians, practitioners and entrepreneurs is today particularly felt in management studies [1, 2, 16]. The cause of this gap can be found, on the one hand, in the lack of relationships, i.e. the inability to build proper networks between the academic world and professionals, and, on the other hand, in communication problems determined by a too technical and one-way academic language [17]. Through collaborative research, the action research, which fully embraces the concept of a bottom-up approach already tried out within integrated territorial development policies, goes beyond the consolidated concept of the third mission [2].

In the framework of the fourth mission, the university contributes to sustainable development not only through technological transfer but also by identifying and implementing sustainability-advancing knowledge in collaboration with the local society. This is because "sustainability issues need to be dealt within each context, because challenges and solutions are place-bound and involve different stakeholders able to understand and address them" [4]. Several European research calls, including, for example, those by Horizon 2020, Heritage Plus, and, in Italy, the National Operative Program, reaffirm the need to strengthen the link between research and enterprises. In particular, the Heritage Plus call was designed to generate new "research-based knowledge to promote the sustainable use and promotion of the cultural heritage", which advocates the birth and strengthening of collaborative networks between University, enterprises, associations and civil society. In this context, Universities and researchers are invited to develop concrete, participatory, shared and specific solutions to address the problems faced by society and the research output, part of a wider development process, results "from an involvement with members of an organization or individuals or groups of professionals over a matter that is of genuine concern to them" [18].

Table 2. Results of the decision-aid model

Thematic areas	Politicians	Technicians	Associations and Companies	Average	Weighted average
Legality and social promotion	14.12%	13.46%	12.66%	13.41%	13.39%
Care and protection of landscape, land use and biodiversity	12.13%	12.98%	9.85%	11.65%	11.72%
Valorization of cultural heritage and artistic patrimony	11.10%	10.90%	9.83%	10.61%	10.61%
Valorization and management of environmental and natural resources	10.08%	10.93%	10.19%	10.40%	10.44%
Access to essential public services	10.96%	8.30%	12.22%	10.50%	10.34%
Social inclusion	7.62%	12.68%	7.83%	9.38%	9.66%
Production chains	8.11%	9.92%	9.39%	9.14%	9.23%
Renewable energy development	8.76%	6.97%	11.05%	8.93%	8.81%
Sustainable tourism	9.32%	5.79%	9.22%	8.11%	7.91%
Intelligent networks and community	7.79%	8.08%	7.76%	7.88%	7.89%

Weights (w)	Politicians	Technicians	Associations and Companies	Average
Politicians	42.02%	27.10%	16.40%	28.51%
Technicians	38.67%	35.97%	42.00%	38.88%
Associations and Companies	19.31%	36.92%	41.61%	32.61%

Source: Our elaboration

The research question of this study was born in this context and, at the basis of the Local Action Plan, is placed the principle of consultation and cooperation between public and private actors, in order to identify common paths of development which can activate the latent potential of the territories and leverage the real knowledge of local actors on the issues and on the choice of the intervention areas. The strategic goals of development for the area were defined through a participatory process that allowed a wide variety of perspectives to be taken into account. The socioeconomic partnership, responsible of the entire development process is composed of the representatives of local public and private interests. The decision-aid model facilitated conflict management, and the phased convergence process revealed itself once again to be well-suited for effective participation and conflict resolution. As previously pointed out, "the effectiveness of the process is not only understood as the degree of the attainment of the prefixed objectives, but also as the attainment of an adequate integration level in the governance system" [14], where the transparency of the decisional process directly affects the level of trust in the capacity of the Partnership and the technicians, and also

Table 3. The Local Action Plan of the LAG "Terre Locridee"

Strategic goal	Thematic areas	Actions	Interventions	Financial resources (public quota)
SG1. Social context	Legality and social promotion in areas of high social exclusion	*1.1. Promotion of the expression of local creativity and placing the centrality of young people and women in growth policies*	1.1.1. *LAB 1.* The Rural Center of Locride. Workshop of participation and sharing of development choices	200,000
			1.1.2. The factory of ideas. Innovative solutions for the socio-economic growth of rural areas	150,000
			1.1.3.The common good. Practices and experiments of shared management of online heritage	150,000
			1.1.4. Coordinated information and training actions, with the aim of raising awareness of issues, promoting initiatives, involving young people and women	90,000
			1.1.5. Memory and identity. Projects oriented to the re-appropriation of places by the inhabitants and to the defense of soil and biodiversity	50,000
		1.2. Hospitality and solidarity. Promoting the establishment of networks and services for integration and inclusion	1.2.1. "Social agricultural" companies. Implementation and strengthening of educational and social farms	650,000
			1.2.2. Rural hospitality. Widespread hospitality for rural	250,000

(*continued*)

Table 3. (*continued*)

Strategic goal	Thematic areas	Actions	Interventions	Financial resources (public quota)
			tourism and network with farms	
SG2. Agri-food and artisanal productions	Development and innovation of local supply chains and production systems (agro-food, artisanal and manufacturing)	*2.1. Promotion of the recovery of local tradition productions*	2.1.1. Cultures of tradition. Creation and development of micro-chains and business networks of typical local products	900,000
			2.1.2. Support for the breeding of animals linked to the tradition of places, in the logic of the micro-chain	150,000
			2.1.3. Support for the creation of product processing plants, in line with typical local processes and supporting process innovation	400,000
		2.2. Promotion of a common and integrated territorial marketing action	2.2.1. *LAB 2.* Territorial Marketing Laboratory. Priority to the promotion and sale in national areas with a strong presence of Calabria. Coordination with the other Gal of Calabria for actions to promote highly typical products	250,000
			2.2.2. "Terre della Locride". Territorial brand of Locride products	54,741

Table 3. (*continued*)

Strategic goal	Thematic areas	Actions	Interventions	Financial resources (public quota)
		2.3. Knowledge and marketing promotion	2.3.1. *LAB 3.* Markets of the Lag "Terre della Locride". Solutions for the marketing and consumer education of local products. Coordination with other Lags of Calabria for the promotion and sale of highly typical products	200,000
			2.3.2. Network of farmers' markets	100,000
SG3. The historical and environmental landscape	Care and protection of the landscape, land use and animal and plant biodiversity	*3.1. Re-appropriation of places and awareness of values*	3.1.1. *LAB 4.* "A road to the Locride". Transhumance and paths Network of paths for the use and the contrast to the abandonment of the territory	200,000
		3.2. Restoration and care of the landscape and the environment	3.2.1. Historical landscape. Recovery of historical corporate and public infrastructures	200,000
SG4. Interterritorial and transnational cooperation	Preparation and implementation of LAG cooperation activities	*4.1. Enterprises' competitiveness and environmental protection*	4.1.1. THE GARDEN OF VINEYARDS. Historical cultures, landscape, economy of places	95,000
			4.1.2. GOOD FOOD LOOP. Bergamot and garden of forgotten fruits	63,018

(*continued*)

Table 3. (*continued*)

Strategic goal	Thematic areas	Actions	Interventions	Financial resources (public quota)
		4.2. Strengthening of institutions' capacity and development cooperation	4.2.1. SOCIAL FARM. The social role of agricultural companies	75,000
			4.2.2. RURALSCAPES. Evolution of the rural landscape	60,000
SG5. Strategies for the development of internal areas	PAL sub-program: Integration of the "Internal areas" strategy	*5.1. Promotion and development of assistance services*	5.1.1. LAB 5. Social laboratory for actions aimed at promoting active aging	70,316
		5.2. Care and enhancement of the territory and its resources	5.2.1. Ecosystems. Land maintenance	40,000
			5.2.2. The fruits of the forest. Microchains of non timber products	60,000
SG6. Implementation of the Local Action Plan	Management and animation costs	*6.1. Management and animation*	6.1.1. Management and animation of the LAP	841,371
				5,299,447

Source: Our elaboration

motivates the various partners to participate in the following phase of the implementation of the plan. That will not be easy and calls for a further deepening and continuous reinforcement of the synergies and relationships that started to develop in the planning phase.

Acknowledgements. This study has been supported by GASTROCERT, a Project on Gastronomy and Creative Entrepreneurship in Rural Tourism, funded by JPI-HERITAGE PLUS (2015–2018). The authors thank their collegues of the Technical Team: Luigi Autiello, Rosario Condarcuri, Vincenzo Crea, Carmela Franzese, Guido Mignolli, Monica Mollo, Carlo Murdolo, Nicola Ritorto, Sorana Russo, Stefano Zirilli, Alessandro Zito.

References

1. Trencher, G.P., Yarime, M., Kharrazi, A.: Co-creating sustainability: cross-sector university collaborations for driving sustainable urban transformations. J. Clean. Prod. **50**, 40–55 (2013)
2. Trencher, G.P., et al.: University partnerships for co-designing and co-producing urban sustainability. Glob. Environ. Change **28**, 153–165 (2014)
3. Trencher, G.P., et al.: Implementing sustainability co-creation between universities and society: a typology-based understanding. Sustainability **9**(594), 1–28 (2017)
4. Rinaldi, C., Cavicchi, A.: Universities' emerging roles to co-create sustainable innovation paths: some evidences from the Marche Region. Aestimum **69**, 221–224 (2016)
5. Romeo, G., Marcianò, C.: Performance evaluation of rural governance using an integrated AHP-VIKOR methodology. In: Zopounidis, C., Kalogeras, N., Mattas, K., van Dijk, G., Baourakis, G. (eds.) Agricultural Cooperative Management and Policy, pp. 109–134. Springer International Publishing, Switzerland (2014)
6. Romeo, G., Marcianò, C.: Governance assessment of the leader approach in Calabria using an integrated AHP–fuzzy TOPSIS methodology. Adv. Eng. Forum **11**, 566–572 (2014)
7. Marcianò, C., Romeo, G., Cozzupoli, F.: An integrated methodological framework for the definition of local development strategies for fisheries local action groups: an application to the stretto coast FLAG in South Italy. In: Proceedings of the XXII Conference of the European Association of Fisheries Economist, Salerno, Italy (2015)
8. Budzich-Tabor, U.: Area-based local development - a new opportunity for European fisheries areas. In: Urquhart, J., Acott, T., Symes, D., Zhao, M. (eds.) Social Issues in Sustainable Fisheries Management. MARE Publication, Series 9, pp. 183–197. Springer, Dordrecht (2014)
9. Tola, A.: Strategie, metodi e strumenti per lo sviluppo dei territori rurali. Il modello del GAL dell'Ogliastra (Sardegna) per la valorizzazione delle risorse agro-alimentari e ambientali, p. 176. FrancoAngeli, Milano (2010)
10. Calabrò, T., De Luca, A.I., Gulisano, G., Marcianò, C.: The rural governance system in leader plus: the application of an integrated planning methodology in Calabria (South Italy). New Medit **3**, 38–46 (2005)
11. Stöhr, C., Chabay, I.: Science and participation in governance of the Baltic Sea fisheries. Environ. Policy Gov. **20**, 350–363 (2010)
12. De Montis, A., Nijkamp, P.: Tourism and the political agenda: towards and integrated web-based multicriteria framework for conflict resolution. In: Giaoutzi, M., Nijkamp, P. (eds.) Tourism and Regional Development, pp. 177–200. Ashgate Publishing, England (2006)
13. Marcianò, C. (ed.) Participatory Rural Development Experiences in Calabria, Interreg IVC-Robinwood Plus, Mediterranea University of Reggio Calabria 92 (2012)
14. Marcianò, C., Romeo, G.: Integrated local development in coastal areas: the case of the "Stretto" coast FLAG in Southern Italy. Procedia Soc. Behav. Sci. **223**, 379–385 (2016)
15. Saaty, T.L.: Mathematical Methods of Operations Research. Dover Pub, New York (1988)
16. Cavicchi, A., Santini, C., Bailetti, L.: Mind the "academician-practitioner" gap: an experience based model in the food and beverage sector. Qual. Market Res. Int. J. **17**, 319–335 (2014)
17. Santini, C., et al.: Reducing the distance between thinkers and doers in the entrepreneurial discovery process: an exploratory study. J. Bus. Res. **69**, 1840–1844 (2016)
18. Eden, C., Huxham, C.: Action research for management research. Br. J. Manag. **7**, 75–86 (1996)

Actors, Roles and Interactions in Agricultural Innovation Networks: The Case of the Portuguese Cluster of Small Fruits

Lívia Madureira[1,2(✉)], Artur Cristóvão[1,2], Dora Ferreira[1], and Timothy Koehnen[1,2]

[1] Centro de Estudos Transdisciplinares para o Desenvolvimento, 5000-801 Vila Real, Portugal
[2] Universidade de Trás-os-Montes e Alto Douro, 5000-801 Vila Real, Portugal
`lmadurei@utad.pt`

Abstract. The idea underpinning EIP-AGRI for linking producers and users of knowledge and promoting their interaction around problem-solving is well grounded on the evidence provided by the 'innovation systems' and related literature. Evidence gaps that matter to the implementation of the EIP-AGRI activities comprise the lack of knowledge regarding the best-fit network configuration for different farming systems and farming styles, and the nature and effectiveness of a facilitator function and role to bridge communication between researchers and farmers. This paper contributes with empirical evidence regarding the networks configuration best-fit for different farming system and farming styles, and provide insights on the facilitator relevance and its desirable profile, built on the study of a particular network: the Portuguese Cluster of small fruits (CSF). The small fruit sector is a novel sector in Portugal that has attracted in recent years a large number of new investors, in particular newly-established small-scale inexperienced producers. The insights provided by the CSF analysis emphasises that agglomeration economies based networks, which are very important in some agricultural sectors (e.g. fruit, wine) and in countries or regions where small-scale farms are significant, can in fact be the ground for knowledge and innovation networks in the sense wanted by the EIP-AGRI, since inclusiveness and facilitation functions are accounted for properly.

Keywords: Agriculture · Knowledge and innovation networks
Agricultural knowledge and innovation systems (AKIS)
Clusters · EIP-AGRI

1 Introduction

Innovation has been placed at the heart of the Europe 2020 strategy for growth and jobs [1]. The European Innovation Partnerships (EIPs) are an innovative tool launched recently by the European Union (EU) to tackle major societal challenges that look for solutions by building on the networking and interaction between actors from the research chain and the innovation players. The EIP on agricultural sustainability and

© Springer International Publishing AG, part of Springer Nature 2019
F. Calabrò et al. (Eds.): ISHT 2018, SIST 101, pp. 42–49, 2019.
https://doi.org/10.1007/978-3-319-92102-0_5

productivity (EIP-AGRI) is one of the five EIP launched by the EU. The EIP-AGRI [2] relies on the innovation systems theoretical approach [3–6] that envisages innovation as a part and the result of interactive learning processes involving multiple actors. Within this approach, multi-actors knowledge networks are the ground for innovation processes taking place at the territorial level. EIP-AGRI activities focus on enhancing the network of producers and users of knowledge, that includes farmers, researchers, advisors, business and other individual and collective actors whose interaction generates 'new insights and ideas, and mobilise existing tacit knowledge into focused solutions' [2]. Hall [7] endorses the importance of incremental innovation focused on problem solving or the constant minor adjustments and improvements that farmers make to be successful.

The link between networking behaviour of firms from all economic sectors, including the primary sector, and their innovative performance has been established by the literature [8–10]. In addition, the research on 'innovation systems' highlights the importance of the partner diversity to the innovative capacity of the networks. Research on agricultural innovation systems emphasizes as well the critical role of actor's heterogeneity [11, 12], and defines innovation as an outcome of open-ended interactions among various actors combining knowledge from many different sources [8, 12–14].

Therefore, the idea underpinning EIP-AGRI for linking producers and users of knowledge and promoting their interaction around problem-solving is well grounded on the evidence provided by the 'innovation systems' and related literature. Evidence gaps that matter to the implementation of the EIP-AGRI activities comprise the lack of knowledge regarding the best-fit network configuration for different farming systems and farming styles, and the nature and effectiveness of a facilitator function and role to bridge communication between researchers and farmers.

The FP7 EU project PRO-AKIS[1] encompassed among their goals exploring and identifying the possibilities, conditions and requirements of agricultural and rural innovation networks that constitute examples for the EIP-AGRI. A set of four case studies for in-depth analysis was selected across different European countries. This paper focuses on the Portuguese Cluster of small fruits (CSF), one of the networks selected by the PRO-AKIS project, given that it offered useful insights on how to design and to develop knowledge and innovation networks able to cope with inclusiveness challenges.

2 The Case Study

The introduction and expansion of the small fruit (berries) sector in Portugal is quite recent. It was launched in the 90's, while its overwhelming expansion occurred in recent years: the sector grew from a few hectares in 2009 to 2,656 in 2016. Its recent explosion is largely due to the investment by new farmers, supported by EU funding to help young farmers settle into their production chain. Unemployment and the lack of

---

[1] PRO-AKIS (Prospects for Farmers' Support: Advisory Services in European AKIS. Additional information available at http://www.proakis.eu/).

opportunities in other areas that resulted from the economic crisis in the Southern European countries attracted hundreds of young (under forty years old) farmers, often searching for a new life-style as well. They are mostly highly educated individuals, but with little or no experience in the farming sector.

The Portuguese Cluster of small fruits (CSF) was launched in 2013, and it is a horizontal nationwide network; its coordination structure comprises the main facilitators of knowledge sharing and diffusion processes. It is composed of both experienced and inexperienced producers and a diversified set of other actors, such as: private agricultural advice companies, independent consultants, several FBOs (cooperatives, farmers' groups and associations), up and downstream industry firms, among others. It is a multipurpose network focused on the berry sector organization at national level. Its major concern is to ensure the sector's competitiveness and sustainability. Knowledge and innovation are key factors to achieve these goals, given the huge knowledge and information gaps in the sector caused by its novelty and lack of tradition along with the entry into the sector of hundreds of small and inexperienced producers.

Therefore, the CSF is mainly a knowledge and innovation multi-actor network with a singular configuration: it tries to benefit from the know-how and expertise of experienced pioneer producers, while transferring it to the less or no-experience producers in the central-northern sub-region. The former producers, which are mostly located in the Southern sub-region of the country, have already well-established informal and transnational knowledge exchange networks. Hence, the study of the CSF offers an opportunity to understand the role of clusters as a tool for clustering knowledge generation and diffusion beyond a localised level [15]. In addition, this network presents the opportunity to understand how extra-cluster knowledge exchange and learning can determine the success of intra-cluster flows [16].

By being an export commodity (its domestic consumption is recent and still residual), berry production needed to be concentrated to attain export scale. This situation has originated as a dynamics for agglomeration economy, although it also entailed the central-northern sub-region an increasingly large number of small and fragmented organizations, such as producer groups, farm-based small firms and other business models in general that are also offering technical advice to their members and/or selling advisory services to other producers. Hence, this network offers a good example for EPI-AGRI, given that it has been created to overcome the sector challenges regarding its productivity and sustainability, which underpin its competitiveness within the global markets, and to address simultaneously the risks faced by the sector competitiveness due to the massive entrance of the small-scale and inexperienced producers.

Regarding the information, knowledge and skill activities the CFS includes the implementation of training actions, best practices and innovation dissemination through seminars and workshops, enhancing of growers experience opportunity exchange with best orchard field days and an annual blueberry fair. Among the main outcomes of the CSF to be highlighted is the support to inexperienced growers and the opportunity to create and to aggregate knowledge respecting the local-specific aspects of growing berries in different regions of Portugal. Given the novelty of this crop in Portugal, this knowledge is very scarce and it is crucial to increase the productivity and sustainability of the orchards which are sensitive to these conditions.

3 Methods and Data Collection

The methodological approach adopted for data collection encompassed three phases. In the first phase, aimed at understanding and mapping the actors of the Portuguese small fruit sector, an exploratory study was carried out that included different steps from the collection and systematization of the latest news and events taking place in the sector of small fruits, to the direct observation and participation of the researching team in some of those events, such as: meetings of producer groups, the Blueberry National Fair, and sectoral workshops related to production and harvesting techniques.

During the second phase, the team of researchers participated in meetings involving the facilitators of the CSF. These meetings allowed for the collecting of information on the CSF foundation, previous informal networks and ties, understanding the facilitator's role and to identify and map all the actors involved in the network.

In the third phase, an exploratory-descriptive approach was chosen to gather information about the structure, content and dynamics of the network. For this phase, two different interview guides were constructed, and applied through questionnaires: one for the participants and other for the facilitators. This script was also applied using a matrix to record relationships and interactions between actors and the flow of knowledge/information, whether in the process of creating, sharing or storing knowledge. These interviews took place on site, lasting on average 60 min and were set up by prior contact via email and telephone so that respondents were aware of the objectives of the study and the type of information to be collected. The sample selection distinguished the actors involved, according to the criterion 'role in the network' and considered the following groups: 'facilitators', 'suppliers of knowledge' and 'knowledge searchers'. The interviewed sample included 3 interviews with 'facilitators', 9 interviews with 'suppliers of knowledge', and 24 interviews with small scale farmers ('knowledge demanders'). The data matrix to record the interactions between actors for social network analysis (SNA) was collected through personal face-to-face interviews, by previously identifying the actors belonging to the clusters and adding the external actors in the interviews. The large dimension of the network encompassed by the cluster where members were only partially listed was a limitation in the delimitation of the network.

4 Results

The CSF players can be grouped into four major categories. There is a core group with four organizations that coordinates the network and includes their key facilitators. A second group of members is a larger and more diverse one, encompassing independent producers, producer groups, small and medium firms of producers and others, cooperatives, farmer associations, private advisors, project developers and up and downstream firms, among others, which are direct or indirectly responsible for the knowledge, expertise and information supply. This group also includes regional agencies of the Ministry of Agriculture and, to a lesser extent, researchers and universities. The experienced pioneer producers (independent or members of profit and non-profit producer groups) stand out within this supply-side group. Our estimates suggest them to be around 15 people/organizations. The latter are vital to CSF insofar as they are the main knowledge and expertise suppliers, while being simultaneously innovation-led producers, ergo fundamental to encourage innovation processes within the network. The third category of actors comprises the inexperienced producers. This is the largest group, with hundreds of producers, although not all of them participate in CSF activities. Among this group there are some active knowledge searchers, whereas the majority are apparently passive recipients of information. The local governments and local development associations of the Central-northern region constitute the fourth group of actors in the SFC network. They act mainly as enablers and supporters, promoting the settlement of new producers, and acting as lobbyists in favour of the sector.

The analysis of the sociogram depicted by Fig. 1 illustrates four important features defining the configuration of the CSF network. Firstly, the dichotomous relationship between experienced and inexperienced producers, in which the later are clearly the active searchers for knowledge, expertise and information, and the former play the role of active supply-side exchangers. Secondly, the polarization of the network around 2 central actors, the sectoral association coordinating the CSF and a medium independent producer located in the Southern sub-region, reflecting the geographical fragmentation of the network. Thirdly, the importance of the sectoral association, producer groups and the regional public advisory agency, all located in the Central-northern sub-region as facilitators, intermediating the interaction between inexperienced producers and the

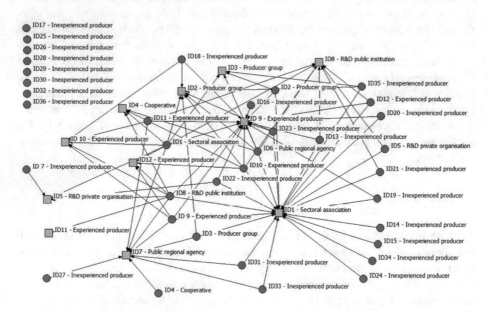

Fig. 1. Interactions among the CSF network interviewed actors (Source: [18])

researchers and the experienced producers, and as well overcoming the geographical fragmentation. Fourth, although the CSF shows to be an inclusive network by bringing on board the small inexperienced and mostly new-established producers, the sociogram suggests there are probably a relevant number of isolated inexperienced producers. These producers' isolation is probably due to lack of time (in view of their status as part-time farmers) and/or their passive attitude towards seeking knowledge and expertise.

The network centralisation index is 8.55%, suggesting a low centrality and an asymmetric configuration, however when the actors centrality is measured (Bonacich power) the sectoral association, the public R&D agency, and the producer groups show to have a central position in relation to the other actors (0.875), closely followed by the experienced producers (0.750). The overall density of the network is around 50%, while the network is asymmetric due to the presence of a significant number of isolated actors (inexperienced growers). Betweenness actor's measure[2] also highlights the central role of the sectoral association and the producer groups, but other institutional actors closely located to the inexperienced growers, the regional public agency and the private R&D agency show to be important as bridging actors. This result underlines the importance of proximity in the case of facilitator bridging role.

The network facilitators have a significant role in mobilising experienced producers, regarding whom inter-personal liaisons appear to be essential to ensure their participation. In fact, producers are motivated to participate in the network by their concern with safeguarding their (good) reputation as Portuguese berries enjoy many export market destinations. The inexperienced farmers are mobilised to participate by the intermediate facilitators, whose job is made easier by the former's acute needs for knowledge and skills to plant, cultivate and harvest the berries.

The learning activities promoted by the CSF are mainly directed to the inexperienced and newly-established producers that include thematic workshops, orchard days, fairs and technical visits to 'best orchards', involving experienced pioneer producers as trainers, that are very much valued by the former group of producers. This confirms that learning from/with the 'peers' who communicate knowledge and information on a common ground, built on empirical experience, facilitates the reduction of the cognitive distances [19], and enhances innovative behaviour by mimicking the innovation-led producers from the Southern sub-region.

5   Conclusions

The study of the Portuguese CSF offers interesting insights on how to design and develop inclusive knowledge and innovative agricultural networks. It contributes to research gaps respecting the configuration and the facilitation role on this type of network to enhance the farmer's innovative behaviour. The lessons learned are also relevant at the policy recommendations level, in particular respecting the operationalisation of the EIP-AGRI regional and local actions to be funded by the national RDP in cases, such as Portugal, where the knowledge and advisory infrastructure is poor and fragmented [18, 20] and where there might emerge the temptation to use

---

[2] To fully understand the social network analysis reported we suggest readers to see [17].

temporary funding of networks to fill structural holes. It appears to be also useful in contexts where the farmers knowledge demand is high, for instance due to the presence of a substantial number of new-entrants in the sector and/or the investment in novel sectors, a situation that will tend to happen across Europe due to the dynamic consumers preferences, markets and climate change.

The lessons learned can be summarised with the following:

- The CSF illustrates that the facilitator function is important and that it might assume different configurations such as: (a) the bridging communication facilitator, analytical brokering and synthetically knowledge delivered by researchers; and, (b) the facilitators that enable the bridge between different actors.
- Pioneers, best farmers or innovation-led farmers appear to be good bridging communication facilitators, in particular when networks are uneven in respect to the knowledge needs and demands, such as the case of networks addressing the needs and demands of newly-established, novel sectors, small-scale and other farmer populations or groups with limitations to access and to mobilise directly the scientific and technical knowledge.
- The inclusiveness can be a critical feature of knowledge and innovation networks focused on productivity and sustainability gains, such as the case of the OG designed under the EIP-AGRI framework, because when there is a segregation between farmers with more access to scientific and technical knowledge and the ones that have less ability to do this, the goals of EIP-AGRI will be attained only in a limited level.

Acknowledgments. We acknowledge funding from the European Commission, 7th Framework Programme through the project PRO AKIS (Prospects for Farmers' Support: Advisory Services in European AKIS, http://www.proakis.eu/). This work does not necessarily reflect the view of the European Union and in no way anticipates the Commission's future policy in this area.

References

1. Commission of the European Communities (CEC), State of the Innovation Union 2012 - Accelerating change. Commission of the European Communities, Brussels, COM (2013), 149 final (2013)
2. EU SCAR, Agricultural knowledge and innovation systems towards 2020 - an orientation paper on linking innovation and research, Brussels (2013)
3. Asheim, B.: Interactive learning and localised knowledge in globalising learning economies. Geo J. **49**, 345–352 (1999)
4. Audretsch, D.: Agglomeration and the location of innovative activity. Oxf. Rev. Econ. Policy **14**, 18–29 (1998)
5. Cooke, P., Uranga, M., Etxebarria, G.: Regional innovation systems: institutional and organisational dimension. J. Res. Pol. **26**, 475–491 (1997)
6. Lundvall, B.: National Innovation Systems: A Comparative Analysis. Oxford University Press, Oxford (1992)

7. Hall, A.: Challenges to strengthening agricultural innovation systems: where do we go from here? In: Scoones, I., Thompson, J. (eds.) Farmer First Revisited: Innovation for Agricultural Research and Development. Practical Action Publishing, Rugby (2009)
8. Klerkx, L., Leeuwis, C.: Matching demand and supply in the agricultural knowledge infrastructure: experiences with innovation intermediaries. Food Policy 33, 260–276 (2008)
9. Pittaway, L., Robertson, M., Munir, K., Denyer, D., Neely, A.: Networking and innovation: a systematic review of the evidence. Int. J. Manag. Rev. 5–6(3–4), 137–168 (2004)
10. Ritter, T., Gemuenden, H.: Interorganizational relationships and networks: an overview. J. Bus. Res. 56, 691–697 (2003)
11. Oreszczyn, S., Lane, A., Carr, S.: The role of networks of practice and webs of influencers on farmers' engagement with and learning about agricultural innovations. J. Rural Stud. 26, 404–417 (2010)
12. Wood, B.A., Blair, H.T., Gray, D.I., Kemp, P.D., Kenyon, P.R., Morris, S.T., Sewell, A.M.: Agricultural science in the wild: a social network analysis of farmer knowledge. PLoS ONE 9(8), e105203 (2014)
13. Conroy, C.: The nature of agricultural innovation. In: Snapp, S., Pound, B. (eds.) Agricultural Systems: Agroecology and Rural Innovation for Development. Academic Press, San Diego (2008)
14. Klerkx, L., Aarts, N., Leeuwis, C.: Adaptive management in agricultural innovation systems: the interactions between innovation networks and their environment. Agr. Sys. 103, 390–400 (2010)
15. Porter, M.E.: Location, competition, and economic development: local clusters in a global economy. Econ. Dev. Q. 14, 15–34 (2000)
16. Giuliani, E., Bell, M.: The micro-determinants of meso-level learning and innovation: evidence from a Chilean wine cluster. Res. Policy 34, 47–68 (2005)
17. Hanneman, R.A., Riddle, M.: Introduction to Social Network Methods. University of California, Riverside (2005)
18. Madureira, L., Ferreira, D., Pires, M.: Designing, implementing and maintaining (rural) innovation networks to enhance farmers' ability to innovate in cooperation with other rural actors. The berry networks in Portugal. Report for AKIS on the ground: focusing knowledge flow systems (WP4) of the PRO AKIS project, December 2014 (2014)
19. Nooteboom, B.: Institutions and forms of coordination in innovation systems. Org. Stud. 21, 915–939 (2000)
20. Knierim, A., Boenning, K., Caggiano, M., Cristóvão, A., Dirimanova, V., Koehnen, T., Labarthe, P., Prager, K.: The AKIS concept and its relevance in selected EU member states. Outlook Agr. 44, 29–36 (2015)

From Binarism to Polarism: On Rural Knowledge Outflows' Role in Fostering Rural-Urban Linkages

Yapeng Ou$^{(\boxtimes)}$ ⒾD and Carmelina Bevilacqua

Mediterranea University of Reggio Calabria, 89100 Reggio Calabria, Italy
yapeng.ou@unirc.it

Abstract. As a worldwide phenomenon, rural-urban divide is the fundamental reason for socioeconomic disparities between rural and urban areas within the same region. To achieve sustainable regional development, it is important to foster rural-urban linkages so as to coordinate rural and urban development. This research is aimed to investigating the reasons for rural-urban divide and suggesting how to address it. To this end, based on literature review and the American context, the research first analyzes the ideology and the political economic dynamics behind rural-urban divide. Then, it looks into how a shift from binarism towards polarism can address rural-urban divide and at the meantime foster rural-urban linkages. Finally, with reference to empirical examples from Boston, the research discusses how rural knowledge outflows in the form of urban farming can contribute to the formation of rural-urban linkages.

Keywords: Binarism · Rural-urban divide · Polarism · Rural-urban linkages
Rural knowledge outflows

1 Introduction

At present, globally there seems to be a predominant urban discourse both at the academic and institutional levels, whereas rural issues are not well coordinated and integrated into urban agendas. At the geopolitical level, the political partiality for urban development has led to a deep-rooted rural-urban divide. Rural-urban divide manifests itself spatially in the form of increasing socioeconomic disparities between the rural and the urban. Rural-urban divide suggests also a "passive interconnection" between the rural and the urban where the rural is the subordinate of the urban. This explains the pervasive chaotic and fragmented urban expansions that have sprawled from the historic centers diachronically into suburbs, semi-rural and rural areas since the postwar period in many European cities [1]. Generally speaking, in contrast to the urban primacy, attitudes of nation states and regions, regardless of their state of development, diverge with regard to the rural. The European Union (EU) and quite many other EU member states' regional policies and local initiatives have created incentives that seek to re-establish sustainable rural economies, successful communities and unique rural space [2].

© Springer International Publishing AG, part of Springer Nature 2019
F. Calabrò et al. (Eds.): ISHT 2018, SIST 101, pp. 50–57, 2019.
https://doi.org/10.1007/978-3-319-92102-0_6

These rural incentives go abreast with urban ones, both aiming at sustainable development. In the US (similar countries like Australia and Canada), by contrast, sustainable urban, not rural, development initiatives are guiding national priorities for change. Consequently, the rural in the US remains "dominated by corporate agriculture, increasing farm size and decreasing work force, and rural decline" (ibid, 2). This research is aimed to investigating the reasons for rural-urban divide and suggesting how to address it. To this end, based on literature review and the American context, the research first analyzes the ideology and the political economic dynamics behind rural-urban divide. Then, it looks into how a shift from binarism towards polarism can address rural-urban divide and at the meantime foster rural-urban linkages. Finally, with reference to empirical examples from Boston, the research discusses how rural knowledge outflows in the form of urban farming can contribute to the formation of rural-urban linkages.

2 Binarism: Rural-Urban Divide as a Problem

As mentioned above, rural America has long been neglected. This causes various socioeconomic and environmental problems in rural America, just to name a few, poverty, food insecurity, environmental pollution, substandard housing and high unemployment [3, 4]. However, the concerns and problems of rural America seldom seem to matter to urban Democrats (Ikerd 2016). As a result, worsening rural economic landscape together with political indifference have paved way for widening rural-urban economic and political divide in the US.

Two reasons account for the rural-urban divide in the US, namely, neo-liberal political economic policies and classic binarist ideology. While the former is superficial and exogenous, the latter is fundamental and endogenous; therefore, it has an overwhelming impact on rural-urban divide. First, political-economically speaking, in the US, both at federal and state levels, neo-liberal economic policies "have favored the market over government intervention" [5] while rigidly complying with the prevailing capitalist value orientation: efficiency and profit maximization. John Ikerd (2016), Professor Emeritus of University of Missouri, refers to the rural America as "economic colonies" of corporations (often multinational), which becomes a victim of economic extraction and exploitation of rural natural and human resources[1]. Besides economic colonization, rural America is unable to resist excessive suburbanization that is a pervading development model in the US. Consequently, rural America suffers the inevitable social and ecological consequences of economic colonization while fails to capture value locally as the economic benefits go to global corporations. This economic subordination makes rural America vulnerable facing major socioeconomic changes such as migration and economic restructuring and enlarges rural-urban disparities. The long-term sustainability of America's natural resource base affects and is affected by its institutions and social organization [6].

---

[1] See John Ikerd, The Real Cause of the Rural-Urban Divide, http://johnikerd.com/2016/08/the-real-cause-of-the-rural-urban-divide/, last accessed 2017/10/22.

Second, ideologically speaking, rural-urban divide in the US and other countries is a manifestation of the classic binarism that dominates theory and practice of regional development. In essence, rural-urban divide reflects a dichotomic thinking which often proves to be "the outcome of an urban-centric, industrialized-economy-geared development model" [7]. Due to rural-urban binarism, the urban and the urban rural prone to be disconnected and segregated academically and institutionally [8]. This is more often than not detrimental both to rural development and urban development, therefore undermines regional development, especially on a sustainable term. Due to urban-centrism, the rural tends to be marginalized in regional development, while the urban is forced to absorb the pressure originated from the rural. Indeed, ignoring the rural can lead to a backlash to the urban, causing deepening rural-urban disparities which tend to become the breeding ground for socioeconomic problems. A typical manifestation is the increasing urban slums following continuous rural migration. Binarism seemingly suggests unpredictability in the US. This is primarily due to a spatial mismatch in the US: while America is a predominantly urban society where eight out of every ten Americans live in urban areas, the future of urban America, however, largely depends on rural-based natural resources, since most of the American land area is non-metropolitan and predominantly rural. For this reason, the urban society and the rural natural resources are mutually interrelated [6].

3 Polarism: Rural-Urban Linkages as a Remedy

The fact that the rural and the urban in the US should have been inalienable demands that there be an ideological shift from rural-urban binarism towards rural-urban polarism that can foster synergizing rural-urban linkages critical to building up rural and urban resiliency. The ideology of rural-urban polarism resonates *yin yang*, a Chinese polarist worldview by which the rural and the urban can be understood as two entities being opposite yet meanwhile complementary to each other. Rural-urban polarism is a promising remedy for addressing rural and urban problems resulting from binarist rural-urban divide. This is primarily because it buttresses rural-urban linkages that are critical to achieving sustainable regional development.

In the first place, rural-urban linkages make it explicit that rural and urban areas coexist as a space-time continuum rather than as discrete geographical territories with boundaries [9]. Within this continuum, what is embedded is relational interdependences and relationality that have been appealed to account for economic action and outcomes; therefore, there needs to be a "relational lens" that coordinates associative, cooperative and collaborative forces [10]. This recognition of rural-urban continuum will stimulate the academia and public institutions to interconnect and aggregate the rural and the urban both in theory and in practice. In so doing, what is most likely to happen is a revolution in regional development schemes. Second, rural-urban linkages help foster a synergy between the rural and the urban. This very synergy can pave the way for 3Cs, namely cooperation, coordination and collaboration that bridge rural development and urban development on the one hand. On the other hand, it can generate co-benefits that are prone to be shared by the two poles given the interconnectedness between them. Already it starts to be recognized that, "Urban areas must embrace their periurban and rural surroundings for their own survival and to make cities work better, and in harmony, with nature's ecological processes [11]." Last but not least, rural-urban linkages are able to mitigate negative externalities caused by rural-urban divide. It is expected that rural-urban linkages help achieve two principal goals: first curtailing economic disparities that undermine regional stability and cohesion, and second reconciling the needs of development and natural preservation as is required by rural-urban sustainability in the long run.

Indeed, rural-urban linkages bolstered by rural-urban polarism convey a crucial message that sustainable regional development is barely achievable if there is no coordinated rural and urban development. This suggests that recognizing rural-urban linkages that mutually strengthen the rural and urban sectors becomes a bad need both in theory and in practice [12–14]. Therefore, rural-urban linkages resulting from rural-urban polarism rather than rural-urban divide stemming from rural-urban binarism are not only what American cities, but all 21st century cities badly need.

4   Rural Knowledge Outflows: A Booster of Rural-Urban Linkages

Today, the future of cities and regions are more than before determined by their competencies and skills to learn and develop themselves in a continuous process to cultivate some specific, differentiated and locally rooted knowledge, and to foster linkages with other knowledge pools [15]. Indeed, local economic development (LED) requires internal and external knowledge flows based on internal and external networks and connections, so that localism due to overly strong internal networks will not decrease the ability of the entire area to acquire external knowledge for innovation and competitiveness [16]. According to van Leeuwen [17], new knowledge and innovations are brought into the local economy by forming connections across a certain system. This means that by fostering internal and external connections, new knowledge and innovations that can drive LED are most likely to emerge.

Based on the above discussions and in response to the conventional, prevailing knowledge flows from the urban to the rural, this research pays a special attention to the opposite ones from the rural to the urban, namely, rural knowledge outflows into the urban. Rural knowledge outflows into the urban is a relevant topic especially at a time when "the cities do not fulfil the promise of a better life that made people leave their rural villages [18]", and there is a bad need for the integration of nature and culture for future sustainable urban and rural landscapes [19]. In this regard, rural knowledge outflows possess the very potentiality to help address urban problems such as poor urban livability, increasing urban risks facing climate change, and deteriorating urban ecosystem.

To illustrate how rural knowledge outflows manifest in the real world, especially how rural values and experiences are transplanted and reproduced in pure urban settings which "act as cultural bridges, potentially linking traditional ecologies, traditional environments, community support networks and informal economies with new global ideas, industrial economies and access to developing urban services" [20], empirical examples from Boston are studied. The major focus is on the spatial representations of rural knowledge outflows in the form of urban farming that brings about emerging dynamic rural-urban spaces amidst the predominantly urban environments[2].

In Boston, rural-urban linkages are recovered mainly through revitalization of agricultural activities in three forms, namely, commercial urban farming, community gardens, and urban agricultural landscaping design. In this linkage recovery process, rural knowledge outflows play a major role, wherein rural agriculture and related traditional values have largely inspired and fueled urban regeneration and social innovation. Within the urban farming ecosystem, the public sector has promoted urban farming by passing in 2013 the new urban agricultural rezoning ordinance Article 89. The private sector has been ever since playing a very active role in translating Article 89 into practice, spurring thriving social entrepreneurship that is economically viable and socially responsible. As for the civil society, community members have partici- pated in the legislative process like that of Article 89 and engaged in urban farming practices in various settings, ranging from community gardens, home gardens to urban farms. They have also helped foster the urban farming ecosystem by becoming con- sumers conscious of their individual impact on environment.

As community gardens such as Fenway Victory Gardens, Berkeley Community Garden, Worcester Street Community Garden and Harvard Community Garden (student-run), urban agricultural landscaping design like Dewey Demonstration Gar- dens, and commercial urban farms like Boston Medical Center Rooftop Farm, Higher Ground Farm atop the Boston Design Center, Fenway Farms in Fenway Park demonstrate, urban farming helps recover rural-urban linkages mainly by:

(1) *playing a significant role in restoring urban ecosystem.* Article 89 legitimizes a great variety of agricultural practices in quite innovative forms, such as aqua- culture, hydroponics and aquaponics. This complex and diverse agricultural pattern can help build up a resilient urban agricultural system. Besides, urban farming provides habitats for beneficial species such as honey bees and con- tributes to rainwater collection by absorbing rain runoff;

(2) *restoring a traditional linkage between agriculture and local communities.* Article 89 encourages the establishment of farmers' markets and farm stands at public and open spaces, which helps create a sense of place and invite community members to socialize and relate to each other. In this way, Article 89 contributes to a considerable extent to the regeneration urban physical and socioeconomic fabrics, which improves the form and the quality of the built environment;

(3) *regenerating urban spaces while generating positive environmental and aesthetic externalities.* Rooftop farming, transforming idling rooftops into thriving urban farms, is especially representative in this regard. It buttresses green urban development by helping cool down buildings which helps reduce energy con- sumption and carbon emissions. Besides, community gardens and urban agri- cultural landscaping design are creating livable, inclusive and attractive public spaces within communities.

---

[2] It is worth noting that, due to space limitation, this research focuses only on the rural traditions and value system embedded in agriculture when referring to the rural knowledge flow towards the urban.

(4) *spurring social entrepreneurship.* Urban farming can promote social entrepreneurship with a community-based and relationship-based approach, stimulate proactive governance, balance efficiency-oriented capitalism and equity-oriented social constructivism, and trigger local value creation by resorting to local, site-specific, collective knowledge and interdisciplinary networks. All in all, by regenerating urban spaces, commercial urban farming is able to bring about both socioeconomic and environmental benefits that are critical to achieving urban sustainability. On the one hand, commercial urban farming can "meet non-production goals of revitalizing an ailing community through job creation, social engagement, education and beautification [21]", therefore driving social entrepreneurship; and on the other hand, it promotes green urban development by being environmentally sustainable.

The strength of rural knowledge outflows in the form of urban farming is, it is largely local knowledge hence highly compatible with adjacent urban context. Rural knowledge outflows in the form of urban farming are also very cost-efficient and beneficial to urban regeneration. Their weakness lies in that their ability to grow as a new economy of scale and create community-based jobs is limited. Public legislative support and social innovation led by the private sector bring considerable opportunities for urban farming to further develop and innovate. In addition, the need to mitigate climate change also brings opportunities to rural knowledge outflows. The threats that rural knowledge outflows face is the deep-rooted "urban-centric, industrialized-economy-geared development model" [7].

5 Conclusion

As a worldwide phenomenon, rural-urban divide is the fundamental reason for socioeconomic disparities between rural and urban areas within the same region. In the US, worsening rural economic landscape together with political indifference have paved way for widening rural-urban economic and political divide. Neo-liberal political economic policies and classic binarist ideology are two major reasons for rural-urban divide in the US. To address socioeconomic and economic problems caused by rural-urban divide, there is a bad need for an ideological shift from rural-urban binarism towards rural-urban polarism that underpins the formation of rural-urban linkages. Rural-urban linkages are critical to sustainable regional development because, first, they will stimulate rural-urban interconnectedness and aggregation both at the academic and institutional levels. Second, rural-urban linkages help foster a rural-urban synergy. Last but not least, rural-urban linkages are able to mitigate the negative externalities caused by rural-urban divide by helping curtail rural-urban economic disparities and reconcile the needs of development and natural preservation as is required by rural-urban sustainability.

Rural knowledge outflows, a major contributor to recovering rural-urban linkages, possess the very potentiality to help address urban problems. In Boston, rural-urban linkages are recovered mainly through urban farming that revitalizes rural agricultural knowledge (and community values embedded in agricultural traditions) in three forms, namely, commercial urban farming, community gardens, and urban agricultural landscaping design. Urban farming as a form of rural knowledge outflows can help recover rural-urban linkages mainly by playing a significant role in restoring urban ecosystem, restoring a traditional linkage between agriculture and local communities, regenerating urban spaces while generating positive environmental and aesthetic externalities, and spurring social entrepreneurship.

It is hoped that this research will evidence the need for a new form of rural-urban interdependence shown by the emergence of urban "rural" spaces characterized by the merging of a rural landscape form with urban economic function [22].

Acknowledgement. This work is part of the MAPS-LED research project, which has received funding from the European Union's Horizon 2020 research and innovation programme under the Marie Skłodowska-Curie grant agreement No. 645651.

References

1. Bencardino, M., Nesticò, A.: Urban sprawl, labor incomes and real estate values. In: Borruso, G. et al. (eds.) ICCSA 2017, LNCS, 10405, pp. 17–30, Springer, Cham (2017)
2. Winchell, D.G., Koster, R.: Introduction: the dynamics of rural change: a multinational approach. In: Winchell, D.G., et al. (eds.) Geographical Perspectives on Sustainable Rural Change, pp. 1–23. Rural Development Institute (Brandon University), Brandon (2010)
3. Cason, K.: Poverty in rural America. In: Moore, R.M., (ed.): The Hidden America: Social Problems in Rural America for the Twenty-first Century, III, pp. 27–41. Susquehanna University Press, Selinsgrove and Associated University Presses, London (2001)
4. Fitchen, J.M.: Endangered Spaces, Enduring Places: Change, Identity, and Survival in Rural America. Westview Press, Boulder (1991)
5. Cocklin, C., Bowler, I., Bryant, C.: Introduction: sustainability and rural systems. In: Bowler, I., Bryant, C.R., Cocklin, C. (eds.) The Sustainability of Rural Systems: Geographical Interpretations, pp. 1–12. Springer, Medford (2013)
6. Brown, D.L., Swanson, L.E.: Rural America enters the New Millennium. In: Brown, D.L., Swanson, L.E. (eds.) Challenges for Rural America in the Twenty-First Century. Penn State Press, University Park, Pennsylvania (2004)
7. Lee, C.C.M.: The Countryside as a City. In: Lee, C.C.M. (ed.) Common Frameworks: Rethinking the Developmental City in China Part 3 – Taiqian: The Countryside as a City, pp. 8–22. Harvard University Graduate School of Design, Boston (2015)
8. Lichter, D.T., Brown, D.L.: Rural America in an urban society: changing spatial and social boundaries. Ann. Rev. Sociol. **37**, 565–592 (2011)
9. Rajagopalan, S.: Rural-urban Dynamics: Perspectives and Experiences. ICFAI University Press, Hyderabad (2006)
10. Kasabov, E.: Rural Cooperation in Europe: In Search of the 'Relational Rurals'. Palgrave MaCmillan, New York (2014)
11. Barbut, M.: Restoring Lost Land. In Sustainable Development Goals: from Promise to Practice, pp. 56–58. UNA-UK (2017)
12. Verdini, G.: The rural fringe in China: existing conflicts and prospective urban-rural synergies. In: Verdini, G., Wang, Y., Zhang, X. (eds.) Urban China's Fringe: Actors, Dimensions and Management Challenges, pp. 1–15. Routledge, London-New York (2016)
13. Lynch, K.: Rural-Urban Interaction in the Developing World. Presbyterian Publishing Corporation, Louisville (2005)

14. Bryant, C.: Urban and rural interactions and rural community renewal. In: Bowler I., Bryant, C.R., Cocklin, C. (eds.) The Sustainability of Rural Systems: Geographical Interpretations, pp. 247–270. Springer, Medford (2013)
15. Cabrita, M.R. et al.: Entrepreneurship capital and regional development: a perspective based on intellectual capital. In: Baptista R., Leitão J. (eds.) Entrepreneurship, Human Capital, and Regional Development: Labor Networks, Knowledge Flows, and Industry Growth, pp. 15–28. Springer, Cham (2015)
16. Fratesi, U., Senn, L.: Regional growth, connections and economic modeling: an introduction. In: Fratesi U., Senn, L. (eds.) Growth and Innovation of Competitive Regions, the Role of Internal and External Connections, pp. 3–28. Springer, Heidelberg (2009)
17. Van Leeuwen, E.S: Urban-Rural Interactions: Towns as Focus Points in Rural Development. Springer, Heidelberg (2010)
18. Van Bueren, E.: Introduction. In: van Bueren, E. et al. (eds.): Sustainable Urban Environments: An Ecosystem Approach, pp. 1–14. Springer, Medford (2012)
19. Van Bohemen, H.: (Eco)System thinking: ecological principles for buildings, roads and industrial and urban areas. In: van Bueren, E., et al. (eds.) Sustainable Urban Environments: An Ecosystem Approach, pp. 15–70. Springer, Medford (2012)
20. Hawken, S.: The urban village and megaproject: linking vernacular urban heritage and human rights-based development in emerging megacities of southeast asia. In: Durbach, A., Lixinski, L. (eds.) Heritage, Culture and Rights: Challenging Legal Discourses, pp. 91–199. Bloomsbury Publishing, Oxford and Portland, Oregon (2017)
21. Pfeiffer, A., Silva, E., Colquhoun, J.: Innovation in urban agricultural practices: responding to diverse production environments. Renew. Agric. Food Syst. **30**(1), 79–91 (2014)
22. Irwin, E.G., et al.: The economics of urban-rural space. Ann. Rev. Resour. Econ. **1**, 435–459 (2009)

Servitisation and Territorial Self Reinforcing Mechanisms: A New Approach to Regional Competitiveness

Domenico Marino[1](✉) [iD] and Raffaele Trapasso[2]

[1] Mediterranea University of Reggio Calabria, 89100 Reggio Calabria, Italy
dmarino@unirc.it
[2] Organisation for Economic Co-operation and Development (OECD),
75775 Paris, France

Abstract. The present paper puts forward a theoretical model to explain the link between servitization and territorial competitiveness based on the situation in Italy. A key assumption of the model is that once the link between manufacturing and KIBS is established within a TES, there is a positive feedback between the increasing productivity (competitiveness) and the link between firms and KIBS, which becomes stronger and stronger triggering a self-reinforcing dynamic. This means that every evolutionary step of the system influences the next and thus the evolution of the entire system, so generating *path dependence*. Such a system has a high number of asymptotic states, and the initial state (time zero), unforeseen shocks, or other kinds of fluctuations, can lead the system into any of the different domains of the asymptotic states [1]. In other words, both the theoretical assumptions and the empirical model outlined in this paper demonstrate that when a functional relationship between manufacturing and services is established (servitization), economic performance is positive or very positive.

Keywords: Servitisation · Regional competitiveness
Self reinforcing mechanisms

1 Introduction

We estimated the contribution of servitisation to the performance of Territorial Economic Systems (TESs) in Italy. We found a high and positive correlation between the specialisation of a given TES in knowledge-intensive business service (KIBS) and productivity, measured as per capita value added. We also found that path dependence strongly influences the capacity of manufacturing firms located in a given TES to diversify their products in order to embed a service component.

The connections between firms and manufacturing and knowledge-intensive business services KIBS) are important in explaining the differences in competitiveness at local level. There is, however, very little literature on the subject. Therefore, the paper by Lafuente et al. [2] is very important as it shows how the growth of employment in a specific territory interacts strongly with servitization and how this functional link can generate virtuous cycles.

© Springer International Publishing AG, part of Springer Nature 2019
F. Calabrò et al. (Eds.): ISHT 2018, SIST 101, pp. 58–67, 2019.
https://doi.org/10.1007/978-3-319-92102-0_7

A comprehensive survey of the literature on servitization can be found in [3]. In this paper, the authors build an interesting taxonomy of the key contributions on servitization, by dividing the different approaches into four quadrants, where the relationship between internal analysis and external analysis is shown on the horizontal axis, and the relationship between mainstream and alternative approaches is shown on the vertical axis.

Quadrant IV, focusing on the KIBS, is of particular interest and is where the present paper is ideally positioned, albeit with a different approach. The present paper puts forward a theoretical model to explain the link between servitization and territorial competitiveness based on the situation in Italy.

A key assumption of the model is that once the link between manufacturing and KIBS is established within a TES, there is a positive feedback between the increasing productivity (competitiveness) and the link between firms and KIBS, which becomes stronger and stronger triggering a self-reinforcing dynamic. This means that every evolutionary step of the system influences the next and thus the evolution of the entire system, so generating *path dependence*. Such a system has a high number of asymptotic states, and the initial state (time zero), unforeseen shocks, or other kinds of fluctuations, can lead the system into any of the different domains of the asymptotic states [1].

In other words, both the theoretical assumptions and the empirical model outlined in this paper demonstrate that when a functional relationship between manufacturing and services is established (servitisation), economic performance is positive or very positive.

However, try to stimulate development in lagging regions by relying on "traditional" policies may not be a good policy choice. Indeed, the paper shows that, due to path dependence and poor response function, in weak TESs, traditional regional policies that focus on compensating the scarce factors of production (for example capital to stimulate production investment) risk creating a Dutch disease effect, because the TES is unable to effectively absorb the additional (traditional) factor of production. Consequently, "compensatory" or "additional" regional policies end up accentuating the differences between regions due to the different response functions and which are manifested in multiple, resilient equilibriums (similar to fitness landscapes). Instead of fostering convergence, the traditional policies create underdevelopment traps (the lowest points in the fitness landscape) from which TESs struggle to escape.

2 The Territorial Economic System

The TES is the physical space in which economic agents interact; the equilibrium properties of this system depend on its structure and, if the space is complex, on the particular attraction basin in which the system stays. The increasing returns, the multiple equilibria, the history dependence can found a meaning in the complex space [4].

By introducing the notion of Territorial Economic System (TES) (Fig. 1) as unit of analysis, it is possible to move towards the increase of interpretative capability when a synthesis among production system, technological knowledge at territorial level and local institution is searched. A TES then consist of interconnection among production system, technological knowledge and *social capabilities*. Each of these dimensions encompasses some factors which determine the performance of the TES (see Table 1).

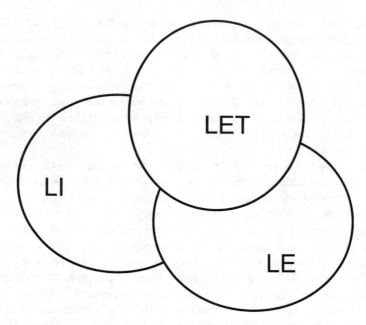

Fig. 1. The elements of the TES.

Table 1. Performance of the TES [5].

Firm Level (FL)	Extra Territorial Level (ETL)	Territorial Level (TL)
Access **to:**	Codified knowledge: technological,	*Intangible elements:*
– Contextual and codified knowledge	organisational and communication codes	Available knowledge and social capabilities
– Local and regional infrastructure networks		
Receptiveness:		*Physical elements:*
– Size		– Available infrastructure
– Organisational structure		– Production system
– Constitutive structure		– Material resources
– Innovative experiences		
– Business networks		

TES is a multidimensional concept that encompasses economic and social dimensions. Whereas the production system has a mainly material connotation, technical knowledge and social capability have a mainly immaterial nature [5]. It is important for a description of the TES to define two dimensions: the proximity and the resiliency. Each territory shows first of all a different degree of proximity which does not necessarily mean contiguity, but can have a functional meaning. There is, in fact, an industrial organization, cultural and temporal proximity.

The resiliency shows the problem of the spatial evolution in the forms of the production, which leads to the question of the historic dynamics and the evolutive trajectories of each TES. It is the capability of the system in the self-organization and in the metabolizing of the change in the external environment. Proximity and resiliency are a way to express the concepts of local interaction and self-organization.

Within TESs economic dynamic takes the form of a *self-reinforcing mechanism*: a positive (or negative) feedback that characterizes the evolution of a dynamic system.

3 TES and Self-reinforcing Mechanism

TES are characterised by a high degree of complexity that select development trajectories stochastically, but then follows a given trajectory based on a self-reinforcing mechanism. The concept of self-reinforcing mechanism can be expressed as a dynamic system, with path dependence and a positive feedback, which tends to a large variety of asymptotic states. Every evolutionary step of the system influences the next one and then the evolution of the entire system, thus generating *path dependence.*

This system has a high number of asymptotic states, and the initial state (Time zero), unpredicted shocks, or other kind of fluctuations, can all conduct the system in any of the different domains of the asymptotic states [1].

Furthermore, the system selects the state in which placing itself. Such dynamics are well known in physics, in chemistry as well as in biology and the final asymptotic state it is called the *emergent structure*. The concept of positive feedback in fact is relatively new for the economic science. The latter generally deals with problems of optimal allocation of scarce/insufficient resources, thus the feedback is usually considered to be negative (decreasing utility and decreasing productivity).

Self-reinforcing mechanism dynamic can be used to assess many different economic problems with different origins: from those related to the international dimension, to those typical of the industrial economy, as well as problems related to regional economics.

Many scholars have assessed multiple equilibria and their inefficiency. Multiple equilibria depend on the existence of increasing returns to scale. If the mechanism of self-reinforcing is not counterbalanced by any opposite force, the output is a local positive feedback. The latter, in turn, amplifies the deviation from some states. Since these states derive from a local positive feedback, they are unstable by definition, so multiple equilibria exist and are efficient.

I the *vector field* related to a given dynamic system is regular and its critical points follow some particular rules, then the existence of other critical points or of stable cycles (also called *attractors*) turns out [6, 7]. The multi-attractors systems have some particular properties that are very useful for our research [6, 7]. Strict path dependence is therefore manifested, and the final state of the system will depend on the particular trail it has been covering during its dynamic evolution from an (instable) equilibrium towards another (instable) equilibrium, and so on. Accordingly, the system's dynamic is a non-ergodic one.

4 Servitization and Self-reinforcing Mechanisms

4.1 New Productive Factors Generating Growth in TES

Description of the evolution of spatialised economies emphasizes the role of a new economic paradigm. The latter is based on a series of different features. For instance, new productive factors seem to have replaced land, work and physical capital. Natural and environmental resources, human resources (skills and human capital) and technology are beginning to get the upper hand following the "technological revolution". Another feature is that co-operation within businesses and between businesses and business systems takes place on a vertical and horizontal scale in which the local dimension and the territorial variables constitute the catalyst for processes of development. In addition, technological expertise and social capabilities [4] are the basic elements capable of explaining the different levels of development seen in different territorial contexts.

Territorial variables, in other words, are decisive factors in explaining development differentials, especially when they are associated with the idea of the market conceived as a social construction. This new market requires rules that will guarantee its smooth running given that access rights, exchange mechanisms and opportunities for distribution of the wealth generated not only do not re-assemble uniformly and autonomously in time and space (Sen, 1984 and 1985), but almost always require outside intervention to achieve the objectives set for development policies. Re-equilibrium policies thus appear necessary to guarantee a more equitable development process. Within the market it is necessary to define collective rules ensuring that positive dynamics (increasing returns) can develop through the interaction of the agents operating in it. Therefore, the territorial dimension and the systemic nature of the production process are fundamental elements to understanding and governing development processes.

4.2 The Impact of Policy Actions Depends on the Response Function that Characterises a Given TES

The collective properties of a given TES in relation to the link existing between productivity growth and information could be represented in terms of response function.

It is possible to create a generalised function - an interpretative model - to describe the propagation mechanism of economic policy in a situation of complexity. The description of the transmission mechanism logically completes the previous observations regarding objectives and instruments. Single economic policy decisions, aimed at achieving the *j-th* objective through the use of the *i-th* instrument, can be represented as an outside stimulus which superimposes itself on interactions between agents.

Agents in this approach are thought of as being spatially distributed and linked to each other by local mutual interactions (of a nearest neighbour type). We use H to indicate the effect of the economic policy. We can thus define an effective *Heff* stimulus which includes both outside stimulus and agent interaction[1]. Obviously, without agent interaction *H* and *Heff* are equal. *Heff* therefore assumes the form:

---

[1] Heff represents the actual output of the implemented policy.

$$Heff = H + \int dr'c(r - r')\delta\gamma(r') \qquad (1)$$

Where $c(r-r')$ is a function of correlation between agents which can constitute an acceptable means of modelling the concept of proximity, $\delta\gamma(r')$ is a variation in the behaviour of agents induced by the policy applied, the integral can be linked to the concept of resilience. This type of behaviour arises in the area of a linear response model for systems with collective properties. The effect of an economic policy on a complex system made up of many agents interacting with each other can therefore be described in this way and modelled by means of the response properties of the system itself. Therefore, in the area of linear response theory we have a cause-effect relationship of the type:

$$E(X) = G(X) \otimes H(X) \qquad (2)$$

where $E(X)$ represents the generalized effect, $G(X)$ the response function, and $H(X)$ the generalized cause. Therefore it is possible to study the generalised transmission mechanism of economic policy by describing the response function as a sort of susceptivity which comes to depend on the distribution of agents within the market. Obviously the type of response depends not only on distribution, but also on the type of interaction between agents.

5 A Two-Stage Least Squares (2SLS) Model to Explain the Linkage Between Innovation, Servitization and Local Development

The relationship between competitiveness, servitization and development can be examined econometrically through a 2SLS regression model. The basic idea analysed is that there is a relationship between added value, intensity of capital accumulation and the capacity for business services development. The equation that must be calculated, therefore, is expressed as follows:

$$VA_t = VA_{t-1} + intaccap + capsviser \qquad (3)$$

It was decided to use a 2SLS regression model to avoid the impact of the autocorrelations which exist between the variables to be estimated. 2SLS regression models succeed in doing this through instrumental variables. The database used to estimate and define the variables used, both predictive and instrumental, is shown in the appendix. The model not only serves to highlight the link between the variables; being based on territorialised data, it also takes regional differentials into account. In short, if the coefficients are significant, this model will not only highlight a link between added value, capital accumulation and the capacity to develop business services, but it will also explain the territorial differentials.

The model was populated with data from Italian regions. The tables with the values of the variables are given in the appendix. The results of the model are outlined in the following tables (Tables 2, 3, 4 and 5):

Table 2. Model description.

		Type of variable
Equation 1	vapercapita2014	Dependent
	vapercapita2009	Predictor
	Intensity of capital accumulation	Predictor
	Ability to develop business services	Predictor
	Perceived risk of crime	Instrumental
	Work regularity rate	Instrumental
	Wealth index	Instrumental
	Share of employees in high-intensity knowledge sectors	Instrumental
	VA rate of change	instrumental
	Share of technical and scientific degree	instrumental

Table 3. Coefficients

		Unstandardized coefficients		Beta	t	Sig.
		B	Std. Error			
Equation 1	(Constant)	−89.064	42.304		−2.105	.051
	vapercapita2009	.762	.152	.738	5.000	.000
	Intensity of capital accumulation	.428	.195	.331	2.197	.043
	Ability to develop business services	.696	.398	.289	1.749	.099

Table 4. Anova

		Sum of squares	df	Mean square	F	Sig.
Equation 1	Regression	12013.559	3	4004.520	60.575	.000
	Residual	1057.730	16	66.108		
	Total	13071.289	19			

Table 5. Model Summary.

Equation 1	Multiple R	.959
	R Square	.919
	Adjusted R square	.904
	Std. Error of the estimate	8.131

As can be seen from the data, the regression has a good level of significance, both overall with a high F value, and for the individual coefficients which have very good Student's T-values. The sign of the coefficients correlates an incremental contribution of the individual variables to added value at time t and highlights a relationship between added value at time t and the "delayed" added value (of the previous period) and two fundamental variables in regional development theory, namely capital accumulation intensity and business services development capacity.

These results are perfectly in line with the theoretical analyses previously developed in defining the territorial economic system. In particular, the link between business services development capacity, which is an excellent proxy for the concept of servitization and added value at time t, highlights the active role of servitization in regional development processes, also in relation to territorial differences. In particular, the model identifies the three variables 'delayed added value', 'capital accumulation' and 'business services development capacity' as the incremental factors determining territorial differences in added value at time t.

By using the data of the Italian regions, a very strong relationship can be established between the 'business services development capacity' variable, which is a proxy for servitization, and 'added value per capita' at time t, taking delayed value added per capita as the baseline.

The territorial value differentials in value added per capita are determined by the initial value added per capita, by servitization and by the intensity of capital accumulation which is an investment-linked variable. Thus, by using the TES as a key to interpretation, it is possible to identify the Territorial Level (TL) in servitization and the Firm Level (FL) in the intensity of capital accumulation. These two dimensions explain most of the differentials in regional development expressed in terms of value added per capita.

6 Policy Considerations and Some Concluding Remarks

An initial consideration concerns the path dependence which characterises regional development trends. A given tool (e.g. regional policy) deployed to promote development in a specific TES, which is characterized by a given response function, may actually create development traps. In other words, both the theoretical assumptions and the empirical model developed in this paper demonstrate, on the one hand, that the functional relationship between manufacturing and services is the basis of positive or very positive economic performance. On the other hand, it is also demonstrated that the weak regions are not equipped to respond in a positive manner (with endogenous growth) to the stimulus represented by "traditional" regional policy, which attempts to compensate for the lack of factors of production, for example by injecting capital to stimulate production investment. This traditional approach risks creating a Dutch disease effect because the TES fails to effectively absorb the additional (traditional) factor of production. It is like trying to fit a piece that does not belong into a puzzle.

This type of "compensatory" or "add-on" regional development policy ends up accentuating the differences between regions, which are due to the different response functions and are manifested in multiple, resilient equilibriums (similar to fitness landscapes). Instead of fostering convergence, traditional policies create underdevelopment traps (the lowest points in the fitness landscape) from which the TES struggle to escape.

Peripheral regions are the ones most exposed to loss of competitiveness since the rules governing the economic system promote the aggregation of factors and "classic" regional policy is unable to counter this trend, despite generous financial compensation.

An effective regional policy should work on two levels: modify the response function of a TES and also provide an investment able to generate a vector (defined as a "generalised cause"). Moreover, interventions should be minimal and aimed at creating stronger connections between economic agents and, in particular, combining production activities with services, to foster the servitization that probably influences "soft" factors inside the TES.

Appendix

See Table 6.

Table 6. Indicators for Italian regions used for 2SLS Model - Standardized and Normalized data (Italy = 100). Source: Our elaboration from ISTAT data.

	Share of employees in high-intensity knowledge sectors	Share of technical and scientific degree	Intensity of capital accumulation	Ability to develop business services	VA per capita 2014	VA per capita 2009	VA Rate of Change	Perceived risk of crime	Work regularity rate	Wealth index
Piemonte	105.65	133.47	126.44	103.72	106.75	104.02	2.62	97.86	99.92	171.67
Valle dAosta	82.35	6.14	139.85	87.99	126.83	128.72	-1.47	43.08	106.38	160.94
Lombardia	118.74	123.55	96.99	115.85	132.69	130.05	2.03	119.31	168.32	257.50
Trentino Alto Adige	75.34	73.48	147.89	84.70	139.82	131.79	6.09	31.10	158.06	271.05
Veneto	80.70	90.76	100.39	95.09	114.04	125.84	-9.37	102.27	137.17	228.89
Friuli Venezia Giulia	90.69	142.17	108.08	109.99	107.77	104.58	3.05	55.25	107.22	130.38
Liguria	123.86	115.00	91.24	100.45	112.55	111.73	0.73	77.13	101.23	132.05
Emilia Romagna	87.63	141.97	100.92	102.94	124.09	119.01	4.27	104.74	143.00	245.24
Toscana	90.00	125.93	90.12	100.63	108.88	107.02	1.74	85.20	128.24	201.96
Umbria	79.99	92.65	100.21	94.90	87.76	92.05	-4.66	118.76	97.35	128.75
Marche	80.45	124.00	97.17	92.82	96.03	97.73	-1.74	88.98	113.56	104.04
Lazio	144.79	136.25	93.61	112.62	114.50	125.77	-8.96	131.76	123.61	177.59
Abruzzo	80.15	74.47	144.40	88.67	89.33	85.14	4.93	81.83	91.13	81.10
Molise	75.51	26.58	124.58	84.75	69.85	78.32	-10.81	30.53	47.54	53.37
Campania	88.68	84.72	82.80	82.99	63.55	67.10	-5.28	116.55	71.32	53.09
Puglia	79.18	50.80	90.67	81.62	63.47	63.43	0.07	107.63	69.32	50.24
Basilicata	77.06	35.58	116.65	91.31	72.00	70.14	2.66	46.16	53.31	40.39
Calabria	79.95	78.16	113.93	80.52	60.89	63.36	-3.90	69.76	41.57	38.29
Sicilia	85.60	60.87	81.79	82.82	63.11	68.87	-8.36	88.31	62.60	40.87
Sardegna	80.85	59.97	95.87	87.90	71.38	66.70	7.02	43.08	54.81	68.21

References

1. Arthur, W.B.: Self-reinforcing mechanisms in economics. In: Anderson, P.W., Arrow, K.J., Pines, D. (eds.) The Economy as an Evolving Complex System. Addison-Wesley, Reading (1988)
2. Lafuente, E., Vaillant, Y., Vendrell-Herrero, F.: Territorial servitization: exploring the virtuous circle connecting knowledge-intensive services and new manufacturing businesses. Int. J. Prod. Econ. **192**, 19–28 (2016)
3. Vendrell-Herrero, F., Wilson, J.R.: Servitization for territorial competitiveness: taxonomy and research agenda, competitiveness review. Int. Bus. J. **27**(1), 2–11 (2017)
4. Krugman, P.: History versus expectations. Q. J. Econ. **106**(2), 651–667 (1991)
5. Latella, F., Marino, D.: Diffusione della conoscenza ed innovazione territoriale: verso la costruzione di un modello, Quaderni di Ricerca di Base dell'Università Bocconi 2, (1996)
6. Marino, D.: Territorial economic systems and artificial interacting agents: models based on neural networks. Int. J. Chaos Theor. Appl. 3(1/2) (1998)
7. Marino, D., Trapasso, R.: The new approach to regional economics dynamics: path dependence and spatial self-reinforcing mechanisms. In: Fratesi, U., Senn, L. (eds.) Growth and Innovation of Competitive Regions, pp. 329–367. Springer, Heidelberg (2009)

Kratos 2020, Strategic Plan Great Valley of Crati River

Ferdinando Verardi[1], Domenico Passarelli[2],
and Andrea Pellegrino[3]([⊠])

[1] Technical and Territorial Planning, Reggio Calabria, Italy
[2] Mediterranea University of Reggio Calabria, Reggio Calabria 89100, Italy
[3] Cultural Tourism Vocation, Reggio Calabria, Italy
lasst.pellegrino@gmail.com

Abstract. Applied experience of democratic participation during a multi-agent integrated planning, organized from the bottom up in a territorial mosaic of Vast Area. The debate was feed through the weaker and local actors of governance (understood as "conversation", landscape), feeding awareness, and improve of endogenous processes of sustainable development. The Strategic Plan as a conscious mode to dilate time, and not like a final goal, a spontaneous harmonious link between hierarchies of the territorial attraction centers. A polycentrism of continuity and proximity to the creation of a territorial identity that leans on the environmental and cultural strategy to is determine. A real experience inspired by non-deterministic theoretical scientific approaches (and subordinated to the result); but it based on democratic ideals, while respecting the role of the institutions, ensuring a high value to the cardinal principle of consultation between the government bodies at various levels and encouraging their participation in the territory in a planned and orderly manner in respect of all local actors. The experience is certainly interesting for today's and future operators who will contribute to enhancing the sustainable development of the area, finding utopian guide and inspiration for democratization processes, with the hope that it can become a concrete best practice for developing good democracy and the determination of a coordinated and strategic planning of territories within the European Economic Space.

Keywords: Strategic planning · Democratic participation
Multi-agent planning

1 The Objectives of the Strategic Plan

The main objective of the strategic plan is to build in a shared way the future of the "Valle del Crati", which encompasses the territory of Kratos. To this end, the strategic planning method is aimed at facilitating the understanding, dialogue and research of solutions through the continuous interaction among the development actors, favoring and facilitating the creation of structured participatory practices-seminars, work tables, projects-in land management. The Strategic Plan is a long-term program, born for the

© Springer International Publishing AG, part of Springer Nature 2019
F. Calabrò et al. (Eds.): ISHT 2018, SIST 101, pp. 68–74, 2019.
https://doi.org/10.1007/978-3-319-92102-0_8

renewal of the old industrial cities, through the analysis of problems, dynamics, tensions and opportunities in the area, identified, monitored and developed, in order to build global solutions and complete. It is useful to reiterate that strategic planning is an organizational process necessary to define a strategy or the direction to take in order to make decisions on the allocation of resources. In this direction lies the great opportunity offered for this area by the work presented here. Thinking about the "Valley of the Crati" through Kratos means elaborating shared "visions" (in the sense of shared vision of Lefebvrist social space), thinking, structuring and managing services not in reference to the single city, but able to go beyond the boundaries of the municipal boundaries, producing economies of scale capable of helping to contrast the crisis (from an economic, social, environmental and institutional point of view), making the territories more efficient and able to offer greater opportunities for development and wellbeing for those who live there. In fact, several studies show that in Europe and in the world, it is precisely the territorial systems (of a large area like *Kratos*), and no longer the single cities or metropolitan area, as happened in the past, to compete and generate development projected over time. The strategic plan is one of the most important innovations concerning the urban and territorial rules, is now one of main approach to challenges for the planning and development of all those territories that focus on enhancing their resources and their own peculiarities. In Europe PS emerged in the 1990s, and the tendence is orienting in the Reticular Structure or global model, like in Turin and Barcelona (it is not the representation of the city that determines the explication of the process, but the network) [1] (Fig. 1).

Fig. 1. Brand mark of KRATOS

2   Kratos 2020 Is a Discourse Between Actors

In Kratos experience the planning work coordinated by the various actors of governance whit a dialogic approach at the different levels of administration, whit ability to be self-determination within a long process of comparison for the construction about a preventive work (not for necessity). It is a widespread area in adjacent to urban core of Cosenza (areas with anthropic intensity that orbits around a well-defined identity system and especially recognized to exogenous operators in the system). The Vast Area

Kratos is in adjacent (to the Urban Area of Cosenza) a political system and managerial relations with a greater formal and substantial structure in relation to the area in question (Kratos), and compared to superordinate governance. Identification that acquired the Urban Area of Cosenza ("Urban Area") in the hierarchy of "attracting centers" finds its natural territorial expression in the *Valle del Crati*, rediscovering spontaneously the potential needed to rebalance with respect to the rest of the relational systems on a provincial and regional scale. *The local institutional relations system of PS Kratos is not a competition* whit other political configuration of the Urban Area, non-induced but spontaneous processes from below, and greater contractual effectiveness with regard to higher-level of governance. In the practice of european PS the analisis of competitors between city is important, but in *Kratos* there isen't city and competivity, but *there is convergence in a territory*. Kratos develops organically as a territorial device capable of being identified globally (natural territorial invariant). With greater clarity because it is free from the human relationships of the local border power. The polycentrism of the *Urban Area* is reinforced by the natural conceptual projection along the Crati river, finding energy to be able to relate to the European economic space, maintaining a natural role in the Mediterranean space and therefore becoming more attractive and competitive the whole area *"Urban Area of Cosenza - Crati Valley"* (all areas guarantees the existence of the other), in the general framework of the hierarchies of the global attractions. In Turin there is network of relations on the global framework, in Kratos the area of existence is related at the national framework, whit a connotation of the infrastructural picture balanced to the diagram of the centralities/economic resources (approach of London PS). The plasticity of the natural ecological corridor along the is expanding the time compared to the constraints of the central power, and is most appealing to "non-economic speculation", but to human relations, viability and substantial sustainability beyond mesh of the economy constrictive. The territorial gadget "Kratos", therefore becomes a process that has produced the shared idea of a Strategic Plan, a functional, permeable, and implementable tool for planning according to good democratic practices, through an ever-open instrument of consultation. Through this systemic of land management, not a clear mission but sequence of Cooperative-Action matured afther process of analisis, it is was easier to identify the different planning activities by putting them into system, with a greater understanding of the different punctual works within the PS, and therefore able to attract funding and further projections for a sustainable tourism development. The spontaneous experience of the Kratos Strategic Plan becomes a substantial link for transnational development scenarios, an example of self-government that can self-perpetuate, demonstrating its capacity to manage its territory. A search of rebalancing beyond the "phenomenon of attraction" like Bilbao or Barcelona goal reached by the Olympics, Kratos is awareness of its local potential and capacity to work in the following phases of design, identifying regulatory, financial instruments, but also the most appropriate technical/administrative forms to define the next design and execution phases, in order to clarify the best performance form of exogenous technical operators the government system, and to ensure proper administrative control over the whole process. In this scenery, it becomes easier to implement the intervention of other stakeholders in the territory, in order to ensure comparison and participation in government decisions, protecting local actors as active, solidarity, and promoting synergies and competitive contexts among stakeholders exogenous, as has already been the case for the most interesting Strategic Plans in Europe with the open to global cultural debate[1].

2.1 Vision

The Strategic Plan can become a means for best practice, and an [2]interpreter of a possible path of sustainable development of cities spread in the "Urban Area". A Territorial Area gadget capable of overturning the concept of *"city to the landscape"*, placing the protagonist role in the territory as compared to densely anthropized areas. Kratos communicates itself with a journalistic [2] but not a publicistic approach (Roger Brunet 2003) with a lever in the ethical/ecological revolution, the PS is simple, permeable instrument that easily fits in the development scenarios and national coordination of metropolitan cities. Following modern regeneration of the territory, with the care of Security and Accessibility to the urban system, in a polycentric framework of densely anthropic sites. Intermodality care in keeping with the authenticity of local values, but also sociality and landscape viable, with a general framework of governance, capable to analyze the processes and interpreting proper administrative control, in the space of actors of Democracy. The "Open Tool Kratos" is so spontaneous contribution of local actors for the harmonious development of places, cultural heritage is effective because dialectical and democratic debate in a scientific framework with deterministic approach. A *sensible instrument*, permeated by human and social values, so as to increase the opportunities of solidarity and equality, ensuring that the instrument remains a means and not an objective, inclusive and in the hands of the Men. Among the new missions of the strategic plan, the missions are pursued to be able to combine the strongly cultural

---

[1] KRATOS is not a compact city (like Barcelona), but a compact management of resources (preventive approach), avoiding high environmental costs that are objected in the mature scientific debate about PS (Bertuglia et al. 2003). KRATOS is an early response to the phenomenon of De-urbanization (with phenomena of hourglass social structure) in relation to single city projections in the urban area of Cosenza. A way of solution for homogeneous redistribution of regular work, and trying to avoid the phenomenon of "haves not" (impossibility in having systems of communication in contemporary cities). Kratos can increase a gentrification (balanced) phenomenon in the historical centers depopulated orbiting the river Crati, and creation of valuable economic activities in the tertiary sector (very high cultural quality). Phenomenon driven by the growth of innovation in transport and communication (Detragiache 1995, 1998). Kratos does not debate on the urban environment but on the territory, as it declaims in the mature scientific debate of the PS (Tomazinis 1985, Bertuglia et al. 1998, Bertuglia et al. 2000), using the most correct supralocal scale (Peano et al. 2002). Kratos intersects urban factors and planning tools (Bertuglia 2002, Champion 2002, Davico et al. 2002, Guerois and Pumain 2002, Haag 2002, Peano et al. 2002). The PS Kratos is analogous to the PS 'Area Nord of Milan and Piacenza (48 municipalities of the province involved) for reticular structuring between institutions (www.asnm/www/futuro/index.HTML/Kratos' approach is similar to the strategic plan of Turin (Torino Internazionale), but don't have a financial speculative approach, because it seeks its own "sincere" approach to the principles of sustainability (Cavallaro 2000).

[2] Partecipate process in communication (Pulsifier e Taylor 2005, Gibelli 1996, Ames 1998, Pumain, Saunders, Saint-Julien 1989, Pareglio 2000), and ability to self-criticism (Bobbio L. 1994) - Like PS of London 2008 in map of net for the green spaces london.gov.uk/thelondonplan; PS of Lione map educational and ecological urban area millenarie3.com; banner in Barcellona.bcn2000.es; brand map in PS of Rimini comune.rimini.it.

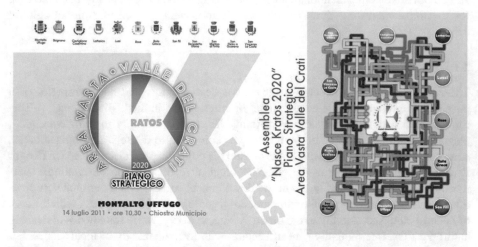

Fig. 2. A concrete bottom-up experience with a multi-agent integrated planning able to ensure endogenous coordination and development for the use of financial resources, as well as being a good practice of addressing the planning and usability of local resources for even exogenous players to the system. A planning of work among local actors in the area concerned, with a clear vision of the territorial framework of territorial management by the superordinate governance, guaranteed the strategic decision-making process an operational coordination that facilitated horizontal and vertical consultation by the local public bodies and actors involved.

Fig. 3. With a clear Large Area framework by local administrators, the spontaneous process of planning It's guaranteed, in a consciousness sustainability key and with an endogenous development path, rebuilding a territorial identity capable of developing its own development idea. The actors of the area find out clearly the *"scenic space"* in which *"to play"*, in respect of their institutional role, proceed to a project of territorial planning, intended to discourse (landscape) in which to create their own conceptual space in respect of the superordinate organs.

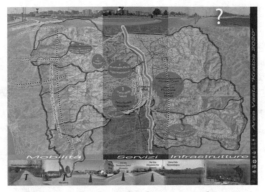

environmental approach inherent in its becoming

Fig. 4. Working on the weaknesses of the integrated system, it was possible to set up a work that involves the assessment, control and environmental management authorities at the various stages of the decision-making process.

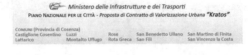

PROPOSTA DI CONTRATTO DI VALORIZZAZIONE URBANA "KRATOS"

Fig. 5. A timely and coordinated design between the different *"dots"* of the territorial mosaic (eleven communities), guaranteed the technical and administrative sustainability of the overall proposal of the Kratos Strategic Plan, an integrated design that is perfectly integrated in the general framework of territorial infrastructures, and ensures the systematic nature of local works within a framework of overall territorial sustainability of public works proposed within the PS. An operational transparency of the administrative machine, which best proposes the whole territory in a national strategic framework. With this work, it becomes easier to intercept new finances for the development of planning at an over-municipal level, and in a Vast Area framework by encouraging permeability to the system for development that is free from the administrative boundary, but at the same time in harmony with it. The created administrative system becomes a *"gadget"* that qualifies the territory, making it attractive for private investment, because it is retrained and well coordinated from a functional point of view. The quality in terms of overall sustainability of the strategic projects set up, guarantees the investor's security because it locates an overview of the local PA, and has a capacity to control and manage decision-making processes when it comes to becoming.

significance of the landscape with its strongly "strategic" re-consideration (consistently with the contents emerging from the European Landscape Convention fully implemented by the regional urban planning law 19/02 of *"Regione Calabria"* and reported in the Regional Territorial Framework), through which landscape and territory become an indispensable resource for the creation of employment opportunities, development and social welfare. Following the "discourse", understood as the still open landscape, so far undertaken by PS Kratos (Figs. 2, 3, 4 and 5).

3   Tale (*Discourse*) of the Strategic Plan with Images

See (Fig. 6)

Fig. 6. The clear knowledge of local administrators of its territory, has allowed us to play the role which is appropriate within the vast area where the Strategic Plan, identifying the strategic interventions needed to develop the Route of Identification some "Urban Area" toward "Valle del Crati", entering into a logic of overall competitiveness of the two systemized areas in the provincial and regional context.

Sustainable Agriculture Development: Lacks on Law and Urban Plots

Susana Campos$^{(\boxtimes)}$, Alexandra Ribeiro, and Micael Santos

Universidade de Trás-os-Montes e Alto Douro (UTAD),
5000-801 Vila Real, Portugal
susanacampos@utad.pt

Abstract. Sustainable development in agriculture is a central theme in the global world. Changes with dietary habits and population growth highlighted the importance of global sustainability as a criterion for achieving a more balanced development, both in rural communities and in cities. Technology and innovative practices will be the solution. So, this paper aims to give a multi-disciplinary perspective about sustainable development and demonstrate how the urban plots can contribute to the sustainable development in the cities. Furthermore, we hope to be able to provide a debate in the scientific community and open new perspectives for future research that can address the gap in legislation on these subjects.

Keywords: Law · Sustainable development · Urban plots

1 Introduction

The global food markets are undergoing significant changes over time [1]. It is amply clear that the world cannot achieve sustainable economic growth with old fashioned consumption and production patterns. Modern civilization now constitutes a dense network of sub-units which has become huge and highly complex and is unable to form a well-functioning system [2].

The literature on the new paradigms of sustainable development in agriculture has highlighted the importance of global sustainability as a criterion for achieving human development, both in local agricultural communities and in cities, nations and the world, and this concept is a decisive factor in the territories development [3]. Sustainability is the engine for the development of agriculture and the right way to reduce environmental impacts. Environmental resources are increasingly limited by excessive use of land due to food production. It causes pressure on society to look for alternative ways of producing more sustainable food. For this all the stakeholders need to join hand and take steps in maintaining the environmental balance and promote sustainable lifestyles [4]. Herrera, 2008 argues that social changes can only be achieved thanks to a main driver: the development of spaces for critical and collective learning.

Thus, sustainable agriculture is a broad concept that seeks to extend its understanding to the preservation of the identity of the social value of agriculture, bridging the production and preservation of agricultural systems [5].

© Springer International Publishing AG, part of Springer Nature 2019
F. Calabrò et al. (Eds.): ISHT 2018, SIST 101, pp. 75–84, 2019.
https://doi.org/10.1007/978-3-319-92102-0_9

The globalization and the multi-dimensionality of poverty and development concepts, rises the adoption of new paradigms of intervention emphasizing their multidimensional nature, like land and people. Countries must define new organizational structures of intervention, meaning that for a certain policy to be successful, it presupposes the involvement of the different social actors in a process of decentralization, giving more (decision) power(s) to local as opposed to central. One must consider, beyond any doubt, the active participation of citizens and other non-governmental entities in the process of defining policies. Only then will we achieve an integrated governance towards sustainability.

The next section presents a review of the relevant literature. It includes consumer perspectives literature, a review about sustainable production, a review about urban plots and how they can contribute to the sustainable development in the cities. Finally, we explore the legal constraints of urban plots in Portugal.

1.1   Consumer Perspectives

At the last two decades have witnessed impressive changes in consumers' perceptions of food. Latest trends like slow food and organic food appeared [6] and well-being and healthy lifestyles have even become a social and economic megatrend [7] in addition to the environmental, ethical, and economic aspects of food consumption that have regional, national, and global impacts [8]. Contemporary trends will affect food demand in the future and this is conditioned by the level of involvement in promotion by retailers as well as the size and the economic value of the segments of concerned consumers [9].

Consumers face several barriers hindering sustainable consumption choices, such as the price, availability, convenience, quality, habit, and lack of belief in product claims, perceived consumer effectiveness, and lack of information [10]. These barriers allow us to explain the gap between consumers' sustainability attitudes and their sustainable consumption behavior [11, 12].

Nowadays, EU consumers emphasize and give preference to food produced and processed according to natural methods. Organic vegetables can be considered foods that meet this need and, at the same time, can provide higher yields to producers. European consumers have greater purchasing power and are willing to pay more for European products. There is, however, a trade-off between the quantity of available product and the quality, because due to increasingly stringent safety standards and shorter value chains, the quantity decreases but with higher quality products [13].

1.2   Sustainable Production

Scientists, organizations and policy makers are emphasizing on sustainable basis. However, massive use of fertilizers and pesticides cause serious threats to the environment and sustainability in agricultural productivity. Through organic farming high quality food could be produced without effecting the environment [14].

Organic farming consists of replacing chemicals and other products with organic fertilizers and non-chemical crops [15]. This type of agriculture receives financial support of European Union, given subsidies to producers to convert conventional production to organic farming. Subsidies are a way of encouraging farmers to practice organic farming and compensate for low production yields during conversion [15].

The area under organic farming has been increasing in the European Union over the years and this is due to the strategy adopted by producers to maximize their profits, receiving subsidies and still being able to sell their products at a higher price because the organic products have a higher value [16].

The motivations for the implementation of more sustainable production practices can be based on internal, external and strategic factors [17, 18]. External factors are related to customer demand, investor, community and public pressure, competitors and regulatory compliance. The internal factors are related to managerial attitudes, concerns about employee safety, company culture, concern about environmental impacts and the state of the environment, land protection and social responsibility [19]. Finally, the strategic factors are related to the competitive advantage, differentiation, marketing benefits, public image, brand reputation, product quality and cost savings [20].

It has been identified that there are barriers to sustainability that may hinder the adoption of various sustainability practices in production, such as lack of knowledge among decision makers, lack of leadership, insufficient stakeholder pressure, inadequate time and ability to implement practices, and high cost of capital [21–23].

1.3   Urban Plots and Sustainability

[24] emphasize the importance of agricultural areas, particularly those that exist in peri-urban areas. They advocate a balance between production and ecological as well as social aspects. The urban plots could be the key to improve a better nutrition, and healthy food consumption. In that's way defines urban gardens such as:

"The urban plots translate a spontaneous way of using the interstitial spaces of cities, allowing self-sufficiency, reducing energy consumption, increasing economic activity by generating jobs and having a multiplier effect on the economy, product availability Organic farming of healthy products".

Urban plots constitute complementary spaces of the family income, the urban population can through these spaces obtain the products they consume in their food, in a straightforward way, fostering the consumption of fresh and healthy products and can provide relaxion moments, encouraging social interaction and facilitating social integration [25].

The plots work, as well as a multifunctional space for sustainable urban development. Agriculture as an activity emerges as well as an articulation between rural culture and the improvement of public services in terms of offering better environmental conditions of life [26]. Likewise, urban agriculture can cover several types of urban structures, gardens, backyard, as well as the use of roofs, balconies, terraces and interior of dwellings and even facades of the buildings. In effect, the diversity of urban agriculture is one of its main attributes and can be adapted a wide range of urban situations and the diverse needs of their actors [27].

To that extent (…), urban agriculture and namely urban plots may bring countless benefits to cities, among which are the strengthening of urban food safety, reduction of urban poverty, better management of urban environment, improvement of citizens health, the development of a more participative administration and the protection of urban biodiversity.

To this purpose, as described by [28], urban policies should/must encourage the implementation of urban agriculture as a way to promote sustainable urban development, argument corroborated by other authors (e.g. [29, 30]). Urban plots may play an important role in the promotion of health, social inclusion, civic participation and sustainable living practices in urban environments [31].

Urban plots may also serve as catalysts of integration. According to [32] (2017) in certain immigrant or needy communities they go well beyond plain plots where to grow vegetables, they serve as alternative homes, as places of food production and feeding, where children are nurtured and protected, where the sick are healed, and people get together in leisure moments, to socialize and for relaxed personal thinking/meditation/ reflection.

Therefore, the advantages of these plots may be of economical type through the promotion of inter-communal trade and diversification of local economies but go well beyond that perspective. We can also find advantages of environmental type, such as the reduction of air pollution, the absorption of carbon dioxide, wastewater re-utilization, reduction of solid waste, improvement of soil quality, re-utilization of waste land and raising population awareness to the importance of the environment, increasing ecological awareness; and of human and social types, namely food safety issues, food accessibility and reduction of its cost, formation of a sustainable society, ecological awareness, food variety (fresh and good quality food), cohesion and well-being of the population and historical and cultural heritage. Biological urban gardens are a model of enhancement of the quality of life of its users, contributing significantly to make people feel happier and more satisfied with life, changing and improving their habits [25, 33].

Urban Plots (community gardening) means a collective cultivation of plants by several people in a shared AREA [34]. Urban plots have been introduced in the US as an urban enrichment source and as a way to bring social harmony to neighbourhoods, increasing the sense of community in the residents [35]. The urban plots concept is usually used to designate a multipurpose space, not only to produce food, but also to provide recreational and leisure spaces, essential for any city that is sustainable and able to respond to the needs of the local inhabitants [36].

The Urban plots are plots of land, mostly for horticultural production, divided into small plots and cultivated by people or groups of people interested in producing their own food (see Fig. 1). In this way, urban plots represent one of the most common strategies for implementing urban agriculture in developed countries.

These plots can be places, not only of planting, but also where people come together to cooperate and to live in this activity. Urban agriculture can enable people to re-think their rural past and re-appreciate the environment and nature [37]. They provide fresh consumer products and helps to relieve the stress of city life as leisure spaces, not just for those who grow it. They are given a key role: to cement social relations, to create a sense of solidarity between neighbors and responsibility for the environment.

Urban Plots appear as a local strategy to try to reverse the situation of food insecurity, as it turns the cities into greener places, creating areas of food production located near the sale areas or for own consumption (see Fig. 2). Food security, as already

Fig. 1. Urban plots in Senhora da Hora, Porto, Portugal.

mentioned and adopted by the FAO, has been bringing renewed academic interest in community plots and urban gardens [38].

Urban plots, besides being green spaces with high biological wealth and possessing several beneficial functions for the city, also represent a way of practicing urban agriculture [37] (2002) refers the key factor in its performance in protecting the

Fig. 2. Urban plots in Oeiras, Lisbon, Portugal.

environment and Nature. Urban plots promote the development of new knowledge and friendships, exchanges of ideas and experiences among users. In addition, a number of studies carried out in the field highlighted the potential of urban gardens for social cohesion, the positive evolution of the relationships observed among the adherent population [39]. Accordingly, in large cities, it is necessary to consider alternatives which do not compete for urban spaces, such as non-living areas or ceilings that can be easily accessed to provide food with quality and in sufficient quantity.

With the growing interest in urban plots, increase does urban agriculture and a new paradigm of development emerges based on the concepts of sustainable development. This new paradigm results from the inability of policies to have effective and visible action to eradicate poverty and from the fact that malnutrition still shows high rates (even in more developed countries). This strategy is nothing new. In the more recent past, green urban spaces and urban plots were very positively linked with the implementation of Agenda 21 and sustainable development policies while at the same time promoting environmental equity [40]. Although food production is the main purpose of urban agriculture, the advantages associated with this activity are very diversified [41]. The role of urban plots has widened, especially in more developed societies where it has been extended to other aspects, aesthetic, social, environmental, didactic, among others [42].

The participation of citizens in these plots helps to create changes in people and changes inspire and prepare them to foster participation in society in a broader way [43, 44] (2013, p. 3) *point out that development centered on people and local territories is a path that has attracted the attention of politicians, technicians and academics/scholars, which is based on, on the one hand, on a democratic vision of society and considers citizens and civil society as active stakeholders in governance of local affairs and, on the other hand, in an integrated, cross-sectoral, endogenous, specific and supported in research approach* ...... Effectively, voluntary citizens are the ones who work the plots, although the result of this work is for their own benefit. In this way, urban plots play a crucial role in food security, well-being, quality of life and if instituted early in all schools, can become a fundamental part of a systemic education that supports the understanding of local and global food issues.

Several studies (e.g., [45, 46]) argue that experiences with nature since childhood positively influence the adoption of more sustainable behaviours, especially with regard to food. It is intended that fruits and vegetables are grown in or around school areas, providing a source of small-scale basic foods as well as other activities [47]. In particular, a study by [48] points to the fact that these gardens might facilitate positive links between school and home, and, simultaneously, enable the community to deepen their knowledge of the foods they eat and greater civic responsibility.

It is therefore urgent to assume that urban plots, as agricultural spaces in and out of cities, can contribute significantly to their sustainable development.

1.4   Legal Constraints of Urban Gardens in Portugal

If this approach to urban plots is maintained, the scenario is adverse to the sustainability of urban agriculture, at least in the configuration in which it has been developed, and favourable of a sharp separation between the city and rural areas. The sustainability of urban agriculture also seems in doubt if we measure its academic and legislative treatment as a sign of its socio-economic and cultural dignity. The fragmentation and legal and dogmatic non-systematization stand out in the urban agriculture problematic, thus undermining its sustainability.

A study carried out in Portugal that aims to assess/demonstrate that the Community Project, called "Horta à Porta", contributes effectively to sustainable urban development and consequent improvements in quality of life for inhabitants of Porto [49]. This study reveals that overall, the benefits most cited by garden users are: "Better food" (82%), "Food safety/product ingestion of better quality "(77%)", "Strengthening social cohesion/Sharing good times" (48%), "Physical and mental well-being" (22%), "Complement in the family budget" (18%) and "Occupation" (12%). These motives and contributions are corroborating by [39] (2000). So, it is important to solve some constraints in terms of the legal framework in domestic law. For this purpose, we find the constitutional grounds (d) and (e) of article 9 and 66 and 93). However apart from this constitutional framework, there are no further laws in the legal system that regulate urban agriculture and urban plots. We only have a local treatment of the issue, through the municipal regulations, currently existing in multiple municipalities, mostly urban but also more rural. Therefore, one can easily conclude that there is a lack of a systematic, integrated, coherent and strategic vision of this matter, a legal instrument with a general framework. The law laying the foundations of environmental policy, the Law on Soil Bases, Spatial Planning and Urbanism (Law no. 31/2014 of May 30 [50]) and the Law regarding wasteland (Law no. 68/93, of September 4 modified by law no. 72/2014 of September 2 [51]) may try to help frame the treatment of urban agriculture, and the plots in particular, but do not deal directly with this matter. On the other hand, urban plots are not usually provided for in Municipal planning director. The only national diploma that addresses the issue of urban agriculture and urban plots is Resolution of the Council of Ministers no. 61/2015 of 11 August, approving the strategy "sustainable cities" and refers to in some of its points urban - rural integration.

2 Future Research

The global challenges of growth, urbanization, scarcity and environmental change become the key drivers to develop a sustainable model. Meanwhile, people need to change their standard of life. To promote a sustainable development in the cities, innovative strategies redefine a more ecological and healthy lifestyle. In this way, urban plots can provide a model of sustainability in action. The next question is: Will policy makers take the example of urban plots as a model for improved sustainable development strategies in the cities?

Understanding which the best way is to achieve the linkages between rural and urban areas is a new challenge that we must reinstated. Furthermore, urban agriculture is emerging in major centres, and it is consensual that creativity and specific skills move freely in the global society.

Further studies are needed to evaluate social and environmental impacts in space and time. On the other hand, it is necessary to adapt the environmental legislation to the real context and provides tools to replicate the well succeed cases from other countries.

3 Conclusions and Policy Implications

For a more sustainable food consumption and production, consumers must be informed about how their food is produced and in the other hand ensure that the types production/processing methods used are in line with consumers' expectations. Furthermore, is necessary integrate sustainability elements in all food production standards. If consumers are informed about their options, improve consumer access to more sustainable and healthier products and allow consumers to make informed choices and create an environment that supports healthy and sustainable food choices, for example by increasing availability and range of sustainable food products. Food trends is a topic of interest because nowadays with globalization are exposes several complexity issues and it turns a topic to explore. Urban plots can provide a model that improve sustainability and give more nutritional food. Besides allowing self-sufficiency.

Urban plots must be preserved and disseminated, considering the potentiality of their multiple functions, beneficial to citizens and the city. In this way, local entities should look to the urban production of food, as an activity capable of improving the quality of life of the community and should be integrated into the urban planning police [49].

Urban plots have a major impact on the economic, social, (…) environmental, pedagogical and aesthetic dimensions [45, 48]. Reinforcing [52], that the concept of urban plots and urban agriculture is associated with a multifunctional character that can be associated with a recreational occupation, a means of overcoming economic difficulties, a restoration and/or recovery of the landscape, or a remnant of past times and the rural world. In this way, urban agriculture, in addition to providing a food production function, increases food security and income savings, and contributes to an environmentally healthy environment [29].

Policymakers must improve the further development and the adoption of more sustainable and innovative practices in driving the uptake of improved technologies and increased production and enhance a healthy and conscious consumption.

References

1. Beghin, J.C., Aksoy, M.A.: Global agricultural trade and developing countries, 44 (2005)
2. Barrientos, S., Gereffi, G., Rossi, A.: Economic and social upgrading in global production networks: a new paradigm for a changing world. Int. Labour Rev. **150**(3–4), 319–340 (2011)
3. Folke, C., Hahn, T., Olsson, P., Norberg, J.: Adpative governance of social-ecological systems. Annu. Rev. Environ. Resour. **30**, 441–473 (2005)
4. Jeppesen, S., Hansen, M.W.: Environmental upgrading of third world enterprises through linkages to transnational corporations. Theoretical perspectives and preliminary evidence. Bus. Strateg. Environ. **13**(4), 261–274 (2004)
5. Lamine, C.: Sustainability and resilience in agrifood systems: reconnecting agriculture, food and the environment. Sociol. Ruralis **55**(1), 41–61 (2015)
6. Tischner, U., Kjaernes, U.: Sustainable consumption research exchanges. In: SCP Cases in the Field of Food, Mobility, and Housing, pp. 201–237 (2007)
7. Reisch, L.A., Gwozdz, W., Barba, G., De Henauw, S., Lascorz, N., Pigeot, I.: Experimental evidence on the impact of food advertising on children's knowledge about and preferences for healthful food. J. Obes. (2013). Article No. 408582

8. Garnett, T.: Food sustainability: problems, perspectives and solutions. Proc. Nutr. Soc. **72**, 29–39 (2013)
9. Gereffi, G., Lee, J.: Why the world suddenly cares about global supply chains. J. Supply Chain Manag. **48**(3), 24–32 (2012)
10. Watkins, L., Aitken, R., Mather, D.: Conscientious consumers: a relationship between moral foundations, political orientation and sustainable consumption. Clean. J. Prod. **134**, 137–146 (2016)
11. Follows, S.B., Jobber, D.: Environmentally responsible purchase behaviour: a test of a consumer model. Eur. Mark. J. **34**(5–6), 723–746 (2000)
12. Carrigan, M., Attalla, A.: The myth of the ethical consumer – do ethics matter in purchase behaviour? Consum. Mark. J. **18**(7), 560–578 (2001)
13. W.B.C. for S. Development, Vision 2050: The new agenda for business, p. 80. WBCSD (2010)
14. Ashraf, I., Ahmad, I., Nafees, M., Muhammad, M., Ahmad, B.: A review on organic farming for sustainable agricultural production. Pure Appl. Biol. **5**(2), 277–286 (2016)
15. Guesmi, B., Serra, T., Kallas, Z., Gil Roig, J.M.: The productive efficiency of organic farming: the case of grape sector in Catalonia. Span. J. Agric. Res. **10**(3), 552 (2012)
16. Willer, H., et al.: A survey of community gardens in upstate New York: implications for health promotion and community development. Can. J. Diet. Pract. Res. **16**(4), 14–21 (2016)
17. Dodds, R., Graci, S., Ko, S., Walker, L.: What drives environmental sustainability in the New Zealand wine industry? Int. J. Wine Bus. Res. **25**(3), 164–184 (2013)
18. Santini, C., Cavicchi, A., Casini, L.: Sustainability in the wine industry: key questions and research trends. Agric. Food Econ. **1**(9), 1–14 (2013)
19. Silverman, M., Marshall, R.S., Gordano, M.: The greening of the California wine industry: implications for regulators and industry associations. J. Wine Res. **16**(2), 151–169 (2005)
20. Duarte, A.: How 'green' are small wineries? Western Australia's case. Br. Food J. **112**(2), 155–170 (2010)
21. Lynes, J.K., Andrachuk, M.: Motivations for corporate social and environmental responsibility: a case study of Scandinavian Airlines. J. Int. Manag. **14**(4), 377–390 (2008)
22. Moon, S.G., De Leon, P.: Contexts and corporate voluntary environmental behaviors: examining the EPA's green lights voluntary program. Organ. Environ. **20**(4), 480–496 (2007)
23. Revell, A., Blackburn, R.: The business case for sustainability? An examination of small firms in the UK's construction and restaurant sectors. Bus. Strateg. Environ. **16**(6), 404–420 (2007)
24. Olsson, E., Kerselaers, E., Søderkvist Kristensen, L., Primdahl, J., Rogge, E., Wästfelt, A.: Peri-urban food production and its relation to urban resilience. Sustainability **8**(12), 1340 (2016)
25. Mourão, I.: Horticultura Social e Terapêutica: contexto, no (2013)
26. Borec, A., Turk, J.: Sustainable Rural Development (2010)
27. Gomes, C.A., Saraiva, R.: SERÁ A AGRICULTURA (2016)
28. Tornaghi, C.: Critical geography of urban agriculture. Prog. Hum. Geogr. **38**(4), 551–567 (2014)
29. Barthel, S., Parker, J., Ernstson, H.: Food and green space in cities: a resilience lens on gardens and urban environmental movements. Urban Stud. **52**(7), 1321–1338 (2015)
30. Moragues-Faus, A., Morgan, K.: Reframing the foodscape: the emergent world of urban food policy. Environ. Plan. A **47**(7), 1558–1573 (2015)
31. Turner, B., Henryks, J., Pearson, D.: Community gardens: sustainability, health and inclusion in the city. Local Environ. **16**(6), 489–492 (2011)

32. Hondagneu-Sotelo, P.: At home in inner-city immigrant community gardens. Hous. J Built Environ. **32**(1), 13–28 (2017)
33. Dos, E., Sociais, B., Ambientais, E., Biológica, A., Doutora, P., De Maria, I.: Económicos Das Hortas Sociais Biológicas Do Município Da Póvoa De Lanhoso, pp. 14–22 (2015)
34. Eigenbrod, C., Gruda, N.: Urban vegetable for food security in cities. A review. Agron. Sustain. Dev. **35**(2), 483–498 (2015)
35. Shacham, E., Donovan, M.F., Connolly, S., Mayrose, A., Scheuermann, M., Overton, E.T.: Urban farming: a non-traditional intervention for HIV-related distress. AIDS Behav. **16**(5), 1238–1242 (2012)
36. Pinto, R.F., De Engenharia, E., De Gualtar, C.: Apresentação Oral: Planejamento Sustentável II 436 VIABILIDADE AMBIENTAL DE HORTAS URBANAS: O CASO DE BRAGA, PORTUGAL, Viabilidade Ambient. Hortas Urbanas O Caso Braga, Port., 13 (2008)
37. Madaleno, I.M.: Cities of the future. Food Nutr. Agric. **29**, 14–21 (2001)
38. Guitart, D., Pickering, C., Byrne, J.: Past results and future directions in urban community gardens research. Urban For. Urban Green. **11**(4), 364–373 (2012)
39. Armstrong, D.: A survey of community gardens in upstate New York: implications for health promotion and community development. Health Place **6**(4), 319–327 (2000)
40. Ferris, J., Norman, C., Sempik, J.: People, land and sustainability: community gardens and the social dimension of sustainable development. Soc. Policy Adm. **35**(5), 559–568 (2001)
41. Freire, I.J., Ramos, C., Lucas, M.R.: Agricultura urbana: impactos económicos. sociais e ecológicos **1**, 1–20 (2016)
42. Prefeitura, A cidade, 2008(1) (2008)
43. Levkoe, C.Z.: Learning democracy through food justice movements. Agric. Hum. Values **23** (1), 89–98 (2006)
44. Júlia, H., Fernandes, D.C., Marta-costa, A.A., Cristóvão, A.: Empoderamento De Comunidade Rurais Como Prática De Revitalização De Aldeias, pp. 86–99 (2013)
45. Truong, S., Gray, T., Ward, K.: 'Sowing and Growing' life skills through garden-based learning to reengage disengaged youth. Learn. Landsc. **10**(1), 361–385 (2016)
46. Carlsson, F., Frykblom, P., Lagerkvist, C.J.: Consumer preferences for food product quality attributes from Swedish agriculture. Ambio **34**(4), 366–370 (2005)
47. Sottile, F., Fiorito, D., Tecco, N., Girgenti, V., Peano, C.: An interpretive framework for assessing and monitoring the sustainability of school gardens. Sustainability **8**(8), 801 (2016)
48. Carlsson, L., Williams, P.L., Hayes-Conroy, J.S., Lordly, D., Callaghan, E.: School gardens: cultivating food security in Nova Scotia Public Schools? Can. J. Diet. Pract. Res. **77**(3), 119–124 (2016)
49. Fernandes, A.L.P.: Agricultura Urbana e Sustentabilidade das cidades _Projeto 'horta à porta' no Grande Porto (2014)
50. Lei n.º 31/2014 de 30 de maio que aprova a lei de bases gerais da politica pública de solos, de ordenamento do território e de urbanismo (2014)
51. DR: lei n.º 68/93 que aprova a lei dos Baldios, publicada na I série do DR de 04 de setembro de (1993)
52. Doutor, O., Carlos, L.: Ana Paula Monteiro Pereira Co-orientador : Engenheira Teresa Alexandre Figueiredo Serranos dos Santos (2017)

Religious Fruition of the Territories: Ancient Traditions and New Trends in Aspromonte

Donatella Di Gregorio^(✉), Alfonso Picone Chiodo,
and Agata Nicolosi

Mediterranea University of Reggio Calabria, 89100 Reggio Calabria, Italy
donatella.digregorio@unirc.it

Abstract. The proposed work is intended to highlight certain routes born from a religious background that have become interesting from an environmental and naturalistic point of view. These routes can offer themselves within a framework of recovery and enhancement of local resources aimed at giving back to the communities concerned often forgotten testimonies of history and shared memory, fragments of life, literature, and art. The particular path of faith examined is that of the historical destination of the Sanctuary of the Madonna of the Mountain at Polsi in Aspromonte, long a place of devotion and faith. The objective of the work is to examine the conditions by which the pilgrimage can contribute even further to the creation of value and socio-economic development for the territories concerned, in a dual path that links together faith and the re-appropriation by the local community of places that have been contaminated and subjugated by the presence of the *'ndràngheta,* in order to promote virtuous processes of economic and social legality. This sees the Aspromonte as generator of sustainable development, of fair trade, of responsible tourism, and of ethical agri-food production.

Keywords: Religious landscape tourism · Pilgrimage
Sustainable development

1 Introduction and Literature Review

Religious tourism represents a global market of approximately 300–330 million people who visit religious sites every year [1], with an estimated turnover of around 18 billion dollars (IS.NA.RT - National Institute for Research on Tourism). According to the World Tourism Organization [2], 40% of this religious tourism takes place in Europe. It is indeed a very complex phenomenon that does not simply combine tourism and religion, but stems from motivations that many authors identify as inner expression [3] and that represents an important element in the life of us all [4, 5].

Liutukas focuses in particular on the search for identity and values that the traveller/pilgrim seeks in his experience, on the traveller's motivation, and on his inner disposition. The categories that he identifies (valuistic journeys, spiritual tourism, holistic tourism, personal heritage tourism, and pilgrimages) provide a useful framework for understanding values-based tourism.

© Springer International Publishing AG, part of Springer Nature 2019
F. Calabrò et al. (Eds.): ISHT 2018, SIST 101, pp. 85–93, 2019.
https://doi.org/10.1007/978-3-319-92102-0_10

The pilgrimage is defined as a journey undertaken by a person in search of a place and of an individual or collective "state of well-being". It represents a precious ideal, and the destinations embody the ideals and values that the pilgrim cannot attain from home.

The International Conference of the World Tourism Organization (UNWTO), organised in 2013 on the theme of spiritual tourism and sustainable development (International conference: Spiritual tourism for sustainable development), recognised a considerable importance to this topic [6, 7] above all in relation to the role that it can have for understanding between peoples and peace in the world, arguing that (p. 2): "the cultural exchange and dialogue evoked by spiritual tourism are the very corner-stones of mutual understanding, tolerance and respect, the fundamental building blocks of sustainability..."

The pilgrimage is one of the oldest forms of devotion of human populations and exists in all the major religions of the world [8–11]. In fact, travels deriving from religious reasons are considered as one of the most ancient forms of tourism, a pre-cursor of modern tourism [12, 13].

The pilgrimage has a positive economic impact [14, 15] and can contribute to the promotion of a region or of a country [8]. For example, the Camiño de Santiago has become an important tourism product to promote Galicia. Through a strong publicity campaign aimed at pilgrims throughout the world [13], it has contributed to improve the routes, to create new hostels for pilgrims, rural houses and hotels. However, some authors raise the problem of the functional conservation of religious heritage [16] and of the excessive anthropic pressure that a phenomenon of far-reaching devotion can cause in fragile territories. The pilgrimage relates itself to the environment, that con-ditions it and in turn is conditioned by it, with both positive and negative impacts [17].

It is well known that territory serves different purposes and has diverse roles in the socio-economic development of an area: imagining the development of an area without respect for its landscape or environmental dimensions seems impossible. In fact, very often, it is precisely the recovery of the environment that activates and powers mechanisms of improvement for the economic conditions of an area. It is important to emphasise that in a context in which the environment and the territory play a funda-mental role, the concept of development is not unconditional. Without a doubt, it must be a concept of development that assures the continuation of human presence in the territory and also ensures the exercise of economic activities.

In this regard, as many authors highlight [18, 19], the need to intervene in the management of the pilgrimage seems evident, in order to encourage behaviours that contribute to a path of responsible and conscious faith, oriented towards the respect for and conservation of the environments and the nature.

In the Bruntland Report of 1987, sustainable development was defined as "de-velopment that meets the needs of the present, without compromising the capacity of future generations to meet their needs". In 1991 the essential aspect of the ecosystem was introduced. In fact, in "Caring for the Earth: A Strategy for Sustainable Living", Sustainable Development was defined as "the satisfaction of the quality of life, keeping within the limits of the capacity of the ecosystems that sustain us". However, it was only at the Rio de Janeiro Earth Summit in 1992 and, even more so, at the World Summit on Sustainable Development (WSSD) in Johannesburg (2002), that the concept of Sustainable Development was extended as a three-dimension integration in which the economic, the social, and the environmental are closely interlinked (www. arpa.fvg.it). It is, essentially, about achieving those conditions which for decades have been assimilated to the concept of "sustainable development" and of considering the importance attributed to the participation of local communities, as the European Landscape Convention observed.

The study aims to examine instruments and strategies that could lead to the better use of a historical destination of pilgrimage in Aspromonte: The Sanctuary of the Madonna of Polsi, long a place of faith and devotion. The objective of the work is to focus on the role the pilgrimage can play in the creation of value and identity in the context of sustainable development of a site that is extremely interesting from an environmental and landscape perspective.

The path that is to be followed is twofold, intending to link faith and the re-appropriation by local communities of places that have for too long been contaminated and subjugated by the presence of the 'ndrangheta, in order to promote virtuous processes of economic and social legality. This is to utilise itineraries of faith and hope for the recovery and enhancement of local resources, to give back to the local communities often forgotten testimonies of history and of shared memory, fragments of life, literature, and art, that are able to act as catalysers for developing economic activities, a solidarity economy, and responsible tourism.

2   The Values and Identity of Religious Routes in Aspromonte

Those who know the Aspromonte know that, for centuries, the many tracks and trails have been carved by the passage of the shepherds who, moved by necessity, were not deterred by the difficult and arduous paths. Their presence contributed to render the mountain alive and rich in the resources that have constituted the economy of the area: wood and livestock. But along with the shepherds, the Aspromonte was also a destination for pilgrims who, no less than the shepherds, were willing to accept fatigue and sufferance in order to reach the various places of worship, of devotion, and of the consignment of hopes and prayers.

This work draws the attention of scholars to certain routes that were born from a religious background but that have become interesting from an environmental and naturalistic point of view, which can offer themselves within a framework of recovery and enhancement of local resources that is aimed at giving back to the communities concerned often forgotten testimonies of history and shared memory, fragments of life, literature, and art. In this regard, it was decided to highlight the historical destination of the Sanctuary of Polsi, long a place of devotion and faith of an entire people on account of the Madonna of the Mountain. It is difficult, today, to trace the religious origin of the site. What we know for certain is that between the seventh and eighth centuries A.D., many Byzantine monks chose the green ravines of Calabria as places to hold elections, due to their solitude and remoteness. It was exactly during this period that there was a great flourishing of Basilian monasteries throughout the region, but it is more secure that this church was erected in the first century of the one thousand years.

To this end, two routes have been considered on the basis of the paths traditionally followed by the pilgrims. To these routes, certain activities found along the trails have been linked that could represent attractions for the tourist, such as agri-food and/or artisanal enterprises and diffuse forms of hospitality, with the scope of creating relationships that are not casual but are instead stable and durable, as well as useful for the common good. This requires, in our view, a strong collaboration between internal and local resources that are not always disposed to believe in the social and economic potential of the religious fruition of an area.

The paths of faith may, beyond their strictly religious context, also have a highly symbolic value [2, 3, 19]: the re-appropriation by the local communities of the Aspromonte in order to promote the beauty and richness of a territory that, for far too long, has been identified only with the 'ndràngheta. This is to see the Aspromonte as generator of shared interests, of virtuous processes of sustainable development, of responsible tourism, and of ethical agri-food production, and at the same time the Aspromonte as bearer of opportunities for development connected to the growth of sharing networks, to experience the reuse of confiscated lands, to a regenerating welfare that points to redefining a network of services in the rural areas in a manner consistent with the resources, specificities, and local needs in order to strengthen those communities and make them vital, attractive, and consistent with the new demand for rurality, for promoting the image, the history, the culture, and the traditions of a place [20, 21].

Becoming a healthy system for the legal economic and social growth of the communities also means strengthening the role of the institutions along the route. The intention is to tie the religious fruition incisively to the trails and mountain roads of the Aspromonte, weaving relationships with the diffused forms of hospitality already partly existing and operating in the area, as well as with the multiplicity of enterprises and producers that operate in many Aspromonte realities who are creating genuine and remarkably traditional products, often according to methods that are the fruit of ancient knowledge and traditions, and made, above all, in compliance with the law.

The path of the Madonna of the Mountain at Polsi is considered particularly important by the Diocese of Locri-Gerace which, together with the groups "Libera, Associazioni, nomi e numeri contro le mafie" (literally "Free Associations, names, and numbers against mafias"), the "Policoro Project", and "Caritas", aims to raise awareness of the use of the Sanctuary not only as a place of worship, of prayer, and of pilgrimage, but also as a place for the formation of the conscience, kept protected from any other interest: "Our land nurtures the dream of becoming a land of hope and a place of beauty, and knows that it has to commit to purifying itself from every mafia drift" (Mons. Francesco Oliva, Bishop of the Diocese of Locri-Gerace). This is to see the Sanctuary, therefore, as a space for human, social, and religious growth.

3 Paths and Roads to the Madonna of the Mountain

Until a few decades ago, Polsi could only be reached by means of trails and mule tracks. This, however, was not an obstacle for the pilgrims, accustomed as they were, like all Calabrians, to walking. In fact, the walk itself was a prelude to the feast that would be celebrated in Polsi [22].

The sanctuary is to be found in an inner place, almost at the heart of the Aspromonte, not on high ground like many other abbeys or holy sites. Of the various routes that reach Polsi some have a particular charm due, perhaps, to the role they have held in the past or, perhaps, to the extreme environmental significance they have today.

One of these is the trail to Polsi that leaves from the village of San Luca which, in 1990, the CAI (Italian Alpine Club) included in the Grand Italian Trail - the longest trekking route in the world - that connects the Italian peninsula and its main islands with the Alps. The trail, which leaves from the native village of the famous author Corrado Alvaro, initially unfolds along the banks of the Bonamico river reaching, before the walls of the valley narrow, the habitation of San Luca that was founded in 1592 by refugees from the ancient town of Potamia, which was destroyed by a landslide and whose Greek name recalled its proximity to the river. The trail then passes through a pleasant landscape, reaching the base of the landslide that damned the river in 1973 and briefly created the lake of Costantino before it almost disappeared again due to the accumulation of further debris. From here the route then climbs two rocky plateaus overlooking the lake before finally descending to the pebbly shore where the stream flows into the lake to create some small beaches. When the walls of the valley narrow, the trail leaves the river and ascends up the ridge and connects to the ancient track and, crossing streams, bracken, and gorse, gains a panoramic viewing point over the valley of Bonamico, the lake, and the crags of Pietra Castello. Continuing on, it enters a majestic holm oak wood (Santu Štefanu), so thick that even the sun struggles to penetrate. The trail continues with slight ups and downs until it joins the track stretching from the refuge of Cano down to Polsi, only a few hundred metres from the Sanctuary itself.

Another interesting trail is the one used in the past by pilgrims coming from the villages of the Piana di Gioia Tauro. The trail reaches Polsi from the forest refuge in the locality of "Vocale"[1] that is in the municipality of San Luca, at a height of 1,286 m above sea level [23–25]. The route follows along from the refuge of Vocale, running slightly downhill through a large beech tree forest, before arriving at Puntone la Croce (with its spring and water fountain), where the panorama opens onto Pietra Castello and onto Polsi directly behind it (Fig. 1). Here the pilgrims, delighted at the sight of their destination, erupted in screams and began shooting festively into the air, or dropped the rock they had been carrying as atonement for their sins, as one can note, in fact, from the enormous mounds of stone lying all around [26, 27][2].

---

[1] The forest refuges in the province of Reggio Calabria, located for the most part in the Aspromonte Park, are structures that were built in different eras, using different styles and materials, and that are currently in varying states of conservation ranging from bad to good, which have significant potential to receive hikers, pilgrims, and travellers. The refuge at Vocale can be reached from San Luca, from Gambarie, and from the Tyrrhenian area, by trekking to Delianuova and continuing on up to the Piani di Carmelia.

[2] Corrado Alvaro, native writer of San Luca, talks about the Sanctuary as a place of devotion and popular faith.

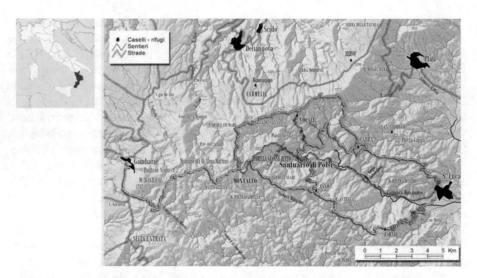

Fig. 1. Map of the trails connecting to the Sanctuary of Polsi (Dr. Roberto Lombi).

4 Towards a Qualification Strategy of Redevelopment: Tools and Objectives

The complex relationships between pilgrimage and the natural environment is arousing a growing interest from the scientific and institutional world, and if on the one hand the scientific world contributes with a wide and articulated bibliography [28, 29], in many cases, the tools and management of sites at a local level are inadequate. The re-appropriation of the territory can be achieved through routes that are able to generate virtuous processes of economic and social legality (Fig. 2) that revolve around the use of religious places of worship [30]. Among the possible hypotheses we can, for example, list: (i) the necessity of a re-appropriation of the Aspromonte through the social reuse of the confiscated assets from organised crime; (ii) a recognition and definition of the forms of hospitality along the routes; (iii) an adequate recognition of historical-cultural sites of high interest, often forgotten or underestimated, and the identification of thematic routes; (iv) the recognition of places of particular landscape/naturalistic value; (v) the recognition of valuable agri-food products typical of the territory; (vi) the recognition of valuable artisanal products; (vii) the paths of memory, of culture, and of the anti-mafia commitment, such as that dedicated to Lollò Cartisano, an innocent victim kidnapped and killed by the 'ndrangheta in 2003.

These possible strategies require a continuous interaction and involvement with the local communities, with the associations, and with the entities which work towards a structural and socio-economic redevelopment of the territory in order to develop actions of valorisation in tandem, also in relation to the religious fruition of the site.

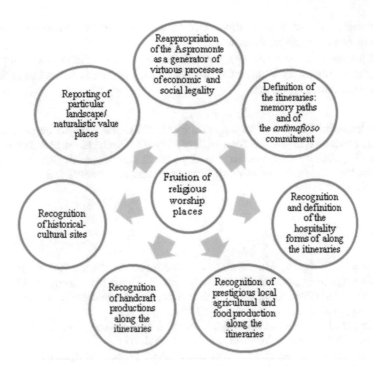

Fruition of religious worship places

Reappropriation of the Aspromonte as a generator of virtuous processes of economic and social legality

Definition of the itineraries: memory paths and of the *antimafioso* commitment

Reporting of particular landscape/ naturalistic value places

Recognition and definition of the hospitality forms of along the itineraries

Recognition of historical-cultural sites

Recognition of prestigious local agricultural and food production along the itineraries

Recognition of handcraft productions along the itineraries

Fig. 2. Hypothesis of connection between the various interventions on the Aspromonte territory [30]

5   Conclusions

The experience of pilgrimage embodies an existential desire to seek or manifest the identity and the values of a person. The internal disposition to choose a travel desti-nation can be seen as a spiritual awareness of life's journey (religious and non-religious), and can be a means to express ones' personal or social identity, or for searching for or reaffirming that identity [3].

The historical destination of the Sanctuary of Our Lady of Polsi in Aspromonte, long a place of devotion and faith for an entire people due to the Madonna of the Mountain may represent, at the same time, both a path of faith and a path of awareness and shared memory. It is territory/laboratory as a place for the formation of the con-science, for the consolidation of local webs and networks, and for promoting:

- virtuous economic systems based on the rule of law and social justice, and on ethical, economic and sustainable development;
- productive processes of development linked to multifunctional and social agricul-ture, involving agri-food production, agritourism, and the reorganisation of the management of production activities, that are aimed at obtaining quality products through methods that are respectful of human dignity;
- systems of welfare in rural areas that are adequate for local needs, specific and consistent with available resources, that provide employment opportunities to the

communities, increase per-capita income, and improve the dynamics of networks that can unlock local know-how and skills;
– social and productive recovery of assets confiscated from the mafia in areas with a strong mafia presence, in order to return them to the community and transform them into shared common goods.

Acknowledgement. The study was supported by GASTROCERT Project on Gastronomy and Creative Entrepreneurship in Rural Tourism, funded by JPI Cultural Heritage and Global Change - Heritage Plus, ERA-NET Plus action "Development of new methodologies, technologies and products for the assessment, protection and management of historical and modern artefacts, buildings and sites".

References

1. De Lucia, C., Fragassi, M., Pazienza, P., Vecchione, V.: Indicazioni di "Policy" per il turismo religioso in provincia di Foggia. Rivista di Scienze del Turismo **2**, 23–33 (2014)
2. UNWTO: International Conference: Spiritual Tourism for Sustainable Development, Ninh Binh City, Madrid, 21–22 November 2013 (2013)
3. Liutukas, D.: The manifestation of values and identity in travelling: the social engagement of pilgrimage. Tour. Manag. Perspect. **24**, 217–224 (2017)
4. Gilli, M., Ferrari, S.: Marginal places and tourism: the role of Buddhist centres in Italy. J. Tour. Cult. Change **15**(5), 422–438 (2017)
5. Nicolosi, A., Pezzino, V., Sciacca, L., Strazzulla, G.: Educare alla pace: la questione Palestinese. Pellegrinaggio di Giustizia nei Territori Occupati, Algra Editore (2017)
6. D'Amore, L.J.: Spirituality in tourism: a millennium challenge for the travel and tourism industry. Tour. Recreat. Res. **23**(1), 87–89 (1998)
7. Cheer, J.M., Belhassen, Y., Kujawa, J.: The search for spirituality in tourism: toward a conceptual framework for spiritual tourism. Tour. Manag. Perspect. **24**, 252–256 (2017)
8. Collins-Kreiner, N.: Researching pilgrimage. Ann. Tour. Res. **37**(2), 440–456 (2010)
9. Amaro, S., Antunes, A., Henriques, C.: A closer look at Santiago de Compostela's pilgrims through the lens of motivations. Tour. Manag. **64**, 271–280 (2018)
10. Jerryson, M.: Monks with guns: Westerners think that Buddhism is about peace and non-violence. So how come Buddhist monks are in arms against Islam? AEON (2017). https://aeon.co/essays/buddhism-can-be-as-violent-as-any-otherreligion. Accessed 01 May 2017
11. Nicolosi, A., Bandini, C.: Creative resistance through creative agricultural practice: the case of the tent of nations in the Bethlehem area. In: Laven, D., Skoglund, W. (eds.) Proceedings of the Valuing and Evaluating Creativity for Sustainable Regional Development, pp. 147–149. UNESCO, Mid Sweden University (2016)
12. Digance, J.: Pilgrimage at contested sites. Ann. Tour. Res. **30**(1), 143–159 (2003)
13. Timothy, D.J., Olsen, D.H. (eds.): Tourism, Religion and Spiritual Journeys. Routledge, London (2006)
14. Norman, A.: Spiritual Tourism: Travel and Religious Practice in Western Society. Continuum, London (2011)
15. Kouchi, A.N., Nezhad, M.Z., Kiani, P.: A study of the relationship between the growth in the number of Hajj pilgrims and economic growth in Saudi Arabia. J. Hosp. Tour. Manag. (2016)

16. Baraldi, L., Pignatti, A.: Il patrimonio culturale di interesse religioso: Sfide e opportunità tra scena italiana e contesto internazionale. FrancoAngeli, Milano (2017)
17. Privitera, D.: I Parchi e il cicloturismo: integrazione strategica per lo sviluppo locale. Agribusiness Paesaggio & Ambiente **XIV**(3) (2011)
18. Pritchard, A., Morgan, N., Ateljevic, I.: Hopeful tourism: a new transformative perspective. Ann. Tour. Res. **38**(3), 941–963 (2011)
19. Kato, K., Progano, R.N.: Spiritual (walking) tourism as a foundation for sustainable destination development: Kumano-kodo pilgrimage, Wakayama, Japan. Tour. Manag. Perspect. **24**, 243–251 (2017)
20. Lu, D., Liu, Y., Lai, I., Yang, L.: Awe: an important emotional experience in sustainable tourism. Sustainability **9**(12), 2189 (2017)
21. Nicolosi, A.: Agricoltura sociale e terreni confiscate alla criminalità organizzata. In: Privitera, D., Nicolosi, A. (eds.) Comunità, Luoghi e condivisione: esplorazione di modelli alternative di consumo, FrancoAngeli, pp. 99–118 (2017)
22. Picone Chiodo, A., Raso, D.: San Luca e Polsi, Storie e Paesaggi d'Aspromonte, Città del Sole (2008)
23. Picone Chiodo, A. (ed.): Guida ai caselli forestali della provinciale di Reggio Calabria. Club Alpino Italiano sezione Aspromonte, Stampa A&S Promotion (2006)
24. Gemelli, S.: Storia, tradizioni e leggende a Polsi d'Aspromonte. Gangemi Editore, Roma (1992)
25. Giampaolo, D.: Un viaggio al Santuario di Polsi in Aspromonte, ristampa anastatica, Città del Sole (2011)
26. Grado, S.: A garden or a laboratory? Conflicting views about nature preservation and local development in the Aspromonte National Park, Italy. Plann. Stud. **12**(4), 327–342 (2007)
27. Alvaro, C.: Polsi nell'arte, nella leggenda e nella storia. Iriti editore, reprint (2005)
28. Salerno, S. (ed.): Domenico Zappone. Il pane della Sibilla. Viaggio nei luoghi di Corrado Alvaro. Rubbettino editore (2014)
29. Kujawa, J.: Spiritual tourism as a quest. Tour. Manag. Perspect. **24**, 193–200 (2017)
30. Di Gregorio, D., Picone Chiodo, A., Nicolosi, A.: Paesaggi culturali e sentieri della memoria: strumenti di sviluppo per territori e comunità. In: 21st International Proceedings Interdisciplinary Conference, The Paradise Lost of the Cultural-Landscape Mosaic. Attractiveness, Harmony, Ata-raxy Venice, Italy, 6–7 July (2017, in press)

The Network of the Villages of the Metropolitan City of Reggio Calabria, a Complex Attraction in the Design of Quality and Safety of the Territory

Natalina Carrà[(⊠)]

Mediterranean University of Reggio Calabria, 89100 Reggio Calabria, Italy
ncarra@unirc.it

Abstract. The territory of the metropolitan city of Reggio Calabria is complex and fragile; a complexity derived from a territorial and social structure, the result of the millennial history of man's work on a difficult nature, the result of which is a contamination between the natural environment and the anthropic elements that shape and structure the places. The fragility referred to the physical-urban dimension and to the aspects of the structure of the social structure, also concerns the management of services and public spaces, connections and urban and territorial infrastructures, as well as demographic aspects: many of these urban centers in the last decade have undergone abandonment and depopulation with values often above the national average.

Interventions on a territory with these characteristics are therefore linked to a need to adapt to new needs and therefore in this specific case, to a change in the concept of efficiency closely linked to the notion of use, function and quality of space and places. Imagining the network of villages, the beating heart of this context, means thinking of a metropolitan area aimed at increasing the efficiency of services, spaces and commons. It means strengthening the cultural level of the community to create awareness and commitment towards the protection and enhancement of the identities of the places.

Networking also means facilitating physical and social accessibility to places and services; this plays a fundamental role in the equilibrium of the metropolitan area because it affects the propensity of users to experience the metropolitan space in conditions of safety.

Keywords: Villages · Urban quality · Territory project

1 Identity and Fragility of a Complex Territory

All territorial contexts have peculiarities that make them unique, they testify to their uniqueness, made of environment, men and activities that over the centuries have deposited their tracks. Every territory must therefore be seen in its complex and articulate *attraction* - a system in which the context - as a bearer and holder of history and culture - is embedded in an unique political, social, economic relational fabric. *There are new relationships between culture and territory: the territory continues to be*

© Springer International Publishing AG, part of Springer Nature 2019
F. Calabrò et al. (Eds.): ISHT 2018, SIST 101, pp. 94–102, 2019.
https://doi.org/10.1007/978-3-319-92102-0_11

*the only container of the culture of which the group is living (...). It would therefore be appropriate to consider the continuity that a territory invokes; in fact, parts of culture - of every culture - are exposed to the risk of silence and death, as well as continuous modification and transformation*s [1].

The territory of the metropolitan city of Reggio Calabria (former provincial territory) is complex and fragile; a complexity derived from a territorial and social structure, the result of the millennial history of man's work on a difficult nature, the result of which is a contamination between natural environment and man-made elements that conform and structure the places. The connections between anthropic activities and environmental/natural features have historically structured the territory, according to forms that local people perceive as an expression of their peculiar, cultural and identity characters. A territorial and environmental context in which man has lived for millennia, leaving visible and less visible traces, shaping the environment for its own needs, exploiting its resources, but above all keeping it alive and structuring its potential in a different way in different historical periods. The course of history, through the signs, traces and fragments that often remain intact, even in marginal parts marked by abandonment and economic weakness, is noticeably marked in this context. Demonstrating that even an essentially rural area, such as the one in the metropolitan city of Reggio Calabria, has history to communicate and resources to be valued as bearers of a profound local identity and possible generators of new economic and social dynamics.

The settlement structure of this territory also has a physical and socio-economic fragility, in fact there is a significant presence of urban areas with high hydrogeological and seismic risk, as well as a strong local dispersion on the territory with consequent consumption of agricultural land. The urban fabric often has the characteristic duality of the center and the periphery: in planned areas and with historical-identity valor is contrasted with an informal, distinct periphery characterized by phenomena of abuse and sometimes degradation.

The fragility of sites related to the physical-urban dimension and the aspects of the social structure also concern the management of public services and public spaces, connections and urban and territorial infrastructures, as well as demographic aspects: many of these urban centers in the last decade have experienced phenomena of abandonment and depopulation with values often above the national average.

The physical layout of the territory, orographic by the Aspromonte massif, constitutes a very special structure of the metropolitan city of Reggio Calabria. It has a marginal location in relation to the vast territory, and this leads to an atypical metropolitan configuration on which few urban centers are gravitating to other metropolitan contexts. That is, a good part of the former provincial territory does not have significant dependency relationships with the city, and many of these centers have no daily economic and social relations with Reggio Calabria.

We are faced with a pattern in which the emerging urban armor is composed by a large number of small and tiny municipalities, whose expansion involves widespread conurbation phenomena, a unique large city, and a few other mid-size centers.

The new administrative identity thus leads to the newly established metropolitan city, to assume more articulated territorial functions. In addition to the delicate task of guaranteeing and enhancing a territorial and urban structure characterized by great contrasts and complexity. It is spread in a very discontinuous, morphologically wide area, where the requirements of protection in the broad sense come frequently in conflict with the pace of transformation and the dynamics of often unfavorable and/or adverse socio-economic trends.

These characteristics of a complex territorial and social system, which has many components or subsystems in interaction between them, are precisely the behaviors of the individual subsystems along with their reciprocal interactions that feed the complexity.

Consequently, it is necessary to introduce forms of articulated and composite transformation linked to the adaptation of cognitive frameworks and institutional structures, which in the metropolitan city are an active subject in the government of the territory, within which it is necessary to re-declare the principles and governance goals from a new perspective, based on system and policy integration.

The transformations to be introduced to address and design the metropolitan dimension pass thus through its physical and cultural complexity. And the changes and transformations are directed towards one of the greatest resources of this territory: the variety of its landscapes (coastal, mountainous, hilly, historic and archaeological landscapes, ancient villages). The landscape, which changes according to the physical, natural and anthropic characteristics of the territory, is a resource that tells the identity, and as such, a true wealth for design.

The villages, the protected areas, the historical and archaeological heritage have become a key part of the development of new policies and the creation of *special productions chain for the territory and urban spaces* (in this case metropolitan areas) increase in the use of new technologies and, of course, improvements in infrastructure [2]. They may be the basis for a planning model where cities, urban network of the cities and the policies of urban network, are the most active part of urban and territorial development processes and urban cohesion, considering city and territory as an integrated system that allows you to achieve the metropolitan dimension more effectively.

2   The Force of Fragility. The Italy of the "Hundred Cities" and the Universe of Small Councils, Territorial Processes Underway

The fragile word has often indicated the precariousness of a territory, but sometimes the places we define fragile are, in fact in a different way, sometimes rare or occasional bearers and heirs of traditions to be passed on to younger generations. They are often places of some sort of disturbing beauty, blended with perfect harmony [3].

In addition to being the strongholds of identity and guardians of our historical, artistic, natural and cultural heritage, these small towns are also the privileged place to experiment with good practices of urban quality and life. Their fragility when well-interpreted and used, represent a value for the territory. These sites are a true garrison for the territory, especially for activities to counteract hydrogeological disaster and for the activities of widespread maintenance and protection of commons. They are fragile urban contexts, but of high naturalistic and historical-identity value, and it is precisely on this peculiarity that one can aim to find advanced forms of transformation; a sort of way out of the crisis that points to the interweaving of identity, innovation and strength of the territory.

The dynamics of the last century, stress that, in addition to the reality of the *"cento città"*, theorized by Cattaneo [4], that was, and is interesting and unique for its physical, historical, social and cultural differences, exists and is very large, that of the capillary network of villages and small towns, where most of the national population resides. Realities characterized by a strong identity, as well as by a good quality of life, but often shocked by sudden transformations.

The implementation of the National Strategy for the Internal Areas and the recent text "Measures for the support and enhancement of small municipalities, as well as provisions for the upgrading and recovery of the historic centers of the same municipalities" (Law No. 158/2017) allow to address the main issues related to the revival of the villages/small municipalities, starting from systems for enhancing cultural heritage, urban planning and territorial development. The goal is to stop the depopulation, recognizing also the specificity of the places, these centers need differentiated policies and specific support to their peculiarities. These measures seem to protect the peculiarities of the country, the urban and territorial armor, the true essence of Italy in the hundreds of cities, a microcosm of local realities that today lives a new dimension.

The eighth *National Small Towns Association* of *Legambiente* (presented in June 2017) [5], draws attention to three opportunities to pursue: *the residential, agricultural and tourist one*, to hinder, according to criteria of a circular economy, depopulation, aging and the denaturing of much of the national territory. The report starts from a different reading key, which does not resemble, but overturns, the current situation with numbers in perspective, trace development hypotheses and actions that could materialize on these three possible opportunities.

The report highlights that in small municipalities, we count about one empty house every two occupied. Housing that represents an opportunity for social re-use, building and tourism recovery that could be utilized for new housing, including on the reception of migrants. Considering also the return of young people to agriculture and the recognition of Italian excellence products on the international market, the union between the agro-food chain and the tourist-food and wine attraction can become a coordinated system of site development. As for the growth of the tourist offer, the report highlights the need to build network, on-line and on-the-spot, thanks to the broadband and the linkage with industry operators. Work, extending the townships to business certification, trusts in this process as a competitive accelerator and performance.

The potentials of these territories are many and substantial, and in recent years there have been many small municipalities that have shown that it is possible to counteract the settlement disadvantage by focusing on tradition and innovation, on the protection of the environment, on the regeneration of the housing stock, starting up processes of physical and socio-economic transformation that start from the bottom by involving communities and all local operators.

3   The Roots of the Settlement and Infrastructural System of the Metropolitan City of Reggio Calabria in the Network of Small Villages. A Territory Project

The territory of the metropolitan city of Reggio Calabria has an extension of 3,183 sq km (about 1/5 of the regional area) and comprises 97 municipalities with a total population of about 28% of the population of Calabria, of which 26 are those with a population of more than 5,000 inhabitants, while 71 are those with less than 5,000 inhabitants, or 73% of the metropolitan municipalities have a population of less than 5,000 inhabitants, higher than the national level, which accounts for around 70%.

The population dynamics of the last decades have led to the concentration of the population predominantly in the city of Reggio and the coastal centers and have caused a gradual abandonment of mountain villages and small villages (Fig. 1). Apart from Reggio Calabria, no other center of the area exceeds 20,000 inhabitants (2017); the most populous municipalities on the Tyrrhenian side are Palmi (about 19,000 inhabitants), Gioia Tauro (about 20,000 inhabitants), Taurianova (about 16,000 inhabitants) and Rosarno (about 15,000 inhabitants); on the Ionian Sea, the most important center in this sense is Siderno (about 18,000 inhabitants).

To imagine, therefore, that small villages in the metropolitan city can represent beyond the roots of identities, also the network of a strategic vision for the metropolitan territory project, in terms of complementarity, emergencies, identity and cultural-social capital, is valued even by numbers. In this *vision*, however, the capacity for cohesion, sharing, vision and convergence towards the integrated action of a metropolitan system must be put in place, exceeding the objective limits, resulting from articulated, discontinuous, unmanageable and small communities for demographic presence.

The metropolitan city territory has a wealth of cultural, environmental and historical resources and cultural attractors, often located in these villages, which are, but can become, more and more territorial competitiveness elements not only national but also international. Adding to this are extraordinary territorial features and values between the landscape contest of the Aspromonte National Park and the beaches of the ionic slope, the Tyrrhenian cliffs, the rivers, towers and castles, and all the historical remains, guardians of the roots of the places.

These small towns/villages must therefore be understood as *new* competitive sites, which on the one hand safeguard the forms of historical identity and individual characteristics, and on the other hand, emphasizing new modes of engagement, activate processes capable to preserve small communities and project them into the future.

The new administrative level, the metropolitan city, is therefore a key milestone for re-occupying its territory, as well as contributing to the revitalization of the villages themselves, to their safe disposal, also to contain hydrogeological phenomena and to prevent seismic ones, all in accordance with the principles of environmental, social and economic sustainability.

Involving development processes by investing in the valorization of local identities, on the environment and on the cultural and human heritage, that is to put the territory and its identities into play, leads to the creation of *new networks of relationships* that go

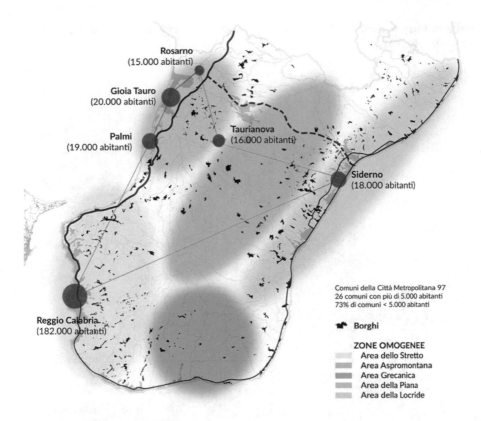

Fig. 1. The villages in the settlement system of the metropolitan city. Homogeneous zones and settlement weights

beyond places, they go towards local development models that need social cohesion and authenticity for the promotion of development.

4 A Vision for the Project

Interventions in the area are often linked to the need to adapt to new needs and therefore, in this specific case, a change in the concept of *efficiency* closely related to the notion of *use, function and quality* of space and places. To imagine the *network of villages* means in this context an organization of the metropolitan area aimed at increasing the efficiency of services, spaces and commons. It means strengthening the cultural level of the community to create awareness and commitment protecting and enhancing the identities and specifics of places. Networking also means facilitating physical and social accessibility to places and services; this plays a key role in the metropolitan area's balance because it affects the user's propensity to live metropolitan space in safe conditions. In this process of reorganizing the territory, the components of the environment have to be re-read through the pattern of use that citizens give back and according to their needs, not ignoring the instances of protecting the identity of the territory (Fig. 2).

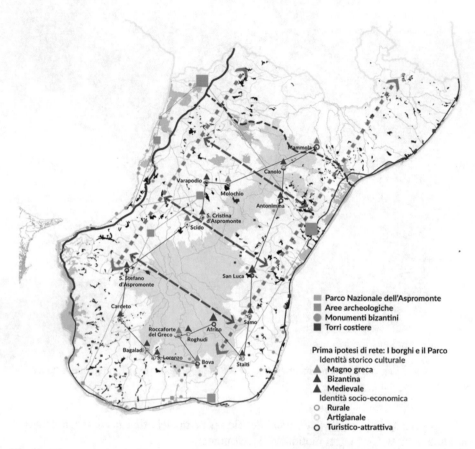

Fig. 2. The network of villages and complex attractors. First network hypothesis: the villages and the park

The territorial design thus conceived is above all an elaboration of individual, public and collective knowledge. Three are the main aspects of relevance in this approach: the full assumption of uncertainty (in context conditions and in defining the outcomes of the actions) in the territorial planning processes; the search for methods of reducing complexity through the selection of relevant themes and the attempt to represent them, though inevitably simplified; the full recognition, and not just ritual and rhetoric, of the process in the construction of the plan/territory project, understood as a continuous opportunity of "collective" learning [6].

By focusing on the methods of reducing complexity through the selection of major themes in the design of the territory as a complex attraction, one can find the identification and definition of special uses of urban (metropolitan) spaces. Special use chains are tools for enhancing the quality of metropolitan territory and are understood as the articulated network set comprising the main activities and their main material and information flows, technologies, resources and organizations which contribute to the operation of a given sector/area, among those that are the cornerstones of the new metropolitan design. Defining a chain in all areas where it is declining, touching on technical-scientific and strategic issues means defining areas that, by their characteristics, can make the difference, that is to offer opportunities for original innovation in the metropolitan organization, giving space to the specificities of the territories, a process of local development is more efficient/lasting when is fond by peculiarities of a particular context: the resources of the places become a strength of the quality and the difference in the supply of products and services that result from the enhancement of peculiar natural values and anthropic.

Among the transverse principles or interpretative paradigms in the construction of urban metropolitan image and in the production of services and infrastructures can be defined as:

- *Sustainability:* strategies, interventions and actions must contribute to systematically strengthening the sustainability of interventions, in its various contexts relating to environment, territory, heritage conservation, socioeconomic system, culture and citizenship.
- *Innovation*: strategies, interventions and actions must help systematically innovate products, processes, technologies and organization of activities, innovate ways of using them, create new and more advanced skills.
- *Accessibility/physical and cultural permeability:* strategies, interventions and actions must help to enhance the accessibility of both places and services (cultural) to activate valorization processes. Accessibility is broadly understood here: it means ensuring the enjoyment of all categories of people without any physical or social distinction; means physical accessibility of the territory through sustainable mobility systems; means giving users the opportunity to understand and interpret the history, complexity and variety of the patrimony visited.

Spreading the awareness of the cultural and social value of a place, it carries an attitude of respect and protection of places and common spaces, which consequently increases the sense of security and well-being of those who live and inhabit live in those places, removing vandalism, degradation and crime. At the same time, knowledge of the cultural value of places is the basic prerequisite for designing actions that bring social benefits, as well as economic, to the entire community. It is a dimension that requires *new visions* of good urbanism and of a territorial planning based on a network of sites and the protection and enhancement of the richness and variety of resources and natural and cultural landscapes.

References

1. Bonato, L. (ed.): Portatori di cultura e costruttori di memorie. Ed. dell'Orso, Alessandria (2009)
2. Tarpino, A.: Il paesaggio fragile: L'Italia vista dai margini. Einaudi, Torino (2016)
3. Faccioli, M. (ed.): Processi territoriali e nuove filiere urbane. Franco Angeli, Milano (2009)

4. Cattaneo, C.: Considerazioni sulle cose d'Italia nel 1848. Einaudi, Torino (1946)
5. Polci, S., Gambassi, R. (eds.): 8 Rapporto nazionale PICCOLI COMUNI, Indagine realizzata con Legambiente, Supporto SM Gianfranco Imperatori Onlus, giugno (2017)
6. Magnaghi, A. (ed.): Scenari strategici. Visioni identitarie per il progetto di territorio, Alinea editrice, Firenze (2005)

Consumers' Preferences for Local Fish Products in Catalonia, Calabria and Sicily

Agata Nicolosi[1], Nadia Fava[2], and Claudio Marcianò[1(✉)]

[1] Mediterranea University of Reggio Calabria, 89100 Reggio Calabria, Italy
claudio.marciano@unirc.it
[2] Politecnica III, 17003 Girona, Spain

Abstract. Research on public markets in small provincial towns is scarce, particularly on the role they play in maintaining a relationship with the local culture, environment and production. This paper examines consumers' habits and preferences for food shopping in three European regions with respect to the purchase of fish products. The goal is to investigate consumers' preferences for local fish to highlight the motivations that lead to different choices. A multiple correspondence analysis explores the motivations behind purchasing preferences, showing the complex process that drives individual consumer choices. Based on 504 interviews conducted in cities and areas adjacent to the cities of Girona, Reggio Calabria, and Lipari, we found no evidence of converging habits and homogenization on preferences. It supports the perspective in which the interplay between local culture and consumption of local products is strictly associated.

Keywords: Consumers' preferences · Food markets · MCA

1 Introduction

Global fish demand continues to grow, and people around the world appreciate the health benefits of regular fish consumption by reaching a per capita consumption of 20.5 kg/year in 2016, according to FAO data. However, the disparities in consumption between more developed and less developed countries, although high, are tending tend to shrink, and, while in developing countries fish consumption tends to be based on consumption of locally available and seasonal products, in developed countries there is an average annual consumption per capita of 27–28 kg [1], with a considerable and increasing share of product imported from other countries. In the face of these trends in the three Mediterranean localities investigated, however consumers have a tendency to prefer local fish products.

This study looks at the preference of consumers of fish products in three European Mediterranean regions, Catalonia in Spain, Calabria and Sicily in Italy. The choice of the favorite places of purchase by consumers (specialized shop, City market or super/hyper market) has made it possible to examine the attractiveness of the different points of sale, in relation to the frequency of purchase, the means of transport used to reach them, and motivations that push people to choose one or the other of the different points of sale [2, 3].

© Springer International Publishing AG, part of Springer Nature 2019
F. Calabrò et al. (Eds.): ISHT 2018, SIST 101, pp. 103–112, 2019.
https://doi.org/10.1007/978-3-319-92102-0_12

2 Methods and Data Collection

In order to examine the ways in which consumers organize their purchases, the survey used direct face-to-face interviews with consumers by means of a semi-structured questionnaire with both open-ended and/or closed questions.

The questionnaire was administered in the following locations:

- Girona (Catalonia/Spain), interviews with people intercepted mainly in the city market in the period between April and May 2017;
- Reggio Calabria (Calabria/Italy), interviews with people intercepted in the city and in neighboring coastal areas, in city markets and in other particularly crowded places, in the period between July and September 2017;
- Lipari, in the Archipelago of the Aeolian Islands (Sicily/Italy), interviews with people intercepted in the city markets and in other particularly crowded places, in the period between July and September 2017.

The questions in the questionnaire were aimed at identifying the following characteristics: the consumption of different agricultural products and fish products and the sale points in which they are purchased; the frequency of purchase; and the reasons for purchase. It was also asked which type of transport was used to reach the places of purchase, the proximity to the place of residence, the place of purchase being seen as a space for socializing and sharing. Having been asked questions about preferences for local fish and local food products, their attention was then directed to health food, a wide range, traceability, labeling, reputation, and ethical and environmental and social sustainability aspects of purchased food products.

The questionnaire also identified the socio-demographic characteristics, sex, age, education, employment and family income of the interviewed consumers. In addition, consumers were presented with a grid of four motivational questions, prepared ad hoc, to mark, from 1 to 10, their motivations for purchasing local food in the vicinity [4]. Based on the responses collected, a database was created, with the help of SPSS software (version 20) in order to analyze consumers' preferences and the reasons for their choice. The database of collected data was processed, analyzed, and initially interpreted through descriptive analysis to highlight the principal characteristics, and then through Multiple Correspondence Analysis (MCA). From a technical point of view, MCA is used to analyze a set of observations described by a set of nominal variables. We refer to literature for more detailed information on MCA properties and objectives [5, 6]. MCA aims to attribute factor scores to each observation and to each category in order to represent relative frequencies in terms of the distances between individual rows and/or columns in a low dimensional space [3, 7].

In this study, analysis should allow us to find evidence on the relationship between the seven individuated motivations that lead consumer choices for fish products. Through a representation in a low-dimensional space – designed on the basis of few principal components we aimed to define some clusters (profiles). MCA identifies the factors underlying the structure of the data, making it possible to construct the principal components which optimally summarize the data, hence highlighting the interdependent relationships between a small number of variables [6, 8]. The results are therefore interpreted on the basis of relative positions of the points and their distribution along the dimensions. As for the study of individuals, two individuals are close to each other if they answered the questions in the same way. The focus is more on populations rather than on individuals.

In order to implement the MCA, seven dichotomous variables were selected, as shown in Table 1. The variables identified, from among those that showed greater connectivity with quality attributes and purchasing motivations, related to social, cultural, and buying habits.

Table 1. The variables chosen for the MCA.

Question	Variable name	Modalities	
Do you prefer buying local fish products or industrial brands/products?	Local fish	Yes	No
Where do you buy it? You can select more than one answer			
– Specialized shop/retail shop (the favorite answers in Reggio Calabria and Lipari)	Retail shop	Yes	No
– City market (the favorite answer in Girona)	City market	Yes	No
Please indicate how these conditions affect your food shopping			
– Food safety, quality and health benefits	Health	Yes	No
– Organoleptic qualities (flavor, color, odor, freshness, crunchiness, etc.)	Qualities	Yes	No
– Wide range of choice	Wide range	Yes	No
– Price	Price	Yes	No
– Possibility to buy food products with zero km transportation	Km0	Yes	No

3   Results

3.1   Description of the Sample

In order to provide a characterization of the sample, Table 2 reports some general characteristics of the interview participants. The consumers of local fish product interviewed in the survey have a medium-high cultural level in all of the localities investigated, and they are more or less equally divided between men and women. Most fall within the 18–39 age group (Reggio Calabria and Lipari) and the over 65 age group (31% of the Girona sample). With respect to occupation, we can see in all cities that the greatest percentage of our consumers are principally employed or self-employed, and in Girona the two main groups are the employed and the retired. It is important to note that a high percentage of the respondents gave no answer about their occupation in the case of Reggio Calabria and Lipari. There is the same situation for the data about family income, in particular in the case of Reggio Calabria (26.3%). Consumers in Lipari and Girona have a lower middle/upper middle family income. On average, the number of family members in the three localities investigated was around four members (only in the Girona sample we observe the percentage of 34.5% in the case of two family members, mainly because the average age is higher). Most of the consumers interviewed live in the place where the survey was administered or they are simply tourists on holiday (this applies in particular for Reggio Calabria and Lipari).

Table 2. Descriptive statistics of the sample

		Girona		Reggio Calabria		Lipari	
		n.	%	n.	%	n.	%
Gender	Male	61	43.0	120	53.6	70	50.7
	Female	81	57.0	104	46.4	68	49.3
Age	18–39	30	21.1	75	33.5	46	33.3
	40–54	39	27.5	61	27.2	32	23.2
	55–65	29	20.4	62	27.7	31	22.5
	>65	44	31.0	26	11.6	29	21.0
Education	Primary school	29	20.4	40	17.9	2	1.4
	Middle school	39	27.5	26	11.6	7	5.1
	High school	27	19.0	104	46.4	73	52.9
	University Degree	47	33.1	54	24.1	38	27.5
	Not answered	-	-	-	-	18	13.0
Occupation	Employee	53	37.3	58	25.9	40	28.2
	Self employee	14	9.9	63	28.1	42	29.6
	Retired	54	38.0	30	13.4	11	7.7
	Unemployed	13	9,2	15	6.7	13	9.2
	Other/not answered	8	5.6	58	25.9	36	25.4
Family income	High	3	2,1	3	1.3	3	2.2
	Upper middle	59	41.5	88	39.3	36	26.3
	Lower middle	59	41.5	94	42.0	54	39.4
	Low	21	14.8	20	8.9	8	5.8
	Not answered	-	-	19	8.5	36	26.3
Family members	1	32	22.5	34	15.2	12	8.7
	2	49	34.5	43	19.2	28	20.3
	3	26	18.3	56	25.0	37	26.8
	4	24	16.9	61	27.2	40	29.0
	5	5	3.5	28	12.5	20	14.5
	>5	6	4.2	2	0.9	1	0.7
Why place survey	I live here	108	76.1	161	71.9	79	57.2
	Work here	8	5.6	24	10.7	24	17.4
	Holiday	7	4.9	29	12.9	26	18.8
	Other	19	13.4	10	4.5	9	6.5

3.2   Analysis of MCA Results

Concerning the MCA results, Figs. 1 and 2 show how consumers differs in the three regions. It is necessary to examine separately the positions assumed by the variables considered in the two dimensions and quadrants of the two-dimensional space identified by the MCA. Projecting each variable in the four quadrants allows us to identify the contribution of each (variable). The analysis also calculates the inertia and eigenvalue for each dimension in order to examine how much of the total data variability is explained [9, 10]. In the present study the two-dimensional solution was considered the most appropriate.

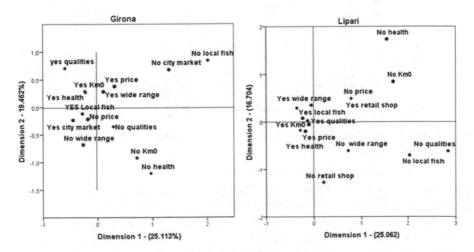

Fig. 1. Results taken from MCA data in Girona and Lipari; position of the original modalities in two-dimensional space.

In Girona, the two dimensions explain 44.6% of the variance. In this case, the two-dimensional semi-axis separates consumers who express their preference for fish consumption in terms of safety, quality, price, Km0, and a wide choice of products, from those who are indifferent to those requirements. The first and second dimensions have eigenvalues of 1.758 and 1.362, respectively, and inertia of 0.251 (25.1% of unfolded variance) and 0.195 (19.5% variance explained). In Girona, it is the first dimension that mostly explains consumers' behavior (Fig. 1). It separates those who give importance to price and to a wide variety of choice from those who prefer food safety, quality, and Km0, and who prefer to buy fish products at the city market. In particular, in the negative quadrant for Dimension 1, and, in the positive quadrant for the Dimension 2, we find the consumers who consider the quality of the purchased fish products to be important, are attentive to food safety, and buy Km0 products. Their profile can be defined as *"healthy consumers, attentive and sensitive to new trends"*. In this group, we find the greatest consensus among the consumers interviewed in Girona (33.8%): they are mostly women (62.5% of the segment), graduates (35.4%), mainly

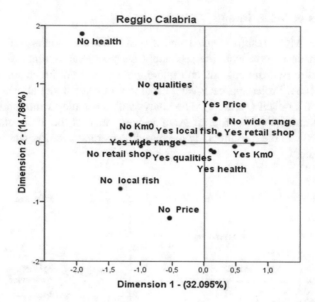

Fig. 2. Results taken from MCA data in Reggio Calabria; position of the original modalities in two-dimensional space

aged between 40 and 54 years (35.4%), and public employees (37.5%), but there are also retirees (31.3%) and people with a low average income (47.9%).

From Fig. 1, we also see that the variables "price" and "wide range of choice" are in the positive quadrant for both dimensions. 16.2% of the consumers interviewed are in this quadrant. The different socioeconomic characteristics of the subjects that are placed in this quadrant are as follows: mainly young (39% of the positive segment for the two dimensions), employees (45% of the cases), slight prevalence of males (57%), high education degree (48% graduates), average high income (48%), four household members (39.1%). The purchase of fish products from the choice available, is mainly dictated by the convenience of the price and the wide range of offer. Their profile can be defined as "saver consumers", they seek solutions in which will save money, and are interested in the assortment/range of fish products.

In the positive quadrant for Dimension 1 and in the negative quadrant for Dimension 2 we find 21.8% of the consumers: with a slight majority of women (54.8%), an age over 65 years (51.6%), a low level of schooling (32.3%), retired (54.8%), an average high income (51.6%), and one to two family members (64%). Their main characteristic is that they do not pay particular attention to safety, quality and Km0, these are "habit consumers" who buy the local fish products they know.

We find 28.2% of the interviewed consumers in the negative quadrant for both dimensions. They buy fresh local fish at the city market because of their loyalty to dealers, and the reputation and trust in the store are important. They are mainly women (60%); the prevalent age group is greater than 65 years (35% of the segment' consumers); they are graduates (37.5%), retirees (37.5%), with a low average income (45%), and are in households with one or two members (57.5%). As for their profile, these consumers can be considered *"loyal consumers"*, passionate and curious, looking for what the city market can offer.

In Lipari, the two dimensions explain 41.8% of the variance. In this Sicilian town, consumer preferences are located along the negative axis for Dimension 1 (accounting for 25.1% of the explained variance). The second dimension has an inertia of 0.167 explaining 16.7% of the variance. In particular, the first dimension separates those who express a clear preference for purchasing fish in retail shops from those who attach relevance to all the other variables. The quality, the wide range of choice and the purchase of local fish products are considered relevant (in the positive quadrant for Dimension 2 and negative for Dimension 1) (Fig. 1). Consumers express a marked preference for buying in retail stores and specialized stores from trusted salesman, so in the opposite quadrant (negative for both dimensions) we find attention to food security and price and the catch at Km0 which are important but also taken for granted by the islanders.

Examining in detail the four quadrants to identify the profile of consumers, it emerges that in Lipari 43.5% of the consumers are placed in the negative quadrant for Dimension 1 and positive for the Dimension 2. The quality, the wide choice and the purchase of local fish products are considered relevant. Their socioeconomic characteristics are: there is a slight prevalence of women (55%); age is more or less equitably distributed in the segment (the different age groups fluctuate between 20% and 28%); the consumers in this segment have high-school education, are employed (33%), are housewives and students (27%), do not answer the question on income level (45%), and their family is composed of three to four members (55%). Their profile can be defined as *"traditional consumers with a special bond with the territory"*.

In the positive quadrant for both dimensions we find consumers who are characterized by the purchase of fish products at retail shops. They are not concerned about food security (which records the greatest variability) and are not particularly price-sensitive. Their socioeconomic characteristics are: they are mostly males (73.3), mainly aged 18–39 (in 50% of the segment), have higher education (43.3%), are professionals (33.3%), students or engaged in other activities (33.3%), have average or high levels of income (43.3%), and their family is made up of four to five members (56.7%). Their profile can be defined as *"young loyal and inexperienced consumers"* who buy because they have confidence in of the seller, and are guided in their choices both by the seller and by relatives and friends.

In the positive quadrant for Dimension 1 and negative for Dimension 2, we find consumers who are characterized by not having preferences for local fish products, and are indifferent to quality, variety of choice and place of purchase. In this quadrant we find 15.2% of the consumers surveyed. There is a slight majority of males (52.4%); their prevailing age is between 18 and 39 years (47.6%); they have a high education (high school (43.3%) and a university degree (33.3%)); they are professionals (33.3%) or do other activities (students, housewives, and Other 33.3%), and in 59.3% of the cases do not respond to the questions on household income. In this case we can indicate the profile as *"distracted and indifferent consumers"*. In the negative quadrant for both dimensions are the remaining 28.2% of consumers who are attentive to food safety and Km0 products and, in the choice of purchases, also care about the price. Their profile can be defined as *"health-conscious consumers"*.

In Reggio Calabria the two dimensions explain 46.9% of the variance. The first dimension clearly separates consumers who prefer local and Km0 fish products, and who are attentive to food safety and quality. As shown in Fig. 2, in the positive quadrant for both dimensions, we find the variables related to prices and the place where purchases are made (retail shops), in which express the greatest interest is expressed by the consumers interviewed. We find 48.7% of respondents in Reggio Calabria in the positive quadrant for both dimensions. There is a small prevalence of women (53.2%), of young people between 18–39 years (35.8%) and between persons of 40–54 years (29%), and of those who have a high-school education (47%) and a university degree (28.4%). They are housewives, students, people engaged in other activities (24.8%) and also employees (22.9%); they declare an average high income (51.4%). Their profile is that of *"loyal consumers and connoisseurs/enthusiasts of local fish products"*.

In the positive quadrant for Dimension 1 and negative for Dimension 2 there are those who appreciate other variables, such as safety, and the quality of Km0 products. This is the case of 8% of the consumers interviewed, who are within the age range 40 to 54 years (38.9%), and who are men in 55.6% of the cases. They have high education levels (50% high-school educations and 27.8% with a university degree). They declare an average high income (44% of the segment). They represent a small minority of *"health-conscious consumers, attentive and sensitive to new trends"*.

In the negative quadrant for both dimensions, we find 21.4% of the consumers who claim not to prefer local fish products and are indifferent to the price and the place of purchase. They are mainly men (68.7%), young people aged 18-39 years (40%); have a medium education (47.9% high-school and 22% middle-school), are professionals, traders, and employees; and they claim to have a low average income (in 46% of cases). Their profile can be defined as *"distracted and disinterested consumers"*.

In the last negative quadrant for Dimension 1 one and the last positive quadrant for Dimension 2 we find the remaining 22% of the consumers surveyed: They do not show concern for food safety (which records the greatest variability), and are not particularly attentive to the organoleptic quality, and to the Km0 products; they are interested in finding a wide range of choices. Their socioeconomic characteristics are: a slight prevalence of males (53.7%), the prevalent age range is 55-65 years (35%); they have an average education (42.8% high-school and 20.4% middle-school), and are retirees, teachers, traders, and people who perform Other activities, and claim to have a low average income (in 63.3% of the cases in the segment). Their profile can be defined as *"occasional consumers who love change"*.

4   Conclusions

The attractiveness and reputation of the food market is enjoying renewed vigor in local territorial systems [11] and in the food systems, but it is a topic which has been neglected by policymakers and researchers since the onset of the "green revolution" [12, 13] and the industrialization of the fisheries. This paper has focused on comparative research conducted in three Mediterranean areas in order to examine consumer behaviors and preferences for choosing sale points. The consistent quality of the products, and the freshness and the variety of the fish, all serve to create trust in the buyers and demonstrate the attractiveness that the food markets and local retail outlets are able to generate.

The local food retail sector, both the shops and the public market, are the basic elements of the convivial urban space [14], which is not only more democratic, inclusive, and secure, but also more economically attractive. Furthermore, the local retail system is also connected to the idea of the 'city of short distances', which is typical of Mediterranean cities, such as Reggio Calabria, Girona, and Lipari, that are walkable and have a relatively high residential density, and offer a mixture of diverse activities, cultures, and economies [15]. The current interest in gastronomy presents not only a new possibility for the local economy but also a new perspective for the city along with the food policy studies. More specifically, the analysis of the consumer behavior for fish products shows that choosing local fish and seafood is linked not simply to the need for quality, health and nutrition, but is also linked to the territory, to environmental sustainability and to the improvement of coastal communities [16, 17].

These reconnections provide a starting point on which to build fisheries by strengthening existing local food and economic systems and relationships that include ethics and moral values [18, 19]. Multiple agents are participating and need to be committed to the transitions taking place in the food system, because changes in the cultural, social, economic and logistic competence of mass production are difficult and slow to revert. The present comparative research could represent a tool for reflecting and acting on the food territorial system, by increasing its awareness through the production of new knowledge, aggregating the existing one and proposing it as a tool to support decisions and actions. It can be a support in designing new products, services, and, more generally, can activate a process of cultural innovation, an area in which new ideas and sustainable business models can grow.

Acknowledgements. This study has been supported by GASTROCERT, a Project on Gastronomy and Creative Entrepreneurship in Rural Tourism, funded by JPI Cultural Heritage and Global Change - Heritage Plus, ERA-NET Plus (2015–2018). The authors thank Romà Garrido, Marta Carrasco and Lorenzo Cortese, who administered the questionnaires, respectively, in Girona and in the Aeolian Islands. We are also grateful to Valentina Rosa Laganà who, in addition to administering the questionnaires in Reggio Calabria, has also edited and uploaded the data on the SPSS v.20 database.

References

1. FAO: Food Outlook October Biannual report on global food markets (2016). http://www.fao.org/3/a-i6198e.pdf. Accessed 14 Oct 2017
2. Chen, Y.: Neighborhood form and residents' walking and biking distance to food markets: Evidence from Beijing, China, Transport Policy, pp. 1–10 (2017)
3. Nicolosi, A., et al.: Fresh local fish, stockfish and best practices for creative gastronomy in Calabria: the consumers' point of view. In: Laven, D., Skoglund, W. (eds.) Valuing and Evaluating Creativity for Sustainable Regional Development. UNESCO-Mid Sweden University (2016)

4. Kim, Y.G., Eves, A.: Construction and validation of a scale to measure tourist motivation to consume local food. Tour. Manag. **33**, 1458–1467 (2017)
5. Abdi, H., Valentin, D.: Multiple correspondence analysis. In: Salkind, N. (ed.) Encyclopedia of Measurement and Statistics. Sage, Thousand Oaks (2007)
6. Husson, F., Josse, J., Pages, J.: Principal component methods-hierarchical clustering-partitional clustering: why would we need to choose for visualizing data? Technical Report, Agrocampus (2010
7. Madau, F., Idda, L., Pulina, P.: Capacity and economic efficiency in small-scale fisheries: evidence from the Mediterranean Sea. Mar. Policy **33**, 860–867 (2009)
8. Guinot, C., et al.: Use of multiple correspondence analysis and cluster analysis to study dietary behaviour: food consumption questionnaire in the SU.VI.MAX. Eur. J. Epidemiol. **17**, 505–516 (2001)
9. Hair, J.F., Black, W.C., Babin, J.B., Anderson, R.E., Tatham, R.L.: Multivariate Data Analysis, 7th edn. Prentice-Hall, Upper Saddle River (2009)
10. Costa, P.S., Santos, N.C., Cunha, P., Cotter, J., Sousa, N.: The use of multiple correspondence analysis to explore associations between categories of qualitative variables in healthy ageing. J. Aging Res. 1–12 (2013)
11. Pedroza-Gutiérrez, C., Hernández, J.M.: Social networks, market transactions, and reputation as a central resource. The Mercado del Mar, a fish market in central Mexico. PLoS One **12**(10), 1–21 (2017)
12. Stell, C.: The Hungry City. How Food Shapes Our Life. Chatto & Windus, London (2008)
13. Fava, N., Guàrdia, M.: Territori del cibo e nuovi spazi di socialità. In: Fontana, L: (ed.): Food and the City. Marsilio, Padova (2017)
14. Parham, S.: Food and Urbanism: The Convivial City and a Sustainable Future. Bloomsbury, London (2015)
15. Nicolosi, A., Sapone, N., Cortese, L., Marcianò, C.: Fisheries-related Tourism in Southern Tyrrhenian Coastline. Procedia Soc. Behav. Sci. **223**, 416–421 (2016)
16. Hilborn, R., Costello, C.: The potential for blue growth in marine fish yield, profit and abundance of fish in the ocean. Mar. Policy **87**, 350–355 (2018)
17. Romeo, G., Marcianò, C.: Integrated local development in coastal areas: the case of the "Stretto" Coast FLAG in Southern Italy. Procedia Soc. Behav. Sci. **223**, 379–385 (2016)
18. Palladino, M., Cafiero, C., Marcianò, C.: Relational capital in fishing communities: the case of the "Stretto" Coast FLAG area in Southern Italy. Procedia Soc. Behav. Sci. **223**, 193–200 (2016)
19. Poitevin Des Rivières, C., Chuenpagdee, R., Mather, C.: Reconnecting people, place, and nature: examining alternative food networks in Newfoundland's fisheries. Agric. Food Secur. **6**, 33 (2017)

Campania Region Metropolitan Area. Planning Tools to Redevelop the Aversana Conurbation

Salvatore Losco[✉] and Gianfranca Pagano

University of Campania Luigi Vanvitelli, 81100 Caserta, Italy
salvatore.losco@unicampania.it

Abstract. The substantial modifications of cities and regions in recent history have raised important questions regarding their formation and organization. Such questions render it imperative to reread their form and structure, especially because they are so different from those inherited from the modern movement in architecture and urban planning. Contemporary cities and territories undergo processes of transformation of their urban organization in morphological terms as well as in terms of the social, political, economic and symbolic relationships determined by their constitution. These new urban models have caused traditional physical planning to be ineffective, with the consequential loss of identity of cities and regions. To a considerable extent, this phenomenon can be attributed to a lack of differentiation with their surrounding contexts, historically considered to be complementary to the city (also in a cultural sense). New rationalities govern such processes and as a result, new planning tools will be needed to address the new demand for territory. After outlining some distinctive features of metropolitan areas, this article proposes to identify homogeneous functional sub-areas within a broader metropolitan area. Such sub-areas could be instrumental in defining a structural plan that can act as an interface among the strategies of the Metropolitan Territorial Plan, the sub-regional area, and the planning/operational choices made by local municipal urban plans. The article presents the case-study of the Aversana area to the north of Naples, framed within the more complex planning of metropolitan areas. The proposal is coordinated also with Kipar's plan set forth in the Regi Felix [1] project for the environmental recovery of the Regi Lagni which is the north border towards Casertana conurbation.

Keywords: Metropolitan areas · Planning tools · Redevelopment

1 Introduction

The process of functional specialization of interdependent centers determines the configuration of conurbations and metropolitan areas: each urban system is characterized by a hierarchical [2] process that develops between centers producing rarer goods and general public equipments and centers producing more common and local ones. The presence of centers with dominant functions in the territory implies that parts of the latter acquire, in turn, a dominant function with respect to other parts of the same

© Springer International Publishing AG, part of Springer Nature 2019
F. Calabrò et al. (Eds.): ISHT 2018, SIST 101, pp. 113–123, 2019.
https://doi.org/10.1007/978-3-319-92102-0_13

according to relationships of subordination. The analysis of spatial concentration phenomena can take advantage of different methodological approaches contained in the theoretical elaborations produced starting from the decade before the Second World War. The identification of territorial hierarchies may represent an effective interpretative key for these areas, their knowledge, sufficiently detailed, is instrumental for spatial planning.

The paper will summarize a structural plan hypothesis of the Aversana conurbation, within the pseudo-metropolitan area of Naples, the idea force is the planning of a polycentric region [3]. The paper explains only two of the five systems of the plan: central places system and ecological and environmental system [4]. Infact it is organized in five interrelated systems, the principal aims are the description of a methodology for the hierarchization of settlements belonging to a conurbation and the reading and understanding the results related to a specific territory.

2 Metropolitan Areas, Conurbations, Sprawl and Cities

Most of today's urban population lives in a new city, whose limits - resulting from the location of places of work, study, production, trade, and leisure - no longer coincide with the administrative ones (municipal, metropolitan and regional). A new city has developed, taking over the countryside, producing new territories that have not yet been adequately explored. The physical growth of urban settlements has expanded beyond their administrative limits; populations and economic activities have been redistributed throughout regions by calling into play places surrounding the central core and beyond [5]. This vast entity is full of energy but still lacks order and harmony. It is a peri-urban area, an indefinite fringe zone commonly referred to as hinterland. This hinterland englobes the large city, which subjects it to its social and economic influence and to which it provides the goods and services resulting from its activities. This new city is a human settlement in which the areas surrounding the main urban core have become increasingly attractive. The emergence of these extensive, agglomerative and aggregational processes has given rise to utterly new models of urban form in which the relationship between urban activities that use space and spatial configuration has been reformulated.

The material signs, expressions of this new conformation, can be found in emblematic ways in the metropolitan peri-urban area, consisting of settlements built at different times, of unstructured or interstitial spaces still used for agricultural purposes, connected by infrastructure systems for the communication and mobility of people and things. These conditions have led to multiple consequences in urban planning. On the one hand, local-scale municipal planning can no longer manage such complex, dispersed, extensive, and overlapping areas (especially when small municipalities gravitate around larger urban poles). On the other, provincial-scale planning, while under discussion after the establishment of the metropolitan area [6] is too far removed from issues regarding land use, collective services and housing. A contradiction also emerges with the excessive sectorial fragmentation of regional policy, which clearly separates, and does not interact with, agricultural development measures (with EU origins) and planning tools (essentially having a local jurisdiction). Particular attention should focus on peri-urban metropolitan zones, which represent the most unstable urban fabrics at greater risk because they lack precise identities and roles defined within a regional structure. They can be rediscovered as environmental and social resources for the spaces of the new city. In this perspective, the peri-urban agricultural zone can provide a means for understanding, interpreting and rethinking the space of the dispersed or sprawl [7], city through a new cognitive approach that:

- provides a descriptive framework for peri-urban diffusion and its bio-physical characteristics;
- seeks to overcome the traditional city/country dichotomy by proposing an interpretative strategy that can address new and technically relevant land uses that underline the potential for the renewed role of agricultural practices [8] within development processes;
- highlights the multiple functions of rural areas in the contemporary city as places of production, environmental education and protection, and ecological reconnection with urban areas;
- attempts to identify models (including socio-economic ones) that can support these transformation processes within a market logic that is a consequence of the environmental economy.

Therefore, peri-urban agriculture observed in non-urbanized areas can become a potential generator of externalities driven by the concept of multi-functionality. Such processes should be based on the idea that other social, cultural and environmental functions can be associated with the production of the essential elements for human and animal nutrition through the pursuit of environmental protection, landscape enhancement, biodiversity conservation, recreational activities, and the production of secondary goods and services. This would come about by virtue of proximity to the city [9, 10].

Urban planning takes two opposing views of these issues:

- one oscillates between the acceptance of these models as contemporary urban landscapes and indifference to problems such as sprawl and land consumption [11, 12];
- the other is critical of urban sprawl and seeks tools and spaces that can redefine contemporary urban form as an artifact that, if not complete, is at least structured. In this second case, the operative tool coincides with the planning and design of the periurban landscape, meaning strategically important voids to be filled not with new buildings, but with meaning, functions and forms that preserve and enhance their specific characteristics.

The key role of these spaces, which are flexible and open to transformation, derives from at least three factors:

- they represent decisive opportunities for reweaving fragments and fringe areas within a structured urban design;
- they can become a system of ecological corridors [13] and natural spaces, important for the environmental quality of the city;
- they offer an opportunity for the contemporary city if they are treated, starting in the initial planning phase, as a network of public open space.

Within this logic, the relationship between solid and void is modified. Solids no longer determine the form of the voids, but rather empty space penetrates the built context, redefining it, restoring it, and returning form to the urban settlement. In this intricate play between artifice and nature, residual and interstitial spaces can become new public places and centers within a renewed urban scenario. A new infrastructure [14] shapes the landscape both morphologically and functionally. In fact, continuity of agricultural land render the historical settlements legible and preserves a formal and visual balance between solid and void (especially corresponding to the directions of expansion of the metropolis). At the same time, wooded areas, linear plant formations along watercourses or accompanying agricultural subdivisions are elements of land-scape diversification and, if they maintain some degree of internal connectivity, can become parts of ecological networks. The networks of agricultural and natural open space define form and function on a regional scale on the one hand; and on the other, they configure the limits [15] of the metropolitan area [16].

3   Metropolitan Areas and Homogeneous Functional Sub-areas: The Aversana Conurbation

The first step in the planning process is the identification of homogeneous sub-areas [17] within a broader and highly urbanized region; this is a preparatory action for any metropolitan plan hypothesis whether in its strategic or structural/operational form. A possible geographic, functional and administrative sub-organization within apparent disorder generated by the merger of multiple centers having no common plan and/or design can reveal the role of the sub-area within the metropolitan continuum. The study of the sub-area highlights significant questions regarding its relationships within the entire metropolitan area and the specificities of the sub-area itself. The logic allowing this passage consists in highlighting discriminating factors that evidence regional complexity with respect to some important homogeneous elements regarding metropolitan areas (for example, the identification and recognition of regional poles).

The next step consists of geo-referencing all abstract data by defining the perimeters of the sub-area having common characteristics. Application to a case study seeks to identify valid procedures for the definition of perimeters of sub-areas for which it is possible to recognize significant factors for a hierarchical reading of urban centers belonging to homogeneous spatial domains.

From the recognition of this kind of organization of the urban settlement, which until today has been the result of spontaneous processes and projects that have not been coordinated by planning or development programs, the next step is the identification of possible strategic development goals, while rendering recognizable the organized parts of a continuous system in which chaos and chance still represent its essential features.

This consolidation and development toward a polycentric structure, reinforcing internal poles, might help contain the growing phenomena of urban sprawl and congestion, characteristic of all metropolitan areas.

The Campania metropolitan area, whose borders were defined in Law 56/2014 to coincide with only those of the province of Naples (Metropolitan City of Naples) is much more extensive and includes the Aversa area to the north and Nocera-Sarno agricultural area to the south.

The first area lies entirely within the Caserta province; the second partly in the province of Salerno. All studies conducted over the past thirty years agree that a proper perimeter of the Naples metropolitan area cannot and should not coincide with that of the province alone [18]. Even within the most advanced hypotheses that recognize not a single limit to metropolitan areas but numerous ones relating to specific spatial domains and functions (limits with variable [19] geometries), the limit of the province of Naples is never significant. In light of such an important premise, this study refers to the limit of the Naples metropolitan area, which also includes the Aversana conurbation to the north. Highly disorganized and uneven urban and infrastructural conditions lie within this perimeter are, so the recognition of homogeneous functional sub-areas (Aversa, the Nocera-Sarno agricultural area, etc.) could make strategic-structural metropolitan planning even more effective by identifying an intermediate plan scale somewhere between the overall metropolitan area and the single municipality. It is important to point out that the study and identification of homogeneous functional sub-areas within broader urban areas can become opportunities for intermediate-scale planning (between the province and/or metropolitan city and municipality). It can be useful both in defining the sub-portions of the metropolitan city on a smaller scale as well as consolidating, on a larger one, a number of municipalities that, while not included in the metropolitan city, necessitate coordinated planning. A strategic-structural plan for the entire metropolitan area would have contents and goals that almost entirely coincide with the Provincial Territorial Coordination Plan (PTCP); while an intermediate-scale plan for the metropolitan conurbation could attribute regional plan goals and contents that are more pertinent than the provincial PTCP and less stringent than the local Municipal Urban Plan (PUC - Piano Urbanistico Comunale in Italian).

The Aversana conurbation, located to the northwest of Naples, includes 19 municipalities. The study of its spatial configuration reveals a core, a secondary nucleus, and a satellite [2] area. The Aversana area extends over a flat landscape with population densities ranging from 5970 per sq km in Aversa to 173 per sq km in Villa Literno. Only in Aversa is there a high population density. Villa Literno is one of the largest municipalities in the Province of Caserta and alone makes up 35% of the entire urban area; together with Casal di Principe, it occupies about half the area of the entire conurbation.

4   The Aversana Conurbation: A Structural Plan Hypothesis

The Aversana conurbation should be considered not only a sub-area for analysis having a series of homogeneous features but also as a design/plan area, a minimum planning unit, an urban area for integrated development that can become a magnet on the metropolitan and regional scales. It should not be considered that the conurbation so identified is determined in a univocal way but rather it is conceived as having a variable limit.

The attempt is to define urban realities that, during regional planning and programming processes, can constitute intermediate references, areas that can be rebalanced, homogeneous territorial areas structured between the conurbation and the rest of the region. This type of approach can help to:

- test new knowledge regarding metropolitan areas and identify the form and content of the Metropolitan Territorial Plan - PTM (Piano Territoriale Metropolitano in Italian)
- identify possible strategic goals for improving peripheral metropolitan urban systems with satisfactory proposals on the qualitative and quantitative levels
- identify tools for controlling and managing transformations.

For the preparation of the Plan, the priority goals are:

- environmental sustainability: for the preservation and planning of natural and historic-cultural resources
- urban regeneration: the promotion of historic urban fabrics and cultural heritage; the diffusion of high quality in more recently-built or declining areas; identification of new natural and urban environmental balances
- overcoming territorial imbalances: in terms of population density, provision of services and infrastructure, and quality of life through the creation of a polycentric system
- socio-economic development: through development of local characteristics and identification of innovative strategies for the steady increase of jobs.

An urban planning project that addresses these issues involves systems [20] rather than zones, thus breaking with traditional urban planning. A system is defined in terms of the identity and integration of its parts, the role that each plays, the functions it offers and the relations between the parts rather than in terms of the uses each can accommodate.

Each system is endowed with a specific perimeter and organization. The ecological-environmental system usually extends beyond the administrative limits of a municipality, its confines dictated by the nature of the soils and morphology. The commercial/industrial system extends as far as the district. The residential system coincides with the housing market. The mobility system is as extensive as the accessibility afforded by the networks that serve the area involved. The system of central places identifies certain locations that enjoy special status insofar as they embody the identity of the city and are obviously the most utilized/visited places.

Natural features, connections, filters, and anchoring elements are the main components of an environmental system. Thoroughfares and agricultural roads, crossings and connections, protected pedestrian squares and areas compose a mobility system. Each system suggests specific themes: preservation, recovery, reclamation and compensation are, for example, the main themes of the environmental system.

Traditional urbanism had been dominated by the idea of classifying, separating and removing. Systems-based planning is dominated by the search for relationships, identity and integration. For brevity's sake, this paper summarizes only Central places facilities and the Ecological-Environmental system, leaving aside the mobility, residential and commerce systems.

4.1   Central Places Facilities System

The most important works in the Aversana conurbation were built in the past by subjects who were well aware of the roles of these works in creating the physical, social and symbolic space of the city; roles that often superseded original functions when, for example, convents were transformed into schools, libraries and hospitals. Driven by recent expansion, the public administration has spread throughout the city, occupying apartments, residential blocks, and empty spaces. Public buildings have become addresses rather than references in the mental maps of citizens. Similarly, driven by financial needs or opportunities, families and institutions have left their domiciles and headquarters, which have been taken over indifferently by banks, offices and commercial activities. The Aversana conurbation has a great need for new places for civic centers and new parks. It is not just a question of locating functions, but rather of creating meaningful spaces. The Aversana region should be endowed with public equipments of both metropolitan (general) and neighborhood (urban standards) interest. This last issue is a very complicated and complex one from a technical and legal point of view, insofar as in most of the Aversa municipalities, it is impossible to proceed with the improvement of urban standards due to the fact that many municipal urban plans are outdated and, as a result, the constraints levied for property expropriation have lost their validity (urban standard/plan constraints/five year decree/white areas) (see Fig. 1).

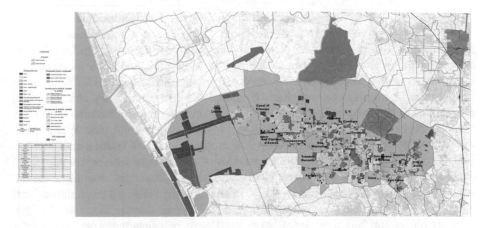

Fig. 1. The Aversana Conurbation thematic map: zoning collage of the nineteen municipal plans, the black areas represent the so-called white areas.

4.2   Ecological and Environmental System

The search for connections that integrate the different parts of the city with its surrounding region and improve the functioning of its ecological system is the cornerstone of the environmental system. This means recognizing that connectivity, meaning the necessity to allow continuous flows and exchanges among the various ecosystem components, is the most significant function of the environmental system. The resulting image is that of a network: in different parts of the environmental system (reservoirs of naturalness; territorial, inter-environmental, and local connections; protection and compensation provisions), it is therefore necessary to identify the configurations of ecological corridors, selected from among the various possibilities on the basis of their current natural value as well as the feasibility of undertaking design and planning initiatives.

The characteristics of a connection depend on:

- its position and ability to connect different contexts (diversity stems from differences in physical-morphological conditions and vegetation and fauna ecosystems). Often, its natural geographical position determines the hierarchical importance of a connection (greenway) and defines its environmental role:
- the type of habitat it contains: the presence of plant and animal species as an indicator of the state of ecosystem health;
- man-made modifications: the analysis of anthropogenic impacts is crucial from the point of view of the identification of ecological corridors that can fulfill environmental functions as well as of the planning of the necessary restoration interventions.

To verify the functionality of a connection, it is useful to examine all existing network infrastructure (roads, sewage systems, pipelines and aqueducts) as potential elements that can interrupt biological continuity; and then to analyze all areas showing strong potential to be integrated with an ecological corridor. Connectivity and the search for biodiversity therefore structure the project for the environmental system. In light of all of the above premises, the draft plan for the Aversana conurbation proposes a multifunctional agricultural park on the northern confine toward the Regi Lagni. Such a park would resolve the problem of the lack of local public equipments and could become a fundamental element in the environmental/ecological network of the entire conurbation. It might become a font of natural resources and green infrastructure, functioning as environmental compensation for the entire Aversana conurbation. In this way, it could also become an ecological network that can improve biodiversity throughout the area. The idea is to physically connect a vast, less built-up area with the built contexts in urban centers through a series of penetrations (ecological corridors) that would allow it to be used more widely, countering phenomena resulting from the vast area of impermeable surfaces that characterize heavily built-up areas (causing heat island phenomenon and low evapotranspiration in urban areas). The plan for this well-appointed and multifunctional agricultural park will determine the preferential and/or prescriptive sites in which the single municipal plans (PUC) will have to allocate neighborhood public equipments and/or urban standards. It should be pointed out that, in order to avoid the forfeiture of the preordained expropriation constraints, a compensation and/or equalization system should be adopted throughout the multifunctional agricultural park.

This implementation mechanism would involve all the systems in the structural plan (see Fig. 2).

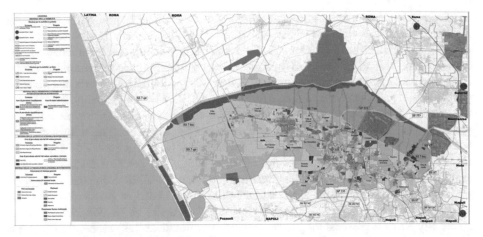

Fig. 2. The Aversana Conurbation thematic map: structural plan, the medium and dark green areas to the north represent the ecological and environmental system.

- the residential system: allowing a share of housing in exchange for relinquishing areas to the public administration for the construction of neighborhood public equipments;
- a mobility system: with areas relinquished for primary urbanization, network infrastructure and in particular the integrated water cycle;
- a system of central locations: with areas relinquished for new public squares, green spaces, social and civic centers;
- the ecological-environmental system: also representing places of naturalness;
- the economic system: insofar as quality agriculture bound to the characteristics of the region will be allowed in the multifunctional agricultural park.

5   Conclusive Remarks

The identified hierarchies are preliminary to any hypothesis of a metropolitan plan, both in the strategic or structural form. They may represent a possible criterion of analysis to put order within an apparent disorder generated by the merger of several centers without any common project. The experiences of planning, planning and local design in recent years show the need to activate development processes capable of achieving and supporting strategic choices towards a new territory of the economy: the Metropolitan City, as the engine of national economies [21]. The city itself, in this context, becomes an economic actor and protagonist of development, even social, which does not underestimate or neglect the consequences of its initiatives. Economic growth can't be separated from a strategic plan for metropolitan areas for the proper development of the territory in order to share not only problems related to economic changes, but also opportunities for redevelopment and regeneration, quality urban life high and economic competitiveness.

Pursuant to Regional Law 16/2004, Regulation N° 5 of 2011 and the Manual, every municipality is required to draw up a PUC having a structural portion and programmatic/operational portion. A structural plan for the Aversana conurbation could involve the single municipal plans (PUC) by replacing their structural portions [22]. The strategic-structural plan and related regulations of the Aversana conurbation plan would replace the structural content of the PUC in each municipality. The intent, through local urban planning, is to recognize the Aversana conurbation as a homogeneous functional sub-area of the Naples metropolitan area extending from Aversa at least to Salerno (north-south) and from the coast to Nola (east-west) or even from Capua to Pontecagnano and from the coast to Nola. The goals of the Aversa area structural plan should also relate and respond to 2014–2020 European economic planning to ensure feasibility and coordination of physical and economic development within a coordinated vision of complex planning. A mid-size city is considered to be a smart city when it presents long-term development that takes into account six factors: economy, people, governability, mobility, environment, living, all referenced in the 2014–2020 EU program. The European Commission itself has determined that the most important goal of our century is to coordinate the plans and strategies of individual municipalities in order to reach shared solutions. With the indications in the Aversa area plan, the goal would be to contribute to creating a smart city and community. Its new urban design should give rise to the new city of Aversa: an intelligent city that can organize itself, optimize its resource use, and change to improve the quality of life of people, achieving overall sustainability. In other words, it will be a city that can organize and use its territory in a different way.

Attributions

Within this contribution, the result of common elaboration of the authors, personal contributions can be identified as specified below: *Introduction, Metropolitan areas, conurbations, sprawls and cities, Metropolitan areas and homogeneous functional sub-areas: the Aversana conurbation* (Salvatore Losco), The *Aversana conurbation: a structural plan hypothesis* (Gianfranca Pagano), *Abstract* and *Conclusive remarks* (collaborative effort).

References

1. De Nardo, A.: Storie di Lagni. Dalla Campania Felix alla terra dei fuochi. Contributi alla storia della non trasformazione di un territorio, Clean Edizioni, Napoli (2017)
2. Losco, S.: Per la definizione del ruolo della conurbazione aversana nell'ambito dell'area metropolitana centrale campana. In: Moccia, F.D., Sepe, M. (a cura di) Metropoli IN-transizione. Innovazioni, pianificazione e governance per lo sviluppo delle grandi aree urbane del Mezzogiorno. Giornata annuale di studi 2004. Atti del convegno: Urbanistica Dossier, vol. 75, no. 201, pp. 387–394 (2005)
3. Parr, J.B.: The polycentric urban region: a closer inspection. Reg. Stud. **38**(3), 231–240 (2004)
4. Jianhua, X., Changlin, F., Wenze, Y.: An analysis of the mosaic structure of regional landscape using GIS and remote sensing. Acta Ecol. Sin. **23**(2), 365–375 (2003)

5. Losco, S.: Le aree metropolitane. In: Colombo, L., Losco, S., Bernasconi, F., Pacella, C. (eds.) Pianificazione Urbanistica e valutazione ambientale. Nuove metodologie per l'eficacia, pp. 119–138. Edizioni Le Penseur, Brienza (2012)
6. Law 56/2015 and similar legislation (2015)
7. Ingersoll, R.: Sprawltown. Meltemi, Roma (2003)
8. Donadieu, P.: Campagne urbane. Una nuova proposta di paesaggio della città, Donzelli, Roma (2004)
9. Losco, S.: Campagna urbanizzata. In: Colombo, L., Losco, S., Bernasconi, F., Pacella, C. (eds.) Pianificazione Urbanistica e valutazione ambientale, pp. 138–148. Edizioni Le Penseur, Brienza (2012)
10. Losco, S.: Campagne urbanizzate e periferie metropolitane: un parco agricolo-urbano nell'area nord della provincia di Napoli. In: Paysage Topscape n. 9, Paysage Editore, Milano (2012)
11. Losco, S., Macchia, L.: Il consumo di suolo nella Conurbazione Aversana e Casertana. Urbanistica Informazioni **257**, 69–74 (2014)
12. Losco, S., Macchia, L.: Problemi di metodo nella quantificazione del consumo di suolo, La Conurbazione Aversana: PLANUM, vol. 29, pp. 1–9 (2014)
13. Jongman, R., Kamphorst, D.: Ecological corridors in land use planning and development policies, pp. 38–41. Council of Europe Publishing (2002)
14. Austin, G.: Green Infrastructure for Landscape Planning. Integrating Human and Natural System, pp. 128–148. Routledge, London, New York (2014)
15. Gaeta, L.: Questioni di metodo sullo studio del confine, Territorio 79/2016, FrancoAngeli, Milano (2016)
16. Fanfani, D. (ed.): Pianificare fra città e campagna. Firenze University Press, Firenze (2009)
17. Losco, S.: La conurbazione Pseudo-Metropolitana di Napoli. Elementi per il riconoscimento degli ambiti territoriali omogenei. In: A.a.V.v.: Il rischio Vesuvio strategie di prevenzione e di intervento, Università di Napoli Federico II, pp. 212–221 (2003)
18. Colombo, C., Giordano, R., Rostirolla, P.: Città metropolitana. L'occasione per riparare il territorio. Rocco Giordano Editore, Napoli (2015)
19. Tira, M.: Verso un territorio a geometria variabile: Ingenio n. 19/2014. http://www.ingenioweb.it/Articolo/1538/Verso_un_territorio_a_geometria_variabile.html. Accessed 10 Mar 2017
20. Secchi, B., Viganò, P.: Sistemi. In: Secchi, B., Viganò, P. (a cura di) Pesaro il progetto preliminare del nuovo Piano Regolatore, Comune di Pesaro, Pesaro, pp. 83–134 (1997)
21. Gibelli, M.C.: Dal modello gerarchico alla governance: nuovi approcci alla pianificazione e gestione delle aree metropolitane. In: Camagni, R., Lombardo, S. (eds.) La Città Metropolitana: Strategie per il Governo e la Pianificazione, Alinea, Firenze (1999)
22. Caserta Province, PTCP - Norme Tecniche di Attuazione, Caserta (2012)

The Role of Social Relations in Promoting Effective Policies to Support Diversification Within a Fishing Community in Southern Italy

Monica Palladino[1](✉), Carlo Cafiero[2], and Claudio Marcianò[1]

[1] Mediterranea University of Reggio Calabria, 89100 Reggio Calabria, Italy
monicapalladino@hotmail.com
[2] Statistics Division, Food and Agriculture Organization of the United Nations, 00100 Rome, Italy

Abstract. Small-scale fishery in Southern Italy is exposed to increasing pressure from changing economic, institutional and environmental conditions. Diversification strategies to accompany the changes will require the support from effective institutions, where fishermen are given a central position and an active role. To explore the strength of the social fabric of the fishing community and its ability to sustain the needed institutional innovation, this article analyses the type of relationships that fishermen declare to entertain with their colleagues. Data collected by interviewing twenty-five fishermen, distributed across the five ports of the area, are used to create social network representations of the revealed acquaintance, information sharing and trust relationships. The discussion on the results is contextualized in the evolving institutional and legal setting associated with the new European Common Fishery Policy. While the current level of sharing managerial information is low, we find that there is sufficient trust among fishermen across the five ports to support increased cooperation, an essential requirement – in our opinion – to enable a small-scale fishery community to respond to the emerging sustainability challenges.

Keywords: Small-scale fishery · Diversification strategies
Network analysis of social relations

1 Introduction

The analysis presented in the article is part of a broader study [1] aimed at analysing in some detail the fishery system in the Southernmost part of the Tyrrhenian coast of Southern Italy (Fig. 1), comprising the ports of *Cannitello* of *Villa San Giovanni*, *Scilla*, *Bagnara Calabra*, *Palmi* and *Gioia Tauro*. The previous study, funded by the "Stretto" Coast FLAG and commissioned to the Agraria Department of the Mediterranea University of Reggio Calabria, was framed in the context of an analysis of fishery development policies, as defined by the new EU Common Fishery Policy through the five-priority axes European Fishery Fund (EFF), which introduced important measures in support of fishery and aquaculture throughout Europe.

One of the innovations in the EFF was indeed the creation of Fishery Local Action Groups (FLAGs) [2] as partnerships between fishermen and other local private and

© Springer International Publishing AG, part of Springer Nature 2019
F. Calabrò et al. (Eds.): ISHT 2018, SIST 101, pp. 124–133, 2019.
https://doi.org/10.1007/978-3-319-92102-0_14

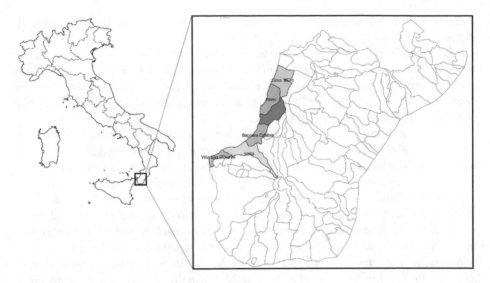

Fig. 1. The study area

public stakeholders. Acting as a tight "community", FLAG partners should contribute to design and implement the strategies needed to address the area's economic, social and/or environmental developmental needs. In such contexts, fishermen are – or should be considered as – key actors, supported in their planning efforts by a number of public and private institutions. Therefore, the way in which they evaluate, participate in, and relate to various type of institutional arrangements becomes crucial to assess the potential for the kind of governance envisaged in the new European development policy to be effective.

Through the analysis of primary data collected by interviewing fishermen and other key stakeholders in the area, the broader study provides a review of the variety of stakeholders in the small-scale fishing sector in the study area and of the relationships that exist among them. By highlighting the main problems and the current prospects of a sector confronted with the need to evolve and adapt to changing economic conditions and new regulations dictated by environmental sustainability concerns, it aims at suggesting feasible practical actions that could contribute to improve the effectiveness of local development plans.

A previous article already explored the extent of exchanges of information and the level of trust between fishermen in the study areas, focusing on the analysis of the network made-up by 26 interviewed fishermen [2]. It revealed interesting, though somehow conflicting results: fishermen know each other, as it might be expected in such a relatively small community, and they also "trust" each other (or at least so declare), which means that there should be a potential for the creation of strong professional links, both formally and informally. When explicitly asked about their sharing information with others, however, they mostly say no, that they do not engage in information exchange to any relevant extent. One of the conclusions from the previous article is that this limits the possibility that relevant information for local development policies may circulate effectively, and therefore that actions explicitly aimed at facilitating information flows might be highly beneficial.

What remained to be determined is how much current institutional settings are to be blamed for not permitting adequate information flows, and how much, instead, the problem rests with the cultural milieu of individuals who, while declaring that they trust each other, do so only or mostly on a personal basis, keeping, at the same time, a fundamental lack of trust in institutions.

In this study, supported by a project funded by JPI Cultural Heritage and Global Change, in which the LAGS and FLAGS of the Province of Reggio Calabria are partners of the Mediterranea University, we conduct further elaborations on the larger set of data to shed light on the issue. The objective here is to explore the strength of the social fabric of the fishing community and its ability to sustain the needed institutional innovation. As in the previous research [1, 2], this is obtained through an analysis of social networks among fishermen. In these networks, each fisherman is treated as a node and "links" are created whenever acquaintance, information sharing, and trust relationships with the other fishermen are declared. The importance of these kind of relations for the development of a small-scale production sector (which shares many features with the small-scale fisheries studied here) is explored by Morone et al. [3] and the literature therein.

While the previous article focused on the results from the analysis of 'one-mode' networks created by the reciprocal links within a group of 26 fishermen, each one of them asked to report on their relationship with the other 25, in this article we extend the analysis to the description of the 'two-mode' networks created with the information provided by 25 of the 26 interviewed on the relationship and information exchanges they entertain with all other fishermen active in the area, including those who were not interviewed directly.[1]

Special attention is devoted to the opinions that the interviewed fishermen expressed regarding the opportunities offered by potential diversification activities as a response to the rising environmental concerns, and the resulting pressures in terms of social and economic sustainability of the fishing activity in the area [4–6]. After a brief summary description of the research methods used in the study, two distinct sections present the opinions of the interviewed fishermen on the potential diversification activities and the explored relations among fishermen, respectively. A concluding section links the two and summarizes the main findings.

2   Methods

The analysis in this article is based on data collected through structured questionnaires, where a convenience sample of fishermen has been interviewed by one of the authors between May and August 2015. Based on perusal of public fleet register data and on a preliminary survey of the five ports in the area, a universe of 128 active vessels was

---

[1] As one of the 26 selected fishermen could only be interviewed by phone, the length of the interview did not allow the collection of all information needed to inform the two-mode network analyzed here.

identified, among which a convenience sample of 26 fishermen to be interviewed was extracted, based on their presence in the area during the study period and their willingness to participate in the study[2].

In the following sections of the article, two aspects are studied separately.

First, the opinions of the interviewed on a list of possible diversification activities are analysed through simple frequency tables of responses.

Next, to analyse the structure of the social relations, each interviewed fisherman was asked to report on the existence and strength[3] of specific kinds of relationships with the skippers of each of the other 127 vessels. This allowed for the creation of two-mode, 25 × 128, networks with directed, possibly valued, links. An analysis is conducted on the network graphs, where each node identifies a fisherman. To facilitate the reading of the graphs, different symbols are used to identify the 25 interviewed (spheres) from the other fishermen (stars), and different colours are used to identify the ports.[4] The links between interviewed fishermen are depicted in red, making it possible to study the extent of reciprocity of the declared relations which applies only to the 25 × 25 sub-network. In total, five networks are created representing: (1) acquaintance, (2) exchange of technical information, (3) exchange of managerial information, (4) declaration of the existence of positive (i.e., "medium high" or "high" levels) or (5) negative (i.e., "low" or "medium low") trust. Conclusions are drawn from visual inspection of the structures and density of the graphs.

3 Diversification

The need to diversify the economic activities of small-scale fisheries in Southern Italy has been explored by Carrà et al. [7]. We borrowed their list of potential diversification activities and asked each interviewed to express a judgment on whether they already engaged in or whether they might consider it, and on how useful they think it might be[5].

Only "fish-tourism" (that is, hosting tourists on board during fishing expeditions), already operated by 7 of the 22 fisherman who expressed an opinion, and the direct sale of fish, operated by 6 out of the 18 who expressed themselves, are activities that already play a certain role in the economy of the small-scale fishing sector in the area. Both are considered useful (15 positive out of 16 opinions expressed for fish-tourism and 9 out of 11 for the direct sale). The interviewed fishermen, however, demonstrated interest towards almost all activities listed. The creation of a brand or a product certification would be taken in consideration by all the interviewed who expressed an opinion (20 out of 20), with 16 of them stating that it may be useful or very useful (Table 1).

One conclusion from this is that promoting diversification might be a successful strategy, if the activities were sufficiently supported financially or otherwise.

---

[2] For a detailed description of the study design, see [2].

[3] For the existence of trust relations, values were expressed on a five-level Likert scale as: "low", "medium-low", "medium", "medium-high" and "high".

[4] To create the network graphs, we used the "Igraph" package [8] under R [9].

[5] Not all interviewed expressed an opinion on each of the listed activities, a reason why the total number of respondents may be less than 25.

Table 1. Evaluation of potential diversification activities

	Already operated	Would consider it	Total	Useful	Quite useful	Very useful
Cleaning and anti-pollution services	4	21	25	1	8	8
Fish tourism	7	15	22	1	6	9
Support to research activities	4	17	21	2	4	9
Branding and product certification	0	20	20	2	8	8
Direct sale	6	12	18	2	2	7
Fish product transformation	4	14	18	1	6	6
Organizing cultural events on fishing and sea	2	13	15	2	6	3
Itti-tourism	1	13	14	1	2	7
Sea tourism (diving, snorkelling and dolphin watching)	1	13	14	3	4	4
Support services to other enterprises	3	6	9	0	2	0

For the list of diversification activities considered, see [3].

A question remains on why some of them, which are deemed useful by the fishermen and that appear to be feasible, have not been started already. We note that some of the most promising ones, such as the creation of brands and other forms of product certification, would require collective action and strong institutional support.

4 Acquaintance, Information Sharing and Trust Relationship Among Fishermen

Acquaintance among the operators of the fishing fleet in the area is quite extensive: skippers of 47 vessels, mostly fishing out of *Bagnara Calabra*, have been declared to be known by colleagues from 3 or 4 different ports (Fig. 2(a), when links reaching a node come from nodes of 3 or 4 different colors). 21 vessels are known only by respondents from the home port form where they operate and the remainders are known by respondents in two ports (either *Bagnara* and *Gioia*, *Palmi* and *Gioia*, *Bagnara* and *Scilla* or *Bagnara* and *Scilla*).

To explore whether the diffuse acquaintance determines also an active exchange of information, each interviewed was asked to declare whether (s)he exchanged technical[6] or managerial information[7] with the others. The resulting "technical information exchange" network (Fig. 2(b)) is somewhat less dense than the "acquaintance" one,

---

[6] By technical information we mean information on the use of fishing gears, on meteo conditions, on navigation courses, etc.

[7] As examples of managerial information, we mentioned information on applications to financial subsidies, participation to development plans, etc.

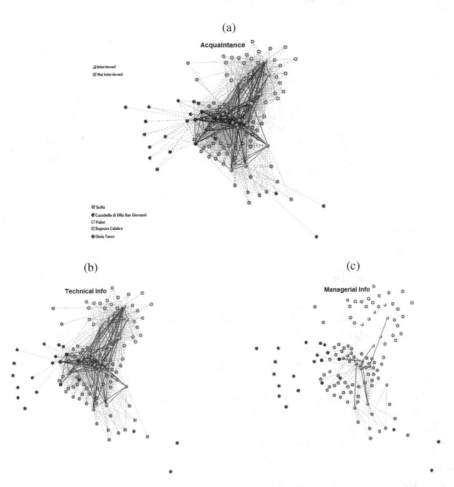

Fig. 2. Comparison between networks.

particularly in *Gioia Tauro* and *Cannitello* of *Villa San Giovanni*, as witnessed by the appearance of isolated nodes.

Sparseness become dramatic when looking at the "management information exchange" network (Fig. 2(c)). Very few fishermen are active in this type of information exchange, though Scilla is a notable exception: management information exchanges there appear more frequent than elsewhere in the rest of the study area.

Each interviewed was also asked to rank – on a 5-point scale from "low" to "high" – the level of trust towards each of the other fishermen. We analysed separately the expression of "positive" trust (ranked 4 or 5) and those of "negative" trust (ranked 1 or 2), excluding the links ranked as 3, to emphasize the quality of the relationship.

Even though there are isolated nodes (once again, more in Gioia Tauro than in other ports), the number of "positive trust" links is quite high, including across different ports (Fig. 3(a)). Expressions of negative trust are much rarer (even though not absent!), seem to prevail more between fishermen of different ports and are expressed by

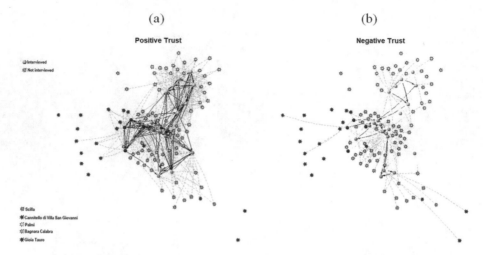

Fig. 3. Two-mode networks of (a) "positive" and (b) "negative" trust.

relatively few respondents. An interesting observation is that, in a few cases, the trust relationship expressed by a pair of interviewed is not reciprocal, so that a link between the same two nodes appears in both networks. Other than in Bagnara, it is rare that two or more different interviewed have expressed a "negative" trust towards a same colleague.

The combined analysis of all these characteristics allows drawing conclusions pointing to strong human relationships among fishermen in the area. While technical information is exchanged freely, either very limited information or very limited cooperation exists with respect to management issues.

5   Discussion and Conclusions

The results of the analysis of the relationships in the entire universe of active fishermen in the Costa Viola confirms and to a certain extent strengthen the main conclusions presented in a previous study [2], as we know now that the result is not limited to the interactions among the 25 fishermen who have been selected for the interview, but extends to the whole community. Fishermen know each other, as it may be expected in such a relatively small, traditional fishing community. They also "trust" each other, or at least so declare, which means that there should be potential for strong professional links. When explicitly asked, however, they do not report to engage in active exchanges of information, either technical or managerial, to any relevant extent.

This apparent paradox can be explained in different ways. First, it may be the case that the extent of trust in the community is overestimated due to self-selection of the fishermen who agreed to be interviewed. It is possible that those who refused to be interviewed are also those who, if interviewed, would have declared lower levels of trust. However, this hypothesis would imply the existence of a high degree of inconsistency in the expression of trust, as the non-interviewed have been reported to be trusted by the others. In any case, even if it were limited to the 25 interviewed, the problem remains. Another explanation might be that fishermen get the necessary technical information independently, or that there is little technical information to be exchanged to start with, as it may well be the case for a well consolidated, traditional activity such as fishing in this area, and in both cases, the limited reported exchange of information would not be a problem.

Where the results is truly problematic if when applied to the exchange of managerial information. Here, we know for a fact that managerial information, such as how to participate in funding opportunity or to be aware of recent regulations, is important and should circulate effectively. New rules are being imposed by the full implementation of European Common Fishery Policy environmental restrictions[8]. There are clear signs that the fishing activity as it was conducted in the past is no longer viable. Diversification is therefore necessary, but seems to be taking off very slowly if at all.

One of the reasons why feasible diversification activities that are considered useful by the fishermen in the area and yet are not already developed, seem to be linked to the fact that their operation would require strong cooperation among fishermen and an active supporting role by institutions; exemplary in this sense is the case of product branding and certification. The point is that such information should probably be generated and diffused not by fishermen individually, but by the various institutions that are created to support the local economy. In the personal interviews that have accompanied the fieldwork, fishermen reported that information on such opportunities have typically reached them quite late, if at all. Fisherfolk feel, for example, they have not been deemed the main actors in the definition of local development plans and programs, which should instead be explicitly customised to their needs.

The various networks of acquaintance, information exchange and trust presented above depict a tight community, with stronger technical links among vessels operating out of the same ports or among those that specialize in the same type of fishing activity, but densely connected also across different ports. This could be evidence that fishermen in the area indeed are interested in discussing common problems (something confirmed also by the informal exchanges one of the authors entertained with them during the study period).

Diffused declared statements of reciprocal positive trust would also suggest a solid basis to promote cooperation beyond what is currently seen, including – and perhaps starting with – making common pleas to access available financial resources from the EU. Such resources, as implemented through Local Development Plans, including the new Development Plan of the FLAG, which covers the period up to 2023, might be effectively destined to promote innovation in marketing of local catches, adding value to it through product and are branding and certification of origin and/or linking it to tourism. Adding value to the limited catches will also have the important benefit of reducing the pressure toward intensification of the fishing efforts, with positive impacts in terms of environmental sustainability.

---

[8] The aspects related to the implications of the EU imposed restrictions to the fishing activity in the area are explored in more depth in [10].

Information collected by interviewing other actors in the area, to explore the extent to which the problem might reside with institutions such as cooperatives and professional associations, is presented in a companion paper [7]. There, we highlight how fishermen knowledge of, and participation in, the activities of various institutions in this area is limited, and perhaps still conditioned by an atmosphere of distrust, generated by events linked to the issuing and enforcement on regulations that have been perceived, by some, as imposed from above, serving partial interests and without giving proper attention to the positions of fishermen as opposed to those of environmentalists. It seems that - rather than among fishermen - the key issue is of still insufficient trust between fishermen and institutions. While we are left without a definitive answer on why such a state of affairs persists and are only partially interested in understanding who is to blame, the fact remains that the future development of the sector, facing growing economic, social and environmental challenges will require a revamped attention to effective forms of public/private partnerships in which fishermen will have a respected role.

Acknowledgements. This study has been supported by GASTROCERT, a Project on Gastronomy and Creative Entrepreneurship in Rural Tourism, funded by JPI Cultural Heritage and Global Change - Heritage Plus, ERA-NET Plus action "Development of new methodologies, technologies and products for the assessment, protection and management of historical and modern artefacts, buildings and sites", 2015–2018. It provides further elaboration of the data collected and the results obtained in a project supported by the "Stretto" Coast FLAG, Calabria Region - European Fisheries Fund (2007–2013), Priority Axis 4: Sustainable Development of Fisheries Areas - Measure 4.1. Strengthening the Competitiveness of Fisheries Areas.

References

1. Palladino, M.: Social Network Analysis del sistema pesca in Calabria: un'analisi sul capitale relazionale nell'area del "GAC dello Stretto". Unpublished Report, Dipartimento di Agraria, Università degli Studi Mediterranea di Reggio Calabria, Italy (2015)
2. Palladino, M., Cafiero, C., Marcianò, C.: Relational capital in fishing communities: the case of the "Stretto" Coast FLAG area in Southern Italy. Procedia Soc. Behav. Sci. **223**, 193–200 (2016)
3. Morone, P., Sisto, R., Taylor, R.: What lies beyond geographical proximity: an investigation into multirelational networks. Centre for Policy Modelling discussion papers, Manchester Metropolitan University Business School 04-139 (2004)
4. Romeo, G., Marcianò, C.: Integrated local development in coastal areas: the case of the "Stretto" Coast FLAG in Southern Italy. Procedia Soc. Behav. Sci. **223**, 379–385 (2016)
5. Nicolosi, A., Sapone, N., Cortese, L., Marcianò, C.: Fisheries-related tourism in Southern Tyrrhenian Coastline. Procedia Soc. Behav. Sci. **223**, 416–421 (2016)
6. Romeo, G., Careri, P., Marcianò, C.: Socioeconomic performance of fisheries in the "Stretto" Coast FLAG in Southern Italy. Procedia Soc. Behav. Sci. **223**, 448–455 (2016)
7. Carrà, G., Peri, I., Vindigni, G.A.: Diversification strategies for sustaining small-scale fisheries activity: a multidimensional integrated approach. Rivista di Studi sulla Sostenibilità **1**, 79–99 (2014)

8. Csárdi, G.: IGRAPH. Network Analysis and Visualization. http://igraph.org/. Accessed 21 May 2017
9. R Core Team: R: a language and environment for statistical computing. R Foundation for Statistical Computing, Vienna, Austria. https://www.R-project.org/. Accessed 21 May 2017
10. Palladino, M., Cafiero C., Marcianò, C.: Institutional relations in the small-scale fisheries sector and impact of regulation in an area of Southern Italy (2018, Forthcoming)

Measuring the Tourism: A Synthetic and Autocorrelate Index for Italy

Domenico Tebala[1]([⊠]) [iD] and Domenico Marino[2] [iD]

[1] National Institute of Statistics, 88100 Catanzaro, Italy
tebala@istat.it
[2] Mediterranea University of Reggio Calabria, 89100 Reggio Calabria, Italy

Abstract. Tourism is a phenomenon that in many Italian provinces, especially in southern Italy, is particularly important in the contribution to collective well-being. Moreover, data measured in a specific area (or province) can be influenced by what happens in nearby areas, generating what is commonly called "spatial autocorrelation" or "spatial interdependence". This study is aimed at identifying a composite "systemic" index to measure the impact of tourist goods and others determinants can make to collective well-being in the provincial context through a composite index through the BES methodology (Equitable and Sustainable Well-being) and to analyze possible spatial autocorrelations between Italian provinces.

The method of construction of the index followed these steps:

(1) analysis of the theoretical framework, methodology used and indicators;
(2) choice of the statistical methodology
(3) statistical analysis: in order to assess the robustness of the identified method and, therefore, improve decision-making, we also completed an influence analysis to analyze the most significant indicators (software COMIC - COMposite Indices Creator)
(4) discussion of the results with the help of a georeferenced map of the synthetic touristic index of Italian provinces and a Cluster Map LISA which shows the provinces with statistically significant values of the LISA index, classified by five categories: (a) Not significant (white); (b) High-High (red); (c) Low-Low (blue); (d) Low-High (light blue); (e) High-Low (light red) (software GeoDa).

Keywords: Tourism · Index · Autocorrelation · Provinces

1 Introduction

Tourism is a phenomenon that in many Italian provinces, especially in southern Italy, is particularly important in the contribution to collective wellbeing. In recent years, tourism has changed a lot, transforming itself into a very complex social and economic phenomenon, capable of having positive effects on the socio-economic well-being of a territory [1].

It is therefore an important added value to the economy of a city, of a region and hence of a whole nation. As a result, the development of this sector can be crucial to

© Springer International Publishing AG, part of Springer Nature 2019
F. Calabrò et al. (Eds.): ISHT 2018, SIST 101, pp. 134–141, 2019.
https://doi.org/10.1007/978-3-319-92102-0_15

improving existing conditions in a place. But, as has already been emphasized, it is not just an economic factor, but much more. It is no coincidence that it is studied from a multiple point of view (economic, social, geographic, psychological, etc. ...). Moreover, data measured in a specific area (or province) can be influenced by what happens in nearby areas, generating what is commonly called "spatial auto-correlation" or "spatial interdependence". In this regard, the LISA indicators (Local Indicator of Spatial Association) provide a local dimension to the measure of autocorrelation, enabling each spatial unit (e.g. the province) to assess the degree of spatial association and similarity with the elements surrounding it. In the case of positive autocorrelation these associations can be of the High-High type (high values observed in a territorial unit and high values even in its vicinity) or Low-Low (low values observed in a territorial unit and low values even in its own around). Conversely, in the case of negative self-correlation, the associations will be of the High-Low or Low-High type. In all other cases, there will be no autocorrelation or non-significant autocorrelation.

However, despite you talk a lot of this importance, it is difficult to measure the extent of their contribution. So this study is aimed at identifying a composite "systemic" index to measure the impact of tourist goods and others determinants can make to collective well-being in the provincial context through a composite index through the BES methodology (Equitable and Sustainable Well-being) and to analyze possible spatial auto-correlations between Italian provinces. The province, that is the territorial sphere where the man is placed at the center of a wide range of environmental, cultural and artistic stresses, is the ideal place to build a synthetic and autocorrelate BES index.

2 Process to Calculate the Indicator

Description of the Theoretical Framework
Tourism is one of the main sectors influencing socio-economic development of the territories and is influenced by various determinants. The approach used involves the construction of macro areas (pillars) by aggregating elementary indicators. Both pillars and elementary indicators have been considered non-replaceable. To construct synthetic index, we adopted the following indicators and polarity [2] (Table 1):

Methodology
The matrix relating to data on Italian provinces was divided into four progressive steps:

(a) Selection of a set of basic indicators on the basis of an ad hoc evaluation model hinging upon the existence of quality requirements;
(b) Further selection aimed at balancing the set of indicators within the theoretical framework of the structure. Outcome indicators are impact indicators as the ultimate result of an action as a result of a stakeholder activity or process;
(c) Calculation of synthetic indices (pillars), by making use of the methodology proved more appropriate to obtain usable analytical information on the tourism of Italian provinces;
(d) Processing of a final synthetic index as a rapid empirical reference concerning the degree of tourism of Italian provinces.

Table 1. Indicators and polarity

Macro areas	Indicators	Polarity
Environment	Surface intended for urban gardens (square meters per 100 inhabitants)	+
	Total density of the green areas (protected natural areas and urban green areas) (percentage on municipal surface)	
	Green historic density and urban parks of significant public interest (m^2 to 100 m^2 of built-up areas)	
Social distress	Rate of homicides reported by police to the court for the province (per 100,000 inhabitants)	−
	Thefts in homes reported by the police to the court (per 100,000 inhabitants)	
	Robberies reported by the police to the court (per 100,000 inhabitants)	
Culture	Number of public libraries (per 100,000 inhabitants)	+
	Number of museums, archaeological sites and monuments (per 100,000 inhabitants)	
	Visitors of non-state institutes of antiquities and art institute for (number per 1,000 visitors)	
	Number of users of public libraries (per 100 inhabitants)	
Infrastructure [3]	Number of tourist ports, airports and railway stations per 100,000 inhabitants	+
Tourist facilities [4]	Number of hotels and number of places bed, number hotel extra exercises and sets bed per 100,000 inhabitants	+
Tourism demand [4]	Number days of presence (Italian and foreigners) in the complex of the receptive exercises per inhabitant	+
	Number of presences (Italian and foreigners) in the complex of the receptive exercises in the non-summer months (days per inhabitant)	
	Total spends in goods and services sustained by a traveller	
	Per-capita GDP [5]	

Missing values were attributed via the *hot-deck* imputation and, where not possible, with Italy's average value.

The choice of the synthesis method is based on the assumption of a formative measurement model, in which it is believed that the elementary indicators are not replaceable, which is to say, cannot compensate each other.

The exploratory analysis of input data was performed by calculating the mean, average standard deviation and frequency, as well as correlation matrix and principal component analysis. Since this is a non-compensatory approach, the simple aggregation of elementary indicators was carried out using the correct arithmetic average with a penalty proportional to the "horizontal" variability.

Normalization of primary indicators took place by conversion into relative indexes compared to the variation range (min-max).

Attribution of weights to each elementary indicator has followed a subjective approach, opting for the same weight for each of them. Since, in some cases, the elementary indicators showed different polarity, it was necessary to reverse the sign of negative polarities by linear transformation.

For the synthetic indicator calculation, we used the *Adjusted Mazziotta-Pareto Index* (AMPI), which is used for the min-max standardization of elementary indicators and aggregate with the mathematical average penalized by the "horizontal" variability of the indicators themselves. In practice, the compensatory effect of the arithmetic mean (average effect) is corrected by adding a factor to the average (penalty coefficient) which depends on the variability of the normalized values of each unit (called horizontal variability), or by the variability of the indicators compared to the values of reference used for the normalization.

The synthetic index of the i-th unit, which varies between 70 and 130, is obtained by applying, with negative penalty, the correct version of the penalty method for variation coefficient (AMPI +/–), where:

$$AMPIi\text{-}=Mri\text{-}Sricvi \tag{1}$$

where Mri e Sri are, respectively, the arithmetic mean and the standard deviation of the normalized values of the indicators of the i unit, and $cvi= Sri / Mri$ is the coefficient of variation of the normalized values of the indicators of the i unit.

The correction factor is a direct function of the variation coefficient of the normalized values of the indicators for each unit and, having the same arithmetic mean, it is possible to penalize units that have an increased imbalance between the indicators, pushing down the index value (the lower the index value, the lower the level of health).

This method satisfies all requirements for the wellbeing synthesis:

– Spatial and temporal comparison
– Irreplaceability of elementary indicators
– Simplicity and transparency of computation
– Immediate use and interpretation of the obtained results
– Strength of the obtained results

An influence analysis was also performed to assess the robustness of the method and to verify if and with which intensity the composite index rankings change following elimination from the starting set of a primary indicator. This process has also permitted us to analyze the most significant indicators.

The analysis was conducted using the COMIC (Composite Indices Creator) software, developed by ISTAT. The software allows calculating synthetic indices and building rankings, as well as easily comparing different synthesis methods to select the most suitable among them, and write an effective report based upon results.

3 Description of the Results

Table 2 reveals a moderate variability, especially for the infrastructure index (σ = 7.653) while Table 3 doesn't show significant correlations between the analyzed indicators of the macro areas except those most closely related to tourism: direct correlation between the tourist facilities index and tourist demand index (r = 0.779), between the tourist facilities index and infrastructure index (r = 0.515) and between tourist demand index and cultural index (r = 0.489).

Table 2. Mean, σ and frequency – Tourism macro areas

	Environment	Social distress	Culture	Infrastructure	Tourist facilities	Tourism demand
Mean	99.544	99.242	99.586	99.339	99.828	99.643
σ	4.379	7.049	5.754	7.653	6.165	5.989
Frequency	110	110	110	110	110	110

Table 3. Correlation matrix of the macro areas

Macro areas	Environment	Social distress	Culture	Infrastructure	Tourist facilities	Tourism demand
Environment	1.000					
Social distress	−0.091	1.000				
Culture	0.085	0.192	1.000			
Infrastructure	0.119	−0.326	−0.099	1.000		
Tourist facilities	0.148	−0.069	0.285	0.515	1.000	
Tourism demand	0.208	−0.162	0.489	0.407	0.779	1.000

Table 4. Influence analysis: mean and σ of the shifts for basis indicator of macro areas

Macro areas	Mean	σ
Environment	10.370	9.240
Social distress	9.060	7.690
Culture	6.390	6.120
Infrastructure	8.770	6.470
Tourist facilities	3.440	3.680
Tourism demand	5.010	4.700
Mean	**7.170**	**6.320**
σ	**2.437**	**1.827**

The influence analysis describes the indicators that most influence the composition of rosters in tourism of provinces. In analyzing Table 4, we can see that most significant macro areas are not tourist facilities and tourist demand, but environment ($\sigma = 9.240$), social distress ($\sigma = 7.690$) and infrastructure ($\sigma = 6.470$).

4 Discussion and Conclusions

The cartographic representation of the final composite index value, alongside with the descriptive analysis of data, yields the usual dualistic pattern South/Center-North of Italy as in other domains of BES, except for the case of Naples (Fig. 1), and shows how the health of the Italian population is closely related to the socio-economic components. This is a further demonstration of the well-known social, cultural and financial disparities within our Country.

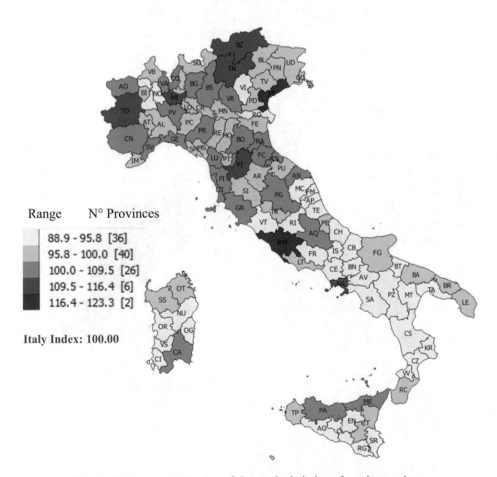

Fig. 1. Territorial distribution of the synthetic index of tourist provinces

In particular, the best performances are grouped in the Lazio (Rome), Trentino-Alto Adige (Trento and Bolzano), Piedmont-Valle d'Aosta (Turin and Aosta), Lombardy (Milan), Tuscany (Florence and Pisa) regions, but the most "touristic" province is Venice (123.03 Index – Italia Index 100), thanks, primarily, to the tourist facilities index (136.6) (27439 hotel extra exercises and 268105 sets bed) and the tourism demand index (116.9) (about 40 days of Italian and foreign presences in the total of resorts per inhabitant).

The province of Benevento (Campania) occupies the last place in the ranking (index 88.9), preceded by Avellino (89.8) although, as has already been said, Campania is well represented by Naples (index 111.1). However all the South is heavily penalized (50% of the provinces below the average are southern), except some isolated cases.

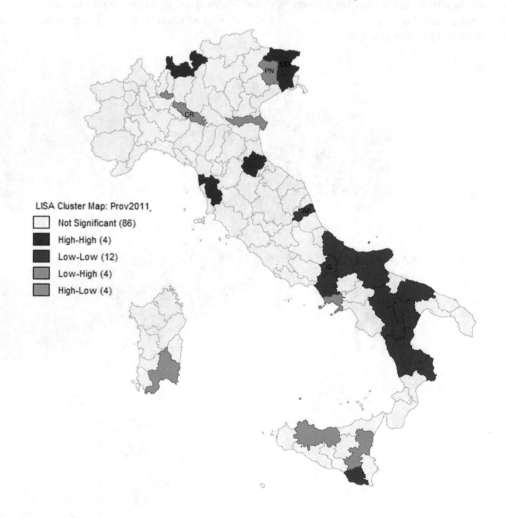

Fig. 2. Spatial autocorrelation - Lisa Cluster map Index

What is the role of autocorrelation [6] or space interdependence? In other terms, is the index of a province influenced by what happens in neighboring provinces? Our results show a good spatial interaction (Moran's index 0.39) and in particular in the South there is a certain positive autocorrelation between 12 provinces, with low values observed in a province and low values even in the surrounding areas (Low-Low Blue Color) (Fig. 2). Therefore, it was possible to identify spatial patterns able to describe areas of "multidirectional dependence", where contiguous areas show similar levels of the same phenomenon.

This again confirms the gap between the well-being of the South and the North with regard to the tourist sector, which is more influenced by tourist determinants (environment and social distress) than by indigenous variables (tourist services and tourism demand).

Our systemic index would be useful to have an idea of the general status of tourism as determined even by social and financial factors and to guide government actions.

References

1. Pernicola, C.: Il Community Empowerment sulle Comunità Locali a vocazione turistica (2007)
2. Istat: UrBes 2015 - Il benessere equo e sostenibile in Italia (2015)
3. Il portale Italiano dell'Open Data: http://datiopen.it, Accessed 25 May 2017
4. Istat: http://dati.istat.it. Accessed 25 May 2017
5. Eurostat - European Commission: http://ec.europa.eu/eurostat/data/database. Accessed 25 May 2017
6. Istat: http://www.istat.it/it/files/2014/10/M.-Mucciardi-E.-Otranto_Presentazione.pdf. Accessed 25 May 2017

A Flexible and at "Variable-Geometry" Planning for Italian Metropolitan Cities: The Case of Reggio Calabria and the "Area dello Stretto"

Giuseppe Fera(✉)

Mediterranea University of Reggio Calabria, 89100 Reggio Calabria, Italy
gfera@unirc.it

Abstract. The Law No. 56/2014 provided for the constitution of 14 metropolitan cities in Italy and established that the boundaries of these should coincide with those of the old Provinces. This choice has generated various perplexities since an institutional fixed border risks not to correspond with the geographical and territorial realities; in fact, today there are Metropolitan Cities under-dimensioned with respect to the real extension of the metropolitan area and others whose perimeter goes far beyond the metropolitan area, including also large rural territories and park areas. The thesis we intend to develop in the paper is that the boundaries of a metropolitan area may vary depending on the point of view, i.e. the role and meaning assigned to the metropolitan city and the goals that its institution must pursue; that is, if the metropolitan city is an instrument to optimize the relational flows within the territory between the central urban area and its hinterland (as it was conceived in the past) or, on the contrary, as the urban policy of the European Union seems to indicate, if its main role is to be a "development engine", a privileged place for research and innovation. However, the need to keep the two aspects together, to foster economic development and to reorganize the territory in functional terms, requires the construction of flexible planning systems, capable of responding to the complexity of the required objectives and taking into account that different goals may correspond to different territorial boundaries; a plan that the paper indicates as at "variable geometry" and that concerns both territorial and strategic planning.

Keywords: Metropolitan cities · Polycentric city
Flexible - "variable geometry" planning · Reggio Calabria

1 Metropolitan Areas and Metropolitan Cities[1]

I recognize having had strong doubts on the events that led to the approval of the Minister Delrio law n. 56/2014 and on some choices that were made. It is the sign of the times, of a phase in which Politics proves to be absolutely unable of serious and thoughtful choices in the interests of citizens and of the nation, but shows a large

---

[1] While *Metropolitan area* in this paper is used to indicate a territorial general phenomenon, we are using Metropolitan city referring to the italian institutional body set by the 2014 Minister Delrio act.

© Springer International Publishing AG, part of Springer Nature 2019
F. Calabrò et al. (Eds.): ISHT 2018, SIST 101, pp. 142–152, 2019.
https://doi.org/10.1007/978-3-319-92102-0_16

amount of nearness and improvisation. Apart from the sudden cancellation of the Provinces, then denied by the referendum of December 4th 2016 (don't sell the bear's skin before killing it), I was quite puzzled by the reductionism and superficiality used to decide the coincidence of metropolitan cities with the old provincial territories. It was a choice that did not take into account the profound differences that characterize the single territorial realities in terms of size, settlement models, economic development, etc. So today we are faced with metropolitan cities undersized with respect to their real extension (the metropolitan areas of Milan and Naples extend far beyond their old provincial boundaries) and against Metropolitan Cities that include entire mountain territories and natural parks (Reggio Calabria and Messina)[2].

A more thoughtful and careful choice should have been considered the rich amount of studies and researches elaborated on the topic, both nationally and internationally, and the different methods proposed for the perimeter of metropolitan areas; or still to be referred to the many studies developed by Regions and Provinces in the drafting of their territorial plans.

Also in the case of the Metropolitan City of Reggio Calabria there were several studies, developed in recent years, that have suggested a perimeter of the metropolitan area. The first idea, for instance, was the *conurbation* identified by Gambi [3], including the centers of Reggio Calabria, Villa San Giovanni and Campo Calabro. This hypothesis was resumed by two of the major Italian planners of the time, Giuseppe Samonà and Ludovico Quaroni; the first, on the occasion of the competition for the General plan of Messina at the beginning of the 60s [4], the second on the occasion of the *International Strait Bridge Competition* of 1969 [5], as part of a proposed layout scheme for the Metropolitan area of the Strait. More recently, this hypothesis has been reconsidered within the studies for the drafting of the *Piano Territoriale di Coordinamento* of the Province of Reggio Calabria (see Fig. 1).

On the other hand, more recent studies[3], considering the processes of sprawl that have taken place since the 80s of the last century, have worked towards extending the boundaries of the gravitational area around the core capital city; for example, the QTR/P, the *Quadro Territoriale Regionale*, identified a *Metropolitan area of the Strait* composed of 15 municipalities for a total of about 280.000 inhabitants. This metropolitan area consisted of three sub-areas:

- a Core area including the centers of Reggio, Villa S. Giovanni and Campo Calabro;
- the Tirrenian coast up to the Port of Gioia Tauro;
- a belt of 7 minor hill towns.

A study by the University of Barcelona [8] on the Spanish and Italian *Functional Urban Regions (FUR)* identified as a metropolitan area a territory composed of 21 municipalities, for a total of about 270,000 inhabitants, a linear system extended along

---

[2] To know about the current situation of Metropolitan cities in Italy and the implementation of the Delrio act see: De Luca, Moccia [1]; Sbetti, Giannino [2].

[3] At an international level one of the main texts about metropolitan areas is the one of OECD (Organization for Economic Cooperation and Development), developed starting from the methodology of the Functional Urban Areas [6], based on working commuting fluxes. See: OECD 2013 [7].

144   G. Fera

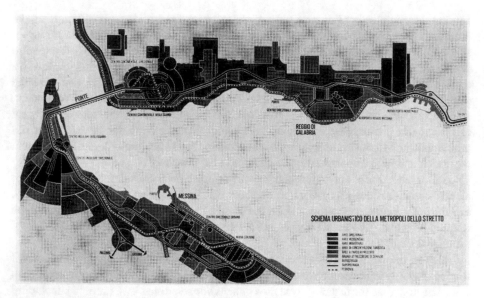

Fig. 1. Ludovico Quaroni, layout scheme of the Metropolis of the Strait of Messina, including Messina, Reggio Calabria and Villa San Giovanni.

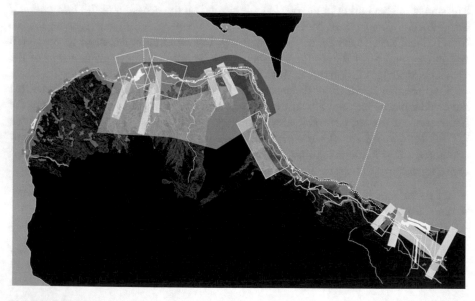

Fig. 2. The Metropolitan area of Reggio Calabria, according to the Regional Territorial Plan (QTR).

the coast of the Strait from Bagnara Calabra to Melito Porto Salvo, a layout roughly resumed in other planning documents, including the Strategic Plan of the city of Reggio Calabria.

Of all these studies, as mentioned before, no one was considered, and today the Metropolitan city of Reggio Calabria includes the territory of the former entire province, a territory comprising 96 municipalities with approximately 550.000 inhabitants (Fig. 2).

The choice immediately raised some doubts: Does a metropolitan city make sense including a 65.000-hectare natural park, within which there are numerous small towns with an exquisitely rural character and with poor functional relations with the main center? After a first impulse to give a negative answer, it is necessary here to think over the role and significance of metropolitan areas today and on the goals that inspired the Delrio Law, linked to the recent directives in urban and territorial policies of the European Union (see Fig. 1).

2 The Metropolitan Areas: From a System of Relational Flows to an Engine for Sustainable Development

In a study of some ten years ago, the OECD (2006) [9] highlighted the role of large urban concentrations as drivers of economic development processes, as they allow:

- to make the most of the so-called "agglomeration economies", deriving from the concentration, in a limited space, of complementary and integrated activities;
- advantages coming from high levels of specialization of service activities and diversification of the offer;
- high capacity to concentrate research and development (R & D) activities and generate, consequently, innovation;
- high level of human and social capital;
- high level of services and physical infrastructures (transportation, production spaces, etc.) usable by companies.

According to the above, the role of metropolitan areas has been the topic of some careful and deep reflections by the institutions of the European Union [10], which has rethought its Development and Cohesion Policies by focusing on the large metropolitan areas, the place of technological innovation, social capital, infrastructures, etc.

But this strategy has some limitations and critical issues, admitted by the European Union itself, such as the costs deriving from social conflicts linked to concentration of poverty and marginalization, the environmental costs of congestion and pollution, the stress of infrastructures, etc. But the main danger of large metropolitan areas is that they tend to grow and to feed themselves at the expenses of the surrounding territory, causing depletion and depopulation, and fueling territorial imbalances.

In other words, we can state that the government of metropolitan areas, in these years, is switching from a tool for the coordination of local transport policies and the management and localization of services, to a device for the promotion of economic and social development. It is evident that this represents a genetic change, requiring new methodologies of definition and consequently of the perimeter setting. In addition, we move from a concept linked to the acknowledgment of an existing state of affair (the metropolitan area encompassed on the basis of the functional relationships already detected, such as commuting flows) to a new conception linked to a possible and desirable development, to a future vision of sustainability, progress, equity, well-being.

If so, the criteria for identifying metropolitan areas shift from the mere consideration of functional relationships (commuting flows, work statistic systems, etc.) to the identification of natural, landscape, human and infrastructural *resources that can promote development*, often extending the area of interest to integrate new "resources" and ultimately to create "critical mass", indispensable for the competition with other territories in the global market. And we must also consider all those peripheral resources, located in a geographical and economic marginality, but which, if properly integrated into the pulp of the central areas, can contribute to the overall development. Otherwise, as we said before, these areas can be affected by demographic and economic depletion; exactly what happened during these years to the "internal areas" of our region and to the small mountain centers.

Due to the above, the choice to identify the area of the Metropolitan city with the former provincial territory makes sense and reason, being the fortunes of the weak and internal rural areas integrated with those of the strong urban areas. But another important innovation should be considered in the way of conceiving metropolitan areas, that is the emergence of a *polycentric urban vision*[4], alternative to a conception of the metropolitan area focused on the central urban core and on its relations with its hinterland. Indeed urban polycentrism is not such a recent idea, if we think that it was born within Organic Urbanism in the first half of the last century and was inspired by Howard's Garden City. The latter, even before the Abercrombie's Plan for the County of London and the Greater London Plan, had imagined the great London metropolis surrounded by Garden cities; an idea that in the middle of the century inspired the politics of the New towns.

More recently, the polycentric organic vision has evolved to imagine the territory as the place of a system of reticular relations, a network in which each urban center represents a node of the network [14]; a concept that, as we mentioned, has inspired the urban and regional policies of the European Union that have identified urban polycentrism as a tool to guarantee a "balanced territorial development", to foster a new and more balanced relationship between urban territories and rural territories and promote urban densification processes to face environmental and social criticalities (land consumption, commuting, lack of relationships, etc.) resulting from a process of urban sprawl nowdays out of control [10].

Therefore, once the general hypothesis of a Metropolitan city has been accepted as a vast, polycentric and inclusive territory, which can therefore incorporate territories with strong rural characteristics, it is necessary to ask what should be the objectives of the territorial planning and management and which are the possible critical issues to remove; knowing that the first possible risk (and I would also say probable) is to assume a line of intervention that privileges the issues related to the central core, where obviously some

---

[4] The bibliography on Polycentric city is wide, starting from the first text of Munford [11] up to the most recent contributions that we can suggest, the models of Ecopolis by Magnaghi [12] and Biopoli by Saragosa [13].

fundamental resources are concentrated in terms of innovation capacity and infrastructure, but above all the decision-making political system of the metropolitan city, whose Mayor coincides, as required by law, with the Mayor of the capital city.

In light of the considerations made previously, with particular reference to the Metropolitan City of Reggio Calabria, we can at this point carry out some reflection on the planning and programming tools that the national law entrusts to Metropolitan cities, i.e. the Strategic Plan and the Territorial Plan.

3 Strategic and Territorial Planning for Metropolitan Cities

The *Territorial Plan of the Metropolitan City* (PTCM), provided by art. 18bis of the Regional Planning law of Calabria, is conceived as an instrument of territorial planning and coordination, and replaces the PTCP, the former territorial plan of the Provinces, which remains in force until the first one is approved. Beyond the differences between the two instruments (the responsibilities of the new plan seem to go beyond those of the old Ptcp) we are still in the presence of a "Wide area" planning tool with general coordination functions, with two fundamental objectives that I feel we can foresee/suggest:

- building the reference framework for the environmental compatibility of the strategic plan in order to ensure that the strategies and actions envisaged by the same, are consistent with the overall conservation needs of the natural areas, landscape protection, mitigation of environmental risks and pollution;
- building the general framework of the urban structure of the Metropolitan city, which should represent the spatial reference model for guiding the economic and social choices of the strategic plan.
- However, it remains to be clarified an aspect that does not appear well defined by the law, namely the relationship between the PTC and municipal planning.

The construction of the territorial model of the PTCM is entrusted to the Strategic Plan, which the Delrio law sets with a three-year deadline. Also in this case some perplexities must be raised; in fact, the forecast of a strategic plan with a "three-year" deadline, connotes it not properly as a "strategic" plan but more similar to an operational plan, like the "Three-year program of public works" of the Municipalities. To be effective this short term plan needs to have upstream a real Strategic Plan, with a medium-long term perspective and strongly integrated with the Territorial Plan, up to the point where the two could coincide. In fact, it is very difficult to build up a model of spatial planning not deriving from a more comprehensive economic strategy, in other words from a model of economic and social structure, of which the territorial model is the spatial realization. At the same time, we know that a model of sustainable economic development cannot be separated from territorial and landscape resources and that the strategies envisaged, for example, in terms of large infrastructures of the territory, must be compatible with environmental conditions and the level of risk affecting the territory. In fact, the urban planning law of Calabria (art.18 bis c.4), in setting the goals and tasks of the PTCM, states that it must pursue the fundamental objective of assuring a territorial structure compatible with the environmental conditions and consistent with the need to maximize the specificity and potential of the different territories.

It is evident that the warranty of a territorial structure, compatible with the environmental conditions, requires a medium-long term time frame, incompatible with the three-year period set by law. This goal can be pursued, above all, through the parallel *Strategic Environmental Assessment* supporting the plan. The idea of imagining a strategic plan and a territorial plan as two strongly integrated instruments, or rather as a single planning tool, that is, a strategic plan with territorial value, would have the advantage of having to produce a single process of environmental assessment.

The second aspect, the maximum enhancement of the specificity and potential of the individual territories brings us back to the title of this contribution, namely the need for flexible planning and/or "variable geometry".

A Territory with Variable Geometry. Starting from this last point, we can state that the former province of Reggio Calabria, due mainly to its geographical position and its geomorphological characteristics, which heavily influenced its history, culture and social and economic conditions, presents high differentiation among the different parts. In a very synthetic way, in relation to the numerous studies produced in these years, as can be deduced from the instruments of Territorial Planning (Qtr and Ptcp), but above all to the experiences of Community strategic planning, a subdivision of the former provincial territory in five sub areas is commonly recognized; areas that essentially follow the articulations decided by the territoril communities with the *Progetti Integrati territoriali* (PIT)[5] and subsequently resumed by the *Quadro Territoriale Regionale* (QTR) with the *Territori Regionali di Sviluppo* (Trs)[6]:

- *Pit 19: Piana di Gioia Tauro*, including 10 municipalities with 86,500 inhabitants; the system historically with the highest agricultural vocation in which the leading role of the large commercial port tends to be increasingly affirmed today;
- *Pit 20: Community of the Aspromonte Park*, including 23 municipalities with a total of approximately 82,000 inhabitants; the great green lung at the center of the entire Metropolitan City;
- *Pit 21: Locride*, 39 municipalities for a total of about 130,000 inhabitants; perhaps the portion of territory richest in cultural resources and traditions, with the highest potential for tourism development;
- *Pit 22, Area dello Stretto*, 13 municipalities for a total of about 230,000 inhabitants; the core area, the urban - metropolitan system of Reggio Calabria, we have already mentioned, extended along the coast from Bagnara to Melito (excluded);

---

[5] The *Progetti integrati territoriali* (Pit) were launched by the European Program 2000–2006, regarding in Italy all the Mezzogiorno regions; in Calabria 20 Pit were produced, of which 5 in the former Province of Reggio Calabria. See: http://www.regione.calabria.it/archiviopor/progetti-integrati/10-programmazione-2000-2006/188-progetti-integrati-2000-2006/1924-menu-progetti-integrati-2000-2006.

[6] The *Territori Regionali di Sviluppo -TRS-* were the basic units for the allocation of the Development Funds of European Union ad for the Regional Development Policies. The whole Calabria was divided in 16 Trs, 3 metropolitan areas (Reggio, Catanzaro – Lamezia, Cosenza), 7 Urban Territories and 6 Rural Areas.

– *Pit 23, Area Grecanica*, 12 municipalities for a total of about 43,000 inhabitants; a small settlement system with strong connotations of a cultural character and identity.

This territorial structure has been incorporated into the Statute of the Metropolitan City, through the establishment of the *Zone omogenee* (Article 39), which "constitute the operational articulation of the Metropolitan Conference and articulation on the territory of the Metropolitan City's activities and decentralized services. They represent the favorable framework for the joint organization of municipal services and for the delegated exercise of functions of metropolitan competence". But, beyond this main articulation, the territory of the Metropolitan city appears as a "moving mosaic", a kaleidoscope in which every card moves at each rotation, returning a different overall image. Out of metaphor, the internal partition of the metropolitan territory must be considered in a variable way depending on the many different sectoral plans and/or institutional aggregations envisaged or that could be born as a *Unione di Comuni*: Health agencies, School districts, Strategic Plans, Territorial Pacts, Gal Leader, etc. Many of these aggregations arise from a project shared by several administrations, established traditions, a common will, that the Strategic Plan of the Metropolitan City will certainly have to take into account (see Fig. 3).

But at the same time the consideration and accounting of the individual territorial realities has to consider the need to build a common strategy and path, in which each territorial reality can be recognized with mutual benefit. From here arises the need for a "variable geometry" planning [15, 16] able, within an absolutely unitary project, to indicate diversified objectives and strategies in relation to the different themes (transport, tourism development, production chains, etc.) or different territorial articulations, homogeneous zones or unions of communes; considering, furthermore, that all the environmental or landscape themes (water basins, natural areas, hydrogeological instability, etc.) have characteristics that do not consider and do not refer to any administrative boundaries.

In the specific case of the Metropolitan city of Reggio Calabria, there is a widely shared position, both in the technical and in the political world, which maintains that a real development perspective for the Metropolitan City is only possible within a framework that includes also the Metropolitan City of Messina, through the creation of an integrated metropolitan area of the Strait of Messina [17]. Such a perspective would not only create a system with a critical mass of 1, 2 million inhabitants, but would create an undeniable added value by means of a common union and management of existing resources (Universities, cultural services, landscape, historical assets, infrastructures). This idea has been positively implemented within the strategies of the Calabria region that in the Urban Planning Law provides that the Strategic Plan of the Metropolitan City is elaborated after coordination with the *Conferenza permanente interregionale per il coordinamento delle politiche nell'Area dello Stretto* (art. 4 regional law of 27 April 2015, n. 12).

The idea of planning coinciding with an administrative boundary does not seem to be real and effective. The principles of autonomy and subsidiarity are endorsing the definitive sunset of a hierarchical and authoritative planning, in which the higher institutional level, that imposes its choices to the lower one, is going to be replaced by

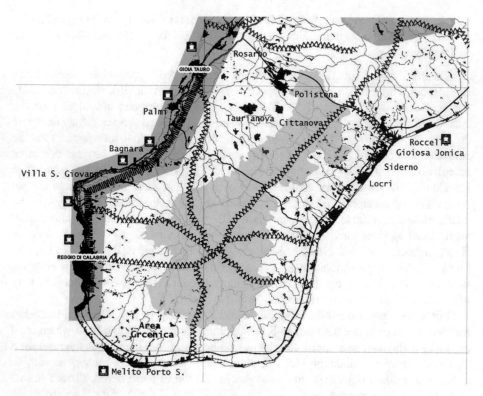

Fig. 3. The Zone Omogenee of the Città Metropolitana: in the middle in light grey the Area of the Aspromonte park, in dark grey the Metropolitan area of Reggio Calabria, including the port of Gioia Tauro; then the rural area of the Piana around Polistena and Taurianova; on the upper right side the urban linear system of the coast of Locride and at the bottom the Area grecanica.

some cooperative and reticular forms, where the choice to join to a project don't arise from an administrative or institutional affiliation, but from a common interest or benefit. All this leads to focus our attention on the process of making up and management of the strategic - territorial plan or, in other terms, on the governance of the plan itself. Apart from all the technical aspects, strategic planning is characterized for being a form of participatory and cooperative planning; therefore it is necessary to focus our attention on the organization of the interaction among the different actors, also for getting the necessary flexibility for the plan to be consistent with the variable geometry of its internal and external borders.

Apart from the need for further studies on the topic, it seems to me that the question of the "variable geometry" of the plan is mainly referred to the composition and organization of the "interactive tables" set up for designing the plan. These tables (or forums or whatever you want to call them) will be able to involve specific territorial or thematic areas and public and private subjects.

It is precisely on the government of the decision-making process that researchers are working elsewhere in Europe; an interesting example comes from France and from the *Schémas de cohérence territoriale*[7], a tool specifically designed to foster the processes of strategic inter-municipal and/or intersectoral planning, with the aim of making coherent with each other or with respect to a single project, the different sectoral plans (urban policies, construction, mobility, environmental protection) or plans referring to different municipalities. Municipalities that intend (or must) to develop a SCoT are constituted in a Union, and they are entrusted with the task of elaborating, approving and managing the Scheme. In short, to return to the Italian contest and in particular to the Strait area, we believe it is necessary to develop a more deep reflection on how we can structure an organic planning process that faces the needs and demands for development that may arise from single territories in which the vast area of the Metropolitan City is historically articulated. To this end we can imagine some work and research paths related to the following objectives emerging from the above:

- To ensure maximum integration between the Strategic Plan and the Metropolitan City's territorial plan, so as to realize maximum consistency between the "design" of the territorial planning envisaged by the Territorial Plan, with its actual implementation through the actions and measures activated by the strategic plan;
- Elaboration of a single Strategic Environmental Assessment that defines the framework of environmental and landscape compatibility and the consistency of this framework with the planned spatial model and the actions necessary to achieve it;
- To assign to this integrated strategic territorial plan, in the first draft a medium-long term vision and to provide specific triennial implementation plans;
- Structuring and organizing the decision-making process, guaranteeing the maximum participation of municipalities, local institutions, companies and associations, providing interaction tables organized for homogeneous areas or specific issues;
- In this framework, to provide a planning step considering the widest dimension relative to the Integrated Area of the Strait and therefore of a specific joint interactive table with the metropolitan city of Messina;
- Provide an adequate "control room" or a coordination structure capable of governing the various moments of participation, incorporating the guidelines of these elaborated to bring them back to a unitary design and strategy.

In conclusion, if the planning of metropolitan cities requires a flexible and "variable geometry" planning process, this should focus on the governance of the plan. This also for a more general reason, since I have seen too many strategic and territorial plans fail because they are incapable of really involving, if not in a reduced way in the elaboration phase, the territorial community, entrepreneurs, professions, workers' organizations, etc. It is my firm belief, in fact, that the strategic plans should not be made by municipalities or metropolitan cities; at most these can assume a coordination function and make available spaces and structures to work; a real strategic plan must come from the bottom, from the active push of the world of production, of work, of research, which will also be called to implement it, through appropriate partnership structures.

---

[7] See the web site of the *Ministère de la Cohésion des territoires* http://www.cohesion-territoires.gouv. fr/schema-de-coherence-territoriale-scot.

In such a complex and territorially complex reality as the metropolitan metropolitan city is, it will not be easy but it is still necessary to try.

References

1. De Luca, G., Moccia, D. (eds.): Pianificare le città metropolitane in Italia. Interpretazioni, approcci, prospettive, INU Edizioni, Roma (2017)
2. Sbetti, F., Giannino, C. (eds.): Città metropolitane, territori competitivi e progetti di rete. Urbanistica Dossier, INU Edizioni, Roma (2017)
3. Gambi, L.: Atlante delle regioni d'Italia: la Calabria. Utet, Torino (1965)
4. Cardullo, F.: Giuseppe ed Alberto Samonà e la metropoli dello Stretto di Messina, Officina edizioni, Roma (2006)
5. Ciorra, P.: Sei maniere di pensare l'urbanistica. In: Terranova, A. (ed.) Ludovico Quaroni, architetture per cinquant'anni, Gangemi ed., Roma - Reggio Calabria (1985)
6. OECD (Organization for Economic Cooperation and Development): Functional Urban Areas in OECD Countries (2016). http://www.oecd.org/cfe/regional-policy/functional-urban-areas-all-countries.pdf. Accessed 15 Feb 2018
7. OECD (Organization for Economic Cooperation and Development): Definition of Metropolitan Urban Areas (FUA) for the OECD metropolitan database (2013). https://www.oecd.org/cfe/regional-policy/Definition-of-Functional-Urban-Areas-for-the-OECD-metropolitan-database.pdf. Accessed 15 Feb 2018
8. Boix, R., Veneri, P.: Metropolitan areas in Spain and Italy, Institut d'estudis regionals i metropolitans de Barcelona (IermB) (2009). https://ideas.repec.org/p/esg/wpierm/0901.html. Accessed 15 Feb 2018
9. OECD (Organization for Economic Cooperation and Development): Competitive Cities in the Global Economy (2006). http://www.sourceoecd.org/regionaldevelopment/92640270. Accessed 15 Feb 2018
10. European Union, Regional Policy Dept.: Cities of tomorrow. Challenges, Visions, Ways forward (2011). http://ec.europa.eu/regional_policy/sources/docgener/studies/pdf/citiesoftomorrow/citiesoftomorrow_final.pdf. Accessed 15 Feb 2018
11. Mumford, L.: The City in History, Harcourt, Brace, New York (tr. it., La città nella storia, Ed. di Comunità; ried. Bompiani 1963, Milano, 1977) (1961)
12. Magnaghi, A. (ed.): Il territorio dell'abitare - Lo sviluppo locale come alternativa strategica. Franco Angeli, Milano (1998)
13. Saragosa, C.: Città tra passato e futuro. Un percorso critico sulla via di Biopoli, Donzelli ed., Roma (2011)
14. Dematteis, G.: L'ambiente come contingenza ed il mondo come rete. In: Urbanistica, n. 85 (1986)
15. Piroddi, E.: Le forme del piano urbanistico. Franco Angeli, Milano (1999)
16. Tira, M.: Verso un territorio a geometria variabile. In: Ingenio Web (2014). https://www.ingenio-web.it/3106-verso-un-territorio-a-geometria-variabile. Accessed 15 Feb 2018
17. Fera, G.: L'Area Metropolitana dello Stretto: storia, presente, prospettive. In: Fera, G., Ziparo, A. (eds.) Lo Stretto in lungo ed in largo. Prime esplorazioni sulle ragioni di un'area metropolitana integrata nello Stretto di Messina, Università Mediterranea- Centro Stampa di Ateneo, Reggio Calabria (2016)

Mobility, Accessibility, Infrastructures

The Rehabilitation of Some Historical Urban Port Areas in Sardinia

Alessandra Casu

University of Sassari, 07041 Alghero, Italy
casual@uniss.it

Abstract. Due to the approval and implementation of the Regional Landscape Regulatory Plan (PPR), in Sardinia most of the attempts to transform the shorelines with settlements have stopped. The only places in which these transformations are allowed are urban areas, often with ports: so urban waterfronts have become a place for projects and programs, which not always are real outcomes of urban policies or strategic plans. Nor these projects and programs privilege the intervention on public open spaces rather than on buildings, and buildings are often reduced to mere 'containers' or sometimes, exhibitors of the *griffe* by a *starchitect*. Another effect of the PPR is that it has imposed to restart the process of municipal urban plans, which is still in progress, and the redesign of the urban waterfronts is one of the most important opportunities to be seized.

The essay analyses some urban regeneration projects related to urban historical ports in Sardinia (Cagliari, Olbia, Porto Torres, La Maddalena, Alghero), from both the viewpoints of urban policies and design.

Keywords: Urban waterfront · Urban policies · Urban design

1 Introduction

Sardinia coastal territories, in recent years, appear as the places to which the most important interventions and regulations are dedicated. Actually, the Regional Law 8/2004 (the so-called 'savecoasts Decree') and the Regional Landscape Regulatory Plan (PPR) have stopped most of the attempts to transform the coastal areas for a depth of 2 km, allowing these transformations only in already urbanized areas: so urban waterfronts have become a place of attention and of projects, that almost never are real outcomes of urban policies, or are preceded by moments of strategic planning, aimed to identify the most appropriate functions or the instruments for implementing programs that privilege the intervention on public open spaces, or that don't reduce buildings to mere 'containers' or sometimes, exhibitors of the *griffe* by a *starchitect*.

The PPR imposed to restart the process of municipal urban plans, adapting them to it: in most cases, this process is still in progress, and the redesign of the urban waterfronts is one of the most important opportunities to be seized. In some cases, proposals for urban waterfronts have been already included in the preliminary schemes of the municipal urban planning, anticipating the process of rehabilitation.

In general, the urban waterfronts renewal involves the port areas: e.g., in 2005, the Provincial Council of Sassari submitted a proposal called 'Harbor Park of the County

© Springer International Publishing AG, part of Springer Nature 2019
F. Calabrò et al. (Eds.): ISHT 2018, SIST 101, pp. 155–163, 2019.
https://doi.org/10.1007/978-3-319-92102-0_17

area' which, in fact, is a reinterpretation of the port system in North Sardinia (Fig. 1), in which the intention is 'to integrate the hinterland economy with the coastal one'. The 'port park' is 'a system based on some key infrastructures, aimed at rebalancing', and therefore it needs intermodality and connections between ports and their hinterland. Not among the tangible and intangible actions of the project, more themes related to urban ports and waterfronts, as currently meant [1, 2], can be recognized: 'the integration of the port with the fabric' using the waterfront as a 'connector [...] among the modern residential districts and attractions of the city and the Old Town' (Alghero) or with the mutual usability and accessibility among vast harbor areas, archaeological resources and commercial districts (Olbia, La Maddalena, Porto Torres).

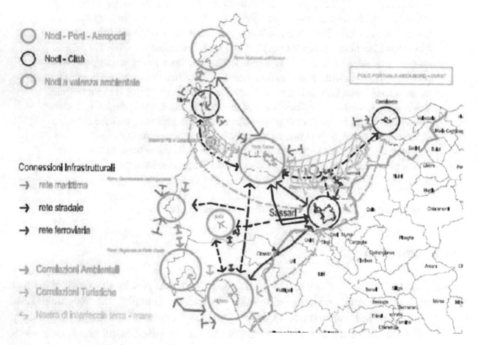

Fig. 1. The network of relationships among inner and coastal territories in North-western Sardinia.

2   Urban Policies at Work?

Policies related to urban ports and waterfronts seem almost related to three themes, or three knots still to be dissolved [3]: the conflict of jurisdiction and ownership rights over the involved areas; the conversion of disused industrial areas [4]; the imitation of *à la page* examples (Barcelona and Bilbao, needles to say), with the run-up to a *griffe* or to a 'rare' function to localize.

2.1   This Land Is My Land

The main problem to be addressed is the one of rights on the areas, primarily the competence to plan: Sardinian urban waterfronts almost belong to the maritime domain, and the transfer of powers on public lands by the State to the Region has not yet produced a simplification or an easier integration of plans, programs and projects, subject to port authorities and their planning instruments and government.

The most striking case takes place in Cagliari, where several projects for the waterfront intersect. The urban planning has been limited by two frames: the Regulatory Port Plan (PRP) and the wide-area Plan by the Consortium of industrial development area. This second Plan includes the construction of the port channel, and thanks to which it has been possible to move the commercial traffic from the urban front—in which there are the Town Hall, the bus station and the facades of the XIX century Via Roma, built with the demolition of the medieval walls.

For the eastern urban waterfront the frame was the old Landscape Regulatory Plan made by a Superintendent. Up to 2000 the reference was the Masterplan approved in 1965, which provided an axis parallel to the shoreline, extending Via Roma and not according to building types that resume the arcaded facades of the existing urban waterfront: they were high, with elevated pedestrian platforms, reproduced and proposed towards the housing in St. Elia area, insisting on public lands obtained —as the areas occupied by the Trade Fair, the Monte Mixi sports facilities and the stadium reclaiming the marshes and the mouth of the port channel that fed the salt factories.

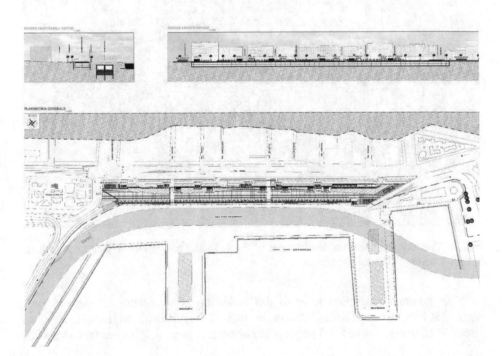

Fig. 2. The proposal for the redevelopment of the historical waterfront of via Roma in Cagliari.

As the Regulatory Port Plan dated forty years, the Port Authority predisposed a new Port Regulatory Plan, dusting off an old proposal included in the '60s urban master-plan, which saw the closing of the bypass of the town shifted from urban waterfront, at elevated or underground level. The first hypothesis, advanced in the old urban mas-terplan, in fact deprive a portion of the waterfront to any public use that was not the mobility; the last hypothesis, although less feasible for financial and technical diffi-culties, would stretch out the XIX century facades as far as the docks, using collectively these public spaces, that once were fenced (Fig. 2).

2.2   The Abandoned Land

Following the divestment of the first production units, and with a view to the urban renewal of the industrial port and its integration with the commercial port, the orien-tation documents for the urban planning in Porto Torres have provided for the redesign of the 'industrial' waterfront and port, configuring it as a long linear park that integrates the traffic at different altitudes for different categories of traffic (including cycles) and joins with the historical urban port area (Fig. 3), which should match with the meeting space, the access to the archaeological area of Turris Lybissonis Roman city, and the new maritime intermodal passenger station (the intermodal center for goods is adjacent to the industrial area, but far from the sea).

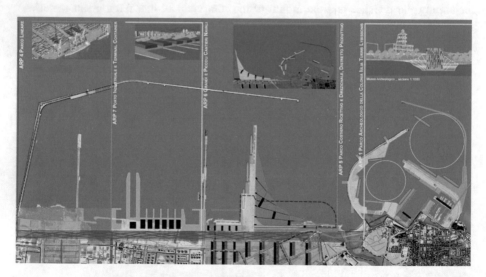

Fig. 3. The proposal for the redevelopment of the industrial and historical waterfront in Porto Torres. Courtesy: Giovanni Maciocco

This program guidelines have to deal with the new structure of the old port, determined regardless of the indications in the Port Regulatory Plan. Despite the clear intention of integrating sectoral projects in planning in general, it is clear that a strategic planning process, not carried out *a priori*, leaves too much uncertainty and does not allow to define choices by more predictable and, therefore, more aware outcomes.

A similar situation occurred in La Maddalena: here the divestiture of the U.S. base on the island of St. Stefano and of the Navy Arsenal released large urban areas in prime waterfront location. A strategic plan was funded in this case, but it started when the most significant choices had already been made.

The most striking example is made up by the former Naval Arsenal: location of shipyard production, repository of specific knowledge that would find an interesting market among boaters who frequent the region, it was transformed in a five-star hotel (Fig. 4) for the G8, regardless of the obvious vocation of places. A contested tender was experienced for its management but, due to the discovery of a severe seawater pollution, the contractor that won the tender had to give it up.

Fig. 4. The former Arsenal of La Maddalena, refurbished and converted in hotel.

Other unrealized works are referred to public works *strictu sensu*: it is mainly the redevelopment of public spaces along the waterfront (Fig. 5), with an undermined financial and technical viability.

2.3   Pictures at an Exhibition

In the case of Alghero the Town Council, anxious to show its Catalan origins, committed the redesign of the northern seafront by architect Joan Busquets assisted by local practitioners: a sort of *rambla* running from the Aragonese walls overlooking the harbor to the Lido (the coastal area most frequented by tourists, with bathing establishments and hotels). Although urban strategies and policies are credited to 'open' the city in locating the tourist facilities beyond the walls, urban design is problematic: in dealing with the sea front in the same way a Rambla in Barcelona, with a pedestrian platform in the middle of four car lanes, it overlooks that the rambla flows to the sea between two rows of buildings, while in Alghero it is facing the sea: the result is a symmetry driven away like a highway, while pedestrians don't go through their 'island' (Fig. 6).

A project that could be developed distinguishing spaces of places from those of flows [5] is reduced to a plan in which mobility marks the partitions, decorated with redundant signs. In a waterfront divided into three sections (the sandy coast to the north of the ancient city, the rock on which it stands, the rocky coast south of the walls),

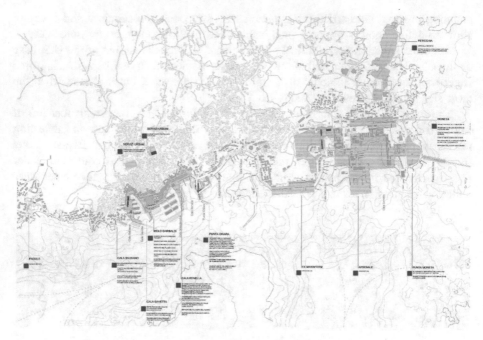

Fig. 5. The project for La Maddalena waterfront redevelopment.

instead of designing a unitary line surrounding the built city, a too broad catalog of street furniture items, dissonant each other and with the urban legacy, was used.

However, it should be recognized that a strategic purpose of the project is pursued: as it happened with the coastal walk along the ramparts of the walled city (but with better outcomes for this project than for the new promenade), the redevelopment of the public spaces is now retinue of facilities, which could help rid the old town center to a share of congestion, and make it more livable.

2.4   The Winner Takes It All

The ultimate synthetic example is located in the capital, Cagliari: the former Durum Wheat Semolina factory overlooking the edge of the harbor, an interesting example of industrial heritage, which is one of the few applications of the Hennebique building system imported from Piedmont by a manufacturer of Sardinian origin.

A project (Fig. 7) entrusted to the Brazilian Pritzker prize Paulo Mendes da Rocha, in proposing a meritorious function (a students' campus in an optimal location for its proximity to major transport infrastructures), faces a former proposal that erases most of the traces of the physical memory of a XX century place of work and innovation.

Not a voice among the scholars was raised, so that at least one legacy of this important piece of the collective memory would remain: some of the people active in some industrial heritage associations were called to design the redevelopment of the area, proposing a huge building as a dormitory, attached to only one of the old silos, and to demolish all the other witnesses of that industrial past.

Fig. 6. The new 'Busquets promenade' at Alghero. To the right, the cycling lane and the port.

Fig. 7. A *maquette* of Paulo Mendes da Rocha's project in Cagliari.

3 Urban Design Strands

In the projects currently proposed, the attitudes that emerge seem affected by a sort of 'Spanish Influenza', which could be summarized in three main points, which find their reference archetypes in some of the most famous recent accomplishments[1]:

- The waterfront as an interface between a town and its port, primarily through the reuse of brownfield or dismissed sites for the relocation of economic activities [4]. The mutual independence remains, as an edge between town and port: one of the problems in Olbia, partially addressed in Cagliari and looming in Porto Torres;
- The construction 'in the harbor', as the example by Herzog and de Meuron in Tenerife, attempting to overcome this traditional separation through a sort of 'rooting': attempted in Olbia with the construction of the Archaeological Museum at the root of the Isola Bianca pier; a hypothesis that would be interesting to apply in Porto Torres, with the integration of the Roman city of Turris Lybissonis with the harbor; part of Mendes da Rocha's idea for the students' dormitory in Cagliari;
- The 'square by the sea', according to the terminal in Yokohama made by FOA, separating a level for flows and another for parks or public spaces. It is the answer given by Zaha Hadid in *Betile* design competition for a museum in Cagliari: a 'fluid' space leading to a topology *à la* Escher, in which interiors spill outward, opening with a large public space on water. It is a sort of large-scale sculpture to be exhibited together with its designer, as the urban context was a gallery.

4 Final Remarks

As also Gordon [6] and Savino [3] stressed, beyond the operations of institutional promotion, neither a real integrated project nor a really strategic approach have emerged in this context yet, perhaps due to the many knots still to be dissolved: nor an occasion for debate - either *a priori* or *ex post* - has been launched on the matter, neither proposal was - in fact - the subject of public discussion.

These are some emblematic examples of how the construction of the waterfront, like still many current urban projects, does not follow the policy-plans-projects-programs sequence but advances without an order, designed by piecemeal projects, by financial opportunities and events, by conflicts of competence and availability of areas, according to an alternation of *grandeur* and local conveniences.

---

[1] This taxonomy arose discussing with Giovanni Maciocco during the preliminary works for the new Masterplan of Porto Torres.

References

1. Bruttomesso, R. (ed.): Waterfronts. A New Frontier for Cities on Water. Cities on Water Editions, Venice (1993)
2. Giovinazzi, O. (ed.): Città portuali e waterfront urbani. Ricerca bibliografica. Centro Internazionale Città d'Acqua, Venezia (2007)
3. Savino, M.: Città e waterfront tra piani, progetti, politiche ed immancabili retoriche. In: Ead. (ed.) Waterfront d'Italia. Piani politiche progetti, Angeli, Milano, pp. 36–69 (2010)
4. Chaline, C.: Le réaménagement des espaces portuaires délaissés. Les Annales de la Recherche Urbaine 55(56), 43–47 (1993)
5. Castells, M.: The Informational City: Economic Restructuring, and Urban Development. Blackwell, Oxford and Cambridge (1989)
6. Gordon, D.: Transforming the Urban Waterfront: why does it take so long? Aquapolis 3(4), 13–16 (1999)

3D Mapping of Pavement Distresses Using an Unmanned Aerial Vehicle (UAV) System

Giovanni Leonardi[✉], Vincenzo Barrile, Rocco Palamara,
Federica Suraci, and Gabriele Candela

Mediterranea University of Reggio Calabria, 89100 Reggio Calabria, Italy
giovanni.leonardi@unirc.it

Abstract. The aim of road surface monitoring is to detect the distress on paved or unpaved road surfaces. Depending on the types of surface rupture, required parameters are measured on-site to determine the severity level of the road damage. Local infrastructure engineers and supervisors must therefore optimize their resources when monitoring road conditions and scheduling maintenance activities. Automation of road surface monitoring process may result in great monetary savings and can lead to more frequent inspection cycles, for this reasons departments of road maintenance, repair and transportations have become more interested in using automatic systems for pavement assessment. The scope of the presented work is the performance evaluation of a UAV system that was built to rapidly and autonomously acquire mobile three-dimensional (3D) mapping data to identify pavement distresses.

Keywords: UAV · Pavement management system · Road distresses

1 Introduction

Monitoring and maintenance of paved road such is an expensive but essential task in maintaining a safe operation. The road surface condition assessment is an essential piece for recent pavement management systems methods as rehabilitation strategies are planned based upon its outcomes. Traditional road surface monitoring methods use visual inspection for proper evaluation of existing pavements that must be continuously and effectively monitored using practical means [1]. Conventional visual inspection of road distress assessment requires manual visual inspection by personnel in-charge. The truck-based pavement monitoring systems represent the common methods have been in-use in assessing the remaining life of in-service pavements and the detection of road distress. This system produces accurate results, but their use can be expensive and data processing can be time consuming, which make them difficult to use considering the demand for quick pavement evaluation. Because of the small funding base of local government, the human and financial resources available for maintaining roads are often inadequate. In the last decades the rate of making and utilization of computer vision methods for pavement engineering applications have been exponentially increased [2] with the help of developments high-level processing techniques to extract information from the images and sensing technologies to capture images efficiently and accurately under various lighting conditions.

© Springer International Publishing AG, part of Springer Nature 2019
F. Calabrò et al. (Eds.): ISHT 2018, SIST 101, pp. 164–171, 2019.
https://doi.org/10.1007/978-3-319-92102-0_18

Satellite images can be able to monitoring road distress. However, due to the cost and the limited spatial resolution quality of the image, it is not preferred. So Images from different platforms such as Unmanned Aerial Vehicles (UAVs) can be cost effective compared to satellite's imagery or traditional aerial photography.

In this paper a road pavement condition was monitored in Reggio Calabria sing an Unmanned Aerial Vehicle (UAV) to inspection of road distresses, such as road surface cracks and potholes, like through image processing algorithms for the detection of cracks in building facades.

2   Methods for Paved Road Condition Evaluation

The flexible pavement surface due to repeated vehicle loading, environmental inter-action and insufficient structural strength can realize pavement failures as rutting, fatigue cracking, and migration of fine particles (pumping) [3, 4]. Rehabilitation and road maintenance work is not only time-consuming, but it also creates traffic distur-bances and is also very costly. Pavement condition inspection can be realized in various ways. The First and the oldest way, yet in-use, consist to walk on the roads and carrying out visual analyses and measuring the required parameters through manual inspection (Fig. 1). In this procedure, a pavement expert travels along the road and collects both visual and quantitative data from the pavement surface by investigating the cracks with bare eye. Visually inspecting the infrastructure and evaluating them by subjective human experts is the simplest method.

Fig. 1. Visual inspection on road surface (www.fordsix.com).

Depending on the types of surface rupture, required parameters are measured on-site to determine the severity level of the road damage, furthermore this method is time consuming and does not produce satisfyingly accurate results.

Problems such as lacking of data reading consistency and time-consuming have made the visual inspection method to be less competent. Furthermore, the main drawbacks posed by this conventional manual visual inspection are the lacking of measurement consistency and longer duration of traffic disturbance to give way to the inspection work [1].

So, the second way and one of the most popular data collection systems, are camera-mounted trucks (Fig. 2). To speed up the pavement condition inspection, cameras attached on modern vehicles are operated rather than walking on the roads as described in the first method. Using this method, continuous and high-resolution data can be collected at high speeds, so they are widely used in pavement inspection systems but the initial cost of developing such truck-based systems is relatively expensive.

Fig. 2. Visual inspection on road surface.

Pavement inspection systems can be categorized into two parts as data collection, and data processing. In data collection part, the information from the pavement surface is gathered as 2D images or 3D surface data by using moving vehicles or aerial platforms. After performing field experiments and collecting data, in general, the gathered data are processed using sophisticated image processing algorithms and machine learning techniques. The integration of more automated and semi-automated remote sensing methods is a relative new concept and may provide valuable results when generated and interpreted in a systematic manner. The application of these methods is likely to be less expensive, more consistent, and conducive to faster inspections that cover wider areas [5]. Non-destructive evaluation techniques, such as Digital Image Processing (DIP), Ground Penetration Radar (GPR), fiber optic sensors, laser systems (LS) or Hybrid systems (HS) are emerging procedures for health monitoring. Furthermore, in the last decade Unmanned Aerial Vehicles (UAVs) have become extremely popular especially in civilian applications because of its low cost and practicality. In fact, they are able to offer high resolution and almost real-time images often cheaper than other space born platforms. A UAV, commonly known as a drone (Fig. 3), is an air vehicle which can fly without a human pilot and with the advancements of its technology, newer commercial UAVs come to the market more frequently. Compared to traditional evaluation methods just described, remote sensing techniques offer non-destructive methods for road condition assessment and image-processing techniques to assess road conditions are considered a good nondestructive method to quantify pavement distress like cracking. Autonomous inspections of paved road infrastructure remain a relatively unexplored area of application, however, their quick response times, maneuverability, and resolution, make them important alternatives for pavement assessment [6].

In this paper, the authors propose a UAV based pavement crack identification system for monitoring flexible pavements' existing conditions based on the combination of image processing techniques and machine learning. In the following section, the method developed is presented.

Fig. 3. Unmanned Aerial Vehicles (UAVs).

3   Review of UAV in Monitoring Road Condition

A UAV, so called a drone, is an air vehicle which can fly without a human pilot. UAVs started to be used by military reconnaissance purposes in the 1950s.

With the advancements of its technology, newer commercial UAVs come to the market more frequently. Thanks to potential of high flexibility and efficiency of UAV system, pavement managers can assess large, often inaccessible areas, in little time, providing valuable up-to-date information on pavement deterioration.

The first road condition assessment using UAV was introduced by Zhang [7]. The flights were done by a radio controlled low cost airframe helicopter equipped with position and velocity detection sensors, then was using a number of image processing algorithms are under development to calculate some features such as length and size of corrugation, geometry of cross section, rutting, potholes etc. In addition, the acquired imagery and developed algorithms may also be useful for the extraction and measurement of other road properties, such as road way width, curvatures, etc. which are also important components in road way management. Zhang and Elaksher [8] published a study where an improved form of UAV-based road condition assessment was performed. The new system included a low-priced model helicopter armed with a 10.1-megapixel resolution digital camera, a GPS receiver and an Inertial Navigation System (INS) and Ground Control Station (GCS). The ground resolution up to 5 mm was obtained from a UAV flight at 45 m altitude. To obtain more accurate measurement for the road surface distress, new set of algorithms were developed for generation of 3D surface model and orthoimages. The difference between the 3D model and field survey data that author has individuated indicated that maximum accuracy could be achieved up to 0.5 cm ground accuracy using the UAVs. The results show that the proposed system is practical for local transportation agencies to gather surface conditions of unpaved roads.

4   Evaluating Monitoring Condition Using UAV

4.1   Mission Description

The aerial survey was performed on 03/12/2017 in one area in the harbour of Reggio Calabria (Fig. 4) to avoid traffic during the inspection. The pictures were shot around the solar noon to minimize the shadows using a commercial UAV DJI Mavic Pro (DJI,

Shenzhen, China) and Pix4d software (Pix4d SA, Lausanne, Switzerland) for automatic mapping area and image acquisition through waypoint. Mavic pro camera specs are reported in Table 1.

Fig. 4. Street object of the aerial survey.

To obtain a high quality ordered dataset and produce a model with defined characteristics, an image acquisition plan was created, defining also camera settings and geo information acquisition type. The image acquisition plan was divided into three steps: (1) definition of image acquisition plan type, (2) definition of Ground Sampling Distance (GSD), and (3) definition of image overlap. The Image acquisition type was set to automatic waypoint flight mission to create 3d object, so the UAV perform an automatic flight as shown in Fig. 5. The GSD calculation [10] define the height of flight according to the required definition (cm/pixel) of the model. To obtain centimetre precision (GSD < 1) at comfortable flight height, considering Mavic Pro camera specs (Sw = real sensor width = 6,17 mm, FR = real focal length = 5 mm) the height flight was set to 30 meters for vertical flight. The overlap between two consecutive images, that determined the image acquisition rate at fixed UAV speed, was set to 80%. The image rate acquisition calculation for front overlap, is made automatically by Pix4d software.

Moreover, the software calculates the image acquisition plan and mission settings automatically, defined the following parameters: flight eight (and consequentially GSD), overlap (%), and area to be mapped as shown in Fig. 6.

The overall survey to map 1 km^2 was performed in almost 2 min. Mavic pro camera setting were set to automatic and geo tag information were captured using the UAV built in GPS and store in the images EXIF data.

4.2   3d Reconstruction Workflow

3d reconstruction process was performed using Agisoft Photoscan software. The core of the software is made by the two algorithms: SIFT algorithm [9] is a method for image feature generation that transforms an image into a large collection of local feature vectors, each of which is invariant to image translation, scaling, and rotation, and partially invariant to illumination changes and affine or 3D projection; SFM algorithm [10] is used to match the common feature on each photo, and using the photogrammetry principles of triangulation, generate the camera poses and the

Table 1. Mavic pro camera specs.

	Mavic pro camera
Sensor	Sony 1/2.3" (CMOS)
Effective pixel	12.35 M
Lens	FOV 78.8°, 28 mm f/2.2
ISO range	100–1600
Electronic shutter speed	8 s–1/8000 s
Image size (pixel)	4000 × 3000
GPS	Built-in

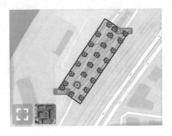

Fig. 5. Waypoint flight mission and overlap scheme for automatic mapping area.

Drone	Date	Time	Type
Mavic Pro	03 Dic 2017	12:25:30 PM	Grid

Location	Dimensions	Overlap	Camera Angle
38,130344°, 15,653241°	27 m x 83 m	80% (72%)	70°

Altitude	Images	Path	Flight time
30 m	24	306 m	2min:17s

Fig. 6. Mission information and setting.

three-dimensional point cloud of the object. Then, starting from point cloud, the software reconstructs a 3d mesh. The output DEM (Digital Elevation Model), reconstructed using WGS84 geotag information, is represented in the following figure (Fig. 7):

Fig. 7. DEM of the area.

Fig. 8. Cracks on pavement.

Starting from DEM the crack on the pavement were highlighted applying a filter on 3d point cloud under the pavement level using 3dReshaper software (Fig. 9 and Table 2). Four different potholes were individuated and measured directly on DEM as shown in Fig. 8:

Fig. 9. Real measurement and DEM measurement.

Table 2. Comparison between the two measurements.

Potholes		Real measurement (m)	DEM measurement (m)	Error
1	a	1,51	1,507	0,003
	b	0,46	0,47	0,010
	Depth	0,52	0,54	0,020
2	a	0,88	0,86	0,020
	b	0,46	0,45	0,010
	Depth	0,54	0,57	0,030
3	a	1,24	1,22	0,020
	b	1,12	1,1	0,020
	Depth	6	5,98	0,020
4	a	0,81	0,8	0,010
	b	0,7	0,72	0,020
	Depth	6,1	6,18	0,020

The difference between error and DEM measurement are <0,03.

5 Conclusion

The pavement road degradation analysis is fundamental for the planning of maintenance interventions. Innovative methods, such as the use of drones, allow savings both in terms of time and maintenance costs. The analysis carried out verified the drone use for the identification of the potholes on the pavement surface and their geometric characteristics, in particular of the depth. Through the comparison between the measurements made in site and those obtained from the DEM model, it is clear the validity of drone use. In order to reduce the images process time, a real time procedure will be developed to analyse the images obtained by drone and identify the distresses with its geometric characteristics.

References

1. Liq Yee Tiong, P., Mustaffar, M., Rosli Haini, M.: Road surface assessment of pothole severity by close range digital photogrammetry method. World Appl. Sci. J. **19**(6), 867–873 (2012)
2. Cet, K., Georgieva, K., Kasireddy, V., Akinci, B., Fieguthd, P.: A review on computer vision based defect detection and condition assessment of concrete and asphalt civil infrastructure. Adv. Eng. Inform. **29**(2), 196–210 (2015)
3. Buonsanti, M., Cirianni, F., Leonardi, G., Scopelliti, F.: Dynamic behavior of granular mixture solids. Key Eng. Mater. **488–489**, 541–544 (2012)
4. Leonardi, G.: Finite element analysis for airfield asphalt pavements rutting prediction. Bull. Pol. Acad. Sci. Tech. Sci. **63**(2), 397–403 (2015)
5. Schnebele, E., Tanyu, B.F., Cervone, G., Waters, N.: Review of remote sensing methodologies for pavement management and assessment. Eur. Transp. Res. (2015)
6. Leachtenauer, J.C., Driggers, R.G.: Surveillance and Reconnaissance Imaging Systems: Modeling and Performance Prediction. Artech House, Boston (2011)
7. Zhang, C.: An UAV-based photogrammetric mapping system for road condition assessment, The International Archives of the Photogrammetry, Remote Sensing and Spatial Information Sciences, ISPRS Congress, XXXVII. Part B (2008)
8. Zhang, C., Elaksher, A.: An unmanned aerial vehicle-based imaging system for 3D measurement of unpaved road surface distresses. Comput. Aided Civ. Infrastruct. Eng. **27**(2), 118–129 (2012)
9. Lowe, D.G.: University of British Columbia, Canada - Object Recognition from Local Scale-Invariant Features (1999)
10. Wu, C.: University of Washington - Towards Linear-time Incremental Structure from Motion (2012)

The Myth of Pedestrianisation or the Reasons of Hierarchy

Alfonso Annunziata[1]([✉]) and Carlo Pisano[2]

[1] University of Cagliari, 09124 Cagliari, Italy
tutoralfannunziata@gmail.com
[2] University of Florence, 50121 Florence, Italy

Abstract. The rediscovery of the functional and symbolic role of environmental networks, cycling and pedestrian paths, and open spaces, particularly in the urban realm, leads to an increasing sensitivity to the discontinuity of territories generated by the main mobility networks. Thus, an instance of domestication emerges, resulting in transformation strategies of large mobility infrastructures - or their portions - into multi-functional pedestrian spaces. However, pedestrianisation interventions have not been completely defined within a theoretical system that clearly establishes modes and conditions of implementation. This article aims to point out how the conditions of hierarchy constitute a supporting tool for controlling and verifying the project of pedestrianisation. The article focuses on a concrete case of modification of an urban road network: the transformation of a portion of an important distributor road in the urban area of Cagliari into a pedestrian space. This analysis uses the fundamental conditions of hierarchy as a tool to assess to what extent the modification of the road network articulation has resulted in conditions of lesser inter-connectivity, legibility and functionality.

Keywords: Hierarchy · Arteriality · Constitution · Configuration
Network

1 Introduction

Contemporaneity is characterised by the rediscovery of the role of environmental networks, cycling and pedestrian paths, open spaces and mass transport, as a condition for redistributing the spatial capital, increasing resilience of urbanised territories, restoring continuity between different ecological systems and increasing the plurality of modes of practising spaces [1–5].

A question emerges, related to the discontinuities determined by the primary distribution networks - and by their constitutive discrete elements [5–7]. This question meets the criticism to some figures on which the network articulation is based and the search for alternative metaphors and paradigms [8, 9]. As a consequence, an instance of domestication of main transport infrastructures arises: it often results in transformation policies of large mobility spaces routes - or their portions - into multi-functional pedestrian spaces.

© Springer International Publishing AG, part of Springer Nature 2019
F. Calabrò et al. (Eds.): ISHT 2018, SIST 101, pp. 172–181, 2019.
https://doi.org/10.1007/978-3-319-92102-0_19

When these interventions modify the hierarchies that establish road patterns and relations between different modes of movement, they lead to conditions of lesser functionality of transport networks: pedestrianisation strategies have not been fully defined within a coherent theoretical system meant to establish fundamental criteria, conditions and modalities of implementation.

This article aims to point out how hierarchy is a fundamental condition of some desired structural properties of transportation networks: coherence, legibility, inter-connectivity of road patterns, and safety and fluidity of circulation [4, 10–14]. Moreover, it infers a qualitative method for control and appraisal of the project of pedestrianisation from the structural conditions of hierarchy. The article consists of three sections; the first one defines the notion of hierarchy, related to transport networks, and identifies its fundamental conditions; its purpose is to show that hierarchy is the condition of some desirable networks structure properties. The second section identifies the cause of the discontinuity of minor networks; it proposes a hierarchy of modes of movement, based on speed bands, as a criterion for defining the relationships between different modes of interaction simultaneously present along a route. The third section describes a qualitative analytical method, based on the conditions of the hierarchy, and describes its application to a case study: the restoration of an arterial street as a pedestrian space. The conclusions reflect on the findings of the previous sections and consider their potential as criteria for the definition and implementation of regeneration actions in road networks links.

2 Hierarchy

As stated by Marshall [11], in order to define hierarchy, a preliminary consideration concerning the distinction between the different dimensions of structures is required: *composition*, which refers to the metric characteristics of a system; *configuration*, related to its topology; and *constitution*, which identifies distinct types of elements and determines their relations of interdependence. This distinction results in the definition of hierarchy as a structure of types [11] and, more precisely, as a particular form of *constitution*, identified by four specific structural conditions: (1) differentiation of components; (2) ordered ranking of elements; (3), *arteriality* or necessary connections and (4) *access constraint* that restricts admissible connections only to those established between elements of the same type or complementary, i.e. whose rank diverges of just one position (or degree). *Arteriality*, in particular, is a fundamental feature of road networks: the contiguity of routes that constitute strategic itineraries at a specific level of scale. Thus, a network could be conceptualized as a structure of contiguous and complete sub-networks of different scales. *Arteriality* is the implicit fundament of any relevant functional classification of roads, since it is a category by which the specific function of each arc can be recognised in relation to the network, and referred to the size of the geographic realm served. Some measurable properties have been identified to infer the function of each route from the analysis of the network configuration: cardinality, determined by observing the conditions of continuity and termination in the nodes of each route [12], and betweenness, that measures the significance of each arc as a bridging element between distinct topological shorter paths [10].

2.1 Favourable Properties of a Network

A road pattern that satisfies the structural conditions of hierarchy also possesses two fundamental desirable properties of networks [11, 13]: legibility and inter-connectivity. These properties are related to the user experience and determine individual behaviors and route choices, whose interaction defines the distribution of flows [13].

Legibility determines to what extent it is simple to gather the structure of a network. *Arteriality* and *access constraint* are fundamental features that allow users to determine their position within the network, according to the status of the route along which they proceed [11]. Thus, this condition implies the recognisability of the type and of the rank of each route.

Inter-connectivity depends on the number, type and pattern of routes and intersections. It can be measured as a function of the perceived continuity of movement across a network:: the latter is determined by measuring the number of transfers between roads of different classes [13]: as *arteriality* implies contiguity of strategic routes, it also ensures continuity of movement within the sub-network of primary distribution, thus resulting in an increased inter-connection of the entire system. However, this formulation doesn't consider that the crossing of an intersection in case of vehicular mobility – even if between routes of the same class – results in an increased path discontinuity, due to the unavoidable variations of speed and direction. Salingaros [4] observes that, beyond a certain dimension, in an isotropic network of local-scale routes, the number of nodes traversed during a generic trip increases to the point of resulting in a perception of greater discontinuity. The introduction of larger scale elements and their configuration within a contiguous sub-network result in an increase of inter-connectivity and path continuity. Thus, the requirement of ensuring continuity implies the reduction in the number of accesses to the main routes and the spacing of consecutive intersections: it results in the configuration of arterial roads as discrete elements [15] and in the requirement of ensuring conditions of *access constraint*.

Hierarchy is also an emerging property of complex systems [14, 16]: an isotropic network tends to modify and self-organize according to a hierarchical *constitution*. Jiang's analyses [10] demonstrate the validity of this hypothesis by verifying the power laws informing the distribution of size and frequency of some fundamental variables related to the form or function of the routes in a network: for example, a rank order rule is recognized in the distribution of street length values and the cumulative distribution of paths connectivity values and betweenness values resembles the 80/20 principle. Likewise, considering the distribution of flows within the road layout of the urban area of Gavle, Jiang concludes that the mobility function of a route descends from its morphology and its structural role within the configuration of the network.

3 A Hierarchy of Modes of Movement

Having defined hierarchy as a specific type of *constitution*, indeed, as a structure of types, it can be concluded that the discontinuity of urbanized territories, pedestrian and cycling paths, and open spaces depends on specific functional and morphological features of types; thus, on parameters selected as a basis for route classification. The discontinuity of minor networks is implicit in conventional route typologies where the system function is derived from the mobility function, according to an imposed inverse relationship between the distribution function and the access function: thus an exclusionary relationship – hence a separation – is introduced between the transit function and any form of minute scale interaction that unfolds within the road space or between roads and buildings. Consequently, cycling and pedestrian surfaces occupy the last tier of the hierarchy and, consistently with the condition of *arteriality*, they constitute fragmented and discontinuous fabrics, dispersed among arterial routes exclusively designed for vehicular transit. Therefore, the restoration of continuity of territorial structures requires to overcome the inverse relationship between transit function and urban function and to interpret them as distinct but compatible dimensions of infrastructural spaces; this paradigm shift results in the possibility of conceptualizing strategic routes as multi-modal and multi-functional spaces; it is also consistent with the definition of route types in constitutional terms, according to the principle of *arteriality*.

The fundamental question arising from this first conclusion concerns the definition of the conditions of different modes of movement coexistence in the same road section. This question can be expressed in terms of determination of admissible connections between elements of a hierarchy: the speed differential is considered as the most relevant criterion for determining the degree of separation between distinct modes of movement [11]. Thus, speed is identified as the parameter for a modal hierarchy; according to this, consistently with the principle of *access constraint*, the conditions of contiguity and separation between different modes of transport and their relative surfaces are determined. Moreover, these conditions are a criterion for establishing admissible connections between roads engaged by different modes of movement and for ensuring fluidity and safety of circulation, as these conditions imply a substantial reduction of conflicts among modes of movement characterized by significantly different speeds. In addition, since the categories of traffic admitted along a road depend on its function within the transport network, the degree of separation among surfaces intended for specific modes of movement constitutes a morphological parameter that determines the consistency between the typological definition of a route and its geometry. This coherence is implicit in the first principle of hierarchy: the distinction of types.

4   Methodology

This section proposes a qualitative method based on the structural conditions of hierarchy and explores its application as a tool for verifying whether the transformations of arterial routes alter the functionality of mobility networks. The proposed analytic method does not consider the principle of *arteriality* as an explicit criterion: the adherence of the network to this condition is implicit in the typological definition of each arc, classified according to the principle of *arteriality*, precisely because topologically contiguous within a complete and continuous sub-network that encompasses a specific geographic area. Consequently, only the conditions of *access constraint* and differentiation of components are individuated as pertinent parameters. The first condition is verified if non-complementary routes, attributable to types whose ranks differ of more than one position, do not connect contiguously. The second condition is verified if the morphological characters of an element are appropriate to its system function. The variable considered is the degree of separation between lanes and surfaces serving different uses or modes of movement, classified according to their speed values. Therefore, 8 categories, according to Marshall stratification by speed [11], are identified: Very high speed, associated with train, fast motor movement on motorway (S5); High speed, considered as the highest speed for a carriageway associated with a footway (S4); Medium-high speed motor transport movement (S3.5); Medium speed (S3); running, cycling, medium-slow motor movement (S2.5); jogging; slow cycling or very slow motor movement (2); walking pace, cycling or parking at walking pace (S1.5) and walking speed (S1). With reference to these categories, three conditions are imposed: (1) modes of movement attributable to the same class can be accommodated by the same lane; (2) modes of movement of different ranks unfold along contiguous lanes if, among their coded types, there is a rank difference equal to or less than one integer value (S3 and S2); (3) if the difference between their relative coded types exceeds one integer value, the different modes of movement are segregated, each one within a confined surface, or within surfaces separated by lanes intended for middle-tier movement modes.

The analysis is conducted by reconstructing the *configuration* of the considered portion of the road layout, represented as a route structure [11, 12]. The network function of each element, hence its type and rank, is determined according to the road classifications, available in specific databases. Consistently with this criterion, the typology derived from the open street map database was chosen. Here, some types are defined according to several parameters and express the different dimensions of a road space. These denominations specify the fundamental definition related to the significance of the route within the network. Nevertheless, these denominations have been rejected when implying an inverse relationship between the transit status and the place status of a route.

4.1   Case Study

The evaluation method defined previously has been applied to a concrete case of modification of an urban road network: the pedestrianisation of a portion of a main distributor in the urban area of Cagliari.

Opened in 1883, Via Roma consists of two carriageways - "Lato Portici" towards the city and "Lato Porto" towards the sea - separated by a central tree-lined promenade. It is one of the main arteries crossing the center of Cagliari: some of the major buildings of the political power, such as the Civic Palace and the Regional Council, align along its margins. After the Second World War, via Roma assumed the role of fundamental urban distributor, intensely frequented by pedestrians and engaged by large flows of private vehicles, trolleybuses and buses.

Since the nineties, urban policies aimed at a global requalification of the historic districts of Cagliari as restricted traffic zones and at the pedestrianisation of main commercial streets have determined a subdivision of the urban fabric into two portions connected by two links only: via Roma to the south and Porta Cristina to the northern extremity of the hill of Castello. This partial fracture and the consequent concentration of different flows along Via Roma have led to several proposals for the re-arrangement of the road surface: the re-configurations of the road sections, the concentration of public transport lines within a dedicated carriageway and the transformation of the central promenade into a parking area; these interventions add to the radical hypothesis of removal and confinement of vehicle lanes within entrenchments or tunnels, in order to release Via Roma from urban and metropolitan scale traffic flows. In this context lies the decision taken by the City Council to undertake the temporary pedestrianisation of Via Roma Lato Portici from August, 11 to September, 17, 2017. This experimentation, never conducted before for such a prolonged period, resulted in a significant modification of the network *configuration*.

4.2 Analysis

The analysis of configurational and constitutional features of the road network reveals the existence of three circuits, or rings, that constitute contiguous subsystems of strategic itineraries, consistently with the condition of *arteriality*, at different scales. Hence, a metropolitan arterial sub-network is identified, consisting of trunk roads and primary distributors, and an inner sub-network contiguous at the scale of the compact urban settlement; these two circuits include the strategic itinerary unfolding along Via Roma Lato Porto, Via Riva di Ponente and Viale La Playa.

Then, a contiguous sub-network of secondary and tertiary distributors is identified; this inter-district circuit encompasses Via Roma Lato Portici. This route carries out a local distribution function towards Marina district and it connects to the finest scale district network composed of local streets and pedestrian routes (Fig. 1).

From this analysis emerges that Via Roma constitutes a multi-modal and multi-functional space, since it is part of contiguous circuits of strategic itineraries both at the metropolitan and at the district scale. Moreover, it is a fundamental arc within the public transport network. This condition is pointed out by the road classification derived from the Open Street Map database; this defines Via Roma Lato Porto as a primary road and Via Roma Lato Portici as a tertiary route.

In this sense, the pedestrianisation of Via Roma Lato Portici introduces a modification of the road pattern that invests different levels of scale: it determines the overlapping - along the strategic itinerary composed of Via Roma Lato Porto, Viale La Playa and Via Riva di Ponente -between the inter-district/local sub-network and the primary network: therefore, a tertiary route, the Largo Carlo Felice, and a local street, via Campidano, connect directly to a primary distributor, Via Roma Lato Porto; moreover routes such as Via Sassari, identifiable as links between the primary and the inter-district subnetworks, evolve into strategic routes within the inter-district sub-network, canalising movement of secondary distribution and penetration towards the primary distributors. As a consequence, Via Roma Lato Porto, Via Riva di Ponente and Viale La Playa - in addition to a function of primary distribution - perform functions of penetration and secondary distribution. Consequently, the structure of routes does not verify the *access constraint* condition.

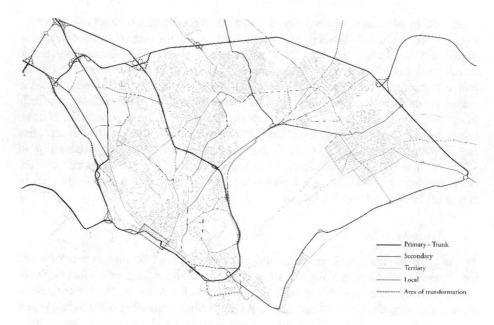

Fig. 1. Configuration of the road network and classification of its routes

Moreover, it is observed that, along certain routes, the relationship between the modes of movement does not verify the condition of *access constraint* derived from a modal hierarchy funded on speed values (see Table 1). Particularly, in Viale La Playa there is no transition space between medium-high speed motor movement (S3) and parking (S1.5) along road-sides; Likewise, along the first segment of Via Sassari, promiscuity arises between distribution (associated with medium-high speed motor movement, S3) and access (S2) of private vehicles, public transport bus access (S2.5), parking (S1.5), and walking (S1) (Fig. 2).

These criticalities, partially pre-existing, are exasperated, in consequence of the pedestrianisation intervention, by the concentration, along adjacent routes, of functions of secondary distribution and penetration previously served by via Roma Lato Portici. The intervention of pedestrianisation thus modifies the link status of adjacent routes [11, 17]: it increases their significance as bridging elements between portions of the network weakly connected to each other. Therefore routes adjacent to Via Roma Lato Portici are affected by a loss of coherence between network function and morphology: Therefore, the first principle of hierarchy, the distinction of components is not satisfied. It can therefore be concluded that the transformation of Via Roma affects the fluidity of the circulation, the recognition of the role - and type - of the paths and the readability of the system. These considerations are confirmed by situations of congestion observed during the period of the experimentation, at the extremities of Via Roma, particularly in via Sassari, Viale la Playa and Traversa prima La Playa, and admitted by the Municipal Administration [18].

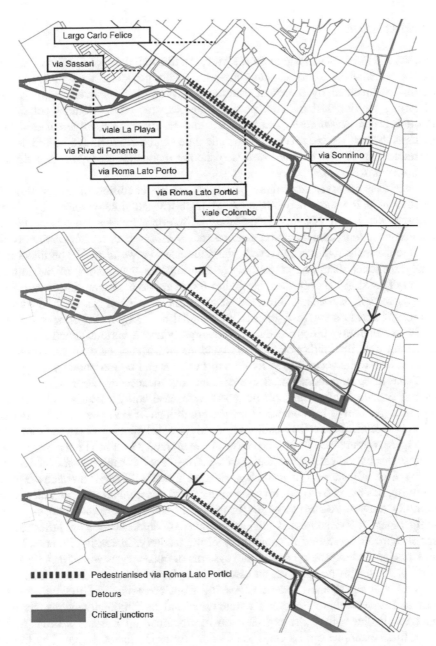

Fig. 2. Effects of the overlapping between the primary and the inter-district sub-networks

5 Conclusions

Hierarchy is a fundamental figure for the interpretation and the organization of mobility systems; nonetheless, as it is often identified with particular configurations, it is increasingly invested by a radical critique that prefigures its overcoming [5, 11]. Reaffirming the conclusions from previous studies, this article pointed out the distinction between *constitution, configuration* and *composition* as different aspects of structures, defined the hierarchy as a specific type of *constitution*, and investigated its fundamental conditions, referred to road patterns: distinction of types, rank determination, *arteriality* and *access constraint*.

It was then observed how these principles are the condition of certain desirable properties of a transport network, such as legibility and inter-connectivity. Subsequently, it was noted that the discontinuity of minor networks depends on specific configurational and compositional features of networks and of infrastructures that constitute the main routes. Conversely, the reference to *arteriality* as a pertinent basis for the functional definition of routes, by enabling the overcoming of the imposed inverse relationship between the distribution function and the access function, enables the *configuration* of roads as multi-modal and multi-functional spaces. Likewise, the condition of *access constraint*, referred to a modal hierarchy based on speed classes, permits to coherently articulate the relationships between various functions along a road. In this way, the fluidity of circulation and the consistency between the typological definition of a route according to its system function and its morphology are guaranteed. From these considerations it emerges the significance of hierarchy as the fundamental condition of a transport network. This case study analysis, conducted by applying a qualitative method based on the conditions of *access constraint* and distinction of types, confirms that modifying hierarchical relationships between routes and modes of movement results in a decrease in functionality of a road network.

These results indicate the necessity of verifying that the *constitution* of a network, resulting from adjustments of its arches, satisfies the conditions of hierarchy. Finally, by pointing out the need to preserve the hierarchical *constitution* of road patterns, this case study analysis re-introduces the centrality of the *composition* of roads, as the specific aspect to be addressed by interventions of domestication of infrastructural elements aimed at restoring the continuity of the territory, increasing its porosity and reconstructing and expanding the minute systems of open spaces and of pedestrian and cycling paths. In this perspective, numerous examples demonstrate the potential of the project as a comprehensive research spacing from corrections of the infrastructure layout to the re-configurations of its transverse and longitudinal sections, from the connection to the soil to the organization of margins and residual spaces [6, 19]. Among these examples are La Gran Via de Les Cortes Catalanes designed by Carmen Fiol and Andreu Arriola, Jordi Henrich's Ronda del Mig, the Moll de la Fusta, by Manuel Solà-Morales, in Barcelona; the Atlantic Passeo in Porto, also designed by Manuel de Sola-Morales, and the rest area in Nîmes - Caissargues, designed by Bernard Lassus. It therefore emerges the necessity of the project as a device for investigating strategies that do not alter the function of a route within the mobility system – hence its constitutional and configurational properties – but pursue the integration of different movement functions and practices in the space of the infrastructure, by modifying its compositional features.

References

1. Alexander, C., Ishikawa, S., Silverstein, M., Jacobson, M., Fiksdahl-King, I., Ange, S.A.: Pattern Language. Towns - Building - Construction. Oxford Press, New York (1977)
2. Jacobs, J.: The Death and Life of Great American Cities. Random House, New York (1961)
3. Gehl, J.: Life Between Buildings - Using Public Space. Van Nostrand Reinhold, New York (1987)
4. Salingaros, N.: Principles of Urban structure. Techne Press, Amsterdam (2005)
5. Secchi, B.: La città dei ricchi e la città dei poveri. Editori Laterza, Bari (2014)
6. Secchi, B.: Figure della mobilità. Casabella **739–740**, 81–83 (2005)
7. Bianchetti, C.: I vantaggi della continuità. Anfione & Zeto **25**, 111–116 (2014)
8. Magnani, C., Val, P.A.: Appunti per una tassonomia. Casabella **553–554**, 28–37 (1989)
9. Magnani, C.: Per una genealogia delle tecniche del progetto. Casabella **739–740**, 60–61 (2005)
10. Jiang, B.: Street hierarchies: a minority of streets account for a majority of traffic flow. Int. J. Geogr. Inf. Sci. **23**(8), 1033–1048 (2009)
11. Marshall, S.: Streets and Patterns. Spon Press, New York (2005)
12. Marshall, S.: Line structure representation for road network analysis. J. Transp. Land Use **1–2016**(9), 29–64 (2016)
13. Xie, F., Levinson, D.M.: Measuring the structure of road networks. Geogr. Anal. **39**(2007), 336–356 (2007)
14. Yerra, B.M., Levinson, D.M.: The emergence of hierarchy in transportation networks. Ann. Reg. Sci. **39**(3), 541–553 (2005)
15. Pisano, C.: Patchwork Metropolis. Progetto di città contemporanea. Lettera Ventidue, Siracusa (2018)
16. Barabasi, A.L., Albert, R.: Emergence of scaling in random networks. Science **286**, 509–512 (1999)
17. Annunziata, A., Annunziata, F.: Reasons for a cultural renovation of the road infrastructure design. J. Civ. Eng. Archit. **9**(2015), 961–969 (2015)
18. Ansa: Via Roma pedonale: Zedda, esperimento ok. http://www.ansa.it/sardegna/notizie. Accessed 24 Oct 2017
19. Donini, G.: Margini della mobilità. Meltemi Editore, Milano (2008)

Tourist Flows and Transport Infrastructures: Development Policies for the Strait Airport

Claudio Zavaglia$^{(\boxtimes)}$, Jusy Calabrò, and Raffaele Scrivo

Mediterranea University of Reggio Calabria, 89100 Reggio Calabria, Italy
claudio.zavaglia@unirc.it

Abstract. This research study, after having focused on the importance of an airport structure and the opportunities that this could generate especially in the context of the tourist accessibility of an entire area, falls in the case of the airport of the "Tito Minniti" Strait of Reggio Calabria, for some years now the subject of a drastic decrease in turnout and loss of interest from a political point of view. Through a series of analyzes, focused on tourist users coming from the surroundings of the Milanese capital which has the Aeolian Islands as their destination, we will propose an evaluation of scenarios focused on the Milan-Lipari section. The goal is to demonstrate how, among the potential of the airport structure of Reggio Calabria, there is to capture the tourist flows that call at other airports (Naples, Catania and Palermo on all) being able to offer very competitive itineraries both in terms of costs but above all extremely convenient from the point of view of travel time.

Keywords: Airport · Tourist flows · Backdrop analysis

1 Introduction

Airports are an essential element for the strategies and policies of economic development of a territory, both at a national, regional and local level. They represent an important competitive advantage for the promotion of an area. The relationship with the territory, however, is both conflictive, but also and above all, synergistic. The airport infrastructure is itself an economic activity in its own right, but its survival needs policics of accompaniment and enhancement by public and private bodies.

The subject of the research object of this contribution has observed the airport of the Metropolitan City of Reggio Calabria, which was the first airport of the Region. Unfortunately, despite the potentials related to the resources of the territory in which it is located, the Area of the Strait represents a catchment area of about 1,500,000.00 inhabitants [1], after a long agony, from the end of 2016, the management of SoGAS SpA, is declared invalid by ENAC, remaining in temporary management until July

The paper is the result of the joint work of the authors. Although scientific responsibility is equally attributable, the abstract and the paragraph nn. 4, 5 were written by C. Zavaglia; the paragraph nn. 6 was written by J. Calabrò; the paragraphs nn. 1, 2, were written by R. Scrivo.

© Springer International Publishing AG, part of Springer Nature 2019
F. Calabrò et al. (Eds.): ISHT 2018, SIST 101, pp. 182–191, 2019.
https://doi.org/10.1007/978-3-319-92102-0_20

2017; it is currently managed by SACAL, which manages the international airport of Lametia Terme, awaiting a new call for tenders for its management.

This paper illustrates a part of the research activity that intends to identify the conditions that can improve the competitiveness of the airport infrastructure, both nationally and internationally, using accessibility indicators.

2   Tourist Circulation and Potential Demand

To tackle the topic, the research has carried out surveys on the circulation of tourism and air transport, identifying in the latter the origin of the extraordinary increase in international tourism in Italy. In addition, a research carried out by Confturismo in collaboration with Ciset on international tourism, confirmed the importance for Italy of the economic sector as evidenced by the incidence of 7% in the Gross Domestic Product, with an influx of two million employed people and with 80 million people in accommodation facilities with about 350 million overnight stays [2].

The period analyzed, from 2001 to 2015, showed that arrivals in our country have increased by 50%, with a threshold of over 50 million units. However, these major arrivals were not translated into proportional increases in income due to a lower average stay (from 4.1 to 3.6 days between 2001 and 2015) and the consequent 35% reduction in expenditure per capita real (from 1.035 to 670 euros). This means that, since 2001, our country has "lost" 38 billion of valuation revenues deriving from international tourism [1]. Analyzing arrivals by areas of evidence, international tourism in Italy is 70% of European origin - Germans in the lead that are confirmed as the first market in many regions - increasing the weight of non-EU countries with a contribution of over 35% to the growth of the period [3]. The growth of Chinese tourism, which in just a few years became Italy's 5th incoming market and destined to grow further in the coming years, should be highlighted. Although the forecasts for the three-year period 2016–18 see growth in arrivals from all countries, especially China and the USA. The Italian regions that absorb over 60% of international arrivals are Veneto, Lombardy, Tuscany and Lazio, which also represent the main entrances to our country. The regions of the Mezzogiorno, on the other hand, attract only 12% of arrivals and 14% of presences. The southern islands are more attractive than the continental part, showing that the poor infrastructure is only a partial problem, the lack of attractiveness is mainly due to the provision and average quality of services, hospitality, catering, cultural and entertainment [4]. On the other hand, the "EU Regional Competitiveness Index", prepared by the European Commission in 2013, highlights how the infrastructural endowment of the regions of the Mezzogiorno is actually among the lowest in Europe, however, Sicily and Sardinia, despite being at the bottom to the ranking of Italian regions, show a better ability to attract international tourism [5].

With reference to the foregoing it should also be considered that the development of air transport has generated a considerable expansion of the tourist space. Destinations considered distant and expensive (such as Bangkok, Honolulu, Palma, Tenerife, Faro), today, not only are offered at affordable prices but, some of these, have become specialized bases for charter traffic, such as Rhodes, Kos, Olbia, Djerba, Palma.

Therefore, if in the past railway stations represented an element of tourist attraction, today the opening, growth and diversification of flows are conditioned by the presence of the airport and the creation of direct lines, so much so that sometimes opening of a stopover can be a preliminary condition for tourist investments (for example, the Mediterranean Club Village in Senegal).

The insufficiency of the air service or its specialization in a single starting market can, on the other hand, make the tourist market lose its share.

3 Objectives and Methodology

As anticipated in the introduction, the objective is therefore to identify strategies that can improve the competitiveness of the Reggio Calabria infrastructure. In this study, the focus is particularly on the possibility of capturing the airport tourist flows that have as a destination the Aeolian Islands and that base their itinerary on the airports of Naples or Catania and then ferry to the archipelago [6].

The analyzes were focused on the catchment area of Northern Italy, especially around Milanese. For tourists who are in fact in these areas, and have as their goal the achievement of the Aeolian Islands, the need to make an airport stop in one of the structures closest to the islands is essential. This necessity requires comparison of the alternatives available.

Through an analysis of scenarios, which envisage the use of quantitative reference parameters such as travel times and travel costs for the Milan - Lipa-ri section, the objective is to demonstrate how the most convenient route is to which provides for the use of the airport of Reggio Calabria.

The systematization of the expected results and of the existing and hypothetical dynamics through a SWOT analysis allows to outline a conclusive picture [7].

4 Scenario Analysis

The plan highlights the connection solutions that can be used by those wishing to reach the islands; every possible alternative takes as a reference the port of Lipari as desti-nation, as it is the main port of the archipelago, also known as "Port of the Aeolian Islands".

4.1 Scenario 1: Milano Lipari – Stopover in Napoli

The first scenario takes the Milan - Lipari route into consideration by opting to ferry to the Aeolian islands from the port of Naples (Fig. 1). After having anticipated the various alternatives offered by Neapolitan local transport in advance, the most con-venient solution is to use the bus to move from the Naples Capodichino airport to the port of Naples Margellina and then embark on the ferry towards Lipari (Table 1).

The itinerary would therefore have journey times of 8 h and 38 min for an expense ranging between € 70 and € 252 *. (* 'The price increases nearing the peak season, in August).

Fig. 1. Connection Milano - Lipari through Napoli (Image processing: C. Zavaglia).

Table 1. Detail of the connection through Napoli; analysis of the means of transport, timing and costs [Source: Alitalia, SNAV: dec 2017 - Data processing: C. Zavaglia].

Stopover Napoli	Time	Cost	Freq.	Transp.
Aerop. Mil. Linate	2 h 8 min	40–190€	11 + 4	Airplane + Bus
Napoli	6 h 30 min	30–60€	3	Ferry
Lipari	–	–		
	8 h 38 min	*70–250€*		

4.2   Scenario 2: Milano Lipari – Stopover in Catania

The second alternative considered considering the solution that leads from Milan to Lipari using the international airport of Catania. Once in Catania by plane it is necessary to move by bus to the central station of the city and then take a train to Messina, and then take the ferry to the Aeolian Islands. Taking this route into consideration, the route would therefore have an overall schedule of 6 h and 20 min for a total cost of between € 65 and € 230 (Fig. 2 and Table 2).

Fig. 2. Connection Milano - Lipari through Catania (Image processing C. Zavaglia).

Table 2. Detail of the connection through Catania; analysis of the means of transport, timing and costs. (Source: Alitalia, Liberty Lines: dec 2017 - Data processing C. Zavaglia).

Stopover Catania	Time	Cost	Freq.	Transp.
Aerop. Mil. Linate	1 h 45 m	45–210€	6 + 10	Airplane
Catania	2 h 50 min	10 €	6	Bus + Train
Messina	1 h 45 min	25–30€	3	Ferry
Lipari	–	–		
	6 h 20 min	*80–250€*		

4.3   Scenario 3: Milano Lipari – Stopover in Palermo

In the third scenario, instead, the timing and costs of an itinerary that envisages reaching the island of Lipari by exploiting the other sicilian international airport, the "Falcone Borsellino" of Palermo-Punta Raisi. The most efficient combination then provides to move by bus and ferry from the port of the Sicilian capital to the Aeolian Islands (Fig. 3 and Table 3).

Fig. 3. Connection Milano - Lipari through Palermo (Image processing C. Zavaglia).

Difficult in terms of time, about 4 h of navigation, it appears the move by ferry from Palermo to Lipari. The only alternative that in this case could be valid would be to make a further move by train from Palermo to the port of Milazzo to take advantage of the greater proximity to the Aeolian Islands. Although the ferry from Milazzo reaches Lipari in just 1 h and 5 min, to move from Palermo to Milazzo itself takes about 4 h and 30 min arriving at a total journey time of 8 h against 6 h and 32 m previously expressed. In the overall budget this alternative appears therefore, although presenting costs in line with the other solutions, not very convenient in terms of travel time.

Table 3. Detail of the connection through Palermo; analysis of the means of transport, timing and costs (Source: Alitalia, Liberty Lines: dec 2017 - Data processing C. Zavaglia).

Stopover Palermo	Time	Cost	Freq.	Transp.
Aerop. Mil. Linate	2 h 32	40–230€	5 + 7	Airplane
Palermo	4 h	45–65€	2	Ferry
Lipari	–	–		
	6 h 32 m	*85–295€*		

4.4   Scenario 4: Milano Lipari – Stopover in Reggio Calabria

The last scenario taken in analysis is the Milan-Lipari section using the Strait airport to sail from the port of Reggio Calabria, via Messina, to the Aeolian Islands.

Going into detail, the scenario provides, once landed at the airport "Tito Minniti" of Reggio Calabria, to reach the port of the capital of the metropolitan city through public transport and then ferry to Messina and then to the Aeolian Islands. The itinerary would therefore have a total journey time of 4 h and 30 min compared to costs ranging between 85 and 290 € (Fig. 4 and Table 4).

Fig. 4. Connection Milano - Lipari through Reggio Calabria (Image processing C. Zavaglia).

Table 4. Detail of the connection through Reggio Cal; analysis of timing and costs (Source: Alitalia, Liberty Lines: dec 2017 - Data proc. C. Zavaglia).

Stopover Reggio Cal.	Time	Cost	Freq.	Transp.
Aerop. Mil. Linate	1 h 35	60–250€	2 + 3	Airplane
Reggio Calabria	1 h 5 min	3–10€	16	Hydrofoil
Messina	1 h 50 min	23–30€	3	Ferry
Lipari	–	–		
	4 h 30 min	*86–290€*		

5 Evaluation of the Scenarios and Expected Results

From the analysis of the possible scenarios, summarized in the Table 5, it emerges that the Reggio Calabria airport is potentially the cheapest in terms of economic and travel times for its use as a terminal from which to reach the Aeolian Islands.

Table 5. Summary table and comparison of the analyzed scenarios.

Milano - Lipari		
Airport stopover	Travel time	Cost
Napoli	8 h 38 min	70–250€
Catania	6 h 20 min	80–250€
Palermo	6 h 32 min	85–295€
Reggio Calabria	4 h 30 min	86–290€

However, although it is therefore advantageous under these two aspects, analyzing the itinerary in detail, a criticality emerges that is not important. The weak point of this solution is in fact the frequency of flights to reach Reggio Calabria.

Analyzing the frequencies of the means of transport in the scenarios highlighted in the previous paragraph we can see how the airport of the strait offers, under this point of view, very few alternatives. In fact, if the Sicilian airports provide 5–6 daily direct connections with the Milan airport, not to mention the Falcone Borsellino airport of

PARTENZA		ARRIVO		PARTENZA		ARRIVO
LIN 10:00	→	REG 11:35		REG 16:15	→	LIN 17:55
ⓘ Volo diretto | 01H:35' | Operato da: Alitalia				ⓘ Volo diretto | 01H:40' | Operato da: Alitalia		
LIN 12:00	→	REG 15:30		REG 12:20	→	LIN 16:10
ⓘ 1 Scalo | 03H:30' | Operato da: Alitalia Cityliner				ⓘ 1 Scalo | 03H:50' | Operato da: Alitalia		
LIN 11:00	→	REG 15:30		REG 12:20	→	LIN 17:10
ⓘ 1 Scalo | 04H:30' | Operato da: Alitalia Cityliner				ⓘ 1 Scalo | 04H:50' | Operato da: Alitalia, Alitalia Cityliner		
LIN 10:00	→	REG 15:30		REG 12:20	→	LIN 18:10
ⓘ 1 Scalo | 05H:30' | Operato da: Alitalia Cityliner				ⓘ 1 Scalo | 05H:50' | Operato da: Alitalia		

Fig. 5. Offer of the Alitalia company for the connection between Reggio Calabria and Milano, 1 direct flight and 3 alternatives with stopover in Rome (Source: Alitalia dec 2017 - Image processing C. Zavaglia).

14:10 - 15:55   Blue Panorama   1 h 45 min   LIN-REG   11:40 - 13:25   Blue Panorama   1 h 45 min   REG-LIN

Fig. 6. Offer of the Blue Panorama company for the connection between Reggio Calabria and Milano, 1 direct flight (Source: Blue Panorama dec 2017 - Image processing C. Zavaglia).

Naples, which makes no less than 11, that of Reggio Calabria stops only 2 flights offered (Figs. 5 and 6).

Alitalia and Blue Panorama are in fact the only companies that currently operate in this sense by providing each a direct flight. The presence of only two direct means that the proposed itinerary proves to be very constrained thus making this solution absolutely not very flexible with respect to tourism needs. The images 5 also show how flights with a stopover in Rome can not be taken into consideration because of the long waiting times in the change of flight at Rome airport, thus making the solution inconvenient as regards travel times.

6 S.W.O.T. Analysis and Conclusions

If on the one hand the route that sees the airport of Reggio Calabria as the fulcrum of the movements towards the Aeolian islands offers competitive prices and the shortest travel times, on the other, the little flexibility of the itinerary is a limitation decisive action [8]. To better understand the dynamics of the current situation and analyze the possible evolutions that could derive from it, we will cross the data taken into consideration, referring to strictly internal contextual aspects, with the potential

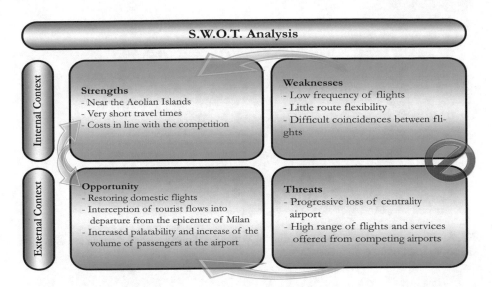

Fig. 7. SWOT analysis in the context of the "Tito Minniti" airport of Reggio Calabria (Data and image processing C. Zavaglia).

opportunities and possible threats coming from an external context, thus going to complete what is defined as a SWOT matrix (Fig. 7).

From the analysis of the strengths and weaknesses it is quite evident that the Reggio Calabria airport has a great opportunity. The proposal on the market of a route that exploits the proximity to the place of destination to generate an offer based on unparalleled travel times, keeping competitive in costs and ensuring an important efficiency in terms of flexibility, choice in schedules and coincidences, would represent for those who want to visit the Aeolian Islands, the choice certainly cheaper from every point of view.

For an infrastructure that in recent years has seen a collapse in the number of passengers (Fig. 8), contextualized to the difficult political situation in business [9] which saw the number of flights drastically reduced, the more immediate translation of the realization of this structured service would be to pick up and divert to Reggio Calabria those tourist flows that today take advantage of the other itineraries proposed and examined in the previous paragraphs [10].

Although tourism to the Aeolian Islands has a seasonal character, prevalently summer-spring, this would result in a major increase in the airport's movements of the strait giving new life to a structure that, despite being in a stratified geographical position and is a of the main connection structures of the metropolitan city, in recent years it has undergone a radical decline [11–14].

190   C. Zavaglia et al.

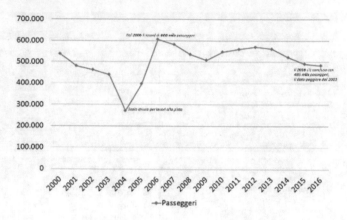

Fig. 8. Passenger airport of the strait from 2000 to 2016 (Source: Italian Association of Airport Managers Dec 2016).

References

1. Istat. http://www.istat.it/it/. Accessed 02 Oct 2017
2. Unive. http://www.unive.it/pag/18630/. Accessed 02 Oct 2017
3. Union Camere Sicilia. http://www.unioncameresicilia.it/. Accessed 02 Oct 2017
4. Osservatorio Turistico Sicilia. https://osservatorioturistico.sicilia.it/geoportale/. Accessed 02 Oct 2017
5. Confcommercio Sicilia. http://www.confcommerciosicilia.it/. Accessed 02 Oct 2017
6. Gabrielli, L., Giuffrida, S., Trovato, M.R.: From surface to core: a multi-layer approach for the real estate market analysis of a central area in Catania. In: Gervasi, O., Murgante, B., Misra, S., et al. (eds.) ICCSA 2015. LCNS 9157, pp. 284–300. Springer, London (2015)
7. Camagni, R.: Economia urbana. Principi e modelli teorici. Carocci (1992)
8. Della Spina, L.: The integrated evaluation as a driving tool for cultural-heritage enhancement strategies. In: Bisello, A., Vettorato, D., Laconte, P., Costa, S. (eds.) Smart and Sustainable Planning for Cities and Regions. Results of SSPCR 2017. Green Energy and Technology. Springer (2018). https://doi.org/10.1007/978-3-319-75774-2_40. ISSN 1865-3537
9. Della Spina, L., Scrivo, R., Ventura, C., Viglianisi, A.: Urban renewal: negotiation procedures and evaluation models. In: Gervasi, O., et al. (eds.) Computational Science and Its Applications – ICCSA 2015. Lecture Notes in Computer Science, vol. 9157. Springer, Cham (2015). https://doi.org/10.1007/978-3-319-21470-2_7. ISBN 978-3-319-21469-6 (Print) 978-3-319-21470-2 (Online)
10. Calabrò, F., Della Spina, L.: The public-private partnerships in buildings regeneration: a model appraisal of the benefits and for land value capture. In: Advanced Materials Research, vol. 931–932, p. 555–559 (2014). https://doi.org/10.4028/www.scientific.net/AMR.931-932.555. ISSN 1662-8985
11. Camagni, R.: La teoria dello sviluppo regionale. CUSL Nuovavita, Padua (1999)
12. Della Spina, L.: Integrated evaluation and multi-methodological approaches for the enhancement of the cultural landscape. In: Gervasi, O., et al. (eds.) Computational Science and Its Applications – ICCSA 2017. Lecture Notes in Computer Science, vol. 10404. Springer, Cham (2017). https://doi.org/10.1007/978-3-319-62392-4_35

13. Della Spina, L., Lorè, I., Scrivo, R., Viglianisi, A.: An integrated assessment approach as a decision support system for urban planning and urban regeneration policies. Buildings **7**, 85 (2017)
14. Della Spina, L., Ventura, C., Viglianisi, A.: A multicriteria assessment model for selecting strategic projects in urban areas. In: Gervasi, O., et al. (eds.) Computational Science and Its Applications – ICCSA 2016. ICCSA 2016. Lecture Notes in Computer Science, vol. 9788. Springer, Cham (2016)

The Port-City Interface

Jusy Calabrò[(⊠)], Alessandro Rugolo, and Angela Viglianisi

Mediterranea University of Reggio Calabria, 89100 Reggio Calabria, Italy
jusy.cal@gmail.com

Abstract. Assuming that waterfront have always been special places "where land and water meet" [1] nowadays they are mostly places for urban renewal, where conflicts and debates, about what their role should be within the urban context, emerge. The "port-city interface" aims at explaining the spatial, institutional and socio-economic relations occurring between cities and ports; evaluating new approaches to take into consideration urban regeneration initiatives where different functions compete causing often urban decay. Particularly, factors of success will be highlighted to address urban regeneration initiatives, toward sustainable approaches, in those sensitive, complex urban areas, for development. The "interface" [2], where port and city communicate, is often a ground characterized by competition, but it could also be a possibility for both city and port to grow. The comparison among selected case studies highlights the dynamics involved into those places, from the spatial, institutional and socio-economic points of view, in order to provide a useful framework for the decision-making process performing between port and city related areas. The possible interaction could enhance the innovation and sustainability attempts of regeneration initiatives in those peculiar places, providing added value for the whole urban system [3].

Keywords: Port-city · Interface · Urban regeneration · Sustainability
Innovation

1 Introduction

The first part of this paper introduces a scientific literature review to understand the historical reasons behind the city-port areas detachment; then, a contextualization of waterfront regeneration initiatives is provided. The second part aims to be a synopsis of the case studies, analyzed along the PhD research activity, which here are considered as international best practices of urban regeneration linked to port-city dynamics, belonging to the same urban policy ground and strictly linked to similar strategies for local development. Then, the main factors involved into the urban regeneration performance for each case study analyzed are highlighted. Moreover, governance processes, led by port and city authorities, are investigated to underline dynamics and

This is the result of the joint work of the authors. Although scientific responsibility is equally attributable, the abstract and the paragraphs nn. 3,4 were written by J.Calabrò; the paragraphs nn. 1, 2.1 were written by A. Viglianisi; the paragraphs nn. 2, 5 were written by A. Rugolo.

© Springer International Publishing AG, part of Springer Nature 2019
F. Calabrò et al. (Eds.): ISHT 2018, SIST 101, pp. 192–199, 2019.
https://doi.org/10.1007/978-3-319-92102-0_21

accountability matters, determining the ongoing spatial changes in the port-city areas. Conclusions will provide a general view of the case study comparison, including those key-factors which have been assessed into the port-city regeneration performances.

2 The Port-City Interaction

Nowadays the urban crisis is mostly perceived as a product of social disease, a weakness, arising from the change in the relationship between urban spaces and their functions. That crisis can be traced especially within those transition spaces, without any specific distinctiveness, such as those between city and port related areas.

Interactions between port and city are expressed in a variety of ways: "some ports are set within urban areas, others have no city associated with them" [2]. There are towns with a strong relationship with their ports, because of territorial contiguity or rather their historical importance for the city; instead, to some extent, "economically and geographically, port and cities have grown apart" [2]. Besides, different ways of understanding planning dimensions, could be traced back to historical reasons of development or influenced by global economic processes spreading out all over, with increase in traffic volumes and the consequent higher demand for space. All factors which brought to a progressive detachment from the city, by shaping the interface as a marginal "place", often a "non-valuable urban producer" [4]. Although many scholars agree on the importance of ports for the city economy, few of them have provided conceptual models to explain the reasons. Hyauth and Hoyle, among the first theorists of the "port-city relationship", established its importance in the context of urban and territorial development, considering the different interactive dynamics between ports and their cities [5]. Hoyle, in particular, described them in six phases, from the close association to the progressive separation, until the current phase of regeneration that aims at sustainable local economic development. "Changes in these systems led to the growing spatial and functional segregation of city and port and the changing landscape of the city-waterfront" [6].

2.1 The Waterfront Transformation

The last decades have displayed the "dramatic ongoing transformation of downtown waterfronts in both North America and Europe as a result of myriad issues, including environmental, economic, social and political factors" [7].

It should be remembered that port sector is extremely unstable, due to political and global economic factors, and the international trade situation as well.

In such context the planning process is extremely complex because of local rules and institutional competences regarding the port area boundaries, thus, the communication among planning tools is often unsolved. Moreover, we can see as in Europe Port Authorities and City-planning authorities are more likely to compete for space and power, rather than working together. Port Authorities' role, within the planning process, does not only address development and assets within the port areas, but it strongly influences those spaces of interaction between city and port, by determining conditions and degrees of relation with respect to their status, public or private, or both.

Accordingly, several differences occurred in the government responses, mainly between United States and Europe, in considering the port-city areas as key factors in their challenges of regeneration: from land uses and planning codes implementation, to environmental matters and sustainability issues. Most European river-port cities, as well as seaport cities, reviewed their policies towards service- centered metropolis, offering a renewed quality of the urban environment in the interface. A new way of intending a place not so "urban" neither port-related, rather characterized by a non-homogeneous urban frame, lacking specific functions: here we can often find big transportation infrastructures, old rails, physical barriers that contribute to the contemporary status of blighted area. The research activity, described here, aims to demonstrate how a proactive city-port relationship is a key factor in the pursuit of a balanced urban development of the "liquid city" [8]. Here "urban regeneration" is considered as a holistic process of development which, following a place-based approach, includes local needs, socio-economic attitudes, environmental issues, blighted areas redevelopment, under an overall strategy for local economic development.

3   Case Studies Analysis

Case studies were used to categorize the main issues interacting within those transition places from the port to the city, highlighting approaches, methods and practices used throughout the investigated urban regeneration processes.

The Case studies analyzed have been grouped into two clusters in order to show differences and opportunities to be explored within diverse international contexts. The categorization allows to analyze the case studies by following the common spatial and institutional background they belong to: a first comparison is conducted among the EU case studies, a second one among the US case studies. A final comparison of methods, practices and tools is provided, according to the different policy areas investigated, referring to spatial, socio- economic and institutional dimensions. The most performing outcomes were stressed to understand the ways through which a urban system can be affected positively by the interface transformation and if it can bring new, flexible meanings and functions strictly connected to real community needs.

Fort Point, Boston (USA) [9]: The 100 Acres Master Plan within Fort Point South is the result of a common effort among public authorities, agencies, neighborhood representatives, owners and associations of the area, that contributed to give rise a participatory urban process. The BRA, Boston's Planning and Economic Development Agency, involved community and stakeholders to draw up a plan for growth and development within the 100 Acres, a Planned Development Area, taking into account the pre-existent facilities and infrastructures capacity, in order to encourage a lively urban district [10]. A shared process of urban regeneration that gave back to the city a district in which innovation and maintenance work together, being example of development and sustainability throughout the Boston Area: not only by attracting capitals and businesses, thanks to its location, but thanks to its renewed potential, it is also appealing to professional studies, artistic galleries and an active social community.

North Embarcadero, San Diego (USA): it is an area breaking with the urban mesh, not properly harmonized with the context of the close Down-Town, rather heavily characterized by large "parking lots" - big parking areas that contribute to the discontinuity and the lack of urban identity of this huge interface strip with the city. The Plan for the northern part, the Embarcadero, is the outcome of the institutional cooperation effort among the San Diego Unified Coast District, the Municipality and the local community, including that of Little Italy. The goal was pursued with good standards of sustainability and "liveability" for this area, by providing the community with a large linear pedestrian park, completely opened to the sea and its activities, but also with a desirable urban area, with great development opportunities for the entire waterfront.

Titanic Quarter, Belfast (UK): represents one of the more extensive urban regeneration projects on water in Europe [11], with its 18 acres on the Lagan River in Belfast. The regeneration process has preserved its historic value, enhancing it through the symbols of a famous past. The district indeed is considered as example of the new English revival, transformed from a blighted area into a multi-functional space, appealing for its innovation attempts. It is a mixed-use neighborhood with a strong historical maritime character, having been the basin within which the Titanic was built up. Its redevelopment is carried out within the framework of a strategic vision shared among the major partners of the project: the City Council, NITB, Titanic Quarter Limited and the Titanic Forum. The project preserved the original identity of the neighborhood while evolving into a strategic pole in the urban context, being a residential and commercial area, but even on a larger scale, thanks to the numbers of art galleries and the valuable leisure activities attracting tourists and artists from all over the region.

Kop Van Zuid area, Rotterdam (NL): it is a urban regeneration project under the guide of national policies, with high flexibility standards, both of planning tools and of negotiation, among the different administrative levels, towards the compact-city main goal [12]. A flexibility which moves under a public participation that guides, but does not constrain, providing services and infrastructures, opportunities for locating, through incentives, encouraging the implementation of projects and pushing towards concrete chances for development, with the involvement of local stakeholders and the community. The motto of the development strategy was "cooperation-coordination-consultation-consent-compromise" which represented the common thread of the project interventions. The project, supported by a strong marketing strategy, aimed to relocate the Kop Van Zuid area as the glorious site of the docklands in Rotterdam. This transformation process can be considered as example which matches the public role, driving toward collective benefits, and the private one, as promoter, in synergy under the common objective; even if it sometimes came out as an experiment of "privatization urban policies" [12].

Hafen City, Hamburg (DEU): Here the port area has progressively detached from the city, because of the increasing request for specialization to transhipment, becoming a separate entity. The former industrial area is now part of the historic center, representing its extension towards the sea, promoter of new life styles. The urban regeneration of the interface keeps its historical character, though strengthening and implementing new urban functions, with a mix of public and residential spaces that enhances the relationship with water. The importance of the integration between the city and its port provides new urban connotations to an area that seemed deprived by its original sense, and that now it represents a positive example far beyond that town. The partnership consisted of the City of Hamburg and the authority for the implementation process management and the development of the Masterplan, (the Hafen City Hamburg GmbH), also responsible for public funding, under national laws, which was created for the purpose to maintain the peculiar structures of both city and the port [13].

The different practices investigated on the port-city relation brought to highlight port and city policies in Europe, mostly centered around the institutional dialog between Port Authorities and City planning ones. Here can be seen how the institutional issues are the proper ground of the interface, and how such level seems to be stronger in the final spatial outcome then in other countries. In US instead the comparison among the case studies is concentrated more on the spatial level, since the institutional context is not homogenous in all States, and the role played by Port Authorities in the planning process as well. It may change according to laws, procedures and importance of the port considered, and their role may be perceived weaker than European contexts, where the interface dynamics depend mainly on competences and jurisdictional matters.

4   Urban Regeneration Initiatives Between Port and the City

Since "a successful implementation of a regeneration project requires public and private funding, provision of high quality services, administrative effectiveness, promotion of tourism and public participation" [13], in the port-city areas such goals could be pursued through the development of balanced "mixture of land and sea uses whilst preserving the distinctive identity of the waterfront [13]. It can be seen as urban regeneration tools, which employ public-private partnerships to be implemented, have been used in many city-port areas, often solving the gap in communication among institutions: thanks to their attitude, all the actors involved can establish roles, skills, advantages, and spatial-functional determination as well, in order to overcome the static nature of planning tools, still inconsistent with the real needs of dynamism which those places need.

Significantly, in most of US cities, approaches to revitalize such urban areas through economic initiatives are mostly focused on the mixed-use of the functions, considering local features and preexistent facilities, consistently with the pillars of sustainability principles.

In Europe institutional and spatial contexts are different from the USA ports-city zones: if we report the important role played by port authorities (public bodies) in Europe for almost all ports, on the contrary, in the USA port areas are ruled under the same planning code of the city, with particular measures in coherence with the Coastal Zone Management (CZM).

A different way of intending planning dimensions which could be explained by historical reasons of development: if in Europe the waterfront development was originally seen as "maritime in character", because of the maritime technological transformation that forced ports to expand towards more specialized areas (Rotterdam,

Amsterdam), in US it is considered as part of the same city's renewal planning process, which applied new development strategies, often very far from port activities. The case studies analysis shows the main dynamics elapsing in city-port relations and some important issues to be faced: institutional dialog, jurisdictional competences, spatial integration, competition for space and functions, are the emerging ones.

Particularly, as for historic ports, or neighborhoods strictly connected to port and water activities, surveys and interviews[1] conducted showed as the involvement and the strong participation of the local community [14] ensured the high sustainability levels of the whole urban transformation process (Fort Point), matching many interests rooted in the area. On the contrary, in those cases where the port area embraces and dominates the city, the so-called work-class port, with the industrial activity strongly shaping the environment, (Titanic Quarter - Kop Van Zuid and North Embarcadero) the role of public-private partnerships was crucial for the development strategy implementation. The integration among different functions was achieved by linking spatial meanings and economic dynamics involved in the interface areas. Urban policies within the port, therefore, cannot fail to consider and use those factors that should become the "invariants" of a urban regeneration strategy that aims to be sustainable for the economic-social-environmental and cultural components.

5 Conclusions

It can be seen as sometimes "laws and regulations dominate the spatial outcomes of governance processes between city and port, and that these tend to frustrate experimental efforts towards truly sustainable results" [15].

Indeed, "development orientations that foresee an on-going port migration process away from the urban core are still common among urban planning and policy makers, which impedes on the joint governance processes needed for building renewed, sustainable port-city relations and spatial projects" [15].

Evidences coming from the port-city relations studied show how different in meanings the interface can be: according with the historical character of cities on water, the interface may reinforce their maritime attitude, or rather becoming an opportunity for that urban area to grow, by catching new meanings within the urban context. Furthermore, where the city and port are perceived as institutions, as in the European contexts explored, and roles and skills go beyond any spatial results, the image of the interface may reflect the struggle for power among all forces involved. Again, where the interface area is considered as the same as urban area for re-development, as in the US cases, the spatial outcome is often the result of negotiation processes, guided by tools as flexible as reliable, sharing both responsibilities and profits among the actors involved, with a strongly market-oriented approach [16].

From the case studies comparison, it is found how the most effective interventions turned out to be those ones linked by some common factors which acted in the same way along the urban transformation process, sorts of potential paradigmatic principles.

---

[1] Interviews and surveys have been conducted by the authors.

198   J. Calabrò et al.

Among them, the possible role that the public-private partnerships [17] could take emerges: it may allow to overcome that formal firmness which, thanks to their moldable nature might catalyze private interests and public objectives through an "agreement" with duties and tasks, shared responsibilities and benefits [18].

The participation of the public sector then, as a guide of the urban transformation process [19], in cooperation with management Authorities created ad hoc, proved to be pivotal for the success of the initiatives: a widespread community consensus, reached by working on their awareness, restored huge parts of the city to its users. Nevertheless it must be said that the risk for those transition areas, working among powers, functions and forms, is likely to become ground for political struggle, which can halt the completion of the initiatives and influence the urban environment [20]. The international examples discussed here represent an attempt to put into practice some urban planning theories and policies based on the need for flexibility. We can see as "successful waterfront redevelopment requires an understanding of global processes and an appreciation of the distinctiveness of port-city locations. Waterfront revitalization occurs at the problematic and controversial interface between port functions and the broader urban environment" [21]. On the other hand, short-term actions are often preferred, supported by political affirmation needs. It can be argued that a possible answer to recognize those place of interface as crucial for both city and the port, could be found in some European contexts where "responsible authorities in seaport cities have started to reconsider the transformation-oriented planning and development approach to their respective waterfront zones" [22]. That seemed to be an important step forward to consider the port-city interface in a more tangible way. As the "cycle of dereliction, neglect, planning, implementation and revitalization of old harbor areas as well as the necessary construction of port infrastructures are part of a complex network of stakeholders and interests" [22], the most important issue to deal with is certainly the right use of urban planning practice and tools. The analyzed cases suggest how urban regeneration initiatives, thanks to their holistic nature, may allow to enhance the interface multifunctional attitude, by using an integrated approach, able to face: trans-scalar dynamics, being port both important for the local and the national economy; inter-institutional matters, by supporting the institutional cooperation toward complementary objectives of growth; social issues, preserving local needs and the liquid and moldable attitude of cities on water [23].

Those adaptive urban tools might provide the path to be followed in order to reach the development objectives, both for urban and port related areas [24].

References

1. Bunce, S., Desfor, G.: Introduction to political ecologies of urban waterfront transformations. Cities **24**(4), 251–258 (2007)
2. Hoyle, B.S.: The port-city interface: trends, problems and examples. Geoforum **20**(4), 429–435 (1989)
3. Carta, M.: Creative City. Dynamics, Innovations, Actions, Barcelona (2007)
4. Daamen, T.: Sustainable development of the European port-city interface. In: ENHR-conference, pp. 25–28, June 2007

5. Wiegmans, B.W., Louw, E.: Changing port–city relations at Amsterdam: a new phase at the interface? J. Trans. Geogr. **19**(4), 575–583 (2010)
6. Hayuth, Y.: The port-urban interface: an area in transition. Area **14**(3), 219–224 (1982)
7. Kotval, Z., Mullin, J.R.: The changing port city: Sustainable waterfront revitalization. J. Town City Manag. **1**(1), 23–38 (2010)
8. Carta M.: Dal waterfront alla città liquida. Waterfront d'Italia. Piani, politiche, progetti. FrancoAngeli, Milano (2010)
9. Cavelli, C.: The Fort Point District Plan. In: Schacter, Busca, Hellman, Ziparo, Boston in the 1990's, pp. 166, Gangemi (1994)
10. Boston Redevelopment Authority, Fort Point Channel Working Group, The Fort Point District: 100 Acres Master Plan (2006)
11. Smith, M.: Titanic Quarter: Building the future from the past, Urbanistica, 148 (2011)
12. McCarthy, J.: Reconstruction, regeneration and re-imaging: the case of Rotterdam. Cities **15** (5), 337–344 (1998)
13. Papatheochari, D.: Examination of best practices for waterfront regeneration. In: Littoral 2010-Adapting to Global Change at the Coast: Leadership, Innovation, and Investment, p. 02003, EDP Sciences (2011)
14. Calabrò, F.: Local communities and management of cultural heritage of the inner areas. an application of break-even analysis. In: Gervasi, O., et al. (eds.): Computational Science and Its Applications, ICCSA 2017. Lecture Notes in Computer Science, vol. 10406. Springer, Cham (2017). https://doi.org/10.1007/978-3-319-09147
15. Daamen, T., Vries, I.: Governing the European port-city interface: institutional impacts on spatial projects between city and port. J. Trans. Geogr. **27**, 4–13 (2012)
16. Hodge, G.A., Greve, C.: PPPs: the passage of time permits a sober reflection. J. Compil. Inst. Econ. Aff. **29**(1), 33–39 (2009)
17. Calabrò, F., Della Spina, L.: The public-private partnerships in buildings regeneration: a model appraisal of the benefits and for land value capture. Adv. Mater. Res. **931–932**, 555–559 (2014). https://doi.org/10.4028/www.scientific.net/AMR.931-932.555
18. Reuschke, D.: Public- Private Partnerships in Urban Development in the United States, NEURUS-Network of European and US Regional and Urban Studies (2001)
19. Calabrò, F., Cassalia, G.: Territorial cohesion: evaluating the urban-rural linkage through the lens of public investments. In: Bisello, A., Vettorato, D., Laconte, P., Costa, S., (eds.): Smart and Sustainable Planning for Cities and Regions. Results of SSPCR 2017. Green Energy and Technology. Springer, Cham (2018). https://doi.org/10.1007/978-3-319-75774-2
20. Ciulla, V., De Capua, A.: La Nuova Forma Urbana. LaborEst **12**, 85–88 (2016). https://doi.org/10.19254/LaborEst.12.14
21. Daamen, T., Louw, E.: Sustainable development of the European port-city interface: evolving insights in research and practice. In: 24th Annual European Real Estate Society Conference. ERES: Conference. Delft, Netherlands (2017)
22. Schubert, D.: Waterfront transformations and city/port interface areas in Hamburg. Dimensión Empresarial **13**(1), 9–20 (2015)
23. Della Spina, L., Lorè, I., Scrivo, R., Viglianisi, A.: An integrated assessment approach as a decision support system for urban planning and urban regeneration policies. Buildings **7**, 85 (2017). https://doi.org/10.3390/buildings7040085
24. Della Spina, L.: The Integrated Evaluation as a Driving Tool for Cultural-Heritage Enhancement Strategies. In: Bisello, A., Vettorato, D., Laconte, P., Costa, S., (eds.): Smart and Sustainable Planning for Cities and Regions. Results of SSPCR 2017. Green Energy and Technology, Springer, Cham (2018). https://doi.org/10.1007/978-3-319-75774-2_40

From 'Highway into Greenway': How Public Spaces Change Zoning Regulations

The Case of Northend Park Rose Fitzgerald Kennedy, Boston, USA

Israa H. Mahmoud[1]([✉]) [iD], Bruce Appleyard[2] [iD],
and Carmelina Bevilacqua[1] [iD]

[1] Mediterranea University of Reggio Calabria, 89100 Reggio Calabria, Italy
israa.hanafi@unirc.it
[2] School of Public Affairs, San Diego State University, San Diego 92182, USA

Abstract. The case study presented in this paper is a manifestation for an urban regeneration project that transformed a Highway into a Greenway. **The first part** aims to understand the contextual background of the highway regeneration Project, and it analyses the key factors of the long-debated land use and how the public authorities mandated the development of open public places as a policy. **The second part,** entails the rebirth of the Public Space as part of the Rose Kennedy Greenway where the role goes beyond the semantics from just a park towards being considered the front porch of the city oldest Neighborhood, the Northend. **The last part** analyses in depth the cultural programming of the Public Place and its character as a livable destination in the heart of the Downtown district. The methodological approach uses a public Life Matrix of evaluation to identify users' behavioral patterns through intercept surveys, frequency of social activities through intensive three months long physical observation analysis, and lastly in-depth interviews with local Stakeholders, related Governmental bodies and Boston development and planning authorities. The findings highlighted a tendency that community involvement in the planning and placemaking process helped inform the Public Policy about the needs of surrounding neighborhood residents; as well as, emphasize the Public Private Partnerships in successful urban regeneration projects such as the case of the Northend Park.

Keywords: Public space · Urban regeneration
Northend Park of Rose Kennedy Greenway

1 From a Highway to a Greenway: A Land Use Dilemma

The Northend Park takes part from the -so called- Boston's ribbon of contemporary parks. The Rose Kennedy Greenway is a mile and a half of contemporary parks in the heart of the Boston city. The greenway as a roof garden topping a highway tunnel connects people, city scape and fun; it connects a series of 7 parks, whereas there are public art installations, water fountains, historical sites (such as: the freedom trail),

© Springer International Publishing AG, part of Springer Nature 2019
F. Calabrò et al. (Eds.): ISHT 2018, SIST 101, pp. 200–210, 2019.
https://doi.org/10.1007/978-3-319-92102-0_22

public transit and bike sharing stations (such as: Hubway), food trucks vending locations and public restrooms [1].

In this light, this paper aims to highlight the role of urban regeneration projects that involves distressed urban areas, through actions, cultural programs, and public policies on a larger scale, to improve the living conditions; by the development of public spaces, parks, squares, etc. and mobilization of cultural capital. This includes, as well, the role of economic and human behavioral facts about public spaces, which is the focus of economic regeneration process to a certain extent. The reason for the selection of the Northend Park in the city of Boston is that it fulfills the criteria of being a public space formed after a regeneration project that involved different stakeholders as well as transformed an eyesore to a vital livable destination in the heart of the city.

Following the removal of the elevated highway (Interstate 93 and 90) as mandated by the Boston Redevelopment Authority [2] a big debated land use dilemma paved the way towards the development of only 75% of the land "as series of parks and urban plazas". That was adopted later in the "Air-Rights Park plan" mandating the 25% surface development only while the rest remained as public open space [3, 4].

The buried sections of Interstate 93 and 90, are now topped with the 1.5-mile-long Rose Fitzgerald Kennedy Greenway, see Fig. 1. The Project have been federally funded and overseen by the state where the Bostonians recall it as "the battlefield of Menino". Somehow, though, Mayor Tom Menino managed to sell investors and ordinary citizens alike on his vision for the clean, efficient, and business-friendly city that would emerge from the project's dust [5]. 'The Big Dig' project ballooned into a $22 billion boondoggle, but Boston came out of it as a better city. Property values have doubled, streets are safer, and economy is more robust than ever.

"The Rose Kennedy Greenway will transform the heart and character of Boston in so many ways. We have taken a space that previously hosted an outdated elevated steel highway and turned it into a vibrant park. Today's dedication of the North End Parks is a significant step forward in the overall development of this Greenway." Governor Patrick: opening ceremony of the Northend Park [6]).

Context Overview. The Central Artery Master plan developed in 2001 by the Massachusetts Turnpike Authority provided a framework designating eight acres for open spaces and parks in Bullfinch triangle and Northend. The plan stipulated the parcels development focusing on reconnecting districts with a seam of futuristic-yet historical-open space that knits its' neighborhoods together [7, 8].

In 2003, the mapped central artery corridor master plan showed designated specific parcels for open space development; amongst which parcels 19, 21, 22 were assigned to Massachusetts Horticultural Society to develop outdoor gardens and an enclosed winter garden. Then, the Massachusetts Turnpike authority delegated the assignment of developing parcels 6, 12 and 18 to the Greenway Conservancy which created the Dewey Square Park and Fort Point channel parks. The "Big Dig" plan nonetheless called the creation of a pair of one-way surface road extending the length of the Greenway, now called "John Fitzgerald surface road", to handle local traffic that existed underneath most of the replaced Central Artery. In the meantime, the Greenway parks and development parcels were bordered as well by other crossing streets (North, Cross, and Sudbury streets) separating the developed parcels. Both parcels 8 and 10

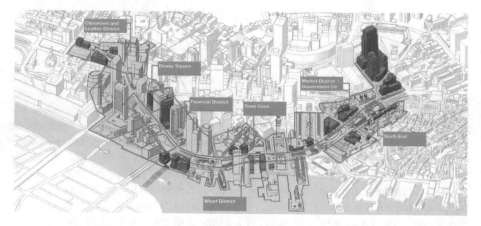

Fig. 1. Rose Fitzgerald Kennedy Greenway sub-districts and seam of connected parks.

gave birth to the Northend parks in 2005[1]. Northend parks, described as a "city hinge" to connect the old-Italian neighborhood to the financial district and Downtown of Boston is a major milestone for the recognition to the role of public spaces to generate social capital and foster communities' development.

2   The Rebirth of a Public Space

The Northend Parks were designed by Gustafson Guthrie Nichol and Crosby Schlessinger Smallridge in 2005. Both firms commissioned by the city of Boston, specifically by the Massachusetts Turnpike authority [9], to design a park that reflects the link between the historic neighborhood of Northend and the rest of the city. The Northend park and plaza opened in November 5th, 2007, have been replacing an area that was formerly an eyesore, the long awaited controversial sixteen-year-long Central Artery/tunnel project of the "Big Dig"; One of the largest and most expensive public works project in the American History, replacing an elevated highway that bludgeoned the city for nearly half century, dividing the city waterfront and historic Northend from the adjacent Downtown.

The Northend Park and Plaza takes part of The Emerald Network, in a parallel way, a 200 miles of greenway networks that is an initiative under the livable streets alliance. The Emerald Network is a seamless shared-use greenway paths in the urban core of the city of Boston and adjacent cities that provides walking, biking connections through parks. The major aim of the greenway project is to connect people to jobs, to neighborhoods' assets by foot, bikes and any non-motorized means [10]. The later one is an initiative of the Massachusetts Government for "rethinking urban transportation", nonetheless, the fundamental aim of the planned shared-use paths across the greenway is to connect neighborhoods with open spaces, transit stations and jobs therefore

---

[1] See http://www.rosekennedygreenway.org/files/7713/0084/3916/North-End.gif

increase mobility, promote active recreation, improve climate change resilience, and enhance the city's competitiveness in the global economy.

Physical Design Concept. Geographically located in coordinates 42.362358 N, 71.055875 W; the Northend parks and plaza are physically encompassed with a size of 2.83 acres and approximately 12.000 m^2. The design of the open space and park plaza (identified in Boston Maps geospatially as ID 570) in which the open spaces layer of data is defined as: the open spaces under conservation and recreation interest in Boston, Massachusetts regardless the ownership [11] frames the entry to Northend neighborhood from one side and to downtown on the other.

The notion of design for the Northend park and plaza is that it works as a significant "hinge", between the grand civic spaces of Quincy market, Government Center, and Haymarket; leading the way to approach the intimate Northend, Boston's oldest neighborhood. While located at the threshold between downtown and a historic/touristic neighborhood, the design conception has not been easy, creating a critical link between the importance of Northend as a "home" to largely Italian community since 1890s. In fact, the neighborhood - still distinguishably abutting the narrow streets and alleys- containing several historical sites remained till now an appealing attraction; hence, local residents have accepted the resulting influx of tourists and the gradual increase of restaurants and retail shops over the years [12, 13].

A steel pergola lines one side of the site and is the conceptual "front porch" of the North End neighborhood, complete with site furnishings that encourage its use, makes it an "exception" as described by [5]. A reflective water feature separates the porch from a series of lawns and perennial gardens. Through the park design on the circumference by streets and walkways (North Street, Hanover Street, and Salem Street walkways) reconnect the City to the North End; Each cross the gardens, water feature and pergola. The site's rich history is reflected in interpretive elements that include granite marking the edge of the Mill Pond and the water's edge, descriptive quotes and a timeline engraved in leaning rails, and an engraved stone map illustrating the changing landform of the site, see Fig. 2.

Fig. 2. Current view of the Northend Parks. Source: the first author, Summer 2016.

3 The Northend Park as a Successful "Public Place"

The ever-evolving debate about the difference between a "public space" and a "place" goes beyond semantics on the distinction between the two concepts. A place is shaped by the environment in which people invest meaningful times; it has its own history, a unique cultural and social identity that is defined by the way it is used and who are the people using it. In addition, Physical, social, environmental, and economic aspects of communities can be nurtured through the creation of places [14, 15]. The Northend park case in particular is characterized by an unmistakable cultural enrichment due to touristic traffic crossing through following the Boston's historic landmark (the Freedom Trail); that said, the flexible design of The North End Parks features spaces including green landscapes with a path system, plazas with pergolas and water features that run through both parcels and appeal to a wide range of people, including North End residents of all ages and the thousands of tourists and Bostonians who visit each year [9].

The Gustafson Nichol [16] design for these two adjacent parcels restored views and street connections that were severed for decades by the elevated highway. The North End Parks, that had always been a physical and social threshold, are now one of the most popular Greenway destination in Boston. Nonetheless, the community involvement in the design process played an essential role towards creating a distinguished public place. The enthusiastic involvement of the neighborhood community centers and an engaging public process -through local collaboration and public meetings- were fundamental in shaping the design of a new "front porch" for the North End. The residents' traditional, lively street culture is celebrated in the parks' design of furnished terraces, intimate garden spaces, and interactive water features. A large pergola defines the North End neighborhood's gateway and 'front porch' as a place to gather, to stroll, and to be seen.

According to Ken Greenburg an urban design consultant hired by the City to visualize the future of the RFK Greenway, noted the importance of keeping existing businesses in the North End to retain the Italian ambience and help retain a 24-h environment. Whereas, he predicted, the North End's parks could be "one of the great public spaces in America" [17]. Historically, since 1950's, the freedom trail crossed the site in confined space beneath the Central Artery viaducts. Now, the park design weaves high the city's historic freedom trail that now crosses the main path while it used to sit in the dark shadow of the elevated highway.

Cultural Programming. More in depth, The Northend Park as listed by June 2015, with a total budget of $400,000 was subject to face-lifts in lighting fixtures, pathway lights along the freedom trail were returned functional and condensed. Benches have been replaced, along with gliding porch swings under the pergolas and the greenery in the gardens area have been replaced, condensed and redesigned to be more welcoming [18]. In fact, The Park is programmed to get advantage of all spaces, while the primary use is passive activities (such as watching water fountains, relaxing on lawns and sitting around benches, chairs and existing tables); active engagement programs include free fitness yoga and Pilates classes [19] or Berklee seasonal Musical Concert series during July and August yearly [20], food trucks vendors, and Galleries on the fence.

Nonetheless, the continuous effort of the Rose Fitzgerald Kennedy Greenway Conservancy in promoting the cultural programming of the greenway through festivals and special events such as artistic installations, outdoor movies displays and Dogs carnivals [21]. In addition to that, the public art program that displays along with the active placemaking activities of creating and activating open spaces have bolstered a longer term economic value creation in the whole urban Downtown business district and the Northend in particular [22, 23].

4   The Northend Park as a Liveable Social Destination

While the physical analysis is as important for any site, one of the most important measurable qualities of any public place is its attraction for users; people gather in plazas, walk in parks, the relationship between people and their space is an essential component of urban design [24]. Based on this notion, the following in-depth analysis is on the users of the Northend park with the help of a public life matrix toolkit technique. The main outcomes are part of an intensive visual observation analysis, an intercept users' statistical survey, interviews with on-site and in-field experts, and a video camera surveillance records analysis.

The First Phase of the analysis started in December 2015, the physical observation of the plaza, see Fig. 3- left, showed a touristic tendency in the crossing between Hanover street and the Blackstone street whereas the freedom trail path is. In other words, in good weather season, the human flow increases, tourists stop often in specific spots to take pictures, enjoy the view of the waterjet fountains, and maybe grab something to eat from the adjacent bakeries of the Italian neighborhood. In compliance with that, the Northend park is as vibrant as one can imagine a public place; yet, in some crossings as reported in **Vision Zero**[2], pedestrians run the red lights or even cross away from the crosswalk due to longer walk signal time [25].

Following that, an intensive visual observation timeline[3] conducted for three months period from April 2017 till July 2017, see Fig. 3-right, showed an attractiveness factor to the usage of the Northend Park as Cultural-Based Destination due to its' contingency to the Italian Neighborhood food and restaurants cluster with a special frequency occurrence on weekend days. Whereas sociability (defined as a livable street life and diversity in public place use and stewardship [14, 26, 27]) measures a noteworthy feminine presence in the plaza and in a frequent time frame between 5:30 pm and 8 pm that peaks in the weekends days.

The second Phase of the case study analysis treated the statistical side of the users' surveys. In the 70 intercepts users' surveys that have been conducted[4], see Fig. 4, there

---

[2] Vision Zero is a real-time online Platform to report Safety Issues as they occur instantly in the city streets; Supported by the city of Boston and ESRI mapping tool.

[3] The Visual observation timeline was conducted from 10 am to 10 pm for three full months and divided into 2 h slots based on a preliminary analysis that most frequent users do not spend more than 2 h in the Northend Park daily.

[4] The survey was completely anonymous for the 70 users and was tested with local stakeholders and urban experts from the greenway conservancy. Statistical analysis of Survey takers was conducted on site using a paper survey form then by building a. SAV database with STATA software and re-analysed with DataCracker online tool.

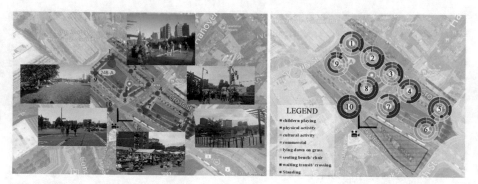

Fig. 3. Left -Observation Points Base map with exemplary views from around the Northend Park, April/July 2017. Right - Mapping Stationary activities in ten Observation points.

has been analyzed a tendency of 37% towards the usage of the public space from the surrounding inhabitants of the Northend district. Considering the strong influence of the park parcels in shaping the cultural aspect of the neighborhood, it is significant that most of the frequent users are from the same zip code of 02113 whereas the park is. Nonetheless, a noteworthy female attraction of 14% of survey takers during the weekends of 16th and 22nd July due to good weather conditions (recorded an average temperature between 19 and 28 °C) and planned Musical Concerts from 5:00 to 7:00 pm. In the interim, the age and gender differences were not substantial, both categories scored between 34 and 38 years as most significant quantiles; a noticeable low correlation of 0.80 between gender and age in the survey results.

Contrariwise, higher correlation was assessed between age and social clusters of users; 57% preferred frequenting the park alone in the average age of 34 and less, while 35% preferred group activities in an age bracket between 35 and 40; at last, only 8% of survey takers were in couples with an average age of 34. While looking at gender in correlation to sociability of the park, 62% of female surveyors were walking alone; and felt "neutral or positive" about it vs. 38% of the opposite gender. Another visible aspect of users' behavioral analysis is by referring to their yearly income, 26% declared having an annual income between $60 k and $90 k (with a majority of 29% of male), while 23% preferred rather not declare their income (with a majority of 24% of female); that however, falls in the median-high percentile wages in Boston MSA.

Lastly, the 70 users were divided into subcategories by latent class analysis, while 'daily' or 'weekly' were the highest percentile of female gender frequency with 38%, 'monthly' was the remarkable sub category of opposite sex with just 14% of sample analysis and 95% confidence interval. On the same scale, the female gender scored 48% in staying 30 min or more up to 1 h in the plaza, that however, coincidences with the female superlative value in the sociability of the public place as confirmed with the visual observation analysis.

The third phase of the case study analysis was the interviews with in-field experts; interestingly, the gained insights from the meetings with Boston Development Planning Agency officials, the Emerald Network, and the Rose Fitzgerald Kennedy Greenway Conservancy were all concerned about the success of the Northend Park due to its'

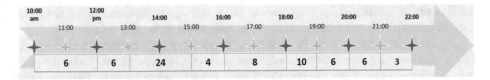

Fig. 4. Timeline of the Surveys conducted during the month of July 2017.

strategic location in the city Downtown area. 6 interviews were conducted with 4 different stakeholders entities and 2 urban planners, most of interviews stressed on conserving the livability of downtown area; whereas the Northend park case study is; in addition to work on equitable cultural programming, inclusiveness and resilience in planning processes, wayfinding, and safety issues.

Furthermore, from the personal interviews it was qualitatively deducted a noticeable governmental approach to conserve the age/gender/social class diversity and encourage people mixing along the RFK greenway. Meanwhile, on the top of the agenda comes safety, security, and accessibility from all colors to the park and to surrounding economic activities. Mostly, those were the planned goals of the blue print document **"Imagine Boston 2030"** - released in 2016 - at its' core as a guiding policy; specifically investing in open spaces, arts and cultural programming together with transportation and infrastructure [28]. In fact, one of the measurable tools to identify the success of the case study of the Northend Park was the Livability performance[5] whereas the transit corridor -where it falls- showed a high ranking in walkability, biking opportunity, accessibility to economic opportunities, social and governmental services, vibrant community and cultural recreation opportunities, healthy and safe neighborhood while scored low in mixed income housing opportunity and that is expected in the Northend case [29, 30].

5   Conclusion

While the Northend park and plaza were in the heart of the regeneration project of a city-wide scale highway; the forefront of the public place had a historical root of Community engagement and a deeper cultural-based territorial neighborhood relationship. That said, founding on the notion of the Northend park as a city hinge and the historical freedom trail path, the physical design and the imagery of the park was a challenging task. The city ballooned the economic debt of the big dig giving birth to an ever-evolving economic urban regeneration project of the Greenway; being, then, on

---

[5] The livability Performance is an index of a calculator designed to numerically describe the characteristics of an individual transit corridor to identify urban policies to implement based on 12 digits FIPS code and Census Block Groups IDs. The results are integrated in a helix of livability indicators such as 1. quality transit, walking, biking 2. Mixed income housing, 3. Accessible economic opportunities, 4. Social and governmental service, 5. Vibrant community, cultural and recreational opportunities, 6. Healthy and safety neighbourhoods.

the last end of a series of open spaces and parks and due to its location, the Northend park is considered one of the greatest parks and plazas in the city of Boston.

By analyzing the Cultural Programming of public events occurrence in the park, Users' behavioral patterns through intercept surveys, intensive physical visual observation analysis, and in-depth interviews with urban anthropologists, Governmental bodies and related urban development authorities, as well as the usage of the livability Calculator, all showed excellent performance keys in the sociability of the park.

Behavioral patterns tracing whether in weekends or weekdays as well as the Stationary activities measurement through imagery or videos recordings unfolded the daily life attractiveness factor of the Northend Park. There is an average of 250 pedestrians crossing an ordinary day of the week and double the count for any weekend. Whatever the time or the day, there will always be someone in the park standing, sitting, walking across, waiting for a friend, riding their bike, and/or children playing in the fountains along a summer breezy day. It is undeniable as well to highlight the activity of the Neighboring Haymarket Public Market held in weekends, that augments the park sociability and brings more users in some peak hours.

Surveys revealed nonetheless the matching sociability patterns in the park; vicinity to two metro stations, a residential neighborhood, and a cluster of restaurants such as the case of the Northend brought a diversity of groups and individuals to the park constantly in remarkable commute times and peak lunch/dinner-times. In addition, Key informants' interviews unfolded the governmental and statuary approaches in the planning development of the case study. However, the microscale of the Northend park is not on the priorities of placemaking in the **"Imagine Boston 2030"** plan as the guideline of development; yet, the inclusiveness, community engagement and safety issues are on the top notch for a variety of stakeholders involved. Whether Boston Planning and development Agency, the Rose Fitzgerald Kennedy Greenway Conservancy, the Livable streets Alliance, the Vision Zero Coalition, etc., there seems to be a certain linkage of collaboration on the wider scale on the networks of parks and open spaces dept. and the transportation dept. in that sense.

In Sum, the Northend Park is a set stage to a successful social public place case study that brought vibrancy and a diversity of users to the downtown core of the City of Boston; the cultural aspects of the regeneration project are distinguishable, and the human attraction plays a turmoil role in shaping the overall park image.

Acknowledgment. The First author would like to thank **Michael Nichols**, Chief of Staff of the Greenway Conservancy as July 2017; **Nidhi Gulati**, the Program Manager for the Emerald Network in Boston; **Katerina Zimmerman**, the urban anthropologist for her insights on how this case study would be developed in April 2017; **John Tad Read**, Senior Deputy Director for Transportation & Infrastructure Planning - Boston Planning & Development Agency (BPDA), **Laura Jasinski,** Associate Director of the Boston trustees of reservations and **Natalia Urtubey,** Director of Engagement for the Imagine Boston 2030 plan and Co-Urbanize platform for the project (See https://courbanize.com/projects/imagineboston/comaps/3?loc=16%2C42.36238887921827%2C-71.05332612991334).

"This work is part of the MAPS-LED research project, which has received funding from the European Union's Horizon 2020 research and innovation programme under the Marie Skłodowska-Curie grant agreement No 645651".

References

1. Kennedy, R.F.: Greenway Conservancy Visiting the greenway. http://www.rosekennedygreenway.org/visit/visiting/. Accessed 08 Apr 2017
2. Boston Redevelopment Authority, Boston 2000: A Plan for the Central Artery, United States of America. https://ia800501.us.archive.org/21/items/boston2000planfo00bost/boston2000 planfo00bost.pdf. Accessed 08 Apr 2017
3. Federal Highway Administration, The Central Artery Environmental Oversight Committee. https://www.environment.fhwa.dot.gov/strmlng/artery/artery_4.asp. Accessed 01 May 2016
4. Turner, R.L.: A spectacular winter garden is the centerpiece of the Massachusetts Artery. But can the society get the job done?, Boston Globe, 1–4 , 30 March2003. http://archive.boston.com/beyond_bigdig/news/artery_033003_magazine_1.htm. Accessed 08 Apr 2017
5. Goldhagen, S.W.: Park Here, New Republic. https://newrepublic.com/article/76951/city-parks-urban-planning. Accessed 08 Apr 2017
6. MTA, North end parks openings, July (2008a). http://web.archive.org/web/20080724170121/, http://www.masspike.com/bigdig/parks/nendpkopen.html. Accessed 08 Apr 2017
7. McCown, J.: Boston Reconnecting, Architecture week, 2–3 (2001). http://www.architectureweek.com/2001/0926/building_2-1.html. Accessed 08 Apr 2017
8. EDRG: Economic Impact of the Massachusetts Turnpike Authority and Central Artery/Third Harbor Tunnel Project, Volume II', Boston: Economic Development Research Group Inc., 24, (2006). http://www.edrgroup.com/pdf/mta-economic-v2.pdf. Accessed 08 Apr 2017
9. MTA, the North End Parks, may (2008b). http://web.archive.org/web/20080514151741/ http://www.masspike.com/bigdig/parks/nendpark.html. Accessed 23 Apr 2017
10. Liveable streets, Emerald Network, In-Progress (2017). http://www.livablestreets.info/emeraldnetwork. Accessed 10 Apr 2017
11. Analyze Boston, Open Space layer BOSTONMAPS: OPEN DATA', Boston: Boston GIS, April (2017). http://gis.cityofboston.gov/arcgis/rest/services/EnvironmentEnergy/OpenData/MapServer/7. Accessed 08 Apr 2017
12. City of Boston Community, Open Space & Recreation Mission: the Neighborhoods: Central Boston (2006). https://www.cityofboston.gov/parks/pdfs/os3c.pdf. Accessed 08 Apr 2017
13. Crosby Schlessinger Smallridge, North End Parks (2016). http://www.cssboston.com/portfolio/nep/1/2. Accessed 25 Apr 2017
14. Carmona, M., et al.: The social dimension. In: Public Places, Urban Spaces: The Dimensions of Urban Design, 2nd edn.,pp. 133–166, Routledge, Abingdon (2010)
15. Mackenzie, A.: Placemaking and Place-Led Development: A New Paradigm for Cities of the Future, PPS (2015). https://www.pps.org/reference/placemaking-and-place-led-development-a-new-paradigm-for-cities-of-the-future/. Accessed 20 Oct 2017
16. Gustafson Guthrie Nichol, North End Parks. http://www.ggnltd.com/north-end-parks. Accessed 25 Apr 2017
17. Reidy, C.: North Enders determined to shape their future, The Boston Globe, (15), 27, June 2004. http://archive.boston.com/business/articles/2004/06/15/north_enders_determined_to_shape_their_future/. Accessed 08 Apr 2017
18. RFKGC, North End Parks Community Meeting: 2015 Improvements, Boston. http://www.rosekennedygreenway.org/files/9814/2533/7168/RFKGC_North_End_Community_Meeting_Presentation_2-25-2015.pdf. Accessed 08 Apr 2017
19. Conti, M.: Pilates on the North End Greenway Attracts Hundreds, NorthEnd Waterfront. Boston, (2012). http://northendwaterfront.com/2012/09/hundreds-turn-out-for-pilates-on-the-north-end-greenway/. Accessed 01 May 2017

20. Berklee, Berklee returns to The Greenway for another summer series of Concerts, Music & Movies on The Greenway. https://www.berklee.edu/events/summer/greenway-series. Accessed 04 Jul 2017
21. RFKGC Special Events, GSA Today (2017)
22. BRA, Rose Kennedy Greenway: Creating Long - Term Value (2010)
23. RFKGC, Rose Fitzgerald Kennedy Greenway Conservancy Inc. NON-PROFIT Report, Boston (2016)
24. Nassar, U.A.E.: Landscape as a Tool to Enhance Behavioural Response and Activities in Historic Urban Parks: An Evaluative Methodology - Alazhar Park, Suez Canal University (2010)
25. City of Boston, vision zero safety issues. http://app01.cityofboston.gov/vzsafety/. Accessed 19 Apr 2017
26. PPS, What Makes a Successful Place? http://www.pps.org/reference/grplacefeat/. Accessed 08 Apr 2017
27. Wortham-Galvin, B.D.: An anthropology of urbanism: how people make places (and what designers and planners might learn from it). Footprint **7**(2), 21–40 (2013). http://footprint.tudelft.nl/index.php/footprint/issue/view/13. Accessed 08 Apr 2017
28. City of Boston, Imagine Boston 2030: A Plan for the Future of Boston, Boston redevelopment Authority. http://imagine.boston.gov/%5Cnpapers3://publication/uuid/A8A094EF-DDA7-4F57-B273-0714CD2B6D58. Accessed 08 Apr 2017
29. Appleyard, B.S. et al.: Livability Calculator for the TCRP H-45 Handbook, Building Livable Transit Corridors: Methods, Metrics, and Strategies, Washington, DC (2016)
30. Oliver, MASSGIS online mapping tool. http://maps.massgis.state.ma.us/map_ol/oliver.php. Accessed 29 Mar 2017

Villa San Giovanni Transport Hub:
A Public-Private Partnership Opportunity

Angela Viglianisi$^{(\boxtimes)}$, Alessandro Rugolo, Jusy Calabrò,
and Lucia Della Spina[iD]

Mediterranea University of Reggio Calabria, 89100 Reggio Calabria, Italy
angela.viglianisi@unirc.it

Abstract. Transport is the foundation of sustainable economic growth and
development. When well balanced, transport greatly facilitates economic and
social development, as it allows the free movement of people, goods and services. This paper estimates the potential opportunities for a Public-Private
Partnership to deliver efficient transport infrastructure and services for a new
transport hub in the city of Villa San Giovanni, which lies in an excellent
geographic position and possesses several other strengths. The area is very
attractive from an economic point of view because of its privileged position on
one of the most scenic natural areas of Italy: the Strait of Messina. Planning
would need to be strategic, and Public-private Partnerships are a special feature
of governance. Urban transformation projects are very complex and have to be
examined from several points of view (socio-cultural, environmental, infrastructural, administrative, and economic-financial) to determine their sustainability, especially in Villa San Giovanni. This paper illustrates a synthetic
evaluation method to assess the renewal potential of the railway station in Villa
San Giovanni and the spatial and urban regeneration opportunities that would
follow. In addition, it aims at building a conceptual framework to support
decision makers in selecting the best Public-Private Partnership model for the
development of a transport hub. This paper presents a working model of participatory management, based on PPP, which serves the purpose of supporting
Public Administrations in managing the urban regeneration processes more
efficiently.

Keywords: Metropolitan cities · Public-Private Partnership
Policy support strategies · Economic sustainability · Financial feasibility

This is the result of the joint work of the authors. Although scientific responsibility is equally
attributable, the abstract and the paragraphs nn. 1, 5 and 6 were written by A. Viglianisi;
paragraph nn. 2, was written by A. Rugolo; paragraph nn. 3, was written by J. Calabrò; paragraph
nn. 4, was written by L. della Spina.

© Springer International Publishing AG, part of Springer Nature 2019
F. Calabrò et al. (Eds.): ISHT 2018, SIST 101, pp. 211–221, 2019.
https://doi.org/10.1007/978-3-319-92102-0_23

1 Introduction

The lack of financial resources in the field of transport infrastructure encouraged the search for alternative financing models, which resulted in a number of relatively recent scientific articles and studies. At present, Italy needs far more financing for transportation infrastructure than what the Government can provide.

Meanwhile, the capacity of the Italian government to provide public services on its own in an effective and efficient way is being questioned and reassessed on various levels. Accordingly, the involvement of private investors in the development of Metropolitan Transportation Systems has been promoted by the Italian public sector by adopting the Public Private Partnership (PPP) model. During the last few years, the processes of urban regeneration in Italy have been stalled due to the economic and political situation of the country. Over the past few decades PPP has become a new way for delivering and financing public sector projects. The lack of public resources that could enable the start of the above-mentioned processes, as well as a complex and incomplete legislative and regulatory framework are the main causes of the crisis affecting the construction industry [1]. As for urban sustainability, mobility and transportation systems as well assume a leading role in guaranteeing high levels of efficiency and livability of urban and metropolitan systems. In many countries, limitations upon the public funds available for infrastructure led governments to invite private sector entities to enter into long-term contractual agreements for the financing, construction and/or operation of capital intensive projects [2]. For this reason, this paper discusses the role of evaluation in supporting the decision-making process in urban planning and urban regeneration processes [3]. Territorial transformations constitute complex decision-making problems and underline the growing need for appropriate tools for evaluating programmes and intervention projects [4]. This paper develops a theoretical framework for the thematic of public-private economic negotiation in the realization of urban regeneration programmes [5], especially for the urban infrastructure.

This paper aims to investigate the role of evaluation in negotiation procedures, and to provide a methodological contribution which can be used to start activities for the proper assessment of Public-Private Partnerships. The recent global economic crisis is favouring the implementation of transport infrastructure projects such as highways, light rails, bridges, seaports and airports through Public-Private Partnerships [6].

To this end, any PPP project decision should be based on a realistic, systematic and comprehensive analysis of benefits, risks and costs, making use of:

- Broader parameters, such as social impact on the general public;
- Qualitative project factors, such as impact on environment and safety;
- Qualitative aspects of bids, such as market response.

In this article, the evaluation methods of the case study of the Villa San Giovanni transport hub will be closely examined from a theoretical point of view. Villa San Giovanni can become a significant component of the metropolitan area of the Strait of Messina, which includes the conurbation of Reggio Calabria and Messina. Villa San Giovanni represents a link between the two cities, and could, as a transport hub, enhance the transport dynamics of the territory. More specifically, the evaluation of the different investment strategies aimed at the hub allows to determine which one would work best for the transport system of Reggio Calabria. Urban transport is a critical component of urban infrastructure and the lifeline of a metropolitan city. For this reason, the Public-Private Partnership [7] could represent an alternative institutional model for implementing private finance initiative in the provision of transport infrastructure. Until the end of the previous century, as the public sector owned and was responsible for the transport infrastructure in almost all countries, it also was, with a few exceptions, directly engaged in the financing and administration of transport infrastructure construction, maintenance and operations. Nevertheless, the 2000s were marked by the significant development of Public Private Partnerships, and from then onwards, PPPs have gained worldwide acceptance. These infrastructures are needed to keep up with living standards and to create the conditions for sustainable development.

2 Public-Private Partnerships in Urban Regeneration Programmes

The European Commission defines PPP as the "transfer of investment projects to the private sector, traditionally executed and financed by the public sector" [8].

Public-Private Partnerships have played a significant role in boosting the undergoing processes of national economic growth, and in developing social infrastructures such as roads, facilities, buildings and public services such as health, utilities, etc. PPPs come into play when the country is going through a period of economic crisis, especially when there is a significant gap between available public funding and expenses, the more so if national debt keeps rising and the country is crippled by budget deficits. In the present context of global financial turmoil, PPPs act as an economic stimulant in developing countries and foster sustainable growth on a global level, though the difficult economic environment has substantially affected the international finance market [9, 10]. In general, the term refers to forms of cooperation between public authorities and the world of business which aim to ensure the funding, construction, renovation, management or maintenance of an infrastructure or the provision of a service [8].

Public-Private Partnerships can be an effective way to build and implement new infrastructures or to renovate, operate, maintain or manage existing transport infrastructure facilities [11].

2.1 Types and Forms of PPP

The "Green Paper on public-private partnerships and Community law on public contracts and concessions" [8] presented by the Commission of the European Communities on 30 April 2004, makes a distinction between:

- PPPs of a purely contractual nature, in which the partnership between the public and private sector is based solely on contractual links;
- PPPs of an institutional nature, involving cooperation between the public and private sector within a distinct entity.

There are different types of PPPs, which can be defined based on the number of stakeholders involved, their contractual agreements, the allocation of risks and responsibilities between public and private partners, and specific project context and content [11]. Stanghellini [12] identified six main types of private engagement:

- Project Financing;
- Concession of building and management;
- Companies for the management and improvement of their real estate;
- URC - Urban Regeneration Companies (STU - *Società di Trasformazione Urbana*);
- Public and private Real estate funds;
- Urban planning framework agreements between public and private partners.

3   The Thinking Behind the Development of the Intermodal Terminal

The development of intermodal terminals is a complex process that includes many actors and functions. The type of actors involved (Public-Private) greatly affects the initiative and logic behind intermodal development. Public actors tend to focus on social and economic benefits, while private developers may see the terminal development as an investment opportunity in itself and aim at selling the site or part of it for profit.

These experiences led to a considerable debate in literature about the effectiveness of new transport hubs and whether investing in them would be worthwhile.

Therefore, it is extremely important to understand the factors that make these systems successful. Cities have begun to function as communities of knowledge, innovation, creativity and learning, thus becoming more dynamic, complex, diverse, open and intangible. In this context, new urban strategies and policies have to be considered.

This paper will present an attempt at clarifying the concept of "Transport Hub" as an outcome of creative urban regeneration processes. There is a strong belief that such systems have indirect socio-economic and urban regeneration benefits, in addition to direct transport and mobility improvements. Besides direct transport and environmental benefits generated by new urban public transport systems, there are other advantages related to urban regeneration, such as city aesthetics, employment creation, social and economic development and cohesion [13].

Thereby, urban regeneration and renewal have become the most common policy for managing spatial changes in a sustainable way [14]. Railway stations function as nodes in transport networks and as places in an urban environment. They impact on accessibility and environment, thus contributing to property value. The literature on the effects of railway stations on property value is mixed in its findings in respect to the impact magnitude and direction, ranging from a negative, or insignificant, to a positive impact [15]. The European Union's White paper of 2001 [16] places emphasis on encouraging intermodal transportation as the solution to the problems seen in the transportation business. Railway stations have assumed a strategic role in urban renewal programmes aimed at providing efficient and liveable urban and metropolitan systems, but, in order to do so, they have to be designed according to an adequate project of urban regeneration [17]. Therefore, the intermodal concept is being applied to railway stations to reflect a new form of service. The essence of efficient intermodal transport lies in the use of transfer between road, rail, and other transport modes [18]. These types of Public and private partnership can be defined as a decision-making process aimed at achieving economic, social, cultural and environmental goals through the development of spatial visions, strategies and programmes and the application of a set of policy principles, tools, institutional and participatory mechanisms and regulatory procedures [19].

The traditional methods of benefit measurement have their roots in economic theory, and thus offer an incomplete understanding of how local communities perceive the value of public transport. This is why the development of a new methodology is needed to measure the socioeconomic effects and impact produced by public transport investments in urban areas. The development of intermodal transport requires the consideration of three of its attributes: transport links, transport nodes, and the provision of efficient services. In this case, urban transport has a significant role but is just part of a wider integrated policy that sees the combination of transport and urban strategies as its core. This proves once again that planning must avail itself of different tools that have to be in constant interaction. In this case study, only Public-Private Partnerships are an innovative procurement model that appeared as an alternative to traditional methods such as public work contracts.

4 Methodology: Realist Evaluation

This study used a realist evaluation methodology informed by a realist synthesis. It is based on qualitative and quantitative research methods for data collection. The purpose of this paper is to develop an evaluation model that can support future Public Administration decisions that could influence urban regeneration processes that need to be implemented with the involvement of private investors.

The introduction of such systems is a costly investment, which also needs a re-organization of the urban network and a harmonized land development strategy in order to maximize the utility of the project. In particular, the transport infrastructure has played only a limited role in the evolution of the urban policy of the Metropolitan City of Reggio Calabria. A number of methodological issues raised by the case study are discussed. The object of this paper is the "evaluation of public convenience" and PPPs represent an alternative for the financing and development of this infrastructure projects [5]. The purpose of this paper is to facilitate a more efficient allocation of resources by demonstrating that developers would find more beneficial to carry out a particular intervention, rather than the possible alternatives. PPPs are the only viable mechanism to fund the transport hub in Villa San Giovanni. The main goal of this analysis is to increase the interest of developers in carrying out this project [5].

The Value Cost (Vc) can be defined as the sum of the Factors Cost on Production Process (Pf), more generally:

$$Vc = \Sigma\, Pf \qquad\qquad (1)$$

The estimation of the Value Cost can be developed by:

- Synthetic procedures;
- Analytical procedures.

The synthetic procedure was preferred because of the quality and quantity of information on the project that we have. In the application of the synthetic method set out in this paper, the purpose is to provide a summary estimate of the project.

The Production Cost is estimated by using a synthetic procedure. It provides an estimate of the likely Production Cost. The following average unit costs were calculated to appraise the cost estimates of the most significant investment components, which were found to be well within the cost range of other comparable projects.

The procedure for developing a Parametric Cost [20] estimation was performed using different Units value of building product (for example: euro/m^3, euro/m^2 and euro/unit). Rough cost estimates are generally understood as being based on unit prices obtained from regional market surveys, quotations from different suppliers or from similar projects in the same regional context. It should be made sure, however, that cost estimates are all-inclusive. The Production Cost estimates have also been confirmed by examining comparable projects in order to cross-check with current market conditions.

This has been done by consulting:

- Operators in the construction companies and/or professionals of the sector (direct sources);
- selected sector literature (indirect sources).

Methodology can be summarised as follows:

1. Analysis of metropolitan areas;
2. Analysis of the metropolitan Strait area;
3. Analysis of context and issues in the project/case study area;
4. Definition of project interventions;
5. Analysis of technical and economic feasibility;
6. Definition of a system of conveniences among the various stakeholders involved.

Special focus has been given to:

1. The salient features of the intervention (with reference to the feasibility study);
2. The current infrastructure context of the area where the project will be carried out;
3. The structure of transport demand in the catchment area, its likely development, the evaluation of the amount of people who will find convenient to use the new railway station;
4. The determination of standard parameters;

5. The analysis of each building type, achieved by breaking up the project into sectors, homogeneous functional areas, and space units, according to a functional logic consistent with the goals of the programming phase and estimate expenditure.

Furthermore, key factors for an urban redevelopment project are also identified from analyses that are considered crucial for the success of the project.

Therefore, the Parametric Cost [20] is the operational proposal and the reference economic data, obtained from technical-economic basic information that refer to previous interventions identified according to the following criteria:

– Recognizability and traceability of Production costs, achieved by indicating the geographic, technical, and time reference contexts;
– Representativeness of the sample according to its typological function and its basic technical and construction characteristics;
– Descriptive value of the parameter in relation to the intrinsic characteristics of the project.

The evaluation of Parametric Costs has been carried out by using the relevant descriptive parameters (surface, volume and other parameters that identify the scope of works such as tracks and switches removal, etc.) and by subdividing the contracted amount for the various phases of the planned project. The Parametric Costs [20] for each sector and type of project refer to completed or in progress building works.

Though project interventions are chosen based on their representativeness for the fields and types considered, economic data are affected by the specificity of the context. When it was not possible to find data on tendered and contracted amounts, the parametric costs were estimated according to the available economic data that provided information on their type. It is clear that a parametric cost calculated for a preliminary estimate of a production cost must refer to the feasibility assessment of the project interventions, characterized by variable elements.

5 The Experience of Public-Private Partnerships in the Villa San Giovanni Transport Hub. Highlights of the Project Intervention

As cities grow and traffic congestion increases, governments are looking for alternatives to encourage a shift from private to public transport. The project area/case study for this research is the Villa San Giovanni railway station, which presents a high development potential, and, if properly approached, will increase its importance in the Strait Area. One of the main reasons behind this choice is that, according to the preliminary studies, Villa San Giovanni can be considered as an "urban infrastructure" in the metropolitan area of Reggio Calabria-Messina, where the goal is to create a wide but integrated urban network on a metropolitan scale. The aim is to add a powerful "missing link" in the core of the Area of the Strait, an infrastructure that will strategically tie together the cities of Messina and Reggio Calabria. This section will present the main results obtained with the application of the assessment methodology described previously. Villa San Giovanni already presents some sort of intermodality, which, however, needs to be reorganized. The same area hosts a functioning railway-ferry station, several platforms used by private bus companies, and quays used by a private company that guarantees sea transport links to Sicily 24/7. The project includes the modernisation of infrastructure and existing terminals and the construction of other transport connections that will ensure the increase of the transport capacity.

The metropolitan area of the Strait needs efficient and competitive sea transport links.

The purpose of the project is:

- To upgrade obsolete structures;
- To rationalize roads in order to improve traffic flow entering and leaving Villa San Giovanni;
- To create an area that can accommodate the traffic flow of the metropolitan area of the Strait;
- To create a network of services for the exclusive use of the metropolitan area of the Strait;
- To provide the currently lacking passenger facilities and amenities;
- To restore the area of Cannitello, a suburb of Villa San Giovanni;
- To transform the quays, currently used by a private company, into a marina.

The new railway station will be built in the former "Piazzale Anas" and it will extend on several floors, thus creating a sheltered and adequate connection with platforms and walkways that lead to the piers, where ferries from the two currently operating companies will dock. Inside, the station will host several facilities, included those for passengers, such as the ticket office, a waiting room, restrooms, luggage storage rooms, tourist and State Railways information offices, business units and a restaurant). Apart from the reconstruction of the main railway station itself, the project includes different measures for the improvement of public transport:

- The creation of an integrated bus and rail station;
- Renovation of the main railway station and optimisation of the interchange between regional and urban public transport.

Estimates include all costs incurred for this project (Table 1):

The project includes the renovation of the main railway station building and its surrounding area with combined land development and infrastructure investments.

After a thorough research on the current parametric prices, the approximate calculation of demolition, building and adaptation costs of the new railway station was carried out. The total Production Cost of the project is estimated at €13,051,062.00. In addition, a surface of 696 m^3 destined to commercial units has been estimated and is intended to be sold. In order to evaluate the parametric value of the commercial units, a study on the current value attributed by real estate to buildings with the same characteristics was carried out. Accordingly, the estimated selling price of the commercial units was estimated at 1,228.00 €/m^3. Therefore, the sale of such units can result, in terms for revenues, in a private investment of 854,688.00 €. Compared to the amount of the investment, the revenue is quite small. This data will be of great importance for the further development of the project.

Table 1. Estimation of the Production Cost (CP) of the new Villa San Giovanni transport hub.

Project intervention dimension			Parametric value		Tot project intervention
Track removal	301	ml	50.00	€/ml	15,050.00 €
Switch removal	4	each	2,600.00	€/each	10,400.00 €
New road	2,368	m^2	64.00	€/m^2	151,552.00 €
Quays upgrade	130	m^2	1,249.00	€/m^2	162,370.00 €
Renovation of passenger facilities	576	m^2	2,000.00	€/m^2	1,152,000.00 €
Underpass upgrade	176	m^2	2,000.00	€/m^2	352,000.00 €
Platform building	3,747	m^2	2,000.00	€/m^2	7,494,000.00 €
Pedestrian underpass connecting buildings B and C	580	m^2	2,000.00	€/m^2	1,160,000.00 €
New railway station for the metropolitan area	5,613	m^3	454.96	€/m^3	2,553,690.00 €
Cost of Production of transport hub	13,051,062.00 €				

6 Conclusions

Public Private Partnerships have been developed and introduced as a new way of project delivery that sees the collaboration between public sector and private sector [7]. Referring to similar works accomplished in the past is the fastest method to obtain a preliminary estimate of the Production costs. At the same time, the application of simple, but logical processes, together with detailed project information, can generate cost estimating tools that are accurate, reliable, up-to-date and adaptable to the needs of public and private users and to the type of building in question [12]. This paper presents an experimental model and any further improvement of the public transport network accessing the transport node could contribute to the maximum efficiency of the project.

In this paper, the term Public-Private Partnership is classified as an institutionalised form of cooperation between Public Administration and one or more private partners in a project with common interests via a distribution of decision rights, costs and risks. Public-Private Partnerships are the most practical approach for the development of the hub [13]. Therefore, the project cost estimates used in the Production Process Evaluation are, in this case, deemed to adequately reflect opportunity costs for developers.

The results obtained indicate that the method applied could make it easier for Public Administrations and private companies involved in the financing of the project to estimate Production costs. Thus, the focus on investment in public transportation is strategically correct: what is required now is that the ability of infrastructure investment to regenerate urban development needs to be recognised and acted upon.

Fortunately, the opportunity is still there: the proposed railway station illustrated above has the scope to be re-developed in a manner that could act as a growth catalyst for urban regeneration [14]. The proposed methodology is a useful and easy-to-use assessment that could provide a synthetic evaluation of the health of policies and actions intended to achieve a better configuration of a developing urban context.

The evaluation model presented in this paper can help both public and private sector to assess whether a potential public project is suited for PPP. In conclusion, the financial feasibility of the project should firstly facilitate the participation of private partners, and secondly, the participation of government bodies, which will provide more substantial guarantees from financial institutions in a context of heavy credit restrictions [10, 21].

References

1. Guarini, M.R., Battisti, F.: Evaluation and management of land-development processes based on the public-private. In: Xu, Q., Li, H., Li, Q., Sustainable Development of Industry and Economy, 3rd International Conference on Energy, Environment and Sustainable Development (EESD 2013), pp. 154–161. Shanghai (2014)
2. Darrin, G., Mervyn, K.L.: Evaluating the risks of public private partnerships for infrastructure projects. Int. J. Project Manage. **20**(2), 107–118 (2002)
3. Copiello, S.: Progetti urbani in partenariato. Studi di Fattibilità e Piano economico finanziario. Alinea, Firenze (2011)
4. Della Spina, L.: Integrated evaluation and multi-methodological approaches for the enhancement of the cultural landscape. In: Gervasi O. et al. (eds.) Computational Science And its Applications, ICCSA 2017, vol. 10404, pp. 478–493. Springer, Cham (2017). https://doi.org/10.1007/978-3-319-62392-4_35
5. Calabrò, F., Della Spina, L.: The public-private partnerships in buildings regeneration: a model appraisal of the benefits and for land value capture. In: Advanced Materials Research, vol. 931, pp. 555–559. Trans Tech Publications, Switzerland (2014). https://doi.org/10.4028/www.scientific.net/AMR.869-870.43
6. Fusco Girard, L., Nijkamp, P.: Le valutazioni per lo sviluppo sostenibile della città e del territorio. Angeli, Milano (1997)
7. Delmon, J.: Public-Private Partnership Projects in Infrastructure. An Essential Guide for Policy Makers. Cambridge University Press, Cambridge (2011)
8. European Commission: Green Paper on Public-Private Partnerships and Community Law on Public Contracts and Concessions, Brussels (2004)
9. Stanghellini, S., Mambelli, T.: La valutazione dei programmi di riqualificazione urbana proposti dai soggetti privati. Scienze Regionali **1**, 77–103 (2003)
10. Stanghellini, S.: Il negoziato pubblico privato nei progetti urbani. Principi, metodi e tecniche di valutazione. Dei, Roma (2012)
11. Campolo, D., Rugolo, A., Scrivo, R.: Lo Spin-Off Universitario "Urban Lab S.R.L." strumenti e metodi per le trasformazioni urbane, LaborEst **9**, 81–85 (2014)
12. Dalla Longa, R.: Criticità e progettualità del PPP in Partnership Pubblico-Privato. Dedalo **30**, 22–25 (2012)
13. Bartik, T.J.: Evaluating the impacts of local economic development policies on local economic outcomes: what has been done and what is doable? In: Organisation for Economic Co-operation and Development (Eds.) Evaluating Local Economic and Employment Development: How to Assess What Works Among Programmes and Policies, c. 5. OECD, Paris (2004)

14. Nesticò, A., De Mare, G.: Government tools for urban regeneration: the cities plan in Italy. a critical analysis of the results and the proposed alternative. In: International Conference on Computational Science and Its Applications, pp. 547–562. Springer, Heidelberg (2014)
15. Cervero, R.: Transit-Oriented Development in the United States: Experiences, Challenges and Prospects, TCRP Report 102 (2004)
16. European Commission: White Paper European transport policy for 2010: time to decide, Brussels (2001)
17. Bertolini, L.: Sustainable urban mobility, an evolutionary approach. Eur. Spat. Res. Policy **12**(1), 109–125 (2005)
18. Stellin, G., Stanghellini, S.: Politiche di riqualificazione delle aree metropolitane: domande di valutazione e contributo delle discipline economico-estimative. Genio rurale-estimo e territorio **7**(8), 47–55 (1997)
19. Calabrò, F.: Local Communities and Management of Cultural Heritage of the Inner Areas. An Application of Break-Even Analysis. In: Gervasi O. et al. (eds) Computational Science and Its Applications, ICCSA 2017, vol. 10406, pp. 516–531. Springer, Cham (2017). https://doi.org/10.1007/978-3-319-62398-6_37
20. Bassi, A.: Costi parametrici per tipologie edilizi. Maggioli S.p.A, Milano (2007)
21. Calabrò, F., Cassalia, G.: Territorial cohesion: Evaluating the Urban-Rural Linkage through the Lens of Public Investments. In: Bisello, A., Vettorato, D., Laconte, P., Costa, S., (eds.): Smart and Sustainable Planning for Cities and Regions. Results of SSPCR 2017. Green Energy and Technology, Springer, Cham (2018). https://doi.org/10.1007/978-3-319-75774-2_39

Road Degradation Survey Through Images by Drone

Giovanni Leonardi$^{(\boxtimes)}$, Vincenzo Barrile, Rocco Palamara,
Federica Suraci, and Gabriele Candela

Mediterranea University of Reggio Calabria, 89100 Reggio Calabria, Italy
giovanni.leonardi@unirc.it

Abstract. Early detection and measurement of the distresses are necessary to keep the pavement functions at an acceptable level and to pledge the safety users. The use of digital photography in order to record pavement images and, later, to identify the surface distresses has undergone continuous improvements during recent years. Image measurement methods are effortless, safe, and can be performed in a short time. In this paper, an image processing measurement is presented to locate potholes and cracks on the pavement surface through pictures by drone.

Keywords: Road degradation · Drone · Pavement management system

1 Introduction

The principal function of a road is to provide a surface with shape and frictional characteristics that permit vehicles to be controllable and, along with their contents, to remain undamaged while travelling at the desired speed. Pavements system are constructed to ensure that roads will fulfill this function reliably and durably. Deterioration and failure of roadway may occur because of overuse, and mismanagement and therefore maintained in good condition requires substantial expenditure furthermore the need for maintenance increases as road infrastructure ages, since it becomes more fragile, less resilient and journeys are more susceptible to disruption. Pavement management systems are widely used to maintain safe, durable and economic road networks. A pavement management system consists of a coordinated set of activities, all directed toward achieving the best value possible for the available capital and must serve different management needs or levels. In fact, the primary benefit of a pavement management system is that it helps users select cost-effective alternatives for pavement maintenance and rehabilitation, however, sometimes due to lack of budget and experience many local agencies do not have a formal pavement management system, but use an informal method for determining which pavements receive a specific maintenance treatment at any time. Pavement distresses information is one of the key elements needed for the pavement management system. In this work it was developed a method of automatically identifying and quantifying distress in pavement road by applying techniques from digital image processing to digital photographs of pavement. The

© Springer International Publishing AG, part of Springer Nature 2019
F. Calabrò et al. (Eds.): ISHT 2018, SIST 101, pp. 222–228, 2019.
https://doi.org/10.1007/978-3-319-92102-0_24

images were collected with an Unmanned Aerial Vehicle and the image enhancement was performed with a specific algorithm.

2 Pavement Road Degradation

The pavement road degradation is caused by environmental factors, traffic loads and materials' properties decline. The degradation types are identified by problems of roughness, friction and lift. The main surface degradations in flexible pavements are manifested by cracks, potholes, rutting and shoving, local deformations or beaten longitudinal tracks. It is possible to distinguish these types of surface cracks: alligator cracking (Fig. 1a), block cracking, longitudinal and transverse cracking (Fig. 1b), reflection cracking.

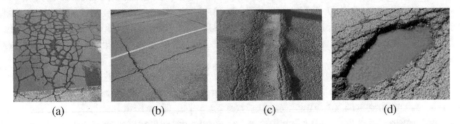

(a)   (b)   (c)   (d)

Fig. 1. (a) Alligator cracking, (b) longitudinal and transverse cracking, (c) longitudinal depression, (d) pothole.

The tensile and shearing stresses at the bottom of the asphalt layer form because of vehicle wheel loads or environmental loads, they increase greatly when there are cracks in the base layer because of the stress concentration at the crack tips. Thus, if there are cracks in the base layer and the asphalt pavement is not sufficiently thick, further cracks will soon form. These cracks, which propagate on the asphalt concrete (AC) layer over the existing cracks, are called reflective cracks. A number of different design strategies have been employed in an effort to slow the development of reflective cracking, including increased overlay thickness [1], joint sawing and sealing [2], reinforcement layers [3], etc. Reflective cracks born at the bottom and progress up to the surface (a bottom-up crack), but this is not the only way. In fact, there is a second mode of crack initiation and propagation: top-down cracking. There are three basic views on the top-down cracking mechanism: high surface horizontal tensile stresses due to truck tires (wide-based tires and high inflation pressures are cited as causing the highest tensile stresses); age hardening of the asphalt binder resulting in high thermal stresses in the AC (most likely a cause of the observed transverse cracks); a low stiffness upper layer caused by high surface temperatures. In thick pavements, the cracks most likely initiate from the top in areas of high localized tensile stresses resulting from tire-pavement interaction and asphalt binder aging. After repeated loading, the longitudinal cracks connect forming many-sided sharp-angled pieces that develop into a pattern resembling the back of an alligator.

Rutting distress is a longitudinal depression (Fig. 1c) - permanent deformation - in the pavements' wheel paths accompanied by bulging of the surface (shoving) at the sides of road generally parallel to the direction of traffic [4]. Increasing axle loading, tire pressure, pavement temperature and loading repetitions are the causes of rutting [5, 6]. Rut decreases the safety on the road because it increases the difficulty of vehicle steering and also, while it deepens, water funneled in them increases the risk of hydroplaning. Therefore, several scientists and highway design agencies consider rutting as a major distress in asphalt pavements. The occurrence of rutting can be associated with problem of asphalt concrete mix-design, instability of subgrade or of other layers, or contribution of all these. In particular, several researchers have determined various factors affecting permanent deformation of a AC mix: volumetric composition, particularly binder content and void content and loading conditions [7]; temperature, bitumen content, viscosity and air voids [8]; number of load cycles [9]; etc. Obviously, these factors have been determined by a number of tests to investigate the potential rutting in a AC layer, for example, uniaxial repeated long creep test, dynamic mechanical analysis test, 2D and 3D imaging, analytical analysis of film thickness and air void, triaxial compressive strength test, uniaxial repeated creep testing, and other more.

It is known that with time and environmental exposition, bitumen degrades chemically and becomes brittle. In addition, moisture may penetrate in the pavement structure, causing loss of adhesion between aggregates and binder [10]. Furthermore, thermal efforts and vehicular loading over aged asphalt pavements are the cause of crack propagation and aggregate losses that lead to formation of potholes (Fig. 1d) [11]. They generally have sharp edges and vertical sides near the top of the hole. Potholes are most likely to occur on roads with thin AC surfaces (25 to 50 mm) and seldom occur on roads with 100 mm or deeper AC surfaces [12]. These are the main cause of reductions in roads service level and the most aggravating pavement distress for traffic safety [13].

3 Maintenance and Pavement Management System (PMS)

Distresses have adverse effect on ride comfort and cause a reduction of safety level. As a matter of fact, that high proportion of traffic accidents is caused by the pavement deterioration. Moreover, it is evident that providing maintenance on road pavements when they are next to the failure point is not a cost-effective pavement management technique. The first step towards maintenance is the identification of pavement distresses and their documentation for further action. Therefore, if the type of surface degradation is identified in advance, an efficient maintenance program could be provide. Assessing the actual pavement condition and forecasting future performance of the asset are the base for an efficient Pavement Management System (PMS). Pavement management is defined like "the process of maintaining the pavement infrastructure cost-effectively" [14], or like "a systematic method for routinely collecting, storing, and retrieving the kind of decision-making information needed to make maximum use of limited maintenance (and construction) dollars" (American Public Works Association - 1993). The PMS is used to plan the best strategies and optimal timing for interventions on road during its service life, depending on updated inventory and database of the actual geometric features, functional and structural conditions of the road. Therefore, the PMS is a program for improving the quality and performance of pavements, minimizing costs through good management practices.

Choosing the optimal timing of preservation efforts can lead to lower life cycle costs. In turn, lower life cycle costs can be one of the outputs of a more sustainable roadway. The fundamental idea is that pavement management will lead to lower overall life cycle costs for a pavement or network of pavements and thus be a more sustainable approach. Thus, there is an indirect relationship between a pavement management system, which can help in determining the best timing of preservation efforts, and sustainability. In general, pavement deterioration, as pictured in Fig. 2, is slow at first and then it increases at an increasing rate. Preservation efforts provide a step increase in pavement condition and essentially reset the deterioration process. Preservation efforts applied too soon do not achieve much improvement in condition for their cost while those applied too late achieve an improvement in condition at substantial cost [15].

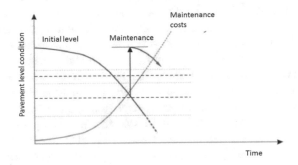

Fig. 2. Pavement deterioration and maintenance.

A basic pavement management system is realized in five principal components: roadway condition surveys, database containing all related roadway structure information, analysis scheme, decision criteria and implementation procedures [16].

Evolution of pavement management system was marked by the domain specific technologies and new materials which appeared in recent years.

4   Use of Digital Imaging to Detection Pavement Distresses

The use of digital photography to record pavement images and subsequent distresses detection and classification has undergone continuous improvements over the past decade. Pavement engineers had long recognized the importance of distress information in quantifying the quality of pavements. Traditionally, pavement condition data are gathered by human inspectors who walk or drive along the road to assess the distresses and subsequently produce report sheets, but it is high cost and time consuming. Worse still, work has to be done along fast-moving traffic. Such condition would endanger the safety of the personnel involved. Finally, large differences will exist between the actual condition and evaluation results because of the subjectivity of the evaluation process. In the wake of tedious manual measurements and safety issues, a variety of types of methods have been devised to identify the distresses on pavements apart from the crude process of manual inspection. Image processing, ultrasonic detection and infrared detection are the common methods to detection distress on pavements. Thanks to advances in electronic sensors and computer systems various systems and methods, named pavement imaging systems, exist or are under development, that assess the pavement surface condition when imaged on video or photography; they are used for the automated pavement surface distress data collection and the processing of the acquired data for the purpose of obtaining information about the roadway pavement surface condition [17]. Principal steps in automated pavement evaluation are: the distress data acquisition and the analysis of distress data acquire. The first step is characterized by image acquisition (the camera, the lens, and the computer hardware). This step includes the recording of the pavement distress data by a moving vehicle that in the last decades can be obtained by the use of a Unmanned Aerial Vehicle (UAV). The distress data analysis represents the software aspect of an automated pavement inspection system, that includes the digital image processing and the image interpretation to identification distress types and their quantification. Digital imaging presents many problems that recent researches is trying to solve, in fact due to rough surface and non-uniform illumination of the pavement image, pavement images have a significant amount of background noises. To resolve this problem have been created methods of image enhancement like: Neighborhood averaging and median filtering, segmentation and threshold, but furthermore there are several limitations concerning the software capabilities of existing pavement imaging systems, especially due to the lack of methodologies for image interpretation. Chan et al. [18] developed a thresholding method based on statistical parameters of the images to detect cracks on pavement images. A set of sample images including linear cracks and alligator cracks were used to test their developed method. Experimental results show that their method could process the pavement images in real time and reduce the negative effects caused a variation of pavement reflection rates. Lin et al. [19] developed an algorithm for pothole detection. In their paper, texture measure based on the histogram is extracted as the features of the image region, and a non-linear support vector machine is built up to identify whether a target region is a pothole. The experimental results showed that the algorithm can achieve a high recognition rate.

5 Application

The aerial survey was performed in an area of Reggio Calabria. Pictures were captured using a commercial drone, UAV DJI Mavic Pro (DJI, Shenzhen, China), and the height flight was set to 25 m for vertical flight. The picture obtained by drone (Fig. 3a) was processed by an algorithm in MATLAB® and it was converted in a black and white image (Fig. 3b).

MATLAB® Image Processing Toolbox™ provides a comprehensive set of reference-standard algorithms for image processing, analysis, visualization and algorithm development. Finding shapes, counting objects, identifying colors, measuring

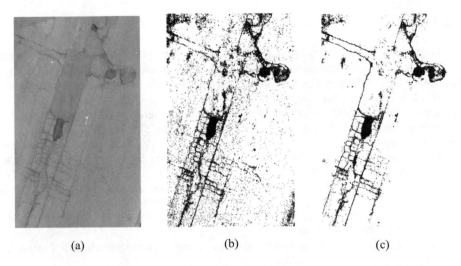

(a)   (b)   (c)

Fig. 3. (a) Picture of road surface degradation captured by drone; (b) black and white image of road surface degradation obtained by MATLAB®; (c) black and white image of potholes and cracks obtained by MATLAB®.

object properties were obtained through the image analysis. This process of extracting meaningful information from images was adopted, in particular image segmentation process was used. Image segmentation is the process of partitioning an image into parts or regions. This division into parts is often based on the characteristics of the pixels in the image. Indeed, different regions in the image were found for abrupt discontinuities in pixel values, which typically indicate edges. These edges defined the regions.

Successively, the black and white image was processed again by MATLAB® in order to obtain a clear image of potholes and cracks (Fig. 3c).

6 Conclusion

The images analysis acquired by innovative and rapid systems such as drones, is a useful method for road infrastructure managers. The goal of these innovative methods is certainly to reduce the time for identifying the surface pavement distresses, but above all to reduce the investigations costs. In this paper, a simple method to individuate distresses such as potholes and cracks from images acquired by drones has been presented. The pictures, obtained by drone from 25 m height, have been processed in Matlab identifying the surface deterioration through different colours. This analysis is sufficient to identify potholes and cracks on the pavement surface without interferences in vehicular traffic during the survey, moreover, this method reduces economic and time costs in order to obtain an accurate output.

References

1. Gulden, W., Brown, D.: Overlays for Plain Jointed Concrete Pavements. Research Project No. 7502. Georgia Department of Transportation, Forest Park, GA (1984)
2. Vyce, J.M.: Reflection Cracking in Bituminous Overlays on Rigid Pavements. Report No. 109. New York State Department of Transportation, New York (1983)
3. Kim, J., Buttlar, W.G.: Analysis of reflective crack control system involving reinforcing grid over base-isolating interlayer mixture. J. Transp. Eng. **128**, 375–384 (2002)
4. Mirzahosseini, M., Aghaeifar, A., Alavi, A., Gandomi, A., Seyednour, R.: Permanent deformation analysis of asphalt mixtures using soft computing techniques. Expert Syst. Appl. **38**, 6081–6100 (2011)
5. Buonsanti, M., Leonardi, G.: A finite element model to evaluate airport flexible pavements response under impact. Appl. Mech. Mater. **138–139**, 257–262 (2012)
6. Buonsanti, M., Leonardi, G., Scopelliti, F.: Modelling micro-damage in granular solids. Key Eng. Mater. **525–526**, 497–500 (2013)
7. Garba, R.: Permanent deformation properties of asphalt concrete mixtures. Norwegian University of Science and Technology (2002)
8. Shenoy, A., Romero, P.: Standardized procedure for analysis of dynamic modulus |E*| data to predict asphalt pavement distresses. Transp. Res. Rec.: J. Transp. Res. Board **1789**, 173–182 (2002)
9. Zheng, Q., Zhao, D., Chen, C., Song, Z., Zheng, S.: Quantitative test technology study on the mesoscopic strength parameters of the mineral aggregate contact surface of bituminous stabilized macadam. Constr. Build. Mater. **94**, 622–631 (2013)
10. Miller, J.S., Bellinger, W.Y.: Distress identification manual for the long-term pavement performance program. Report No. FHWA-RD-03-031, Federal Highway Administration, Washington, D.C., USA (2003)
11. Roberts, F.L., et al.: Hot Mix Asphalt Materials, Mixture Design and Construction, 2nd edn. NAPA Education Foundation, Lanham (1996)
12. Yang, Y., Qian, Z., Song, X.: A pothole patching material for epoxy asphalt pavement on steel bridges: fatigue test and numerical analysis. Constr. Build. Mater. **94**, 299–305 (2015)
13. Archilla, A., Diaz, L.: Effects of asphalt mixture properties on permanent deformation response. Transp. Res. Rec.: J. Transp. Res. Board **2210**, 1–8 (2011)
14. Wolters, A., Zimmerman, K., Schattler, K., Rietgraf, A.: Implementing pavement management systems for local agencies. Illinois Center for Transportation (2011)
15. Stevens, L.B.: Road Surface Management for Local Governments - Resource Notebook. Publication No. DOT - I-85-37. Federal Highway Administration, Washington, D.C. (1985)
16. Peterson, D.E.: National cooperative highway research program synthesis of highway practice 135. In: Pavement Management Practices, NCHRP, TRB, National Research Council, Washington, D.C. (1987)
17. Ouyang, A., Luo, C., Zhou, C.: Surface distresses detection of pavement based on digital image processing. In: 4th Conference on Computer and Computing Technologies in Agriculture (CCTA), Nanchang, China, October 2010. Advances in Information and Communication Technology, AICT-347, Part IV. IFIP, pp. 368–375. Springer (2011)
18. Chan, P., Soetandio, C., Lytton, R.L.: Distress identification by an automatic thresholding technique. In: 1st International Conference on Application of Advanced Technologies in Transportation Engineer, San Diego, California (1989)
19. Lin, J., Liu, Y.: Potholes detection based on SVM in the pavement distress image. In: 9th International Symposium on Distributed Computing and Applications to Business, pp. 544–547. Engineering and Science, DCABES (2010)

Gis Based Multi-criteria Decision Analysis for the Streamlining of the Italian Network of Minor Airports

Maria Rosaria Guarini[(✉)] and Anthea Chiovitti

Sapienza University of Rome, 00185 Rome, Italy
mariarosaria.guarini@uniroma1.it

Abstract. Alongside the general growth of air traffic in Primary-level airports of each country, since the 1990s the popularity of "individual" (private), General Aviation, non-scheduled "point to point" flights all over the world and in Italy increased. They have proved to be a valid alternative to rail and road transport for short-medium-distance journeys (for different classes of business and tourist passengers). Taking into consideration the national and international airport system development scenarios, the paper illustrates the results of in- depth analyses carried out aiming to build up a full integrated GIS-based Multi-criteria Decision Analysis evaluation methodology. This is geared towards formulating strategies for the development and streamlining of the 51 existing Italian minor airport network and for the identification of the right locations for new hubs that could be required for the construction of an efficient second-level air transport network (the "Highway in the Sky"). The methodology is implemented according to different evaluation levels that verify: the suitability of air- port services and infrastructure (status quo) and the attractiveness of airport hubs given the territorial facilities found in their catchment areas.

Keywords: Multi-Criteria Decision Analysis
Multi-Criteria Spatial Decision Analysis · Geographic Information System
Civil aviation infrastructure · Minor airport · Integrated transportation system

1 Introduction

Since the 1990s in relationship with the general growth of air traffic in airports be- longing to the main network (Primary-level including large numbers of passengers and goods on both commercial and/or charter long-haul and medium-haul flights) of each country, we have observed the progressive growing popularity of "individual" [1], or private, non-commercial, non-scheduled "point to point" flights (Secondary-level including small numbers of passengers and goods travelling on private, short-haul flights) all over the world [2, 3], in Europe [4, 5] and in Italy [6]. The latter, operating using General Aviation aircraft [7, 8] out of the secondary airport network, are proved

---

The paper must be attributed in equal parts to the authors.

© Springer International Publishing AG, part of Springer Nature 2019
F. Calabrò et al. (Eds.): ISHT 2018, SIST 101, pp. 229–239, 2019.
https://doi.org/10.1007/978-3-319-92102-0_25

to be a valid alternative to rail and road transport for short- and medium-distance journeys (100–500 km) for a number of different classes of users, including business and tourist passengers. This phenomenon become evident since early 2000s in the USA [9] and in many European countries as well England, France and Germany [8, 10] also as result of the introduction of the latest aircraft - advanced ultra-light aircraft and very light jets - on the general civil aviation market. These have dynamic performance features (greater ease in take-off and landing, they do not require handling operations or even a control tower in many cases, they are easy to move due to their smaller size, etc.) and extremely low fuel consumption.

Though it provides a service for the manufacturing areas and businesses in surrounding territories, Italy's network of minor airports (consisting of 51 airports) has not been particularly developed, is poorly connected to intermodal transport systems [10] and not homogeneously widespread on the territory. If it were to be properly enhanced and expanded, the network of Italian minor airports could take on strategic importance both in the national infrastructural system, playing a role that could support primary airport networks, and in the development of territorial vocations. The factors in favor of the latter development can be summarized as follow [10, 11]: (i) absorbing air traffic, encouraging the streamlining of operations in main airports; (ii) ensuring that the traffic demand from specific sectors is met for individual (business, air-taxi service flights); (iii) supporting activities that run parallel to passenger and goods transport, i.e.: (a) non-aviation activities, which include a wide range of commercial and tourist services designed to satisfy different types of users; (b) aviation activities, such as working flights, civil defense, reconnaissance activities, patrols, training, education and support provided to civil protection forces and more generally all forms of territorial surveillance to avoid damaging alterations and attacks.

Coherently with the national and international airport system development scenarios at global, European and Italian level [2, 6], this paper illustrates an in-depth analyses carried out as part of ongoing research, divided up into a number of steps. This research aims to construct an integrated GIS-based Multi-criteria Decision Analysis evaluation methodology [12–14]. This methodology is geared towards formulating strategies for the development and streamlining of the existing Italian minor airport network and the identification of the right locations for the new hubs that could be required to construct an efficient and homogeneous second-level air transport network (the "Highway in the Sky") on the National territory taking into account several aspects.

The methodology is built considering of a set of evaluation elements (of multi-criterial and spatial nature), chosen for assessing both the suitability of airport services, of the infrastructures (status quo) and the attractiveness of Italian minor airports also taking into account the territorial facilities found in their catchment areas. In detail the set of evaluation elements concern with: (i) infrastructural characteristics intrinsic to airports (tracks, taxiways and aprons); (ii) infrastructural characteristics extrinsic to airports: airport services (airport services, air traffic support services, services to passengers, intermodal connectivity); (iii) manufacturing vocation (companies and the labour market), tourism (the number of state museum institutes and related visitors) and the transport network infrastructure (boasting road, rail and public transport infrastructure) in the areas served by each minor airport; (iv) current urban and territorial redevelopment processes (town planning forecasts and implementation from a vast level to a local scale) so as to ensure the integration and overlapping of systems (urban, environmental and transport systems); (v) economic and financial aspects (cost/benefit ratio for the completion, upgrading and/or conversion of airport hubs).

In the previous phases of research the methodological approach (Multicriteria Decision Analysis - MCDA) was defined aiming at assessing the suitability of airport service and infrastructure facilities (status quo) [15]. Subsequently, a georeferenced return of data (through using the Geographic Information System - GIS) was implemented by the tight coupling integration type between MCDA and GIS [16]. Thus, the evaluation methodology proposed can be defined as a type of Multi-Criteria Spatial Decision Analysis (MCSDA) [17–20]. This paper illustrates an upgrade of the evaluation methodology to the full integration type, which permits both the assessment and the in-depth analysis for evaluating the status quo of Italian minor airports and evaluating (from scratch) the attractiveness of airport hubs in view of the territorial facilities of their catchment areas (see above point iii). The upgrade was carried out in order to implement and manage a greater quantity of spatial/territorial data in an integrated and automated way, data that characterizes the decision-making problem analyzed here. Such an upgrade will also prove an essential basis for the development of further research phases (see above point iv and v).

The paper is structured as follows: definition of the evaluation methodology (Sect. 2) including the structure of the research (Sect. 2.1), the organization and spatial elaboration of evaluation elements (Sect. 2.2), the implementation procedure (Sect. 2.3), the results (Sect. 3), and in the end the conclusions (Sect. 4).

2 Evaluation Methodology

2.1 Structure of the Research

As mentioned earlier, this paper shows the evaluation methodology based on a full integration between Multi-Criteria Decision Analysis (MCDA), which allows us to consider a multi-criterial data set, and Spatial Decision Support Systems (SDSS) that implement the spatial/territorial analysis [12–20] exploiting the potential of Geographical Information Sciences (GISciences) [21]. The Multi-Criteria Spatial Decision Analysis (MCSDA) so defined allows: (i) the construction of databases with a suitable amount of information that can be consulted and easily updated, linked to the national infrastructure system; (ii) the formation of Territorial Information Systems regarding the network of minor airports for the organization and consultation of data supporting decisions; (iii) the combination of the geographic data (map criteria) with assessment indicators (information for decision-makers, stakeholder's preference and uncertainties) [22]. The integrated evaluation methodology is designed to judge airport hubs both individually and taken together as far as the following aspects: (i) The suitability (status quo) of infrastructure and service facilities by the identification of: A, intrinsic infrastructural characteristics (Criteria, Cj.A; Sub-criteria, Sj.A; Indicators, Ij.A) and B, extrinsic characteristics: airport services (Criteria, Cj.B; Sub-criteria, Sj.B; Indicators,

Ij.B) to obtain Judgment Levels (JL) involving A.n classification for: a "structural classification" (JLA), a "services classification" (JLB); a "overall classification of resources" (JLC) combining the JLA and JLB; (ii) The attractiveness given the territorial facilities of their catchment areas by the identification of parameters to define the D, manufacturing, tourism and infrastructure territorial vocation (Criteria, Cj.D; Sub-criteria, Sj.D; Indicators, Ij.D) in order to obtain a "territorial activeness classification" (JLD) and a "airport potential classification" (JLE) based on a combination of the previous evaluation phases JLC and JLD.

The evaluation methodology so defined requires an enormous range of data specially of the spatial kind. For this the Multi-Criterial Spatial Evaluation Model (MCSDA) proposed is based on the creation of a Database Management System (DBMS) [22]. Thanks to the use of computer software, a DBMS allows to create, manage and search data that helps making an assessment by constructing a database arranged in fields and columns. Using the database, the evaluation model is to be implemented using the management and processing of the data (of a multi-criterial kind) by tools (computer software) for geo-referencing and spatial analysis. Such instruments process the database using topographic overlay functions, spatial queries, network analysis, isochrones and heatmaps. Among different software packages available, the open-source QuantumGIS software was chosen to make implementation possible without having to resort to purchasing a license.

In order to conduct the analysis through the MCSDA referring to each airport (evaluation alternatives) operatively (Fig. 1) is to be performed: (i) The hiring (from the data used in previous research development phases) [15] and the definition (ex novo) of a set of evaluation elements made up of alternatives (An), criteria (Cj.n), sub-criteria (Sj.n), indicators (Ij.n), specific resource parameters (Ij.n) for JLA, JLB, JLC, JLD and JLE; (ii) The organization and spatially processing the specific resource parameters, related to each alternative, through the DBMS for JLA, JLB, JLC, JLD and JLE; (iii) The Multi-criteria type evaluation procedure (MCSDA) definition by interrogation of spatial data in order to provide a specific classification and then results extraction for each JLA, JLB, JLC, JLD and JLE. The implementation of these evaluation levels aims to singling out the parts of the country that are "covered" by minor airports and those that are "exposed" and to identify new areas (among those judged to be exposed) to permit that the network will achieve the national coverage.

2.2 Hiring, Organization and Spatial Elaboration of Evaluation Elements for JLA, JLB, JLC, JLD and JLE

Italy's 51 minor airports (A.n alternatives) are geo-referenced in the 1984 World Geodetic System (WGS84) of geographic coordinates (Fig. 2), in accordance with the information provided by the Ente Nazionale per l'Assistenza al Volo (ENAV) [23] and following the Nomenclature of Territorial Units for Statistics (NUTS). A shape-file is created for minor airport system allowing to archive vector data (e.g. location, form, attributes).

The JLA is worked out using the Cj.A, Sj.A, Ij.A and i(Ij.A) designed (in previous phases of research) to describe the intrinsic infrastructural characteristics (Tracks, Taxiways, Apron) of A.n [15].

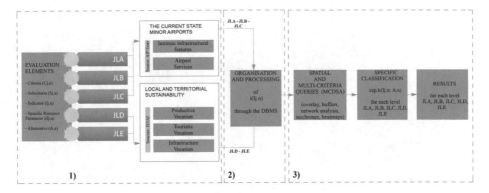

Fig. 1. Framework of the model.

The JLB is worked out using the set of Cj.B, Sj.B, Ij.B, i(Ij.B) and the potential resource parameters (ppd.n) designed to describe the airport services available for each alternative A.n. The JLB refers to data used in previous phases of research development (concerning: airport service, air traffic support service, services to passengers, inter-modal connectivity) [15], combined with the addition of new Cj.B, Sj.B, Ij.B and i (Ij.B) concerning the Intermodal connectivity and related to the proximity of: (i) hubs (or medium- or long-haul main airports); (ii) interports; (iii) ports; (iv) railway stations. The elements for evaluation used were built up on the basis of: (i) *the geographic area of analysis* identified by considering the territories closest to minor airports found in the 50 km buffer area (a spatial query in Sql Buffer); (ii) the *consultation and inclusion of the data* retrieved (2017) *from the Aeronautical Information Publication (AIP)* [23], which helps convey the level of airport services at each analyzed alternative; *the choice of Italian Istituto nazionale di Statistica (ISTAT) statistical data*, which helps convey the level of services linked to intermodal connectivity.

Evaluation elements for the JLC comes from the output obtained in the JLA and JLB according to the procedure described at Sect. 2.3.

For the JLD is used the set of Cj.D, Sj.D, Ij.D, i(Ij.D) and potential resource parameters (ppd.n) for the alternative A.n (Table 1) describing the facilities found in the territory surrounding each airport built up on the basis of: (i) *the definition of Gegrafical areas of analysis* identified by taking into account the territories closest to minor airports and the average time it takes to cross them by road, measured at two levels: 15 min; 30 min; (iii) *the choice of ISTAT statistical data*, which helps build up an area's profile as regards: (i) manufacturing, (ii) tourism, (iii) infrastructure in the territories surrounding minor airports.

The specific resource parameter i(Ij.n) identified for each A.n in each evaluation level have to be arranged in a Microsoft Excel spreadsheet (which included the one created during previous phases of the research), sorted into fields and columns. The spreadsheet is laid out with A.n placed in rows and Cj.n and Sj.n in columns, while the i(Ij.n) ca be placed at the point where they cross. Saving the Excel file in a comma separated value format (.csv) it is possible to link the Excel spreadsheet to the Quantum GIS (QGIS) software.

NUTS 1	NUTS 2: Regions	N°	Minor airport
North-West (ITC)	Piemonte	1	Alessandria
		2	Biella Cerrione
		3	Casale Monferrato
		4	Novi Ligure
		5	Torino Aeritalia
		6	Vercelli
	Lombardia	7	Alzate Brianza
		8	Como Idroscalo
		9	Calcinate del Pesce
		10	Cremona Migliaro
		11	Milano Bresso
		12	Valbrembo
		13	Varese Venegono
		14	Vergiate
		15	Voghera
	Liguria	16	Sarzana Luni
	Val D'Aosta		Region without minor airports

NUTS 1	NUTS 2: Regions	N°	Minor airport
North-East (ITD)	Trentino Veneto	17	Trento Mattarello
		18	Asiago
		19	Belluno
		20	Legnago
		21	Padova
		22	Thiene
		23	Venezia
		24	Verona Boscomantico
	Emilia	25	Carpi Budrione
		26	Ferrara Aguscello
		27	Ferrara
		28	Modena
		29	Pavullo nel Frignano
		30	Lugo di Romagna
		31	Ravenna
		32	Reggio Emilia
	Friuli V Giulia	33	Gorizia
		34	Udine Campoformido

NUTS 1	NUTS 2: Regions	N°	Minor airport
Centre (ITE)	Toscana	35	Arezzo
		36	Lucca Tassignano
		37	Massa Cinquale
	Umbria	38	Foligno
	Marche	39	Fano
	Lazio	40	Aquino
		41	Guidonia
		42	Latina
		43	Rieti
		44	Roma Urbe
		45	Viterbo
South (ITF)	Abruzzo	46	L'Aquila Preturo
	Molise		Region without minor airports
	Campania	47	Capua
		48	Salerno
	Puglia	49	Lecce Lepore
	Basilicata		Region without minor airports
	Calabria		Region without minor airports
Islands	Sicilia	50	Palermo
	Sardegna	51	Oristano Fenosu

Fig. 2. Geolocalization of the Italian Minor Airports.

2.3   MCSDA Procedure for JLA, JLB, JLC, JLD and JLE

For each evaluation level, the procedure is to be implemented by defining spatial queries on the database in Structure query language (Sql) according to the methodology outlined below.

According to the procedure defined in the previous phases of research [16] for each alternative A.n, in order to define: (i) the JLA, specific classification function f(csp.n) is to be implemented in Sql to obtain a geo-referenced specific classification csp.n(A.n) for each indicator I.A.n. The set of csp.n (A.n) is to be aggregated into two sub-levels (JLA1 and JLA2) through the Concise classification functions f(csnt.n) in order to

Table 1. Evaluation elements for JLD.

Criteria (CD.n)	Sub-criteria (SD.n)	Indicators (ID.n)	Specific resource parameter i(ID.n) for alternative (A.n)	Potential resource parameter (ppd.n)	Sources
Territorial activeness (JLD)	SD.1 Tourist Vocation	ID.1.1 Number of visitors (Demand) in state antiquity and art museum institutes (by province)	i(ID.1.1;An)	ppd.n(ID.1.1;A.n)	*ISTAT, Territorial Indicators for Development Policies, Cultural Heritage, indicators n. 076_P, census 2015, http://www.istat.it/it/archivio/16777*
		ID.1.2 Number of institutes (Supply) belonging to state antiquity and art museum institutes (by province)	i(ID.1.2;An)	ppd.n(ID.1.2;A.n)	*ISTAT, Territorial Indicators for Development Policies, Cultural Heritage, indicator n. 077_P, census 2015, http://www.istat.it/it/archivio/16777*
		ID.1.3 Amount of touristic accomodation (Supply) (hotels and similar establishments holiday accommodation and other short-stay accomodation, camping areas and equipped areas for motorhomes and caravans) in Isochrone area 15 minutes	i(ID.1.3;An)	ppd.n(ID.1.3;A.n)	*Elaboration in Structure Query Language from data ISTAT, Tourism Ateco 2007, Indicator: total collective accommodation establishments, 2015, http://dati.istat.it/?lang=it&SubSessionId =39dcb0ba-24c5-4a0e-b637-2b85f16a9f52&themetreeid=-200*
		ID.1.4 Number of touristic accomodation (Supply) (hotels and similar establishments, holiday accommodation and other short-stay accomodation, camping areas and equipped areas for motorhomes and caravans) in Isochrone area 30 minutes	i(ID.1.4;An)	ppd.n(ID.1.4;A.n)	*Elaboration in Structure Query Language from data ISTAT, Tourism Ateco 2007, Indicator: total collective accommodation establishments, 2015, http://dati.istat.it/?lang=it&SubSessionId =39dcb0ba-24c5-4a0e-b637-2b85f16a9f52&themetreeid=-200*
	SD.2 Territorial Equipment	ID.2.1 Area served by infrastructures: Space Interrogation in Structure Query Language - Isochrone area 15 minutes	i(ID.2.1;An)	ppd.n(ID.4;A.n)	*Elaboration in Structure Query Language from data ISTAT, Territorial bases and censorship variables, pubblication 2011, http://www.istat.it/it/archivio/104317*
		ID.2.2 Area served by infrastructures: Space Interrogation in Structure Query Language - Isochrone area 30 minutes	i(ID.2.2;An)	ppd.n(ID.5;A.n)	*Elaboration in Structure Query Language from data ISTAT, Territorial bases and censorship variables, pubblication 2011, http://www.istat.it/it/archivio/104317*
		ID.2.3 Number of passengers (Demand) using public transport in the provincial capitals	i(ID.2.3;An)	ppd.n(ID.6;A.n)	*ISTAT, Territorial Indicators for Development Policies, City, data n. 650_C, census 2015, http://www.istat.it/it/archivio/16777*
		ID.2.4 Urban public transport networks in the provincial capitals	i(ID.2.4;An)	ppd.n(ID.2.4;A.n)	*ISTAT, Territorial Indicators for Development Policies, City, indicator n.138_C, census 2013, http://www.istat.it/it/archivio/16777*
	SD.3 Productive Vocation	ID.3.1 Number of companies (in municipal areas)	i(ID.3.1;An)	ppd.n(ID.3.1;A.n)	*ISTAT and Infocamere Movimprese, Territorial Indicators for Development Policies, Demografia d'impresa data n. 138_P, census 2016, http://www.istat.it/it/archivio/16777*
		ID.3.2 Participation of population in the labour market (in municipal areas)	i(ID.3.2;An)	ppd.n(ID.3.2;A.n)	*ISTAT and Infocamere Movimprese, Territorial Indicators for Development Policies, Work indicator n. 108_P, census 2015, http://www.istat.it/it/archivio/16777*

obtain a concise classification csnt.n (An) in three level of suitability (Suitable, Adaptable, Not Suitable); (ii) the JLB, Threshold satisfaction (ssd.n) is to be defined and considered in the construction of the Value classification function f(cdm.n), implemented in Sql, to obtain a geo-referenced value classification and related score cdm.n(A.n) referred to each indicator I.B.n. Each cdm.n(A.n) is to be implemented by the weighting and standardization of the value classification function f(cdm.n) in order to obtain a service requisites classification (weighted and standardized) cdm.n(A.n) in five level (Very High, High, Average, Low, Very Low).

For the definition of JLCw, a weight of 2 is assigned to each Value Classification Score of the JLA representing the importance of the Structural requisites (JLA) with respect to the Service requisites (JLB). A Weighted Structural Classification JLAw and a Weighted Services Classification JLBw is thus obtained. By adding the values obtained in JLAw and in JLBw for each Alternative A.n a Weighted General Equipment Classification JLCw is carried out. Starting from JLCw, considering that "categorical variables can come from numeric variables by aggregating values" [24], the Univariate Data Analysis, is implemented. Thresholds attractivity (ssd.n) for alternatives A.n are defined through percentile values (St.n), between 0 and 1, for: (i) the JLCw on the Weighted General Equipment Classification in order to define 3 intervals of attraction (St.1 = 1; St.2 = 0.6; St.3 = 0.3; St.4 = 0.0). These are considered in the construction of the Value classification function f(cdm.n) in order to obtain the Classes Cl.C.n(A.n); (ii) the JLD for each indicator ID.n on each range of specific resource parameters i(ID.n) in order to define 5 intervals of attraction (St.1 = 1; St.2 = 0.8; St.3 = 0.6; St.4 = 0.4; St.5 = 0.2; St.6 = 0.0). These are considered in the construction of the Value classification function f(cdm.n) in order to obtain the value classification and related score cdm.n(A.n). Each cdm.n(A.n) is analyzed by concise classification function f(csnt.n) in order to obtain a Territorial attractiveness classification csnt.n(A.n) (Table 2); (iii) the JLE on the airport potential classification Scores, obtained by adding the Weighted General Equipment classification JLCw values and the Territorial attractiveness classification csnt.n(A.n), in order to define 3 intervals of attraction (St.1 = 1; St.2 = 0.6; St.3 = 0.3; St.4 = 0.0) These are considered in the construction of the Value classification function f(cdm.n) in order to obtain Classes Cl.E.n(A.n).

3 Results

The methodology has been built to obtain results that could be useful to qualify and quantify the parameters upon which the construction of the second-level network should be based. The implementation of the proposed procedure for each level of data evaluation, could allow to output results that as regards to: (i) the suitability of an airport's infrastructure and services, could be useful to evaluate the status quo of Italian minor airport System (by the implementation of the JLA, JLB and JLC); (ii) the attractiveness of airport hubs given the territorial facilities found in their catchment areas, could provide an airport potential classification for each alternative A.n (by the implementation of the JLD and JLE). Particularly the results of JLE could represent the synthesis of results obtained from all evaluation elements considered in the previous phases of evaluation procedure proposed.

Table 2. MCSDA JLD implementation procedure.

Criteria (CD.n)	Sub-criteria (SD.n)	Indicators (*ID.n*)	Specific resource parameter i(ID.n) for alternative (A.n)	Potential resource parameter (ppd.n)	Attractivity Thresholds (*ssd.n*)	Value classification function *[f(cdm.n)]*	Value classification and related score (*cdm.n*)	JLD Concise classification function *[f(csnt.n)]*	JLD Territorial attractiveness classification (*csnt.n*)
Territorial attractivenesses (JLD)	SD.1	ID.1.1	i(ID.1.1;An)	ppd.n(ID.1.1;A.n)	St.1(ID.1.1)	if ssd(ID.1.1;St.6) <ppd.n(ID.1.1;A.n) = ≤ssd(ID.1.1;St.5)	cdm.n(ID.1.1;A.n)=0.2		
					St.2(ID.1.1)	if ssd(ID.1.1;St.5) <ppd.n(ID.1.1;A.n) = ≤ssd(ID.1.1;St.4)	cdm.n(ID.1.1;A.n)=0.4		
					St.3(ID.1.1)	if ssd(ID.1.1;St.4) <ppd.n(ID.1.1;A.n) = ≤ssd(ID.1.1;St.3)	cdm.n(ID.1.1;A.n)=0.6		
					St.4(ID.1.1)	if ssd(ID.1.1;St.3) <ppd.n(ID.1.1;A.n) = ≤ssd(ID.1.1;St.2)	cdm.n(ID.1.1;A.n)=0.8	If 0.00 < M[fnl(cdm.n)] ≤ 0.20;	csnt.n = **VL**
					St.5(ID.1.1)	if ssd(ID.1.1;St.2) <ppd.n(ID.1.1;A.n) = ≤ssd(ID.1.1;St.1)	cdm.n(ID.1;A.n)=1		
		...	...	...	...	... ...	...		
	SD.2	ID.2.1	i(ID.2.1;An)	ppd.n(ID.2.1;A.n)	St.1(ID.2.1)	if ssd(ID.2.1;St.6) <ppd.n(ID.2.1;A.n) = ≤ssd(ID.2.1;St.5)	cdm.n(ID.2.1;A.n)=0.2	If 0.20 < M[fnl(cdm.n)] ≤ 0.40;	csnt.n = L
					St.2(ID.2.1)	if ssd(ID.2.1;St.5)<ppd .n(ID.2.1;A.n) = ≤ssd(ID.2.1;St.4)	cdm.n(ID.2.1;A.n)=0.4		
					St.3(ID.2.1)	if ssd(ID.2.1;St.4) <ppd.n(ID.2.1;A.n) = ≤ssd(ID.2.1;St.3)	cdm.n(ID.2.1;A.n)=0.6	If 0.40 < M[fnl(cdm.n)] ≤ 0.60;	csnt.n = A
					St.4(ID.2.1)	if ssd(ID.2.1;St.3) <ppd.n(ID.2.1;A.n) = ≤ssd(ID.2.1;St.2)	cdm.n(ID.2.1;A.n)=0.8		
					St.5(ID.2.1)	if ssd(ID.2.1;St.2) <ppd.n(ID.2.1;A.n) = ≤ssd(ID.2.1;St.1)	cdm.n(ID.2.1;A.n)=1	If 0.60 < M[fnl(cdm.n)] ≤ 0.80;	csnt.n = **H**
		...	...		...	... ...	...		
	SD.3	ID.3.1	i(ID.3.1;An)	ppd.n(ID.3.1;A.n)	St.1(ID.3.1)	if ssd(ID.3.1;St.6) <ppd.n(ID.3.1;A.n) = ≤ssd(ID.3.1;St.5)	cdm.n(ID.3.1;A.n)=0.2		
					St.2(ID.3.1)	if ssd(ID.3.1;St.5) <ppd.n(ID.3.1;A.n) = ≤ssd(ID.3.1;St.4)	cdm.n(ID.3.1;A.n)=0.4	If 0.80 < M[fnl(cdm.n)] ≤ 1.00;	csnt.n = VH
					St.3(ID.3.1)	if ssd(ID.3.1;St.4) <ppd.n(ID.3.1;A.n) = ≤ssd(ID.3.1;St.3)	cdm.n(ID.3.1;A.n)=0.6		
					St.4(ID.3.1)	if ssd(ID.3.1;St.3)<ppd .n(ID.3.1;A.n) = ≤ssd(ID.3.1;St.2)	cdm.n(ID.3.1;A.n)=0.8		
					St.5(ID.3.1)	if ssd(ID.3.1;St.2) <ppd.n(ID.3.1;A.n) = ≤ssd(ID.3.1;St.1)	cdm.n(ID.3.1;A.n)=1		
		...	...		...	... ...	...		

Legend: VL = Very Low; L = Low; A = Average; H = High; VH = Very High; M = aritmetic mean

4 Conclusions

The evaluation methodology proposed for subsequent levels of analysis helps to define the overall performance of Italy's infrastructural system of minor airports.

Obtainable outputs could serve as the basis for future phases of analysis also through the use of Multi-Criteria Decision Analysis (MCDA) integrated with, for example, Fuzzy Analysis, Strategic Planning Tools (SPT), Participation Techniques (PT), Financial, Economic, Risk and Sensitivity Analysis [25–29] built up to evaluate: (i) the amount of investment necessary and the socio-economic, financial and environmental repercussions associated with identifying/constructing each airport hub in the geographic area concerned (following the redevelopment/upgrading of existing infrastructure and/or the planning of new infrastructure); (ii) the inclusion of second-level network hubs within the processes of territorial and urban transformation underway (planning forecasts and implementation programs from the macro-area level to the local level); (iii) the sustainability of upgrading/expansion work and/or relocation (on the basis of the exposed areas identified) and therefore the structure of the network on the basis of the investment time necessary to guarantee that it becomes fully operational. The evaluation model proposed can be integrated with and/or replaced by other descriptive elements of additional and/or supplementary profiles for alternative A.n.

References

1. Becheri, E., Biella, A.: L'intermediazione della filiera del turismo organizzato. Maggioli Editore, Santarcangelo di Romagna (RN), Italy (2013)
2. International Civil Aviation Organization (ICAO): ICAO Annexes Collection, Annex 14, Aerodromes - Aerodrome Design and Operations, vol. 1, 7th edn., July 2016, http://cockpitdata. com/Software/ICAO%20Annex%2014%20Volume%201%20%207th%20Edition%202016. Accessed 29 Nov 2017
3. Airbus, Global Market Forecast, Growing Horizons 2017–2036. http://www.airbus.com/ content/dam/corporate-topics/publications/backgrounders/Airbus_Global_Market_Forecast_ 2017-2036_Growing_Horizons_full_book.pdf. Accessed 29 Nov 2017
4. Eurocontrol. Monthly Network Operations Report Overview, September 2017. http://www. eurocontrol.int/sites/default/files/publication/files/nm-monthly-network-operations-report-overview-september-2017.pdf. Accessed 29 Nov 2017
5. Eurocontrol, Monthly Network Operations Report Analysis, September 2017. http://www. eurocontrol.int/sites/default/files/publication/files/nm-monthly-network-operations-report-analysis-september-2017.pdf. Accessed 29 Nov 2017
6. Ente Nazionale Aviazione civile (ENAC), Traffico commerciale complessivo internazionale e nazionale Servizi di linea e charter (arrivi+partenze) 1 gennaio - 31 marzo 2017. http://www. enac.gov.it/repository/ContentManagement/information/N22169819/Dati_di_Traffico_ 2017_1_trimestre.pdf. Accessed 29 Nov 2017
7. International Civil Aviation Organization (ICAO), Doc 9060/5, Reference Manual on the ICAO Statistics Programme, 7th edn. (2013). https://www.icao.int/MID/Documents/2014/ Aviation%20Data%20Analyses%20Seminar/9060_Manual%20on%20Statistics_en.pdf. Accessed 29 Nov 2017
8. Aircraft Owners and Pilots Association (A.O.P.A.) General Aviation Statistics. https://www. aopa.org/about/general-aviation-statistics. Accessed 29 Nov 2017
9. Aircraft Owners and Pilots Association (A.O.P.A.), The Wide Wings and Rotors of General Aviation, The Industry's Economic and Community Impact on the United States. https:// www.aopa.org/-/media/files/aopa/home/news/all-news/2015/gama_whitepaper_final_mres. pdf?la=en. Accessed 29 Nov 2017
10. Criscuolo, C.: Verbale dell'incontro tra Ente Nazionale Aviazione civile and Italian Light Airport Network (i.LAN), Presentazione di uno studio finalizzato alla codifica di una nuova tipologia di infrastruttura di volo intermedia fra "Avio-superficie" e "Aeroporto", giugno 2007, Roma. https://www.yumpu.com/it/document/view/51080454/scarica-il-verbale-della-riunione-filas. Accessed 29 Nov 2017
11. Ministero delle Infrastrutture e dei trasporti (MIT), Ente nazionale per l'aviazione civile (ENAC): Piano nazionale degli aeroporti, febbraio 2012. https://www.enac.gov.it/La_ Comunicazione/Pubblicazioni/info-1156450804.html. Accessed 29 Nov 2017
12. Chen, J.: GIS-based multi-criteria analysis for land use suitability assessment in City of Regina. Environ. Syst. Res. **3**(13) (2014)

13. Malczewski, J.: GIS-based multicriteria decision analysis: a survey of the literature. Int. J. Geo-Graph. Inf. Sci. **20**(7), 703–726 (2006)
14. Liu, Y., Lv, X., Qin, X., Guo, H., Yu, Y., Wang, J., Mao, G.: An integrated GIS-based analysis system for land-use management of lake areas in urban fringe. Land Scape Urban Plan **82**(4), 233–246 (2007)
15. Guarini, M.R., Battisti, F., Buccarini, C., Chiovitti, A.: A model of multicriteria Analysis to develop Italy's Minor Air-port System. In: Gervasi, O., et al. (eds.) 15th International Conference on Computational Science and its Applications - ICCSA 2015. LNCS, vol. 9157, pp. 162–177. Springer, Heidelberg (2015)
16. Guarini, M.R., Locurcio, M., Battisti, F.: GIS-based multi-criteria decision analysis for the "Highway in the Sky". In: Gervasi, O., et al. (eds.) 15th International Conference on Computational Science And Its Applications - ICCSA 2015. LNCS, vol. 9157, pp. 146–161. Springer, Heidelberg (2015)
17. Sharifi, M., Boerboom, L., Shamsudin, K., Veeramuthu, L.: Spatial multiple criteria decision analysis in integrated planning for public transport and land use development study in Klang Valley, Malaysia. In: Proceedings of the ISPRS Vienna 2006 Symposium, Technical Commission II. In: ISPRS Archives, vol. XXXVI (2), pp. 85–91. http://www.isprs.org/proceedings/XXXVI/part2/pdf/sharifi.pdf. Accessed 29 Nov 2017
18. Joerin, F., Thériault, M., Musy, A.: Using GIS and outranking multicriteria analysis for land-use suitability assessment. Int. J. Geogr. Inf. Sci. **15**(2), 153–174 (2010)
19. Sugumaran, R., DeGroote, J.: Spatial Decision Support Systems. Principles and Practices. CRC Press Taylor and Francis Group, LLC (2011)
20. Boroushaki, S., Malczewski, J.: Using the fuzzy majority approach for GIS-based multicriteria group decision-making. Comput. Geosci. **36**(3), 302–312 (2010)
21. Roche, S.: Geographic Information Science I: Why does a smart city need to be spatially enabled? Prog. Hum. Geogr. **38**(5), 703–711 (2014)
22. Connolly, TM., Begg, CE.: Database Systems: A Practical Approach to Design, Implementation, and Management, 4th edn. Addison-Wesley, Boston (2005)
23. Ente Nazionale per l'Assistenza al Volo (ENAV), Aeronautic Information Publication, https://www.enav.it/sites/private/it/ServiziOnline/AD.html. Accessed 11 Sep 2016
24. Verzani, J.: Using R for Introductory Statistics. CRC Press, Boca Raton (2014)
25. Della Spina, L., Lorè, I., Scrivo, R., Viglianisi, A.: An integrated assessment approach as a decision support system for urban planning and urban regeneration policies. Buildings **7**, 85 (2017)
26. Nesticò, A., Sica, F.: The sustainability of urban renewal projects: a model for economic multi-criteria analysis. J. Prop. Invest. Financ. **35**(4), 397–409 (2017)
27. Guarini, M.R., Chiovitti, A., Battisti, F., Morano, P.: An integrated approach for the assessment of urban transformation proposals in historic and consolidated tissues. In: Borruso, G., et al. (eds.) 17th International Conference on Computational Science and Its Applications - ICCSA 2017. LNCS, vol. 10406, pp. 562–574 (2017)
28. Morano, P., Tajani, F., Locurcio, M.: GIS application and econometric analysis for the verification of the financial feasibility of roof-top wind turbines in the city of Bari (Italy). Renew. Sustain. Energy Rev. **70**, 999–1010 (2017)
29. Del Giudice, V., De Paola, P., Forte, F., Manganelli, B.: Real estate appraisals with Bayesian approach and Markov Chain Hybrid Monte Carlo method: an application to a Central Urban Area of Naples. Sustainability **9**(11), 2138 (2017)

Real-Time Update of the Road Cadastre in GIS Environment from a MMS Rudimentary System

Vincenzo Barrile[1] , Giovanni Leonardi[1] , Antonino Fotia[1] ,
Giuliana Bilotta[2(✉)] , and Giuseppe Ielo[1]

[1] Mediterranea University of Reggio Calabria, 89100 Reggio Calabria, Italy
[2] University IUAV of Venice, 30135 Venice, Italy
giuliana.bilotta@iuav.it

Abstract. The local authorities that are responsible for the management of road infrastructures must have by law a Road Cadastre by realizing one from scratch or updating the existing one. The Italian road network is spread over 160,000 km of state and provincial roads, so in the face of an efficient solution in terms of costs/performance, a rapid timetable is required for the data acquisition and the construction of the system. The elements of the territory represented in the GIS for the management of the Road Cadastre, on the one hand are characterized by the high mutability that affects both their position and the attributes necessary for classification; on the other hand, the integration of the GIS is only possible through a process of continuous and constant updating of the database. This note, taking up the experimental activities presented in [1] concerning the preparation of an MMS for automatic driving, uses only some of the equipment appropriately predisposed (GPS - GIS), integrating them with two low cost cameras. This in order to test a rudimentary MMS for the updating of the Road Cadastre that through the realization and implementation of suitable software allows to obtain appreciable and interesting results in terms of identification acquisition and transposition in real time on GIS system of elements of the territory for the management of the Road Cadastre. In fact, the proposed system allows real-time recognition (via computer vision) of road artefacts (drains, grids) and vertical signs, as well as the subsequent geo-referencing of acquired frames and automatic transposition in GIS environment (thanks to a dedicated software for updating of the Road Cadastre).

Keywords: Mobile mapping systems · Geographical information systems
Mobile computing

1 Introduction

Road management is practically a necessity in most of the world countries.

In this regard, it is essential to carry out periodic surveys in order to obtain, on the road axes of competence, the identification of degradations and the knowledge of the evolution of road characteristics.

© Springer International Publishing AG, part of Springer Nature 2019
F. Calabrò et al. (Eds.): ISHT 2018, SIST 101, pp. 240–247, 2019.
https://doi.org/10.1007/978-3-319-92102-0_26

The perfect integration of geometric and photographic data within an organized and easily interrogated archive allows the extraction and automatic creation of numerous elements present within the detected scenario.

New detection techniques are opening up new possibilities in the field of acquisition, processing and return of metric and thematic information, and consequently stimulate new studies, both experimental (data acquisition) and new modeling - IT studies (treatment of observations). On the other hand, the potential deriving from the use of neural networks is well known and extensively tested.

The quality and usefulness of the services provided by a generic Geographic Information System also depend on the ability of the software component to extrapolate and make explicit the information contained in an implicit form in its database, especially in those areas characterized by a particular dynamism.

MMS is particularly performing in this regard. The term Mobile Mapping System indicates the system of detection techniques from moving vehicles, and the direct georeferencing of the acquired data, expressed in the desired reference system, thanks to a positioning and orientation system on board installed.

The Laboratory of Geomatics, established at the Faculty of Engineering of the Università degli Studi Mediterranea of Reggio Calabria, in view of the creation of an autonomous MMS for automatic driving has already used an appropriately equipped vehicle during the experimental phase, for the first experiments in this regard.

In relation to the activities referred to in this note, the same vehicle was equipped with two color cameras (resolution 720×288 at 12 fps at 72 dpi, IP65 Waterproof) of video capture, suitable GPS instrumentation, a laptop and a UPS unit of control [1].

More in detail, the first camera, oriented along the direction of travel, allows to acquire the data relating to signage and geometric characteristics (for example, width of the roadway or lanes), the second camera instead, located at the rear of the vehicle, is oriented in the opposite direction of travel to record the state of the pavement [2].

The use, integration and implementation of software systems (GIS - Neural networks - Object recognition - Pattern recognition) some of which are made in-house and suitably combined and assembled in a single application (Fig. 1), allows:

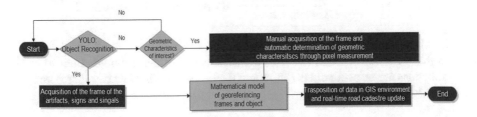

Fig. 1. Flowchart complete system.

- Determine the geometric characteristics through measurements on the pixels [3]
- Acquisition of the frames of the artifacts and of the vertical signs (Object detection - Pattern recognition) [4–6]
- Georeferencing through a specific mathematical model
- Data transposition in GIS environment and real-time Road Cadastre updating.

2 Determination of Geometric Characteristics Through Pixel Measurement

The geometric characteristics of the road, of the artefacts and of the deterioration are obtained directly from the frames having fixed a correction plane and the focal point of the camera [7].

Segments are automatically plotted on frames (Fig. 2a) starting from the reference axes and applied two scaling functions to determine the length and width.

Fig. 2. (a) Correction plan (grid). (b) Example road artifact measures (manhole).

More specifically, in order to determine the length function, different measurements were made in pixels of sample objects of known dimensions at different distances from the pick-up point, the results obtained were then analyzed and approximated in a single function:

$$y = 0.0005167 \cdot x^2 \tag{1}$$

In which is expressed the function of scale for length experimentally obtained.

$$x = \frac{Distance\ between\ the\ axes}{n°pixel\ between\ axes} \cdot n°pixel\ object\ to\ be\ measured \tag{2}$$

In which is the function of scale by width.

In the example shown in Fig. 2(b) the dimensions of the manhole obtained by the two functions are respectively 0.705 m and 0.696 m, reporting an error therefore in the order of half a centimeter.

3 Acquisition of the Frame of the Artifacts, Signs and Signals (Object Detection - Pattern Recognition)

Object Recognition in the Computer Vision is the ability to find a determined object in a sequence of images or videos. You Only Look Once (YOLO) is an advanced object recognition system developed within the Object Recognition [8]. This method, programmed by Joseph Redmond, a mathematician and computer scientist at the University of Washington, is much faster than traditional artificial intelligence systems.

The above-mentioned methodology (used for the implementation of the system proposed in this note), compared to the methodology proposed in [9], uses a single neural network [10]. For the recognition of vertical signage, this neural network provides for bounding boxes and the probabilities of associating the pixels of the image to a class directly from the complete images (avoiding the partition of the image) in a single evaluation for the recognition of vertical signage [11].

Because the entire detection pipeline is a single network, it can be optimized end-to-end directly on detection performance. The unified architecture is extremely fast [12]. The YOLO basic model processes images in real time at 45 frames per second, and is well suited for use on MMS for real time detection, in movement, of vertical signs and artefacts.

In this regard, we report an example of the application of the methodology inherent in the recognition of a single signal (stop) acquired with the method described above acquired with the MMS in circulation along a street of Reggio Calabria (Fig. 3).

Fig. 3. Example of object recognition.

YOLO reframes object detection as a single regression problem, straight from the image.

The immediate restitution of the data is well suited to be used as an integral part of a system dedicated to updating a Road Cadastre [13].

The proposed system identifies the elements of interest, memorizing the frame by acquiring it at a given distance and georeferencing it according to the diagram shown in the Fig. 4.

In particular, the distance is determined in relation to the number of pixels of the recognized object (the signal), having the known dimensions. In the system calibration phase, in fact, several frames of the same signal were acquired at different distances and the respective pixels were measured [14].

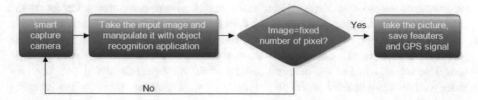

Fig. 4. Flow chart of frame acquisition from smart cam.

4   Mathematical Model of Georeferencing Frames

The mathematical model of georeferencing used is derived from the El-Sheimy and Schwarz 2004 formula (Table 1):

$$r_p^m = r_{IMU}^m(t) + R_{IMU}^m(t) \cdot \left[s_p \cdot R_s^{IMU} \cdot r_p^s + r_{IMU/S}^{IMU} \right] \tag{3}$$

Applying the projection on the street level of the positioning to the recognition of the pattern:

$$R_F = r_p^m + R_v \tag{4}$$

The value of the R_v (Fig. 5) vector is determined in relation to the position of the bounding box of the signal in the frame with respect to the progress of the MMS. The dimensions of the signal are known according to the type of road in which it is installed [14].

Table 1. Diagram of vector's position from the point P.

r_p^m	Vector position point P in the mapping frame	Unknown
$r_{IMU}^m(t)$	Original position of the sensor in the mapping frame at time t	From GPS/INS
$R_{IMU}^m(t)$	Rotation matrix from the IMU body frame to mapping frame at time t	From GPS/INS
s_p	Ratio between the camera-point distance and the vector r_p^s length	From restitution
R_s^{IMU}	Rotation matrix between camera frame and body frame	Calibration of shooting system
r_p^s	Vector of the image coordinates (x, y, −c) of the point P	Measured
$r_{IMU/S}^{IMU}$	Vector position of the camera in the body frame	Calibration of shooting system

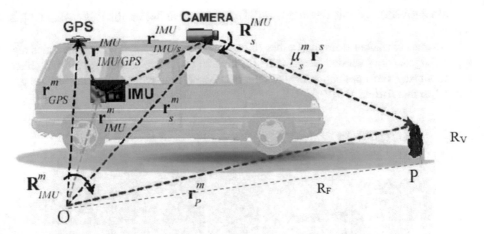

Fig. 5. Georeferencing vector calculation scheme.

5 Transposition of Data in GIS Environment and Real-Time Road Cadastre Update

Remodeling the application implemented by Barrile and Cotroneo [9], the solution proposed and implemented today includes:

(1) The GIS I/O module that manages the interfacing with the GIS software
(2) The plug in for data acquisition from frames
(3) The Kernel, which coordinates the various modules, pre-processes and post processes the I/O data of the modules themselves, and interacts with the user.

As shown in Fig. 6, the GIS I/O Module requests incoming GIS layers representing the roads along which the survey is performed (georeferenced layers) [15].

Fig. 6. Georeferencing vector calculation scheme.

The need to import this data is linked to the fact that the Kernel of the application works of spatial adjustments (Simple and Intelligent Snapping) on the elements detected and recognized. Through this module, the Kernel exports to the GIS one or more specific layers containing the spatial and alphanumeric data (type of signal, etc.) of the identified elements [16, 17].

We must add that this module was developed by leaning on the ESRI Map Object library.

The data extracted from the frames (GPS coordinates, geometric characteristics and type of signal) are included in an XML file created by an appropriate graphic interface to accompany the application itself, which is associated with a specific "language" related to this format [18] (Fig. 7).

Fig. 7. Software interface window.

6 Conclusions

The new software, although still in the experimental phase, has given excellent results in the performance of the functionalities inherent to the implemented features. Real-time identification and automatic acquisition of road signs and deterioration; export and automatic updating of data relating to the detected elements (position and cataloging) in GIS systems; snapping to implement corrections on the positioning of the recognized symbols, identification of points with coordinates close to those of homologous points derived from the geometry of the signs present in the frames.

References

1. Barrile, V., Meduri, G.M., Critelli, M., Bilotta, G.: MMS and GIS for self-driving car and road management. In: Gervasi, O., Murgante, B., Misra, S., Borruso, G., Torre, C.M., Rocha, A.M.A.C., Taniar, D., Apduhan, B.O., Stankova, E., Cuzzocrea, A. (eds.) ICCSA 2017. LNCS, vol. 10407, pp. 68–80. Springer, Cham (2017). https://doi.org/10.1007/978-3-319-62401-3_6
2. Ballard, D.: Computer Vision. Prentice-Hall, Englewood Cliffs (1982)

3. Barrile, V., Meduri, G.M., Bilotta, G.: Comparison between two methods for monitoring deformation with laser scanner. WSEAS Trans. Sign. Proces. **10**, 497–503 (2014)
4. Sonka, M., Hlavac, V., Boyle, R.: Image Processing, Analysis, and Machine Vision. Springer, New York (1998). https://doi.org/10.1007/978-1-4899-3216-7
5. Fu, L.M.: Neural Networks in Computer Intelligence, pp. 101–130. McGraw-Hill, New York (1994)
6. Simpson, P.K.: Artificial Neural Systems-Foundations, Paradigms, Applications, and Implementations, pp. 80–103. Pergamon Press, New York (1990)
7. Barrile, V., Bilotta, G., D'Amore, E., Meduri, G.M., Trovato, S.: Structural modeling of a historic caste using close range photogrammetry. Int. J. Math. Comput. Simul. **10**, 370–380 (2016)
8. Redmon, J., Divvala, S., Girshick, R., Farhadi, A.: You only look once: unified, real-time object detection. In: The IEEE Conference on Computer Vision and Pattern Recognition (CVPR), Las Vegas, NV, USA, pp. 779–788 (2016)
9. Barrile, V., Cotroneo, F.: Un software per l'aggiornamento automatizzato del catasto strade in ambiente GIS. Bollettino SIFET (2005)
10. Barrile, V., Cacciola, M., D'Amico, S., Greco, A., Morabito, F.C., Parrillo, F.: Radial basis function neural networks to foresee aftershocks in seismic sequences related to large earthquakes. In: King, I., Wang, J., Chan, L.-W., Wang, D. (eds.) ICONIP 2006. LNCS, vol. 4233, pp. 909–916. Springer, Heidelberg (2006). https://doi.org/10.1007/11893257_100
11. Gonzalez, R.C., Woods, R.E.: Digital Image Processing. Pearson Education Inc., Englewood Cliffs (2002)
12. Barrile, V., Cacciola, M., Morabito, F.C., Versaci, M.: TEC measurements through GPS and artificial intelligence. J. Electromagn. Waves Appl. **20**, 1211–1220 (2006)
13. Chen, R., Toran-Marti, F., Ventura-Traveset, J.: Access to the EGNOS signal in space over mobile-IP. GPS Solut. **7**(1), 16–22 (2003)
14. Barrile, V., Cacciola, M., Meduri, G.M., Morabito, F.C.: Automatic recognition of road signs by hough transform. In: 5th Symposium on Mobile Mapping Technology ISPRS Archives, vol. XXXVI-5/C55, pp. 62–67 (2008)
15. Barrile, V., Postorino, M.N.: GPS and GIS methods to reproduce vehicle trajectories in urban areas. Proc. Soc. Behav. Sci. **223**, 890–895 (2016)
16. Wolf, P., DeWitt, B.A.: Elements of Photogrammetry with Applications in GIS. McGrawHill, New York (2000)
17. Quddus, M.A., Ochieng, W.Y., Noland, R.B.: Integrity of map matching algorithms. Transp. Res. Part C: Emer. Technol. **14**(4), 283–302 (2006)
18. Al-Shaery, A., Zhang, S., Rizos, C.: An enhanced calibration method of GLONASS interchannel bias for GNSS RTK. GPS Solut. **17**(2), 165–173 (2013)

Heritage, Landscape and Identity

Assessing the Landscape Value: An Integrated Approach to Measure the Attractiveness and Pressures of the Vineyard Landscape of Piedmont (Italy)

Vanessa Assumma$^{(\boxtimes)}$ ⓘ, Marta Bottero ⓘ, Roberto Monaco,
and Giulio Mondini

Politecnico di Torino, 10129 Turin, Italy
`vanessa.assumma@polito.it`

Abstract. The paper deals with an integrated evaluation methodology finalized to evaluate the landscape value of the UNESCO site "Vineyard landscape of Langhe, Roero and Monferrato" (Piedmont, IT). The methodology consists in the employment of a system of landscape indicators with the aim, on one hand, to measure the landscape value in economic terms and on the other hand, to examine the pressures exercised on the landscape and its components. The present evaluation model could be considered a feasible tool in the decision support system for the definition of territorial scenarios of change.

Keywords: Landscape indicators and indexes · Ecosystem services
Decision-making process

1 Introduction

The global market provides generally a price for some goods and services. For other goods and services, as the environmental goods, which are known as externalities, the market price does not exist or only capture a small section of the market [1, 2]. The concept of Total Economic Value (TEV) is normally employed for the economic evaluation of the environmental goods and services. It is usually conceived by economists [3] as a series of attributes that compose any good or service. Disaggregating the TEV value into individual components might be useful to comprehend the different aspects of value that can be described as follows:

- Direct use values, which derive from the use of goods or services by people located in the ecosystem, in terms of consumption (e.g. harvest, timber) and non-consumption (e.g. recreational activities);
- Indirect use values, which refer to the services outside the ecosystem that generate indirect benefits, for instance, the protection of the forests;
- Non-use values, known as existence and bequest values, which refer to the awareness of people about the availability of a service, even if they never use it directly.

© Springer International Publishing AG, part of Springer Nature 2019
F. Calabrò et al. (Eds.): ISHT 2018, SIST 101, pp. 251–259, 2019.
https://doi.org/10.1007/978-3-319-92102-0_27

Evaluating the TEV as economic indicator might be relevant for the decision-making process to analyze the economic value of territories. There are many economic evaluation techniques in literature for supporting the decision-making process and usually employed for the evaluation of environmental goods and services. These are distinguished into:

- Monetary methods: these are always based on the analysis costs and benefits related to the goods under investigation. One of the most innovative techniques is the Choice Experiment (CE), which highlights the user preferences among a series of alternatives toward an environmental good [4]. This tool facilitates the definition of the value of each landscape component;
- Non-Monetary: these are mainly based on Multicriteria Decision Analysis (MCDA), these methods consider the several aspects of a problem, both qualitative and quantitative and include in the evaluation the preferences of the actors involved in the process. The Analytic Hierarchy Process (AHP) is one of the most versatile MCDA tools in the evaluation of complex problems in the decision-making process [5, 6], and many applications are available.

Among the different evaluation techniques for environmental and landscape goods, a very important role is played by systems of indicators for landscape evaluation and management [7, 8], which allow considering the different dimensions of the process such as ecology, history, culture, land cover consumption, economy, society and well-being. Furthermore, it has to be noticed that systems of indicators favor a complete spatial interpretation of landscape components if integrated by Geographic Information Systems (GIS).

The present study proposes the construction of a system of indicators for the analysis of landscape economic aspects [9].

The system of indicators is organized according to two sub-systems: the indicators of Value (V) and the indicators of Pressure (P). The firsts are finalized to assess the landscape value, while the seconds highlight the negative impacts that involve the landscape components [10, 11]. In the present research, the indicators of Value measure landscape quality and economic performance, while the indicators of Pressure consider the risk and its components able to compromise the economic state of a territory. The process can be synthesized as follows:

- Firstly, the set of value and pressure indicators are defined;
- Secondly, the components and the specific attributes are identified;
- Thirdly, standardization and aggregation rules are defined;
- Lastly, synthetic indexes are calculated: the Landscape Economic Value (LEV) and the Landscape Economic Pressure (LEP).

2   Case Study: The Vineyard Landscapes of Piedmont

The Vineyard landscape of Piedmont, Langhe, Roero and Monferrato (LRM) is a particularly attractive context located in South Piedmont, in the provinces of Alessandria, Asti and Cuneo, which has recently been included in the World Heritage

List of Unesco (WHL) for its Outstanding Universal Value (OUV), according with "integrity" and "authenticity" criteria: a significance that overcomes the national borders making the site of common importance for both present and future generations. The LRM context has been defined as "living cultural landscape" for its cultural, anthropic and perceptive components. The Unesco site perimeter is composed of 6 core zones, which include 29 municipalities and 2 buffer zones that protect the conservation factors of this site. The perimeter is inspired by the Units of Landscape (UL) that are a specification of the Ambits of Landscape, provided into the Landscape Regional Plan of Piedmont. In the LRM context, there are 30 UL, submitted to specific lines and prescriptions to assure the landscape continuity and the relations among the elements of the wine-making process [12]. The territory under observation consists of 101 municipalities for over 10.000 hectares, which have been organized in 8 homogeneous territorial clusters, distinguished in 6 core zones (CL1–CL6) and two buffer zones (CL7 and CL8) as described below: the clusters of Diano d'Alba (CL1), Grinzane Cavour (CL2), Neive (CL3), Nizza Monferrato (CL4), Canelli (CL5), Rosignano Monferrato (CL6), Asti (CL7) and Casale Monferrato (CL8).

3 Application of the Evaluation Model

The evaluation model has been structured in a system of landscape indicators, considering 4 categories of economic indicators suitable to estimate the Landscape Economic Value and 4 categories of pressure indicators that exercise negative impacts on landscape and are suitable for assessing the Landscape Economic Pressure. The main aim consists in the monitoring of LRM status in economic terms. The procedure followed in this evaluation can be described as follows:

1. Definition of a system of indicators (Table 1), collection of municipal data and organization in a system of territorial clusters;
2. Standardization of the indicators at municipal scale to facilitate the comparison of the indicators and a subsequent aggregation into partial indexes. The formula below converts the indicators into a non-dimensional indexes, in a range between 0 (minimum of value or pressure) and 1 (maximum level of value or pressure);

$$I_i = x_i / x_i^{max} \tag{1}$$

3. Weighting and aggregation of the indicators, according to a set of weights (w_i) defined by a group of experts and then ranked through the Analytic Hierarchy Process (AHP). The weights used in the model are reported in Table 1;

$$A = \sum_{i=1}^{5} w_i I_i,, \sum_{i=1}^{5} w_i = 1$$
$$T = \sum_{i}^{4} = w_i I_i, \sum_{i=1}^{4} w_i = 1$$
$$M = w_{10}I_{10} + w_{11}I_{11}, w_{10} + w_{11} = 1$$
$$F = w_{12}I_{12} + w_{13}I_{13}, w_{12} + w_{13} = 1$$
$$S = w_{14}I_{14} + w_{15}I_{15}, w_{14} + w_{15} = 1 \tag{2}$$
$$L = \sum_{i=1}^{3} w_i I, \sum_{i=1}^{3} w_i = 1$$
$$F = \sum_{i=1}^{3} w_i I_i, \sum_{i=1}^{3} w_i = 1$$
$$P = w_{22}I_{22} + w_{23}I_{23}w_{22} + w_{23} = 1$$

4. Aggregation of partial indexes and definition of final synthetic indexes:

$$LEV = y_1 A + y_2 T + y_3 M + y_4 F, \sum_{k=1}^{4} y_k = 1$$
$$LEP = y_5 S + y_6 L + y_7 F + y_8 P, \sum_{k=1}^{4} y_k = 1 \tag{3}$$

5. LEV and LEP index have been correlated to the territorial surface (Km^2), from which a Specific Landscape Economic Value (SLEV) and a Specific Landscape Economic Pressure are obtained;

$$SLEV = LEV/Km^2; \quad SLEP = LEP/Km^2 \tag{4}$$

6. Lastly, both LI$_{VP}$ and SLI$_{VP}$ synthetic indexes have been calculated to measure the landscape context status, making a comparison between values and pressures.

$$LI_{VP} = (LEV - LEP)/LEV; \quad SLI_{VP} = (SLEV - SLEP)/SLEV \tag{5}$$

Table 1. The systems of landscape value indicators and landscape pressure indicators.

Value	Indicators	w_i	Pressure	Indicators	w_i
Agriculture (A) y_1 0.570	Farms (No.)	w_1 0.049	Soil (S) y_5 0.530	Soil Consumption (C$_i$)	w_{14} 0.833
	Bio Farms (No.)	w_2 0.245		Urbanized Surface (ha)	w_{15} 0.167
	DOP/PGI Farms (No.)	w_3 0.129	Landslides Risk (L) y_6 0.138	Landslide surface	w_{16} 0.053
	Workers (No.)	w_4 0.401		Residents at risk	w_{17} 0.474
	Agriculture Surface (m^2)	w_5 0.176		Vulnerable elements	w_{18} 0.474
Tourism (T) y_2 0.168	Arrivals (No.)	w_6 0.402	Flood Risk (F) y_7 0.256	Flood risk surface	w_{19} 0.053
	Presences (No.)	w_7 0.281		Residents at risk	w_{20} 0.474

(continued)

Table 1. (*continued*)

Value	Indicators	w_i	Pressure	Indicators	w_i
	Total beds (No.)	w_8 0.064		Vulnerable elements	w_{21} 0.474
	Farmhouse – beds (No.)	w_9 0.253	Pollution (P) y_8 0.075	Polluted sites (No.)	w_{22} 0.875
Real Estate Market (M) y_3 0.075	Real estate value (€/m²)	w_{10} 0.400		Power lines (No.)	w_{23} 0.125
	Agricultural value (€/ha)	w_{11} 0.600			
Forestry (F) y_4 0.187	Forest surface (m²)	w_{12} 0.250			
	Forest farms (No.)	w_{13} 0.750			

For the collection of the data, information about the economic components have been gathered, considering both national and regional data sources (ISTAT, Sistema Piemonte, Osservatorio immobiliare – Agenzia delle Entrate). The territorial and environmental data are available by the catalog of Geoportale Piemonte and #ItaliaSicura [13], which is a kind of ReNDIS-web platform that give an overall vision about the risks from national to local scale.

4   Results

Following the methodology described in the previous sections, the synthetic indexes have been calculated and compared in order to analyze in an integrated way the landscape system.

4.1   Landscape Economic Values

The results calculated for the LEV are illustrated in Fig. 1. As it is possible to see, the highest LEV values refer to the clusters of Asti (CL7) and Diano d'Alba (CL1), equal to 1 and 0,267. The lowest LEV value is recorded by the cluster of Grinzane Cavour (CL2) equal to 0,010. The remaining LEV values are ranged between 0,063 and 0,142. Subsequently, a spatial distribution analysis investigates the importance of the economic indicators in the territory. The results of SLEV in fact overturn LEV values: if the buffer zones (CL7 and CL8) show the highest LEV value and Grinzane Cavour (CL2) the lowest one, in this case the highest SLEV values are recorded by the clusters of Grinzane Cavour (CL2) and Nizza Monferrato (CL3), respectively equal to 1 and 0,823. The lowest SLEV value is recorded by the cluster of Casale Monferrato (CL8), equal to 0,192.

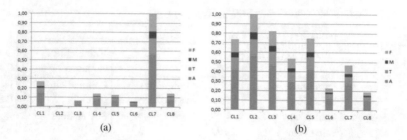

Fig. 1. Results of LEV (a) and SLEV (b) values for the clusters system.

4.2 Landscape Economic Pressures

The results of LEP are illustrated in Fig. 2. The maximum LEP index is recorded by the cluster of Asti (CL7), equal to 1 and followed by the cluster of Diano d'Alba (CL1) equal to 0,156. The lowest LEP value is recorded by the cluster of Grinzane Cavour (CL2) equal to 0,004. The remaining clusters are ranged between 0,022 and 0,114. The SLEP values have been calculated considering the spatial distribution of the pressure indicators in the territory. The maximum SLEP index is the cluster of Canelli (CL5), equal to 1, followed by the cluster of Neive (CL3), equal to 0,844. The clusters of Diano d'Alba (CL1), Nizza Monferrato (CL4) and Rosignano Monferrato (CL6) show medium SLEP indexes ranged between 0,597 and 0,642. The cluster of Grinzane Cavour (CL2) equal to 0,548, while the lowest SLEP indexes are recorded by the cluster of Rosignano Monferrato (CL6) and the cluster of Casale Monferrato (CL8), equal to 0,121 and 0,082.

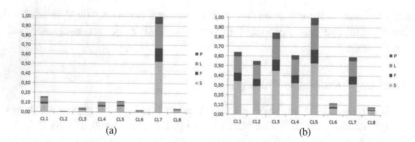

Fig. 2. Results of LEP (a) and SLEP (b) indexes for the clusters system.

4.3 Landscape Value-Pressure Synthetic Indexes

As shown in Sect. 4, both LEV and LEP indexes have been related with a mathematical formula (5), through which new synthetic indexes have been calculated: the Landscape Value-Pressure Index (LI_{VP}) and the Specific Landscape Value-Pressure Index (SLI_{VP}). The maximum LI_{VP} value is reached by the cluster of Casale Monferrato (CL8), equal to 1 and followed by the clusters of Diano d'Alba (CL1), Grinzane Cavour (CL2), Neive (CL3) and Rosignano Monferrato (CL6), with values ranged between 0,549 and

0,892. The clusters of Nizza Monferrato (CL4) and Canelli (CL5) show medium LI_{VP} values, respectively 0,363 and 0,232. Lastly, the cluster of Asti (CL7) equal to 0,075. Figure 3 provides a spatial representation of LEV, LEP and LI_{VP} values for the landscape context under examination.

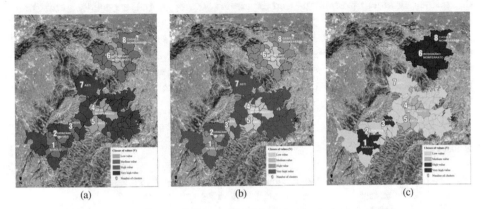

(a)   (b)   (c)

Fig. 3. Spatial representations of LEV(a), LEP(b) and LI_{VP}(c) indexes.

Table 2. Results of the evaluation model in relation to the area (km^2) of the clusters.

Clusters	Area (km²)	LEV	SLEV	LEP	SLEP	LI$_{VP}$	SLI$_{VP}$
CL1	133,38	0,267	0,737	0,156	0,642	0,625	0,625
CL2	3,81	0,010	1,000	0,004	0,548	0,824	0,787
CL3	29,06	0,065	0,823	0,045	0,844	0,549	0,682
CL4	102,99	0,142	0,534	0,109	0,612	0,363	0,417
CL5	906	0,128	0,748	0,114	1,000	0,232	0,000
CL6	101,23	0,063	0,227	0,022	0,121	0,892	0,833
CL7	783	1,000	0,470	1,000	0,597	0,075	0,114
CL8	277,71	0,142	0,192	0,040	0,082	1,000	1,000

5   Discussion of Results and Conclusions

The results in Table 2 give an overall picture of the economic features of landscape under examination. Firstly, it is interesting to notice that all the clusters provide positive results of LI_{VP}. This means that the values of the territories encompass the pressures, ensuring an overall quality. Secondly, it is possible to highlight that both LI_{VP} and SLI_{VP} show low values for the buffer zone 1 (CL7), considering the minor quality of the territory with reference to the related core zones (CL1–CL6) that provide higher values. This evaluation model represents a very important contribution in the decision-making process because allows both a careful analysis of resources located in the territory and the landscape evaluation through innovative support methodologies. This facilitates, on one hand, the definition of territorial scenarios of change and, on the other hand, the cooperation between decision-makers and stakeholders. The present system of landscape indicators is feasible and implementable, finalized to monitor both the landscape economic value in terms of attractiveness and the potential impacts that might compromise the economic status of territories. Furthermore, the current evaluation model also has shown in precedent studies an integration capability with other disciplines, producing promising and expending results [9, 14], especially in the frame of innovative regeneration projects or strategies [15] and energetic and economic performance [16]. However, a sensitivity analysis on the set of weights might be useful to examine how the indicators behave in this territory. We attend as future perspective a concrete application of this evaluation model, through the participation of both public and private actors in the process. We will expect also to involve local communities and tourists to test the reliability of the model.

References

1. Tempesta, T.: Economia del paesaggio rurale. In: Tempesta, T., Thiene M. (eds.) Percezione e valore del paesaggio. Franco Angeli, Milano (2009)
2. Dixon, J., Pagiola, S.: Economic Analysis and Environmental Assessment. In: Environment Department, EA Sourcebook (1998)
3. Pearce, D.W., Warford, J.J.: World without End: Economics, Environment and Sustainable Development. Oxford University Press, New York (1993)
4. Adamowicz, V., Hanley, N., Wright, R.E.: Using choice experiments to value the environment. Environ. Resour. Econ. **11**, 413–428 (1998)
5. Bottero, M., Comino, E., Riggio, V.: Application of the analytic hierarchy process and the analytic network process for the assessment of different wastewater treatment systems. Environ. Model Softw. **26**(10), 1211–1224 (2011)
6. Saaty, T.L.: The Analytic Hierarchy Process. McGraw Hill, New York (1980)
7. Bottero, M.: Assessing the economic aspects of landscape. In: Cassatella, C., Peano, A. (eds.) Landscape Indicators: Assessing and Monitoring Landscape Equality. Springer, Dordrecht (2011)
8. Cassatella, C., Peano, A.: Landscape Indicators: Assessing and Monitoring Landscape Equality. Springer, Dordrecht (2011)
9. Assumma, V., Bottero, M., Monaco, R.: Landscape economic value for territorial scenarios of change: an application for UNESCO site of Langhe, Roero and Monferrato. In: 2nd ISTH 2020 New Metropolitan Perspectives 2016, vol. 223, pp. 549–554, Procedia, Social and Behavioral Sciences (2016)
10. Brunetta, G., Caldarice, O., Pellerey, F.: La Valutazione Integrata Territoriale. Scenari del commercio in Provincia di Trento, SR Scienze Regionali **16**(3), 401–432 (2017)
11. Comino, E., Bottero, M., Pomarico, S., Rosso, M.: Exploring the environmental value of ecosystem services for a river basin through a spatial multicriteria analysis. Land Use Policy **36**, 381–395 (2014). Elsevier
12. UNESCO World Heritage List: Vineyards Landscape of Piedmont: Langhe-Roero and Monferrato, Candidacy Dossier, 1–2 (2014)
13. D'Angelis, E.: #ItaliaSicura per agire sul rischio idrogeologico. In: Ecoscienza, n. 3, p. 100 (2015)

14. Assumma, V., Bottero, M., Monaco, R., Soares, A.J.: La valutazione ecologica-economica del paesaggio:un'applicazione al Monferrato Ovadese. In: XXXVIII Italian Conference of Regional Sciences (AISRE), Cagliari, 20–22 September 2017
15. Mondini, G.: Integrated assessment for the management of new social challenges. Valori e Valutazioni **17**, 15–17 (2017)
16. Barthelmes, V.M., Becchio, C., Bottero, M., Corgnati, S.P.: Cost-optimal analysis for the definition of energy design strategies: the case of a nearly-zero energy building. Valori e Valutazioni **16**, 57–70 (2016)

Intergenerational Discounting in the Economic Evaluation of Projects

Antonio Nesticò$^{(\boxtimes)}$ ⑩ and Gabriella Maselli ⑩

University of Salerno, 84084 Fisciano, Italy
anestico@unisa.it

Abstract. The Social Discount Rate (SDR) is among the most critical parameters of the Cost-Benefit Analysis (CBA), because it strongly conditions the results. In the case of economic evaluations, that is when the analyses are conducted from the point of view of the community, the SDR allows to make financially comparable the costs and the benefits that the investment generates over time. Thus, it influences both the "weighting" of the cash flows temporal distribution and the measure of inter-generational equity associated with the project. Extremely important issues for those interventions that display their effects on a very long time horizon. In these circumstances, the traditional discount procedures show limits because they end up excessively reducing the financial terms that occur over a certain period. A possible solution to this problem is the use of hyperbolic discount procedures through declining discount rate (DDR).

In the present paper we intend to first outline in essential terms the theoretical framework of the approaches proposed in the literature for the estimation of the DDR. It is about the Consumption-Based Approach to DDRs and the Expected Net Present Value (ENPV). In the second part of the study a critical examination of the same approaches is proposed, in order to highlight their limitations and prominent theoretical aspects. These elements are useful to outline research perspectives for the characterization of an innovative model for estimating the declining discount rate, which can reduce at the same time the theoretical problems and the operational difficulties of the estimation methods currently used.

Keywords: Economic evaluation of projects · Social discount rate
Intergenerational discounting · Declining discount rate · Economic model

1 Intergenerational Projects. Conventional Procedures Limitations in Discounting Cash Flows

The estimate of the Social Discount Rate (SDR) is fundamental in the economic evaluation of investment projects. In fact, even small SDR variations significantly influence the analyses' results and, consequently, the order of priority of the

---

The contribution to this paper is the result of the joint work of both authors, to which the paper has to be attributed in equal parts.

© Springer International Publishing AG, part of Springer Nature 2019
F. Calabrò et al. (Eds.): ISHT 2018, SIST 101, pp. 260–268, 2019.
https://doi.org/10.1007/978-3-319-92102-0_28

interventions to finance where it is necessary to select among several initiatives. Therefore, with repercussions on the whole allocation process of public resources.

The question is of particular theoretical and operational commitment when subjected to a cost-benefit test there are projects that exert their effects on a long temporal horizon. In this case, the use of a discount rate taken through traditional estimation procedures, can lead to very low cash flows of the current values, different from the ones of the valuation time, thus strongly contracting the weight on the performance indicators.

Very frequent are those plans, programs and projects whose benefits are perceived by successive generations instead of those that have realized them [1, 2]. Just think of investments with strong social implications, such as those in the human resources field: education, training, research, especially preventive health. Of infrastructural interventions on the territory [3–5] and projects on cultural heritage [6–11]. But also to all the environmental initiatives, for which the beneficiary generations are often different from those that bear the costs [12]. About that, it is worth mentioning, among others: the multi-objective reorganization schemes of water basins, covering the water supply for industrial and civil purposes, the defense of the soil, the production of electricity; reforestation interventions, which reach the regime phase after 15–25 years from planting; investments for nuclear implants, which are characterized by significant environmental risks only after the first 20 years from the entry into operation of the power stations; projects that aim to reduce greenhouse gas emissions, whose initial costs are very high, while the benefits are evident for centuries [13].

Thus, the long "life" of social policies, investments in infrastructures, environmental projects, initiatives on cultural heritage, generating benefits, costs and in some cases risks over a period of time that goes beyond that of the generations that evaluate them, requires recourse to methods and techniques that allow, in the relevant analyses, to take into account such benefits, costs and risks.

The issue has important consequences in the cash flows discount transactions that are estimated for projects subject to economic evaluation according to the Cost-Benefit Analysis (CBA) logic. Indeed, the "conventional" discount procedures, which are the ones conducted through time-invariant discount rates, cause an accentuated and sometimes unacceptable contraction of the cash flows values that are produced for future generations.

A possible solution to the problem is the use of "hyperbolic" discount procedures [14–18]. This leads to the exclusion of the use of time-invariant rates in favor of time-declining rates, able to associate a greater weight to more distant over time events. In spite of rare voices of dissent [19, 20], there are numerous empirical evidences for inter-generational projects that induce to favor the actualization precisely through hyperbolic discount functions [18, 21–25].

While on one hand the use of time-declining rates can allow the correct discounting of long-term cash flows, on the other hand it raises the time inconsistency problem [26–28]. According to Arrow (et al.): «The problem is this: at time 0 the discount rate between period t and t + 1 is a long-term (low) discount rate. But, when period t actually arrives, the individual will apply a short-term (high) discount rate to period t + 1. Therefore, the individual will want to consume more in period t than he had planned to consume when he formulated his plans at time 0. The fact that the individual wishes to change his decision due simply to the passage of time means that his decision is time inconsistent. [...] This would occur simply due to the passage of time: in 2050, the forward rate used to discount benefits and costs from 2051 to 2050 would be higher than the one used in 2012 to discount benefits and costs from 2051 to 2050» [13]. It should however be noted that updating the DDR as soon as new economic information is available, allows to contain significantly the limit of the time inconsistency [12], which however does not have such weight as to undermine the "hyperbolic discounting" [29].

From the literature it emerges that the logic behind the DDR is the most suitable to actualize costs and benefits related to inter-generational projects. This is also confirmed by the actions of government authorities who have adopted declining discount rates for the economic evaluation of projects. In this regard, the United Kingdom imposes the use of a DDR with an initial value of 3.5% and value that in the long term, for t = 300 years, becomes 2% [30]. Similarly, the French Government proposes a constant rate of 4% for the first 30 years of the evaluation, to achieve the value of 2.2% after 300 years [31].

In the present paper, we intend to reconstruct in essential terms the theoretical framework of reference for the DDR, highlighting operational issues and limits to be resolved. The aim is draw up research perspectives useful for the characterization of the estimation protocols of the declining rate, in compliance with the mandatory theoretical principles and operational needs in order to implement the tool with simplicity in the individual case studies.

2   Theoretical Approaches for the Estimation of the Declining Discount Rate

The scientific literature proposes two approaches for the DDR estimation:

(1) the *Consumption-Based Approach*, which uses the Ramsey formula;
(2) the *Expected Net Present Value Approach* (ENPV).

For both of them, the theoretical assumption consists into including an uncertainty factor in the temporal structure of SDR [13]. For the Consumption-Based Approach the uncertainty concerns the growth rate of consumption, while in the ENPV is the same discount rate to be modeled as uncertain. The essential principles of the two approaches are examined below.

2.1   The Consumption-Based Approach to DDRs

In the CBA, the Consumption-Based Approach to DDRs reflects the social planner prospective, which acts to maximize the social welfare [32–34]. The project's cash flows are discounted to the rate of consumption that «is the rate at which a society is willing to postpone a unit of current consumption in exchange for more future consumption» [35]. This rate can be estimated through the Ramsey formula (1928):

$$r_t = \rho + \gamma \cdot g_t \tag{1}$$

The (1) shows that the consumption discount rate r_t depends on the pure temporal preference rate ρ, on the elasticity of the marginal utility of consumption or risk aversion γ and on the growth rate of consumption g_t.

The greatest difficulty in implementing the formula is linked to the estimation of the growth rate of consumption since, no doubt, any g_t estimation for the next century or millennium is subjected to a potentially huge error [36]. Several models are proposed for the g_t forecast, from which it emerges that the DDR value is strongly influenced by the persistence of the shocks in the economic growth process [36, 37]. Indeed, this persistence, if high, tends to amplify the long-term risk and, consequently, the negative "precautionary" effect associated with the social discount rate.

The most common models for the g_t estimation hypothesize a consumption growth process according to a Brownian motion, that is to say random, and with shocks to the consumption growth that are not correlated over time. In other words, the growth rate of consumption is modeled as a sequence of random variables that are normally, independently and identically distributed (i.i.d.) with mean μ and variance σ^2. In this case, it is shown that the discount rate is independent from the time horizon, as shown by the mathematical expression:

$$r_t = \rho + \gamma \cdot \mu - 0,5 \cdot \gamma^2 \cdot \sigma^2 \tag{2}$$

In (2) the term $0,5 \cdot \gamma^2 \cdot \sigma^2$, called "precautionary", summarizes the uncertainty on the growth rate of consumption. This term, in fact, determines a contraction of the discount rate. More complex models assume that shocks on growth are persistent. This assumption determines for r_t a decreasing term structure, according to which the persistence can be explained by a autoregressive model of degree 1, AR (1). This necessarily implies the complex estimation of a shocks correlated parameter.

The future uncertainty of g_t can also be taken into account by modifying the structure of (2) through the introduction of one or more uncertainty parameters. In this way we define a model, called "with parameter uncertainty" and frequently used for g_t prediction, on the basis of which the consumption register is supposed to follow a Brownian motion with trend $\mu(\theta)$ and volatility $\sigma(\theta)$, in which θ represents the probability that the respective parameters have to occur. In particular, there may be three cases:

(a) the trend μ of the growth rate of consumption is uncertain, i.e. $\mu = \mu(\theta)$. In this case, the function of the discount rate becomes

$$r_t = \rho - \frac{1}{t} \cdot \ln \sum\nolimits_{\theta=1}^{n} q_\theta \cdot e^{(-\gamma \mu_\theta)t} - 0,5 \cdot \gamma^2 \cdot \sigma^2 \tag{3}$$

with $\sum q_\theta = 1$, where q_θ is the parameter μ probability associated with the uncertainty;

(b) the volatility σ of the growth rate of consumption is uncertain, $\sigma = \sigma(\theta)$. So

$$r_t = \rho + 0,5 \cdot \gamma \cdot \mu - \frac{1}{t} \ln \sum\nolimits_{\theta=1}^{n} q_\theta \cdot e^{\left(0,5\gamma^2\sigma_\theta^2\right)t} \tag{4}$$

where $\sum q_\theta = 1$, with q_θ probability of the parameter σ associated with the uncertainty;

(c) both the mean μ and the volatility σ of the growth rate of consumption are uncertain, according to the formulations (3) and (4).

2.2   The Expected Net Present Value Approach to DDRs

According to the Expected Net Present Value (ENPV) approach, the discount rate is considered as an uncertain parameter [12, 13, 15, 16, 38–40]. The logic behind the method can be explained by an example. Consider a project that generates net benefits of € 1,000 in 100 years. If you use the mean discount rate of 3.5% constant throughout the analysis period, then the Net Present Value (NPV) of the project is equal to:

$$NPV_1 = \text{€}\,1,000 \cdot e^{-rt} = \text{€}\,1,000 \cdot e^{-0.035 \cdot 100} = \text{€}\,30.20.$$

However, the value 3.5% of the rate is uncertain and there may be the probability that it is 5% or 2%. In this case, the NPV of the €1,000 of future benefits is equivalent to:

$$NPV_2 = 0.5 \cdot \left(\text{€}\,1,000 \cdot e^{-0.05 \cdot 100}\right) + 0.5 \cdot \left(\text{€}\,1,000 \cdot e^{-0.02 \cdot 100}\right) = \text{€}\,71.04.$$

It is evident that the NPV resulting from an uncertain rate is more than twice that the one estimated at the mean discount rate.

From this result Weitzman [16] derives an important corollary: «Computing the expected net present value of a project (ENPV) with an uncertain but constant discount rate is equivalent to computing the NPV with a certain but decreasing "certainty-equivalent" discount rate» [38]. According to this corollary, it is necessary to evaluate the discount factor and the respective "certainty-equivalent" discount rate, defined below.

The discounting of future costs and benefits that occur at time t is done by a P_t discount factor defined as:

$$P_t = \exp \cdot \left(-\sum\nolimits_{i=1}^{t} r_i\right) \tag{5}$$

When r is a stochastic variable, it is necessary to introduce an uncertainty factor E which defines the declining structure of the discount rate. Then, the discount factor E (P_t), called "certainty-equivalent", is written as:

$$E(P_t) = E \cdot \left(\exp \cdot \left(-\sum\nolimits_{i=1}^{t} r_i\right)\right) \tag{6}$$

It follows that the corresponding certainty-equivalent forward rate $\tilde{r}_t$ for discounting between adjacent periods at time t as equal to the rate of change of the expected discount factor:

$$\frac{E(P_t)}{E(P_{t+1})} - 1 = \tilde{r}_t \tag{7}$$

In summary, the fundamental problem consists in predicting the time trend of the discount rate, to determine the values r_i which have to be included in (5). The forecast can be made by using econometric models based on the trend of past data related to market interest rates, often concerning long-term government bonds. However, these are models that are not easy to implement and for which doubts are raised regarding the nature of the data to be used [12, 25, 39].

3 Critical Analysis of Approaches for DDR Estimation. Ideas for Future Researches and Conclusions

In the light of the theoretical framework on the approaches for the Declining Discount Rate estimation, outlined in essential terms in the preceding paragraph, salient elements, analysis protocols and certain critical issues both for the Consumption-Based Approach and for the Expected Net Present Value Approach emerge.

The Consumption-Based Approach is based on a rigorous theoretical approach, the Ramsey formula, also anchored to economic data and demographic indicators that reflect the socio-economic structure of a country [40]. These aspects are appreciated to the point that this approach is now followed by government administrations for the DDR estimation.

However, it should be noted that the forecast of the consumption growth rate g_t that appears in formula (1) is based on certain restrictions:

(1) the reference model for the Ramsey formula extended with non-correlated shocks over time, as it is shown in formula (2), is constructed on the already indicated restrictive hypotheses, which on the one hand do not always allow a satisfactory approximation of historical data; and that, on the other hand, they lead to a constant discount rate [22, 36, 37];

(2) the model for the g_t prediction in which shocks on growth are persistent, leads to discount rate functions characterized by a rapid decline in a few years. Moreover, as has been mentioned, the estimation of the parameter that correlates the shocks is particularly complex;

(3) the model with parameter uncertainty, to which refer the formulas (3) and (4) according to which the consumption follows a Brownian motion with trend $\mu(\theta)$, volatility $\sigma(\theta)$ and where θ is the probability that the parameters have of occurrence, it is the only one able to offer the best compromise between methodological consistency and simplicity of calculation. However, the random variable g_t is modeled through a normal distribution that is not always the one that best approximates the historical data.

Especially the coherence of the general logic is appreciated in the Expected Net Present Value Approach. However, the same method is critical. *In primis*, the rate r_t is estimated on the basis of long-term market interest rates, which sets all the limits linked to the distortions to which the market is subjected [35, 41]. *In secundis*, for the r_t prediction it is necessary to employ rather complex econometric forecasting models and, consequently, of not simple implementation [12].

What are the synthetic considerations that can be drawn after the developed examination?

The Expected Net Present Value Approach shows a solid theoretical framework and is based on the rigorous concepts of the "certainty-equivalent" discount factor and the "certainty-equivalent" discount rate. However, it was said of the critical issues related to both the nature of the data to be used, which in the ENPV are the interest rates of the Government Bonds, both of operational nature in the implementation of the econometric models useful for the prediction of the r_i that appear in the formula (5).

Well, it is believed that the criticalities on the nature of the data to be used for the prediction of the r_i can be solved through the Ramsey formula which, using the economic and demographic informations of a Country, allows to read the meaning of the elaborations and is easy to use.

Moreover, with reference to the project risk components included in the r_i values of the formula (5), an estimation can be based on the future projection with probabilistic laws of the r_t of the Ramsey formula (1). In essence, the consumption growth rate g_t - on which the discount rate r_t value depends - can be modeled as a random variable, to which a probability function is associated on the basis of the corresponding historical data. In other words, starting from the latter function, it is not difficult to derive, for each year, a series of probable values to associate with the rate g_t and, consequently, to the unknown quantity r_t.

These are considerations that clearly lead to the characterization of a new model for the DDR estimation, rigorous from the theoretical point of view because it is based on methods widely recognized in the scientific literature, and at the same time easier to implement than the approaches described in this paper since it can be built on easily retrievable data.

The definition of an innovative estimation model for the declining discount rate, repeatable in its computational *iter* and of non-complex use, can increasingly direct operators to use the DDR in the inter-generational projects' economic evaluation. The discounting through time-declining rates is in fact a procedure as valid as it is still not widely used in operational practice.

References

1. Dolores, L., Macchiaroli, M., De Mare, G.: Sponsorship for the sustainability of historical-architectural heritage: application of a model's original test finalized to maximize the profitability of private investors. Sustainability **9**(10), 1750 (2017)
2. Bencardino, M., Nesticò, A.: Demographic changes and real estate values. A quantitative model for analyzing the urban-rural linkages. Sustainability **9**(4), 536 (2017)

3. Morano, P., Locurcio, M., Tajani, F., Guarini, M.R.: Fuzzy logic and coherence control in multi-criteria evaluation of urban redevelopment projects. Int. J. Bus. Intell. Data Min. **10**(1), 73–93 (2015)
4. Napoli, G., Giuffrida, S., Trovato, M.R., Valenti, A.: Cap rate as the interpretative variable of the urban real estate capital asset: a comparison of different sub-market definitions in Palermo, Italy. Buildings **7**(3), 1–25 (2017)
5. Nesticò, A., Sica, F.: The sustainability of urban renewal projects: a model for economic multi-criteria analysis. J. Prop. Invest. Financ. **35**(4), 397–409 (2017)
6. Navrud, S., Ready, R.C.: Valuing Cultural Heritage: Applying Environmental Valuation Techniques to Historic Buildings, Monuments and Artifacts, 1st edn. Edward Elgar Publishing Ltd., Cheltenham (2002)
7. Tweed, C., Sutherland, M.: Built cultural heritage and sustainable development. Landsc. Urban Plann. **83**(1), 62–69 (2007)
8. Munda, G.: Social Multi-criteria Evaluation for a Sustainable Economy, 1st edn. Springer, Heidelberg (2008). https://doi.org/10.1007/978-3-540-73703-2
9. Ferretti, V., Bottero, M., Mondini, G.: Decision making and cultural heritage: an application of the multi-attribute value theory for the reuse of historical buildings. J. Cult. Herit. **15**(6), 644–655 (2014)
10. Calabrò, F.: Local communities and management of cultural heritage of the inner areas. An application of break-even analysis. In: Gervasi, O., Murgante, B., Misra, S., Borruso, G., Torre, C.M., Rocha, A.M.A.C., Taniar, D., Apduhan, B.O., Stankova, E., Cuzzocrea, A. (eds.) ICCSA 2017. LNCS, vol. 10406, pp. 516–531. Springer, Cham (2017). https://doi.org/10.1007/978-3-319-62398-6_37
11. Della Spina, L.: Integrated evaluation and multi-methodological approaches for the enhancement of the cultural landscape. In: Gervasi, O., Murgante, B., Misra, S., Borruso, G., Torre, C.M., Rocha, A.M.A.C., Taniar, D., Apduhan, B.O., Stankova, E., Cuzzocrea, A. (eds.) ICCSA 2017. LNCS, vol. 10404, pp. 478–493. Springer, Cham (2017). https://doi.org/10.1007/978-3-319-62392-4_35
12. Newell, R.G., Pizer, W.A.: Discounting the distant future: how much do uncertain rates increase valuations? J. Environ. Econ. Manag. **46**(1), 52–71 (2003)
13. Arrow, K.J., Maureen, L., Cropper, C.G., Groom B., Heal, G.M., Newell, R.G., Nordhaus, W.D.: How should benefits and costs be discounted in an intergenerational context? The views of an expert panel. Resources for the future discussion paper, pp. 12–53 (2013)
14. Weitzman, M.: On the environmental discount rate. J. Environ. Econ. Manag. **26**(2), 200–209 (1994)
15. Weitzman, M.: Why the far-distant future should be discounted at its lowest possible rate. J. Environ. Econ. Manag. **36**(3), 201–208 (1998)
16. Weitzman, M.: Gamma discounting. Am. Econ. Rev. **91**(1), 261–271 (2001)
17. Henderson, N., Langford, I.: Cross-disciplinary evidence for hyperbolic social discount rates. Manag. Sci. **44**(11), 1493–1500 (1998)
18. Cropper, M.L., Laibson D.: The Implications of Hyperbolic Discounting for Project Evaluation. World Bank Policy Research Working Paper Series 1943, Washington, D.C. (1998)
19. Read, D.: Is time-discounting hyperbolic or subadditive? J. Risk Uncertain. **23**(1), 5–32 (2001)
20. Rubinstein, A.: Economics and psychology? The case of hyperbolic discounting. Int. Econ. Rev. **44**(4), 1207–1216 (2003)
21. Harris, C., Laibson, D.: Dynamic choices of hyperbolic consumer. Econometrica **69**(4), 935–957 (2001)
22. Gollier, C.: Discounting an uncertain future. J. Public Econ. **85**, 149–166 (2002)

23. Hepburn, C., Koundouri, P., Pasnopoulou, E., Pantelidis, T.: Social discounting under uncertainty: a cross-country comparison. J. Environ. Econ. Manag. **57**, 140–150 (2009)
24. Gustman, A., Steinmer, T.: Policy effects in hyperbolic vs exponential models of consumption and retirement. J. Public Econ. **96**, 465–473 (2012)
25. Freeman, M., Groom, B., Panopoulou, E., Pantelidis, T.: Declining discount rates and the Fisher Effect: inflated past, discounted future? J. Environ. Econ. Manag. **76**, 32–39 (2015)
26. Lesser, J., Zerbe, R.: What can economic analysis contribute to the sustainability debate? Contemp. Econ. Policy **13**(3), 88–100 (1995)
27. Schelling, T.: Intergenerational discounting. Energy Policy **23**(4–5), 395–401 (1995)
28. Groom, B., Hepburn, C., Koundouri, P., Pearce, D.: Declining discount rates: the long and the short of it. Environ. Resour. Econ. **32**, 445–493 (2005)
29. Heal, G.: Valuing the Future: Economic Theory and Sustainability. Columbia University Press, New York (1998)
30. Treasury, H.M.: The Green Book: Appraisal and Evaluation in Central Government. TSO, London (2003)
31. Rapport Lebègue: Révision du taux d'actualisation des investissements publics. Commissariat Général au Plan, Paris (2005)
32. Dasgupta, P.: Discounting climate change. J. Risk Uncertain. **37**(2), 141–169 (2008)
33. Goulder, L.H., Williams, R.: The choice of discount rate for climate change policy evaluation. Clim. Change Econ. **3**(4), 1–18 (2012)
34. De Mare, G., Nesticò, A., Tajani, F.: The rational quantification of social housing. In: Murgante, B., Gervasi, O., Misra, S., Nedjah, N., Rocha, A.M.A.C., Taniar, D., Apduhan, B. O. (eds.) ICCSA 2012. LNCS, vol. 7334, pp. 27–43. Springer, Heidelberg (2012). https://doi.org/10.1007/978-3-642-31075-1_3
35. Zhuang, J., Liang, Z., Lin, T., De Guzman, F.: Theory and practice in the choice of social discount rate for cost-benefit analysis: a survey. ERD, Working Paper No. 94, Asia Development Bank (2007)
36. Gollier, C.: Pricing the Future: The Economics of Discounting and Sustainable Development. Princeton University Press, Princeton (2011)
37. Gollier, C.: Discounting with fat-tailed economic growth. J. Risk Uncertain. **37**, 171–186 (2008)
38. Cropper, M.L., Freeman, M.C., Groom, B., Pizer, W.A.: Declining discount rates. Am. Econ. Rev. **104**(5), 538–543 (2014)
39. Groom, B., Koundouri, P., Panopoulou, E., Pantelidis, T.: Discounting the distant future: how much does model selection affect the certainty equivalent rate? J. Appl. Econom. **22**, 641–656 (2007)
40. Nesticò, A., De Mare, G., Conte, A.: Approcci teorici ed empirici nella stima del saggio sociale di sconto, Valori e valutazioni 14 (2015)
41. Florio, M., Sirtori, E.: The social cost of capital: recent estimates for the EU. Working Paper No. 3, Centre for Industrial Studies, Milano (2013)

Real Estate Landscapes and the Historic City: On How Looking Inside the Market

Laura Gabrielli[1(✉)], Salvatore Giuffrida[2], and Maria Rosa Trovato[2]

[1] Department of Architecture, University of Ferrara, 44121 Ferrara, Italy
laura.gabrielli@unife.it
[2] Department of Civil Engineering and Architecture,
University of Catania, 95124 Catania, Italy
{sgiuffrida,mrtrovato}@dica.unict.it

Abstract. The real estate capital is one of the most resistant forms of the process through which the social surplus product was consolidated, making possible the phenomenon of cities as a gradual layering of the "traces" of a settled community. The complexity and complementarity between homogeneity of the urban fabrics and heterogeneity of the architectural shapes led to the multiplicity of functions that properties play, encouraging the expectations of the players of its enhancement process: administrations, owners, large and small investors. This paper focuses on the interpretation of the urban pattern of the historic city through the analysis of the housing markets. The research deals with the case study of the town of Syracuse, a multifaceted urban context from several points of view. The formal and functional articulation of this real estate market justifies the use of different, layered and structured analysis tools to identify sub-markets, deepening the relationship between value/price.

Keywords: Real estate market · Complex urban context · Sub-markets
Hard clustering · Fuzzy clustering

1 Introduction

The real estate sector is one of the triggers of urban development and refurbishment mostly influenced by the a-synchronicity of the values/prices dynamics. The observation of the real estate market aims at finding out consistent values/prices relation-ships but especially in the historic city, these relationships often results weird.

In such imperfect real estate markets, the spectrum of opportunities depends:

- in static terms, on the ability to capture the unexpressed options of the assets whose expected prices are high enough to justify the re-adaptation costs in the short term;
- in dynamic terms, on the ability to interpret the predictable changes of both real estate capital asset and urban context attractiveness in the medium/long run.

This dialectic allows the potential purchasers and investors to combine and/or alternate the three primary functions of real estate capital - personal use, productive investment and speculative investment.

© Springer International Publishing AG, part of Springer Nature 2019
F. Calabrò et al. (Eds.): ISHT 2018, SIST 101, pp. 269–276, 2019.
https://doi.org/10.1007/978-3-319-92102-0_29

This contribution is the first step of a broader study concerning such dynamics. It reports the results of deepened analysis of this real estate market of the old town of Syracuse trying to represent its sub-markets based on a detailed analysis of a sample covering the three main historic districts. Both the hard and fuzzy cluster analysis performed, highlight some apparent inconsistencies that can be assumed as tips of the "beyondness" of such an urban context.

2 The Case Study

The old town of Syracuse, has been being characterized by a sudden process of enhancement started from Ortigia - the ancient settlement located on an islet connected to the mainland by two bridges, definitely jeopardized district up to '60 - and now involving the whole city.

Since 1990, the Plan for Ortigia, the establishment of the Faculty of Architecture of the University of Catania, the number of urban regeneration projects with national and European funding, and the subsidies for refurbishing private buildings, aroused unexpected real estate performances and the interest of large and small investors [1].

The Umbertino district, dating back to the nineteenth and early twentieth centuries, is formed by buildings of significant size and of good architectural quality, many of them still showing medium-low prices. The functionality of the settlement system based on a grid of large square meshes and the contiguity at Ortigia make it the most favorable area for investment potential even for lower-middle prices.

Borgata is a large, small bourgeois neighborhood, headed by the great square of Santa Lucia, which was built between the end of the nineteenth and early twentieth century. The low prices level and the easy accessibility encourage the settlement of young people and students (Fig. 1).

Fig. 1. Geographical framing of the studied urban context.

The proposed real estate observation addresses a sample of 101 units of which asking prices are reported. The properties are characterized by six primary character-istics [2], articulated in 33 sub-categories (Table 1).

The relationship represented by the chart shows a progressive widening of the price range as the aggregate index grows. The bubble size represents the size of every single

Table 1. Hierarchy of characteristics.

k_{e1} Location and overall accessibility of the building	1. Complexity and urban shape	1. Location and settlement quality; 2. Mix of functions; 3. Social status (professional, income level, etc.) of the area; 4. Public space maintenance level
	2. Urban facilities	1. Public facilities; 2. Public services
	3. Accessibility	1. Mobility from/to the area with private transportation; 2. Mobility from/to the area with public transportation; 3. Mobility within the neighbourhood
k_{e2} Neighbourhood characteristics		1. Functional and 2. Symbolic characteristics
k_i Unit location within the building		1. Panoramic quality; 2. View; 3. Brightness; 4. Orientation; 5. Security
k_t Technological characteristics	1. Building	1. Building structure quality; 2. Installations level; 3. Finish and windows quality; 4. Maintenance levels
	2. Unit	1. Finish; 2. Installations quality; 3. Maintenance levels
k_{a1} Building architectural quality		1. Use suitability; 2. Integration of structure and instalment; 3. Finish and technology; 4. Morphological consistency; 5. Space decorum; 6. Maintenance levels
k_{a2} Unit architectural quality		1. Size, functionality and distribution; 2. Additional surfaces; 3. Quality of finish

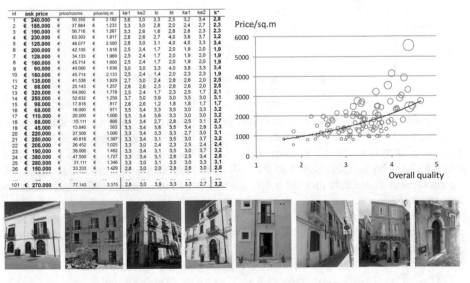

Fig. 2. Sample of the database: early analyses of the data.

property: the smaller real estate matches a lower quality and price, and vice versa, being larger properties located in buildings of greatest historical-architectural value.

Figure 2 provides an early perception of the relationship between unit prices and the aggregate quality index k^*. The same analysis was carried out for each of the six key characteristics. The observation is accompanied by a sample representation on the map (Fig. 3), which allows us to associate each property with the different characteristics and price.

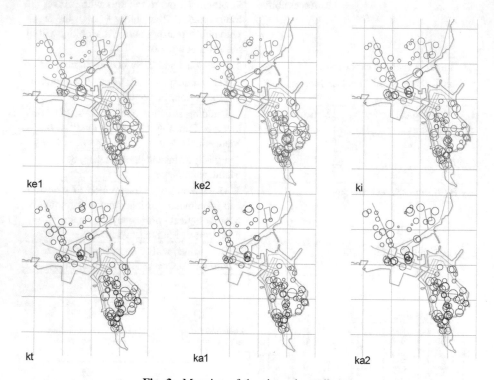

Fig. 3. Mapping of the six main attributes.

Figure 4 shows a summary of the data. In the image on the left, within the ranges of variation of the different characteristics, and the normalized prices from 1 to 5, the gray points indicate the first and third quartiles, while the central one in red is the median. On the right, the sample relative to the different quantitative, qualitative and economic characteristics the charts provide a synoptic view of the asynchrony between features and prices.

The ripples of probability density functions of the qualitative analysis show a certain heterogeneity of the sample. It is possible to appreciate how overall quality level prevails over all aspects and how they are in some way compensated, resulting in a sufficiently compact distribution from the aggregate quality index k^*.

The prices distributions (€/room and €/sq.m) are compact and asymmetrical; the most frequent price occurs on medium-low levels, suggesting the potential of such

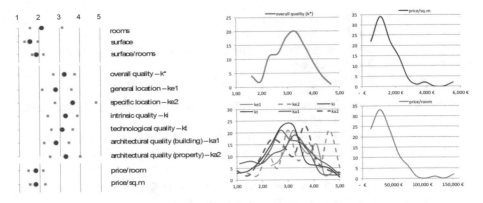

Fig. 4. Variables ranges, characteristics and price distributions

asymmetric market. The long right tail of the distribution indicates the presence of excellent properties, resulting in very high unit prices.

3 Methods

Methods that aggregate spatial units to partition a larger area are mainly based on cluster analysis [3]. Those approaches are useful techniques in which the creation of new areas regarding grouping aims to identify homogeneous parts concerning a set of variables. The defined groups are characterized by being relatively homogeneous within them and heterogeneous among each other. A defined set of variables evaluates homogeneity (similarity within-clusters) and heterogeneity (diversity between clusters).

The Cluster Analysis is a multivariate method, which aims at classifying of observations into some different groups based on a set of measured variables. The degree of association between two objects belonging to the same group is maximal, but if they belong to a different cluster, it is minimal. The cluster analysis helps to identify groups and their structures within the data and analyze those clusters of similar observation rather than individual data. Moreover, cluster analysis portrays relationship not revealed otherwise within the observed data, developing taxonomies.

Cluster analysis has been widely used in real estate analysis in order to segment sub-markets and partition different areas of the cities. The method has significant implications for explaining how the real estate market works [4–6].

3.1 Hard Clustering

In the partitional approaches, we used the K-mean method, which aims at grouping data into K clusters based on how close is an observation to the mean of the data in each cluster. The method segments the data, minimizing the within-cluster variation. The steps in the process are different, consisting in assigning, randomly, each observation to a K cluster, reassign the observations to other clusters to minimize the within-cluster variation, which is the squared distance of each observation from the mean of each cluster. Finally, the last step consists in repeating the process until no data needs to be reassigned. As K-means method does not build a hierarchy (the cluster affiliation of data could change during the process), the approaches belong to the non – hierarchical clustering approaches.

A process for determining the optimal number of clusters is [7]:

- assumed the dataset X, a specific clustering algorithm and a range of some clusters [M_min, M_max], are defined;
- the clustering algorithm is repeated from predefined values of M_min to M_max;
- the clustering results (partitions P and centroids C) are obtained, and then the index value for each of them is calculated;
- the cluster M is selected, for which the partition offers the best outcome according to some criteria (minimum, maximum or knee point).

3.2 Fuzzy Clustering

Fuzzy clustering methods use the theory of fuzzy sets and allow to associate a unit with groups with a certain degree of belonging, expressed by a membership function that assumes values within the range [0, 1] (none or complete similarity). Each data belong to multiple clusters, and the sum of the membership of each point on all clusters must be = 1. The attention in these methods arises from the awareness that there is some degree of imprecision in the data and that such a technique can represent them more than a crisp approach. Fuzzy clustering methods are richer in information as they provide the degree of consistency of a unit with each cluster, enabling a hierarchy of groups to be established (hierarchy is given by the different degree of belonging to the group) to which may belong to unity because groups are seen as fuzzy sets. Non-hierarchical classification methods have the ability to supply a certain number of a priori fixed groups directly, through iterative procedures that try to optimize an objective function.

Several algorithms differ for function adopted and therefore for the different iterative procedure chosen to calculate the degree of membership of the units to the groups.

The objective function determines for each solution a measure of the error based on the distance between the data and the representative elements of the clusters. We try to minimize the following objective function C [8] defined by cluster membership and distances. To verify the goodness of fuzzy clustering, i.e., the number of clusters, NCSS software requires the value of the Average Silhouette to be compared with the Dunn's partition coefficient and Kaufman's partition coefficient.

4 Results and Discussion

One of the most significant aspects of the K-Means cluster analysis is the not well-delineated clusters and not dominant features for clustering [9, 10]. The significant results of the Fuzzy cluster analysis can be observed in the degree of membership in the unit price/k* graphs, where we can appreciate that in the cluster boundary areas the degrees of membership progressively decrease (Fig. 5). The clear distinction between clusters based on the overall quality index suggests that the aggregation of the six criteria is satisfactory.

K-means cluster analysis, which uses binary description domains, represents this ambiguity in the overlapping of different clusters (Fig. 5). Fuzzy cluster analysis, which instead uses blurred domains, avoids this ambiguity by defining clustered groups and represents the reduction of the degree of membership of elements as they approach the next cluster border (Fig. 5).

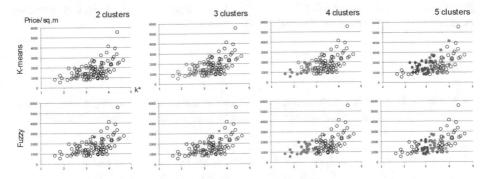

Fig. 5. Comparison between hard and fuzzy clustering results.

Some examples of these ambiguities can be highlighted in:

1. The case of properties in the highest cluster which do not occupy a favorable or valuable location (ke1 and ke2 low), and therefore has a low degree of membership (small bubble);
2. The presence in the same building of two properties belonging to different clusters but one of the two has a lower degree of belonging;
3. The presence in the same micro-urban area of several properties owned by the various groups and with the same degree of membership.
4. The wide range of prices in the highest cluster.

5   Conclusion

The study proposes an analytical platform for the real estate survey providing a multiplicity of queries and their synoptic view in the perspective of an interactive approach supporting the formation of multiple domains of descriptions. Based on 33 indices grouped in six main criteria distinguishing 101 sample properties, two domains of description were defined, aimed at delimiting the possible sub-markets, the first (K-means) of the binary type, the second of the Fuzzy type. The complexity of the context constituted a significant difficulty for implementing the two procedures [11, 12].

K-means procedure, performing binary distinctions, provided a weaker description of the sample, measured by low significance indices, and displayed by the overlap of the clusters in the chart showing the relation between unit price and overall quality.

Fuzzy clustering, performing a more adaptive distinction pattern, provided a stricter delimitation of the clusters that is significant for a sample of properties that meet diversified preferences structures and select agents differently motivated.

While in this first phase of experimentation the definition of the sub-markets concerned the characteristics, in reality, the description of the structure of this real estate market as a whole, needs to be represented taking into account prices as well.

Such integration will make evident the relationship between price map and urban equalization in urban fiscal policies, especially in the perspective of the relevant gentrification processes that is going to extend to marginal districts [13], supported by negotiation pattern based on the real price dynamic [14].

References

1. Gabrielli, L., Giuffrida, S., Trovato, M.R.: Functions and perspectives of public real estate in the urban policies: the sustainable development plan of Syracuse. In: ICCSA 2016, vol. 9798, pp. 13–28 (2016)
2. Forte, C.: Elementi di estimo urbano. Etas Kompass, Milano (1968)
3. Tryon, R.C.: Cluster Analysis: Correlation Profile and Orthometric (Factor) Analysis for the Isolation of Unities in Mind and Personality. Edwards Brothers, Ann Arbor (1939)
4. Hruschka, H.: Market definition and segmentation using fuzzy clustering methods. Int. J. Res. Mark. 3(2), 117–134 (1986)
5. Fletcher, M., Gallimore, P., Mangan, J.: The modelling of housing submarkets. J. Property Invest. Finan. 18(4), 473–487 (2000)
6. Goodman, A.C., Thibodeau, T.G.: Housing market segmentation and hedonic prediction accuracy. J. Hous. Econ. 12(3), 181–201 (2003)
7. Gabrielli, L., Giuffrida, S., Trovato, M.R.: From surface to core: a multi-layer approach for the real estate market analysis of a central area in Catania. In: Gervasi, O. et al. (eds.) Computational Science and Its Applications-ICCSA 2015, vol. III, pp. 284–300. Springer (2015)
8. Kaufman, L., Rousseeuw, P.J.: Finding Groups in Data. Wiley, New York (1990)
9. Gabrielli, L., Giuffrida, S., Trovato, M.R.: Gaps and overlaps of urban housing sub market: a fuzzy clustering approach. In: Stanghellini, S., et al. (eds.) Appraisal: From Theory to Practice, vol. 15, pp. 203–219. Springer, Cham (2017)
10. Giuffrida, S., Ventura, V., Trovato, M.R., Napoli, G.: Assiologia della città storica e saggio di capitalizzazione Il caso del Centro storico di Ragusa Superiore, Valori e valutazioni, 41–55 (2017)
11. Hwang, S., Thill, J.C.: Delineating urban housing submarket with fuzzy clustering. Environ. Plan. B: Plann. Des. 36, 865–882 (2009)
12. Bourassa, S.C., Hoesli, M., Peng, V.S.: Do housing submarkets really matter? J. Hous. Econ. 12(1), 12–28 (2003)
13. Bencardino, M., Nesticò, A.: Demographic changes and real estate values: a quantitative model for analyzing the urban-rural linkages. Sustainability 9(4), 536 (2017)
14. Della Spina, L., Scrivo, R., Ventura, C., Viglianisi, A.: Urban Renewal: Negotiation Procedures and Evaluation Models. In: Gervasi, O. et al. (eds): Computational Science and Its Applications-ICCSA 2015, Lecture Notes in Computer Science, vol. 9157. Springer, Cham (2015)

Regional Development Policies in Italy: How to Combine Cultural Approaches with Social Innovation

Andrea Billi[1] and Luca Tricarico[2(✉)]

[1] La Sapienza University of Rome, 00185 Rome, Italy
[2] Politecnico di Milano, 20133 Milan, Italy
luca.tricarico@polimi.it

Abstract. Over the years, the cultural enterprises sector in Italy has been considered a field of confrontation for regional development policies and institutional reorganizations, with a twofold challenge: experience new entrepreneurial cultures borrowed from other policy fields and enhance a considerable (but hardly recognizable) cultural heritage and assets. This paper intends to analyze the current state of cultural entrepreneurial policy making in Italy by examining critical points in terms of its capacity to (1) produce value, (2) include youth and (3) promote inclusive forms of urban and local developments. By combining these three objectives, the paper aims to outline a general reflection on how innovative organizations and policy tools could be explored to promote a cultural innovation agenda for the country.

Keywords: Cultural policies · Entrepreneurship · Social innovation
Urban and regional development

1 Introduction

Over the last 10 years, both academics and policy makers have highlighted the importance of cultural and creative industries as key drivers in facilitating economic and social development processes [1]. This increasing interest in culture-led strategies, both at a national and international level, also indicated the lack of a clear and unambiguous definition of cultural and creative enterprises/industries (CCIs) and in particular, the interactions between the two sectors [2]. The connection between culture and creativity (recently emphasized by the EU approach to CCis) has always been considered stronger and more direct than the link between culture and social innovation, which might lead to new and unpredictable potentialities[1] [4]. Despite the wide diversity in perspective and research tools, there is a significant convergence of opinion on the connection between cultural production, social cohesion and economic development. This means that this sector may be able to produce a smart, sustainable and

---

[1] In recent years, new and interesting initiatives, often promoted by youth, created new implications in terms of social inclusion and urban regeneration, even in a period of decreasing public subsidies [3].

© Springer International Publishing AG, part of Springer Nature 2019
F. Calabrò et al. (Eds.): ISHT 2018, SIST 101, pp. 277–287, 2019.
https://doi.org/10.1007/978-3-319-92102-0_30

inclusive growth[2] driving other productive sectors in innovative approaches. This link is particularly interesting in the Italian experience, both in the high level of youth unemployment and in the relevance of cultural heritage. The main picture of the evolution of the cultural sector is shown by the Fondazione Symbola, which is considered as the benchmark in the definition and measurement of the CCIs. In the sixth report, "Io sono cultura", the contribution of the "Cultural and Creative System" to the Italian economy is considered to be particularly relevant, especially if we consider the *core* cultural activities and the *culture-driven* activities (tourism, handicraft, etc.). Symbola estimated that the Cultural System creates about 6% of Italian wealth which is €89.7 billion with about 1.15 million people employed in this field, which is 6.1% of the total workforce in Italy. The analysis by Symbola (2016) suggests two interesting points: (1) During the period 2008–2014, most of the core cultural activities went through a marked slowdown also due to the important reduction of public transfers (−8.1% in 2008; [5]), (2) the trend of the System as a whole reveals a substantially positive picture, especially thanks to the "culture-driven" sector, whose performance was generally above the average of other productive fields in a period of general recession. The cultural and creative sector is important, not only in terms of creation of value and employment but also as an element of innovation which is able to produce interesting indirect effects. However, the impact of culture-led strategies follows the traditional economic growth geography: it is more important in northern Italy and especially in the big urban centers or medium-sized cities with a high rate of development and standard of living [6]. This proves that certain local communities are only able to optimize their local *assets* if they are grounded in a good socio-economic and political context. The key to boost bottom-up and community based initiatives is a clear and shared strategy and good administrative governance [3, 7, 8]. The limitation is that cultural and creative industries are more likely to grow in ecosystems already rich in opportunities and capabilities. For this reason, it is important to set an innovative agenda in supporting the economy of the culture and to use creativity as a tool to reduce inequalities between territories, metropolis and small urban settings, outer suburbs and towns, accelerating social innovation experiences[3]. The pressure for innovative culture-led policy making must be linked to include opportunities for young people, who can play a role in bottom-up innovative energy[4] and reduce the high percentage of NEETs (not in education, employment or training) mainly in southern Italy. The experience of Taranto demonstrates that social innovation and culture activities can boost each other [10]. With respect with this theoretical background, the present contribution provides an interpretive framework for the role of social innovation and institutional actors in designing an innovative approach to culture-led development agendas in Italy. Section 2 analyses the recent trends and experiences of youth engagement in Italian enterprises compared to other Europe contexts while Sect. 3

---

[2] In accordance with the objectives of the Treaty of Lisbon available on http://ec.europa.eu/europe2020/pdf/annexii_en.pdf.

[3] New approaches could overcome traditional policies linked to public subsidies which have proved to absorb a significant amount of public resources [9], without reducing inequalities, especially for young people.

[4] The percentage of employed people aged 25–34 years is 22.7% while for other jobs it is 17.9% [6].

describes the main changes in national public policies concerning the cultural sector in Italy; the final section is devoted to the description of possible organizational and policy innovations to maximize cultural and regional developments.

2 Cultural Enterprises in Italy: No Youth Means no Innovation

Culture and creativity are often considered as key resources for Italy, a country "which has culture in its DNA" and could use them to overcome the economic crisis and tackle future social and economic challenges. This scenario depends on the creation of a new "Italian way" for the development of this sector: on the one hand, by proposing an original path which does not follow development models conceived in other contexts and on the other, by going beyond the dichotomy of protecting artistic heritage while promoting contemporary cultural production. In other words, it is necessary to re-evaluate a "performative" dimension of our cultural sector in addition to the contemplative one. We need to harness the energy and know-how which gave rise to our heritage to promote a model diffused and shared by local communities which goes beyond the intensive commercial exploitation of large attractions [11]. This development model must take into account the inextricable link between the culture and creativity sectors with a multi-dimensional declination of sustainability, made from the interwoven aspects of: (1) economic performance: assuring a future to cultural projects in spite of cuts to public transfers; (2) technological innovation: a common willingness to innovate and to explore new value-creating models, exploiting the new opportunities related to the advent of the fourth industrial revolution; and above all, (3) youth engagement: widespread involvement of new generations both in the existing organizations and as promoters of new entrepreneurial projects. The data available on the cultural sector in Europe for the five-year period 2008–2014 [5], shows that in Italy, the 15–29 age bracket has the lowest index for youth participation in the labor market (12%). Furthermore, the percentage regarding the cultural and creative sector is even lower (10%), 8% less than the 28% EU average (Fig. 1). Other surveys included in the young workers' bracket (people aged up to 39), analyzed a wider range of cultural activities. In this case, the surveys show the presence of "young workers" is higher than the average in other sectors. The recent "Italia Creativa" report [12] observed a positive deviation of 4% in the relationship between "young" workers (15–39) compared to all other economic sectors. In the same period, the report noticed the strong presence of freelance professionals and an extensive use of temporary workers as well. Indeed, the report shows a stronger and increasing presence of project contracts and other forms of atypical work but it does not provide specific data on the matter. As the report underlines, the absence of thorough statistical sources explains lack of public interest in the matter[5]. If we want cultural and creative professions to continue producing

---

[5] Specific data on the entrepreneurship of young people in the creative sector are also missing. We can suppose that young people may have the same difficulties of other sectors in creating and strengthening new entrepreneurial activities.

opportunities for development in Italy, it will be necessary to re-assess the work policies for this professional category [13]. Moreover, the report by Global Entrepreneurship Monitor [14] shows this cultural and structural weakness ranks Italy the penultimate out of 60 countries because of the fear of failure in an entrepreneurial initiative. This fear seems to be confirmed by the Eurostat data (2016) on the survival of new cultural enterprises created in Italy in 2008. Indeed, only 30% survived at least five years (Fig. 2). A mixture of social, economic and regulatory issues makes the Italian labor market more complex than ever and favorable to those who already work.

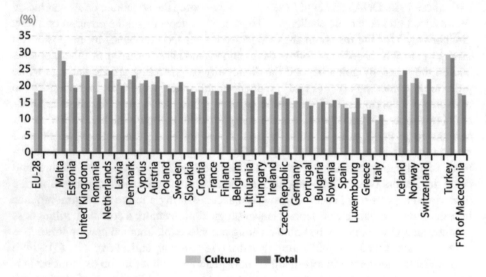

Fig. 1. Employment rate of people aged 15–29 in the cultural sector in Europe compared to the general index. Elaborated on Eurostat data (2016).

The problem is not only related to the Italian context; 25% of young people in middle-income countries and 15% in rich countries are NEETs [15]. The cultural and creative sector, characterized by high accessibility and low necessity of investments, does not avoid the issue of the exclusion of new generations from the labor market [16]. According to ICOM's evaluations (Institute for Competitiveness), in recent years, the missing involvement of NEETs has affected the Italian GDP by about €36 billion with a loss of €15 billion in terms of revenue [17]. Furthermore, the lack of inter-generational labor mobility and the obsolescence of the ruling classes in Italy, affect almost every professional field [18]. This conflicts with the willingness of young people to accept challenges, the incessant calls for new business models, the exhortation to tackle current changes instead of avoiding them, and to adopt new technical and organizational paradigms for value creation. Istituto Toniolo's data [19] is quite clear on the entrepreneurial propensity of young Italians. The sample shows 55% of the interviewees consider the "ability to adapt as the most useful element in finding a job" and 91% agree with considering work "as a direct tool for earning an income". Difficulties and uncertainties affect their vision of the future and social trust: "In particular,

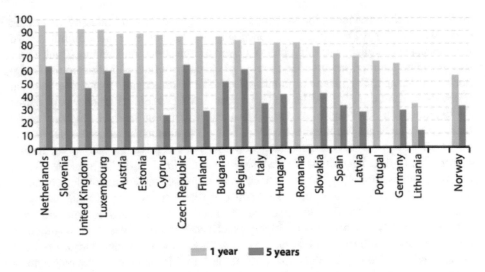

1 year   5 years

Fig. 2. Survival of European cultural enterprises created in 2008, one and five years later. Elaborated on Eurostat data (2016).

78% of Neets perceive the future as risky and unsure while among those who study, the percentage falls to 72%. Less pessimistic are those who have a permanent contract: 65%". So the younger generations can be perceived as "the great absentees" in the ongoing transformation which is recalled more than practiced or, as an extraordinary well of under-utilized potential for the innovation of the cultural and creative sector.

3   The Ongoing Change: Institutional Reform and *Start-up* Culture

The new cycle of the European Funding 2014–2020, brought profound changes to the institutional/political framework which leads territorial development in Italy. The cultural and creative sector is now considered as a key driver in national and regional strategies funded by the EU Structural Funds and some changes can represent an interesting opportunity to innovate local development policies, either combining culture and social innovation approaches or the multilevel governance (national/regional/local) of culture led strategies. The main innovation regarding cultural policies was the reform of MiBACT (Ministry of Cultural Heritage and Tourism) which in 2014, had been reorganized both at central and territorial levels, emphasizing the role of each structure (museum, archeological areas, etc.) as a promoter of local economic development, linking the cultural mandate with tourism. In Italy, the preservation and restoration policies were historically driven by the Ministry through its detailed territorial network, while local development was driven by the regions and municipalities, with the concrete problem of territorial governance of culture-led strategies (and funding). A result of this new approach was the implementation of the National Operational Plan (PON) 2014–2020 Culture and Development, directly managed by the MIBACT, emphasizing the role of the Ministry as booster and coordinator of cultural policies, reducing the timeframes of programs and introducing a specific strategic axis for supporting entrepreneurship in the creative and cultural sector. Similarly, the establishment of an Italian system of museums conferred to 20 museums of overriding national importance, the role of executive office and allowed the designation of managers by means of public selection at international level. Other relevant innovations can be mentioned here: (1) the creation of 17 regional centers (responsible for the other non-autonomous museums) as a real structure of local development rather than simply providing cultural activities; (2) the implementation of new rules on tax credit (*Art Bonus*) which emphasize a new role for the private sector in supporting cultural heritage and activities; (3) a considerable growth of MiBACT's budget, which has been almost doubled. These changes have addressed new cultural developments which previously had never been recorded over the last 20 years. In this changing framework, the MiBACT also increased its reputation, especially in southern Italy where the MiBACT's action is positively perceived because it acts as an institutional facilitator, and some directors of the new structures play a relevant role in symbolic change, which goes beyond their institutional role[6]. In this scenario Young engagement in cultural entrepreneurship must play a significant role, not least for the symbolic value of their activities. Therefor analyzing certain multi-sectorial policies in the support of cultural and creative industries, we can understand the main issues in dealing with reducing the exclusion and difficulties faced by young people in establishing new initiatives. These specifically concern three main kinds of policy: (i) policies for creating cultural enterprises, (ii) youth policies, and (iii) policies for innovative enterprises (*start-ups*). (i). A recent article published by Nuccio and Ponzini [20] highlights the inaccuracy of the assumption that public investment in culture generates local economic development. Research into the policies aimed at promoting the development of "cultural districts" based on the model of the industrial districts [21], showed the measures are barely efficient and elements for evaluation are lacking. On the one hand, we can confidently assert that this moment led the Italian Cultural and Creative Sector to be part of a *Smart Specialisation* strategy. This is due to a relevant availability of resources as well as to a real effort by public administrations to realize institutional structures in order to foster new initiatives[7]. On the other hand, at present, the interest is imbalanced and it tends to foster investment with a strict procedure which is hardly adaptable to the experimental features of creative enterprise: there is no clear distinction

---

[6] The winning 2019 ECoC (European Capital of Culture) bid of Matera represent an effective evidence of this trend. This may not only represent an opportunity in institutional terms, to define and experience stronger coordination between national, regional and local actors but also in structural terms, to give a more defined (and reproducible) aspect to some innovation experience by better identifying the relationship between culture and social innovation.

[7] With regard to the policies supporting innovative enterprises (start-ups), since 2012 the central government and sometimes even the Regions strongly committed in order to promote the establishment of ecosystems fostering new enterprises characterised by a high technology and scalability. The law describes a specific condition for start-ups "with a social impact" that includes those which "provide cultural services", in particular for "enhancing heritage and cultural goods" [22]. The program was created in order to support venture capital through tax- cuts for investments, labour contracts and funding for enterprises' incubation and acceleration programs.

between profit and non profit or even other pre-set legal forms[8]. (ii) Public policies can create a very positive and stimulating environment for youth entrepreneurship, if they are implemented using an appropriate approach and not simply as placement services for young people. Informal education, training on the job for example can be key factors to reach the young as target of those policies, which should allow young people to take a leading role. Only if they feel part of a true changing process, will they feel responsible to develop/implement ideas and projects which can become business opportunities. At current state of the Italian youth policy framework the principal tool *Fondo Nazionale Politiche Giovanili* (National fund for Youth Policy) has been reduced in the last years for the 96% of its original amount, due to a strict judgment of the *Corte dei Conti* (Italian Court of Auditors) on how to use the huge resources put on budget [23]. (iii) Unlike policies supporting cultural enterprises, the policies concerning *start-ups* diffused the culture of experimentation, collaborative and innovative logic, the recognition of the value of a widespread knowledge and a bottom-up innovation. However, the *start-up* culture is strongly oriented to conceive projects characterized by a high scalability and to promote profit opportunities for investors. On the other hand, it neglects the relationship with the territory and cultural identity. This approach is risky because it could create many platforms and apps which do not take into account the territories' resources and consequently, do not involve the several actors necessary to overcome the obstacles of the cultural and creative *business model* in order to assure a widespread and inclusive development. An apparent increased accessibility of the "start-up" culture often corresponded to severe selection criteria according to "picking the winner" logic which has been diffused and experienced for years in other national and international contexts [24]. According to a recent report published by GP Bull-hound [25], if we consider the European start-up ecosystem, the labor market selection is actually lacking young entrepreneurs to explore new innovative models. In the same ecosystem, most successful *start-ups*[9] have been established by people over 30 and more than half, by people over 40. Most of these enterprises rarely have a connection with the existing entrepreneurial ecosystems which causes confusion in the use of different language and meanings and consequently, creates mistrust among actors in the traditional sectors. Moreover, in Italy, the remarkable public investment in business incubators produced minimal income: 14 official incubators were promoted by universities and public institutions out of a total of 30 which in 2015, had right to facilitation provided by the Decree 2.0[10]. According to the European Business and Innovation Centre Network [26], in the three-year period 2010–2012, creating a job in our country through business incubators cost approximately €38,000 with a public investment of 68% compared to 23% in Germany. Germany is the leading economy in Europe regarding start-ups, employing an average of 14.6 workers (compared to 2.9 in Italy). In conclusion, the systems supporting *start-ups* are interesting but should be

---

[8] The lack of clarity in the procedure caused some controversies. An example on the notice Cultura Crea is described at the following article published for the online magazine Vita.it (http://bit.ly/2CU5EuS).

[9] We mean successful models which consider the so-called *Unicorn* as successful. They are start-ups which are worth a billion dollars with obvious problems of inclusion and diffusion.

[10] A decree's overview is available at: http://bit.ly/2dAXv2h.

seen as a part of a wider system and by themselves, cannot support the development of a widespread cultural and creative innovation and entrepreneurship. Entrepreneurship and innovation, in the cultural and creative field are perceived to be the result of a process that is difficult to launch, even with public funding and support to *start-ups*. For cultural enterprises organizations we mean those with few accounting and administrative competences, lacking in capital, dealing with activities whose intangible assets are often *core business* and consequently, whose models are very different from traditional enterprises [27]. According to a recent study by the European Commission [28], cultural enterprises often lack technical abilities in financial relations, consequently causing a considerable level of mistrust by the banks. This mistrust is confirmed by the fact that no credit institution would entrust or finance them because of their difficulty in communicating with an ecosystem unable to transmit its value. At present, social impact investment seems the only possible tool to interact with projects characterized by inadequate definition, high complexity and a functioning different from the logic of ventures usually supporting the *start-up environment*.

4 What Is Missing? Combining Cultural Innovation and Territorial Development

In light of the analysis conducted in this paper, we can elaborate some considerations on the policies able to promote new territorial and cultural innovations. The term "cultural innovation" refers on the one hand, to the introduction of "elements of change" in policies supporting the production and use of cultural goods and services and on the other, to experiences of *social innovation through culture*. These include practices and projects carried out by communities to fulfil their needs by using, producing and exploiting tangible and intangible assets. With regard to the organizations mentioned in Italy, there are increasing experiences of *diffused innovation* [29], born and supporting each other by means of a combination of technology and creative allocation of the resources of the territory. These organizations are hybrid [30], ranging from profit and non profit, established in the middle of the economic crisis, not only to meet emerging social needs but also to create opportunities for self-employment. They have a strong entrepreneurial vocation and are almost militant bearers of demand for change. They act in several sectors of great knowledge, they use new technologies maintaining a strong connection with the territory, they change progressively by adapting themselves to their context which provides a significant capacity to forestall tendencies and needs, and to transform problems into opportunities. In Italy, they are called social and cultural start-ups but they rarely develop *business* models able to meet the interest of ventures. At the same time, they struggle to find public support measures (insufficient economic and financial solidity, hybrid legal forms). The value generated by those experiences does not translate into (moderate) economic aid of activities but rather into the strengthening of the social cohesion they allow. This happens especially in contexts where public policies and private actors are inefficient in the experience and dissemination of new economic practices. We discuss economies and organizations working on different levels of disintermediation in services for communities, individuals and several local actors by overcoming limits through the complex *community-based* process of the establishment of relational, financial and spatial *assets* [31]. European institutions, particularly related to urban contexts, showed a growing interest in the recent European Pact of Amsterdam [32], especially with regard to the *City Makers* matter. The contribution of this "community of practices" is the ability to innovate and organize independently, alternative forms of urban development which are mainly tied to the creation of new social spaces powered by the contexts in which they act. In Italy, research on these matters was carried out by Avanzi with Segnali di Futuro for the metropolitan area of Milan. The subject on a national scale is analyzed in the report *Community Hub* [33]. The mapping of the national context shows that entrepreneurial innovation experiences are diffused and distributed in northern, central and southern Italy as well ranging from large cities to small towns and the outlying areas. The experiences mapped and analyzed, demonstrate a significant difference between *Community Hub* and traditional cultural enterprises. The starting point is not the supply of resources supporting activities but rather the definition of the activities themselves, based on existing territorial resources after having involved and collaborated with the communities. With regard to the policy tools, they could be used to enhance the development of bottom-up cultural innovation initiatives. The tools are different and can cause new challenges which the System should be able to comprehend: (1) the *social impact investments/finance* can be a concrete incentive for cultural innovation projects because it puts the focus on the social impact of investments in addition to the focus on financial sustainability. We need to consider that the measures experienced up until now need a proven demand for investment which is to say, new social impact entrepreneurial projects which would not have arisen without the experiences; (2) the potential of development of projects tied to the *Manufacture 4.0 and to digital handicraft*, the radical transformation of production systems through the merging of technologies and new customization opportunities. The intangible component of production is growing; the power of producers and consumers can become a stimulus for the return of manufacture to urban centers. The FabLab experience teaches us that hybrid artistic and manufacturing experience spaces can promote positive externalities such as the development of a circular economic circuit and sharing between close and virtual communities [34]. Even though our country is always behind in terms of introducing innovation, it is ranked first in the world for the number of FabLabs. According to a recent census published by Fondazione Make in Italy [35], this data also depends on the fact that more than half of the FabLabs rely on volunteer work; (3) The *policies for Urban regeneration and revitalization of the outskirts* can play a fundamental role in promoting bottom-up initiatives. Utilization of unused public[11] and private heritage could contribute to the creation of 73,000 jobs reducing unemployment to 4.8% [37, 38][12]. We need to take into consideration that the quality of initiatives for urban regeneration depends firstly, on the real presence in the territory of a set of resources and ecosystems ready for new "cultural

---

[11] Further information on the potential resulting from the entrepreneurial enhancement of public asset in Italy are available in the article of the Economist "Setting out the store" [36].

[12] To this end, the scope selected by the last *Bando per le Periferie on the "adjustment of infrastructures for social, cultural and educational services, as well as for cultural and educational activities promoted by public and private actors"* seems to be particularly relevant References to the notice are available on http://bit.ly/2Ff2Ldp.

infrastructures" and it cannot be addressed only by public guidelines. "Ecosystems" means the ability of local administrations to assure accessibility to infrastructures and cultural spaces, the advantages of philanthropy and finance to support development, but especially, the education of human capital which has to be able to use technology and informatics as well as to involve communities and territories in cultural projects with a bottom-up approach.

References

1. Sacco, P.L., Ferilli, G., Blessi, G.T., Nuccio, M.: Culture as an engine of local development processes: system-wide cultural districts I: theory. Growth Change **44**(4), 555–570 (2013)
2. NESTA: A manifesto for the creative economy (2013). http://www.nesta.org.uk/publications/manifesto-creative-economy. Accessed 15 Oct 2017
3. Tricarico, L.: Imprese di Comunità nelle Politiche di Rigenerazione Urbana: Definire ed Inquadrare il Contesto Italiano. Euricse Working Papers **68**(14) (2014)
4. Ferilli, G., Sacco, P.L., Tavano Blessi, G., Forbici, S.: Power to the people: when culture works as a social catalyst in urban regeneration processes (and when it does not). Eur. Plan. Stud. **25**(2), 241–258 (2017)
5. Eurostat: Cultural Statistics, Statistical Books (2016). http://bit.ly/2aRbMs4. Accessed 15 Oct 2017
6. Symbola-Unioncamere, Io sono Cultura-Rapporto 2016. L'Italia della qualità e della bellezza sfida la crisi (2016). http://www.symbola.net/. Accessed 15 Oct 2017
7. Tricarico, L., Le Xuan, S.: Le community enterprises, un'esperienza britannica di rigenerazione urbana. Urbanistica Informazioni **249**, 63–65 (2013)
8. Tricarico, L.: Community action: value or instrument? an ethics and planning critical review. J. Architect. Urbanism **41**(3), 221–233 (2017)
9. Falasca, P., Lottieri, C.: Come il federalismo fiscale può salvare il Mezzogiorno. Rubbettino (2008)
10. Sonda, G.: Taranto: a social innovation lab. bottom-up urban regeneration practices. Tafter J. **91**, Novembre-Dicembre 2016. http://bit.ly/2HY8mmT. Accessed 15 Oct 2017
11. Caliandro, C., Sacco, P.L.: Italia reloaded: ripartire con la cultura. Il mulino (2011)
12. Ernst and Young (EY): Italia Creativa (2016). http://www.italiacreativa.eu/pdf/ItaliaCreativa.pdf. Accessed 15 Oct 2017
13. Montalto, V.: Economia culturale a trazione creativa. Nova Sole 24ore. (2016). http://bit.ly/29sdFJq. Accessed 15 Oct 2017
14. Gem Monitor: Global report (2016). http://www.gemconsortium.org/report/49480. Accessed 15 Oct 2017
15. The Economist, Generation Uphill, Special report: the young (2016). http://econ.st/23h1g2k. Accessed 15 Oct 2017
16. Rosina, A.: Neet. Giovani che non studiano e non lavorano. Vita e pensiero (2015)
17. ICOM: Studio I-Com per La Scossa: Giovani chi li ha visti? Il PIL mancato di una generazione fantasma (2011). http://bit.ly/2CSy2Oi. Accessed 15 Oct 2017
18. ISTAT: Rapporto annuale 2016 (2016b). http://www.istat.it/it/files/2016/05/Ra2016.pdf. Accessed 15 Oct 2017
19. Istituto Toniolo: Rapporto Giovani (2016). http://bit.ly/YXBNxw. Accessed 15 Oct 2017
20. Nuccio, M., Ponzini, D.: What does a cultural district actually do? critically reappraising 15 years of cultural district policy in Italy. Eur. Urban Reg. Stud. **24**, 405–424 (2016)

21. Becattini, G.: Distretti industriali e made in Italy. Le basi socioculturali del nostro sviluppo economico. Bollati Boringhieri, Torino (1998)
22. Ministero dello Sviluppo Economico (MISE): Guida per startup innovative a vocazione sociale alla redazione del "Documento di Descrizione dell'Impatto Sociale" (2015). http://bit.ly/2tVcwrU. Accessed 15 Oct 2017
23. Corte dei Conti: Sezione centrale di controllo sulla gestione delle amministrazioni dello Stato, Indagine di controllo su Fondo per le politiche giovanili (2013). http://bit.ly/2FpLfB0. Accessed 15 Oct 2017
24. Cantner, U., Kösters, S.: Picking the winner? empirical evidence on the targeting of R&D subsidies to start-ups. Small Bus. Econ. **39**(4), 921–936 (2012)
25. GP Bullhound: Indipendent Technology Research. European Unicorns: Do They Have Legs? (2015). http://bit.ly/1HJzxh9. Accessed 15 Oct 2017
26. European Business and Innovation Network Centre (EBN): Ec-Bic Observatory 2013 and the last 3-Year trends, (2013). http://bit.ly/2npPbfO. Accessed 15 Oct 2017
27. Tricarico, L., Geissler, J.B.: The food territory: cultural identity as local facilitator in the gastronomy sector, the case of Lyon. City, Territory Architect. **4**(1), 16 (2017)
28. European Commission towards more efficient financial ecosystems Innovative instruments to facilitate access to finance for the cultural and creative sectors (CCS): good practice report (2016). http://bit.ly/1OXmis3. Accessed 15 Oct 2017
29. Bonomi, A., Masiero, R.: Dalla smart city alla smart land. Marsilio Editori (2014)
30. Venturi, P., Zandonai, F.: Ibridi organizzativi. L'innovazione sociale generata dal gruppo cooperativo Cgm. Il Mulino, Bologna (2014)s
31. Tricarico, L.: Imprese di comunità come fattore territoriale: riflessioni a partire dal contesto italiano. CRIOS **11**, 35–50 (2016)
32. European Union (EU): Urban Agenda for the EU, Pact of Amsterdam (2016). http://bit.ly/293tcns. Accessed 15 Oct 2017
33. Avanzi, Dynamoscopio, Kilowatt, Cooperativa Sumisura, Community Hub: I luoghi puri impazziscono (2016). http://bit.ly/2eFw8Zh. Accessed 15 Oct 2017
34. Menichinelli, M.: Business models for fab labs. openp2pdesign. Org (2011). http://bit.ly/2oD2pUB. Accessed 15 Oct 2017
35. Menichinelli, M., Ranellucci A.: Censimento dei Laboratori di Fabbricazione Digitale in Italia. Rapporto di Fondazione Making in Italy CDB (2014). http://bit.ly/2HXz7I5. Accessed 15 Oct 2017
36. The Economist, Setting out the store: Advanced countries have been slow to sell or make better use of their assets. They are missing a big opportunity (2014). http://econ.st/1gslPiA. Accessed 15 Oct 2017
37. Campagnoli, G.: Riusiamo l'Italia: da spazi vuoti a start-up culturali e sociali. Gruppo 24 ore (2014)
38. Tricarico, L., Zandondai, F.: Local Italy, i domini del "settore comunità" in Italia. Fondazione Giangiacomo Feltrinelli, Milano (2018)

Economic Sustainability in the Management of Archaeological Sites: The Case of Bova Marina (Reggio Calabria, Italy)

Carmela Tramontana⬤, Francesco Calabrò(✉)⬤,
Giuseppina Cassalia⬤, and Maria Carlotta Rizzuto⬤

Mediterranea University of Reggio Calabria, 89100 Reggio Calabria, Italy
francesco.calabro@unirc.it

Abstract. This paper deals with the issue of economic sustainability in the management of cultural heritage. The first part is dedicated to theoretical aspects, while the second part presents a case study of an archaeological area in the province of Reggio Calabria (Italy). Regarding the theoretical aspects, the question of economic sustainability is placed in the debate on the so-called New Public Management, and highlights which variables affect the economic balance in the management phase of a site starting from the nature of the management entity and of the activities carried out.

Keywords: Cultural heritage · Management model · Economic sustainability

1 Introduction

Until a few years ago, statistics reported that until a few years ago the share of cultural activities in Europe barely exceeded 0.5% of GDP, while direct employment averaged 0.12% of the working population [1].

Nevertheless, the sector of cultural heritage plays an increasingly important role from an economic point of view, thanks to its effects on the functionally related sectors, more specifically on:

- the conservation supply chain, which is characterized by the high impact of human resources on spending units, not only in terms of workforce but also of the necessary professional skills;
- the cultural tourism supply chain, which in recent years has been characterized by a progressive growth in demand.

Conservation and enhancement of cultural heritage are relevant from a social point of view as well: they are the most contributing factors to the constitution of a community identity. As is known, strengthening local identity contributes to social integration and cohesion, pre-conditions that are essential to any development process.

This is the result of the joint work of the authors. Although scientific responsibility is equally attributable, Sects. 3.1, 3.2, 3.4, 3.5 were written by Carmela Tramontana; Sects. 1, 2.4, 5, were written by Francesco Calabrò; the paragraphs Sects. 2.2, 3.3, 4.1 were written by Giuseppina Cassalia; the paragraphs Sects. 2.1, 2.3 were written by Maria Carlotta Rizzuto.

© Springer International Publishing AG, part of Springer Nature 2019
F. Calabrò et al. (Eds.): ISHT 2018, SIST 101, pp. 288–297, 2019.
https://doi.org/10.1007/978-3-319-92102-0_31

For these reasons, in the last programming periods the lagging regions have invested substantial amounts of the European structural funds on the conservation and enhancement of their cultural heritage.

The several interventions for the physical recovery of artefacts are often not matched with the adequate organizational skills needed to manage such heritage. The result is something of a paradox: once they have been recovered, the artefacts often end up being underused or unused.

The progressive cuts in public spending have prevented both superintendents and local authorities from taking on enough human resources as to ensure the adequate management of the sites. The involvement of private subjects, on the other hand, has often been the result of impromptu approaches rather than of a careful reflection on the management model best suited to the specific situation [2].

This paper presents a case study of the management plan of an archaeological area where significant involvement of the local community is foreseen, in the form of voluntary work provided by members of cultural associations. This type of solution will be more and more common in the future, if we want to guarantee the usability of the assets in a situation of increasingly lacking available economic resources.

More specifically, this contribution aims at thoroughly examining the evaluative and economic-estimative tools able to assess the feasibility and economic sustainability of innovative solutions for cultural heritage management.

2 Cultural Heritage Management

2.1 Cultural Heritage in the Era of the New Public Management

Focus on the protection and conservation of cultural heritage as a legacy for future generations is part of a broader debate on non-renewable resources for sustainable development [3, 4].

Within the subject of sustainability, the economic dimension, especially in relation to asset management, has assumed growing importance since a new concept of Public Administration, known as New Public Management has established itself.

The need for a careful rethinking of the functioning of the public administrative machine is the result of the excessive amount the public debt reached in the 1990s, which made the levels of public spending unsustainable for western economies [5, 6].

However, the new international equilibrium created by the fall of the Berlin Wall and the lack of a need to build consensus through public spending are certainly not foreign to such rethinking [7, 8]. The progressive contraction in spending that affected almost all the public sectors coincided, at least in Italy, with widespread dissatisfaction towards the quality of the services provided [9].

All this generated forms of managerialization of the Public Administration [10, 11] that follow the so-called "3 E rhetoric", where the 3 E's stand for Economics, Effectiveness, and Efficiency. Given the limits of the present contribution it is not possible to judge its results here. In addition to these, other principles have been introduced into the Italian legal system by the process of reform of the Public Administration, which has been implemented mainly during the 1990s [12]:

- autonomy and decentralization;
- planning;
- accountability;
- opening up to the market.

The latter aspect is of particular relevance in the context of this contribution. Different strategies of opening up to the market have been implemented: either by focusing on the needs of the user, with an approach similar to that of private companies or through organizational models of services open to the contribution of private subjects [13, 14].

All this entailed a shift in the focus on the needs of budget balance, a condition essential for involving private individuals in the realization of actions of public interest [15].

2.2 Management Modelling of Cultural Heritage

Machu Picchu is one of the few sites that can generate enough revenue from visitors to cover all management costs, included those for conservation and restoration [16].

These are very particular situations, in which the attractiveness of the site contributes in one way and the economic conditions of the area hosting it in the other. In these cases, an entrance ticket of the same price has a different weight in different economic contexts. In economically disadvantaged countries entrance tickets to touristic sites represent a very significant income which can, as in the Peruvian case, guarantee economic equilibrium.

In the US, cultural institutions, such as museums, survive with very few public subsidies. The Museum of Modern Art in New York operates on a budget which is covered only by 5% by public subsidies, while the remaining 50% and 45% are covered respectively by revenues derived from the activities of the museum, and by revenues from other resources (generally private donations) [17]. In this case revenues directly derived from museum activities could not guarantee the economic sustainability of the structure: budget balance is achieved thanks to the practice, common among the citizens of that context, of financially supporting the services that the community deems relevant.

These cases led us to consider a management model of sites and structures closely linked to the context: every situation requires a model tailored on its specific context, as instructed by the "contextualist approach" [18].

2.3 Heritage Management

In Italy, the forms of management of cultural heritage are regulated by Article 115 of the Code of Cultural Heritage and Landscape [19], which identifies two main forms: direct or indirect. In the economy of this contribution, we will focus on the indirect form.

Indirect management is implemented by using the institution of the concession to third parties of the promotion and development activities through public tender procedures based on a comparative assessment of specific projects.

As specified in Paragraph 4 of the above mentioned article 115, the purpose of assigning the site to an indirect management is to ensure a higher enhancement level of the cultural heritage.

Paragraph 4 introduces a fundamental concept: the choice between direct and indirect management is not made arbitrarily, but through a comparative assessment in terms of economic and financial sustainability and effectiveness based on previously defined objectives.

2.4   A Proposal for Formalization

The need to guarantee economic-financial balance in the processes of cultural heritage enhancement is even more felt in presence of public-private partnerships. The approach used is grounded on a fundamental statement of fact: different cost structures can be associated to different management subjects, while the revenue structure for the specific asset in question remains constant.

Especially during the planning phase of interventions and for the purposes of activating forms of public-private partnership, it is necessary to identify, at a preliminary stage, the type of management subject most suited to the specific asset in question, in relation to its specific cost structure and the capacity of the asset to generate revenue.

Based on the nature of the activities and managing entity, it is possible to hypothesize the following three models [20] (see Table 1):

Table 1. Managing models by subject type.

Managing model	Nature of entity and activities
Model P - Profit	For profit entity, for profit activities
Model NP - Non-Profit	Non-profit entity, non-profit activities
Model M - Mixed	Non-profit entity, for profit activities

The economic-financial balance E_f in the management phase can therefore be considered a function of two variables:

V_1 - type of activities and involved subjects (public, for-profit private, non-profit private with for profit and non-profit activities)

V_2 - attractiveness of the asset (intrinsic attractiveness, extrinsic attractiveness).

$$E_f = f(V_1; V_2) \tag{1}$$

The variable V_1 is mainly related to the costs of human resources and the presence or lack of profit. As for human resources, they may constitute a cost for the non-profit organizations as well, if the activity takes the form of a for profit activity in all respects.

Repeated tests carried out by means of a common Cost and Revenue Analysis, in which the cost items change from time to time while the revenues remain constant, allowed to understand the following:

- the composition of investment capital (public, private, public-private mix);
- the type of management entity;
- the amount of charges that the manager subject may pay, to ensure the economic sustainability of the activity and the maintenance of the asset for future generations. In our previous article [20] a more exhaustive presentation of the model for the verification of the economic-financial sustainability of management models (MoGe) is given.

3 Case Study

3.1 The Archaeological Area of San Pasquale in Bova M. (Reggio C., Italy)

The object of study, the area of the Archaeological Park of Bova Marina, is located in San Pasquale, along the left bank of the creek of the same name. In 1983, during the works to enlarge the roadway of the new trunk road No. 106, a synagogue was found, thus allowing the acknowledgement of the Jewish presence in the area adjacent to the ancient Roman Regium. The site reached its development peak in Roman times, between the 2nd and 4th century AD. The archaeological remains were initially interpreted as belonging to a Roman villa, until, in one of the rooms was found a mosaic flooring with the image of the menorah, the Jewish seven-branched candelabrum that allowed scholars to identify the site as a synagogue. The synagogue went through at least two phases: the first during the 4th century AD, and the second in the 6th century, when it seems it was violently destroyed and the local Jewish community left the area.

The archaeological park is located 43 km south of Reggio Calabria and covers an area of about 6 ha. It extends in an area of luxuriant Mediterranean scrub, among olive and bergamot trees, and consists of the actual excavations, the Documentation Centre for Cultural Heritage and Judaism in the Greek-Calabrian Area, a café and the archaeological museum. The Superintendence for the Archaeological Heritage of Calabria owns the land where the archaeological park lies and the Antiquarium, while the Municipality owns the Research Centre for Cultural Heritage and Judaism in the Greek-Calabrian Area (see Fig. 1).

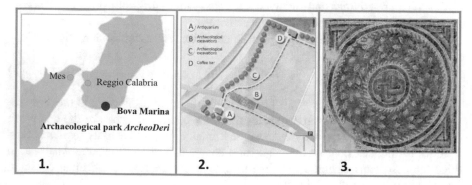

Fig. 1. 1. Location; **2.** Archeological area ArcheoDeri **3.** Synagogue's floor mosaic.

3.2   The Management Model

In the case in question, the preliminary considerations on variable V_2 of the MoGe model (intrinsic and extrinsic attractiveness of the asset) took into account the existing and potential flows of visitors, however limited. These considerations prompted to carry out the initial verification of the economic and financial sustainability of the management model in the hypothesis of a non-profit management subject that carries out non-profit activities.

The first verification was therefore carried out in the case that the provision of cultural services related to the Archaeological Park of Bova Marina would mainly be entrusted to the voluntary activities of local cultural associations and voluntary organizations in general.

At this time of economic crisis local administrations hardly cope with the most compelling needs of the community, and do even more so with those services, such as cultural services, that are not considered essential. The direct management of initiatives with a high cultural content becomes feasible only in those areas endowed with a good infrastructural apparatus and placed at the centre of large flows of trade and tourism that are not limited to seasonal periods [21, 22].

The main issue that could arise in the management phase, in which the fixed costs are relevant for the cash flows of a voluntary association, concerns the respect of the financial equilibrium. This is why such associations, especially during the start-up phase, rely on potential annual contributions from public institutions (municipality, metropolitan city and region).

3.3   Analysis of Current and Potential Demand

The key elements of the touristic offer of the province are the sea and the city of Reggio Calabria, which represent factors of competitiveness in the region. More specifically, the destinations on the coast of Reggio Calabria are receptivity basins throughout the year. This allows companies to implement seasonal adjustment strategies and achieve good touristic return (in terms of sales and territory usability) even in low season. In summer, peak season for this product, the growth of sales in the businesses of the coast allows the province to compete with the seaside touristic offer of the rest of the region.

As for multi-product integration, events and festivals organized in the area represent a further push. These moments increase the demand for tourism, support the local economy, and represent a good opportunity for publicity and the expansion of the catchment area of the territory, overcoming local borders and supporting businesses in the tourism sector.

The starting point is the particular interest of companies for cultural events, fairs and exhibitions of various kinds (e.g. music festivals such as Paleariza).

Besides the sea, which attracts 89.3% of tourists, the most visited places are historical centres (84.4%) and cathedrals and other religious places (41.7%). Moreover, the interest in the cultural offer of the area, mainly shown by foreigners, also determines a turnout of tourists in museums and art galleries (36.8%) and in places where traditional craft still has a strong presence (35.8%) [23].

Counting both presences in accommodation facilities and the estimate of those in second homes and rented accommodation, the total number of tourists in the province of Reggio Calabria almost reaches 1.6 million.

Bova Marina is part of this general framework of tourism demand and supply, as it occupies a strategic position in the heart of the Greek-Calabrian area, halfway between Reggio Calabria and the Locride area.

3.4   Current Offer Analysis

The intervention aims at improving the usability of the archaeological area and its surroundings, thus allowing to cater to different needs.

It aims at increasing the attractiveness of the area, improving the usability of the territory from a cultural and touristic point of view, and at including the visit to the archaeological park in local itineraries within the regional and supra-regional networks.

In the last year the following events were organized: five art exhibitions (two of ceramics, two of photography and one of painting); a concert; some conferences that have been attended by personalities of the national and international academic world. The park has also served as venue for the Week of the Neo-Hellenic Culture, carried out under the patronage of the Embassy of Greece in Rome, and for the 2011 Spring Day of the FAI (Fondo Ambiente Italiano - Italian Trust for Places of Historic Interest or Natural Beauty), besides for the presentations of books, films, etc. Finally, the celebration of the European Day of Jewish Culture is of particular importance, for which the Municipality is accredited by the UCEI (Unione delle Comunità Ebraiche Italiane - Union of Italian Jewish Communities), which has European significance.

Last year the total number of attendees has been of about 3,500.

3.5   Description of Activities and Cultural Events Planned

The planned activities and cultural events are as follows: Guided tours; Workshops; Internships for university students; Conferences; Library and multimedia archive; Sale of gadgets, souvenirs, catalogues, books, multimedia products, etc.; Day of Jewish Culture; Week of Neo-Hellenic culture; Art and photography exhibitions.

The activities inside the Park are all linked to the visits, specifically the visit to the excavations, Antiquarium and Documentation Centre.

4   Financial Profile

4.1   Costs

The following cost items have been identified: Purchase of materials for organizing activities, events and services; Promotion and marketing; Staff; Utilities; Overheads (see Table 2).

Costs for activities, events and services to the public are represented only by the chip offered to guides as a flat-rate reimbursement. The general operation costs of the park are the weekly cleaning of the premises and the annual update of the website (if it will prove to be an activity that volunteers cannot provide). Park maintenance will be

Table 2. Operating costs.

Outputs	
Type of cost	Full year
Purchase of materials for activities, events and services	€ 1,700.00
Promotion and marketing	€ 680.00
Staff	€ 50,630.00
Utilities	€ 9,800.00
Routine maintenance and major refurbishment (provision)	€ 16,000.00
Overheads	€ 7,900.00
Total	**€ 86,710.00**

entrusted to AFOR (Agenzia Forestale Regionale - Regional Forestry Company), which is already affiliated with the Municipality of Bova Marina and an integral part of the future management of the San Pasquale Park.

4.2   Revenues

The following revenue sources have been identified: Income from activities, events and services fees; Rental incomes; Membership fees and offers; Public financing; Other forms of co-financing (donations) (see Table 3).

Table 3. Revenues.

Type of revenue	Full year
Income from activities, events and services fees	€ 16,800.00
Rental incomes	€ 17,200.00
Membership fees and offers	€ 1,200.00
Public financing	€ 4,000.00
Other forms of co-financing (donations)	€ 50,000.00
Total	**€ 89,200.00**

The fees have been estimated considering a unit rate of EUR3.00 for guided tours and EUR10.00 for enrolment in laboratory activities, which will cover the costs of purchase of educational materials. Laboratory activities are limited in number (3 designed for children, 2 designed for youth and 2 designed for children with special needs and disabilities) and are included in the summer programme, which coincides with the tourist season. The internship, is aimed at university students in the archaeological sector (a group of 40 students), is offered annually and costs EUR60.00 per person. The expected total number of paying people is about 700; the induced flow of non-paying visitors, present both during workshops and the various events, amount to 3,500 people (with an annual increase of about 5%): it is just enough to generate the minimum income needed for the management of the refreshment spot.

The rental fee for the refreshment activity is EUR 1,000.00 per month for the first year, and EUR 1,250.00 per month for the following years. The rent was calculated on the basis of a market investigation on market values of commercial activities and restaurants recorded last year in the town of Bova Marina. These values range between EUR 1,200.00 and EUR 3,000.00 per month. The definition of the fee is therefore aligned with market prices and is entirely precautionary for the purposes of sustainability over time of the activity. The rent of the conference room ranges from EUR 200.00 to EUR 500.00 depending on the importance and duration of the event to be hosted.

The association is expected to count 50 members in total. Each member will pay an annual enrolment fee of EUR5.00 for youth and of EUR 15.00 for adults. The latter could pay, for the first year only, an additional fee for the start of the association. In addition, an additional money inflow is expected, thanks to possible private donations and offers from visitors.

5 Conclusions

In this case, the application of the MoGe model suggested to initially verify the hypothesis of creating a non-profit entity with non-profit activities.

The costs and revenues analysis confirmed the sustainability of the hypothesis, demonstrating how, even in the case of a cultural asset visited by a very limited number of people, it is possible to achieve an economic balance in the management phase.

The possibility of charging the managing operator for the costs of ordinary and extraordinary maintenance is especially relevant: in this way the sustainability of the management model is guaranteed, even if this requires a strong component of voluntary work and a significant fundraising activity.

This activity could initially be directed towards the numerous emigrants of Bova Marina who still show in various ways attachment to their place of origin.

The involvement of the associations is however an essential element to guarantee the usability of the asset and the sustainability of its management.

The research perspectives lean towards an improvement of the knowledge of the relationship between the Economic and Financial Plan of a single project and the budget of the management entity [24].

References

1. Greffe, X.: La gestione del Patrimonio Culturale. FrancoAngeli, Milan (2003)
2. Canesi, R., Antoniucci, V., Marella, G.: Impact of socio-economic variables on property construction cost: evidence from Italy. Int. J. Appl. Bus. Econ. Res. **14**(13), 9407–9420 (2016)
3. Della Spina, L.: The integrated evaluation as a driving tool for cultural-heritage enhancement strategies. In: Bisello, A., Vettorato, D., Laconte, P., Costa, S., (eds.) Smart and Sustainable Planning for Cities and Regions. Results of SSPCR 2017. Green Energy and Technology. Springer (2018). https://doi.org/10.1007/978-3-319-75774-2_40
4. Bandarin, F.: Patrimonio Culturale e Naturale e sviluppo economico. La dimensione internazionale. In: Greffe, X. (ed.) La gestione del Patrimonio Culturale, pp. 9–12. FrancoAngeli, Milan (2003)
5. Della Spina, L., Lorè, I., Scrivo, R., Viglianisi, A.: An integrated assessment approach as a decision support system for urban planning and urban regeneration policies. Buildings **7**, 85 (2017). https://doi.org/10.3390/buildings7040085

6. Panozzo, F.: Management by decree. Paradoxes in the reform of the Italian public sector. Scand. J. Manag. **16**(4), 357–373 (2000)
7. Rose, N., Miller, P.: Political power beyond the state: problematics of government. Br. J. Sociol. **43**(2), 271–303 (1992)
8. Hood, C.: New public management in the 80's: variation on a theme. Account Organ. Soc. **20**(2–3), 93–110 (1995)
9. Della Spina, L., Scrivo, R., Ventura, C., Viglianisi, A.: Urban renewal: negotiation procedures and evaluation models. In: Gervasi, O. et al. (eds.) Computational Science and Its Applications, ICCSA 2015. LNCS, vol. 9157. Springer, Cham (2015). https://doi.org/10.1007/978-3-319-21470-2_7
10. Ferlie, E.: Quasi strategy: strategic management in the contemporary public sector. In: Pettigrew, A.M., Whittington, R. (eds.) The Handbook of Strategy and Management, pp. 279–298. Sage, London (2002)
11. Flynn, N.: Public Sector Management. Sage, London (2007)
12. Zan, L.: La gestione del patrimonio culturale. il Mulino, Bologna (2014)
13. Della Spina, L.: Integrated evaluation and multi-methodological approaches for the enhancement of the cultural landscape. In: Gervasi, O., et al. (eds.) Computational Science and Its Applications, ICCSA 2017. LNCS, vol. 10404. Springer, Cham (2017). https://doi.org/10.1007/978-3-319-62392-4_35
14. Morano, P., Tajani, F.: Decision support methods for public-private partnerships: an application to the territorial context of the Apulia region (Italy). In: Stanghellini, S., Morano, P., Bottero, M., Oppio, A. (eds.) Appraisal: From Theory to Practice. Green Energy and Technology, pp. 317–326. Springer, Cham (2017)
15. De Mare, G., Manganelli, B., Nesticò, A.: Dynamic analysis of the property market in the city of Avellino (Italy). The Wheaton-Di Pasquale model applied to the residential segment. In: Murgante, B., Misra, S., Carlini, M., Torre, C., Nguyen, H.Q., Taniar, D., Apduhan, B. O., Gervasi, O. (eds.) ICCSA 2013, LNCS, Part III, vol. 7973, pp. 509–523. Springer, Heidelberg (2013). https://doi.org/10.1007/978-3-642-39646-5_37
16. Lusiani, M., Zan, L.: Assetti istituzionali e business model: prospettive su autonomia e sostenibilità. In: Zan, L. (ed.) La gestione del patrimonio culturale, pp. 89–96. il Mulino, Bologna (2014)
17. Frey, B.: Arts & Economics. Analysis & Cultural Policy. Springer, Heidelberg (2000)
18. Pettigrew, A.M.: Context and action in the transformation of the firm. J. Manag. Stud. **24**(6), 649–670 (1987)
19. Code of Cultural Heritage and Landscape, Legislative Decree No. 42, 22 January 2004
20. Calabrò, F.: Local communities and management of cultural heritage of the inner areas. An application of break-even analysis. In: Gervasi, O., et al. (eds.) Computational Science and Its Applications, ICCSA 2017. LNCS, vol. 10406. Springer, Cham (2017). https://doi.org/10.1007/978-3-319-62398-6_37
21. Morano, P., Tajani, F.: The break-even analysis applied to urban renewal investments: a model to evaluate the share of social housing financially sustainable for private investors. Habitat Int. **59**, 10–20 (2017). https://doi.org/10.1016/j.habitatint.2016.11.004
22. Campanella, R.: Un progetto di territorio per il turismo sostenibile. LaborEst **10**, 17–22 (2015)
23. Calabria 2014. Thirteenth report on tourism (2014)
24. Carbonara, S.: La stima dei costi del patrimonio edilizio privato nella ricostruzione post-sismica abruzzese: un'analisi critica delle procedure utilizzate. TERRITORIO, fascicolo 70, Franco Angeli (2014)

PLUS Hub: A Cultural Co-creative Enterprise for Local Urban/Rural Regeneration

Gaia Daldanise[1](✉)[iD] and Maria Cerreta[2][iD]

[1] National Research Council of Italy (CNR), Institute of Research on Innovation
and Services for Development (IRISS), 80134 Naples, Italy
g.daldanise@iriss.cnr.it
[2] Federico II University of Naples, 80134 Naples, Italy

Abstract. At European, national, regional and local level, several approaches
are related to the Cultural and Creative Production System: new cultural districts, creative reuse of buildings and industrial sites, cooperation for common
goods are becoming economic trends. This process is building the conditions
necessary to encourage new cultural and creative industries especially in
developing alternative forms of governance and management of resources for
local regeneration. Considering this context, the research aims at responding to a
yet open question in place-based regeneration policies and strategies: how the
evaluation and urban planning process together with Cultural and Creative
Production could convert economic austerity and cultural diversity in new local
opportunities? The paper explains the "Community Branding (Co-Bra)"
methodological approach for a learning and negotiation process that combines
management models and multi-criteria/multi-group evaluation methods. Within
the framework of Matera ECoC 2019, the case study of Pisticci (MT), the
third-largest town in Basilicata (Italy), tested the Co-Bra method and started a
cultural co-creative enterprise for urban/rural regeneration. The multidimensional approach together with a financial analysis, focused on the recognition of
social, economic and cultural opportunities, provides strategies for both valorising cultural heritage and strengthening places network by a "community
hub" with a multilevel governance: so-called "PLUS - Pisticci Laboratorio
Urbano Sostenibile" (Pisticci Sustainable Urban Lab).

Keywords: Co-evaluation · Place branding · Cultural creative enterprise

1 Introduction

In Europe, some processes and practices demonstrate that financing culture increases
benefits in exploiting local resources (human, social, relational, territorial capital) for
real "tailor-made" policies at the local level in response to the specific needs.

In a progressive "liquefaction" of the social structures, new communication technologies [1, 2] and new cultural creative production systems are catalysts of a local
development in valorising cultural heritage.

Since 2007, in the regulation on the ERDF - European Regional Development Fund
(article 4), some investment priorities concern the protection and conservation of
cultural heritage, the development of cultural infrastructures and cultural services, with

© Springer International Publishing AG, part of Springer Nature 2019
F. Calabrò et al. (Eds.): ISHT 2018, SIST 101, pp. 298–307, 2019.
https://doi.org/10.1007/978-3-319-92102-0_32

a total of 434 operational programs in the European Union. The funds allocated by the ERDF for culture (about 6 billion euro), represent 1.7% of the total, in which 2.2 billion euro are allocated for the development of cultural infrastructure, 2.9 billion euro for the protection and conservation of cultural heritage and 797 million euro for cultural services [3], such as the National Operational Program (PON) Culture and Development - Programma Operativo Nazionale (PON) Cultura e Sviluppo - ERDF 2014–2020 (about 490,9 million euro).

At European, national, regional and local level, several approaches are related to the Cultural and Creative Production System as example: European and regional programs (such as Creative Europe, Green Paper on cultural and creative industries etc.); the transparent financial statement for a sustainable budget; the PPPP (Public Private People Partnership) partnership; collaborative platforms; and bottom-up initiatives as social street; community hub, fab lab. Each approach takes into account the principles of the Charter for Multilevel Governance in Europe (2014), which recognizes as essential the need to work in partnership for achieving the goal of greater economic, social and territorial cohesion in Europe.

In Italy recent researches (I am culture - *Io sono cultura* - Reports 2013 and 2014, Symbola Foundation and Unioncamere) estimated in 2012 about 5.7% of added value produced by the cultural production system, supported by the entrepreneurial system (5.4%), and a positive impact in terms of employment (5.8% of the total employed) in this cluster, despite still relevant differences in Southern Italy. In this context, it is also significant to highlight that "historical and artistic heritage" and "performing arts and visual arts" have the best performance in this production in spite of a number of companies and added value quantitatively lesser within the sector (as shown by PON Culture and Development ERDF 2014–2020). If these values could increase by building a widespread national network of small and medium creative enterprises strongly linked to local cultural production with services and products related to the territorial boundaries? In this perspective, probably culture could really be a key driver in responding to local development needs because the revenues of these local businesses, in proximity welfare logic, would be oriented directly to the community on which they have social, economic and environmental impacts.

New cultural districts [4, 5], creative reuse of buildings and industrial sites, abandoned areas and cooperation become trends, which are assuming also a quantitative and qualitative dimension from an occupational point of view. In many of these initiatives related to the re-use/valorisation of "common goods", culture represents the starting point for projects with positive social impacts, within collaboration processes among citizens, private organizations and public institutions [6].

In particular, cultural and creative enterprises are progressively consolidating their role in the promotion of tangible and intangible resources both nationally and internationally. The key factor is the involvement of creative figures in the governance and management processes of the cultural production sectors. These figures, as catalysts of resources responding to the needs of inhabitants and city-users, are able to build resilient and tailor-made development models on the territory. This kind of models often shows a hybridisation in processes or products based on perception and experiences strongly linked to *genius loci* and to local communities [7] and this could be useful to an efficient, effective and equitable regeneration process.

Since the end of the 20th century, several countries have developed policies from this cultural point of view [8]. First of all, UK has the role of launching the term "creative industries" for providing economic legitimacy to cultural policies and for implementing creativity in an economic perspective. From 1998 to 2001 the Department for Culture, Media and Sport (DCMS) developed this approach based on a definition of creative industries as activities having "their origin in individual creativity, skill and talent, and which have a potential for wealth and job creation through the exploitation of their intellectual property" [9].

In Italy, the regulatory framework includes a legislative proposal entitled "Disciplina e promozione delle imprese culturali e creative" (Discipline and promotion of cultural and creative enterprises) - in progress - which defines what is a cultural and creative enterprise and which are its requirements. Furthermore, the report "Io Sono Cultura 2017" (I am Culture 2017) shows alternative models of cultural and creative enterprises and its specific economic impacts in the whole country. The 6% of the wealth in Italy has produced thanks to the Cultural and Creative Production System (cultural industries, creative industries, artistic historical heritage, performing arts and visual arts, creative-driven productions): 89.9 billion euros. Data are growing by 1.8% compared to the previous year [10].

Several cultural and creative industries (ICC) are growing for the developing of alternative forms of governance as in the case of: cultural foundations; public service company; public companies and institutions; other service companies; consortiums and associative forms; dependent bodies. These models are often accompanied by any forms of tax relief able at supporting the economic sustainability and management of the ICC. In Italy, in particular, the following tax reliefs are developing: tax relief for intangible assets (so-called "patent box"); ZES or urban free zones for cultural and creative districts; Differentiated VAT; super amortization.

In line with this discussion, the research question focuses on a yet open question concerning place-based regeneration policies and strategies for valorising cultural heritage and environmental resources in cross-scale dimension. How the evaluation/urban planning process together with Cultural and Creative Production could convert economic austerity and cultural diversity into new local opportunities?

The research focus is oriented to demonstrate how a creative co-design process of local identity is able to extract the local development needs by local communities. The paper attempts responding to the above mentioned research question through the following structure: the first part (Sect. 2) defines the methodological approach of Community Branding (Co-Bra); the second one (Sect. 3) explains the PLUS hub case study and the results of this emerging cultural co-creative enterprise; the third (Sect. 4) shows discussion and conclusions about the whole process and scenario.

2 The Community Branding (Co-Bra) Approach

In recent times, a new discipline emerged between strategic planning and destination management, so-called place branding, which links elements characterizing the brand with places' peculiarities. Despite it isn't a new phenomenon, a strong growth of cities in the global international context [11] emphasizes the competitiveness within urban policies through the concept of city branding [12, 13]. In order to implement new processes or products for a healthy competition (innovation) and to adapt production continuously to market demand (flexibility), the brand is the new institution of the "information economy" as the factory was in the industrial economy. In this sense, place branding represents a culture for the governance of the territorial supply, especially in building a system of innovative principles, rules and procedures for managing a territory and enabling its society to action.

The process of creative co-design of place identity can be activated by the convergence of interests among economic operators and community in order to enrich the value of the project actions and make effective their feasibility. This value could be able to overcome traditional approaches in the process of urban and social capital production for ensuring a sustainable local development.

In a time of economic austerity, hybrid approaches are crucial for local resources enhancement, combining both management aspects and both community planning in a circular process of conservation/innovation [14]. The effectiveness of this kind of process is in building a place common vision and its efficient perception by city-users (linked to place branding and relational marketing) and both the management of resources oriented to an economic valorisation (linked to place marketing, resource-based theory and value chain). Innovation and production become key issues of this process that starts from the assimilation and creation of new knowledge highlighting the connections among places, organizations and people. In this process, resources, skills and social environment are complementary to the development of key urban functions. This strategic connectivity [15] could improve the active and participatory protection of cultural heritage by implementing the management skills of city-prosumers [16].

An experimental field consists in the methodological approach of this research: the "community branding". This approach is intended as a strategic process with the potential of building people awareness and community self-organization. It is a glocal process that uses physical and digital connections for extracting perceptions from territorial realities and for building operational links between forms of local tradition and tools of global innovation.

In fact, the circular methodological approach proposed (Fig. 1) combines the Community Impact Evaluation [17] with a set of evaluation and management tools: so called Community Branding (Co-Bra) [18]. By Co-Bra, we mean a strategic, "glocal" process for redefining territories, communities and their economies combining:

– Place Branding for the governance of the cultural local supply;
– Place Marketing for the management of the cultural local demand;
– Community Planning and Community Evaluation to identify cultural, social and economic opportunities from local creative production.

In this process, resources, skills and social environment are related thanks to the tangible and intangible "links" that are complementary to all territorial functions.

Through this physical and digital strategic connectivity, resources, social aggregation, active and participatory protection of the local assets could be re-activated by implementing the management skills of all the actors within the territorial system.

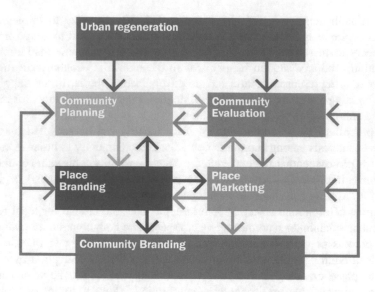

Fig. 1. Community branding (Co-Bra) approach [18].

The various steps according to the proposal are structured for building a new supply and related territorial demand starting from the perception of the community's wishes and from relationships established between places and people.

With the aim of defining appropriate monitoring and action tools, the 12 phases of the CIE are integrated with the 5 steps of the place branding process [19].

The whole process, intended as a Deliberative Multi-Criteria Evaluation - DMCE [20], uses transdisciplinary evaluation tools and approaches (multi-criteria methods such as PROMETHEE), management (Canvas approach of the Business model) and communication (Walkabout - storytelling) that bring results in terms of analysis of project interventions but also in terms of action on the territory through experiences.

This approach was tested in South of Italy (in Basilicata Region), through the lived experiences in Pisticci community (near Matera) with the identification and evaluation of the local priorities [21, 22] and creative bottom-up initiatives.

3   A Cultural Co-creative Enterprise: The PLUS Hub Case Study

The historical-cultural and landscape richness in urban/rural environment as Pisticci is an evidence demonstrated by projects and researchers on field (as Green Lucania project and PhD researchers) but also by the sensitivity of such local creative and social initiatives (as an example "Imbianchini di Bellezza" - Painters of Beauty, "Legambiente circolo Pisticci", Lucania Film Festival, "Teatro Lab", P-stories Walkabout, etc.). Consequently, identifying cultural values has been a spontaneous process induced by the know-how and experience of the community sectors who have enriched the process with elements that characterize territorial dimension [23] also in line with the vision of Matera European Capital of Culture 2019.

Especially, the engine of this place-based regeneration process consists in co-designing and co-evaluating a community hub [24] so-called PLUS –"Pisticci Laboratorio Urbano Sostenibile" (Pisticci Sustainable Urban Lab), able to build a common vision of the territory (PLUS community brand) in a local network of cultural diversity coming from community instances.

Starting from the data analysis of some questionnaires submitted in an earlier phase of the research, the territorial development proposal has pursued the following objectives: brand identity (PLUS); interactions among values (Pisticci network of sustainable urban organizations); the local experiences (Pisticci place of sustainable labs).

In line with this perspective, PLUS hub pursues the following objectives: building shared cultural landmarks (construction); activating processes of PLUS community (cooperation); communicate the PLUS brand (communication).

The focus is demonstrating the strength of a place network through the construction of a community hub intended as an experiment of co-creative community enterprise: a physical and digital platform of productive network for the territory implemented by the links between human and creative resources.

The main approaches and tools have been selected with the aim of both building a better interaction among stakeholders and both obtaining concrete results in terms of actions and services designed for the different sectors of the community. The two approaches used during the co-design roundtables on governance, activity and eco-nomic sustainability of PLUS hub are: the World Café method and the Business Model Canvas method. The World café method [25] is used here for improving the interaction during the co-design round tables and it is based on the theoretical assumption that the participants' contribution can be maximized by the dynamics of the action, by the informality of the dialogue and by the freedom of expression. The technique concerns an "incremental and circular discussion" completed and enriched thanks to the rotation of all participants at regular intervals. In addition to this approach, the Business Model Canvas (BMC) is included as a method for evaluating the strategic choices of a creative cultural enterprise. The Canvas method [26] is easily adaptable to the logic of com-munity planning, because of its visual language that enables understanding, discussing and analysing possible opportunities for an innovative business. This method allows the comprehension of complex elements concerning business models to a wide public. In particular, the "Value Proposition" of the BMC has allowed establishing the needs of specific community groups and actions/services that increase the benefits and decrease the disadvantages for each group.

Starting from the alternative vocations defined together with the community and during a performing media storytelling so-called walkabout ("Luoghi di Zonzo#1" - Places of Wandering), and thanks to the collaboration of participants in co-design process, we identified the project actions structured on four "experiential" variables: (1) recovery of tangible and intangible assets; (2) digital platforms; (3) services for resident and temporary citizens; (4) "urban contract".

The fourth variable attempts at defining models of co-governance through urban pacts ("urban contract") among stakeholders [27].

For each project action, the direct and indirect impacts (D, I) were classified on the various community sectors according to economic, social and cultural criteria (E, S, C) related to the macro-criteria of hardware, software, orgware and virtualware (investment categories for place branding strategies).

From the collected data and from the evaluation of 34 economic, social and cultural indicators it emerged that vocation 4 (Artisan and Creative Density) is a priority [18] for activating urban regeneration in Pisticci, engaging new local economies also in line with the cultural path of Matera ECoC 2019.

We used a costs/opportunities approach for defining sectoral preferences of stakeholders and for establishing coalitions to be consolidated in starting the regeneration process. Through a detailed analysis of the possible costs related to the physical interventions, and managing costs for the activities, different types of revenues are identified that fit the proposal in its cultural creative variety.

Starting from main financing models for cultural and creative enterprises [28], different categories for this co-creative community are included in the proposal: self-financing; fundraising; collaborative platform; ticketing; provision of services; private investments.

The costs/opportunities calculations include a total investment cost of € 1,519,000 and a total/year revenue of € 389,428. Performing a long-term evaluation (for about 20 years), the Net Present Value (NPV) is 956,277.62, while the Internal Rate of Return (IRR) is 13.63% (as shown in Fig. 2).

Fig. 2. Pisticci historical centre (source PLUS hub A.P.S.).

Consequently, the financial analysis demonstrates how the PLUS hub proposal in vocation 4 could become a strategic priority for the territory because the benefits are higher than the resources used (Fig. 3) elaborated by the authors.

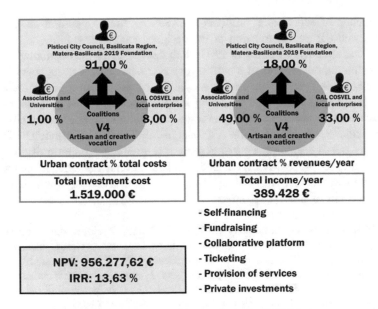

Fig. 3. Costs/opportunities analysis for PLUS hub.

4 Discussion and Conclusions

At the national level, the two development leverage concern both the strengthening of facilitated taxation and both the building up of networks (Artlab'17 "Il Sud in trasformazione" - The South in transformation - Taranto and Matera). The objectives pursued at national level are: 1. to enhance the diversification of economic activities; 2. to implement attractiveness through human capital, identifying attractive scenarios for external and internal entrepreneurs; 3. to strength the supply and at the same time the demand; 4. to make responsible and active the community and fortifying its cultural background; 5. to create a virtuous GDP through cultural heritage; 6. to strength the relationship between public and private actors; 7. to stimulate creativity in the process and not just in production.

In this sense, cooperation is a key tool for identifying innovative and inclusive actions aimed at developing productive networks. The activation of multi-actor partnerships involving social partners, universities, NGOs and representative groups of civil society, in line with the subsidiarity principle, recognizes solutions that reflect needs of citizens and territorial resources.

In this context, place branding is crucial for the development of real specific experiential products and forms of collaboration to activate partnerships. One of the most important interaction is about the link between culture and sustainability. Through a process called "capability building", culture produces consumption patterns linked to experiential, symbolic and participatory values allowing the construction of positive environments for institutional and human resources development but also for the strengthening of management systems [29]. In this sense, three phases are fundamental:

the construction of experiential products and services, the cooperation of the involved stakeholder and the communication of the designed experiences [30].

The financial analysis results of PLUS hub demonstrates the economic and social, as well as cultural, advantages of a multilevel governance for urban regeneration [31] if the institutions are oriented to have a medium-long term vision and a collaborative approach [32, 33] listening and engaging people involved in the process. Multilevel governance models are forms of operational and institutionalized cooperation, innovative and efficient, based on skills and responsibilities of each individual. Therefore, cultural and creative enterprises can represent a significant opportunity, capable of triggering virtuous processes suited to the Italian cultural heritage, in which public, private and private-social stakeholders can interact constructively towards the synergistic valorisation of present resources.

Acknowledgments. Within the unitary work, Maria Cerreta developed the first part (Sect. 1) about the national and international background of the research issues; Gaia Daldanise developed the third section (Sect. 3) related to the development of PLUS hub case study. All authors developed the research methodological approach (Sect. 2) and the discussion and conclusions on the whole process (Sect. 4). The authors are very grateful to "Imbianchini di bellezza", PLUS hub association, P-stories, Open story lab, Lucania Film Festival, Teatro Lab, AVIS, University of Basilicata, Pisticci Municipality, "Fondazione Matera-Basilicata 2019" and Basilicata region.

References

1. Bauman, Z.: Culture in a Liquid Modern World. Polity Press, London (2011)
2. Bauman, Z.: Glocalization and hybridity, glocalism. J. Cult. Polit. Innov. **1**(1), 1–5 (2013)
3. Biclazio: TURAS Transitioning Towards Urban Resilience and Sustainability (2011)
4. Sacco, P.L., Pedrini, S.: Il distretto culturale: mito o opportunità. Il Risparmio **51**(3), 101–155 (2003)
5. Chiarvesio, M., Di Maria, E., Micelli, S.: Global value chains and open networks: the case of Italian industrial districts. Eur. Plan. Stud. **18**(3), 333–350 (2010)
6. Legacoop, Legambiente: Rigenerare le città (2011)
7. Esposito De Vita, G., Trillo, C., Oppido, S.: Urban regeneration and civic economics: a community-led approach in Boston and Naples. J. Comp. Cult. Stud. Archit. **9**, 28–40 (2016)
8. Sacco, P.L., Crociata, A.: A conceptual regulatory framework for the design and evaluation of complex, participative cultural planning strategies. Int. J. Urban Reg. Res. **37**(5), 1688–1706 (2013)
9. ESSnet: European Statistical System Network on Culture. Final report, European Commission - Eurostat (2012)
10. Unioncamere-Fondazione Symbola: L'Italia della qualità e della bellezza sfida la crisi, Technical report (2017)
11. Sassen, S.: Cities in a World Economy. Sage Publications, Thousand Oaks (2011)
12. Dinnie, K.: City Branding: Theory and Cases. Palgrave Macmillan, Basingstoke (2011)
13. Patteeuw, V.: City branding: image building & building images. NAi (2002)
14. Fusco Girard, L.: Creativity and the human sustainable city: principles and approaches for nurturing city resilience. In: Sustainable City and Creativity: Promoting Creative Urban Initiatives. Ashgate Publishing, Farnham (2011)

15. Daldanise, G.: Innovative strategies of urban heritage management for sustainable local development. Procedia Soc. Behav. Sci. **223**, 101–107 (2016)
16. Rifkin, J.: La nuova società a costo marginale zero. Mondadori, Milan (2014)
17. Lichfield, N.: Community Impact Evaluation. University College Press, London (1996)
18. Cerreta, M., Daldanise, G.: Community branding as a collaborative decision making process. In: 17th ICCSA 2017. LNCS, vol. 10406, pp. 1–13. Springer, Heidelberg (2016)
19. Place Brand Observer: The Five Steps of Successful Place Branding Initiatives (2016)
20. Proctor, W., Drechsler, M.: Deliberative multicriteria evaluation. Environ. Plan. Gov. Policy **24**(2), 169–190 (2006)
21. Montrone, S., Perchinunno, P., Torre, C.M.: Analysis of positional aspects in the variation of real estate values in an Italian Southern metropolitan area. In: 10th ICCSA 2010. LNCS, vol. 6016, pp. 17–31. Springer, Heidelberg (2010)
22. Brigato, M.V., Coscia, C., Curto, R., Fregonara, E.: Valutazioni per strategie di sviluppo turistico sostenibile. Il caso del Bacino Metallifero dell'Iglesiente, Territorio, pp. 1–11 (2014)
23. Cerreta, M., Inglese, P., Manzi, M.L.: A multi-methodological decision-making process for cultural landscapes evaluation: the green Lucania project. Procedia Soc. Behav. Sci. **216**, 578–590 (2016)
24. Calvaresi, C.: Community Hub, due o tre cose che so di loro, cheFare (2016)
25. World Café Community Homepage. http://www.theworldcafe.com/. Accessed 3 Dec 2017
26. Osterwalder, A.: The business model ontology: a proposition in a design science approach (2004)
27. Perulli, P.: The Urban Contract: Community, Governance and Capitalism. Routledge, Abingdon (2016)
28. Bertacchini, E.E., Pazzola, G.: Torino creativa. I centri indipendenti culturali sul territorio torinese. Edizioni GAI, Torino (2015)
29. Federculture: Le industrie culturali e creative in Italia (2013)
30. Buonincontri, P., Micera, R.: The experience co-creation in smart tourism destinations: a multiple case analysis of European destinations. Inf. Technol. Tour. **16**(3), 285–315 (2016)
31. Esposito De Vita, G., Ragozino, S.: Natural commercial centers: regeneration opportunities and urban challenges. Adv. Eng Forum **11**, 392–401 (2014)
32. Clemente, M., Giovene di Girasole, E.: Cultural heritage as a common good for the valorization and regeneration of seaside cities. In: Aveta, A., Amore, R., Marino, B.G. (a cura di) La Baia di Napoli. Strategie integrate per la conservazione e la fruizione del paesaggio culturale. Artstudiopaparo, Napoli (2016)
33. Clemente, M., Daldanise, G., Giovene di Girasole, E., Processi culturali collaborativi per la rigenerazione urbana. In: XX Conferenza Nazionale SIU Urbanistica è/e azione pubblica. La responsabilità della proposta. Planum Publisher (2017)

Social Enterprise and the Development of Cultural Heritage Assets as Catalysts for Urban Placemaking

Deniz Beck[✉] and Samuel Brooks

University of Portsmouth, Portsmouth PO1 2UP, UK
deniz@denizbeck.com

Abstract. This paper aims to establish the value of urban cultural heritage sites to local population centres as a means to strengthen community bonds and form new hubs of social activity. Sympathetic, sustainable regeneration of such shared assets can often create opportunities for new forms of Social Enterprise, and their powerful historical identity can greatly assist with further placemaking within established urban environments.

The Hotwalls Studios comprises thirteen self-contained artist's studios and workshops housed within the historic casemate arches of Point Battery, a defensive fortification situated at the mouth of Portsmouth Harbour. As the centrepiece of the site, the studios have been deliberately designed as passive environments in which artists and visitors are encouraged to interact and engage with one another, with the goal of improving communities through cultural and social endeavours. The site now plays host to regular arts-based events, including exhibitions, seasonal markets and public activities. In the context of this recent project in Portsmouth, UK, we hope to demonstrate that such developments greatly benefit both heritage sites and local populations. Providing financial security, nurturing social and creative pursuits, encouraging sustainable design and reinvigorating communities with a new understanding of a shared historical and cultural identity, are all benefits of a framework which we believe can be successfully applied to many such sites throughout the world.

Keywords: Cultural heritage · Social Enterprise · Regeneration

1 Introduction

Forming effective new social centres in existing urban environments is a constant and evolving challenge to local governments and civic organisations, and whilst such projects can be superficially achieved through standard development practices, cultivating a genuine sense of 'place' that is both meaningful and valued to the community it serves is not only the most critically important component in its long-term success, but also one that is, in our experience, the most elusive to effectively harness.

Consequently, this outcome can often only be achieved by relying on complex, bespoke solutions tailored to an imprecise combination of environment, population and intended usage, with understandably inconsistent results. However, by utilising an appropriate ideological framework, we believe that existing cultural heritage assets and

© Springer International Publishing AG, part of Springer Nature 2019
F. Calabrò et al. (Eds.): ISHT 2018, SIST 101, pp. 308–315, 2019.
https://doi.org/10.1007/978-3-319-92102-0_33

increasingly limited financial resources can be efficiently employed to achieve mutually beneficial and wide-ranging rewards for urban communities.

The Importance of Community Cohesion. The establishment of community hubs focusing on creative, leisure and social activities can be powerful tools in strengthening bonds within urban populations [1], and when appropriately implemented can have immediate and observable effects on community cohesion as well as far-reaching and often intangible benefits for physical and mental well being [2]. This can in turn reduce the burden on conventional primary public services and infrastructure, by promoting capacity for personal and social well-being rather than simply managing the effects of ill health and disease after they occur [3].

Obstacles to Implementation. In December 2010 the UK government announced that funding to local authorities would be cut by 28% over the next four years [4], with funds made available for public services and infrastructure in 2017 forecast to be 22% lower than in 2010 [5]. Population centres are increasingly looking towards alternative sources of finance to bridge this gap, and consequently Social Enterprise funding models have flourished, with the average year on year growth in turnover for organisations over three years old in 2014 standing at 72%, and those studied contributing an estimated £22bn to UK GDP [6]. As a result they have proven to be instrumental in sustaining many ventures that serve social goals.

An additional impediment in realising these projects is a lack of suitable sites to conveniently locate such enterprises relevant to the populations they serve, and this is felt no more acutely than in Portsmouth; as an island city, land is at a premium and vacant or underused property is usually swiftly redeveloped into housing. Available locations often lack adequate space and opportunities for public access, and landowners are justifiably reluctant to pursue community-based ventures that may be unprofitable or problematic to administer when compared to more conventional commercial enterprises.

Utilising Heritage Assets. Fortunately, Portsmouth boasts an abundance of cultural heritage assets such as historic monuments and fortifications that occupy pivotal locations in established population centres. Acting as powerful and identifiable landmarks in residents' collective subconscious, these sites present an inherently authentic identity that we believe can greatly assist in the establishment of new community-based enterprises.

However, when considering these sites, additional problems become apparent, and potential developers and investors are often discouraged by the extensive regulatory frameworks that surround their redevelopment [7]. Already considerable financial expenditure and long development programmes are both factors which can be further exacerbated when dealing with historical structures, with their antiquated construction regularly revealing hidden complications that consistently require significant remedial work.

2 Thesis

2.1 Methodology

Our method of addressing the shortcomings apparent in the processes outlined above centres on the effective combination of two distinct aspects to achieve a result that provides tangible, secondary benefits for each process alongside the primary combined objective of a successful community-focused amenity.

The chart below (Fig. 1) demonstrates how this is achievable by pursuing investment through Social Enterprise funding models in concert with the conservation of heritage assets. Individually, both approaches are proven to achieve their own set of positive results, but our belief is that in combination they can offer a further intangible yet powerful constituent that, if adequately cultivated, can further increase the likelihood of overall success in establishing a meaningful sense of 'place'.

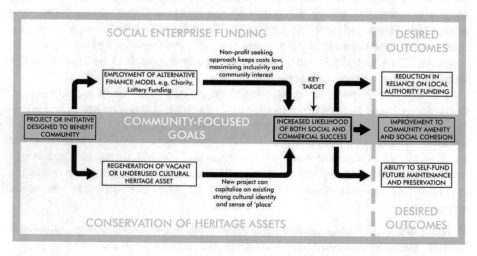

Fig. 1. Chart demonstrating how the combined inputs of alternative funding models and regeneration of heritage assets can achieve mutually beneficial, community-focused outcomes.

With increasing pressure on traditional methods of financing, local authorities are increasingly turning to alternative external sources to fund works to civic services and infrastructure [4]. Whilst this has the immediate benefit of alleviating pressure on public finances, it also provides an opportunity to pursue non-profit ventures and the attendant diversity of potential end-uses that can benefit communities over private interests. In the case of Portsmouth, the city council retains direct ownership of many heritage assets, and the goal of end-use sustaining the self-funding of ongoing maintenance can further incentivise the embracing of such community values as inclusivity and public accessibility through measures usually biased by profitability, such as affordable leasing. Projects instigated by local authorities themselves can also reduce both cost and programme length by utilising in-house resources and streamlining the bureaucratic processes and regulatory frameworks that surround the redevelopment of historic sites.

By harnessing the acknowledged cultural identity and strong sociological origins of urban heritage, we believe a foundation for both functional regeneration and establishment of a new sense of 'place' can be formed. This would have the consequence of engendering a deeper appreciation of the more intangible benefits of heritage assets in their users, particularly around those sites that have become vacant, derelict or otherwise lost their purpose in modern society.

We expect the net result of such endeavours to be a long-term successful solution that simultaneously addresses the need for community amenity and social cohesion, the ability to effectively fund future preservation efforts, and a financial self-sufficiency that simultaneously underpins and benefits from the binding of both these concepts.

2.2   Case Study: The *Hotwalls Studios*, Portsmouth, UK

History and Significance of the Site. For hundreds of years, Portsmouth's importance as a naval base made it one of the most heavily defended cities in the world, and it was on the site of Point Battery that early inhabitants erected the first permanent fortifications. At its peak, the Point was the cultural heart of the city, attracting thousands of sailors and visitors from around the world, and immortalised in works by people such as English satirist Thomas Rowlandson and composer William Walton [8].

Point Battery Fortifications. In the mid-19th Century, the Battery was expanded through the construction of soldiers' accommodation to the north and a series of small buildings on the Western boundary with Broad Street, enclosing a tapered triangular parade ground. Upgrades to gun emplacements and the installation of electric searchlights took place around the turn of the 20th Century, but the final and most significant modifications took place following its acquisition by the City Council in 1958; the Western structures adjoining Broad Street were demolished, the accommodation block reduced in size, and the site and adjacent foreshore finally opened to public access. This combination of modifications carried out over the last 150 years has culminated in the structure taking the form visible today.

Social Origins of 'the Point'. For around 300 years, the Point existed outside the City of Portsmouth, with the adjacent harbour serving as the Royal Dockyard. Although initially a convenient site for locating warehouses, the small peninsula was often the first landfall for many visiting and returning sailors; the high-volume and transitory nature of these patrons, during a time when Britain was building its empire and asserting control over trade routes, created a robust and vibrant social scene.

All land traffic was channelled through the narrow King James's Gate, the nightly closure of which compounded the sense of separation from the city; its isolation, trade in exotic and illegal goods, and high concentration of all-day drinking 'dens' (as many as half of all those in the city) earned it the nickname 'Spice Island'. Brothels, eating houses and pawn shops also proliferated, and its reputation for lawlessness brought with it regular smuggling and press-gang activity.

Advances in shipbuilding allowed maritime traffic to increasingly bypass the area altogether, and by the early 19th Century its popularity was already in decline.

Reintegrated with the city over 150 years ago, the area has nonetheless retained its strong nautical connections; its unique, close-knit origins, and sense of a 'community within a community' are evident in the summer months through the enduring popularity of its modest beach front and public spaces.

2.3 Proposals for Regeneration

The combination of historical significance and unique sociological heritage evident in the Battery and its surroundings can be summarised in the words of Dr Richard Massey, Assistant Inspector of Ancient Monuments for English Heritage:

> *"With commanding views across the Solent and the mouth of Portsmouth Harbour, the battery also occupies a pivotal location within the historic defences of the city which [...] is fundamental to any appreciation of Portsmouth as a naval and maritime centre. This significance has invested Point Battery with a powerful sense of place and considerable communal heritage value, and should emphasise the potential for re-defining this area as a cultural focus and public space".*

Recognising the importance of the continued preservation of this monument, Portsmouth City Council invited local Architects to tender for the conversion of the existing empty casemates and storage areas. By developing appropriate usage of the vacant spaces, revenue could be generated which would self-fund the future conservation of the structure.

The chosen objective of the project was to build on the existing artistic connection by forming an artist's quarter and creative centre in the vacant arches and ancillary buildings. Focusing on the conversion and renovation of the arches themselves, each space would form a versatile working environment for its occupants and would collectively act as a gallery showcasing a variety of work and encouraging socialising and interrelation between visitors and tenants.

Funding and Programme. The Hotwalls Studios project was a direct result of a £1.75m investment from the Coastal Community Fund, a lottery funded organisation established to encourage sustainable economic growth in coastal communities. Additional funding of £40k was provided by the Partnership for Urban South Hampshire, (PUSH) a consortium of local authorities, private stakeholders and government agencies supporting, amongst other beneficiaries, creative industries. Along with a one-time investment of £100k from Portsmouth City Council, management would initially be overseen by the council, with the intention of handing responsibility over to studio tenants and associated community organisations [9]; once funding was approved, at least one studio was required to be operational within a year.

As the project was to be financed entirely by fixed-sum investments, based on past experiences with similar heritage projects we anticipated unforeseen problems arising during the build. As a result, ensuring build cost stayed within the fixed budget required the design team to be innovative and flexible in their approach to dealing with such obstacles, alongside the regular review and restructuring of the costs of individual programme items.

2.4   Development Approach

Architectural Design. Minimal contemporary glass frontages to each of the openings serve both the purpose of creating a cultural space with the desired focus on artist/visitor interaction, and reinforcing a clear definition between modern addition and historic context. Coupled with re-opening the gun embrasures, this allows an improved flow of light, benefitting occupants and visitors alike whilst reconnecting the public space on one side with the significant seaward views and beach front on the other.

The new cafe-bar would see its vaulted brickwork cleaned of paintwork and render and restored to its original finish, with new services channelled beneath a suspended floor. Obviously contemporary but deliberately low-key and simplified, the terrace, along with the renovation of the adjacent searchlight position, serves to reinvigorate the impression of the building when viewed from the busy shipping channel and acts as a 'showcase' for the renewed usage of Point Battery.

New bench seating and selective soft landscaping extend the appeal of the arches into the adjoining open space, and are an important factor in 'binding' the various spaces together in one coherent form. This was essential in encouraging people to feel comfortable using the space for leisure, relaxation and socialising. Our ambition was to ultimately see the parade ground employed as a regular, temporary marketplace, supporting activity associated with the studios and reinforcing its role as a meeting place and hub for the city's creative community.

Heritage Considerations. Inherent in this project was the idea of striking a balance between the provision of modern fixtures and conveniences for a variety of contemporary uses, and the restoration and preservation of the historic fabric of the building. To this end, the design was driven by a focus on sustainability and reversible, non-destructive work rather than material alterations. This would essentially allow the removal of any additions and the reversion of the structure to its pre-development state with minimal visible changes, should the need arise.

3   Conclusion

3.1   Outcome

Although no formal survey of occupant satisfaction has been conducted, figures provided to us by the city council show that current occupancy stands at 100%, following an initial business model based on 80% occupancy. In addition, the original footfall target was achieved early on in its operation, and attendance in the first year reached double the targeted figure. The area continues to enjoy heightened popularity since opening [10], and considering the sensitivity and age of the site and the associated propensity for unforeseeable complications, the project was delivered both on-time and on-budget with no significant disruptions.

Conservation. The design team aimed to impress our scheme upon the structure with minimal impact on the existing fabric; through its utilitarian forms and layout, the building itself dictated how we could efficiently deliver the brief within a minimal

Fig. 2. The *Hotwalls Studios* as they are today; the existing casemate arches now housing 13 artist's studios, historical interpretation centre and waterfront cafe bar.

footprint. This also helped to cultivate a sense of responsible stewardship, rather than ownership, in its new occupants.

Minimalist contemporary glass frontages create a working cultural space whilst clearly defining modern addition and historic context. All modern services have been concealed where possible both within these façades and discreet internal raised floors, removing the need for face-fixed ducting which would undoubtedly compromise the interior spaces' aesthetic simplicity.

3.2   Success as an Exercise in Placemaking

The Hot Walls Studios project has transformed a vacant Victorian fortification into a new community hub and activity space whilst supporting the ongoing conservation of an internationally important Scheduled Ancient Monument. What has been delivered is a working cultural centre where local artists can establish affordable studios, rented from the council, and which subsequently benefit from regular open days and work-shops allowing direct interaction with the public. This unique approach is both inclusive and empowering to both visitors and artists and of all ages, abilities and disciplines, and it would be significantly more difficult to achieve such results through private enterprise.

Tenants are encouraged to open their doors and interact with visitors to showcase local arts, and an appropriate infrastructure supports future events such as markets and exhibitions. Complimentary services such as the bar/restaurant ('The Canteen') and coffee shop, and historical interpretation centre add extra visitor attractions, and along with administrative facilities ensure a regular presence on site, reducing the likelihood of criminal activity and promoting a safe, inclusive atmosphere.

The success of this project ultimately hinged on the integration of appropriate, contemporary amenities with the sensitive restoration of the historic building fabric, and both these objectives have arguably been achieved, along with the primary emphasis being on their combined social function. It is this exercise in meaningful placemaking that has been the most fruitful; the popularity of the project with both residents and visitors through the use of a shared identity and history has both strengthened a community and ensured the ongoing conservation of a treasured monument. It is our belief that these results could be replicated in many other publicly-owned urban heritage sites around the world through a similar approach; utilising the ever-growing variety in the methods of social enterprise, the opportunities and means with which this can be achieved are more accessible and diverse than ever.

References

1. How Arts and Cultural Strategies Enhance Community Engagement and Participation. https://staging.planning.org/research/arts/briefingpapers/engagement.htm. Accessed 2 Dec 2018
2. Cameron, M., Crane, N., Ings. R.: Be Creative Be Well, Arts, Wellbeing and Local Communities: An Evaluation, pp. 30–31. Arts Council England, London (2012)
3. Jones, M., Kimberlee, R., Deave, T., Evans, S.: The role of community centre-based arts, leisure and social activities in promoting adult well-being and healthy lifestyles. Int J Environ. Res. Pub. Health **10**(5), 1948–1962 (2013)
4. The Growing Role of Social Enterprise in Local Government, The Guardian, 11 September 2012. https://www.theguardian.com/social-enterprise-network/2012/nov/09/growing-role-social-enteprrise-local-government. Accessed 2 Dec 2018
5. Britain's Local Councils Face Financial Crisis, The Economist, 28 January 2017. https://www.economist.com/news/britain/21715673-amid-painful-fiscal-squeeze-some-authorities-may-soon-be-unable-meet-their-statutory. Accessed 2 Dec 2018
6. NatWest SE100 Annual Report 2014. https://se100.net/analysis/annual-2014. Accessed 2 Dec 2018
7. Institute of Historic Building Conservation: Incentives for the protection, restoration and maintenance of historic buildings. The 2016 IHBC Yearbook. Cathedral Communications Ltd.
8. Portsmouth Point-Ouverture, for Orchestra (1925). http://www.waltontrust.org/en/16-tooltip/64-portsmouth-point-ouverture-for-orchestra-1925. Accessed 2 Dec 2018
9. Portsmouth City Council-Hotwalls Studios. https://www.portsmouth.gov.uk/ext/development-and-planning/regeneration/hotwalls-studios. Accessed 12 Feb 2018
10. Portsmouth City Council-Hotwalls Studios Celebrates a Year of Creative Success. https://www.portsmouth.gov.uk/ext/news/hotwalls-studios-celebrates-a-year-of-creative-success-. Accessed 12 Feb 2018

Hypothesis for the Development of Identity Resources Surrounding the San Niceto' Castle in the Metropolitan City of Reggio Calabria

Immacolata Lorè[✉], Tiziana Meduri, Roberta Pellicanò,
and Daniele Campolo

Mediterranea University of Reggio Calabria, 89100 Reggio Calabria, Italy
immacolata.lore@unirc.it

Abstract. The study is aimed at developing a valorization proposal for a historically relevant area in the Metropolitan City of Reggio Calabria by the use of the cultural identity of the territory. Specifically, a plan is proposed to manage the cultural heritage of the Barony of Saint Aniceto that starts from local production, and provides an integrated plan for the development of cultural identity. The method consists of a preliminary investigation of the data that led to the identification of effective tools for valorization actions. The next step involved the pre-feasibility study of the proposed project. The paper intends to demonstrate that a development project could be the possible way out of the view of culture as a marginal aspect of economic life but rather is based on the socio-economic policies for the development of a territory.

Keywords: Cultural heritage · Valorization · Development · Management

1 Introduction

In recent years, the cultural heritage sector has been involved in significant management changes thanks to diffusion of the idea that the culture can represent a driving force for the economic and social growth of a territory. This is a significant evolution of a tendency that is gradually shifting the meaning of the purpose of cultural assets; it shifts from a purely conceptual vision in which heritage was understood as a memory of a cultural identity, and therefore a vehicle for education and training, towards a vision that has enriched the conservation strategy through a much more dynamic and economically productive approach.

The local product is playing an increasingly important role in the project for the enhancement of the territories and the attention to this resource has also been underlined by UNESCO, including the Mediterranean Diet within the Intangible Cultural Heritage List, underlining the lifestyle and the cultural, anthropological and productive

---

The paper is the result of the joint work of the authors. Although scientific responsability is equally attributable, the abstract and Sects. 1, 4, 6, 11 were written by I. Lorè, T. Meduri and D. Campolo; Sect. 2, 3, 5, 7, 8, 9, 10 were written by R. Pellicanò.

© Springer International Publishing AG, part of Springer Nature 2019
F. Calabrò et al. (Eds.): ISHT 2018, SIST 101, pp. 316–326, 2019.
https://doi.org/10.1007/978-3-319-92102-0_34

aspects [1]. In this direction, the paper intends to illustrate a valorization proposal of a relevant historical area of Reggio Calabria through the use of cultural heritage that identifies the area of interest: the Motta San Niceto, historical closure of the hilly defensive circle, located to the east of the city of Reggio Calabria about four kilometers from the sea and three kilometers from the municipality of Motta San Giovanni [2]. The reasons that led to the choice of this topic are linked to the recognition of the intrinsic value for the territory under examination of the "motta", as a historical proof, and in particular of the Saint Niceto's castle.

2 Historical Overview of the Motta San Niceto

The Saint Niceto's castle is the only example of late Byzantine architecture that is well preserved in Calabria [3].

The castle belonged to the barony of Saint Niceto, bordered to the west by the Valanidi stream that still today defines the municipal border with Reggio Calabria and to the east by the possessions of Pentedattilo, whose boundaries coincided with the S. Elia stream, the current administrative boundary between the municipalities of Montebello and Melito di Porto Salvo [2].

The territories of Motta San Giovanni, Montebello Ionico, Saline, Bocale, Pellaro and Valanidi belonged to this area. In its historical context, the Saint Niceto's castle not only had a defensive function, but also represented an important source of resources and a point of connection between the territory of Reggio Calabria and the Ionian area [3]. The study is aimed at the intangible cultural resources of the examined area, introducing the distinctive position of its enogastronomic traditions, recognized in the case study as strong points, with reference also to the quality of life; indeed, the analysis of these resources provides a description about the size and attractiveness of the area related to tourist demand [4]. The endowment of resources of the intangible heritage with an enogastronomic character is considered of fundamental importance, as it highlights the relationship between landscape and cultural heritage, that is one of the distinctive features of the territory, as well as an asset of important value in economic competition [5].

3 Identification of the Identity Resources of the Territories Bordering the Motta San Niceto

The cognitive analysis was conducted to identify the identity characteristics of the examined territorial context and has enabled placing attention on the local products that have been part of the barony of Saint Niceto. This is because what the local product contains is connected to a critical knowledge of the territory that involves not only the economic aspects, but also the social ones. These aspects are linked to the historical memory of local citizen on production and its use in gastronomic preparations or social consumer practices, and thus representing an important identity element [6]. Specifically, the products that characterize the various areas under consideration are: the bergamot (PDO) and the wine (PGI) for the territory of Pellaro; licorice and olive oil for the territory of Montebello Ionico; and almonds, honey, prickly pears and the cured meat "azze anca grecanico" for the area of Lazzaro in the municipality of Motta San Giovanni [7–9]. Typical enogastronomic products are strictly linked to their territory because they derive from them the identity (tradition or excellence) and the recognizability in the market. A local economy that aims at enhancing its natural resources can lead to a considerable growth of tourism and can generate benefits, not only for the companies involved in production process, but, more generally, for the environment and the local socio economic system [10, 11] (see Fig. 1).

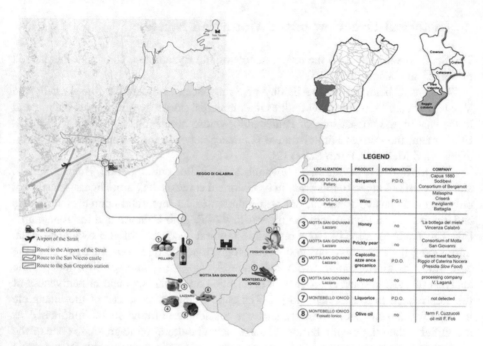

Fig. 1. Identification of route of project, local products and actors operating in the sector.

4   The Proposal of Valorization for the Motta San Niceto

In the awareness that cultural heritage, if properly put to value be it due to historical-artistic assets or linked to typical crops and agri-food products can be a driving force for the economic and social growth of a territory. Therefore, the project proposal is oriented to the realization of a tourist hub that can favor the connection between the resources of the areas bordering the Saint Niceto's castle, and as a consequence, promote cultural activities for the knowledge of the territory's resources, such as the castle and the local products [12].

For the realization of the project idea, a methodological path has been identified that consists of distinct phases, following the bibliographic research and the cognitive analysis of the area and the surrounding context, that are structured as follows:

- Elaboration of the SWOT analysis;
- Analysis of development strategies for the valorization of the identified assets;
- Analysis of best practices;
- Identification of the operative model to be applied to the case study, with particular attention on the definition of the management plan;
- Verification of the sustainability of the identified valorization plan.

5 The SWOT Analysis

The knowledge learned through the preliminary research were later organized according to the SWOT matrix, thus identifying strengths, weaknesses, opportunities and threats, in order to have a clear picture of the different aspects that characterize the examined context. This step is fundamental in order to identify and support the appropriate choices that will be translated into specific actions, and to activate on the territorial enhancement processes for economic and social growth. The purpose of the analysis was also to outline the opportunities for development of the area, through the enhancement of the strengths, and a containment of weaknesses; therefore, the strengths become the references to define the development strategy. The use of SWOT analysis has also solved the problem concerning the choice of the most convenient location for the intervention. To this end, a preliminary survey was carried out on the territory, focusing on the characteristics of use and accessibility of the San Niceto site and the surrounding territories. The analysis of the results led to the identification of the Arenella site in the San Gregorio district of the Metropolitan City of Reggio Calabria (RC), as the most convenient location for the sustainability of the interventions. The choice is related to the presence of a sufficient accessibility system in this area compared to the others, thanks to the connection with the Airport of the Strait and with the railway station of San Gregorio, easily reachable by tourists and citizens [13].

6 Analysis of Development Strategies for the Valorization of Cultural Heritage

The enhancement of cultural heritage resources is achieved through the establishment and stable organization of resources, structures or networks, thus conferring on cultural heritage an increasingly significant role in the framework of development models based on local peculiarities and on the valorization of endogenous resources, thanks to the implications of intangible nature such as the traditions linked to knowledge and creativity that have enriched the notion of heritage [15]. The process of conservation and valorization of cultural heritage if supported by strategies of system and addressed not only to cultural heritage but to resources that represent the distinctive marks that history have left on a territory can play an important role both for conservation of the assets, in this case the Saint Niceto's castle, and the promotion and support of economic development of local communities. Therefore, the present study considers the involvement of local communities, implemented through the networking of the main stakeholders of the territory, in order to increase the knowledge and awareness about cultural heritage, understood as the ability of citizens to recognize their identity in the local heritage, recognizing it as one's own and, consequently, to cooperate for its conservation [16]. The need to solve the problems emerging from the SWOT analysis has as a consequence the definition of the following objectives:

A. guaranteeing accessibility to the site of Saint Niceto's castle;
B. increasing citizens' knowledge and awareness of their heritage through the promotion of activities and territorial animation aimed at citizens and tourists;
C. favoring the participation of local companies in territorial promotion initiatives;
D. increasing knowledge on local productions.

To achieve the objectives, a general intervention strategy has been defined in the construction of a tourist hub, as a pilot action, able to connect the various resources of the territories bordering the Saint Niceto's castle through the promotion of cultural activities aimed at the knowledge of local resources [14]. The integrated supply of resources through the establishment of a tourist hub, has the general objective of generating direct economic impacts, with the outsourcing of activities and services related to its management, as well as indirect impacts. The latter derive not only from the most famous repercussions on the tourism industry, but also from the system that develops around the heritage; it increases the competitiveness of a territory, making it able to attract more human and financial resources, increasing tourist flows, as well as the establishing other productive activities [17].

6.1   Definition of Actions

The identification and choice of actions has as its starting point the accessibility to the site of the Byzantine fortress and then the creation of a network between local companies of the area. The possibility of achieving the objectives has directed the actions illustrated in the following figure (see Fig. 2).

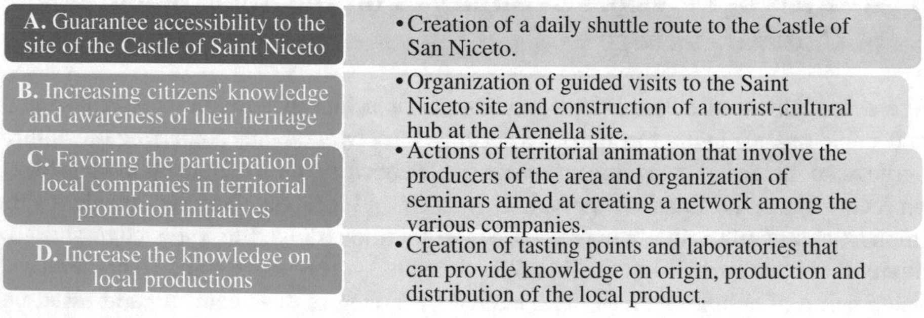

A. Guarantee accessibility to the site of the Castle of Saint Niceto	• Creation of a daily shuttle route to the Castle of San Niceto.
B. Increasing citizens' knowledge and awareness of their heritage	• Organization of guided visits to the Saint Niceto site and construction of a tourist-cultural hub at the Arenella site.
C. Favoring the participation of local companies in territorial promotion initiatives	• Actions of territorial animation that involve the producers of the area and organization of seminars aimed at creating a network among the various companies.
D. Increase the knowledge on local productions	• Creation of tasting points and laboratories that can provide knowledge on origin, production and distribution of the local product.

Fig. 2. Definition of actions.

7 Study of the Best Practice: Slow Food

To propose a line of action that can have a solid basis, it was decided to carry out a survey on the initiatives of the enhancement of tradition and culture gastronomy and in particular of local products [19]. The analysis highlights that an absence of similar initiatives in the Calabrian territory was found, with the exception of the presence of a Slow Food Presidium and Conducts belonging to the Greek area; among the latter, it is considered appropriate to mention the one closest to the area of interest, located in Santo Stefano d'Aspromonte (RC). Slow Food serves to realize these principles: renewing the trust in the right to pleasure for the protection of biodiversity and traditions; educating taste and informed eating by focusing on products, nutrition and local companies active in the area. These principles are consistent with the objectives provided by the hypothesis of intervention described; moreover, the promoted activities, both educational and not, such as Master of Food, are an important reference for the definition of the actions that characterize the project. Interesting examples are the masters organized, in Trentino Alto Adige, on the production of honey at Trento and production of oil in Campania, at Avellino. The master's program includes lectures and tasting experiences at the educational sites; here the participants put themselves to the test and learn about the products and the territory. These events have been very successful and guarantee a large influx of tourists and local producers every year. Another interesting initiative that has as protagonists the consumer and the producer is *Mercati della Terra*, similar to the one made at Villa San Giovanni organized by Slow Food Versante dello Stretto and Costa Viola. The initiative has also proposed, among many events, the laboratory of taste, a moment of special tastings guided by experts in the food and wine sector.

8 The Management Model

The strategic use of resources presupposes a unitary decision-making model that defines its mission and specifies its operational actions and the medium and long term objectives; the aim is to achieve and maintain an advantageous position over the time. It appears that the strategy of a local development process is supported, in addition to the specialization of the designed plan, by an effective management of the planned actions.

For this reason the Management Plan must give consideration to the differences, characteristics and needs of the site, as well as the cultural and natural context in which it is located. It can also incorporate existing planning systems and other traditional ways of organizing and managing the territory.

The management is also the activity aimed at ensuring, with the reorganization of human and material resources, the use of cultural heritage and its valorization. It is therefore referable both to the protection and valorization of cultural heritage and, as required by the Article 115 of the Cultural Heritage Code, the provision of public services (services for visitors, related cultural activities, promotions and sponsorships, etc.).

With regard to the definition of the various actions to achieve the project's key objective, a management model has been defined to illustrate the organizational and management aspects [20].

In this paper, the scenario with a direct type of management is described; the entire organization will be related to the figure of the Consortium of Bergamot, the only owner of the structure, that will represent the entrepreneurial promoter.

The creation of the cultural hub is related to the start of various activities necessary for the involvement not only of local companies that must be collaborate with each other and make their contribution to spread the knowledge of products, but also for citizens and tourists. After defining the location of the intervention and analyzing the actual situation, it was possible to define the use of each building on the site under consideration and its consistency.

The proposed activities and the human resources involved in the project are summarized and illustrated in the following figure (see Fig. 3).

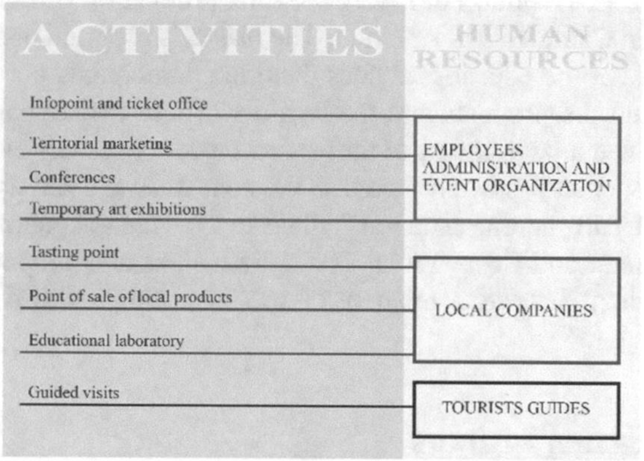

Fig. 3. Activities and the human resources involved in the project.

After defining the location of the intervention and analyzing the actual situation, it was possible to define the intended use of each building on the site in question and its consistency (see Fig. 4).

9   The Stakeholder

The project involves various actors active in the territory who play a key role in achieving the objectives' set. The subjects directly involved are the local companies, to whom participation and active collaboration are required. Through a survey on the territory the producers of the territory were identified, for whom it was possible to promote their products and sell them within the structure. Their participation is essential to ensure the interest of users, interested in knowing the territory and its resources.

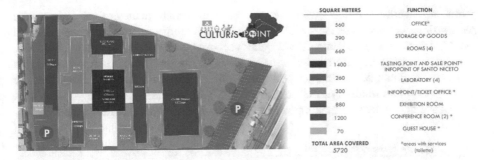

SQUARE METERS	FUNCTION
560	OFFICE*
390	STORAGE OF GOODS
660	ROOMS (4)
1400	TASTING POINT AND SALE POINT* INFOPOINT OF SANTO NICETO
260	LABORATORY (4)
300	INFOPOINT/TICKET OFFICE *
880	EXHIBITION ROOM
1200	CONFERENCE ROOM (2) *
70	GUEST HOUSE *
TOTAL AREA COVERED 5720	*areas with services (toilette)

Fig. 4. Planned activities for the hypothesis of intervention. Functions and consistency of the project areas.

The other activities foreseen by the project require the participation of cultural associations and public institutions, such as schools and universities; these can not only organize events, but also represent potential users. Finally, among the figures involved in the implementation and management of the project is the municipal administration that can represent an important financial support, becoming a partner of the promoter of the initiative (see Fig. 5).

Local territorial bodies
Local companies
Cultural associations
Community
Educational institutions
(schools, universities)

Fig. 5. Figures involved in the implementation and management of the project.

10   The Verification of Pre-feasibility of the Interventions

For the estimation of investment costs, a summary parametric estimate of costs related to building works and external accommodations was carried out; to derive the unit cost of restructuring, a complex of three buildings for offices/laboratories was used as a reference [21, 22]. This choice was dictated by the type of functions envisaged for the structure.

Through the parametric costs obtained, it was possible to calculate the costs of restructuring the buildings of the structure taken into consideration, located in the S. Gregorio district of the Metropolitan City of Reggio Calabria. Subsequently, the investment costs relating to the individual actions were defined. Through the estimate of annual operating costs and revenues, the economic financial plan relating to the profit management model was drawn up. The results that emerged from the pre-feasibility check made it possible to state that an entrepreneur promoter, in this case the Consorzio del Bergamotto, with the proposed management model, could not support the entire project proposal without the contribution of third parties. It could be assumed, as illustrated by the second non-profit scenario that a public entity may be interested in the initiative of the entrepreneurial promoter and decide, as a partner, to co-finance the project by making an annual contribution of €161,710.84 (see Fig. 6).

ESTIMATE OF THE ANNUAL MANAGEMENT COSTS	
ITEM	TOTAL
Human resources	€ 71.760,00
Management fees CULTURIS.POINT	€ 264.000,00
Bus maintenance	€ 3.000,00
Bus operating cost	€ 1.635,84
MARKETING Six-monthly website update	€ 85,00
Domain fee website	€ 150,00
TOTAL	€ 340.630,84

ESTIMATE OF TOTAL REVENUES OF ANNUAL MANAGEMENT			
ITEM	QUANTITY	UNIT PRICE	TOTAL
Rent tasting point	10	€ 1.000,00	€ 120.000,00
Entry ticket tourist and cultural pole			€ 63.600,00
Guided tour of San Niceto castle			€ 58.800,00
Rent conference rooms			€ 57.600,00
Rent exhibition rooms			€ 40.800,00
Revenues of sale point			€ 9.720,00
TOTAL			€ 350.520,00

COSTS	€ 340.630,84		
REVENUE	€ 350.520,00	RESULT	€ 9.889,16

Fig. 6. Estimate of the annual management costs and revenues.

11 Conclusions

Local product is playing an increasingly important role in the project for the enhancement of the territories; the awareness of the importance that the typical product has for the identity and historical memory of the place becomes fundamental for the construction of local development models able to create an active dialogue between consumer (tourist or citizen) and local producer, as well as between community and territorial heritage. The proposed scenarios describe the role of the actors in the territory and the public space, as a container of planned activities, characterizing the project of territorial enhancement and essential for the creation of a network between the areas under consideration [22].

In order to achieve the objective, it is also necessary to define a system of public and private actors that can collaborate for the realization of a sustainable project, engaging in terms of economic and human resources. From this perspective, the actors are identified in municipal administrations, local entrepreneurs and voluntary associations. The complexity of local actors involved becomes a guarantee of sustainability and a self-control of the valorization of common goods [23–25].

References

1. UNESCO: Convention for the safeguarding of the Intangible Cultural Heritage. Parigi (2003)
2. De Lorenzo, A.: Le Quattro Motte estinte presso Reggio di Calabria. Descrizione, memorie e documenti. La motta Anòmeri e la Motta Rossa. Ed. Brenner (1993)
3. Martorano, F.: Calabria Bizantina. Testimonianze d'arte e strutture di territory. La Fortezza bizantina di S. Niceto. Rubbettino editore (2002)
4. Cassalia, G., Tramontana, C.: Un modello applicativo per la valorizzazione del Paesaggio Culturale della Dieta Mediterranea. LaborEst **11**, 78–84 (2015)
5. Mollica, E.: Le aree interne della Calabria: una strategia e un piano quadro per la valorizzazione delle loro risorse endogene. Rubbettino, Soveria Mannelli (1996)
6. Valtieri, S.: Il bene culturale come strategia didattica: conoscenza, tutela, valorizzazione e gestione del territorio calabrese. Falzea Editore, Reggio Calabria (2002)
7. Ministero per le Politiche Agricole. Disciplinare per la denominazione di origine protetta. Bergamotto di Reggio Calabria, olio essenziale (2001)
8. Ministero per le Politiche Agricole. Disciplinare di produzione dei vini ad Indicazione Geografica Tipica «Pellaro» (2013)
9. Ministero per le Politiche Agricole. Disciplinare di produzione DOP. Liquirizia di Calabria (2007)
10. Arfini, F., Belletti, G., Marescotti, A.: Prodotti tipici e denominazioni geografiche. Strumenti di tutela e valorizzazione. Edizioni Tellus (2010)
11. Calabrò, F., Della Spina, L.: The cultural and environmental resources for sustainable development of rural areas in economically disadvantaged contexts. Economic-appraisals issues of a model of management for the valorisation of public assets. In: Advanced Materials Research, vol. 869–870 (2014). https://doi.org/10.4028/www.scientific.net/AMR. 869-870.43, ISSN 1662-8985
12. Pellicanò, A.: Cultura e sviluppo del territorio: risorse, metodologie ed esperienze formative: alta formazione per la valorizzazione dei beni culturali e ambientali nella provincia di Reggio Calabria. Gangemi, Roma (2002)
13. Calabrò, F., Della Spina, L., Viglianisi, A.: Il miglioramento dell'accessibilità per l'incremento della competitività dell'Aeroporto dello Stretto: il contributo della cultura della valutazione, Lo Stretto in lungo e in largo. Prime esplorazioni sulle ragioni di un'area metropolitana integrata dello Stretto di Messina. Edizioni Centro Stampa di Ateneo, CSd'A (2016)
14. Valentino, P.A.: Dossiere generale e guida sulle strategie di valorizzazione integrata delle risorse culturali. Edizioni Formez (2005)
15. Baldacci, V.: Il sistema dei beni culturali in Italia: valorizzazione, progettazione e comunicazione culturale. Giunti editore, Firenze (2004)
16. Bilancia, P.: La valorizzazione dei Beni culturali tra pubblico e privato, studio di modelli di gestione integrata. Franco-Angeli, Milano (2006)
17. Santagata, W.: I distretti culturali. Una formula per lo sviluppo sostenibile. In L'offerta culturale. Valorizzazione, gestione, finanziamento. Roma, Biblink (2001)
18. Andrews, G.: The Slow Food Story. Politics and Pleasure. Pluto Press, London (2008)
19. Guido, M.R.: Tre dimensioni della valorizzazione: l'esperienza, la partecipazione e la gestione. in Primo Colloquio sulla valorizzazione, MiBAC, Roma (2001)
20. XIII Rapporto sul Turismo in Calabria. Regione Calabria, Osservatorio del turismo (2014)
21. Prezzario DEI. Collegio degli Ingegneri e Architetti di Milano. Tipologie edilizie (2004)

22. Calabrò, F.: Local communities and management of cultural heritage of the inner areas. an application of break-even analysis. In: Gervasi, O., et al. (eds.) Computational Science and Its Applications. LNCS, vol. 10406. Springer, Cham (2017). https://doi.org/10.1007/978-3-319-62398-6_37

23. Calabrò, F., Cassalia, G.: Territorial cohesion: evaluating the urban rural linkage through the lens of public investments. In: Bisello, A., Vettorato, D., Laconte, P., Costa, S. (eds.) Smart and Sustainable Planning for Cities and Regions. Results of SSPCR 2017. Green Energy and Technology. Springer (2018). https://doi.org/10.1007/978-3-319-75774-2_39, ISSN 1865-3537

24. Della Spina, L., Lorè, I., Scrivo, R., Viglianisi, A.: An integrated assessment approach as a decision support system for urban planning and urban regeneration policies. Buildings **7**, 85 (2017). https://doi.org/10.3390/buildings7040085

25. Della Spina, L.: The integrated evaluation as a driving tool for cultural-heritage enhancement strategies . In: Bisello, A., Vettorato, D., Laconte, P., Costa, S. (eds.) Smart and Sustainable Planning for Cities and Regions. Results of SSPCR 2017. Green Energy and Technology. Springer (2018). https://doi.org/10.1007/978-3-319-75774-2_40, ISSN 1865-3537

Using an Hybrid AHP-SWOT Method to Build Participatory Ecotourism Development Strategies: The Case Study of the Cupe Valley Natural Reserve in Southern Italy

Salvatore Bianco and Claudio Marcianò[✉]

Mediterranea University of Reggio Calabria, 89100 Reggio Calabria, Italy
claudio.marciano@unirc.it

Abstract. Within Protected Areas, the Natural Reserves are intended to maintain the ecosystems, so environmental protection is the first objective of ecotourism, which is ecological tourism that does not disturb the environment. The purpose of this paper is to evaluate the ecotourism development prospects of the Cupe Valley Reserve in Calabria, Southern Italy. Starting with an explanation of Protected Areas, the paper develops the concept of ecotourism, which allows local communities to benefit from this recreational activity. There follows a description of the area, its tourist flows, and the process that led to the establishment of the Nature Reserve. Then, the methodology section applies the hybrid A'WOT model, whereby various strategies can be formulated in order to weigh the elements that make up the SWOT analysis. The results show how the enhancement of natural resources has changed the economy of the local community, but they also highlight the political problems and obstacles that can impede the development process of the area. Therefore, the strategies are designed to take account of the opinions of the interviewed stakeholders, who represent the main categories of local actors in the entire framework of the Reserve, together with the municipalities and an environmental organization.

Keywords: Protected Areas · Ecotourism · A'WOT
Participatory rural development

1 Introduction

Protected Areas have the function of maintaining the environmental balance of a certain site, by preserving its biodiversity that consists of heterogeneous landscapes in which different species of animals and plants coexist [1]. Alongside the conservation of the natural heritage, protected areas can drive the development of the area by means of ecotourism that, if implemented appropriately, can bring benefits to the local population. Through tourism, public awareness can be raised to become environmental respect, and can also stimulate the emergence of new local economic activities, including those based on crafts, food and wine, which make use of the area's natural resources.

© Springer International Publishing AG, part of Springer Nature 2019
F. Calabrò et al. (Eds.): ISHT 2018, SIST 101, pp. 327–336, 2019.
https://doi.org/10.1007/978-3-319-92102-0_35

In Italy, the first basic law relating to Protected Areas was Law n. 394/1991, the Framework Law on Protected Areas, which laid the foundations for standardizing and coordinating the institutional set-up for the planning, realization, development and management of parks and natural reserves. This Law guarantees and promotes, in an integrated and coordinated way, the preservation and enhancement of natural heritage in all its forms, aiming to mitigate the potential conflicts that can arise between the preservation of natural resources and socioeconomic development [2]. In addition to safeguarding natural resources, Protected Areas can give rise to employment, and thus generate revenue that encourages a positive attitude towards the conservation and protection of nature [3]. From this, a new form of tourism emerges that does not consume resources but is of an educational and adventurous nature, concentrated on natural, cultural and historical sites [4].

So we define ecological tourism or ecotourism as "traveling in relatively undis-turbed or untouched natural areas with the specific objective of studying, admiring and appreciating the scenery, its plants and wildlife, as well as any existing (past and present) cultural aspects of the target areas" [5]. Wunder (2000) states that economic incentives are indispensable for nature conservation, as without the establishment of an economic base for sustainability, the goal of ecotourism cannot be achieved [6]. Moreover, the development of ecotourism will create jobs in tourist services, such as restaurants, souvenir and food shops, campsites, accommodation, transportation, and guide services, and thus provide economic benefits directly to local people [7]. In addition, there will be indirect incentives such as improved infrastructures that will increase the standard of living of local populations. This will encourage respect for different cultures [8], thus increasing environmental awareness and motivation at the participatory level both of which are essential if communities are to be an integral part of the protected area [9].

To protect resources and meet conservation strategies, it is necessary to limit the scope of recreational use, regulate tourism flows, and draft a code of conduct to minimize the negative impact of the growing number of eco-tourists [10]. As well as environmental protection, it is also necessary to provide economic assistance to the communities involved through the diversification of economic activities in ecotourism so that the local population can obtain significant incomes and improve the quality of their lives [11].

Some critics say that the development of tourism is self-destructive, and in the long run contributes to environmental destruction, as the increasing number of tourists threatens the quality of life and the environment, so a functional management approach is required includes respect for social needs and development priorities, protecting biodiversity, and, above all, creating a balance between protection and economic development [12]. In order to ensure the proper implementation of ecotourism, strategies must be established to prevent any negative consequences. One useful method that can be used to do this is the A'WOT hybrid method, which involves applying the Analytic Hierarchy Process (AHP) to the SWOT analysis. This method enables an analytical picture of the situation to be produced and therefore all the qualitative factors present in the SWOT assume a quantitative value by means of the AHP, thus allowing the strategy to be determined more accurately [13, 14]. The present study looks at the Cupe Valley Natural Reserve in Calabria, and the ecotourism strategies that have been proposed by examining the ecotourism potential it possesses, the level of utilization of this potential, and the problems which exist.

2 Methods

2.1 Description of the Study Area

The Cupe Valley Natural Reserve (Fig. 1) is located in Calabria, in the Presila Catanzarese in the Province of Catanzaro, and occupies the municipalities of Sersale and Zagarise with an overall surface area of 647 ha, of which 500 fall within the territory of Sersale, and the rest in that of Zagarise. The Cupe Valley has three geographically distinct territorial areas, all characterized by the presence of natural elements of great ecological and landscape value [15].

Fig. 1. Map of the Cupe Valley natural reserve (shown in green) [15] (Color figure online)

The Cupe Valley Canyon is a rare geological formation deeply embedded in a substrate consisting of polygenic conglomerate, producing on extraordinary scenic effect, resulting from the slow excavation by the stream whose source is at an altitude of 700 m on the Southern slopes of Monte Crozze. The Campanaro waterfall drops down from the slopes of Mount Spineto (1,434 m asl), in the Greco-Carrozzino area, on the border between the communes of Sersale and the communes of Zagarise, right in the heart of the Presilano Catanzarese. The Crocchio Gorges, which comprise the third part of the Protected Area and are located in the middle stretch of the Crocchio river, were mentioned in antiquity by Strabone and Pliny the Elder Arocha. According to the legend, the name of this river derives from that of a nymph who after being raped by a crude shepherd, wept so much that the gods took pity and turned her into a river [16]. The land is also characterized by the rich forests and abundant native flora that constitute a heritage of remarkable scientific value, including in terms of biodiversity. There are also many species known as "officinal herbs". Among the species of greatest scientific interest are: the Woodwardia radicans, a really ancient fern of tropical origin, among others, on the IUCN Red List; the Pteride of Crete, another species of relict fern in the ancient tropical flora; Osmunda regalis, a species of oceanic fern [17]; the hairy

Felcetta (Cheilanthes marantae) reported in Southern Italy only in the Cupe Valley area [18]. Among the vegetation habitats covered by the Annex to Directive 92/43/EEC, are: the evergreen forest; the evergreen cork forest; a rare oriental woodland (Platanus Orientalis) [19]; the equally rare laurel groves; and the native black pine forests of Calabria [17]. The motivation that led the Regional Council of Calabria to establish the Cupe Valley Protected Area is framed in the perspective of sustainable development, referring to a socioeconomic process already in place, and widely known as the "Cupe Valley phenomenon", understood as the process that is affecting this area of the Presilano Catanzarese, and which is increasingly attracting attention at the regional, national and European level. A particularity of this process is that it is a phenomenon that was born and developed from the grass roots level, triggered by the action of a small group of environmental studies and local history enthusiasts, convinced that natural, landscape, historical and cultural resources mainly on part of the Calabrian territory, are an obvious development opportunity [17].

Following the proposal, to make the Cupe Valley a Protected Area, it was followed up by the Environment Commission, and in the Budget Committee, and, in addition, the opinion of the Scientific Committee for Protected Areas of the Calabrian Region was taken into account. The rationale of the law was to recognize the model of sustainable development implemented by the Cooperative "Mediterranean Secrets", as well as the landscape beauty and unique biodiversity of the Cupe Valley. The law was unanimously approved by the Regional Council of Calabria at its meeting on 21 December 2016. It was promulgated through Burc n. 125 of 27 December 2016, and assigned the number 41 in the series of regional laws of 2016 [15]. Although the Reserve was set up in 2016, the activity to enhance its natural resources had already began in 2003 with the action of the "Mediterranean Secrets Cooperative", which established the first trails and has dealt with the management of tourist flows, which have increased from year to year (Table 1), as well as with activities in the catering sector (Table 2) and tourist accommodation (Table 3).

Table 1. Tourist flows in the Cupe Valley area

	2003	2008	2013	2017
Cupe Valley area	3,000	12,000	22,000	27,959

Source: Our elaboration

3 Methodology

As mentioned in the introduction, the methodology used to define priorities and strategies is the SWOT-AHP (A'WOT) hybrid model. This model allows the SWOT analysis to be combined with the analytical hierarchy process (AHP), in order to have an analytical of the situation [20]. There are many applications of this model in the literature on the field of protected areas [13, 14]. The SWOT (acronym of strengths, weaknesses, opportunities and threats) allows the most important internal and external factors to be identified, in order to have a basis for the formulation of the strategy to be implemented [20]. Sometimes, the use of SWOT does not provide a complete picture of the situation,

Table 2. Catering activities in the Cupe Valley area

	2002		2007		2013		2017	
	n°	Places	n°	Places	n°	Places	n°	Places
Farmhouse	1	45	2	75	2	75	3	105
Home restaurant	0	0	0	0	0	0	1	6
Restaurant	2	240	2	240	2	240	4	365
Restaurant pizzeria	1	30	2	50	3	90	3	90
Risto pub	0	0	0	0	1	24	2	52
Cupe Valley area	4	315	6	365	8	429	13	618

Source: Our elaboration

Table 3. Tourist accommodation in the Cupe Valley area

	2002		2007		2013		2017	
	n°	Beds	n°	Beds	n°	Beds	n°	Beds
Farmhouse	1	5	2	12	2	12	3	24
Home restaurant	0	0	0	0	0	0	1	3
B&B	0	0	1	7	3	19	4	27
Housing accomodation	0	0	0	0	1	6	3	17
Cupe Valley area	1	5	3	19	6	37	11	71

Source: Our elaboration

since problems may arise due to the fact that it provides qualitative information, not allowing quantitative comparisons [13]. Furthermore, SWOT does not include any means to analytically determine the importance of factors or to evaluate decision alternatives. To overcome this problem, it can be combined with the AHP, allowing it to reach a more analytical level, making the various factors commensurate [20].

The Analytic Hierarchy Process (AHP) is a multicriteria method that can also be used in the context of strategic planning [21]. This methodology allows the decision maker to analyze and evaluate various alternatives, as well as to adequately manage complex choices, starting from the personal and subjective preferences of the users interviewed. The decision-making process is modeled on the construction of a hierarchical tree on several levels, the number of which grows according to the level of data disaggregation. Compared with other multicriteria methods, the AHP has an advantage, especially in those situations where subjective opinions are a fundamental part of the decision-making process, due to its ability to integrate tangible and non-tangible criteria. Hence, it has been shown that the AHP contributes to making coherent decisions, whose criteria are expressed by subjective measures based on experience. Using it, a pairwise comparison is made between the various factors in order to determine which of them has priority, and to define the importance of each one with respect to the other. Hence, it is clear that, in the hybrid model A'WOT, by applying the AHP technique, the degree of importance of each SWOT factor within the reference category and within each of the other categories can be determined [22, 23].

4   Results

The SWOT analysis was built up thanks to several analyses carried out on the territory (see the previous section) undertaken by several authors, who have thoroughly studied the tourism system. In addition to this, reference was made to the Law establishing the Reserve, and supplemented by a sample of privileged witnesses: the Mediterranean Secrets Cooperative, private actors (restaurateurs, bar managers, B&B managers), tourists and local inhabitants, who were asked to comment on the most significant internal and external factors of the Reserve. The SWOT analysis (Table 4) summarizes of the ecotourism potential of the Cupe Valley Natural Reserve. It has 21 factors divided into 9 Strengths, 6 Weaknesses, 3 Opportunities, and 3 Threats.

Table 4. SWOT summary

Strengths	Weaknesses
Great landscape value (S1)	Local population is not aware of the importance of the Reserve (W1)
Presence of historical-archaeological sites (S2)	Lack of trails in some sites (W2)
Part of the Reserve belongs to the Sila National Park (S3)	Scarcity of financial resources (W3)
High naturalistic value of the sites (S4)	Obsolete web site (W4)
Easy access to the town of Sersale, the main municipality of the Reserve (S5)	Poor training of tourist operators (W5)
The Reserve is very near to the Ionian coast (S6)	Lack of tourist facilities (W6)
Museum network (S7)	
Passion, dedication, and professionalism of the Reserve managers (S8)	
Presence of significant tourist flows (S9)	
Opportunities	Threats
Possibility to use the financial instruments of the European community (O1)	Intervention of the politicians at the regional level (not in harmony with local stakeholders' perspectives) (T1)
High demand for ecotourism and the possibility to create tourist packages with the accommodation facilities of the nearby coastal area (O2)	Conflict between the local authorities belonging to the Reserve (T2)
Generation of employment for the local population (O3)	Growing speculative attention (T3)

Source: Our elaboration

Subsequently, in order to weight the elements of the SWOT analysis, a pairwise comparison of these elements was made by stakeholders identified by non-probabilistic sampling [24]. The final sample consisted of: restaurants (3); bars (2); bed and breakfasts (2); the municipal administration of Sersale; the Reserve as an institution; the Mediterranean Secrets Cooperative which manages the tourist flows; and, finally, a sample of tourists (5). The stakeholders' views were identified by submitting a questionnaire in which there were the comparisons between the SWOT groups and the various elements of each SWOT group. Through these comparisons, the importance of each factor was determined with respect to the others by using the fundamental scale of Saaty [21]. The global SWOT priority values are listed in Table 5, and show how the group that has the highest priority is Strengths, with 37.5%, followed by the Opportunities group, with 28.1%, and the Weaknesses group, with 25%. Last comes the Threats group, with 9.4%.

Table 5. Local and global weights of SWOT factors

	Factors	Local weight (%)		W. Group (%)	W. Global (%)
Sub-criteria strengths	S1	24.57	0.25	37.50	9.21
	S2	8.64	0.09	37.50	3.24
	S3	5.14	0.05	37.50	1.93
	S4	13.27	0.13	37.50	4.97
	S5	20.05	0.20	37.50	7.52
	S6	7.40	0.07	37.50	2.77
	S7	5.25	0.05	37.50	1.97
	S8	10.45	0.10	37.50	3.92
	S9	5.25	0.05	37.50	1.97
Sub-criteria weaknesses	W1	16.65	0.27	25.00	6.66
	W2	21.58	0.22	25.00	5.39
	W3	25.38	0.25	25.00	6.34
	W4	7.36	0.07	25.00	1.84
	W5	11.42	0.11	25.00	2.86
	W6	7.61	0.08	25.00	1.90
Sub-criteria opportunities	O1	42.86	0.43	28.13	12.05
	O2	42.86	0.43	28.13	12.05
	O3	14.29	0.14	28.13	4.02
Sub-criteria threats	T1	33.33	0.33	9.38	3.13
	T2	33.33	0.33	9.38	3.13
	T3	33.33	0.33	9.38	3.13

Source: Our elaboration

In the Strengths group, the highest priority factor is great "landscape value" (S1) (24.6%), followed by the "easy access to Sersale town", the main municipality of the Reserve (S5) (20%), and the "high naturalistic value" (13.3%). The least important strengths were "the presence of major tourist flows" (S9) (5.3%), the "museum network" (S7) (5.2%) and the "part of the Reserve belongs to the Sila National Park" (S3) (5.1%). In the Weaknesses group, the weakest factor is "the local population is not aware of the importance of the Reserve" (W1) (26.7%), followed by "scarcity of financial resources" (W3) (25.4%) and "lack of trails on some sites" (W2) (21.6%). The weaknesses which weigh less were the "poor training of tour operators" (W5) (11.4%), the "lack of tourist facilities" (W6) (7.6%) and the "obsolete web site" (W4) (7.4%). In the Opportunities group, the first two elements (O1 and O2) are equally important, while the third element "generation of employment for the local population" (O3) is considered less important. Lastly, in the Threats group all three factors (T1: "Intervention of the politicians at the regional level"; T2: "conflict between local authorities belonging to the Reserve"; and T3: "growing speculative attention" are equally important. However, T1 and T2 they have emerged because there has been a controversial proposal by some regional politicians. Indeed, regional authorities wanted to create a governance with a Board of Directors, where the politicians would play an important role in influencing decisions while the local actors would prefer a participative bottom-up approach.

Following this participatory method, to determine the global weights of the various factors, the local weights are weighted with the weights of the respective groups. On the basis of the results obtained from the application of the AHP to the SWOT synthesis, some strategies were proposed which are listed below:

- *Diversification of tourism.* This strategy emerges from the analysis, where a seasonal peak of tourists occurs in the summer months, thus enabling the promotion of ecotourism potential which could also stimulate cultural and historical tourism through the museum network of the Reserve (the ethnobotanical museum, the ethnofauna museum, the carolingian museum and the monastery) in the winter, a time of enogastronomic and artistic activity, through the increasing involvement of the local population.
- *Creation of tourism development corridors.* This strategy presupposes cooperation with the existing local amenities, such as the "Orme nel Parco" adventure park, the "Adventure Village of Sellia", and the Taverna Civic Museum which houses the paintings of Mattia Preti, the great Calabrian painter. This cooperation is partly started and, if implemented, could stimulate the emergence of new commercial activities that are geared to the enhancement of natural and artistic resources.
- *Creating an Appropriate Image for the Reserve.* This strategy is linked to the growth of the Reserve and includes collaboration with Legambiente, an Italian environmental organization, which ensures transparency regarding all the proposals for the safeguarding and protection of natural ecosystems.
- *Strengthening co-operation with travel agents and industry operators*: This strategy requires closer cooperation with travel agents and industry operators, thus encompassing much of the regional and national tourist itineraries. Doing so could increase the share of foreign tourists who, in most cases, do not have sufficient information about the area, and, in particular, cannot reach it because, sometimes, transport links are absent.
- *Organization of training programs to develop the concept of ecotourism.* This strategy involves the organization of training programs to develop the concept of ecotourism. A significant fact resulting from the data processing is the local population's ignorance of the importance of the nature reserve. Conferences could then

be organized to make it clear what a nature reserve is, explain what benefits it can bring, and, above all, increase the sense of environmental responsibility in the population.

- *Utilizing the advantages offered by the strategic geographic position.* This strategy concerns the utilization of the advantages offered by the strategic position: in fact, the proximity to the mountains and the sea creates enormous possibilities for tourism; it could increase cooperation with coastal communities, in order to offer visitors various types of tourism, whereby the sea and the mountains can be enjoyed together by enhancing the environmental heritage.

5   Conclusions

The key objective of Protected Areas is to preserve and safeguard biodiversity. In the light of this, we have analyzed the valorization of the Cupe Valley Natural Reserve by the local population, who could become the engine of its future development by means of ecotourism that could bring prosperity to the local community. To do this, it is necessary to build a common vision of for the area and, above all, be aware of the factors that are important in planning and implementing effective development strategies.

The SWOT-AHP hybrid model (A'WOT) enabled a quantitative analysis of the SWOT factors, which made it possible to highlight the convergence and divergence points among the various stakeholders, and therefore it had the advantage of combining the advantages of SWOT qualitative analysis (SWOT analysis) with quantitative analysis (AHP).

As evidenced by the results, the factors on which to focus in order to enhance development are those that affect the implementation of interventions aimed at increasing the services offered to visitors. However, some of the critical issues that may hinder the process have emerged, and there is a distinct lack of interest in the Reserve by the population of the area. Therefore, it is important to carry out a continuous awareness-raising action with regard to this problem, including seminars that can highlight the centrality of the Reserve in the territory.

Acknowledgements. This study has been supported by GASTROCERT, a Project on Gastronomy and Creative Entrepreneurship in Rural Tourism, funded by JPI Cultural Heritage and Global Change - Heritage Plus, ERA-NET Plus (2015-2018).

References

1. Dudley, N.: Guidelines for Applying Protected Area Management Categories, p. 8. IUCN, Gland (2008)
2. Palmieri, N.: The system of protected areas for the conservation of biodiversity. SILVAE **12**, 15–36 (2009)
3. Chen, Z., Yang, J., Xie, Z.: Economic development of local communities and biodiversity conservation: a case study from Shennongjia National Nature Reserve, China. Biodivers. Conserv. **14**, 2095–2108 (2005)

4. Lenao, M., Basupi, B.: Ecotourism development and female empowerment in Botswana: a review. Tour. Manag. Perspect. **18**, 51–58 (2016)
5. Ceballos-Lascuráin, H.: Tourism, Ecotourism and Protected Areas: The State of Nature Based Tourism Around the World and Guidelines for Its Development. IUCN. The World Conservation Union, Gland, Switzerland and Cambridge (UK), (1996)
6. Wunder, S.: Ecotourism and economic incentives - an empirical approach. Ecol. Econ. **32**, 465–479 (2000)
7. Das, M., Chatterjee, B.: Ecotourism: a panacea or a predicament? Tour. Manag. Perspect. **14**, 3–16 (2012)
8. Nyuapane, G.P., Poudel, S.: Linkages among biodiversity, livelihood and tourism. Ann. Tour. Res. **38**, 1344–1366 (2011)
9. De Lange, E., Woodhouse, E., Milner-Gulland, E.J.: Approaches used to evaluate the social impacts of protected areas. J. Soc. Conserv. Biol. 1–7 (2016)
10. Tsaur, S., Lin, Y., Lin, J.: Evaluating ecotourism sustainability from the integrated perspective of resource, community and tourism. Tour. Manag. **27**, 640–653 (2006)
11. Hunta, C.A., Durhamb, W.H., Driscoll, L., Honey, M.: Can ecotourism deliver real economic, social, and environmental benefits? a study of the Osa Peninsula, Costa Rica. J. Sustain. Tour. **23**, 339–357 (2015)
12. Chan, R., Bhatta, K.: Ecotourism planning and sustainable community development: theoretical perspectives for Nepal. South Asian J. Tour. Herit. **6**, 69–96 (2013)
13. Kurttila, M., Pesonen, M., Kangas, J., Kajanus, M.: Utilizing the analytic hierarchy process AHP in SWOT analysis, a hybrid method and its application to a forest-certification case. For. Policy Econ. **1**, 41–52 (2000)
14. Akbulak, C., Cengiz, T.: Determining ecotourism strategies using A'WOT hybrid method: case study of Troia historical national park, Çanakkale, Turkey. Int. J. Sustain. Dev. World Ecol. **21**, 380–388 (2014)
15. Regional Law 27 December 2016 n. 41, Establishment of the Regional Nature Reserve of the Cupe Valley, Calabria Region, Italy
16. Proposed Law 25 November 2015 n. 104 for the establishment of the Regional Nature Reserve of the Cupe Valley, Calabria Region, Italy
17. Lupia, C., Lupia, R.: Le Valli Cupe. Nature and Hiking Guide. Rubbettino (2010)
18. Caruso, G., Lupia, C., Uzunov, D., Pignotti, L.: Italian Botany Society Newspaper. New Report of Cheilanthes Marantae (2004)
19. Caruso, G., Gangale, C., Uzunov, D., Pignotti, L.: Chorology of Platanus orientalis (Platanaceae) in Calabria (S Italy). Phytol. Balc. **14**, 51–56 (2008)
20. Kangas, J., et al.: A'WOT: Integrating the AHP with SWOT analysis. Finnish Forest Research Institute, Kannus Research Station, pp. 189–198 (2001)
21. Saaty, T.L.: The analytic hierarchy process, what it is and how it is used. Math. Model. **9**, 161–176 (1987)
22. Romeo, G., Marcianò, C., Performance evaluation of rural governance using an integrated AHP-VIKOR methodology. In: Zopounidis, C., Kalogeras, N., Mattas, K., van Dijk, G., Baourakis, G. (eds.) Agricultural Cooperative Management and Policy. Cooperative Management, pp. 109–134. Springer International Publishing, Switzerland (2014)
23. Nikodinoska, N., et al.: SWOT-AHP as an inclusive analytical tool of the forest-wood-energy chain: the case study of the Sarntal (South Tylor). Ital. Soc. Silvic. For. Ecol. **12**, 1–15 (2015)
24. Cedrola, E.: Notes on market research, pp. 89–91. Milan (2001)

The Cultural Landscape of the Rocky Settlements of Calabrian Greek Monk

Daniele Campolo[✉], Tiziana Meduri, and Immacolata Lorè

Mediterranea University of Reggio Calabria, 89100 Reggio Calabria, Italy
daniele.campolo@unirc.it

Abstract. The research project starts from a survey of the hypogean sites, on the Ionian and Tyrrhenian coasts of the Province of Reggio Calabria, where traces of "Italian-Greek" monks can be found and which have contributed to the realization of a Cultural Landscape since the times of the first immigrations.

Monks deeply influenced the culture of the places in which they settled due to their knowledge. They devoted themselves to manual work and working in the fields, teaching local populations the new agronomic techniques imported from the East.

Consequently, in the vicinity of the monasteries, urban agglomerations were created, which reflected specific techniques of sustainable land-use, which took into account the characteristics and limits of the natural environment they were established in, and a specific spiritual relation to nature. The methodology used for the historical survey, starting from the bibliographic research, also focuses on research in the area through exploration on the territory and comparing the hypogeal and epigeal sites present in the provincial territory.

Keywords: Cultural landscape · Greek monasticism · Rocky settlements

1 Introduction

The cultural landscape represents the "combined works of nature and of man" designated by the World Heritage Convention.

They illustrate the evolution of human society and settlements over time, under the influence of the physical constraints and/or opportunities presented by their natural environment and of successive social, economic and cultural forces, both external and internal [1].

Cultural landscapes are landscapes that have been affected, influenced, or shaped by human involvement. A cultural landscape can be associated with people or historical events.

The concept of cultural landscape can be summarized in the definition of Carl O. Sauer: "The cultural landscape is molded from a natural landscape by a cultural

---

This is the result of the joint work of the authors. Although scientific responsibility is equally attributable, the abstract and Sects. 1, 2 and 2.1 were written by Daniele Campolo; Sect. 2.2, were written by Tiziana Meduri; Sects. 2.3 and 3 were written by Immacolata Lorè.

© Springer International Publishing AG, part of Springer Nature 2019
F. Calabrò et al. (Eds.): ISHT 2018, SIST 101, pp. 337–345, 2019.
https://doi.org/10.1007/978-3-319-92102-0_36

group. Culture is the agent, the natural area is the medium, the cultural landscape is the result" [2].

Ever since the VI–VII century AD, the Calabrian territory has been profoundly influenced by the migrations of Eastern monks escaping Arab and Persian invasions and iconoclastic persecutions. Calabria was a refuge for monothelite monks fleeing the eastern provinces of the Empire. They found favorable conditions to establish their settlements in Calabria.

The Calabrian coasts offered a cultural network fabric of Greek origin that merged well with the cultural origin of the monks and a geological conformation rich in hypogeal settlements dating back to the Neolithic period that could be used as dwellings.

In the following centuries we can identify at least three migratory flows: Hermits devoted to the isolated quest for God had inhabited the slopes of the Aspromonte and build their settlements. Subsequently, groups of anchorites introduced cenobitic monasticism characterized by the balance between the individual and the community, the respect for diversity and harmony between man and nature.

This valuable wealth of settlements, not yet adequately studied and protected, could open a perspective of historical-archaeological and touristic valorization of this immense cultural heritage.

The territory involved meets all the criteria of cultural landscape and could be enhanced through the creation of a tourist-cultural district promoting the heritage of the rupestrian in the province of Reggio Calabria. A sustainable development project could improve the identity resources of the Greek culture of Calabria, now disappearing [3].

The phenomenon reached such large proportions as to characterise this area. Consequently Byzantine Calabria underwent a slow process of orientalisation of all forms of religious life (rites, cults and liturgy), which accompanied the remarkable spread of churches and monasteries, founded by Eastern monks, that preserved and transmitted the Greek and Hellenistic tradition.

The monks were also import vectors of culture from the East and at the same time, through their movements in Europe, influenced Western countries with their knowledge. They spread out through Italy and Europe, especially developing the local economy with technological innovations in agriculture and animal husbandry.

Therefore, the greek monasticism represents an excellent example of cultural interchanges and coexistence in the world history, reflecting and preserving the greek culture in the south of Italy.

The caves, the hermitages, the remains of the monasteries, the Byzantine churches, the rock settlements are evidence of the presence of Greek monks over seven centuries of Calabrian history.

2 Evolution of Monastic Settlement Systems

The different periods of immigration in the Calabrian territory also correspond to the different types of settlements: the first flows of the 7th–8th century were concentrated in residences in caves; in the 9th–10th century monks built "laure" and "cenobi"; after the XI–XII century they develop monasterial complexes.

The chosen areas for italo-Greek monks settlements were generally in inland areas, far from the coast and the dangers of Muslim incursions, along transhumance routes or near ancient settlements of rock civilizations, in inaccessible areas in order to control the sea routes, with availability of land where monks could experiment new farming methods [4].

The early monks were mainly hermits devoted to ascetic life, meditation and prayer. For this reason, the monks initially adapted to living in handmade caves; they are cavities of different sizes, mainly circular and communicating through tunnels: the caves, used as rooms for monks' cells often have engravings, crosses, niches for lamps and icons, seats carved out of the rock.

A second type of settlement is Laura (Λαύρα), a Greek word that originally meant "narrow path", "gorge", "ravine". The typical laura was made inside some narrow crevasses, on desolate slopes or scarcely covered by vegetation [5].

The central buildings leaned against the steep rock, sometimes carved in the same stone; constructions were usually organized in terraces. The cells of the solitary hermits stood out all around. In caves monks prayed, read, meditated, practiced ascesis and worked. The central core consisted of a church, a meeting room, an oven, a warehouse and sometimes a stable. Sometimes the church was a large cave. Pilgrims and travelers were hosted in a guest house.

All the settlements received the influence of the East: even the urban settlements were built according to the type of "city-kastron", fortified centers with a wide view of the sea to protect the territory from Arab incursions; while the "chorion" (many historic centres in the Grecanic Area in the province of Reggio Calabria still have the name "Chorio") was the fortified agricultural center, built inside a natural or artificial road system, that could connect it with the other nearby centers.

The monks also maintained the right balance of manual work alongside the contemplative - and often eremitic - life, giving rise to authentic centres of production, which left their mark on the economy of the places where the monasteries where located. The rock settlement originates from the fundamental necessity of camouflaging itself in the territory and in the nature, to escape the persecutions and dangers of the coasts [6].

In the territory of the province of Reggio Calabria, along the Ionian and Tyrrhenian coasts, there are several rock complexes. They are part of historical and artistic heritage, some of them date back to the Bronze Age, others were built in more recent times and were inhabited starting from the 5th–6th century by hermit monks.

Four of these settlements, closely linked to the Byzantine monks, should be studied more in-depth: the caves of "Brancaleone Vetus", the caves in the territory of Palmi (Grotte della Pietrosa and of "Macello-Pignarelle"), the caves of Melicuccà (cave of S. Elia Speleota)

2.1   Settlement of Brancaleone Vetus

The historic centre of Brancaleone, now abandoned, rises 4 km away from the coast, on a hill 300 m above sea level. It contains a hypogeum complex, consisting of about twenty units partly incorporated into brick houses. The inhabited nucleus, as well as all the centers of the Greek area of Calabria, was connected to the coast and to neighboring countries, through a system of paths.

In ancient times the land of Brancaleone was called "Sperlinga or Sperlonga", from the latin word "Spelunca" and from the Greek word "σπήλυγξ", meaning cave or cavern.

The origins of the village, datable between the 5th and 6th centuries, probably derive from a first settlement of Greek-Byzantine monks. The entire territory, district of the ancient diocese of Bovesia, was strongly influenced by the presence of the monks: many of the historic centres, now part of the so-called "Grecanica Area", originated from the works of these monks, who contributed to enriching the economic activities and cultural heritage. The village is a classic example of Byzantine settlement where the architecture had to have, because of defensive needs, a complete view of the valley below. The road system was closely connected to the Vallone Monaca that provided water supply.

The housing units are small and circular, often consisting of a single room, with the presence of niches for oil lamps and seats carved into the rock. In later periods the cells were turned into shelters by the first inhabitants of the place, who escaped from the continuous attacks on the coasts. Others units were annexed to homes and were used as service areas, such as silos or cellars. In some of them there are tanks and structures used for processing and for the storage of foodstuffs.

The Hypogeum Complex of Brancaleone is made in a silty arenaceous turbidites rock characterized by layers of silty clay alternating with quartz sandstone layers, originated 20 million years ago (source: Bonfà I.) (See Fig. 1).

Fig. 1. Particular of Brancaleone Vetus. (by Marco Colonna).

In many caves there are graffiti of engraved crosses, niches for oil lamps and seats carved out of the rock. The caves far from the centre and with frescoes have a probable cult destination.

One of the units, known as the Church-Cave of the "Tree of Life" is a circular "hypogeum", constituted and characterized by a central pillar. The name probably derives from this stone element similar to the rock churches of Armenia and Cappadocia. This church-cave was built by the monks of the East between the 8th and 9th century A.D. (source: Stranges S.) (See Fig. 2). The entry hole also presents graffiti of sacralization of the space: an astile cross flanked by a peacock, symbol of resurrection and eternal life.

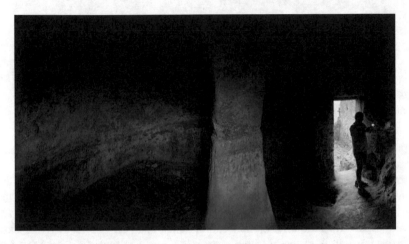

Fig. 2. Church-Cave of the "Tree of Life" (by Marco Colonna).

In the hypogeum complex there is also a rupestrian church with an ellipsoidal plan, now dedicated to the "Our Lady of Rest". It still preserves traces of frescoes both on the walls and on the vault. On the right side two parades of monks ascending are depicted contemplating Christ (See Fig. 3).

2.2   Caves in the Municipality of Palmi

In the territory of Piana di Gioia Tauro there are many testimonies of Greek monk monasteries: over 40 including the Monastery of Sant'Elia and the oldest Mercurion of Tauriana with the early Christian crypt dedicated to San Fantino after his death in 975 in Thessalonica (Thessaloniki).

Cave of Pietrosa (or of Tràchina)

The cave of Pietrosa is located in the municipal area of Palmi, of particular archaeological interest. In the excavations carried out in the 1990s, archeological finds from the Bronze Age and the Helladic period were found. These archeological finds highlighted the relationship between ancient Greece and local populations of the time.

Fig. 3. Astile cross flanked by a peacock. (by Marco Colonna).

The potsherd, from Aegean production, lead to the hypothesis that the Calabrian Tyrrhenian coasts were a landing place for the Mycenaean trade routes [7].

From the archeological finds it is clear that the cave was used as early as the third millennium B.C. and later from the medieval age on as a shelter for flocks. The cave located at 90 m above sea level, on a high cliff of the Costa Viola, in an area of particular value from the environmental and landscape point of view, overlooking of the Tyrrhenian Sea from the Messina's strait to Capo Vaticano (VV) and the Aeolian Islands (See Fig. 4).

The cave has two entrances: one facing west and one facing north. Further down, just a few meters below the level of the cave, there is a disused section of the Southern Tyrrhenian Railway.

Caves of "Macello-Pignarelle" or of "Tarditi"

The caves of "Macello-Pignarelle", located in the "Macello" (boucher's) district of the Municipality of Palmi, form a rock settlement dug into the sandstone by Byzantine monks, probably starting from the 6th century A.D., when numerous Byzantine monks found refuge in Calabria from iconoclastic persecutions.

The settlement, located in a steep rocky ridge a few meters from the town, is composed of several caves excavated on different levels with the entrance facing north overlooking the Tyrrhenian Sea. Moreover, this settlement has a huge importance under the landscape and cultural aspect for the entire territory because in addition to the scenic beauty, it is built within a system of terraces, typical of the Costa Viola. In the past centuries the terraces allowed local farmers to obtain arable even on the steepest and most inaccessible slopes.

Next to the settlement a small stream flowed. The brook in the gully dug in the rock, guaranteed the water supply for the monastery and for daily activities but also for the work activities of the monks.

Fig. 4. The Strait of Messina. (by Daniele Campolo).

The settlement, well known to the inhabitants, was used as a shelter during the bombings of the Second World War. The first cave is made up of a single room with a wide opening, from which the tunnels depart for a few meters inside. The second cave has a depth of 17 m and ends with an apse. The third and largest cave is called "Basilica", it is formed by one nave and two aisles: the nave is 6 m high and 3 m wide with an elliptical apse; the two aisles are 2 m wide and 1.5 m high, because of filling material brought from the outside by rains or by small local landslides; inside, the lateral corridors merge with the central corridor forming a Greek cross.

Caves have numerous Byzantine crosses engraved, some placed on the entrance, other in the vaults; there are also numerous niches for lamps and icons, steps and beds for a large community of monks (See Fig. 5).

2.3   Caves of Sant'Elia

The monastic complex of Sant'Elia lo Speleota (meaning "cave dweller") is located along the provincial road which connects Melicuccà to Bagnara, in a landscape of particular interest, in a valley between the northern slopes of the Aspromonte, among woods of centuries-old olive trees, bridges and tunnels of the evocative and now abandoned Calabrian-Lucan railway line.

Not far from the rocky wall of the original settlement, there are the remains of the eighteenth-century monastery, dedicated to Saints Peter and Paul. Only the cave used as a church remains and it is the place of worship of the saint. The site also presents a cemetery area.

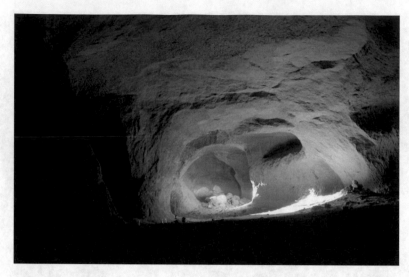

Fig. 5. Particular of the cave "Macello-Pignarelle" (by Daniele Campolo).

In 2005 the Archaeological Superintendence of Calabria discovered eleven pit graves, dating back to the late Middle Ages, entirely carved into the rock. The "bios" of Sant'Elia says the Saint personally dug his grave in the cave where he lived.

The complex is also characterized by the remains of some structures which were used in the monastery for the conservation of foodstuffs or for the production of wine, including an ancient millstone used until the eighteenth century. These remains are important in order to reconstruct the customs of the monastic community.

3   Conclusions: The Cultural Value of the Rocky Settlements

Fitting harmoniously into the natural environment, the settlements offer a model of coexistence, sustainable development and conservation of the environment, providing a valuable message for contemporary society and for future generations, who can draw the essential values for their human and intellectual formation [8].

The Greek monastic experience has provided an extraordinary model of cultural, social and spiritual life, because the monks, dedicated to interior research and to the elevation of man's condition, were carriers for the transmission of innovative knowledge. They have contributed to the improvement of the social and economic conditions of the Calabrian society.

A phenomenon that, radiating from Italy and beyond, had a deep effect on the intellectual and cultural development, shaping the medieval Mediterranean civilizations and contributing to the intellectual, political and economic development of modern Europe [9, 10].

The rocky settlements also represented important cultural centers, in which the universal heritage of knowledge was recovered, preserved and spread through exchanges and acquisitions as the primary investments for the construction of the future [11, 12].

These special sites reveal aspects of Calabria's origins and development as well as the evolving relationships with the natural world. They provide scenic, economic, ecological, social, recreational, and educational opportunities helping communities to understand who they really are.

References

1. U.N.E.S.C.O.: Operational Guidelines for the Implementation of the World Heritage Convention, p. 19 (2017)
2. Campolo, D.: M.C. Escher and the Calabrian cultural landscape of linguistic Greek minorities. In: Advanced Engineering Forum, vol. 11, pp. 458–463. Trans Tech Publications, Switzerland (2014)
3. Marinelli, E.: Il monachesimo bizantino in Calabria. In: Rogerius, Bollettino dell'Istituto della Biblioteca Calabrese, periodico di Cultura e bibliografia, Soriano Calabro, anno IV, n. 2°, pp. 55–61, Luglio-Dicembre 2001
4. Martorano, F.: La rupe come risorsa. Esempi di insediamenti nella Calabria ionica meridionale. In: Insediamenti rupestri di età medievale: abitazioni e strutture produttive. Italia centrale e Meridionale, a c. di De Minicis, E., I, pp. 217–228 (2008)
5. Tiné, V.: La grotta Petrosa di Palmi: i livelli dell'età del bronzo. In: AA.VV.: Palmi, un territorio riscoperto, Rubbettino Editore (2001)
6. Il Paesaggio Culturale dei Monasteri Benedettini. http://www.treccani.it/monasteri_benedettini/. Accessed 08 June 2017
7. Campolo, D., Bombino, G., Meduri, T.: Cultural landscape and cultural routes: infrastructure role and indigenous knowledge for a sustainable development of inland areas. In: 2nd International Symposium New Metropolitan Perspectives, Reggio Calabria, Procedia - Social and Behavioral Sciences, vol. 223, pp. 576–582 (2016)
8. Menestò, E. (a cura di): Le aree rupestri dell'Italia centro-meridionale nell'ambito delle civiltà italiche: conoscenza, salvaguardia, tutela. In: Atti del IV Convegno internazionale sulla civiltà rupestre, Savelletri di Fasano (BR) (2009)
9. Calabrò, F., Della Spina, L.: The cultural and environmental resources for sustainable development of rural areas in economically disadvantaged contexts. Economic-appraisals issues of a model of management for the valorization of public assets. Adv. Mater. Res. **869–870**, 43–48 (2014). https://doi.org/10.4028/www.scientific.net/AMR.869-870.43
10. Della Spina, L.: The integrated evaluation as a driving tool for cultural-heritage enhancement strategies. In: Bisello, A., Vettorato, D., Laconte, P., Costa S. (eds.) Smart and Sustainable Planning for Cities and Regions. SSPCR 2017. Green Energy And Technology. Springer, Cham (2018). https://doi.org/10.1007/978-3-319-75774-2_40. ISSN 1865-3537
11. Della Spina, L.: Integrated evaluation and multi-methodological approaches for the enhancement of the cultural landscape. In: Gervasi, O. et al. (eds.) Computational Science and Its Applications - ICCSA 2017. LNCS, vol. 10404. Springer, Cham (2017). https://doi.org/10.1007/978-3-319-62392-4_35
12. Della Spina, L., Lorè, I., Scrivo, R., Viglianisi, A.: An integrated assessment approach as a decision support system for urban planning and urban regeneration policies. Buildings **7**, 85 (2017). https://doi.org/10.3390/buildings7040085

The Rediscovery of the via *Annia - Popilia* from Capua to Reggio Calabria for Knowledge and Enhancement of the Cultural Route

Rosa Anna Genovese[✉]

Federico II University of Naples, 80134 Naples, Italy
rosaanna.genovese@unina.it

Abstract. The Mediterranean is witnessing the re-emergence of its role as the pulsating heart of the encounter and clash of cultures. The creation of a *cultural path* capable of highlighting the common cultural roots of the regions, surrounding it of exalting their positive values, can come to constitute an essential element for dialogue in the horizon of *Mare nostrum*.

The Via *ab Regio ad Capuam* (known as Via Popilia or Via Annia) is the historic road built by Roman magistrates to join Rome to the *Civitas foederata Regium*, located at the furthermost tip of the Italian peninsula. Its course starts from the Via Appia, a few kilometres South-East of the ancient Capua, branching out through Nuceria, Salernum, the Vallo di Diano, the territories of Calabria and finally Reggio Calabria.

The Via Annia Popilia may therefore represent an axis for sustainable development, the driver of cultural, social and economic growth of the centres it crosses and the territories connected to it in the Regions of Southern Italy (Campania, Basilicata and Calabria, facing the Tyrrhenian Sea, which flows into the Mediterranean), a *Cultural Route* of European breadth through which historic urban landscape, archaeology, architecture and arts can become the meeting point of popular culture and traditions, of oenology-gastronomy and music, of cultural tourism and active participation, both public and private. The Route mentioned should also relate to territorial excellences and, thus, to the cultural and natural Sites inscribed onto the UNESCO World Heritage List and to their respective Management Plans.

Keywords: Mediterranean basin · Integrated conservation · Cultural Routes
Historic roads · Historic Paths · Cultural tourism

1 Cultural Routes, Historic Paths, Historic Roads: Meaning, Conservation and Enhancement

"Cultural Routes often reveal - as I recalled during the 2nd International Symposium New Metropolitan Perspectives ISTH 2020 (Reggio Calabria, Italy, 18–20/05/2016) - the encounter of The East and West, enhancing the contributions of peoples and pas-sing on to younger generations such values as solidarity, freedom, sharing, peace multicultural integration and tolerance. They also represent the irreplaceable narrative keys to establish a relation between man and cultural and natural heritage, both tangible

© Springer International Publishing AG, part of Springer Nature 2019
F. Calabrò et al. (Eds.): ISHT 2018, SIST 101, pp. 346–358, 2019.
https://doi.org/10.1007/978-3-319-92102-0_37

and intangible [1]. The International universe of organisations for the conservation of cultural heritage (UNESCO, ICOMOS and Council of Europe) has not yet reached a unanimous agreement over the meaning and essence of Cultural Routes, disorienting at times the process of perfecting appropriate legislative strategies and the coordination between the institutions responsible for their popularisation, enhancement and implementation.

The Council of Europe maintains that they should unfold around a theme (Resolution CM/Res, 2007) and that they are representative of the memory, history and heritage of Europe. This approach implies they are not meaningful because of their intrinsic value but because they constitute the connection of cultural and touristic interest between elements of heritage, the tool that stimulates multilateral cooperation for intercultural dialogue and European Identity" [2].

Currently, the criteria determining whether an artefact or site should be considered part of cultural heritage have been extended to include the notion of ancient – not interpreted the same way by different cultures – and have come to highlight the meaning and value attributed to such heritage, both tangible and intangible, by the community in which it is located or by the one or the ones to which it is connected historically and/or culturally [3]. The conceptual vision and denominations have similarly been broadened, coming to include Cultural Routes, which in some cases can cover continental or intercontinental distances and host within them various kinds of activities, or landscapes of great variety or distant from one another but united by the significance arising from their joint belonging to the function of the Route itself and its specific purpose [4, 5].

The International *Charter on Cultural Routes*, elaborated by the scientific committee responsible (CIIC) and ratified in Quebec in 2008, in occasion of the 16th ICOMOS General Assembly, defines such types of heritage, the defining elements and a methodological base for their identification, conservation and enhancement.

In short, it can be said that the essential features that distinguish a Cultural Route from other kinds of historic roads or paths of heritage value are those of constituting unique historic phenomena respondent to a precise and practical purpose, predetermined o evolutionarily reflecting part of the community, and different from mere transportation or communication. As the *Charter on Cultural Routes* of the ICOMOS CIIC indeed stresses "... Beyond its character as a way of communication or transport, its existence and significance as a Cultural Route can only be explained by its use for such specific purpose throughout a long period of history...", and then:

"Therefore, Cultural Routes are not simple ways of communication and transport which may include cultural properties and connect different peoples, but special historic phenomena" [...] "Cultural Routes have sometimes arisen as a project planned a priori by the human will which had sufficient power to undertake a specific purpose... On other occasions, they are the result of a long evolutionary process in which the collective interventions of different human factors coincide and are channeled towards a common purpose..." [6].

In accordance with what stated above and with what has already been analysed and approved by the ICOMOS International Committee on Cultural Routes (CIIC), 'Historic Paths' may be included among the various properties that compose them and are part of their functional dynamics and their overall significance. Indeed, they, in turn, correspond to a heritage category by now universally accepted, such as 'Historic Cities', or 'Cultural Landscapes' etc....

A Historic Path is a communication route, usually also used for transport of goods and commodities, connecting two or more geographic locations, traditionally linked to a human community that attributes it a special heritage significance within its culture. Such Paths are the result of the changes of time and, differently from Cultural Routes, do not respond to a certain specific and practical objective, but simply to a need or a human impulse to move physically from one place to another and, if required, to transport goods and wares. Historic Paths must have a texture materially identifiable along their entire course. Nevertheless, if some sections are missing, reference to their historic existence should be demonstrable through both documents and scientific evidence or well attested oral traditions. As mentioned earlier, in some cases there may be Historic Paths or sections of them that, like other properties of different nature, are integrated into one or more Cultural Routes, although in many other cases they exist independently from them. Historic Paths can be of different types, different courses, specific uses and dimensions and they consequently acquire different de-nominations, also according to language and culture, for instance Route, Path, Road, Corridor, etc....

The CIIC Charter therefore establishes the new concept of a Cultural Route, its definition and methods for documenting, conserving and promoting these important global heritage resources. Since the adoption of the Charter there has been lively discussion about the scope of Cultural Routes and how emerging concepts in heritage identification and conservation, for example, historic roads, cultural corridors and cultural landscapes, are acknowledged within the CIIC Charter.

At the CIIC Committee Meeting and Symposium at Ise, Japan in 2009, two working groups were given the task of providing greater clarity about the relationships between historic roads, cultural corridors and Cultural Routes. The CIIC recognizes that a historic road may be an important component of a cultural route.

As many existing Cultural Routes already contain historic roads, it is appropriate to draw on the growing body of international research to gain a better understanding of how historic roads are related to Cultural Routes. The specific working group reached the following conclusions:

- A historic road is a land-based route that allows for individual or group travel along a recognized and established route that has a defined origin and destination.
- A historic road is recognised as a route that serves an important transportation purpose.
- A historic road is distinguished from other land-based routes, such as canals and railways, due to its universal adaptability and non-reliance on scientific engineering (though many historic roads do possess highly developed engineering, a road, by definition does not need possess such attributes).
- A historic road demonstrates regular use and maintenance, and is identified as a non-natural feature in the landscape through compaction, wear or construction.
- A historic road is recognised by a period of regular use and may also be recognised as an example of exceptional design or technology.
- A historic road may facilitate a cultural exchange, or through design and influence, define a cultural exchange.

- A historic road may be layered on much older indigenous pathways and travel routes and so may demonstrate a layering of different cultural activities over time.
- A historic road may be a component of a cultural itinerary; not all historic roads are cultural routes as defined by the CIIC.

Furthermore, "Cultural Routes help to confirm to us that universal civilization is a heritage that belongs to us all, resulting as it does from a historical process to which all of the world's peoples have contributed through their reciprocal cultural influences. By recognising and respecting cultural diversity, Cultural Routes contribute to the enhancement of intercultural dialogue and sustainable development. They may also provide conservation policy with a territorial breadth, cultural integrity and harmonization of actions and contents that has not been accomplished before" [2].

2 A Cooperation Network for the Integrated Conservation of the Mediterranean Cultural Heritage

"There are places", wrote the poet Iosif Brodskij, "that, examined on a map, make you feel connected to providence for a short instant, places where history is inevitable, where geography makes history". The Mediterranean Basin, which is studded with such places and has produced some of the most astounding historic and cultural events of the planet for millennia, possesses all the unique requirements that account for its precocious development. Indeed, it hosts the largest inland sea in the world, close to the fluvial nucleus that the first civilisations developed around, and is the guardian of the richest and most precious sources for the study of ancient cultures.

"Intercultural dialogue represents one of the challenges of the contemporary world and at the same time one of the fundamental human values to build a world of peace and prosperity. In our current historical times the Mediterranean is witnessing the re-emergence of its role as the pulsating heart of the encounter and clash of cultures, and the creation of a 'cultural path' capable of highlighting the common cultural roots of the regions surrounding it, of exalting their positive values. It can come to constitute an essential element for dialogue in the horizon of *Mare nostrum* and offer further possibilities for the aggregation of local communities, bind by shared historical roots, and once also linked by prolific commercial exchanges" [2].

Culture constitutes, in its multiple aspects, an extraordinary resource for the improvement of society and economics, and it implies, as recalled in the *Universal Declaration on Cultural Diversity* of UNESCO (2001), a set of spiritual, tangible, intellectual and emotional features for a society or a community, including also, besides art and literature, lifestyles, values, traditions and beliefs. It thus plays an essential role in human development and in the complex fabric of identity and custom, of individuals and community.

The International Convention of UNESCO for the *Safeguard of Intangible Cultural Heritage* (2003) was conceived to favour the transmission of intangible heritage to future generations through the identification, conservation, protection and enhancement of languages, dialects, music, oral tradition, performance art, religious and folk festivals, knowledge and practices concerning nature and the universe, craftsmanship knowledge, expressions of ancestral culture. With the Convention on *Protection and promotion of diversity of cultural expressions* (2005), UNESCO has completed measures for the conservation, protection and enhancement of the Intangible World Heritage necessary for the harmonious mutual development of man and nature.

The Faro Convention (2005) later introduced the notion of 'social value' of cultural heritage. This social approach offers the interpretation of heritage as a resource for living together, for improving social cohesiveness, and thus for urban regeneration. In order to achieve it, the contribution of local communities for carrying out conservation, protection and enhancement of cultural heritage through a greater participation of inhabitants in decision processes, must be strengthened, so that a 'heritage community' capable of making the city more inclusive may emerge [7].

3 The Cultural Route of the via Annia - Popilia

The Via *ab Regio ad Capuam* (known as Via Popilia or Via Annia) is the historic road built by Roman magistrates to join Rome to the *Civitas foederata Regium,* located at the furthermost tip of the Italian peninsula. It is also known as 'Via Popilia' from the consul Publio Popilio Lenate who would have commissioned it in 132 b.C..

A different interpretative hypothesis 'Via Annia' claims the road was started by Popilio and completed the year after by Tito Annio Rufo in 131 b.C. [8]. Its course starts from the Via Appia, a few kilometres South-East of the ancient Capua (currently Santa Maria Capua Vetere), branching out through Suessola, Nola, the strip of foothill settlements of Somma Vesuvio, the site 'Ad Teglanum' (acknowledged in the Peutinger Table), Sarno, Nuceria, Salernum, the Vallo di Diano, part of Basilicata, the territories of Calabria including Cosentia and finally Reggio Calabria (see Fig. 1).

The archaeology, the landscape and city constitute a highly evocative triad. Mapping finds, reading the continuous transformation process that determined the current layout of inhabited centres and their surroundings, creating a reference system capable of making a dialogue or exchange of information between different subjects with institutional responsibilities possible, therefore creating homogenized documents, updatable in real time, available to the public, means obtaining an irreplaceable tool for the planning of historic cities, and also the promotion of a new public conscience, a renewed sense of collective responsibility.

The research group 'Adopt the Via Annia Popilia' formed by archaeologists, architects and experts of the field, of which I have been the scientific Coordinator for Region Campania and the entire Lions International District 108 YA (2016/2017) since 2015, has operated with a multidisciplinary approach focused on the in depth study of the course the Roman road, knowledge of its historic-territorial and historic-architectural stratification, using the inventory of the most significant cultural emergencies present along it. The classification of tangible (archaeological, historic-artistic, architectural and landscape) and intangible heritage was carried out in order to identify the current historic-cultural features of the territory and to complete proposals for the enhancement of sites and territories involved (Tables 1, 2, 3 and 4).

The Via Annia Popilia may therefore represent an axis for sustainable development, the driver of cultural, social and economic growth of the centres it crosses and the

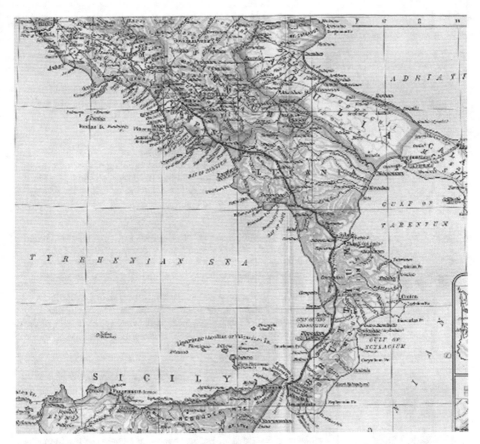

Fig. 1. Ancient Roman Capua-Rhegium (Popilia) Road ('Historical Atlas', William R. Shepherd, New York, Henry Holt and Company, 1923).

territories connected to it in the Regions of Southern Italy (Campania, Basilicata and Calabria, facing the Tyrrhenian Sea, which flows into the Mediterranean), a *Cultural Route* of European breadth through which historic urban landscape, archaeology, architecture and arts can become the meeting point of popular culture and traditions, of oenology-gastronomy and music, of cultural tourism and active participation, both public and private. The Route mentioned should also relate to territorial excellences and, thus, to the cultural and natural Sites inscribed onto the UNESCO World Heritage List (see Fig. 2) and to their respective Management Plans. You may recall that for region Campania they are: (1) The Archaeological Areas of Pompeii, Herculaneum and Torre Annunziata; (2) The Historic Centre of Naples; (3) Cilento and Vallo di Diano, with the Archaeological Sites of Paestum, Velia and the Certosa di Padula; (4) The Amalfi Coast; (5) The Longobards in Italy. Places of power, the church of S. Sofia in Benevento; (6) The 18th century Royal Palace of Caserta with the Park, Aqueduct by Vanvitelli and the complex of San Leucio; (7) The Celebrations of great shoulder-borne procession structures (Intangible Heritage); (8) The Mediterranean Diet (Intangible Heritage);

Table 1. Some significant testimonies of the tangible and intangible heritage concerning the modern town centres found along the *Via Annia – Popilia* in Region Campania (Italy).

Cultural property: tangible, intangible	
	Abbey of Sant'Angelo in Formis, Capua The abbey, which lies in an evocative position was rebuilt by Desiderio, the abbot of Montecassino, between 1072 and 1087. The inside of the church, in three naves, is covered with frescoes from the late 11th century.
	Amphitheatre, Santa Maria Capua Vetere Built between the end of the 1st and the beginning of the 2nd century it is the second largest public building after the Coliseum. Next to it an *antiquarium* holds important testimonies including the inscriptions of the *magistri campani*.
	Carnival, Capua The hundredth anniversary of the Carnival of Capua was celebrated in 1985. Its rich program counts many events: parades of allegoric floats with popular masks, exhibitions, concerts, festivals enliven the city together with characteristic lights, up to the Mardi Gras.
	Benedictine Sanctuary of Maria SS. del Castello, Formicola The Sanctuary stands in the town of Formicola. The hill of the Holy Mary is where the community gathers to worship the Virgin of the Castle, who constantly animates the spiritual and religious life of the inhabitants of the area.
	Palace of the Bishop Seat of the Diocesan Museum, Nola The Diocesan Museum opened in 2000, has been conceived as an integrated venue lively an present throughout the territory, intended to give visibility to works of art present in shrines. The core of the collections exhibited is hosted in spaces adjacent to the cathedral: the structures of S. Giovanni Battista church, the chapel of the Immacolata and the rooms of the Bishop's palace.
	The Gigli of Nola, Nola The Festa dei Gigli di Nola, dedicated to the Patron Saint, together with the 'Varia' of Palmi, the 'Macchina di Santa Rosa' and the 'Candelieri' of Sassari, has been inscribed, since 2013, onto the UNESCO WHL as intangible property in the section "Celebrations of great shoulder borne parade procession structures".
	Basilica of San Felice, Cimitile The complex of the paleochristian basilicas is formed by various buildings dedicated to the saints Felice, Stefano, Tommaso, Calionio, Giovanni, to the Martyrs and the Madonna of Angels, and bears witness, with its historic stratification to the passage from the late empire to the Middle ages, from paganism to Christianity. The original structure of the Basilica of San Felice dates back to the 3rd century.
	Basilica of Santo Stefano, Cimitile Located between the paleochristian basilicas, that dedicated to Santo Stefano presents an original nucleus from the 6th century.

Table 2. Some significant testimonies of the tangible and intangible heritage concerning the modern town centres found along the Via Annia – Popilia in Region Campania (Italy).

Cultural property: tangible, intangible, natural	
	BasilicaVetus Antiquarium, Cimitile The basilica Vetus constitutes the heart of the paleochristian complex of the basilicas, having been built over the tomb of the bishop Paolino, guardian saint and thaumaturgist..The first nucleus dates back to the 4th century.
	Regional Forest, Roccarainola The Forest can be subdivided into four typologies according to elevation. The flora and vertebrate fauna found displays great biodiversity. The Vivaio Costa Grande is part of it, being a site for the cultivation of essences destined to reforestation and the safeguard of biodiversity of regional plant species.
	Parc of Partenio and Roman Amphitheatre, Avella The Parc of Partenio is a very important site for its natural, environmental, historic-religious and cultural resources. The Roman Amphitheatre, a monument that has become the symbol of Avella, lies within the protected area.
	The Viae Publicae, Santa Anastasia The *Viae Publicae* cross municipal territory. The two distinct roman-time roads come from the coast and joining together, heading towards Somma where they used to join onto the *ab Regio ad Capuam*. Under environmentally protective restrictions in compliance with the former Law 1497/39.
	Festa of the Madonna dell'Arco, Santa Anastasia The rites of the Madonna dell'Arco take place on Easter Monday. The long race of the 'fujenti' ends when it reaches the Virgin. The jostling pilgrimage brings a thick never ending crowd of barefoot devotees who follow an antique itinerary of pain with repeated gestures of archaic rituality every year.
	Augustan Aqueduct, Palma Campania Roman aqueduct with two adjacent ducts relative to two different construction and utilisation stages, one dating back to the Augustan Period in *opus reticulatum* and the other of the Constantine Period in *opus latericium*. Chronologic phase: 1st cen. B.C.- 1st cen. A.C.
	Itinerant Museum of Memory and Peace, Campagna The Itinerant Museum of Memory and Peace is located inside the Dominican convent of San Bartolomeo, which was used as prison camp for civilian Jews during the Second World War. The Museum includes a permanent exhibition of photographic panels with documents and pictures of the Shoah, and a reconstruction of a room of the camp and of the synagogue.
	Greek - Roman Theatre, Sarno The theatre, found at Foce in 1965, displays the layout of Greek theatres of 2nd-3rd cen. B.C. with a stone seat cavea. The archaeological digs executed have led to hypothesise the presence of a sanctuary of which the theatre structure was part.

Table 3. Some significant testimonies of the tangible and intangible heritage concerning the modern town centres found along the *Via Annia – Popilia* in Region Campania (Italy).

Cultural property: tangible, intangible	
	Church and Convent of Materdomini, Nocera Superiore First nucleus dating back to 11th cen. Under monument protective restriction art. L.D. 42/2004.
	Monumental Necropolis of Pizzone, Nocera Superiore The monumental necropolis lies along the Cavaiola brook, which separates it from the road 18. Chronologic reference: 2nd century B.C. - 1st century A.C. Cultural reference: late republican and imperial age. Under archaeological protective restriction F. with M.D. 26.11.94.
	Festival and International Madonnari Contest, Nocera Super. The Contest takes place in May, in honour of S. Maria di Costantinopoli and S. Pasquale Baylon. It is linked to the National Madonnari Centre (MN); the Kalos Centre, (Mexico); Santa Barbara Mission, (California USA); Italian Street Painting Festival, all FIEM federates (International Madonnari Events Federation). Enhancement proposal: request to UNESCO to recognize the four contests as 'intangible heritage' properties.
	Church of S. Maria maggiore or *della Rotonda*, Nocera Super. The church of Santa Maria maggiore or of the *Rotonda* (late 5th- 6th century) is a paleochristian baptistery on a circular plant with three apses and an ambulatory. It was rebuilt after the 1944 eruption of Vesuvius.
	Church and Congregation of S. Maria al Quadruviale, Cava de' Tirreni It is located in San Pietro di Cava. The original nucleus dates back to the 4th century. Under environmental and monumental protective restriction, respectively art.136 L. D. 42/04 and art. 10 L. D. 42/2004.
	Festa of Montecastello, Cava de' Tirreni The Festa takes place every year in remembrance of how the plague epidemic of 1656 was escaped. The Protagonist is the piston or arquebus, ancient weapon of the 16th century, which the *Pistonieri* still fire with blanks during the traditional *Giro*, to celebrate the Festa.
	The challenge of the Trombonieri, Cava de' Tirreni Historical re-enactment of the consignment of the *White parchment* (kept in the archives of the Palazzo of the city), celebrated every year with the parade of the Trombonieri and a folkloristic event in commemoration of the historical event in which the city had to defend its liberties: the battle of Sarno.
	Via Torquato Tasso, Salerno Section of road paved with polygonal basalt flagstones discovered in 1879 about one metre under via Tasso, between the junctions of via dei Canali and via Madonnna della Lama, it has been recognised by various sources as the urban section of the Roman consular road.

Table 4. Some significant testimonies of the tangible and intangible heritage concerning the modern town centres found along the *Via Annia – Popilia* in Region Campania (Italy).

Cultural property: tangible, intangible	
	**Garden of Minerva, Salerno** During the Middle Ages it was used as *Giardino dei semplici* for didactic purposes for the students of the Medicine School of Salerno. In the heart of the antique centre in the area called *Plaium montis* in the Middle Ages, it is halfway along an ideal route following the axis of the walled and terraced vegetable gardens that from the Villa Comunale go up towards the Castle of Arechi. Protective restriction: Law 1089/39.
	**The procession of San Matteo, Salerno** The procession of San Matteo, patron saint of the city, takes place on 21st September every year. It starts from the cathedral and moves along the main streets of the centre with great participation of the population and a parade of the statue of the Saint, of S. Gregorio VII, S. Giuseppe and the Martyr Saints Gaio, Ante and Fortunato.
	**The Festa of the Madonna of Costantinopoli, Salerno** On the first Sunday of August a procession over sea and land commemorates the voyage of the 15th century byzantine picture. A boat carrying the picture, accompanied by a spectacular cortege boats, cheered along by chants and fireworks, follows the coastline to land in the place where the picture was found in 1453. After mass celebrated by the Archbishop, the devotees and citizens accompany the Madonna in her return to the church of S. Agostino, made sanctuary of the Holy Mary in 1972, amidst moments of great emotion and feeling.
	**The Martyrdom of Sant'Orsola, Eboli** One of Caravaggio's masterpieces, the *Martirio di Sant'Orsola*, was found in Eboli 1954. It was the last work painted by Michelangelo Merisi in 1610 a few weeks before his death, and exhibited today in Naples at Palazzo Zevallos Stigliano, while a reproduction is on display at the palazzo di Città di Eboli.
	**Baratta Factory, Battipaglia** Important agricultural and food hub of Battipaglia, no longer in activity. Example of post war industrial architecture. Only the office area is currently in use (private property).
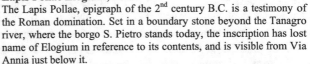	**Lapis Pollae Elogium, Polla** The Lapis Pollae, epigraph of the 2nd century B.C. is a testimony of the Roman domination. Set in a boundary stone beyond the Tanagro river, where the borgo S. Pietro stands today, the inscription has lost name of Elogium in reference to its contents, and is visible from Via Annia just below it.
	**Certosa di San Lorenzo, Padula** Known also as *Certosa di Padula*, it is the largest of the Carthusian monasteries in Italy, and is located in the Vallo di Diano. Rich in stratification, it has been inscribed onto the UNESCO World Heritage List since 1998.

Fig. 2. Territorial vision of the UNESCO Sites in Region Campania (Italy) inscribed onto the World Heritage List.

The art of Neapolitan pizza makers (Intangible Heritage); and in Basilicata The Sassi and the Park of the Rupestrian Churches of Matera (see Fig. 3) [9, 10].

"In this broader perspective it will be possible to create a system connecting the territories of the ancient Roman Road to the sites of the three Italian Regions inscribed onto the World Heritage List, and define the objectives, the actions for the protection and management of the possible *Cultural Route of Via Annia-Popilia* supporting and upholding the outstanding tangible and intangible values rooted in a fabric so strongly stratified." [11].

Because of the interest aroused by the documentation illustrated by me on the theme, in occasion of the ICOMOS International Scientific Committee (CIIC) Meeting that took place at the 'Universidad Politécnica' (Madrid, 16–19 November 2015), the Via Annia - Popilia was inserted, also for its possible future developments for conservation, protection, enhancement and management of the tangible and intangible heritage stratified upon it, into the 'Internal tentative List of Cultural Routes of CIIC'.

Currently this research is focused on deepening knowledge on the Via Annia-Popilia and, in parallel, with the working group I am coordinating to incentivise its protection and enhancement to offer "...the Authorities in charge a programmatic tool (also through the promotion of cultural tourism and eco-tourism) capable of pairing integrated conservation with the innovation of the cultural heritage stratified along its course, offering an opportunity of social and economic development of the territories it crosses, to be carried out for future generations and with the involvement of youth." [11].

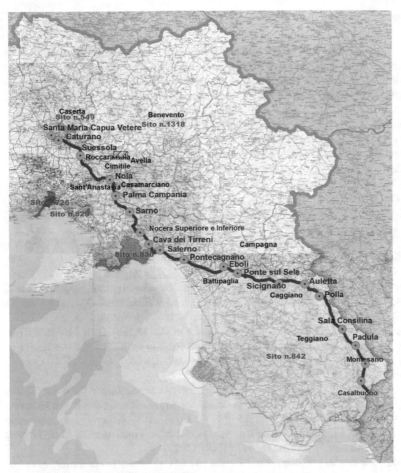

Sites inscribed onto the UNESCO World Heritage List

    The Archaeological Areas of Pompeii, Herculaneum and Torre Annunziata (Site n.829);

    The Historic Centre of Naples (Site n.726);

    Cilento and Vallo di Diano, with the Archaeological Sites of Paestum, Velia and the Certosa di Padula (Site n.842);

    The Amalfi Coast (Site n.830);

    The Longobards in Italy. Places of power, the church of S. Sofia in Benevento (Site n.1318);

    The 18th century Royal Palace of Caserta with the Park, Aqueduct by Vanvitelli and the complex of San Leucio (Site n.549);

    The Celebrations of great shoulder-borne processional structures (Intangible Heritage);

    The Mediterranean Diet (Intangible Heritage).

Fig. 3. Major centres and modern municipal territories identified along the *Via Annia - Popilia* (on cartography IGM 1:100,000) and UNESCO Sites in Region Campania (Italy).

References

1. Genovese, R.A., Dal Paesaggio agli Itinerari culturali: conservazione, tutela e valoriz-zazione tra Oriente e Occidente. In: RICerca/REStauro, coordinated by Donatella Fiorani; Sezione 1C, Questioni teoriche: storia e geografia del restauro, pp. 315–327, Società Ita-liana per il Restauro dell'Architettura (SIRA), Edizioni Quasar di S. Tognon srl, Rome (2017)
2. Genovese, R.A.: Cultural Routes between East and West: a network for cooperation between mediterranean cities. In: Procedia-Social and Behavioral Sciences, Proceedings of 2nd International Symposium "NEW METROPOLITAN PERSPECTIVES". Strategic Planning, Spatial Planning, Economic Programs and Decision Support Tools, through the Implemen-tation of Horizon/Europe2020, ISTH2020, Reggio Calabria, Italy, vol. 223, pp. 619–625. Elsevier, 18–20 May 2016
3. Aa. Vv.: Itineraires culturels. In: Thema & Collecta, n.4. ICOMOS Wallonie-Bruxelles asbl (2015)
4. Genovese, R.A.: Pilgrimage Cultural Routes: The Cammini di San Michele. In: Chaudhary, P. (ed.) Heritage and Cultural Routes: An Anthology, pp. 239–264. Centre for Studies in Museology, University of Jammu, India, Shubi Publications, Gurgoan (2012)
5. Genovese, R.A.: Landscapes as significant element in European Pilgrimage cultural routes. In: Salinas, V.F. (ed.) Landscape as Context and Substantive Element of Cultural Routes, of Historic Towns they Pass through and of Traditional Paths of Communication, pp. 129–146. International Scientific Committee of Cultural Routes of ICOMOS (CIIC), Madrid (2013)
6. CIIC (International Scientific Committee on Cultural Routes) of ICOMOS. The ICOMOS Charter on Cultural Routes, Ratified by the 16th General Assembly of ICOMOS, Québec, Canada, 4 October 2008
7. Genovese, R.A.: Intercultural dialogue and participation of communities to the conservation, protection and enhancement of cultural heritage. In: Smart Travel, Smart Architecture, Heritage Conservation and its Fruition for Dialogue, 19th General Assembly and International Experts Conference of the Romualdo Del Bianco Foundation and Life Beyond Tourism, 11th–13th March, Florence, pp. 103–109 (2017)
8. Merckx, B.: Les anciennes voies de l'Empire romain en tant qu'itinéraires culturels. Perspectives et exemples, In: Aa.Vv.: Itineraires culturels. ICOMOS Wallonie-Bruxelles asbl, Thema & Collecta, n.4 (2015)
9. Genovese, R.A.: Conoscenza, tutela e valorizzazione della Via Annia/Popilia e dei territori attraversati in Campania. In: Caruso, L., Lazzari, M. (eds.) La Via ab Regio ad Capuam, un itinerario culturale come motore dello sviluppo economico e turistico del territorio, pp. 21–37. Distretto Lions International 108YA and CNR IBAM, Lagonegro (2015)
10. Genovese, R.A.: Un GIS per la conservazione e la valorizzazione del patrimonio culturale in Campania: la Via Annia-Popilia, In: Sessa, S., Di Martino, F., Cardone, B. (eds.) GIS DAY 2015. Il GIS per il governo e la gestione del territorio, pp. 51–59. Aracne Editrice, Ariccia (2016)
11. Genovese, R.A.: La Via 'ab Regio ad Capuam': conservazione integrata della strada storica e dell'Itinerario culturale, In: Belli, G., Capano, F., Pascariello, M.I. (eds.) La città, il viaggio, il turismo. Percezione, produzione e trasformazione The City, the Travel, the Tourism. Perception, Production and Processing, VIII Congresso AISU, 7–9 September, pp. 499–505. CIRICE Edizioni, Napoli (2017)

Strategic Collaborative Process
for Cultural Heritage

Eleonora Giovene di Girasole$^{(\boxtimes)}$ ⓘ, Gaia Daldanise ⓘ,
and Massimo Clemente ⓘ

National Research Council, Institute for Research on Innovation
and Services for Development, Naples 80134, Italy
e.giovenedigirasole@iriss.cnr.it

Abstract. Recent developments within national and supranational policies in Europe have come to understand cultural heritage as a "common good" that is pivotal to the sustainable development of territories and communities. In doing so, EU policies nudge member states to adopt participative approaches to development, to invest on citizens' and stakeholders' identification with the local cultural heritage, and to work together for its conservation.

This article starts with a purview of the current debate, both in academia and in policy-making circles, about the interpretation of cultural heritage as common. In this context the article highlights the importance of the use the collaborative processes and defining/adopting approaches and tools for activating a strategic chain of "knowledge and planning".

The paper defines a framework - so called Strategic Value Chain for Cultural Collaborative Process - that highlights the potentials of using *IAD Framework* (by Elinor Ostrom) of collaborative processes together with *Place branding, Place marketing* and *Community planning* for sustainable and context-based urban transformation plans based on a shared cultural identity.

This framework has been utilized to analyze the case study on the historic town of Faenza, in Italy, to highlight the successful points in the urban regeneration process based on cultural identity but also points to work on. For this reason the paper ends with a research follow up, highlighting a possible upgrade of the local process, thanks to approaches and tools' implementation.

Keywords: Cultural heritage · Collaborative process · Strategic Value Chain

1 New Approaches for the Enhancement of Cultural Heritage

In order to fully understand the nature of cultural heritage in historic cities, our usual methods of inquiry must be accompanied by a reading of culture- and identity-based values. With regards to the production of concrete plans, the enhancement of cultural heritage - both in its material and its immaterial dimension - is instrumental to overcoming the semantic flattening that characterizes many projects and interventions that take place in the time of globalization.

This paper presents a progress report and offers some preliminary findings of an ongoing research about shared cultural identity as a driver of urban regeneration and

© Springer International Publishing AG, part of Springer Nature 2019
F. Calabrò et al. (Eds.): ISHT 2018, SIST 101, pp. 359–368, 2019.
https://doi.org/10.1007/978-3-319-92102-0_38

sustainable local development. Identity construction begins with a community recognizing its material and immaterial heritage. This heritage is rooted in the architecture but also in the skills, functions, and traditions of a community.

This interpretation of cultural heritage is mirrored in recent European policies that look at it as a common good and as a fundamental ingredient for sustainable development and social innovation. In doing so, such policies seek to promote participative development processes. In this scenario, new methods for cooperation and horizontal participation have been introduced, as alternatives to vertical models, to implement territorial transformations. On the ground, this change has been attempted by engaging and dialoguing with different actors with the aim of building shared knowledge as well as social and political capital [1].

Against this background, the authors highlights the importance of Place branding, Place marketing and Community planning as tools in the processes for cultural heritage enhancement and for urban regeneration [2]. Indeed, these tools enable experts to better understand the needs of a community, and thus to guarantee a long-lasting urban transformation that is rooted in the territorial context. Furthermore this approach enables a coordinated and efficient use of resources, which are here understood as the opportunities to govern a territory through strategy and by simultaneously pursuing economic growth, environmental protection, and the monitoring of how the affected communities perceive the transformation.

The analysis of the case of Faenza reveals that a collaborative process has been built in a rather implicit fashion. This paper also argues that, in order to improve the results of this process, it would be worthwhile to implement innovative tools to activate and engage all the actors affected by the process.

2 Cultural Heritage for Urban Regeneration: Values, Approaches and Processes

2.1 Cultural Heritage as "Common" for Local Cultural Identity

In the last decades, the concept of cultural heritage and the elaboration of strategies for its conservation have undergone a significant evolution. Meanwhile, its importance for urban sustainable development has been increasingly recognized.

Recent developments within national and super-national policies in Europe regard cultural heritage as a "common good" that is pivotal to the sustainable development of territories and communities. In doing so, European policies nudge member states to adopt participative approaches to local development, based on cooperation among public institutions, citizen, associations, etc. [3]. The inclusion of local communities, indeed, is seen as a means to increase their awareness about the cultural heritage, understood as the capacity of citizens and stakeholders to recognize it as a mirror of their identity and to cooperate for its conservation [4].

The "commons" nature of cultural goods derives from their intimate linkage to the identity, culture, and traditions of a territory, and for their importance for communal life. Cultural goods can be regarded as a particular type of common goods, which «refer to culture expressed and shared by a community» [5]. These are characterized by shared values and attitudes which facilitate cooperation. In other words, «culture and cultural goods as common goods engage users in reproducing them and passing them on to the next generations» [6]. By doing so, cultural goods contribute to sustainable development [7].

Against this background, the conceptualization of cultural heritage, as a common good, and the importance of citizens' participation in enhancing it are found among the conventions and recommendations on cultural heritage that have been issued by the Council of Europe in its work on social and territorial sustainable development.

The "Council conclusions on participatory governance of cultural heritage [8] define cultural heritage as «a shared resource and [as] a common good» (art.1). This definition stresses the idea that the resources composing the cultural heritage, regardless of who formally owns them, bring value to all members of the community and are, as such, "common goods".

The Council of Europe, furthermore, recognizes that cultural goods may become a common good if, once they are recognized, they become relevant in their context [9] and in their community. When communities become active in using it and protecting it, cultural goods create social capital [10].

The "Council of Europe Framework Convention on the Value of Cultural Heritage for Society" (Council of Europe, 2005), signed in Faro, Portugal, in 2005 and valid as of 2011 "outlines the framework of rights and responsibilities of citizens in participating in the creation and preservation of cultural heritage, and lists the possible meanings of its 'value'. It does by proposing a multidimensional approach that regards the contribution of cultural heritage to human and societal development" [11]. In this document, communities are expected to actively identify, study, interpret, protect, preserve, and present cultural heritage, whereas states are invited to promote collaborative development processes based upon synergy between institutions, citizens, and associations that recognize it as a common good.

Recognizing cultural heritage as common goods may also help reconstruct the *common ground* conditions that Elinor Ostrom [12] deemed necessary to create trust, trustworthiness, and reciprocity among members of a community. This happens when they agree on shared rules to access the good, a *social* commons or, in other words, the place where a collaborative spirit is produced, which enables societies to behave as a culturally coherent entity [13].

In her studies, Ostrom [14] proposed the "IAD Framework" to analyze the main components of collective systems. These are posited to develop around an "action arena" that includes "actors" that act in a social space, which she calls "action situation". The IAD Framework identifies the factors that affect the action arena by analyzing the modes of interaction among individuals ("patterns of interaction") within the arena. The structure and the functioning of the action arena are influenced by three classes of external factors: physical factors, the community structure, and the set of rules regulating the usage of the relevant resource [12].

The IAD framework could be considered a starting point to define a new framework for a collaborative cultural process able to build a cultural identity. It could be a driver of urban regeneration and sustainable local development.

2.2 The Strategic Chain of "Knowledge and Planning" for Local Shared Identity

Shared identities can be built though *Place Branding* [14–17], a potentially innovative approach within the "knowledge and planning" [18] sphere. *Place Branding* is defined as a process involving the discovery, the development and the implementation of ideas and actions to "rebuild local identities, the distinctive features and the "sense of place" [19].

The goal of *Place Branding* is precisely the activation of the organizational abilities of citizens and to transform them into actors that pursue a shared vision through diversified modes of communication and interaction. It comes from *Corporate Branding* discipline, as a set of techniques elaborated in the sphere of business administration, that helps building these abilities by promoting citizens' emotional attachment to and identification with the cultural heritage.

Marketing strategies exerts a significant influence over the development of economic as well as urban strategic plans. Indeed, scholarship in this field offers the conceptual and operative tools to define strategies, tactics, and actions. Marketing studies may help elaborating the single competitive choices of individuals and groups, as well as the elaboration of strategic visions guiding them. In these circumstances, marketing serves as a set of tools to develop the relations and linkages between an enterprise (defined as a set of goals and resources) and the incentive structure that is prevalent in a given context [20].

These approaches to cultural heritage management may be accompanied by more consolidated ones, such as *Community Planning* [21–23]. Broadly speaking, the term is used here to designate a continuous and lengthy process of planning, production, and revision carried out together with the relevant community. Since the 1960's, urban planning was expected to deliver on social innovation [24], which planners sought to deliver through *Community Planning* [25]. Indeed, in the United States, Community Planning carried with it a reformulation of planning methods, with more attention being paid to local needs and to engagement with the affected communities.

In this context, *Place Branding* (as a "knowledge" approach), *Place Marketing* and *Community Planning* (the latter two being more closely connected to the realm of planning) might help create a balanced synergy (the UNESCO's "knowledge and planning" tools) in which locally produced and locally rooted knowledge goes hand in hand with informed, place-based [26], and productive planning measures.

This strategic chain, based upon cultural heritage, may be able to outperform the traditional top-down approaches in the production of cultural urban and social capital and bring about sustainable local development through a truly collaborative process.

This strategic chain, based upon cultural heritage values, trough *Place branding, Place marketing* and *Community planning* toolkit, could be useful for building cultural urban capital, economic and social capital in order to make a collaborative process.

3 Strategic Value Chain for Cultural Collaborative Process

Shared cultural identity may thus be considered as a driver of urban regeneration and local development. The analysis of background identified innovative and interdisciplinary methodologies to support Cultural Collaborative Processes.

Started on the IAD [27] framework by Elinor Ostrom, it is possible to imagine a "Cultural Collaborative Process" (Fig. 1) in which the community, associations, cultural entrepreneurs and public institutions build an "Action Arena", highlighting the importance of their cultural heritage, and on its identification as a common good.

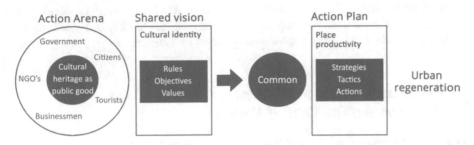

Fig. 1. Framework – cultural collaborative process [28]

Recognizing a common vision based on cultural identity allows the definition of shared rules, goals, and values, thus transforming the cultural heritage from a "common good" to a "common". By developing a *Common Action Plan*, this may allow the construction of strategies, tactics and actions aimed at increasing the productivity of places, for the goal of the urban regeneration.

In order to be able to realize the different steps of this process we could use the tools included in *Place Branding*, *Place Marketing* and *Community Planning* disciplines to define a "Strategic Value Chain for Cultural Collaborative Process" (Fig. 2).

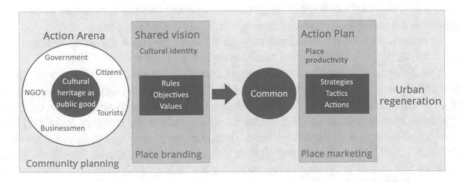

Fig. 2. Framework strategic value chain for cultural collaborative process [28]

The latter is understood here as a strategic cultural value chain that seeks:

1. Community Planning may be of help to increase participation and the sense of sharing within participative processes. The tools of Community Planning can help actors to recognize their cultural heritage and identify it as a "common good" or, in other words, to build the *Action Arena;*
2. *Place Branding* configures itself as a system to build the cultural identity into a *Shared Vision,* defining the values, goals, and rules for the usage of th*e relevant good,* necessary to build a common vision.
3. *Place Marketing* may also be useful in defining a *Common Cultural Action Plan* to define shared strategies, tactics, and actions aimed at urban regeneration and local sustainable development. In that it may *Place Marketing* help to develop and implement a market-conscious philosophy that enables places to communicate their identity strategically, thus implementing their productivity.

4   Collaborative Enhancement of Cultural Heritage as a Process: The Case of Faenza

The case of Faenza was analyzed through the above mentioned interpretation of the *Strategic Value Chain for Cultural Collaborative Process* (Table 1) for highlighting approaches and key tools activated.

The approach adopted in Faenza, implicitly enhanced the city's cultural heritage by building a shared identity that both reflects the local specificities and connects material and immaterial resources. The Action Arena formed through the recognition of the historic centre and its traditions as common good by local community.

Fundamental role had City Council, as the first "heritage actor". The key tool activated by the local administration consists in a "social masterplan" (*Piano Regolatore Sociale*) that involved citizens and cultural NGOs in an urban regeneration process within thematic focus groups. This approach allowed the dialogue among City council, citizens, cultural NGOs, on the Historical Centre value. Relevant research, conducted by the municipality of Faenza together with 1to11lab of Milan, shows the actions of local government allowed a coherent urban development and that the satisfaction of citizens regarding the perception of historic centre quality is quite high [29].

Community found a "shared vision", started from the city's intuition, and later the certainty, that "the inhabitants are the best experts of the territory" [30].

The dialogue with the city administration shows a strong pressure from associations to become part of an "advanced cultural district" [31] or, in other words, to become part of a system that not only looks at culture-led tourism, but which aimed at the creation of a local system able to use culture as a social platform for capacity building [32]. The city administration adopted the requests of the community and prepared a detailed program, called "Rigenera Faenza" [33], which established a network among the local stakeholders. In this program, culture becomes the heart of the governance platform as tactic for cultural heritage that is no longer interpreted merely as a set of objects and festivals, but also as a community asset and as capabilities and competences of the citizens to create local development.

Table 1. Faenza: strategic value chain for cultural collaborative process

Collaborative cultural process	Strategic value chain components	Key tools activated
Action arena		
Heritage actors	City council, citizens, cultural NGOs	City council social masterplan and research tools Focus groups
Cultural heritage as public good	Historical centre Traditional handicraft and art	1to1lab Events International network
Shared vision		
Rules, objectives, values	Shared democratic rules Shared objectives for iconic places, Art/craft and social values	"Rigenera Faenza" projects (as MAP open air museum, Fornarina street etc)
Action plan		
Strategies, tactics, actions	Strategy of social masterplan for capacity building, Tactic of governance platform for cultural heritage, Shared actions among stakeholders and citizens for iconic places	OST (Open Space Technology) and 1to1lab research tool of biofeedback for emotional touch point map

The driver towards regeneration has been operationalized through projects and activities that were in turn based on a set of indicators detailing the potentials of the community.

The main tools of the "Rigenera Faenza" [33] are included the following projects:

- "Fornarina street gets renewed. Are you in?" for the "participative requalification" of buildings in via Fornarina, in the "Borgo" quarter.
- "The quarter I would like to have", which aimed at fostering participation on culture and integration matters. The project was introduced in some blocks in the historic center, around the San Francesco square.
- Open-air museum (or MAP, *Museo all'aperto*), for which the artist collective "Team Ginko" realized a large street art work (about 1000 sqm).

Thanks to art/artisanship and social values, the first objective consists in identifying "shared goals" as driver for change. These shared objectives became shared democratic rules in the networks among people and activities.

Based on cultural identity of historical centre, the City council, citizens, cultural NGOs become a "common" trough the definition of shared rules, goals, and values.

The third step consisted in building a *Common Action Plan*, for iconic places of historical centre, for which City Council provided strategies, tactics and actions for the valorization.

This process aimed at enhancing Faenza's heritage by accruing the welfare of individuals, without at the same time decreasing the community's welfare or depleting the shared resources. The strategy of City council social masterplan, the tactic of a governance platform for cultural heritage and the actions among stakeholders and citizens within the OST or the emotional touch point map (tool for investigate the local demand from citizens needs) [29] activated the community's shared values [34].

Using Open Space Technology (OST) and focus groups, the actions of the project were elaborated starting from the analysis that had already been carried out. With this model, the city administration seeks to draft the above mentioned "social masterplan" (*Piano Regolatore Sociale*).

The focus was to foster participation starting from the identification of the strategic elements and the iconic places of the city (as recognized by its inhabitants), such as the Parco Azzurro ("blue park") in the Borgo quarter.

The participation plan adopted the OST technique, which allows citizens to express his or her point of view on the discussed subject. The use of this methodology was aimed at rearranging the local welfare and to transform it in a "proximity welfare" scheme based on participation and on the provision of "light" services at the neighborhood level. Also thanks to the recourse to OST, stakeholders could effectively cooperate with the city administration in different projects and initiatives. Goals and rules enabled actors to share democratic values through a visible bottom-up drive towards independent thought and critical reflection.

5 Discussion and Conclusions

The regeneration of Faenza may be read as a participative culture-led process in which projects, as well as public and private investments, converged to contrast urban decay and to foster the creation of new public spaces and small-scale production chains in the culture industry.

Participation and social activation tools used within "Regenera Faenza Projects", contributed considerably to build a "shared vision", but represent a traditional participation model, starting from public administration.

With the help of place branding tools, integrated in collaborative process, could be possible to innovate the process and involve horizontally the community, building a greater awareness on heritage identity value as well as a trust in the actors of change for active participation in collaborative process.

"1to1 lab research", using a tool of biofeedback for emotional touch point map, could be considered a place marketing tool for investigating sensorial perception related to the historic centre for emerging current local demand.

It would also be useful to integrate these analyzes with tools that can also highlight the overlap of interests and to capture citizens' ability/willingness to cooperate with several public and private actors involved. The focus is to build a real collaboration able to product new potential local demand.

OST technique represents an efficient community planning tool, to build network among stakeholder, but it could be implemented with other community engagement instruments (such as Planning for Real etc.). The goal is helping actors to define common targets, actively involving them in the decision-making process.

In attempting to promote this kind of processes, actions for the enhancement of the heritage could have to draw upon the toolbox of management and marketing disciplines. In addition, this approach could have to couple the management and organizational aspects on the one side, and participation measures derived from community planning, in order to achieve a process of conservation and innovation [34], on the other. Conservation, innovation and cooperation are the keywords of this process, which starts from the acquisition and from the production of new "knowledge" for evolving in "planning" through linkages between places, organizations, economic actors, public institutions and citizens.

Acknowledgments. Within the unitary work, the following contributions can be individuated in the paper: Sect. 1 has been carried out by Massimo Clemente; Sect. 2.1 by Eleonora Giovene di Girasole; Sect. 2.2 and 4 by Gaia Daldanise; Sect. 3 has been developed by Eleonora Giovene di Girasole and Gaia Daldanise. The conclusions have been shared by all authors.

References

1. Clemente, M., Arcidiacono, C., Giovene di Girasole, E., Procentese, F.: Trans-disciplinary approach to maritime-urban regeneration in the case study friends of Molo San Vincenzo. In Santos Cruz, S., Brandão Alves, F., Pinho, P. (eds.) Book of Proceedings Joint Conference Citta 8th Conference, vol. 2, pp. 701–718. Clássica - Artes Gráficas, Porto (2015)
2. Cerreta, M., Daldanise, G.: Community branding as a collaborative decision making process. In: 17th International Conference on ICCSA 2017, Trieste. Springer, Cham (2017)
3. Zhang, Y.: Heritage as cultural commons: towards an institutional approach of self-governance. In: Bertacchini, E., Bravo, G., Marrelli, M., Santagata, W. (eds.) Cultural Commons. A New Perspective on the Production and Evolution of Cultures, p. 153. Edward Elgar Publishing, Mass (2012)
4. Arcidiacono, C.: Urban Regeneration and Participatory Action Research. Psychology at Portacapuana. Junior Press, Milano (2015)
5. Bertacchini, E., Bravo, G., Marrelli, M., Santagata, W.: Defining cultural commons. In: Bertacchini, E., Bravo, G., Marrelli, M., Santagata, W. (eds.) Cultural Commons. A New Perspective on the Production and Evolution of Cultures, p. 3. Edward Elgar Publishing, Cheltenham (2012)
6. Mariotti, A.: Beni comuni, patrimonio culturale e turismo. Introduzione. In: Aa.Vv.: Commons/Comune, Società di studi geografici. Memorie geografiche 14, p. 437 (2016)
7. Nijkamp, P., Riganti, P.: Assesing cultural heritage benefits for urban sustainable development. Int. J. Serv. Technol. Manag. **10**, 29–38 (2008)
8. European Commission, Council conclusions on participatory governance of cultural heritage. http://eur-lex.europa.eu. Accessed 27 Nov 2017
9. Mattei, U.: Beni comuni, un manifesto. Laterza, Bari (2011)
10. Putnam, R.D., Leonard, R., Nanetti, R.Y.: Making Democracy Work. Making Democracy Work. Civic Traditions in Modern Italy. Princeton University Press, Princeton (1993)
11. Direzione Generale per la Valorizzazione del Patrimonio Culturale, Ricerca e sperimentazione. http://www.valorizzazione.beniculturali.it/. Accessed 27 Nov 2017
12. Ostrom, E.: Governare i beni collettivi. Marsilio, Venezia (2006)
13. Rifkin, J.: La società a costo marginale zero. Mondadori, Milano (2014)
14. Anholt, S.: The Anholt-GMI city brands index how the world sees the world's cities. Place Brand. **2**(1), 18–31 (2006)

15. Baker, B.: Destination Branding for Small Cities: The Essentials for Successful Place Branding. Creative Leaps Books, Portland (2007)
16. Dinnie, K.: City Branding: Theory and Cases. Palgrave Macmillan, New York (2011)
17. Kavaratzis, M., Warnaby, G., Ashworth, G.J.: Rethinking Place Branding. Springer, New York (2015)
18. UNESCO: UNESCO Recommendation on the Historic Urban Landscape. http://whc.unesco.org/en/activities/638. Accessed 27 Nov 2017
19. Govers, R., Go, F.: Place Branding-Glocal, Physical and Virtual Identities Constructed, Imagined or Experienced. Palgrave Macmillan, New York (2009)
20. Caroli, M.G.: Il marketing territoriale. FrancoAngeli, Milano (1999)
21. Forester, J.: Beyond dialogue to transformative learning: how deliberative rituals encourage political judgment in community planning processes. Philos. Sci. Humanit. **46**, 295–334 (1996)
22. Sadan, E.: Empowerment and Community Planning: Theory and Practice of People-Focused Social Solutions. Hakibbutz Hameuchad Publishers, Tel Aviv (1997)
23. Wates, N.: The Community Planning Handbook: How People can Shape their Cities, Towns & Villages in any Part of the World. Earthscan from Routledge, London (2014)
24. Sadan, E.: Empowerment and community practice. http://www.mpow.org/elisheva_sadan_empowerment.pdf. Accessed 2017/11/27
25. Hague, C.: Reflections on community planning. In: Critical Readings, Planning Theory: Urban and Regional Planning, p. 227 (2013)
26. Barca, F.: An agenda for a reformed cohesion policy. A place-based approach to meeting European Union challenges and expectations. http://ec.europa.eu/regional_policy/archive/policy/future/pdf/report_barca_v0306.pdf. Accessed 27 Nov 2017
27. Ostrom, E., Gardner, R., Walker, J.: Rules, Games, and Common Pool Resources. The University of Michigan Press, Ann Arbor (1994)
28. Clemente, M., Daldanise, G., Giovene di Girasole E.: Processi culturali collaborativi per la rigenerazione urbana. In: Planum Atti XX, Conferenza Nazionale SIU, pp. 494–500 (2017)
29. Gallucci, F., Poponessi, P.: Il marketing dei luoghi e delle emozioni. EGEA spa (2010)
30. Comune di Faenza, Rigenerare il sociale. Apparato metodologico ed analisi (2013)
31. Sacco, P.: Cultura e sviluppo locale: il distretto culturale evoluto. Sinergie **82**, 115–119 (2011)
32. Sacco, P.: Sviluppo locale come shock cultural. In: Putignano, F. (ed.): Learning Districts. Patrimonio culturale, conoscenza e sviluppo locale, pp. 47–58. Maggioli, Firenze (2009)
33. Comune di Faenza, Rigenera Faenza (2013)
34. Clemente, M., Giovene di Girasole, E.: La rigenerazione collaborativa della Costa Metropolitana di Napoli: verso un piano condiviso. In: Guida, G. (ed.): Città Meridiane. La questione metropolitana al Sud, pp. 149–160. La Scuola di Pitagora, Napoli (2015)

The Role of Cultural Heritage in Urban Resilience Enhancement

Roberta Iavarone[1] (iD), Ines Alberico[2] (iD), Antonia Gravagnuolo[3(✉)] (iD),
and Gabriella Esposito De Vita[3] (iD)

[1] CNR - National Council of Research, IBAF Naples, 80131 Naples, Italy
[2] CNR - National Council of Research, IAMC Napoli, 80133 Naples, Italy
[3] CNR - National Council of Research, IRISS Napoli, 80134 Naples, Italy
a.gravagnuolo@iriss.cnr.it

Abstract. Place-based urban regeneration suggests the need of shift from a fragmented to a systemic model to understand the interrelation between sub-systems and the effects that a changing sub-system can pose on the others. In this context, urban regeneration strategies resilience oriented may include actions able to improve the economic, physical, social and environmental conditions and thus the human well-being of urban areas.

The great quantity of information involved in the resilience assessment of urban systems require the use of Spatial Decision Support Systems (SDSS) that make possible to record, analyze and summarize data with different spatial and temporal resolution. In the present work, we described the structure of a framework for the Resilience and Disaster Risk Management. Particularly, we focused our attention on the identification of indices expressing the contribution of cultural heritage in making cities resilience toward natural hazards. These tools allow identifying replicable and scaling-up successful practices and converting the impalpable values of cultural heritage in measurable ones.

Keywords: Cultural heritage · Urban resilience · Urban regeneration
Risk management

1 City Regeneration Strategies Resilience Oriented

City resilience is a key concept to manage the complexity of regeneration processes within a context characterized by relevant cultural heritage preservation issues. Since the seventies, the term resilience was used to describe the aptitude of a natural system to adsorb shocks from disturbances and to find an equilibrium state [1]. Afterward, it was adopted in urban planning [2–4] because cities have been theorized as highly complex, adaptive systems [5–7]. The urban resilience is a controversial concept [8] because different opinions exist about the elements making an urban area [9, 10]. Several works define the cities as "complex systems" [11–14] and others theorize urban systems as composed by "networks" or by the combination of systems and networks [5, 15]. Furthermore, resilience concept frequently overlaps with sustainability notion [16, 17]. Several authors consider the resilience enhancement as a means to improve the system sustainability [18–21], others invert this relation and describe the sustainability as a

© Springer International Publishing AG, part of Springer Nature 2019
F. Calabrò et al. (Eds.): ISHT 2018, SIST 101, pp. 369–377, 2019.
https://doi.org/10.1007/978-3-319-92102-0_39

factor contributing to resilience [22–28]. Moreover, resilience and sustainability are considered to have different aims that can be complementary or competitive [29–33].

Recently, the resilience was considered as the capacity of urban systems to prepare, withstand, respond to and adapt more readily to shocks and stresses to emerge stronger and live better in good times. These resilience goals were widely described in ARUP [34] by the term Resourcefulness, Robustness, Redundancy, Flexibility, Inclusiveness, Integration and Reflectiveness.

Resourcefulness systems imply that people and institutions are able to rapidly find different ways to achieve their goals or meet their needs during a shock or under stress. This concept may include investing in capacity to anticipate future conditions, set priorities, and respond, for example, by mobilizing and coordinating wider human, financial and physical resources. *Resourcefulness is instrumental to a city's ability to restore functionality of critical systems, potentially under severely constrained conditions.*

Robustness systems include well-conceived, constructed and managed physical assets, so that they can withstand the impacts of hazard events without significant damage or loss of function. *Robust design anticipates potential failures in systems, making provision to ensure failure is predictable, safe, and not disproportionate to the cause.*

Redundancy systems express the spare capacity purposely created within systems so that they can accommodate disruption, extreme pressures or surges in demand. It includes diversity: the presence of multiple ways to achieve a given need or fulfil a particular function. Examples include distributed infrastructure networks and resource reserves. *Redundancies should be intentional, cost-effective and prioritized at a city-wide scale, and should not be an externality of inefficient design.*

Flexibility systems implies that systems can change, evolve and adapt in response to changing circumstances. It may favor the decentralization and modular approaches of infrastructure or ecosystem management. *Flexibility can be achieved through the introduction of new knowledge and technologies, as needed. It also means considering and incorporating indigenous or traditional knowledge and practices in new ways.*

Inclusive systems empathize the need of a broad consultation to create a sense of shared ownership or a joint vision to build city resilience.

Integrated systems bring together citizens and institutions to achieve greater results.

Reflectiveness systems can accept the ever-increasing uncertainty and changes in today's world. They have mechanisms to continuously evolve, and will modify standards or norms based on emerging evidence, rather than seeking permanent solutions based on the status quo. *As a result, people and institutions examine and systematically learn from their past experiences, and leverage this learning to inform future decision-makers.* In the light of this consideration, several international document [35–37] highlighted the important role of cultural heritage in sustainable development and regeneration process. The United Nation [37] stated that culture allows revitalizing urban areas, strengthening the social participation (point 38) and contributes in developing vibrant, sustainable and inclusive urban economies (points 45 and 60). Cultural heritage symbolizes the historic, aesthetic, social, scientific or spiritual value for past, present, and future generations [38]. Gathering the physical manifestation of past human activities and the modes used by men to interact with the environment, cultural heritage enriches people's lives and provides an inspirational sense of connection between community and landscape [39]. The latter is an "historic layering of cultural, natural, tangible and intangible values" [35].

As suggested by [40], this definition moves the attention from the single "monument" to the context and recognizes the landscape as an "organism" made of complex economic, social, environmental and cultural characters and of relationships between them. The high quality of the landscape contributes to the enhancement of city attractiveness through circular and synergistic processes. Therefore, it is able to stimulate a new demand and a new development perspective [41]. To date, there are few researches [42, 43] about tools able to support the relationship between cultural heritage conservation/regeneration and resilience growth. These tools are necessary to quantify the contribution that cultural heritage conservation/regeneration can provide in any dimension (economic, social, cultural, environmental) and to identify replicable and scaling-up successful practices. They are also required to convert the impalpable values of cultural heritage in measurable ones.

In the present work, we presented a framework structure named "Resilience and Disaster Risk Management" (hereafter, RDRM) implemented for the resilience assessment of urban system exposed to dangerous events. Moreover, the role of cultural heritage for resilience improving was illustrated, it enhances the places identity, wellbeing and quality of life of citizens, contributes to tourism and diverse economic activities growth and preserves and improves environmental quality and reduces the soil consumption.

The RDRM structure allows to manage great quantity of data regarding the economic, environmental and institutional dimensions of resilience. At this aim, drivers, sub-drivers and indicators were identified to quantify the resilience status of urbanized zones. These tools were also classified according to the disaster risk management phases and thus they can be used to manage not only the recovery phase but also to reduce the risk in ordinary planning tools (Regional Territorial Planning, Provincial Territorial Planning, Municipal territorial planning).

2 The *RDRM* Framework for Urban Resilience Assessment

Place-based urban regeneration suggests the need of shift from a fragmented to a system model. It includes actions able to improve the economic, physical, social and environmental components of an area that could change [44]. Resilience assessment is a complex problem because the coexistence and interaction of multiple characteristics [41] modify the dynamic equilibrium of urban systems. From this concept emerged that the analysis of urban systems implies the management of great quantity of data varying in spatial resolutions, precision and accuracy. In this context, Multi Criteria Spatial Decision Support Systems (MC-SDSS) are often used [45].

In the present work we defined the structure of the *RDRM* framework commonly used to help decision-makers in analyzing and synthesize data in thematic maps, graphics, to identify the problem more accurately and to make decisions in shorter time [46–50].

The *"RDRM"* framework is a computer-based system that combines the potentiality of Geographic Information System (GIS) and Decision Support System (DSS), allowing the application of decision logic on conventional information and georeferenced data [51]. The system's root represents the urban resilience and the four branches named *economic, social, environmental and institutional* are the drivers (Fig. 1); they are macro-categories of city system characteristics in the present time (for detail see definition in De Vita et al., this volume). From the single driver, several sub-drivers (e.g. the entrepreneurial ecosystem is vital; the heritage community takes care and valorizes cultural heritage/landscape; natural heritage is preserved; government is open and citizens' participation takes place) start.

These elements describe the characteristics of urban systems resilient to natural hazards. Moreover, one or more phases of disaster risk management (mitigation, preparedness, response, recovery) are associated to the single sub-driver (Fig. 1). From the single driver four branches, representing the phases (mitigation, preparedness, response and recovery) of disaster risk management starts.

This tree structure was repeated for sub-drivers and indicators. In detail from the single phases of disaster risk management start one or more sub-driver and for the single sub-driver several indicators branch off. The latter is a measure, generally quantitative, that can be used to illustrate and communicate complex environmental phenomena simply, including trends and progress over time and thus helps provide insight into the state of the environment [52]. They give the most detailed information of entire framework quantifying the resilient level of cities. Particularly, our attention was focused on the social and environmental indicators expressing the contribution of cultural heritage for improving the urban resilience in both mitigation and recovery phases.

According to [53], we considered the social sub-driver expressing the capacity that resources inherited from the past have in promoting both an individual and a communal sense of identity (Fig. 1). Indeed, people often demonstrates passion and commitment when actively conserving a heritage representing the unique characteristic of their communities [54]. Therefore, this information is important because help to identify areas were citizens may feel more motivated to quickly recover in case of dangerous events.

Two social sub-drivers named *"People recognize and feel proud of their city's identity"* and *"The Heritage Community takes care and valorizes cultural heritage/landscape"* were described and populated with several indicators (Fig. 1). Particularly, the indicator (from SP1 to SP7) associated to the first sub-driver express the importance that citizen attributes to place where they live, while the indicator (from SC1 to SC4) of the second sub-driver show the interest of citizen for the heritage conservation and valorization (Fig. 1).

Similarly to social dimension, two environmental sub-drivers named *"Cultural heritage/landscape is safeguarded from natural hazards"* and *"Natural heritage is preserved"* and several indicator were identified (Fig. 1). These indicators evaluate natural and cultural heritage preservation level towards natural hazards (Fig. 1). In detail, the indicator from "EC1" to "EC4" express the physical characteristic of cultural and landscape heritage while those from "EN1" to "EN7" take into account the measures of equitable and sustainable well-being.

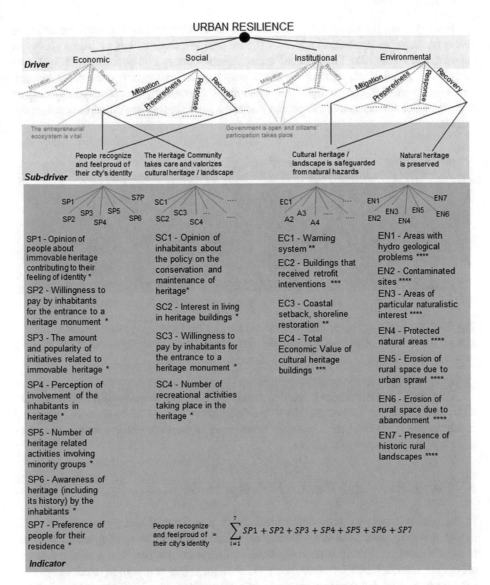

Fig. 1. The tree structure of the evaluation framework. A list of social and environmental indicators defining the contribution posed by cultural heritage in resilience enhancement and thus in urban regeneration processes was furnished. The stars symbolize their references: * [43]; ** [55, 56]; *** proposed in the present work; **** [57]. The formula in the bottom of figure illustrates an example of sub-driver construction.

The paramount vision of social, economic, environmental and institutional components of resilience provided by the "*RDRM*" framework could favor the reduction of natural hazard consequences on urban system.

For example, the existence of such frame could have limited the negative effect of the earthquake with a magnitude of 4.0 Mw [58] occurred at Ischia Island (Italy) in August 2017 (Fig. 2). The lack of citizen preparedness compromised the ordinary activities, spread panic among residents and tourists that interrupting their holiday causing severe damage to the local economy. In addition, the poor application of building rules for the seismic zone and the dense urbanization amplified the earthquake effects.

(a)   (b)

Fig. 2. (a) The effect of earthquake occurred at Ischia Island in August 2017, (photo from [59]) and (b) a wall write of population testifying their preparedness to the arrival of hurricane Irma (photo from [60]).

Differently the Hurricane Irma, that hit the Southern Carolina in 2017 (Fig. 2), had a reduced impact on the coast tanks to the declaration of emergency state that prepared residents and businesses for a potential emergency. About 6.3 million people and more than 400 shelters were put in safe.

3   Conclusions and Research Perspectives

In the present work, we provided a preliminary list of social and environmental indicators useful to detect the contribution of cultural heritage in making cities resilient toward natural hazards in both mitigation and recovery phases. The social indicators allow the identification of areas were citizens may feel more motivated to quickly recover in case of dangerous events whereas the environmental indicators contribute to assess the natural and cultural heritage preservation level. They are part of a wider tree framework structure named "Resilience and Disaster Risk Management" that considers the city as a complex, dynamics, self-organizing systems, continuously changing under the pressure of perturbing factors caused by internal processes or external factors. This framework can be applied in any areas and at different scales (local, regional, national), it is flexible and new data or workflow can be implemented at any time.

Future research will focus on the improvement of the SDSS proposed in the present work through more dynamic and open mapping tools able to make citizens directly involved into maps drawing. A possible solution may be the use of Open Street Map (OSM) that is the world map entirely built by voluntary citizens on open software basis.

Collaborative mapping stimulates the perception of public elements and of cultural and natural heritage as "common good", thus the use of this tool could be seen as an indicator of civic responsibility. In this context, OSM allows not only to assess the level of resilience characterizing the city in the present time but also to identify regeneration activities shared by the population.

References

1. Holling, C.S.: Resilience and stability of ecological systems. Annu. Rev. Ecol. Syst. **4**, 1–23 (1973)
2. Ahern, J.: From fail-safe to safe-to-fail: sustainability and resilience in the new urban world. Landscape Urban Plann. **100**(4), 341–343 (2011)
3. Wilkinson, C.: Social-ecological resilience: insights and issues for planning theory. Plann. Theory **11**(2), 148–169 (2011)
4. Davoudi, S., Shaw, K., Haider, L.J., Quinlan, A.E., Peterson, G.D., Wilkinson, C., Fünfgeld, H., McEvoy, D., Porter, L., Davoudi, S.: Resilience: a bridging concept or a dead end? "Reframing" resilience: challenges for planning theory and practice, interacting traps: resilience assessment of a pasture management system in northern Afghanistan, urban resilience: what does it mean in planning practice? Resilience as a useful concept for climate change adaptation? The politics of resilience for planning: a cautionary note. Plann. Theory Pract. **13**(2), 299–333 (2012)
5. Godschalk, D.R.: Urban hazard mitigation: creating resilient cities. Nat. Hazards Rev. **4**(3), 136–143 (2003)
6. Folke, C.: Resilience: the emergence of a perspective for social-ecological systems analyses. Glob. Environ. Change **16**(3), 253–267 (2006)
7. Batty, M.: The size, scale, and shape of cities. Science **319**(584), 769–771 (2008)
8. Meerow, S., Newell, J.P., Stults, M.: Defining urban resilience: a review. Landscape Urban Plann. **147**, 38–49 (2016)
9. Campanella, T.J.: Urban resilience and the recovery of New Orleans. J. Am. Plann. Assoc. **72**(2), 141–146 (2006)
10. Lu, P., Stead, D.: Understanding the notion of resilience in spatial planning: a case study of Rotterdam, The Netherlands. Cities **35**, 200–212 (2013)
11. Brugmann, J.: Financing the resilient city. Environ. Urban. **24**(1), 215–232 (2012)
12. Da Silva, J., Kernaghan, S., Luque, A.: A systems approach to meeting the challenges of urban climate change. Int. J. Urban Sustain. Dev. **4**(2), 125–145 (2012)
13. Cruz, S.S., Costa, J.P.T.A., de Sousa, S.Á., Pinho, P.: Urban resilience and spatial dynamics. In: Eraydin, A., Taşan-Kok, T. (eds.) Resilience Thinking in Urban Planning, vol. 106, pp. 53–69. Springer, Dordrecht (2013)
14. Lhomme, S., Serre, D., Diab, Y., Laganier, R.: Urban technical networks resilience assessment. In: Laganier, R. (ed.) Resilience and Urban Risk Management, pp. 109–117. CRC Press, London (2013)
15. Desouza, K.C., Flanery, T.H.: Designing, planning, and managing resilient cities: a conceptual framework. Cities **35**, 89–99 (2013)
16. Pizzo, B.: Problematizing resilience: implications for planning theory and practice. Cities **43**, 133–140 (2015)
17. Marchese, D., Reynolds, E., Bates, M.E., Morgan, H., Clark, S.S., Linkov, I.: Resilience and sustainability: similarities and differences in environmental management applications. Sci. Total Environ. **613–614**, 1275–1283 (2018)

18. Milman, A., Short, A.: Incorporating resilience into sustainability indicators: an example for the urban water sector. Glob. Environ. Change **18**, 758–767 (2008)
19. Walker, B., Pearson, L., Harris, M., Maler, K.-G., Li, C.Z., Biggs, R., Baynes, T.: Incorporating resilience in the assessment of inclusive wealth: an example from South East Australia. Environ. Resour. Econ. **45**, 183–202 (2010)
20. Saunders, W.S.A., Becker, J.S.: A discussion of resilience and sustainability: land use planning recovery from the Canterbury earthquake sequence, New Zealand. Int. J. Disaster Risk Reduct. **14**, 73–81 (2015)
21. Jarzebski, M.P., Tumilba, V., Yamamoto, H.: Application of a tri-capital community resilience framework for assessing the social-ecological system sustainability of community based forest management in the Philippines. Sustain. Sci. **11**(2), 307–320 (2016)
22. McEvoy, D., Lindley, S., Handley, J.: Adaptation and mitigation in urban areas: synergies and conflicts. Proc. Inst. Civ. Eng. Munic. Eng. **159**, 185–191 (2006)
23. Chapin, F.S., Kofinas, G.P., Folke, C.: Principles of Ecosystem Stewardship: Resilience Based Natural Resource Management in a Changing World, pp. 1–409. Springer, New York (2009)
24. Avery, G.C., Bergsteiner, H.: Sustainable leadership practices for enhancing business resilience and performance. Strategy Leadersh. **39**(3), 5–15 (2011)
25. Closs, D.J., Speier, C., Meacham, N.: Sustainability to support end-to-end value chains: the role of supply chain management. J. Acad. Mark. Sci. **39**, 101–116 (2011)
26. Ahi, P., Searcy, C.: A comparative literature analysis of definitions for green and sustainable supply chain management. J. Clean. Prod. **52**, 329–341 (2013)
27. Bansal, P., DesJardine, M.R.: Business sustainability: it is about time. Strateg. Organ. **12**(1), 70–78 (2014)
28. Saxena, A., Guneralp, B., Bailis, R., Yohe, G., Oliver, C.: Evaluating the resilience of forest dependent communities in Central India by combining the sustainable livelihoods framework and the cross scale resilience analysis. Curr. Sci. **110**, 1195 (2016)
29. Derissen, S., Quaas, M.F., Baumgaertner, S.: The relationship between resilience and sustainability of ecological-economic systems. Ecol. Econ. **70**(6), 1121–1128 (2011)
30. Fiksel, J., Goodman, I., Hecht, A.: Resilience: navigating toward a sustainable future. Solutions, 1–13 (2014)
31. Lizarralde, G., Chmutina, K., Bosher, L., Dainty, A.: Sustainability and resilience in the built environment: the challenges of establishing a turquoise agenda in the UK. Sustain. Cities Soc. **15**, 96–104 (2015)
32. Lew, A.A., Ng, P.T., Ni, C. (Nickel), Wu, T. (Emily).: Community sustainability and resilience: similarities, differences and indicators. Tour. Geogr. **18**, 18–27 (2016)
33. Meacham, B.J.: Sustainability and resiliency objectives in performance building regulations. Build. Res. Inf. **44**, 474–489 (2016)
34. Rockefeller Foundation and ARUP. City Resilience Framework, New York (2015)
35. UNESCO: Recommendation on the Historic Urban Landscape, UNESCO World Heritage Centre, Paris, France (2011)
36. United Nations: Transforming our World: The 2030 Agenda for Sustainable Development. United Nations, New York, NY, USA (2015)
37. United Nations. Draft Outcome Document of the United Nations Conference on Housing and Sustainable Urban Development (Habitat III). United Nations, New York, NY, USA (2016)
38. Mackee, J., Askland, H.H., Askew, L.: Recovering cultural built heritage after natural disasters: a resilience perspective. Int. J. Disaster Resil. Built Environ. **5**(2), 202–212 (2014)
39. ICOMOS.: The Burra Charter. The Australia ICOMOS Charter for Places of Cultural Significance, Australia (1999)

40. Fusco Girard, L., Gravagnuolo, A., Nocca, F., Angrisano, M., Bosone, M.: Towards an economic impact assessment framework for historic urban landscape conservation and regeneration projects. BDC. Boll. Cent. Calza Bini. **15**, 1–29 (2015)
41. Fusco Girard, L., Nijkamp, P.: Le valutazioni per lo sviluppo sostenibile della città e del territorio. Angeli, Milan (1997)
42. Nocca, F.: The role of cultural heritage in sustainable development: multidimensional indicators as decision-making tool. Sustainability **9**(1882), 1–28 (2017)
43. European Union. Cultural Heritage Counts for Europe project - CHCfE. Full report (2015)
44. Roberts, P.: The evolution, definition and purpose of urban regeneration. In: Roberts, P., Sykes, H. (eds.) Urban Regeneration: A Handbook, pp. 9–36. Sage (2000)
45. Cinelli, M., Coles, S.R., Kirwan, K.: Analysis of the potentials of multi criteria decision analysis methods to conduct sustainability asses. Ecol. Indic. **46**, 138–148 (2014)
46. Crossland, M.D.: Individual decision-maker performance with and without a geographic information system: an empirical study. Unpublished doctoral dissertation, Kelley School of Business. Indiana University, Bloomington, IN (1992)
47. Crossland, M.D., Wynne, B.E., Perkins, W.C.: Spatial decision support systems: an overview of technology and a test of efficacy. Decis. Supp. Syst. **14**(3), 219–235 (1995)
48. Chakhar, S., Martel, J.M.: Enhancing geographical information systems capabilities with multi-criteria evaluation functions. J. Geogr. Inf. Dec. Anal. **7**(2), 47–71 (2003)
49. Mennecke, B.E., Crossland, M.D., Killingsworth, B.L.: Is a map more than a picture? The role of SDSS technology, subject characteristics, and problem complexity on map reading and problem solving. MIS Q. **24**(4), 601–629 (2000)
50. Malczewski, J.: A GIS-based multicriteria decision analysis: a survey of the literature. Int. J. Geogr. Inf. Sci. **20**(7), 703–726 (2006)
51. Shekhar, S., Xiong, H., Zhou, X.: Encyclopedia of GIS. Springer, Cham (2017)
52. EEA: Core set of indicators – Guide, Technical report 1 (2005)
53. Graham, B.: Heritage as knowledge: capital or culture? Urban Stud. **39**(5–6), 1003–1017 (2002)
54. Elsorady, D.A.: Heritage conservation in Rosetta (Rashid): a tool for community improvement and development. Cities **29**, 379–388 (2012)
55. Cutter, S.L.: The landscape of disaster resilience indicators in the USA. Nat. Hazards **80**, 741–758 (2016)
56. Normandin, J.M., Therrien, M.C., Tanguay, G.A.: City strength in times of turbulence: strategic resilience indicators. In: Proceedings of the Joint Conference on City Futures, Madrid, 4–6 June 2009
57. ISTAT. http://www.istat.it/en/well-being-and-sustainability/well-being-measures. Accessed 25 Nov 2017
58. INGV. http://comunicazione.ingv.it/index.php/comunicati-e-note-stampa/1632%20TERRE MOTO-DELL-ISOLA-D-ISCHIA-DEL-21-AGOSTO-2017-ELABORAZIONE-DATI-INGV-PRESENTATA-ALLA-COMMISSIONE-GRANDI-RISCHI-CGR-DEL-25-AGOSTO-2017. Accessed 12 Jan 2017
59. La Repubblica newspaper. http://www.repubblica.it/cronaca/2017/08/22/news/terremoto_ischia_morire_per_scossa_di_bassa_magnitudo-173587827/. Accessed 5 Dec 2017
60. The post and couriers newspaper. http://www.postandcourier.com/news/forecast-shifts-west-for-hurricane-irma-as-decision-on-south/article_8c493c32-9371-11e7-a119-fb3bb9d6d988.html. Accessed 5 Dec 2017

The Urban Being Between Environment and Landscape. On the Old Town as an Emerging Subject

Salvatore Giuffrida[1](✉) ⓘ, Grazia Napoli[2] ⓘ,
and Maria Rosa Trovato[1] ⓘ

[1] University of Catania, Catania 95124, Italy
sgiuffrida@dica.unict.it
[2] University of Palermo, Palermo 90133, Italy

Abstract. The landscape units of the Sicilian mountainous inland, as for the case of Petralia Soprana, are marked by the presence of ancient urban centres controlling the agricultural territory, from which they derived their own wealth, and to which they conferred landscape significance. The unity between economy and landscape has been interrupted by the radical transformation of the socio-economic structure and the technologic progress, which have eroded the consistency between structures and superstructures. We propose an assessment approach based on a synthesis of semiotic and phenomenological view. The approach mainly focuses on the basic concepts and contents of the valuation process that can be assumed as the theoretical premise for the operational tool.

Keywords: Minor old towns · Urban landscape · Landscape semiotics
Landscape phenomenology · Landscape assessment

1 Introduction

Landscape can be defined in different ways, many of which not usual, some others not explicit, but only recognisable within the evaluative or planning approach from which, contingently, an idea of landscape in some way emerges.

In the common behaviors the notion of landscape is even weaker, as it isn't included in the individual system of values, but belongs to a very rarefied and misunderstood aesthetic dimension. This weakness is manifest in landscape policies inconsistency, based predominantly on regulatory constraint rather than on the widespread perception of landscape as an identity matrix.

In some other cases, landscape has overcome its relationship to territory, standing as a brand, then as an abstract and cumbersome surrogate of the basic and original life issues (employment, commerce, administration), a sort of status symbol.

Such a landscape "re-format" is a hope for many people - administrators as well as citizens - especially in those urban towns, once flourishing, rich in history and architecture, now definitely impoverished of workers, youth and prospects.

A similar notion, prevailing in the perceptivist or pure-visibilistic approach, is definitely out of touch with the local material culture, the evolutionary potentialities

© Springer International Publishing AG, part of Springer Nature 2019
F. Calabrò et al. (Eds.): ISHT 2018, SIST 101, pp. 378–386, 2019.
https://doi.org/10.1007/978-3-319-92102-0_40

and vocations, the way in which communities aggregate, the underlying consensus, the economic processes that can be traced back to a spatial context characterized by a resistant "genius loci".

The notion of landscape is at the top of a process of abstraction:

- starting from *environment* – a natural habitat of human acting, a complex of resources, a ruled space – the *reference*, working as a set of constraints;
- passing through *territory* – as a living space organized for the purposes of the broadest and most lasting progress of the human species – the *signifier*, working as a set of opportunities;
- getting to *landscape* – an autonomous manifestation of territory – the *significance*, working as "informing information" [1].

As such, landscape plays the role of resilient matrix of the organization of a settled' community; it works as "generative grammar" [2] of any possible modification (re-combination) of the physical-factual occurrences; it is a "holographic" [3] order-bearer. A landscape unit arises as an "icon", *a fact in itself*.

In the sphere of social communication, facts are worth due to their "resonance" [1, 4]; consequently landscape is one of the codes – likely the most general – of the cultural sub-system; this sub-system uses such filter trying to exclude, limit or rule the most relevant land transformations claimed by: the technological sub-system, due to the efficiency and efficacy of them [5]; the economic sub-system, due to the profitability of them [6, 7]; the political sub-system due to the endorsement achievable from public as a consequence of such transformations.

The cultural sub-system is one of the weakest in the macro-social system, and it can easily be excluded from social communication, so turning into environment.

This contribution proposes theoretical reflections on a possible combination of two approaches, semiotic and phenomenological, from which the contents and a multidi-mensional assessment method can be outlined with reference to the landscape unity centered on the old town of Petralia Soprana in the Province of Palermo, Italy.

Such land context can be assumed as the epitome of the entropic course that has been affecting lots of old town, predominantly located in the hinterland, definitively deprived of their original landscape arrangement, as can be recognized from the *signs* and *phenomena* of a clumsy and asynchronous aspiration to modernity.

2   The Case of Petralia Soprana (PA)

Petralia Soprana is a mountain old town in the Province of Palermo, located nearby the southern limit of the Madonie Park, and connected to the A19 Palermo-Catania highway through SS 120 and SP 129. Together with other important centers of Madonie mount chain it forms a *cultural koinè* of durable identity values and resilient landscape connotations. The urban center lies on the terminal part of the ridge forming the watershed between the river basin of Salso and Imera. Its territory, about 57 km^2 (Fig. 1), is mainly mountainous and devoted to arable ground, with a wooded area of just 21 ha [8]. The result is a discontinuous and "naked" landscape, whose shape and perception is marked by the signs of the human settlement.

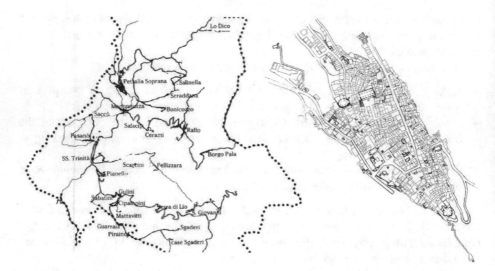

Fig. 1. Territorial frame and the old town of Petralia Soprana [8].

The issues of the conservation of this landscape unit can be reported starting from the transformation of the political-economic and social order in which agriculture and land rent have been replaced by the administrative tertiary and construction sectors; new needs and lifestyles have subverted the topographic hierarchy of the villages territorial network, the age pyramid, the geography of preferences.

The disorganized buildings in Contrada Pinta, born recently just north of the town, the unfinished road above the western slope; the current residential and settlement patterns; the insistent infiltration of ungrammatical building within the older neighborhoods are the main evident signs of maladjustment to modernity.

3 Axiological Approach as Synthesis of Semiotic and Phenomenological Instances

Interpreting reality through a structured system of values means applying an axiological approach, that is assuming value as a cognitive, decision and action pattern, by defining a valorizing substance as the content of a common rationality making us willing to share the subsequent choices. An axiological approach assumes "the true value" [9] as the constitutive substance of social reality.

Value establishes the relationship between interpreted *reality* and *world* to create, and then things exist in so far as they are appreciated in a shared system of preferences outlining a better world. In other words, values constitute the selective/prescriptive code giving form and consistency to things, then things will exist according to the way in which values, intentions, perspectives, preferences and motivations stand in the common life.

As a consequence, landscape can be considered an "unstable" interface between knowledge (evaluation/production capacity) and reality (success of values and products) and its transformations acceptable only if individual (*ethos*) and collective (ethics) values are consistent [10].

The axiological approach overturns the idea of landscape as a "cultural elaboration of a specific natural environment" [11].

The look at landscape through values systematizes, on the one hand, the amazement of those who look at the world "without having their heads" [12], on the other, the bewilderment of those who are not able, "only with their head" [13], to implement strategies in complex and uncertain contexts.

A landscape axiology can cover different approaches such as the semiotic and the phenomenological ones, of which we propose a sort of convergence.

In the field of semiotics, landscape is a generative matrix of territory, working as a semantic chain; it is featured by a "weak" - that is "intensional"- horizontal syntactical bond between signs, related to *sense* rather to *significance*. Then, it needs "intentionality" to be defended. On the contrary, each sign, is characterized by "strong" - that is "estensional" - vertical semantic bonds between signifier and significance, able to bear the necessary modifications so that the horizontal interactions between signs are kept safe, in order to prevent them from getting *inert* [14].

In the field of phenomenology, such a consistent semantic chain assumes a high "landscape responsibility" for being characterized by high external coherence – even at the cost of reducing the internal coherence of each of them – then gives rise to *Landscape* as a phenomenon, an object in itself, unitary and not divisible, raw material of knowledge and the privileged reference for land policy.

Consequently, *Landscape* stands as an "emerging subject" playing the role of "emerging agent"; both status are consistent with the "immanent perception" [15], a cognitive and heuristic activity making landscape, "in person", as the interlocutor of policies, rather then the subject of negotiation from the outside.

Immanent perception is in contrast to pure-visibilism, which performs incomplete territorial analysis – generally divisive. Immanent perception, as opposed to transcendental perception – that refers to consciousness – redefines the unity of landscape as "absolute evidence" through intentionality, which gives form to the objective knowledge of the flow of experiences in the "world of life" [16].

Immanent perception [17] is the perception of intentional objects, in which "consciousness and its object form an individual unity constituted purely of life experiences".

The transformations that lacerate landscape act precisely at this depth. These transformations are the result of a political-economic climate dangerously dominated by:

i. a "scientific-technological apparatus that has become autonomous [and self-referential] due to the inversion of the mean-scope relation" [18] and able to dismantle and reassemble territories overcoming their physiological rhythms;
ii. an economic-financial crisis that, on the private side, does not discourage the compulsive production of new buildings, while on the side of public keeps endorsing large infrastructural redevelopment.

4 The Value of Stones as a Petrified Value

Two types of values influence behaviors and policies: *anthropological*, specific and concrete values, regarding the settled communities, the local economies, the sphere of labor and productive investments; *cultural*, general and abstract values, concerning the global community, endorsed by over-local economic-territorial policies and inspired by the logic of not-reproducible capital.

Landscape takes part to the economic communicative dimension as well, not only because "everything has to come to terms with economy", but because, on the contrary, economy, sooner or later, comes to terms with everything that is worth. The economic dimension enables landscape as a subject of analysis according to the attributes characterizing it as social capital.

The economic nature of landscape - measurable basing on the setting of subsequent behaviors - is the factual recognition of the substance of abstract value:

1. perceived as "extensional" datum, "cognitive unit" resulting from *denotation*;
2. recognized as "intensional" shape, "axiological unit" resulting from *connotation*;
3. supported as "intentional" action by shared decision-making processes.

Accordingly, landscape is (1) social (2) territorial (3) capital enabling the two independent and coalescent abilities of capital value, the explicit liquidity, the *value of the stones*, and the implicit liquidity [1], the *petrified value*.

The petrified value is the "ability to bear" the "landscape predicate" as:

a. *sense*, resulting from the combination of the semantic and syntactic relations of the landscape components (signs), and from the redundancy of the context they form;
b. *essence*, allowing landscape unit to emerge as "novelty", through the intentionality that unifies the "flow of the economic life experiences" and the "immanent perception" connecting conscience and experience.

The relations of *sense* create the axiological fabric of the unintentional "*ethe*"; the transcendence to *essence* defines the invariants of a landscape "ethics".

If the individual interest prevails (2. fails - i.e. the vertical link significance/signifier of signs is loosened), then some parts of the semantic chain break free from (1. fails, that is the horizontal link between the signs in loosened), so that, in the normative defect (3. falls, landscape becomes "inert"), the landscape quality (concrete value) fails as the result of a fall in abstract economic value.

In this sense, 1. *Landscape*, as an abstract value, also founded in the economic sense on "immanent perception", and connected to real hoarding (3.) in the "forms that inform" (2), "emerges in person" [19] and assumes the function of "matrix" of its possible land "footprints". Landscape as a matrix is an "open work" [20] triggering new experiences and perceptions that are still traceable to it.

5 Axiological Contents and Assessment Exemplification

The case of Petralia Soprana is the epiphenomenon of the progressive loss of continuity and unity of the "natural/built/lived" territory. Its shape results from the progressive capitalization (petrifaction) of the "axiological ascension": chance, cause, reason, motivation, legitimation.

The coherence of such ascension "justifies" the adaptation of the social system to environment as "immanent action" of which *Landscape* is phenomenon, and allows us to perceive the maladjustment of this old town and its territory to contemporaneity. The acknowledgement of such gap of value supports preserving/modifying strategies that correct "in reason" the *Image* of *Landscape*.

Based on this ascension, and in a phenomenological sense:

- *Landscape* is the "justification" of the territorial occurrence as a result of the immanence in its *Image*;
- *Image* is the complex of the sensitive occurrences obtained from different perceptions in the "world of life";
- *Immanence* results from axiological positioning (from chance to legitimation) of a territorial occurrence.

The "ascensional" evaluation model is applied to a pattern of "immanent perception". The evaluation consists of:

1. identifying the most dense signs of landscape, according to the immanent perception; *Image* as a whole emerges according two order/disorder declination:
   1.1. the *natural order/disorder*, by identifying the relationship between *structural* elements and the *infrastructural* ones symbolizing the human adaptation:
      1.1.1. the geomorphological (capital) status;
      1.1.2. the vegetational (capital) status;
      1.1.3. the infrastructural (capital) status;
      1.1.4. the architectural-urban (capital) status;
   1.2. the *cultural order/disorder*, by identifying the *superstructural* elements through which the urban settlement has been distinguishing itself as a unit:
      1.2.1. density/compactness of the urban fabric;
      1.2.2. consistency of dimensions and volumes;
      1.2.3. graphic and chromatic consistency;
2. defining the significant characteristics of the immanent perception concerning:
   2.1. *evidence*, in relation to their ability to mark the phenomenon referring to:
      2.1.1. extent (extensional evidence);
      2.1.2. distinction (point impact from colors and materials);
      2.1.3. resilience (reversibility);
   2.2. *justification*, by referring each of them to a different level of the "ascension":
      2.2.1. chance: "environmental irritations", and/or not reasonable human actions;
      2.2.2. causes: natural phenomena under human control, or human predictable actions;
      2.2.3. reasons: consistency of premises and consequences described and verified basing on "standard or dual models of the human rationality" [13];

2.2.4. motivations: result of an underlying pattern of values somehow made explicit;

2.2.5. legitimacies, as the collective acknowledgement (more or less ruled by an institution) of the relationship between motivations and results (Figs. 2 and 3).

signs/quality						immanent perception							assessment	impact
						1. evidence			2. justification						
						1. extension	2. distinction	3. resilience	1. chance	2. cause	3. reason	4. motivation	5. legittimacy		
	natural order	1	geomorphology	a11	curr	-0.1	-0.5	-1.0	0.2	0.8	0.0	0.0	0.0		1.2
	1 order/dis				proj	0.2	-0.1	0.2	0.1	0.1	0.2	0.4	0.2		
	order	2	vegetation	a12	curr	-0.6	-1.0	-0.2	0.4	0.6	0.0	0.0	0.0		1.57
					proj	0.5	0.2	0.2	0.1	0.1	0.3	0.4	0.1		
		3	infrastructure	a13	curr	-0.8	0.1	0.1	0.0	0.8	0.1	0.1	0.0		
					proj	0.1	0.1	0.1	0.0	0.5	0.2	0.2	0.1		
		4	architecture/city	a14	curr	0.6	0.8	0.2	0.0	0.1	0.2	0.4	0.3		
					proj	0.8	0.8	0.5	0.0	0.1	0.2	0.4	0.3		
	cultural	1	density	a21	curr	0.5	0.6	0.2	0.1	0.2	0.3	0.3	0.1		
	2 order/dis				proj	0.6	0.6	0.3	0.0	0.1	0.3	0.3	0.3		
	order	2	extentional coherence	a22	curr	0.5	0.2	-0.2	0.1	0.4	0.3	0.2	0.0		
					proj	0.6	0.2	-0.1	0.0	0.3	0.3	0.2	0.2		
		3	architectural coerence	a23	curr	0.4	0.6	-0.3	0.3	0.3	0.2	0.1	0.1		
					proj	0.8	0.8	0.1	0.2	0.2	0.3	0.2	0.1		

Fig. 2. Impact assessment matrix of *current* status and **project** hypothesis.

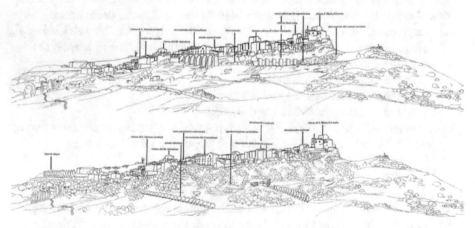

Fig. 3. Comparison of current arrangement and project hypothesis [8].

6 Conclusion

The textual unit of Petralia Soprana is the subject of an axiological approach combining semiotic and phenomenological perspectives to define the contents of a multidimensional approach [21] acknowledging the interactions between the properties of the syntactic unity and its "immanent perception" concerning "facts" and "justification".

The ontology of territory, due to its multiform abundance, is unknowable if not for different and opposite [22] observations which sometimes exclude to each other.

Landscape, as the "phenomenon of territory", is given as a general cognitive unit that reduces the conflict [23, 24] of those visions and as such it is the item of the best and most adequate adaptive knowledge of the underlying ontologies.

The "immanent perception" is the heuristic result overcoming such oppositions; it is the perspective of novelty constituted by the landscape phenomenon.

References

1. Rizzo, F.: Valore e valutazioni. La scienza dell'economia o l'economia della scienza. FrancoAngeli, Milano (1999)
2. Chomsky, N.: Syntactic Structures. Mouton, Den Haag (1957)
3. Morin, E.: Le vie della complessità. In: Bocchi, G., Ceruti, M. (eds.) La sfida della complessità, pp. 49–60 (1995)
4. Luhmann, N.: Sistemi sociali. Il Mulino, Bologna (1990)
5. Napoli, G., Gabrielli, L., Barbaro, S.: The efficiency of the incentives for the public buildings energy retrofit. The case of the Italian Regions of the "Objective Convergence". Valori e Valutazioni **18**, 25–39 (2017)
6. Giuffrida, S., Ventura, V., Trovato, M.R., Napoli, G.: Axiology of the historical city and the cap rate. The case of the old town of Ragusa Superiore. Valori e Valutazioni **18**, 41–55 (2017)
7. Napoli, G., Giuffrida, S., Valenti, A.: Forms and functions of the real estate market of Palermo. Science and knowledge in the cluster analysis approach. In: Stanghellini, S., et al. (eds.) Appraisal: From Theory to Practice. SIEV 2015. Green Energy and Technology, pp. 191–202. Springer, Cham (2017)
8. Boscarino, S., Federico, A., Giuffrida, S., Prescia, R., Rizzo, F. (eds.): Petralia Soprana. Ipotesi di restauro urbano e studi di analisi multicriteriale. Medina, Palermo (1994)
9. Giuffrida, S.: The True value. On understanding something. In: Stanghellini, S., et al. (eds.): Appraisal: From Theory to Practice, pp. 1–14 (2017)
10. De Monticelli, R.: La questione morale. Raffaello Cortina, Milano (2010)
11. Sereni, E.: Storia del paesaggio agrario italiano. Laterza, Roma (1961)
12. Harding, D.E.: Del non avere la testa. In: Hoftadter, D.R., Dennet D.C. (eds.): L'io della mente, pp. 34–40. Adelphi, Milano (1993)
13. Labinaz, P.: La Razionalità. Carocci, Roma (2013)
14. Turri, E.: La conoscenza del territorio. Metodologia per un'analisi storico-geografica. Marsilio, Venezia (2002)
15. Abbagnano, N.: Fenomenologia. Voce del Dizionario di Filosofia. Utet, Torino (1998)
16. Husserl, E.: La crisi delle scienze europee e la fenomenologia trascendentale. Il Saggiatore, Milano (2008)
17. Merleau-Ponty, M.: Phenomenology of Perception. Routledge, London (2013)
18. Severino, E.: La Filosofia Futura. Oltre il dominio del divenire. BUR, Milano (2006)
19. De Monticelli, R.: La novità di ognuno. Persona e libertà. Garzanti, Milano (2009)
20. Eco, U.: Opera Aperta. Bompiani, Milano (1962)
21. Nesticò, A., Sica, F.: The sustainability of urban renewal projects: a model for economic multi-criteria analysis. J. Prop. Invest. Financ. **35**(4), 397–409 (2017)

22. Oppio, A., Bottero, M.: Conflicting values in designing adaptive reuse for cultural heritage. A case study of social multicriteria evaluation. In: ICCSA 2017, pp. 607–623. Springer, Cham (2017)

23. Della Spina, L., Scrivo, R., Ventura, C., Viglianisi, A.: Urban renewal: negotiation procedures and evaluation models. In: Gervasi, O., et al. (eds.) ICCSA 2015. LNCS, vol. 9157. Springer, Cham (2015)

24. Trovato, M.R., Giuffrida, S.: The choice problem of the urban performances to support the Pachino's redevelopment plan. IJBIDM **9**(4), 330–355 (2014)

Cultural Strategies for Urban Regeneration: The Effects of Policies Implemented by European Capitals of Culture

Marialuce Stanganelli[✉]

University of Naples Federico II, 80125 Naples, Italy
stangane@unina.it

Abstract. The paper analyzes the role of art and culture in urban regeneration processes. The effect of cultural strategies upon thirty European cities previously designated European Capitals of Culture is appraised. The analysis was carried out by comparing two kinds of city profiles created by using data from social media (social profile) and data from travel guides (erudite profile). The results highlight a new approach to cultural strategies.

Keywords: Urban regeneration · Cultural strategies
European capital of culture

1 Culture and Cities

Starting from the 1990s, there has been a growing convergence in studies of urban planners, sociologists and economists concerning the role of culture in promoting urban development. Many essays, books and papers from different branches of knowledge have contributed to a reversal of perspective whereby culture shifts from a passive to an active role: "Culture used to be a by-product of wealth. Now, culture is seen as a generator of wealth" [1].

In 1995, Hall [2] explicitly compared the great urban cultures of the past, such as those of Athens and Florence, with innovative urban environments of the past and present (e.g. Glasgow and Silicon Valley), giving rise to a different narrative of urban development. The aim was to demonstrate that cultural networks which made Classical Greek or Renaissance culture possible were analogous to innovative networks on which the industrial revolution and Silicon Valley were structured. It was an important step forward, showing that global cities were not only made of finance [3] but were also determined by other decisive forces, primarily culture.

At the same time, recent economic theories highlighted new emerging forms of the urban economy centered on culture: i.e. a creative or symbolic economy. According to Howkins [4], value in the Creative Economy is produced by creative and imaginative capacity. The Creative Economy is structured into 15 classes of activities (cultural industries) including art, science and technology. Cultural industries are identified in a very broad vision including not only culture, but also games and the entire field of research and development. Notwithstanding, it is significant to have placed cultural activities at the center of what is considered a powerful new economy destined to grow

© Springer International Publishing AG, part of Springer Nature 2019
F. Calabrò et al. (Eds.): ISHT 2018, SIST 101, pp. 387–397, 2019.
https://doi.org/10.1007/978-3-319-92102-0_41

at a rate of 5% per year [5]. Moreover, the Creative Economy reinforces the explicit acknowledgement [1] of the existence of a direct link between culture, innovative capacity, and economic development. Howkins's theory was the starting point of a strand of studies that directly involved cities through the "creative class" of Florida [6] and the "creative city" of Landry [13] and Yencken (1988), giving rise, according to some authors [7, 8], to an elitist and exclusive narrative of urban development.

Narratives also play an important role in the Symbolic Economy, defined by Zukin [7] through three characteristics: it is urban; it is based on the production of symbols; it is also based on the production of spaces, considered both as places and as symbols of the city and culture. Therefore, the Symbolic Economy combines the production of symbols (culture, food, fashion, crafts) with the production of the new urban narrative [7].

While the Creative Economy had an elitist characterization, the Symbolic Economy underlined the necessity of democracy, tolerance and inclusiveness to establish cultural processes able to drive urban regeneration. Indeed, culture is not only an important driver of economic development, it is also an important resource to accomplish broader goals of social improvement: in a world where communities are becoming increasingly multiethnic, culture is an inclusive practice fostering intercultural dialogue and reinforcing social cohesion (UNDP, 2013).

2 Methodology

In many cases, it is difficult to distinguish spontaneous and policy-driven processes in examples of culture-driven urban regeneration. Indeed, many of the "spontaneous" processes in question are actually supported by public action or public capital or they are hosted in public areas. Public support can also intervene in a second phase, to endorse and reinforce a spontaneous initiative. Therefore, it may be presumed that in most cases of success, cultural development is due to the virtuous and complex superimposition and interaction of spontaneous bottom-up initiative and public action.

There are many descriptions of culture-driven regeneration processes as spontaneous and bottom-up phenomena. From the study of Williamsburg (Brooklyn) and other New York neighborhoods, Zukin [8] extracted a sequence of five phases of a culture-based spontaneous regeneration process. That said, few studies have sought to ascertain the long-term effects of public cultural policies on urban regeneration.

In order to clearly understand the role and effectiveness of public policies in this domain, this paper analyzes a specific set of 30 cities which held the title of European Capital of Culture (ECoC) from 1990 until 2016. After the launch of the ECoC initiative in 1985, selection criteria and bidding deadlines were outlined from 1998 onward. In 2005 the Programme had its first legislative framework - for the years 2005 to 2019 - establishing a rotational system of designation, in which one Member State would be allocated the title each year (Decision 1419/1999/EC). The evaluation criteria included "contribution to the development of economic activity, particularly in terms of employment and tourism" and "to encourage the development of links between the architectural heritage and strategies for new urban development". A new Decision - for the years 2007 to 2019 - enacted that the program shall "be sustainable and be an integral part of the long-term cultural and social development of the city". In 2014 a new legislative framework (Decision 445/2014/EC) established rules for the forthcoming period (2020–2033), commenting once again on the importance of a long-term cultural strategy. In detail, a city's candidature must clearly explain "the plans to strengthen the capacity of the cultural and creative sectors, including developing long-term links between the cultural, economic and social sectors in the candidate city". With regard to cultural and artistic contents, "a clear and coherent artistic vision and strategy for the cultural program and the capacity to combine local cultural heritage and traditional art forms with new, innovative and experimental cultural expressions" are specifically requested (ibidem). Therefore, these cities constitute a privileged test bench to ascertain the effectiveness of public cultural strategies, in other words top-down cultural strategies.

In order to evaluate the effects of cultural strategies, for each city two descriptions were identified and compared: before and after ECoC. The initial description was based on the following: the population trend prior to ECoC (Eurostat data); tourism trends before ECoC (TourMIS Data); the main motivations for candidature expressed in official dossiers; the presence of UNESCO heritage sites (UNESCO heritage list). After ECOC, the description was based on: the population trend in subsequent years (Eurostat data); tourism trends in subsequent years (TourMIS Data); city perception after ECOC.

The change in city perception was inferred by analyzing social media reviews and highlights from travel guides as described in Sect. 4. Social media reflect perception of a widespread, differentiated audience giving informal opinions and non-structured opinions on places, while travel guides reflect a more cultured and refined approach to the perception of cities. As regards the former, for each city an analysis of the number of reviews of activities in TripAdvisor was carried out and four indexes were created, describing the following: cultural and natural assets; cultural vibrancy; the tourist industry; and leisure time activities as perceived by reviewers worldwide.

3 Thirty European Capitals of Culture

Given the aims of this study, attention was focused only on cities that were not already considered famous cultural centers and/or had incorporated the designation into a long-term economic and urban regeneration strategy. The cities selected are listed in Table 1. The cities were organized into the following three groups based on population: Metropolitan areas, Large cities (more than 150,000 inhabitants); Medium-size cities (population below 150,000). As regards pre-existing cultural attractions, the cities can be organized into two groups: cities that already had a rich, albeit perhaps not well-known, cultural heritage, and cities that had no previous acknowledged cultural visibility.

The main motivations for ECoC candidature were as follows:

To Overcome an Economic Crisis. This objective directly links the economic and cultural development of a city. The first city which tried to leverage its economy by using its ECoC designation was Glasgow in 1990. Its success was an important path to

Table 1. The ECoC Cities analyzed

Glasgow 1990 Scotland	Sibiu 2007 Romania	Guimaraes 2012 Portugal
Thessaloniki 1997 Greece	Luxembourg 2007	Maribor 2012 Slovenia
Weimar 1999 Germany	Liverpool 2008 England	Marseille 2013 France
Salamanca 2002 Spain	Stavanger 2008 Norway	Kosice 2013 Slovakia
Bruges 2002 Belgium	Vilnius 2009 Lithuania	Riga 2014 Latvia
Graz 2003 Austria	Linz 2009 Austria	Umea 2014 Sweden
Lille 2004 France	Essen 2010 Germany	Pilsen 2015 Czech Rep.
Genoa 2004 Italy	Pécs 2010 Hungary	Mons 2015 Belgium
Cork 2005 Ireland	Tallin 2011 Estonia	Wroclaw 2016 Poland
Patras 2006 Greece	Turku 2011 Finland	San Sebastian 2016 Spain

follow for several cities and regions such as Genoa 2004, Liverpool 2008, Pécs 2010, and Essen for the Ruhr the same year. Initially, the economic benefit identified was cultural tourism; starting with Liverpool 2008 and more explicitly with Essen 2010, creation and development of creative industries became a new target. Both cities with pre-existing cultural heritage and cities without former cultural visibility expressed this motivation.

To Gain New Visibility. Many cities of the former Soviet area (i.e. Weimar, Tallin, Riga, Vilnius, Wroclav) sought to promote their cultural assets through a special showcase, namely ECoC. These cities already possessed an important cultural heritage that was almost unknown to a Western public due to isolation experienced during the Soviet period. Other cities, already well known, tried to use ECoC as an opportunity to add a new aspect to their consolidated international profile, namely Bruges and Luxembourg. In Western Europe, Graz, San Sebastian, Guimaraes and Salamanca, notwithstanding their solid heritage, wished to upgrade their international visibility.

To Create a New Image of the City. Cities with little previous cultural visibility sought to create a new cultural image. In some cases, culture is considered as a complement of other strengths, such as for Lille and Stavanger.

To Confront the Dark Side of a City's History. Several cities have used ECoC as "an opportunity to explore openly and critically the darker side of their history" [9], such as San Sebastian and the former conflict between communities, Liverpool with its major role in the slave trade, and Riga exploring its German and Soviet occupations. The hardest challenge was borne by cities tackling their links with a Nazi past, namely Graz, Weimar and Linz. Graz is adjudged to have been successful in overcoming its past, even if there was some opposition to the "idea by a young Graz-based artist to create a life-size black shadow of the 14th-century clock tower. Some members of the city council had concerns in particular about evoking the "black past" of the city during the Nazi era" [10]. It was more difficult for Weimar to overcome its past. "Weimar has a difficult Janus-faced history: on the one hand, Weimar is the heart and soul of the German Classical Period - home to Goethe, Schiller, and a plethora of other writers, musicians, artists, and architects. On the other hand, Weimar represents the lowest of low points of German history and the assault on civilization by the Nazis" [10].

Weimar did not set up an innovative program but targeted 'high and exclusive culture'. Yet a new design for the Rollplatz by Daniel Buren was rejected after public protests. Ten years later Linz, "Hitler's former *Patenstadt* (favorite town) [10], measured up by how it addresses and confronts its past against the wider backdrop of European history".

Urban Regeneration. Several cities, especially historic cities, presented a wide and structured physical development plan as a key objective of their ECoC (Thessaloniki 1997, Genoa 2004, Pécs 2010); Nordic cities such as Stavanger did not draw up significant physical development plans [11]. Such developments included: Transport infrastructures; Renovation of historic centers; Urban regeneration of some districts; Renovation of historic buildings; New Architecture; Recovery of derelict spaces. Over the years, a slow but progressive change in approach to urban interventions may be noted. While the trend in infrastructure interventions in the 1990s and 2000s was to create new iconic buildings to host cultural venues[1], more recent interventions target the recovery of abandoned industrial spaces and their reuse as venues for cultural and creative activities[2]. Furthermore, in the first period some cities tried to create "cultural districts" or specific zones endowed with cultural venues (Salamanca 02, Graz 03) inspired by the modernist 'zoning' principle. In more recent years a preference was given to widespread cultural infrastructures such as the *Maisons Folies* of Lille 2004, renovating 12 buildings across the region into arts centers. "We decided that we didn't want to build yet another cultural temple, and instead we took over various abandoned industrial spaces and called them *Maisons Folies*" [10] Changes in urban interventions are linked to changes in cultural approach. Whereas the early ECoC cities were inspired by 'high' culture, a different cultural openness slowly established a new broader vision of culture. Glasgow was the first city to put on a community events program for a wide-ranging audience [11]. Starting with Graz 2003, "culture is [considered] a matter of everyday life" [10], a concept that was widely adopted by later ECoC Cities. This shift led to greater interest in folk and street art and in local artists. Indeed, the enhancement of local artistic abilities bodes well for long-term cultural development.

4   City Perception: Social and Erudite Profile of Cities

To understand how city perceptions changed after ECoC, two analyses were carried out: one on social media and the other on a major international travel guide, Lonely Planet. For each city an analysis of the number of reviews of activities on TripAdvisor was carried out. It should be stressed that reviews on TripAdvisor can be posted both by visitors and inhabitants. In less well-known contexts the majority of reviews are written by local inhabitants, or else by those who are moved by pride in their own city heritage, using this international showcase to let foreign visitors know the best places to see. Of course, there is always the danger of some self-absorbed reviews aimed at publicizing a particular hotel or restaurant rather than singing the praises of monuments

---

[1] Glasgow 90, Thessaloniki 97; Graz 03, Genoa 04.

[2] Essen 2010, Pecs 2010, Turku 2011, Tallin 2011, Kosice 2013, Pilsen 2015.

or public spaces. In the full knowledge of the dangers that are intrinsic to any data from social networks, some activities were discarded in order to consider reviews from TripAdvisor a reliable source of information on how a city is perceived by its inhabitants and visitors, a useful source to define a "social" profile of the city. That said, it should be pointed out that this is a very dynamic profile, changing every day on the basis of reviews added by users which could change the rating attribute for each element commented. The data were collected in November 2017.

Another analysis was carried out on a travel guide, comparing the reviews of "places to see and activities to do" in Lonely Planet with TripAdvisor's "social" profile of cities. Lonely Planet reports a cultured approach to places and could be construed as an "erudite" profile of the city. Four indexes were created using data extracted from TripAdvisor:

Cultural Asset Index including the number of museums, monuments, sites and landmarks, parks and natural areas recommended by TripAdvisor users; this index shows the supply of cultural heritage and describes the cultural infrastructure.

Cultural Vibrancy Index including the number of nightlife attractions, concerts, shows, food and drink attractions recommended; this index points out how many places and opportunities there are to experience live, vibrant culture.

Tourist Industry Index grouping data on tourist services such as the number of tours, outdoor activities, shopping attractions, classes and workshops (e.g. on local cuisine or local crafts), travel resources highlighted by users; this index highlights the existence or otherwise of a structured tourist industry.

Leisure Time Activities Index gathers data on elements that are not necessarily linked to tourist use but, in many cases, constitute facilities for residents also used by visitors such as the number of zoos and aquariums, amusement parks, fun and games facilities, spas and wellness centers, transportation.

The results are collected in the following social perception performance matrix (Table 2). Unexpectedly, medium-size cities performed better than large cities, overturning what seemed to be a direct correlation between the number of reviews and number of recommended elements with the relative population. Of course, metropolitan areas and capital cities benefit from having always been a center for the cultural life of surrounding regions and states. What is surprising is that "small is better". It may be supposed that medium-size cities find it easier to manage development projects than big cities. Patras, Umea, Kosice, Pécs, Mons, and Essen are perceived as the cities least endowed with cultural and natural heritage and rank lowest in all indexes, while those perceived as the most endowed are the majority of metropolitan cities (except for Essen, Wroclaw and Lille), with Bruges, Cork, San Sebastian and Graz belonging to other groups. Importantly, almost all the cities designated ECoC over the last 10 years are in the lower part of the rankings. Only those cities which were already renowned for their cultural heritage and tradition occupy the higher ratings. This means that for city ratings, being ECoC from 2009 onwards must be considered "in progress" since the cultural strategy adopted needs a more extended period to show its results.

Table 2. Social perception performance matrix.

Metropolitan Areas	Review Index	Places to visit	Cultural asset	Cul Vibrancy	Touristic Ind.	Leisure time	Heritage List	Medium-size Cities	Review Index	Places to visit	Cultural asset	Cul Vibrancy	Touristic Ind.	Leisure time	Heritage List
Lille	•	•	•	•	•	•		Salamanca	●	•	•	•	•	•	▪
Marseille	•	•	•	•	•	•		Pécs	°	°	°	°	°	°	▪
Riga	•	•	•	•	•	•	▪	Stavanger	•	•	•	•	•	•	
Wroclaw	•	•	•	•	•	•	▪	Cork	●	●	•	●	•	•	
Glasgow	•	•	•	•	•	•		Bruges	●	•	•	•	•	°	▪
Genoa	•	•	•	•	•	°	▪	Umea	°	°	°	°	°	°	
Essen	°	°	°	°	°	°	▪	Maribor	°	•	•	°	•	•	
Vilnius	•	•	•	•	•	•	▪	Luxembourg	●	•	•	•	•	•	▪
Liverpool	●	•	•	•	•	•	▪	Mons	•	°	°	°	°	°	●
Tallin	•	•	•	•	•	•	▪	Weimar	•	•	•	°	•	°	▪

Legend: ●High • Good • Medium ° Scarce

Review Index = % Reviews/Inhabitants

Source: TripAdvisor November 2017

Large Cities	Review Index	Places to visit	Cultural asset	Cul Vibrancy	Touristic Ind.	Leisure time	Heritage List
Thessaloniki	•	•	•	•	•	°	▪
Graz	•	•	•	•	•	•	▪
Kosice	°	°	°	•	°	•	
Linz	•	•	•	•	•	•	
San Sebastian	●	•	°	•	•	•	
Turku	•	•	°	•	•	•	
Patras	°	°	°	°	°	°	
Sibiu	•	•	•	°	•	°	
Pilsen	°	•	•	°	°	•	
Guimaraes	•	•	•	•	•	°	▪

5 Conclusions

There was a considerable change in cultural approach over the years: from an elitist approach centered on high culture, inspired by the creative economy, to a more democratic and inclusive approach more geared to the symbolic economy. Cities that captured this fundamental change had a greater chance of success.

Comparison of pre-ECoC characteristics of cities and present conditions including the social and erudite profile led to several conditions for success being identified. These are discussed below through the exemplification of some city profiles. The conditions for success were identified as follows:

- a long-term, multi-objective cultural strategy;
- a pre-existing cultural habit;
- an inclusive notion of culture;

- the ability to assimilate and rework conflict and contradiction;
- propensity of inhabitants;
- an urban regeneration strategy where "small, widespread and recycle" is better than "big, concentrated and new".

A Long-Term, Multi-objective Cultural Strategy. Culture is time-dependent: it is a cumulative process and needs a long time to take root, grow and expand. One of the main factors detected behind success is conceiving the ECoC year as a step in a process rather than an event. This was achieved by successful cities such as Glasgow, Liverpool, and Graz. A long-term strategy is necessary mainly for cities which start with a low endowment of cultural resources. From the results of this study, a cultural strategy needs about 8–10 years to show its effects. This also means that the EcoC findings can only be evaluated 10 years after designation of the city in question. A successful cultural strategy also has a multi-objective approach, combining culture with other urban aspects: urban regeneration, social inclusion, economic development, and resilience improvement.

Glasgow is a metropolitan area which was not renowned as a cultural centre prior to ECoC. Even if it is not a UNESCO World Heritage Site, Glasgow has accumulated a substantial cultural heritage during its long history, partly due to its major role in the Industrial Revolution. Lonely Planet does not rank Glasgow in the top ten cities of interest in Scotland. That said, noting its importance on tourist routes, a specific travel guide is devoted to this city. Social and Erudite profiles correspond in listing historical, modern and contemporary sites as well as urban spaces among major attractions. Furthermore, TripAdvisor users also highlighted nightlife attractions, reporting the perception of a lively city with many different assets ranging from the past to the contemporary period, and from the urban environment to nightlife attractions. In addition, Lonely Planet highlights the modern heritage of architect Charles Rennie Mackintosh. Even if Glasgow in 1990 had no explicit cultural strategy, over time it has continued to invest in culture and infrastructures, blending new iconic architecture and derelict land recovery, such as Clyde Waterfront Regeneration including Zaha Hadid's 2011 Riverside Museum (listed among the top ten attractions both by TripAdvisor and Lonely Planet), SEC Armadillo1997 and Hydro2013 by Norman Foster.

A Pre-existing Cultural Habit. Culture is facilitated by the presence of a pre-existing cultural habit. The latter does not necessarily originate from the presence of 'high culture' and heritage resources. Cities may also possess a specific cultural habit due to the historical development of a particular activity or ability stemming from their being industrial, mining or multi-religious communities. Glasgow's success story encouraged the candidature of cities that were not regarded as already established cultural centers, namely Liverpool, Cork, Essen, Patras, Umea, Mons, Pècs, Linz, Kosice, Maribor, Pilsen, and Stavanger. Some of them, albeit lying outside major cultural circuits, already had a heritage base in history, nature and culture: some had been ancient cities (Cork, Linz, Pecs); others were major centers of industrial development (Liverpool, Essen, Pilsen); others again (e.g. Stavanger) were notable for their natural resources. Cities which started from a low resource level failed to achieve their goals. This led to the consideration that culture is a kind of energy, and like energy none can be created or destroyed.

Patras is a large port city not renowned as a cultural center prior to becoming ECoC. Lonely Planet does not list it among the top ten cities of Greece in terms of interest. Both Social and Erudite profiles are very poor: the Social profile includes local shops and a local carnival with few (6) reviews. Lonely Planet does not go beyond listing six things to do. Ex-post evaluations stressed that Patras 2006 included plans for development that would come to fruition in the longer term. After more than ten years these long-term effects do not seem to have been realized.

An Inclusive Notion of Culture and the Ability to Assimilate and Rework Conflicts and Contradictions. A pre-existing cultural habit is not enough to guarantee the success of a cultural strategy. What is necessary is an inclusive notion of culture where culture is for everybody and not restricted to a few people of learning. Live culture is nourished by the contradictions and discards of contemporary life and is fed by the mix between new cultures grafted onto previous ones. In order to emerge, it needs democracy and tolerance and to be based on a new urban narrative produced by common people. Among ECoC cities, successful experiences like those of Graz or Liverpool were inspired by a wider cultural concept, while an inclination to high culture, as in Weimar, did not give the expected results.

Graz is a large city and a heritage site (Historic Center and Schloss Eggenberg). Lonely Planet ranks it third among the ten most interesting cities of Austria. Historical and contemporary culture are highlighted both by TripAdvisor and Lonely Planet. Many new venues built during ECoC are still an attraction for visitors, such as Murinsel by Vito Acconci and Kunsthaus by Peter Cook and Colin Fournier. Factors behind the success of the Graz experience included a program based on a wide notion of culture involving events for all inhabitants and an audacious choice to mix contemporary art, performances and architecture in historical spaces that are part of the World Heritage List. Some brave events carried on, such as the Homeless Street Soccer Cup continued in subsequent years.

Weimar is the smallest city to be designated as ECoC. It is a heritage site (Classical Weimar) but needed new visibility after its long isolation as part of East Germany. Lonely Planet does not list it among the ten most interesting cities in Germany. Both TripAdvisor and Lonely Planet underline its historical sites and the concentration camp of Buchenwald. With no contemporary art or traces of contemporary culture, Weimar is presented as a conservative historic city. The literature reports a divide between the ECoC philosophy and organization and the community. The artistic program targeted 'high culture' "as it was argued that such events would be more attractive to international tourists" [11] and local residents were worried that "the event might have a negative impact on the city (higher prices and rents for example), and felt marginalized and not adequately involved" [12].

Propensity of Inhabitants. About all cities declared to promote community engagement with open submission of projects, However, some cities stated that despite community engagement during the events of the years in question, there was a resistance to pursuing this route due to a conservative attitude.

Bruges is a medium-size city which was renowned for its heritage (Historic Center) prior to becoming ECoC. Lonely Planet rank it third of the ten most interesting cities in Belgium. Like Weimar, Bruges is described by TripAdvisor and Lonely Planet as a historical city but more lively and vibrant. TripAdvisor highlights its main attractions as its urban spaces, breweries and leisure time spaces. Bruges has always been a popular tourist destination; its objective was to be "on the map" for contemporary culture. The ECoC year served to endow the city with new cultural venues, such as the Concertgebouw (designed by Paul Robbrecht and Hilde Daem) or such events as the Dance Festival. That said, there is still a weak perception of Bruges as a city of contemporary culture. The organizer noted that Bruges is a very conservative city and that "the general public isn't at all open to installations and contemporary works" [10].

An Urban Regeneration Strategy Where "Small, Widespread and Recycle" Is Better Than "Big, Concentrated and New". From cases of success, a broader concept of culture emerges strictly tied to urban spaces and urban life. Live culture spreads from heritage sites and monumental areas to public spaces, streets, and meeting places where people can meet, gather, and express themselves. It arises from the streets, free from borders and fences, spreading to everyday spaces, assimilating and re-elaborating conflict, contradictions, discards of the contemporary city. It is a culture depending on urban spaces and strictly connected to them. The wide distribution of small cultural structures in the urban area guarantees a greater involvement of many different parts of the city, while large concentrated new cultural districts risk being without enough users after ECoC years. Furthermore, recycling of existing abandoned spaces is highly appreciated by users: they allow new things to be done in old spaces, or the "new" can be created by "reinterpreting the past". After all, this is the way culture grows.

References

1. Zukin: Whose Culture? Whose City? The Paradoxical Growth of a Culture Capital. Hong Kong Conference Cultures (2001)
2. Hall, P.: Cities in Civilization. Culture, Innovation and Urban Order. Weidenfeld & Nicholson, London (1996)
3. Saskia, S.: Le città globali. UTET, Turin (1997)
4. Howkins, J.: The Creative Economy: How People Make Money from Ideas. Penguin, London (2001)
5. UNESCO, United Nations Development Programme (UNDP), Creative Economy Report, NY (2013)
6. Florida, R.: City and the creative class. In: City & Community, vol. 2(1). Blackwell Publishing Inc., USA (2003)
7. Zukin, S.: The Culture of Cities. Blackwell Publishing, Malden and Oxford (1995)
8. Zukin, S.: L'altra New York. il Mulino, Bologna (2015)
9. European Commission, European Capitals of Culture. 2020 to 2033 a Guide for Cities Preparing the Bid. http://ec.europa.eu/culture/tools/actions/capitals-culture_en.htm. Accessed 5 Oct 2017
10. European Communities 2009, European Capitals of Culture: the Road to Success. From 1985 to 2010. http://ec.europa.eu/culture/tools/actions/capitals-culture_en.htm. Accessed 5 Oct 2017

11. Garcia, B., Cox, T.: European Capitals of Culture. Success Strategies and Long-Term Effects (2013). http://ec.europa.eu/culture/tools/actions/capitals-culture_en.htm. Accessed 5 Oct 2017
12. Palmer Rae Associates, European Cities and Capitals of Culture. City Report (2004). http://ec.europa.eu/culture/tools/actions/capitals-culture_en.htm. Accessed 5 Oct 2017
13. Landry, C.: The Creative City: A Toolkit for Urban Innovators. Earthscan, London (2000)

Utilizing Culture and Creativity for Sustainable Development: Reflections on the City of Östersund's Membership in the UNESCO Creative Cities Network

Wilhelm Skoglund[(✉)] and Daniel Laven

Mid Sweden University, 831 25 Östersund, Sweden
wilhelm.skoglund@miun.se

Abstract. Within the last few years, cities around the world have promoted creativity as a new resource for driving future development. As a result, a number of networks have emerged around this theme. The UNESCO Creative Cities Network (UCCN) is one such network that attempts to use creativity as a mechanism to achieve sustainable growth and development. The network has grown rapidly since its inception in 2004 and now has 180 members worldwide, all of which have adopted the UCCN guidelines and directives. In this paper, the authors explore if and how cities use their membership to implement sustainable development goals. The paper uses the northern Swedish city of Östersund as a case study, which has been an active member of the UCCN since 2010. Study findings indicate that membership in the UCCN has enabled Östersund to advance sustainability discourse at a regional level, as well as improve practice in a limited sense. At the same time, findings also identify a number of challenges for integrating sustainability objectives into the UCCN.

Keywords: Creative cities · Sustainable development · UNESCO
Östersund

1 Introduction

Within the last few years, cities around the world have promoted creativity as a new resource for driving future development. As a result, a number of networks have emerged around this theme. The UNESCO Creative Cities Network (UCCN) is one example that attempts to use creativity as a mechanism to achieve sustainable growth and development. The network has grown rapidly since its inception in 2004 and now has 180 members worldwide, all of which have adopted the UCCN guidelines and directives. The mission of the UCCN is to provide networking and knowledge exchange opportunities that support creative development for its member cities. At the same time, membership in the network demands that cities adopt sustainable development strategies according to the United Nations Sustainable Development Goals (i.e., Agenda 2030). Hence, creativity has become a potential mechanism to encourage cities to become more sustainable.

© Springer International Publishing AG, part of Springer Nature 2019
F. Calabrò et al. (Eds.): ISHT 2018, SIST 101, pp. 398–405, 2019.
https://doi.org/10.1007/978-3-319-92102-0_42

This paper explores one city's journey of integrating sustainability into its process of creative development through membership in the UCCN. The paper uses the northern Swedish city of Östersund as a case study, which has been an active member of the UCCN since 2010. The next section (section two) introduces key concepts that frame the case study, and section three offers a description of the methods. The remaining sections discuss the findings and implications of the study.

2   From Experiential Businesses to Creative City Networks

In recent decades, the global economy has undergone substantial changes, which includes dramatic shifts from large-scale industrial production to the need for regions and cities to compete and position themselves. The pioneering article by Pine and Gilmore [1] marked this change by highlighting global economic shifts away from agrarian, industrial or service economies towards a new economic model. They called this new economy the *experience economy*, which is characterized by competition around qualitative, symbolic and cultural aspects rather than cost and standardization. Other scholars have also highlighted this change. For example, O'Connor [2], Power [3], and Pratt [4] referred to this phenomenon as the *cultural industry*, whereas Caves [5] and Hartley [6] have termed it the *creative industry*. Regardless of the definition, the general idea behind this new terminology indicates a change in the competitive paradigm for many industrial sectors, whereby the production and selling of goods and services is increasingly based on other values in order to attract successively more demanding consumers.

As this new approach continued to influence the commercial business sector, other sectors of society soon followed. Academics such as Florida [7, 8], Hospers [9], Scott [10], and Sacco [11] extended the impact of creativity to a place based phenomenon, studying this new economy's impact on contexts stretching from nations to cities. In particular, cities have been attempting to compete for new, creative business establishments, while attracting creative people to settle in their areas. In short, cities have been competing in order to attract the growing *creative class*, a concept most often associated with Florida [7, 8], which has significantly influenced city development strategies. The creative class is a class working within the creative industry, and it is a segment of labor that no longer just moves where work is found, but rather decides where it is pleasant to live, and thereafter *creates* their work. According to scholars like Florida [7, 8], one of the major aspects of a city's success has to do with cultural offerings and the associated creative atmosphere.

Cities have accordingly developed strategies for becoming cultural and creative, and rankings and measurements of their performance send signals where it is most attractive for the creative class to settle. In response, networks of cities have developed in order to improve attractiveness. One example is the UCCN, which was established in 2004 in order to support cooperation between cities that are determined to use culture and creativity as a mechanism for sustainable urban development (UNESCO, 2017).

Through the UCCN, UNESCO has become a key player in the efforts for many cities to become creative. The UCCN has expanded rapidly and today it consists of 180 cities worldwide, with over 60 receiving designation during the last year.

UCCN designation allows a city to share ideas and exchange knowledge with other member cities under an umbrella consisting of seven sub-categories: craft & folk art, design, film, gastronomy, literature, media arts, or music (UNESCO, 2017). The UCCN also serves as a laboratory for testing and developing new ways to monitor and advance development at local levels [12].

Many of the studies on the UCCN thus far have focused on the branding aspects of city membership, including the work of Pearson and Pearson [13] and Rosi [14]. These studies also acknowledge other dimensions and opportunities of networks such as the UCCN, including sharing expertise and knowledge. Within this context, it is highly relevant to study how cities manage the use of culture and creativity as resources for sustainable development.

Dr. Jyoti Hosagrahar, Director of the Division of Creativity, Cultural Sector of UNESCO, has emphasized the importance of connecting creativity to sustainable development. For example, she recently noted that the links between cultural, creativity, and sustainable development will be a key factor in global efforts to implement Agenda 2030 [11].

The following section offers a case study of the Swedish city of Östersund. The case study focuses on Östersund's path towards UCCN membership as well as the impact the membership has had to envision a more sustainable future.

3 Methods

The paper is exploratory and uses standard approaches associated with case study research [15, 16]. From this methodological perspective, the City of Östersund represents a critical/crucial case example and serves as the focus of the case study. Purposeful sampling was then used to gather data from key actors associated with Östersund's involvement in the UCCN. Data included confidential interviews, focus group meetings, as well as personal communication with the researchers. In addition, a wide variety of secondary sources were analyzed including official administrative and policy documents (e.g., local development strategies, Östersund's application to join the UCCN as well as UCCN performance and monitoring reports). Participants in the study included 15 key actors (political officials, policy makers, and administrative actors, and local small business owners). Data were then analyzed for potential explanatory patterns [17].

4 Östersund: A UCCN City of Gastronomy Since 2010

The northern Swedish city of Östersund was established in 1786 and was largely characterized by two army bases and an air force base until the 1990s. In the years following the end of the cold war, the army and air force installations were dismantled and the city lost 1400 direct employment opportunities leading to the loss of 3500 indirect employment opportunities [18, 19]. This was a severe loss for a city which at the time consisted of less than 60 000 inhabitants. The Swedish government compensated for this loss through investments in 1100 new employment opportunities (primarily in the governmental sector), 80 million SEK (approximately 7 million Euro) in transformation grants, and a substantial investment in transitioning the local college into a new university. Together, these investments put the city back on its feet [19]. In the following years, the city reinvented and reshaped itself through these new ventures and also made substantial strides in becoming a well-known tourist destination.

Sweden has not been known for its gastronomic traditions – certainly not in the same way as many Mediterranean countries (e.g., France, Italy, Spain, etc.). In fact, some scholars characterized Sweden as a "food desert", which implies that the country is dominated by an industrialized/productionistic discourse in the agricultural and food sectors. This approach emphasizes an export orientation of agricultural and food products in which urban and rural patterns of food consumption are distinctly different, and where urban consumers demand a considerable amount of fast food [20]. Despite this context, Östersund (and the region of Jämtland-Härjedalen in which the city is located) has been recognized for its culinary heritage through two Slow Food Presidia products (cellar matured goat cheese and suovas). Good soils for agricultural production have also given the region a rich heritage of farming that includes agricultural products considered clean and of high quality [21]. Hence, the city's interest in joining the UCCN was based on a strong food tradition, particularly in the Swedish context. In Östersund's application for UCCN membership, it states that the city's representatives *"want to share knowledge, promote its cultural products, and be a model for other regions as to how to combine sustainable development and gastronomy"* [22]. The goal of the membership stresses the wish to *"attract investments in the creative sector, which we believe is a powerful source of social and economic development for Östersund city"* (Ibid). At the time of designation, 68 other cities were part of the UCCN, of which only one was a city of gastronomy. The UCCN was a network where some members where much more active than others, and there were some members completely inactive. Initially, the UCCN designation was used primarily by Östersund to position culture and gastronomy as development resources on a local and regional basis, and representatives took part in the annual UCCN meetings to build closer inter-city cooperation. At the same time, the city understood that membership also required the city engage more actively in the global sustainability agenda, as noted by one study participant, *"UNESCO clarified to us that we have to give something to the network"*.

In 2014, Östersund proposed to UCCN leaders that the city serve as the host for the annual UCCN meeting in 2016. Östersund won the bid, which accelerated local activities tremendously. The 2016 annual meeting was arranged in conjunction with the research conference, "Valuing and Evaluating Creativity for Sustainable Regional Development".[1] Together, these two events brought over 450 people to Östersund to discuss the nexus between culture, creativity, and sustainable development. Furthermore, the "Östersund Declaration" (Table 1) was developed and promoted locally, and then approved by all member cities in the network during the annual meeting.

---

[1] https://www.miun.se/contentassets/3e8e9d091ec14b9fad83ef28ce89dfba/vec2016-proceedings-webb.pdf.

The declaration connects creative development approaches to the UN Habitat III agenda for sustainable urban development.

As suggested by the Ostersund Declaration, awareness of the UN Sustainable Development Goals have increased in Östersund's political discourse. For example, the goals are now directly referenced in the regional food development strategy, which is directly tied to funding support in the sector for the next three years. This suggests that quintuple helix networks have started shaping around creative development in Östersund, even though business involvement is still somewhat weak.

In addition, after hosting the annual meeting and the associated research conference, activity levels as well as in-depth intra-city cooperation has increased. For example, Erasmus projects supporting youth engagement in sustainable gastronomy have started with several of the other UCCN members, and two regional EU-funded gastronomy projects with regional actors involved have also been initiated.

Another example was the UN Sustainable Gastronomy Day on June 18, 2017. Östersund coordinated the event that included 11 out of 18 members UCCN member cities. In Östersund, the event "Creative City in the Park" included a multitude of gastronomy-oriented activities that attracted 7000 people. This event has been a turning point in sparking broader, general interest in the city's UCCN designation. However, despite these activities, Östersund still does not yet use the UCCN designation as an explicit branding or communication tool. Instead, the city continues to primarily identify itself as a winter sports haven.

Study participants also identified a number of additional challenges that need attention. First, how does a city like Östersund obtain support from UCCN leadership in integrating the UNESCO goals into regional development strategies? Several of our study participants reflected that the UNESCO goals are "lofty" and not always translatable or actionable at local scales. Second, how do cities like Östersund directly and meaningfully connect the CCI with sustainability in a broader, more general sense? This issue is highly relevant in Östersund, and one study participant that is directly involved in managing Östersund UCCN activities described it like this, *"there is a potential conflict here, to balance the needs of the members with UNESCOs goals".*

Another challenge are the serious differences in scale in terms of socio-economic activity and size. After several UCCN meetings, one member of Östersund's delegation asked the obvious question, *"how can we cooperate on gastronomy with massive Chinese member cities like Macao"*? This study participant also reflected that the UNESCO mission statement is *"too urban in its current formulation".* In other words, the experience is that UNESCO, and the UCCN leadership, lacks a supporting function for the members to reach the goals of the UCCN. According to study participants, the focus has been on attracting new members that then have to work on their own to establish their own contacts in the network in order to meet the goals of the UCCN. This, in turn, suggests that without more active guidance or support from UNESCO and the UCCN, member cities may not readily connect creativity with UNESCO's sustainability goals

Another challenge is the lack of awareness about the possibilities associated with membership from local and regional stakeholders.

Table 1. The Östersund Declaration.

ÖSTERSUND DECLARATION
X UNESCO CREATIVE CITIES NETWORK ANNUAL MEETING

Östersund, Sweden
14-16 September 2016

We, the participants of the X Annual Meeting of the UNESCO Creative Cities Network (UCCN), held in Östersund, Sweden, from 14 to 16 September 2016, advocate the importance of culture and creativity as vital and transformative drivers of sustainable development.

Together, we have identified creativity as a strategic factor for sustainable urban development, which provides us with an inclusive framework toward the common objective of placing creativity and cultural industries at the heart of our local development plans and cooperating actively at the international level.

Throughout the X Annual Meeting discussions focused on the theme *Fostering the culture and creative sectors as drivers of sustainable development, maximizing the potential of urban-rural connections*, the delegates of the 116 UCCN Cities have reiterated the importance of the "Hangzhou Outcomes" adopted at the International Conference on "Culture for Sustainable Cities" in Hangzhou, People's Republic of China, 10-12 December 2015.

In view of the adoption of the New Urban Agenda at the Third United Nations Conference on Housing and Sustainable Urban Development (Habitat III, Quito, Ecuador, 17 - 20 October 2016), Östersund's meeting was for us the opportunity to reaffirm our commitment to follow the guidelines of the New Urban Agenda including the integration of culture in urban and regional development in the following ways:

1. **PEOPLE-CENTRED CITIES:** Humanizing cities through culture to enhance their liveability and empower people to connect with their communities and shape their urban environments.

2. **SUSTAINABLE URBAN ECONOMIES:** Alleviating poverty and managing economic transitions by enhancing the cultural assets and human potential of cities.

3. **HUMAN SCALE, COMPACT AND MIXED-USE CITIES:** Promoting culture and creativity in urban development, regeneration and adaptive reuse.

4. **INCLUSIVE MULTICULTURAL CITIES:** Recognizing cultural diversity by promoting collaborative partnerships to encourage community participation and reduce inequalities.

5. **PEACEFUL AND TOLERANT SOCIETIES:** Building on the diversity of culture and heritage to foster peace and intercultural dialogue, and counter urban violence.

6. **SUSTAINABLE, GREEN AND RESILIENT CITIES:** Integrating heritage and traditional knowledge into innovative and culture-based solutions to environmental concerns.

7. **INCLUSIVE PUBLIC SPACES:** Leveraging heritage and cultural and creative activities to foster social cohesion and ensure access to well-designed quality public spaces.

8. **ENHANCED RURAL-URBAN LINKAGES:** Fostering respect for the cultural value of small settlements and landscapes, and strengthening their relationship with cities.

9. **IMPROVED URBAN GOVERNANCE:** Strengthening participatory mechanisms, capacity-building, and developing indicators to assess the role and impact of culture on urban development.

We also wish to reinforce the importance of enabling a diversity of cultural expressions and strengthening the connections between all parts of civil society.

We commit to integrate culture in initiatives, policies and projects towards the achievement of the United Nations Sustainable Development Goals, the Agenda 2030, and to continue a close partnership with UNESCO in implementing them.

In summary, despite some encouraging signs, the analysis above suggest that substantial efforts are still needed in order to integrate sustainability into local creative development within the UCCN context.

5   Conclusion

The purpose of this paper was to explore how UCCN designation can encourage the use of creativity as a resource for sustainable development. Several key themes have emerged through the case study of Östersund's membership in the network:

- The development and adoption of the Östersund Declaration illustrates that UCCN membership helped to directly connect culture and creativity with sustainability discourse. However, the declaration is still largely perceived to be rhetoric, and substantial work remains in actual implementation.
- The network has been utilized in order to promote gastronomy in ways that are increasingly tied to sustainability issues. However, there is more to do in terms of involving businesses and the residents of the city and its surrounding region. Good starting points have been several business oriented workshops (organized by the city's Chamber of Commerce) as well as the UN Sustainable Gastronomy Day.
- Constructing local and regional networks around the gastronomy theme has evolved mostly around policy makers. As a result, businesses need more involvement in order to achieve quintuple helix network effects, especially in order to improve sustainability effects.
- The UCCN has established a platform for the exchange of knowledge and research on an international basis. However, such exchange mechanisms are in need of deeper local connections in order for local effects to occur.
- Branding is under-utilized, which prevents the message from the UCCN from spreading to local and regional community residents.
- The rapid growth of the network has positive and negative impacts. The upside is that more cities get involved in sustainable development through creativity, whereas the downside is that the massive number of new members can burden intra-city cooperation. Consequently, the perception of Östersund UCCN representatives is that some member cities perceive the UCCN as is simply about enforcing its agenda, thereafter leaving the members on their own.
- It takes time for cities to fully grasp the opportunities of network membership while also balancing the dynamics of an ever-evolving network.

References

1. Pine, B.J., Gilmore, J.H.: Welcome to the experience economy. Harv. Bus. Rev. **76**, 97–105 (1998)
2. O'Connor, J.: The cultural and creative industries: a review of the literature. A report for Creative Partnerships. London, Creative partnerships Arts Council, England (2007)

3. Power, D.: The Nordic 'Cultural Industries': a cross-national assessment of the place of the cultural industries in Denmark, Finland, Norway and Sweden. Geogr. Ann. Ser. B Hum. Geogr. **85**(3), 167–180 (2003)
4. Pratt, A.C.: The cultural industries production system: a case study of employment change in Britain 1984-1991. Environ. Plann. A **29**, 1953–1974 (1997)
5. Caves, R.: Creative Industries: Contracts Between Arts and Commerce. Harvard University Press, Cambridge (2000)
6. Hartley, J. (ed.): Creative Industries. Blackwell Publisher, Malden (2005)
7. Florida, R.: The Rise of the Creative Class: And How It's Transforming Work, Leisure, Community and Everyday Life. Basic Books, New York (2002)
8. Florida, R.: The Flight of the Creative Class: The New Global Competition for Talent. Harper Business, New York (2007)
9. Hospers, G.: Creative cities in Europe. Intereconomics **38**(5), 260–269 (2003)
10. Scott, A.J.: Creative cities: conceptual issues and policy questions. J. Urban Aff. **28**, 1–17 (2006)
11. Sacco, P.L.: Culture 3.0: A new perspective for the EU 2014-2020 structural funds programming. OMC Working Group on Cultural and Creative Industries (2011)
12. Hosagrahar, J.: Culture and creativity for sustainable urban development. In: Laven, D., Skoglund, W. (eds.) Proceedings of Valuing and Evaluating Creativity for Sustainable Regional Development, pp. 20–22. Mid Sweden University, Östersund (2017)
13. Pearson, D., Pearson, T.: Branding food culture: UNESCO creative cities of gastronomy. J. Food Prod. Mark. **28**, 1–14 (2015)
14. Rosi, M.: Branding or sharing? The dialectics of labeling and cooperation in the UNESCO creative cities network. City Cult. Soc. **5**(2), 107–110 (2014)
15. Flyvbjerg, B.: Making Social Science Matter: Why Social Inquiry Fails and How It Can Succeed Again. Cambridge University Press, Cambridge (2001)
16. Yin, R.K.: Case Study Methods: Design and Methods. Sage Publications, Thousand Oaks (2003)
17. Miles, M.B., Huberman, A.M.: Qualitative Data Analysis: An Expanded Sourcebook. Sage Publications, Thousand Oaks (1994)
18. Östersunds kommun (2004). http://www.ostersund.se/. Accessed 3 Dec 2017
19. Skoglund, W., Westerdahl, S.: Relationell estetik i gammal militärstad. Färgfabriken Norr i Östersund. In: Lindeborg, L., Lindqvist, L. (eds.) Kulturens kraft för regional utveckling. SNS Förlag, Stockholm (2010)
20. Bonow, M., Rytkönen, P.: Gastronomy and tourism as a regional development tool - the case of Jämtland. Adv. Food Hosp. Tour. **2**(1), 2–10 (2012)
21. Slow Food, Presidia. Slow Food Foundation (2007)
22. Östersund UCCN application, Östersund. Östersunds Kommun (2010)

Make Public Spaces Great Again Using Social Innovation Reflections from the Context of Downtown San Diego as a Cultural District

Israa H. Mahmoud$^{(\boxtimes)}$ ⓘ and Carmelina Bevilacqua ⓘ

Mediterranea University of Reggio Calabria, 89100 Reggio Calabria, Italy
israa.hanafi@unirc.it

Abstract. This paper investigates the role of public spaces in spurring innovation and promoting entrepreneurial activities in Downtown San Diego urban context as a distinguished "cultural district". The idea that in creative cities, flourishing human capital, when coupled with incremental quality of life, could be the driving vehicle to social innovation and economic prosperity. On that public spaces are a cross cutting phenomenon in a lifetime cycle, through which their success could be evaluated contextually based on their formation and implementation policies, and how they work-out to be social innovation catalysts. Hence, this paper studies contextually the Downtown San Diego Partnership (DSDP) and their role to activate and regenerate different public spaces in downtown area to foster economic development. Two successful exemplar cases are studied; to better understand the dynamics by which the cultural programming in urban parks through events occurrence in Downtown as vibrant cultural hub; as well as the focus on a co-working and incubator space as a successful case to explain the human capital attraction to the Downtown area. The conclusions draw on an evaluation matrix of analysis that investigates the Catchment area/Sphere of influence falling in San Diego Downtown area, and helps to reach the envisioned opportunities and the policy measures applied to foster social innovation in those public spaces, and evaluate the success or failure of the Downtown San Diego Partnership to boost the innovation ecosystem.

Keywords: Public spaces · Social innovation ecosystem · Co-working spaces

1 Introduction

While the role of public spaces in catalyzing entrepreneurial activities and spurring innovation is undebatable, the correlation between the social innovation and 'terri-torial milieu' remained under investigation for an academic decade [1]. However, the need for a place-based approach to better understand the spatial dimension in perceiving and accelerating the opportunity for social innovation Ecosystem remained significant. Yet, different key-factors interfere when it comes to social innovation and its' territorial milieu such as governance of public spaces, localization of innovation spaces, as well as the space territorial connectedness and network [2]. In that sense, this paper is structured as follows: the first part introduces the social innovation concept into the

© Springer International Publishing AG, part of Springer Nature 2019
F. Calabrò et al. (Eds.): ISHT 2018, SIST 101, pp. 406–415, 2019.
https://doi.org/10.1007/978-3-319-92102-0_43

research of Public spaces and how the locational factors help build the cultural districts profile.

The second part focuses on the Spatial context of downtown San Diego and displays both cases of the urban park "Courtyard" and the Co-working space "Downtown Works". The third and last part discusses the findings and the conclusions based on criterion evaluation matrix and measurement tools.

2 Public Spaces Role as Catalyst for Social Innovation

A promotional approach in the field shows an interest in cultural hubs and Districts where public spaces formation act as a melting pot for creative industries clusters. The role that creative city spaces act behind scientific policy rationales, claims the share in the knowledge economy and the cultural ranking of a city [3]. Meanwhile, those cultural hubs develop themselves in a later phase - in basis of their context, quality of life, identity, uses and programming- to act as catalyst, physically and virtually, for inward investment, business location decisions [4] and most importantly for human capital attraction that shape the social innovation environment and ecosystem (see Fig. 1).

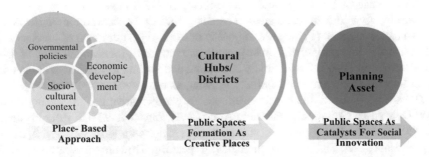

Fig. 1. Cultural hubs formation lifecycle dynamics (Source: the authors, May 2017)

Accordingly, with the role that public spaces should keep tenure of some combined factors of technological innovation, amenities, public services, and a certain level of social inclusion and lifestyle diversity. In a broader sense, those cultural hubs come to be seen as a strategic urban planning asset [5, 6]. Afterwards, those assets distinguish certain public spaces from others based-on context formation policies, cultural programming, sphere of influence/catchment area and stakeholders. Contrariwise, it is undeniable the fact that while some Public spaces might be strategic planning assets, might as well play a role in increasing income inequalities, gentrification, and economic segregation [7, 8].

The Un-Habitat [9] amends the measurement of quality of public spaces by anchoring the urban policies, governmental synergies, and usage of public spaces as a catalyst for economically growing vibrant cities. Quality public spaces-including streets, urban parks, courtyards- are multi-functional joints for social interaction, economic exchange, and theatre for cultural diversity expression.

In that perception, on one hand the role of quality public spaces is pivotal in improving living conditions of urban populations. Locational factors attract knowledge, industry businesses, qualified and creative workforce, and tourism as highlighted in Leipzig Charter for Sustainable cities [10]. Therefore, the interaction between public spaces and planning policies when coupled with infrastructure must ultimately improve in order to create attractive, user-oriented public spaces and achieve a higher standard living environment.

Nonetheless, on the other hand the definition of Social innovation as an approach for individual and collective wellbeing can be elaborated into three interconnected features [11]: first, the satisfaction of human needs (both material and non- material); second, social relations between individuals and groups at different spatial scales; third, empowerment, with micro level initiatives bringing positive macro level change. Spatial scale, therefore, has a strong role in the emergence and effectiveness of socially innovative actions, especially in terms of the level of intervention [12, 13].

3 A Spatial Focus on Downtown San Diego Urban Context

The urban fabric of San Diego- either as a county or MSA-is very diverse and merely, touristically, vivacious in some areas. The contextual study showed the verification of two evaluation criteria: (1) increase in quality of life and (2) attraction of human capital that create the vibrancy that -consequently- attracts the professional pool of labor. Downtown San Diego offers a variety of cultural and social amenities that explains the reason why downtown is exponentially growing as a regional urban core [14].

The spatial focus in this paper goes to the Downtown area; geographically defined as the Downtown Community Planning Area (CPA), Zip Code 92101. Roughly an area of 1,450 acres and encompasses seven thriving neighborhoods, each with its own unique identity (see Fig. 2 - left). Statistically speaking, Downtown is home to 35,000 residents and a growing population of 97% since 2000, notably a dominance of 51% for highly educated residents and 73% high-earning professionals, [15]. A remarkable attribute to downtown San Diego area is the constant physical and economic development in progress, the city efforts to develop a walkable "live, work and play" urban core. Nonetheless, the area has a potential growth, the city government hired one of highly ranked architectural and urban design firm to overlook the redesigning of the downtown skyline from North to South [16].

In a matter of fact, Downtown San Diego has a 90% score in walkability; 78% of residents enjoy its central location [17, 18]. Proximity to different venues of arts and culture as well as other amenities in downtown area makes it attractive to entrepreneurs. The San Diegan Downtown vibrancy is unmistakable; that noted, leads to social innovation; the evidence-based is demonstrated through the two evaluation criteria: the increase in quality of life and the attraction to human capital, as described as "creative class" and Knowledge-based workers.

Fig. 2. Downtown San Diego seven neighborhoods (left) (Source: the authors, 2016) –Arts and cultural organizations Cluster in Downtown San Diego (Right). (Source: [19])

3.1 Downtown San Diego Partnership Focus Area

One of the most prominent cases of those cultural districts in the Downtown area, whereas the Downtown San Diego Partnership (DSDP)-a nonprofit organization-is spatially focused and operating. That adds up to the approach explained earlier of public spaces lifetime cycle where the place-based is transformed into implementation-based through privileged partnerships and associated stakeholders. The following part of this paper examines on the ground scale of the downtown area whereas the evaluation of key factors for the creation of the Cultural Hub/District is observed to verify the role that public spaces are playing in forming creative places that turn to be catalyst for social innovation in San Diego area.

In a study conducted by UC San Diego extension center for research on the regional economy in (2016) [14] about the San Diego Downtown prosperity, the area demonstrated a concentration of 92 arts and cultural destinations and organizations in the urban core. A diversity of venues types between 31 art galleries, 4 museums, 6 live performance theatres, 12 music venues and 10 performing arts groups, including a symphony, an opera, and a professional ballet company. The neighboring Balboa Park boasts upwards of 30 arts and cultural destinations within its boundaries as well (see Fig. 2- right).

While Downtown San Diego demonstrates itself as a cultural hub and performance arts hotspot, 79% of residents enjoyed being to a proximity to that ambient. These individuals attend activities such as musical entertainment (74%), museums (74%), movies (69%), performance arts (67%) and art galleries/events (61%), [14].

The Phenomenon that is occurring in downtown San Diego being transformed in an arts and culture hotspot nowadays in not a laissez-faire. The American planning association (APA) coined the concept in 2015 by highlighting the facts that entre-preneurial activities seek out communities that inspire creativity and push boundaries. That, being correlated by business firms location with artists and cultural facilities together, has a resulting 'multiplier effect', driving further the innovation economy and economic vitality by measurable outcome [20]. In fact, the role that

Downtown San Diego Partnership plays in forming and pushing the cultural scene vitality is un-deniable, relating between cultural-sector firms and creative professionals, along with improving and developing physical facilities deliberates a shared economic advantage to downtown area in that sense.

Urban Parks Model: The Quartyard Case Study. Another example for the spatial concentration of cultural events occurrence and venues clustering in Downtown area is observed in the "urban park" namely "Quartyard". A 25,000-square foot city owned lot at 1102 Market street, constructed in 2014 from repurposed shipping containers in East Village at downtown (see Fig. 3). The public space, with a 1000 persons' minimum capacity, is home to a coffee shop, restaurant, dog park, beer garden, music venue and a rotating assortment of food trucks. Basically, promoting themselves as a venue that brings people together, celebrates community, coffee, food, music, and cultivates the culture of unique social gatherings. Open 7 days a week, Quartyard plays host to numerous cultural events from farmers', pop-up markets to movie nights, film festivals and music outdoor concerts.

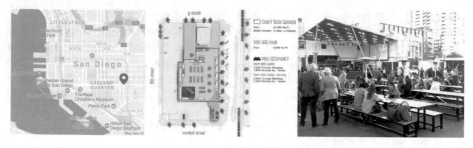

Fig. 3. Quartyard Urban Park location, site design and inside views during an event - from left to right. (Source: the authors, May 2017).

The interesting attribute in the Quartyard urban park example is its' formation as a "communal" public space. As reported by Lothspeich [21], three senior architectural students in 2013 concepted the idea of a "Movable Urban Park". The idea was simple, that is transform vacant city land into a thriving public space that could be assembled quickly, and torn down just as fast, using shipping containers as structural components instead of spending the time, money and effort required to develop an actual building. While missing on financial means, the trio raised funds on kickstarter (an online crowd sourcing platform), $60,000 were campaigned online and in person to prove residents' interest in the project to the city, after gathering investors, partnering with contractors, and receiving legal approval from the city, the urban park was born to light and given the name of "Your City Block".

Co-Working Spaces Model: Downtown Works Case Study. Nonetheless, the same importance of cultural events occurrence is equally measured by successful co-working spaces spreading, the diffusion of these spaces in the San Diego downtown area nowadays is remarkable. One of the highly ranked amongst the entrepreneurial community is Downtown Works, physically located in 550 West B Street 4th Floor San

Diego, CA 92101. The two floors, 5,000 square feet Co-Working space offers tailored services for entrepreneurs and companies located in the venue; plans vary based on startups size and budget. Monthly, daily, permanent, or virtual offices are amongst options; open-desks, exclusive desks, or private offices and meeting rooms as well (see Fig. 4). Members benefit from a variety of amenities, a pet friendly space, on site market and a 24/7-member access [22].

Fig. 4. Downtown works location and views from inside the co-working space – from left to right. (Source: the authors, 2017.)

Besides the physical amenities and the proximity to public transportation, and having the view of the San Diego port, Downtown Works [23] has an in-house accelerator program for startups. Weekly "lunch-n-learn" meetings with tech industry experts to promote the startups exposure to the business community, and to provide hands on experience with advisors and funding to startups.

In an Interview with one of the space founders "Wolf Bielas" conducted May 23rd, 2017, he highlighted the importance of co-working spaces venues to the vibrancy of the entrepreneurial scene in San Diego Downtown that matches with "live, work, play" vision of the city and with entrepreneurs needs above all. Another crucial factor of co-working spaces in his opinion was the collaboration among co-working spaces venues rather than competition in the startups scene.

Within the same concerns of Bielas, he emphasized the fact that the cultural diversity and walkability atmosphere of Downtown San Diego is significantly high in comparison to other cities. Yet, the downtown San Diego area is relatively cheap in office rent prices for newly created startup and companies to get located:

"Nonetheless, there is an attraction of slowly growing investments in the area due to connection of software engineers to labor pool, even if the wage is 30% less than San Francisco, for example, but the cost of living is 50% less than Bay area"

Interestingly, he stressed out the cross-border relationship connection with Tijuana and how this territorial proximity helps the software, hardware, and firmware talented labor pool in common to develop and prosper. Same notion happened around the dynamics of economic activities and local services clusters, (food, art, cultural venues, and shopping, etc.), which affected the local community gatherings scene. On the contrary, he highlighted the missing effect of an educational institution as an anchor in spreading the diffusion of startups, coupled with the missing fact of an anchor company headquarter that would attract other startups types to cluster and develop in downtown area.

Another stimulating fact in this interview was about the public policies and local governmental approach to facilitate the investment in Co-Working spaces and real estate development around Downtown area. Bielas referred to downtown area being exposed to a "perfect storm", whereas the mayor and the city strategic plans are very pro-development by supporting changing in some zoning regulations and giving permits facilitations to have new mixcd-use buildings available for housing, small businesses, and a diversity of arts and cultural venues, together with a variety of retrofitting facilitations for existing buildings in the Downtown area.

In sum, the success of Downtown Works model gives an important retrospective about how the Public-Private interrelation is driving the innovation ecosystem of startups and entrepreneurs to flourish in Downtown area. Whiles the Private sector is a Pushing actor to the Public sector and how this controversial relationship has its' impact over the territorial milieu and social innovation ecosystem.

4 Findings

It is visible to eyesight the uniqueness art and cultural hub of Downtown San Diego; the social demographics data show a certain attraction to art hotspots where the urban vibrancy could be easily measured and perceived such as in urban parks models. While the evidences explain the human capital attraction phenomenon, the closest measurable and tangible criterion was the multiplier of startups in downtown area and how innovation ecosystem is pacing out in Downtown area.

The above-mentioned evaluation matrix (see Table 1), shows the measurement tools used in this research to identify the success or failure based on evaluation criteria that follows the conceptual model of cultural districts introduced earlier. Through the verification of implementation-based approach, the physical attributes in Downtown San Diego area showed a proximity to transportation, high walkability score, vicinity to different amenities; and that, fortifies the fact that a strong cultural District is flourishing, giving way to prosper economic development.

The hurdles to public spaces development are mostly financial; however, the city is tapping into local redevelopment funds, private donations, and economic recovery act to strengthen and transform the downtown area into **a vibrant cultural hub** [24]. Meanwhile, apart from the strengths in the territorial context, the wage multiplier effect for innovative jobs, lower rent prices and competitive advantages for startups are striking reasons, and for that, the downtown area has a wider sphere of influence/catchment area and is "baby-booming" in some specific innovation economy sectors such as software, tech and IT [25, 26]. Nonetheless, the rippling effects of the innovation scene is not limited to downtown area, the latest published **Kauffman Index** about metropolitan areas and city trends in startup activity unveiled a growing drift in San Diego metropolitan region. Among 39 innovative regions, San Diego ranked 4th based on rate of new entrepreneurs in market, opportunity share and startups density; thus looking willingly to join the parade of innovative cities [27].

Table 1. Evaluation analysis matrix for downtown San Diego case study (source: the authors)

Downtown San Diego case study summary			
Concept	Criteria of evaluation	Measurement tool	Assessment results
Place-based approach	• Socio-cultural context • Governmental policies • Economic development	• Demographical context • Arts and cultural hotspots • Open spaces and Parks	• Proximity to transportation, • High walkability score, • Vicinity to amenities
Cultural hubs/districts	• Quality of life increase • Human capital attraction	• Vibrant spaces in area • Startups located in area	• Effective planning strategies and facilities towards mixed-use developments
Implementation – based approach	• Cultural Programming • Stakeholders involvement • Sphere of influence/catchment area	• Cultural Events occurrence • The rippling effects of innovation scene.	• Diversity in implementation techniques • Cross-border labor relations, • Missing Educational Anchor institution

5 Conclusion

Fundamentally, Social innovation is deeply intertwined with privileged public services, public facilities, social inclusion, and lifestyle diversity where the public spaces are proved to be granular catalysts for sharing knowledge and building innovation [28]. While Cultural Hubs/districts are proven to be a driver for cultural-led urban policies [29], it is undeniable that the context diversification in Downtown San Diego area affect positively the two measurable criteria used to verify the implementation-based approach, that are: the increase of quality of life and Human Capital attraction. On one hand, Downtown San Diego Partnership plays a protagonist role as a public-private partnership in terms of public policies and community development through different engagement techniques to regenerate and activate different public spaces in connection with art and culture venues in downtown. On the other hand, the private sector is still pacing out the road while public bodies such as civic san Diego follow out the changes on the ground, yet the cultural vibrancy is undeniable.

Two notable outcomes from the interviews and the physical observation analysis show that: (1) downtown area lacks **an anchor educational institution** that drives a lot of economic activities within neighboring areas, (2) **governmental facilitations** to business development either in urban planning policies or land-use zoning regulations differ on basis of projects. Private businesses have easier trends to obtain permissions for mixed-use developments and retrofitting projects. (3) the Community plan as approached by public administration (e.g. CCDC) is merely tokenism with no ground.

In sum, Downtown san Diego area is a set stage to development of public spaces bringing to renaissance a vibrant urban core; constraints are many, but Public assistances are practical and doable. Even though the CCDC [30] is a "weak link" in the deliverance of Public Space. The cultural district in Downtown area is distinguishable, the human attraction and life quality play a turmoil role in fostering development of knowledge economy and shifting the cultural ranking of San Diego city forward.

Acknowledgment. "This work is part of the MAPS-LED research project, which has received funding from the European Union's Horizon 2020 research and innovation programme under the Marie Skłodowska-Curie grant agreement No 645651".

References

1. Moulaert, F., et al.: General introduction: the return of social innovation as a scientific concept and a social practice. In: The International Handbook on Social Innovation, Collective Action, Social Learning and Transdisciplinary Research, pp. 1–153 (2013)
2. MAPS-LED Project: S3: Cluster Policy And Spatial Planning. Knowledge Dynamics, Spatial Dimension And Entrepreneurial Discovery Process. Second Scientific Report, MAPS-LED Project, Multidisciplinary Approach to Plan Smart specialisation strategies for Local Economic Development, Horizon 2020 - Marie Swlodowska Curie Actions -RISE - 2014 -grant agreement 645651, pp. 9–23 (2017)
3. Evans, G.: Creative cities, creative spaces and urban policy. Urban Stud. **46**, 1003–1040 (2009)
4. Mercer, C.: Cultural planning for urban development and creative cities. In: Shanghai Cultural Planning Conference, 3 (2006). http://www.kulturplan-oresund.dk/pdf/Shanghai_cultural_planning_paper.pdf. Accessed 21 Nov 2017
5. Florida, R.: The Rise of the Creative Class: and How It's Transforming Work, Leisure, Community and Everyday Life. Basic Books, New York (2002). http://www.washingtonmonthly.com/features/2001/0205.florida.html. Accessed 21 Nov 2017
6. Deffner, A., Vlachopoulou, C.: Creative city: a new challenge of strategic urban planning? In: ERSA Conference Papers, pp. 1–14 (2011). http://ideas.repec.org/p/wiw/wiwrsa/ersa11p1584.html. Accessed 21 Nov 2017
7. Florida, R.: The New Urban Crisis. Edited by Martin Prosperity institute, Basic Books (2017)
8. Mallach, A.: Lots of Maps, Little Insight in Richard Florida's Latest, Shelterforce, pp. 20–22, 11 May 2017. https://shelterforce.org/2017/05/11/lots-of-maps-little-insight-in-richard-floridas-latest/. Accessed 21 Nov 2017
9. UN-HABITAT, Global Public Space Toolkit: From Global Principles to Local Policies and Practice. https://unhabitat.org/books/global-public-space-toolkit-from-global-principles-to-local-policies-and-practice/. Accessed 21 Nov 2017
10. European Commission LEIPZIG CHARTER on Sustainable European Cities (2007). http://www.eu2007.de/en/News/download_docs/Mai/0524-AN/075DokumentLeipzigCharta.pdf. Accessed 21 Nov 2017
11. Moulaert, F., et al.: Towards alternative model(s) of local innovation. Urban Stud. **42**(11), 1969–1990 (2005). http://journals.sagepub.com.ezproxy.neu.edu/doi/pdf/10.1080/00420980500279893. Accessed 21 Nov 2017
12. Mehmood, A., Parra, C.: Social innovation in an unsustainable world. In: Moulaert, F., MacCallum, D., Mehmood, A. (eds.) The International Handbook on Social Innovation: Collective Action, Social Learning and Transdisciplinary Research, pp. 53–66. Edward Elgar, Cheltenham (2013)

13. Shockley, G.: Book review: The international handbook on social innovation: collective action, social learning and transdisciplinary research. J. Reg. Sci. 152–153 (2015). Wiley
14. DSDP: Downtown San Diego: the innovation economy's next Frontier A data driven exploration of San Diego's Urban renaissance. San Diego: UC San Diego extension center for research on the regional economy (2016). http://downtownsandiego.org. Accessed 21 Nov 2017
15. US Census Bureau American Community Survey, American FactFinder. http://factfinder2. census.gov. Accessed 18 May 2017
16. Showley, R.: Gensler: Redesigning San Diego, from North to South, The San Diego Union-Tribune, pp. 1–4, May 2017. http://www.sandiegouniontribune.com/business/growth-development/sd-fi-gensler-20170508-story.html. Accessed 21 Nov 2017
17. DSDP Imagine Downtown, San Diego. http://downtownsandiego.org/wp-content/uploads/2015/02/Imagine-Downtown-Presented-by-the-Downtown-San-Diego-Partnership.pdf. Accessed 21 Nov 2017
18. DSDP Update, Imagine Downtown, San Diego. http://downtownsandiego.org/wp-content/uploads/2015/02/Imagine-Downtown-Presented-by-the-Downtown-SanDiego-Partnership.pdf. Accessed 21 Nov 2017
19. DSDP.   https://dsdp2015.carto.com/viz/4bb5a924-fd10-11e5-92ce-0e3ff518bd15/public_map. Accessed 21 Nov 2017
20. Dwyer, M.C., Beavers, K.A.: How the arts and culture sector catalyzes economic vitality, American Planning Association. Edited by Murray, D.J. (2015). https://www.planning.org/research/arts/briefingpapers/vitality.htm. Accessed 21 Nov 2017
21. Lothspeich, D.: Quartyard's New Lease on Life, SoundDiego: Music, Community, Culture (2017).   http://www.nbcsandiego.com/blogs/sounddiego/Quartyard-Lands-New-Location-421784253.html. Accessed 15 May 2017
22. SDtechscene Downtown Works. http://sdtechscene.org/venues/downtown-works/. Accessed: 1 June 2017
23. Downtown Works, CoWork with us! Lightpost Digital. http://www.downtownworks.com/. Accessed 1 June 2017
24. The San Diego Union Tribune, A city of great public spaces. http://www.sandiegouniontribune.com/opinion/editorials/sdut-a-city-of-great-public-spaces-2011jan16-story.html. Accessed 1 June 2017
25. CONNECT, San Diego Innovation Report. http://www.connect.org/innovation-reports. Accessed 1 June 2017
26. Equinox Entrepreneurship How are we doing? Center for Sustainable Enegrgy. https://energycenter.org/equinox/dashboard/entrepreneurship#indicator-idea. Accessed 20 May 2017
27. Morelix, A., Fairlie, R., Tareque, I.: Kauffman Index of Startup Activity: Metropolitan area and City Trends (2017). https://www.KauffmanIndex.org. Accessed 1 June 2017
28. Feldman, M.P.: The character of innovative places: entrepreneurial strategy, economic development, and prosperity. Small Bus. Econ. 43(1), 9–20 (2014)
29. Hesmondhalgh, D., Pratt, A.C.: Cultural industries and cultural policy'. Int. J. Cult. Policy 11(1), 1–13 (2005)
30. CCDC, Centre City Development Corporation: Downtown San Diego Public Open Space Implementation Plan PART 1: Project Initiation, Data Analysis And Synthesis. San Diego. http://civicsd.com/wp-content/uploads/2015/02/POSIP_Part_I_Summary_Report.pdf. Accessed 20 May 2017

Describing a Unique Urban Culture: Ibadi Settlements of North Africa

Beniamino Polimeni[✉]

De Montfort University, Leicester LE1 9BH, UK
beniamino.polimeni@dmu.ac.uk

Abstract. This paper examines the urban structure and architecture of the North African regions, which are characterised by the historical presence of Ibadism. This topic has a definite cultural frame, albeit with some differences across three geographical areas located in the Mediterranean Maghreb: the Island of Djerba in Tunisia, the region of Mzab in Algeria, and the Djabal Nafusa mountains in Western Libya. Although similar configurations can be found in different parts of the Mediterranean territories, in these particular regions, the need for protection and defence, such as the balanced use of natural resources, has played an emblematic role. From the 10th century onwards, in fact, Ibadi communities chose the hard conditions of these lands to preserve their cultural identity, even at the cost of their isolation. Guided by the desert climate and scarcity of natural resources, these peoples developed specific urban solutions, architectural forms and construction traditions.

The objective of this research is to define and compare the main facets of settlements and architectural forms, with the purpose of understanding and verifying the cultural continuity.

Keywords: Djerba · Mzab valley · Djabal Nafusa · Settlements Ibadism

1 Introduction

In their extensive exploration and interpretation of what they call "The Corrupting Sea", Horden and Purcell [1] persuasively define the Mediterranean as a collection of distinctive Micro-Regions. These are regions characterised by discontinuous geographical locations, and differences in scale and territorial configurations, although they are often connected by comparable cultural matrices, which give clues as to their histories, religions and principles, identifiable through similar characters. Regions will often contain cities or be composed of small scattered settlements, in which we find the most direct expressions of the civilisation that created them. Indeed, we recognise the cultural border of the communities who live there, and we understand the relations between the physical spaces and the characteristics which promoted their functions.

The aim of this paper is to analyse the distinctive urban and architectural elements of some of the Mediterranean regions connected by a common cultural feature: the presence of Ibadism. This form of Islam, which is distinct from the Sunni and the Shi, was founded 20 years after the death of the Prophet Muhammad and was spread across

© Springer International Publishing AG, part of Springer Nature 2019
F. Calabrò et al. (Eds.): ISHT 2018, SIST 101, pp. 416–425, 2019.
https://doi.org/10.1007/978-3-319-92102-0_44

North Africa by missionaries who had fled from the Umayyad Caliphate and took refuge in the Nafūsa Mountains at the beginning of the 8th century. Those preachers converted the Berber people into a combat force which battled several times against the different governors of North Africa, dominating a large part of the Maghreb during the 9th century [2].

Their culture and the historical evidence of their presence can be found in three areas of the Mediterranean Maghreb: the Island of Djerba in Tunisia, the region of Mzab in Algeria, and the Djabal Nafusa area in Libya. These are three peculiar geographical spaces which perfectly represent the desire of the aforementioned communities to protect their specific identity.

Of particular note here is the whole territory of a flat island in the Gulf of Gabès, as well as the borders of a valley that cuts a limestone plateau in the Northern Sahara, and the northern limit of a plateau in the Western Tripolitania. Indeed, these are the natural regions where there exist protected values and beliefs which define a specific way of establishing a community. This peculiar condition will be analysed, assuming that "defence" and "Isolation" are the primary needs which have shaped the settlements and influenced their architectural language.

There are three reasons for choosing these case studies. The first one arises from the desire to describe the urban form which allowed the Ibadi and Berber groups to resist both the expansion and the influence of peoples from other Mediterranean areas.

The second reason, and the most compelling, concerns the particular location of the regions and their environmental characteristics: these qualities influenced the architectural solutions which were pioneered, creating urban "organisms" perfectly adapted to the environment.

The third reason is related to the possibility of creating a comparative analysis: this is a structural evaluation, useful for discovering analogies and differences between these case studies, by studying a repertory of forms and solutions which, in local variations, covered the whole western Maghreb.

2 Case Studies

A brief overview of the historical events which generated the urban shapes of our examples is essential in order to understand the existing relationship between physical configuration and the diverse rules of occupation.

The five cities of Mzab were founded ex-novo, inspired by the rigorous principles of the Ibadi doctrine, on the rocky outcrops of a valley almost unoccupied until then.

These peoples, who had inhabited the desert since the 9th century, chose this inhospitable region to found their first settlement (El-Ateuf) in 1012 [3]. Human settlement of the Djabal Nafusa has instead been associated with two different migration processes. The first one relates to the preachers who took refuge in this area, starting from the 8th century, and created an independent Imamate. The second process involves the Berber tribes, converted to Ibadism, who lived in the Jifarah plain: these peoples moved away from the invasions of the Arabs of the Bani Hilal and Bani Sulaym, and occupied the northern border of the mountain. This process gave birth to new urban organisms and peculiar means of territory occupation, often modifying the form and the organisation of the previous settlements.

During the 10th century, a similar displacement involved the Nukkarites, Wahbites and Khalafite tribes, as well as several Berber Kutama groups who took refuge on the island of Djerba to escape persecution from the Fatimids [4]. Here they developed a system of fortified mosques spread throughout the island which allowed them to maintain, for several centuries, a relative cultural independence [5].

In the next paragraphs, the urban, geographical, and architectural aspects of these sites will be described individually using different sources and scales (Fig. 1).

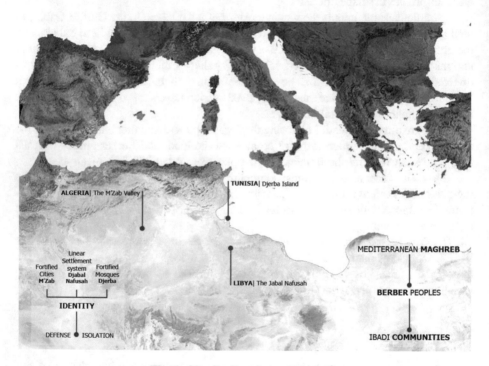

Fig. 1. The Ibadi regions of Maghreb.

3   The Ksour of Mzab Valley

The Mzab is a region of the Algerian Sahara which presently corresponds to the wilaya of Ghardaia. It has the appearance of a vast rocky plain, declining in altitude from 700 m in the west to 300 m in the east, crossed by deep and curved valleys, hence why it is called the Chebka (shabka = network) of the Mzab [3].

This vast area, historically travelled by the nomadic tribes of the Chaanba, Mekhadma and Said Otba, contains five centres of sedentary population [3]. The most important of these is the Pentapolis, the heart of the Mzab, with its five cities founded by the Rustamids in the 11th century. These citadels, spread along the valley of the wadi Mzab over a distance of 12 km, from east to west, are: El Ateuf, Bou Noura, Beni Isguen, Melika and Ghardaia.

These groups, whose capital Tahart was burned down in 909, occupied the Mzab valley because of the defensive possibilities it offered. The occupation of land, organisation of space and source sharing were inspired by rigorous hierarchical principles and are of an exemplary character.

These specific design rules generated sustainable models studied by many modern architects, starting from Le Corbusier, who visited the Mzab region in 1931 and proposed some of the features of the valley for some of his most relevant projects [6].

Each of these small citadels, called Ksours, is dominated by a fortified mosque whose minaret also functions as a watchtower. The position, the structure and the architectural arrangement of these buildings are conceived to protect the community in case of siege. Arsenals and grain stores are generally incorporated into the buildings, and represent the starting point of a structure of houses built in concentric circles. Each dwelling, a square cell defined by a closed or an open courtyard, exemplifies an egalitarian social organisation, based on family-structure respect with the aim of protecting its privacy and autonomy. Outside the walls, this principle of impartiality is exemplified in the cemeteries, where among the many little, unmarked graves, only small mosques and the tombs of sages are distinguishable. The living pattern of the Ibadis of the Pentapolis included a seasonal migration. Each summer, the population journeyed to and settled in palm groves, where their "summer cities" were marked by a less strict organisation, by the highly defensive nature of the houses, the presence of watchtowers and mosques without minarets, similar to those in the cemeteries.

Despite its cultural isolation and the absence of any local resources other than energy, the Ibadi society has been prosperous in finding means of survival in organised and profitable commercial activities; indeed, they have exhibited, over time, an extraordinary aptitude in adapting to the contemporary world [3].

Demographic and urban expansion has partially changed the character of these places, although they still retain the religious cohesion and the traditional social structures which provided their original strength.

3.1   Urban Structure and Architectural Features

These urban organisms, entirely representative of the Ibadi social identity and perfectly adapted to the environment, are examples of balanced community organisation and efficient climate response. Indeed, the clever combination of technologies, materials and architectural solutions has generated a peculiar domestic dimension, whereby the town structure has been developed through successive articulations. This complex system originated from the mosque, and has a "radial", compact organisational structure; in this geometric matrix, the location of the dwellings is parallel to the contour lines and a hierarchic rule between buildings of different dimensions can be seen. Housing units are made of mud, bricks and gypsum, are two floors high, and have covered courtyards with grilled skylights; in addition to this, openings to the alley are limited to the main door and a small window above. All the settlements of the valley are connected to an elaborate hydraulic system which follows the course of the "oued", but which is separated visually by rough open swaths of desert. This system consists of barrages built across the "oued" to create surface and subterranean reservoirs. The Oases are now used as summer towns but were the main centre of the economy of these communities. They are in fact a necessity, and rather than being used just for food production, they are also used for other vital purposes. In particular, date palm cultivation is one of the most significant aspects of the heritage and the built environment of this area, offering shelter, sustenance and building materials [7].

It should be noted that some of the houses of the Ksour have often been redesigned internally to respond to modern requirements of comfort, while some have the appearance of bizarre blocks of concrete. Despite this, however, many have preserved their traditional outward appearance, and become wholly integrated into the hierarchical pyramid, which rises towards the mosque.

The new units which have been developed at the feet of their older counterparts are primarily villas or multistorey buildings, with a garage or even a shop on the ground floor. Although these new types follow the axes created by the original urban tissue, they lack the centrality of the symbolic hierarchy (Fig. 2).

Fig. 2.   The centre of Ghardaia. Picture by Axel Derriks.

4   The Island of Djerba and the Fortified Mosques

Djerba, the largest island in North Africa, is located in the Gulf of Gabes, off the coast of Tunisia. The island consists of a small plateau which reaches its highest point in the south near Sedwikessh (55 m) and slopes down towards the coastal plains in the west of the island. It is one of the rare remaining places in Tunisia where the Berber language is still spoken, and the Ibadi sect is active. Half the population has, in fact, retained the Ibadi faith in its local form of Wahbism [8].

Members of this religious group occupied the island during the middle ages, as a continuation of the process of displacement started in Northern Africa between the 8th and the 10th century. Constructions from this period include numerous small mosques dating back as far as the 12th century. In 1944, René Stablo counted 166 Ibadi and 122 Sunni mosques, which are often converted Ibadi mosques, distributed equally across the whole territory of the island. The reason for this notable distribution is related to the role played in the past by every single mosque.

In Djerba, these buildings were the centre of the spiritual and social life of the city.

They were often used as schools for students of any age, as spaces for the accommodation of travellers, and as places of refuge for the inhabitants in case of danger.

Mosques were also used for different judicial functions, such as the legal confirmation of contracts or as places for celebrating marriages.

It should be noted that the island was extremely vulnerable, mainly due to its flat orographical and geographical position; as such, from the 11th century up to 1560, when it finally passed under Ottoman control, the island suffered a significant number of attacks. These continued threats encouraged the Ibadi community to transform the whole coastline into a structured defensive system. On the frontline, near the sea, there was a belt of several mosques responsible for monitoring the coast (Fig. 3).

These mosques generally had no minaret, so that they were not visible from the sea.

The faithful would take turns, tirelessly dividing their time between prayer and observing the sea. In case of attack, lights or smoke signals were used in the daytime or

Fig. 3. Mosque Sidi Fadhloun near Midoun.

fire at night-time to warn the community. Two examples of this kind of watchtower-mosque are Sidi Jmour in the west and Sidi Yati in the south. In addition to these mosques, many other religious places also allowed the population to take refuge if necessary.

Another line of fortified mosques placed internally also offered refuge. Two of the most remarkable examples of this kind of building are the Jama' Tajdid and the Fadhloun mosques [5]. These fortified edifices, equipped with seats and with spaces for storing victuals, had buttress-like projections and thick walls, sometimes pierced by loop-holes. This double structure defined a consistent geographical system, articulated in the territory by an equally balanced distribution; indeed, the structure also represents an extraordinary example of a defensive system perfectly integrated into the environment.

5   The Urban Settlements of the Nafusa Mountains

The Djabal Nafusa is a semi-circular calcareous escarpment lying on the boundary between the Libyan coastal plain, known as the Jifarah, to the north, and the Tripolitanian Plateau to the south. The height of this chain ranges from 460 to 980 m, while its surface is intersected by several "Uadian", or dry valleys, which cross the Jifarah Plain and flow into the Mediterranean Sea. The "Mountain", which is now referred to as the "Berber heart" of Libya, has been inhabited since Paleolithic times and holds many traces of the peoples who have occupied the territory in the recent past, especially Romans, Vandals and Byzantines. The densest period of occupation is closely related to the Arab diffusion and particularly to the Hilalian invasion. In fact, the migratory process following the attack of the Bani Hilal and Bani Sulaym in the 12th century influenced the structure and the configuration of the settlements located both on the northern and southern parts of the mountain border. Though the Nafusa Berbers had at first accepted Ibadism and continued to occupy the Jifarah plain area, these new historical events resulted in the successive displacements of those communities, which eventually abandoned the plain and settled in internal, less accessible areas [9]. The subsequent process was a slow change in the structure of the cities, with a transformation that continued over three centuries. The towns and the villages located on the borders of Djabal Nafusa became a linear structure: a network of elements linked by similar means of land use, by a common need for defence, and by social connections among the human groups. This new distribution reflects a new need for protection and isolation resolved by the morphological characteristics of the mountain - an ideal place from where to guard the territory against incursions from the north or south. At the same time, the mountain simultaneously maintains a strong link with the Sahara Desert. This new condition can be described schematically in two phases: the first is the proliferation of several new little villages, from east to west, on the edge of the plateau. The second phase involves the fusion of several villages into the first fortified cities such as Kabaw and Jadu, whose existence began with this type of mutation. Of particular note here is the structure of these new towns, as well as several economic changes that encouraged the populations to move toward the border of the plateau. Indeed, all of this reinforced the interdependency between cities and adjacent villages, defining a secure homogeneous network articulated in the territory by a very similar rule of occupation. This was a pattern of source-sharing in which the villages, the paths, the water sources and the technical skills became part of a common vision, creating a cultural sphere defined throughout the ages by a clear hierarchy which is visible in the transformation processes and the solid connections among the settlements. With the passing of time, this physical border became a psychological border and a symbol of a particular social and cultural identity. This strong character, which influences the different examples of the vernacular architecture, arose from the adaptation to the new limited environment and in response to the collective needs of the communities. Indeed, this character itself reflects the environmental, cultural and historical context in which it exists, thus creating a system unique to the Maghreb Territory [9].

5.1 Architectural Types

Cave dwellings, fortified Granaries and above-ground houses on the mountain represent the social values and reflect the need for protection in these communities. Fortified granaries, known as "gasur" [10], are the central elements of most of the border settlements. These protected structures, located in a part of the village which is less accessible from outside, are usually composed of a number of store cells called ghorfa; these are superimposed and exist on two to six levels. The cells are arranged around a central courtyard, their outer walls creating a curtain wall, which can only be crossed by a single protected doorway. One Gasur may consist of several granaries, each of them organised around their courtyard. A family may own several ghorfas handed down from father to son who, while possessing his own keys, nevertheless entrusts the care of the granary as a whole to the custodian of the main entrance. The oral tradition dates these architectures back to the 14th century, although it is possible to suppose that some of them were built earlier. The most significant example of granaries in the west are the Gasur of Nalut, Kabaw and Tirit. A late model is located in Gasr El-Hag – a small village situated on the Jifarah not too far from Jadu.

Besides the communal granary, the region possesses another unique type of architecture: the cave. Artificial cavities were used historically for different functions: in the eastern area of the plateau, they served for human habitation, while the above-ground buildings were restricted to serving as store houses, schools and mosques. In the strictly Berber regions near Iefrin, Jadu, Kabaw and Nalut, a more extensive use was made of above-ground dwellings for habitation and of underground caves for mosques, olive presses, stores, and libraries. Two main factors are responsible for this: firstly, the harder rock and soil made the cave habitations smaller; secondly, the density of the population was far less in the west and building stone was more abundant.

The "taddart" – which means "the place where I live" – is the traditional above-ground house; in its simplest version, it consists of a quadrangular room in which a single opening simultaneously serves to ventilate, illuminate and provide access. The room is covered by a vault and less frequently by a flat roof. This type is generally constructed out of roughly worked limestone boulders bound together by gypsum mortar. The walls which form the rectangular rooms are quite thick to provide a fresh environment in hot weather; the walls sometimes consist of rows of niches and smaller arches, creating a structure which provides the base for construction of vaulted spaces.

Flat roofs are built using a series of small, repetitive, parallel beams made of olive and palm wood which support several layers of a waterproofing mortar. The difference processes involved in the vertical or horizontal multiplication of this unit produce more complex types in which a precise separation between domestic and work-related activities exists. The final stage of this process is the courtyard house: this type of house is built when there is a small space limited by walls in front of a mono-cellular unit which becomes a reference for the addition of different rooms [11] (Fig. 4).

Fig. 4. The old village of Am Safar not far from Kabaw.

6 Conclusions

This study has described the landscape, as well as the urban and architectural elements of a specific area historically occupied by Ibadi Berber groups; the paper has also provided an initial framework which is useful for comparing and defining the relationship between the religious, cultural and formal aspects of these three regions. The major aspects to have emerged from this analysis were both the theme of "defence" and "physical isolation" - two characteristics essential for protecting and maintaining the identity of a culture which is still alive. The traits of this culture were historically determined by the peculiar position of these areas along with the religious and political events which spread the Kharijism and Ibadi doctrines across North Africa. These new schools, which asserted the complete equality of all Muslims regardless of ethnic origin, influenced the urban and architectural forms of these regions, defining a new lexicon in which technical solutions are integral constituents of the buildings and the inhabitants' lifestyle. Studying this language, which entirely embodies the concept of sustainability, could result in a fundamental level of awareness to be used in the construction of a model designed to advance contemporary urban design. The adaptation of these ideas to existing settlements could represent an opportunity to develop new directions in sustainable and original techniques, filtered through local knowledge and industrial innovations.

References

1. Horden, P., Purcell, N.: The Corrupting Sea: A Study of Mediterranean History. Blackwell, Oxford (2000)
2. Lewicki, T.: al-Ibāḍiyya. In: Bearman, P.J., Bianquis, T., Bosworth, C.E., van Donzel, E., Heinrichs, W.P. (eds.) Encyclopaedia of Islam, 2nd edn. Brill, Leiden (1993)
3. Rouvillois-Brigol, M.: Mzāb. In: Bearman, P.J., Bianquis, T., Bosworth, C.E., van Donzel, E., Heinrichs, W.P. (eds.) Encyclopaedia of Islam, 2nd edn. Brill, Leiden (1993)
4. Di Tolla, A.: Ibāḍī 'aqīda-s in Berber. In: Francesca, E. (ed.) Ibadi Theology. Rereading Sources and Scholarly Works. Georg Olms Verlag, Zürich (2015)
5. Prevost, V.: Les mosquées ibadites du Maghreb. Revue du monde musulman et de la Méditerranée **125**, 217–232 (2009)
6. Gerber, A.: Le Corbusier et le mirage de l'Orient. L'influence supposée de l'Algérie sur son œuvre architecturale. Revue du monde musulman et de la Méditerranée **73**(1), 363–378 (1994)
7. Etherton, D.: Concentric towns - the valley of mzab. In: Lewis, D. (ed.) Growth of cities. Wiley-Interscience, New York (1971)
8. Despois, J.: Djarba. In: Bearman, P.J., Bianquis, T., Bosworth, C.E., van Donzel, E., Heinrichs, W.P. (eds.) Encyclopaedia of Islam, 2nd edn. Brill, Leiden (1993)
9. Polimeni, B.: The cities of the libyan nafusah mountain: type of dwellings and urban settlements. In: Dolkart, S., Al-Gohari, O.M., Rab, S. (eds.) Conservation of Architecture, Urban Areas, Nature & Landscape: Towards a Sustainable Survival of Cultural Landscape, vol. 1, pp. 1–16. The Center Csaar for The Study of Architecture, Amman (2011)
10. Norris, T.: Cave habitations and granaries in tripolitania and tunisia. Man **55**, 82–84 (1953)
11. Polimeni, B.: Strutture insediative nella Libia Occidentale: Il caso del Jebel Gharbi. In: Giovannini, M., Prampolini, F. (eds.): Spazi e culture del Mediterraneo 3- Ricerca PRIN 2007-2009, pp. 461–473. Iiriti, Reggio Calabria (2011)

Conservation, Enhancement and Resilience of Historical and Cultural Heritage Exposed to Natural Risks and Social Dynamics

Marco Vona[✉], Benedetto Manganelli, and Sabina Tataranna

University of Basilicata, 85100 Potenza, Italy
marco.vona@unibas.it

Abstract. The recent earthquakes have highlighted a significant economic losses and a low resilience of communities, particularly in historic centers of the town. These effects are mainly due to the high vulnerability of buildings and the correlated economic system. The consequences of the long recovery time on social-economic-management issues are often dramatic. This study proposes an economic and social sustainability approach for the conservation and enhancement of historical and cultural heritage exposed to natural risks and social dynamics. This approach, which develops in three phases, is based on the part that here is detailed on the construction of different quantitative measures of community resilience. In this way, integrated retrofitting interventions and mitigation strategies should be considered and evaluated in order to define the most effective action plan to maximize the resilience.

Keywords: Historical centers · Seismic communities resilience
Integrated retrofitting approach

1   Introduction

The development or the decay of the historical centres is the consequence of natural events or socio-economic phenomena [1, 2]. The vulnerability of historical sites to these events is generally higher than the more recent neighborhoods. In the retrofitting strategies, the requirements of historic centers are often related to social and economic changes (effects of the delocalization of the population, effects of the tourism). In Italy and in other countries with similar real estate assets, several historic centers or their significant part have become ghost towns after natural events. Even in these conditions, the historic centers often remain unique and fascinating places, such as to become a tourist attraction capable of generating significant economic growth in the surrounding area. Their management reuse and development must be based on their vulnerability reduction. The role of the vulnerability of buildings on the resiliency of the cities is the fundamental critical aspect. Consequently, the research will be carried out considering the fundamental role of the resilience of the cities in risk mitigation and government.

The quantification of the expected monetary consequence is the first step in order to define new and more effective risk mitigation strategies [3]. In fact, retrofitting

© Springer International Publishing AG, part of Springer Nature 2019
F. Calabrò et al. (Eds.): ISHT 2018, SIST 101, pp. 426–433, 2019.
https://doi.org/10.1007/978-3-319-92102-0_45

priorities based only on the structural problem of buildings are incorrect. It neglects several important issues such as the environmental, economic, financial losses consequence.

Prioritization of the mitigation strategies is a fundamental topic especially with regard to the limited economic resources and their allocation on national, regional or sub-regional territories [4]. The economic resources allocation should be defined through strategies based on a strongly multidisciplinary framework.

Moreover, the economic impact of vulnerability historic centers should be a matter of primary interest for public administrations, private and insurance companies, banks, owners, professionals, despite operating at different territorial levels, with different objectives. The quantification of the expected environmental and monetary consequence of the vulnerability, in terms of the probable cost of repairing, plays a key role in defining new and more effective risk mitigation strategies.

A new and innovative approach, analysis and evaluation tools will be defined and applied to quantify the direct and indirect economic effects and environmental effects resulting from the conservation and management of the historic centers.

2 Proposed Methodology to Evaluate the Economic Feasibility

The proposed approach would like to become a support tool for decision–makers in planning risk mitigation strategies able to minimize the environmental and economic consequences in spatial and temporal distribution of economic resources. Moreover, it should be noted that the economic value attributed to seismic security already plays an essential role in the process of price formation in the real estate market. If this policy becomes compulsory, in the form of a certain defensive expenditure (insurance premium), it would be a strong incentive for the promotion and the development of private risk mitigation initiatives.

The risk mitigation model develops sustainably the demand for greater safety of historic centers. Moreover, the life's improvement must be satisfied. This must be met without exceeding the load capacity of environmental ecosystems, by introducing environmental issues (ecosystem safeguard) and the economic and social sustainability of activities and investment. In this way historic centers and their communities will be more and more resilient.

First, the evaluation of risks devoted to their mitigation must be based on the collection of available data, processed and homogenized through a Geographic Information System (GIS). Data, procedures and models must be defined and combined using new methods and techniques to manage of studied areas.

Historic centers must be investigated concerning their vulnerability to natural hazards. The definition of the characteristics of the historical buildings must be made through methods able to combine field data collection techniques with imaging techniques. The risk analysis must always converge in the GIS platform following innovative procedures and models. Results must be incorporated into new management models based on the concept of resilience of communities. Specifically, quantitative resilience models must be defined and implemented for the management of areas and buildings.

The proposed approach aims at a significant advancement of procedures and methodologies for risk mitigation and management of the historic centers. The approach allows a deeper comprehension of the critical elements based on past experiences and thus will allow the improvement of mitigation strategies. The multi-disciplinary approach involves several specialization areas: smart technologies for Cultural Heritage and tourism, urban and socio-economic aspects, structural engineering, seismic hazard.

This issue is extremely topical, especially after the latest natural events (for example earthquakes). The damage on cultural heritage has been generating big losses economic, to the regional and national tourism systems and in the long terms the loss of the same identity of communities.

Particularly important is the economic and social sustainability approach. The sustainability is understood as the ability of the proposed approach to generate security, growth, income and work with preventive measures for historic centers rather than wait for the depressive effects of the catastrophic events. Analysis in different historic centers can allow for a much deeper understanding of the phenomena and can therefore allow more appropriate mitigation measures, especially prevention, and management strategies. The proposed methodology is developed in the three phases.

I   Part. Particularly important in this phase is the access to data and information for the next calibration of the proposed models. The investigated areas are firstly characterized using historical data in order to know the past effects of the real hazard. The definition of the structural types of the historical buildings in the investigated centres has been made through methods able to combine field data collection techniques with imaging and satellite techniques.

II Part. The second part develops and makes the methodologies proposed in the project operational and applies them appropriately to the different contexts studied. Historic centers located in the area of the research have been investigated about their vulnerability to all the considered critical events. The vulnerability must be evaluated using innovative simplified methods.

III Part. The third part is dedicated to the application of the models, methods and procedures proposed. They will take in account the degradation condition in an integrated way in order to obtain an integrated model for the ordinary and extraordinary management strategies and evaluation of the risks.

In this study, the first step and part of second step of the proposed approach are applied. To this aim, two case studies are considered and a real application of the developed methodologies are reported. The case studies are: the ghost village of Romagnano al Monte (SA), the small historic center of Miglionico (MT).

The first has an incredible feature. After Irpinia Earthquake the town was widely damaged but not destroyed. It is "frozen" in the 1980 post-earthquake condition. Consequently, it provides the opportunity to investigate about the main goal of the study. The historic center of Romagnano al Monte could be an interesting case study to define new strategies for reusing Italian ghost towns, particularly in the central and south Apennines: cultural heritage has been deserted, historical buildings are considered as ruins, and evidence of the past that is likely to be lost must be saved.

2.1 New Concept for Re-Use of Ghost Towns

Romagnano al Monte [5] currently represents a unique case and could be considered as an interesting open-air laboratory. Re-use does not seem a viable option based on the significant changes in the life style the community of Romagnano al Monte. Moreover, a significant amount of economic resources should be used to retrofitting the buildings to meet modern standards and seismic codes. On the contrary, the strategy for re-use could be based on a new tourism system. In this way, the current state of the historic center should be enhanced as it is. Obviously, this approach considers the maintenance and preservation of the village as it is, however it must guarantee the total safety of users. For this reason, the first part is knowledge of the buildings.

Aerial images, obtained from with Unmanned Aerial Vehicle (UAV), and photographic interpretation techniques are used. In this way, accurate information on building dimensions and their damage state have been obtained using strongly limited economic resources. In this study, the emphasis is given to the accuracy of dimensional characteristics and damage reconstruction considering the improvements that are possible in future development (Fig. 1).

Fig. 1. "Real" view of the Romagnano al Monte village and 3D model from UAV data processing.

For a reliable assessment and retrofitting design of the structures, knowledge of structural-geometrical characteristics, traditional materials and construction techniques is required. The main goal of the re-use proposal is preserve, retain and protect the historic center in its current condition. Structural interventions will be defined only to guarantee the safety of the site but not of the buildings. The buildings and the touristic paths will maintain the integrity and authenticity of ruins. Finally, based on the defined paths, the Economic Feasibility will be measured. The Pay Back Time of the investment one will be is evaluated considering different scenarios and potential investors.

2.2 New Resilience Model for Historic Center

The second case is Miglionico village in the province of Matera, Italy (Fig. 2). It is considered a typical Italian historic center and it is investigated in order to identify the optimal interventions to increase the resilience of community.

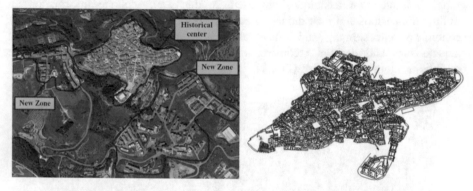

Fig. 2. "Real" view of the Miglionico village and first elaboration of 3D model.

In this case, the development of new strategies for the enhancement of the historic center must be based considering the inhabitants and social-economic issues. For this reason, the proposal is based on the new resilience model. In this proposal, observing on the past earthquakes data, the effects of a seismic risk of the existing buildings on a community's resilience have been evaluated by an integrated approach. The seismic performances of building are evaluated together with the economic and time consequence of seismic vulnerability based respectively on Repair Cost (RC) and Repair Time functions. The approach provides an economic quantification of the real community's resilience, and could be an effective support tool in definition of new effective seismic risk mitigation strategies a synthetic measurement of the community's resilience has defined based on the buildings analysis. To obtain the above goal, the seismic vulnerability has been investigated. Based on typological survey, the first vulnerability distribution has been obtained following the common approach [6].

The vulnerability classes have been defined based on the Damage Probability Matrices (DPMs) [6]. They are based on the vertical and horizontal structural types and their combination and variation inside and along the height of the buildings: the most vulnerable vertical/horizontal combination have been considered. Seismic-resistant and retrofitted buildings have been considered as the lower class of vulnerability. In order to better evaluate the actual vulnerability classes, other work and in field survey are in progress. The statistical distributions of the vulnerability classes are reported in Fig. 3.

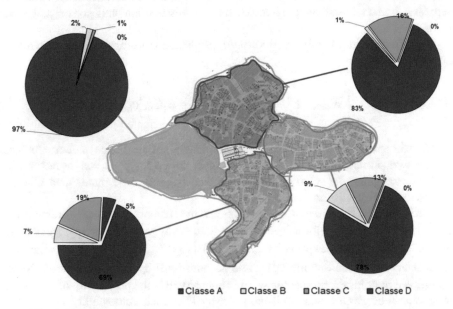

Fig. 3. Statistical distributions of the vulnerability classes in Miglionico village.

The damage scenario can be defined in order to obtain the Repair Cost (RC) and Repair Time functions; then, the resilience index can be defined. In Fig. 4 a qualitative trend of a community's resilience function is reported. The blue area is only referred to the residential buildings stock.

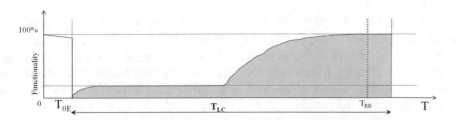

Fig. 4. Qualitative seismic resilience trend for the residential buildings stock.

Three different parts can be identified: A rapid return to functionality in the short term, (a few days immediately after the seismic event), linked to the emotive reaction of the citizens, for the recovery of the functionality of usable and non-damaged buildings and to the positive effects of the reactivation of services on the buildings.

A pseudo-horizontal step linked to the planning and implantation of preliminary activities for the reconstruction process (during this phase, the functionality of the system of residential buildings does not undergo substantial variations). Lastly, an increasing branch based on progressive funding distribution and consequent repair activities.

The resilience index only referred to the residential buildings stock of community can be expressed as:

$$R_{index}(I) = \sum_{class=1}^{n} \left\{ W_{class} \cdot \left(1 - \sum_{i=1}^{n} \frac{E[T_{RB}|C_{r,r|I}]}{T_{LC}} E[C_{r,r}|d_{l,i|I}] P[d_l = d_{l,i}|I] \right) \right\} \quad (1)$$

The resilience index is based on a combination of resilience indexes for each building type presented in a housing system area. For each type, a weight factor W_{type} is used in weighing the relative importance of a building type rather than others in relation to their quantity in a community.

The index is based on a convolution between the state of the damage (probability of occurrence $P[d_l = d_{l,i}|I]$) for a specific seismic intensity, corresponding expected ratio cost $(E[C_{r,r}|d_{l,i|I}])$ and expected ratio time $(E[T_{RB}|C_{r,r|I}]/T_{LC})$. The community's functionality losses are directly taken into account by $P[d_l = d_{l,i}|I]$ based on the vulnerability of buildings for a specific seismic intensity. It is the probability of obtaining a damage level d_l given a macroseismic intensity I for each vulnerability class. Generally, in this study $d_l = L_d$ is the damage levels as provided in the EMS-98, ranging between 0 and 5 ($L_d = 0$ means total absence of damage, while $L_d = 5$ means total destruction of the building). Moreover, in this study $P[d_l = d_{l,i}|I] = DPM[Class, L_d|I]$ are the Damage Probability Matrix adopted in other work in similar case [6]. For scenario analysis, earthquakes are provided in a deterministic form, e.g. by referring to the maximum credible or the most probable earthquake. Therefore, in applying the DPM approach, just one macroseismic intensity value is used to prepare building damage scenarios.

$E[C_{r,r}|d_{l,i|I}]$ is referred to the required economic resources needed to restore the initial conditions and/or any possible improvement. Finally, $E[T_{RB}|C_{r,r|I}]/T_{LC}$ relates to the expected value of the building recovery time (influenced by the level of damage to a building, difficulty in work activities, available economic resource) and the control time. The economic resources available influence both the form of restoration and the time of restoration and control.

3 Conclusion

In this work a new approach has defined in order to improve the conservation and enhancement of historical cultural heritage exposed to natural risks and social dynamics. The proposed method is strongly quantitative. Two case studies are discussed. They can be considered two different and extremely case for historic centers. First is a ghost town which need to developed and use for only touristic goals. Second case is a typical historic center of the mountain village of southern Italy. In this case, the development, use, and re-use of the historic center is based on new quantitative seismic resilience of communities. The procedure is based on three phases. Although limited to the first part, a first application of the procedures has been carried out in order to obtain the base information to apply the procedure in operational way. In fact, the model is based on a numerical evaluation of the performance of buildings, repair or retrofitting cost [3] and related time. Each part could be defined at a different level of accuracy but each part is defined in a quantitative way. For this reason, the reported results play the fundamental role on effectiveness and reliability of the proposal. About this topic, following the proposed approach, the following results can be obtained:

- accurate risk - time - costs curves;
- optimal allocation of resources;
- optimal solution for effective repair and retrofitting strategies.

References

1. Dolce, M., Di Bucci, D.: Comparing recent Italian earthquakes. Bull. Earthq. Eng. **15**(2), 497–533 (2017)
2. Vona, M., Harabaglia, P., Murgante, B.: Thinking about resilience cities studying Italian earthquake. Urban Des. Plann. **169**(4), 185–199 (2016)
3. Vona, M., Mastroberti, M., Manganelli, B.: Novel models and tools to evaluate the economic feasibility of retrofitting intervention. In: COMPDYN 2017, 6th ECCOMAS Thematic Conference on Computational Methods in Structural Dynamics and Earthquake Engineering Rhodes Island, Greece, pp. 15–17, June 2017
4. Vona, M., Anelli, A., Mastroberti, M., Murgante, B., Santa-Cruz, S.: Prioritization strategies to reduce the seismic risk of the public and strategic buildings. Disaster Adv. **10**(4), 21–34 (2017)
5. Vona, M., Cascini, G., Mastroberti, M., Murgante, B., Nolè, G.: Characterization of URM buildings and evaluation of damages in a historical center for the seismic risk mitigation and emergency management. Int. J. Disaster Risk Reduct. **24**, 251–263 (2017)
6. Dolce, M., Kappos, A.J., Masi, A., Penelis, G., Vona, M.: Vulnerability assessment and earthquake scenarios of the building stock of Potenza (Southern Italy) using the Italian and Greek methodologies. Eng. Struct. **28**, 357–371 (2006)

A Multi-criteria Approach to Support the Retraining Plan of the Biancavilla's Old Town

Maria Rosa Trovato[✉] [ID]

University of Catania, 95124 Catania, Italy
mrtrovato@dica.unict.it

Abstract. The historic town centre of our cities may be considered as a stratified and interconnected aggregate of values. Identifying these values for some of the possible configurations of components for a historic town centre is a complex task. This study proposes a cognitive, interpretive and evaluative model for the evaluation and choice of action in the redevelopment plan for a historic town centre, in order to locate amongst all possible redevelopment alternatives, those that best meet the prerequisites of sustainability. In particular, this study analyses the actions of the redevelopment plan for the Biancavilla's historic town centre. The instrumental operational model for evaluation is of a multi-criteria type and in particular uses a MAUT approach. This multi-criteria approach allows for a ranking of the analysed alternatives on the basis of a set of criteria which take into account the sustainability of the interventions.

Keywords: Old town · Retraining actions plan · Theory of the systems
Multi-criteria approach · Multiple Attribute Utility Theory (MAUT)

1 Introduction

The historic town centres of our cities may be considered as a stratified and intercon-nected aggregate of values where it is often difficult to identify. Identifying these values for some of the possible configurations of components of a historic town centre is a complex task. The recognition of these values is instrumental in order to quantify and qualify all the implemented actions in the historic town centre. The values are connected on the one hand, to fairness among the stakeholders of the process, including the socio-economic and political-administrative ones, but also, potential stakeholders, such as future generations; on the other hand, are connected to the historical, cultural, urban, architectural and environment values. These values are related both to the recognition of the objective and subjective values of the historic town centre and to the identification of the present value of knowledge and the historical and cultural identity, as a means of enhancing territorial competitiveness. The different configurations for the historic town centre defined by the implementation of redevelopment processes for it, must be capable of ensuring sustainable development between the all town' components [1, 2]. This study proposes a supporting model for the evaluation and choice of action of the redevelopment plan for a historic town centre, in order to locate amongst all possible

© Springer International Publishing AG, part of Springer Nature 2019
F. Calabrò et al. (Eds.): ISHT 2018, SIST 101, pp. 434–441, 2019.
https://doi.org/10.1007/978-3-319-92102-0_46

redevelopment alternatives, those that best meet these prerequisites. In particular, this study analyses the actions of the redevelopment plan for the Biancavilla's historic town centre [3]. This study proposes the MAUT approach (Multi-Attribute Utility Theory). This multi-criteria approach allows for a ranking of the analysed alternatives on the basis of a set of criteria which take into account the sustainability of the interventions.

2 Methods

2.1 The Cognitive, Interpretive and Evaluative Model to Support the Redevelopment Plan for Historic Town Centre

The historic town centre is a complex and dynamic system which can be represented using a cognitive, interpretative and evaluative model (Table 1).

Table 1. The cognitive, interpretive and evaluative model to support the redevelopment plan for historic town centre.

The old town's sub-systems	Dissipative structures		
	Autopoietic organizations		
Systemic features	AACE	SE	AP
AACE	The matrix of the features of the old town's system and of the relationships between the its subsystems		
SE			
AP			
AACE	Artistic, Architectural, Cultural and Environmental subsystems		
SE	Socio and Economic subsystem		
AP	Political and Administrative subsystem		

This is a general model which can be used to support the evaluation of some plan actions [4]. It is an operational model which may be interfaced with a multi-criteria model, from those, that have been developed in the field of MCDA. The cognitive, interpretative and evaluation model was developed by F. Rizzo, using a specific economic approach, namely, that of New Economy [5]. As a background to this economic, evaluative and sociological approach, the theory of the systems the theories of German sociologist N. Luhmann (his approach to social systems) [6], the theories of the two neurobiologists Maturana and Varela [7, 8], and in particular the genetic epistemology ("every knowledge is action and every action is knowledge") [9], the theories of Prigogine (the dissipative structures and the epistemology of complexity) [10]. According to this approach [11], the city is a "unit difference between the human, built and natural environment". This definition can be extended to each sub system of the city and then to the historic town centre. In particular, the historic town centre can be characterise by the following subsystems: artistic, cultural, architectural, environment and political-administrative. Each sub-system presents the same characteristics of the system. The system's operation as dissipative structure is a prerequisite to define a new order in the direction of sustainable development for the system.

2.2 The MAUT Approach to Support the Process of Regenerating the Historic Town Center

In the MCDA context (Multiple Criteria Decision Analysis) have been proposed several algorithms to support the decision-maker in the identifying the best alternative among a set of alternatives. In particular, in relation to the proposed general model is possible to recall a particular extension of the MAUT [12, 13]. The Multi Attribute Utility Theory is based on the main hypothesis that every decision maker tries to optimize, consciously or implicitly, a function which aggregates all their point of view. So, the decision maker's preferences can be represented by a function, called the utility function U. The utility function is a way of measuring the desirability or the preference of objects, called alternatives. The utility function is composed of various criteria which enable the assessment of the global utility of an alternative [13]. Using an extension of the MAUT approach as well as has been proposed by Matarazzo [14], it is possible to switch from a functional representation to a matrix representation. The element $x_{ii}(x_{ii} > 0 \, for \, i = 1, 2, \ldots, n)$ of the matrix X indicates the evaluation, i.e. the level of the action degree that is attributed to factor (x_i), that corresponding to the i-th row and i-th column. In particular, the action degree of the factor is assumed equal to $x_{ii} = 1$, when it has a level considered "normal", while, is taken between 0 and 1 when has a below level than considered normal and $x_{ii} > 1$ when has a higher level than considered normal. Each element $(x_{ij} > 0 \, for \, i \, different \, from \, j; i = 1, 2, \ldots n \, and \, j = 1, 2, \ldots n)$ indicates the degree of influence exerted by factor (x_i) corresponding to the $i - th$ row on that (x_j) corresponding to the $j - th$ column. The element x_{ij} can assume value equal to 1 when the factor (x_i) has a neutral interaction with the factor (x_j), value > 1 if the interaction is of the exalting type and value < 1 if the interaction is reductive. From the matrix X can be obtained indications concerning the problem under consideration. In fact, it is possible to calculate the complex degree of action of the $j - th$ factor obtained in the following way:

$$V_j = x_{1j} \cdot x_{2j} \cdot \ldots x_{jj} \cdot \ldots x_{nj} = \sum\nolimits_{i=1}^{n} x_{ij} (j = 1, 2, ..n) \qquad (1)$$

Obtained all the values of the complex degrees of action for the n factors $V_1, V_2, \ldots, V_j, \ldots, V_n$ is necessary to establish a weighting factor in following way:

$$\lambda_j (\lambda_j \geq 0 \, for \, j = 1, 2, \ldots n; \sum\nolimits_{j=1}^{n} \lambda_j) \qquad (2)$$

It expresses the relative appreciation of each considered factor of the obtained level V_j previously. By performing the sum of the products of the values V_j for the respective weights λ_j can be achieved V, namely, the degree of total appreciation :

$$V = \lambda_1 V_1 + \lambda_2 V_2 + \ldots + \lambda_n V_n = \sum\nolimits_{j=1}^{n} \lambda_i V_i \qquad (3)$$

This approach is used as a tool to support the decision made about the various design alternatives for the historic town centre redevelopment plan. In particular, all the possible alternatives are considered, including the current state and the future state without the project.

3   Application

The study proposes a model to support the evaluation and selection of different redevelopment plans for the historic town centre. The proposed model to support the evaluation and the choice of alternatives has been implemented for the redevelopment of the Biancavilla's historic town centre, a small town in the province of Catania. In particular, the area which is the object of the redevelopment action, is that of a macro block, which is located right in the heart of the historic town centre. This block consists of an aggregation of buildings from the late 700' [15]. It has developed around an open area which had been allocated as a private botanical garden. The proposal to draft a redevelopment project for the historic town centre has been initiated by the study of the proposed reutilization of the Palazzo Portale. On the one hand the aim of the project was to create a museum to house a collection of archaeological exhibits (400 pieces) and the Hortus Siccus Plantarum Sicularum [16], a garden of some indigenous Sicilian plants, which had been grown by the botanist Salvatore Portal, Bishop of Biancavilla in February 1852. On the other hand, the project wanted to redevelopment the free area, inside the block, where the old botanical garden was allocated (in which more than 2080 plants were present, of the indigenous and exotic type, and that had been grown since 1820, making it older than Catania's public botanical garden, which dates back to 1858). The Palazzo Portale was built in 1908, it is located in the heart of the Biancavilla's historic town centre, namely the main square of the town. This square has always been identified as a privileged place for the social, political and cultural relationships. This building is attributed to the Milanese architect Carlo Sada, who can be considered as a major architect for Catania in the period lasting from the nineteenth to the twentieth century.

The design alternatives provided in the redevelopment plan for Biancavilla's historic town centre are as follows:

the current status alternative or zero map (alternative 1);

the future without the project alternative (alternative 2);

the realization of a multi- storey car park project in the area where the old botanical garden was located, which is now an abandoned area to improve the mobility and parking areas to the historic town center; a separate evaluation for this project has been specifically was conducted in relation to the risk of encouraging the spread of fluoro-edenite, because the Biancavilla's area has been threatened by pollution from these fibers of asbestos, which are carcinogen (alternative 3);

the retraining of the Botanical Garden based on the original project and of the Palace Portal as museum in which to allocate the archaeological collection and the "Hortus Siccus Plantarum Sicularum" (alternative 4);

1. the retraining of the "Palazzo Portale" as above and the realization of the multi-storey car park project in the area of the old botanical garden (alternative 5);
2. the retraining of the "Palazzo Portale" as above and the realization of a project for some research laboratories in herbal and pharmaceutical fields, which are located in the ground floors of some buildings of the block and a new project for area with the old botanical garden as garden laboratory (alternative 6).

4 Results

The family of the considered criteria supporting the MCDA approach are those represented in the following table (Table 2). The chosen criteria take into account the cognitive, interpretive and evaluative model that has been introduced to represent the historic town centre. In particular, a curve of standard transformation has been constructed for each criterion and for each alternative.

Table 2. The criteria.

The criteria	
$AACE_1$	Pollution by fluoroedenite
$AACE_2$	Efficiency of the urban liquid waste disposal
$AACE_3$	Efficiency of the water system
$AACE_4$	Efficiency of the solid waste disposal
$AACE_5$	Architectural quality
$AACE_6$	Functional quality
$AACE_7$	Ability to enhance the architectural quality
$AACE_8$	Ability to increase the quality of the functional space
$AACE_9$	Ability to represent themselves
$AACE_{10}$	Ability to increase the sense of belonging and differentiation
SE_1	Ability to meet new needs, in addition to cultural ones
SE_2	Ability to create new cultural interests
SE_3	Ability to create new economic interests
SE_4	Quality of economic activities value
SE_5	Ability to increase the real estate market
AP_1	Representativeness of the public administration on the territory
AP_2	Ability to collect taxes
AP_3	Reception of the system
AP_4	Involvement of the private sector in the management of the public facilities

The curve of standard transformation allows for the location of the assumed value for each criteria for each considered alternative. Specifically, the evaluation of the criterion is expressed by using a standardized scale, allowing the comparison of the evaluation for different types of criteria (quantity or qualitative type) [17, 18]. In particular, the matrices have been determined for the different considered alternative (for example in the Fig. 1).

	AACE₁	AACE₂	AACE₃	AACE₄	AACE₅	AACE₆	AACE₇	AACE₈	AACE₉	AACE₁₀	SE₁	SE₂	SE₃	SE₄	SE₅	AP₁	AP₂	AP₃	AP₄
AACE₁	1,9	1	1	1	1	1	1	1	1	1	1	1	1	1	0,5	1	1	1	1
AACE₂	1	1,8	1	1	1	1	1	1	1	1	1	1	1	1	1,2	1	1	0,5	1
AACE₃	0,5	1	1,7	1	1	1	1	1	1	1	1	1	1	1	1,2	1	1	1,2	1
AACE₄	1	1	1	1,1	1	1	1	1	1	1	1	1	1	1	1,2	1	1	1,2	1
AACE₅	1	1	1	1	1,1	1	1,2	1	1,2	1,2	1	1	1	1	1,2	1	1	1,2	1
AACE₆	1	1	1	1	1,2	1	1	1,2	1	1	1	1,2	1	1,2	1	1	1	1,2	1
AACE₇	1	1	1	1	1	1	1,6	1	1,2	1,2	1	1	1	1,2	1	1	1	1,2	1
AACE₈	1	1	1	1	1	1	1,2	0,6	1	1	1	1,2	1	1,2	1	1	1	1,2	1
AACE₉	1	1	1	1	1	1	1	1	1,8	1,2	1	1	1	1,2	1	1	1	1,2	1
AACE₁₀	1	1	1	1	1	1	1	1	1	2,0	1	1	1,2	1,2	1	1	1	1,2	1
SE₁	1	1	1	1	1	1	1	1	1,2	1	0,5	1	1,2	1,2	1	1	1	1,2	1
SE₂	1	1	1	1	1	1	1	1	1,2	1	1	0,2	1	1,2	1,2	1	1	1,2	1,2
SE₃	1	1	1	1	1	1	1	1	1,2	1,2	1,2	1	1,7	1	1	1,2	1	1,2	1
SE₄	1	1	1	1	1	1	1	1	1	1	1	1,2	1	1,5	1,2	1,2	1	1	1
SE₅	1	1	1	1	1	1	1	1	1,2	1	1	1,2	1	1	0,6	1,2	1	1	1
AP₁	1	1	1	1	1	1	1	1	1,2	1	1	1,2	1,2	1	1,2	0,4	1	1,2	1
AP₂	1	0,5	0,5	0,5	1	1	1	1	1	1	1	1	1	1	1	1	0,9	1	1
AP₃	1	1	1	1	1	1	1	1	1	1	1	1,2	1,2	1	1,2	1	1	1,7	1
AP₄	1	1	1	1	1	1	1	1	1	1	1	1	1	1	1,2	1	1	1	0,1
Vᵢ	0,95	0,90	0,85	0,55	1,32	1,00	2,30	0,72	6,45	4,15	0,60	0,60	3,53	5,57	1,49	0,69	0,93	8,77	0,12
Weights	0,1	0,05	0,05	0,05	0,09	0,08	0,05	0,05	0,04	0,05	0,01	0,03	0,02	0,05	0,1	0,08	0,05	0,03	0,02
V	1,92																		

Fig. 1. The evaluation matrix for the alternative number 3.

The results show that the present condition, i.e. one in which intervention is absent has a value of the degree of total appreciation that appears to be comparable with that relating to the condition with the project. This is due to the fact that we are evaluating an intervention on a block that represents a fragment of the urban structure, which, although expresses some critical issues related mostly to parking areas, is already among the most qualified in the urban fabric of Biancavilla, in terms of architectural, historical, cultural and social qualification [19, 20].

The alternative that has achieved the highest score of the degree of total appreciation is the number 6 (Fig. 2), that provides the creation of Salvatore Portal's Museum, of a new research structure that is created by implementing a new project for the Botanical Garden and by the integration of this with some building units at ground floor level of the historic block, where the herbal and pharmaceutical research laboratories would be allocated. We can also see, that the alternative which involves the construction of Salvatore Portal's Museum and the Botanical Gardens on the basis of the original project, i.e. alternative 4 (Fig. 2), has recorded a better result than that involves the construction of the same museum and of the project for the multi-storey car park in the area of the old botanical gardens, i.e. alternative 5 (Fig. 2). The alternative which provides only the multi storey car park, representing a priority of the public administration, recorded the lowest rating, i.e. alternative 3 (Fig. 2).

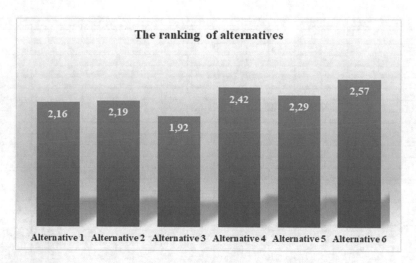

Fig. 2. The ranking of alternatives.

5   Conclusion

The purpose of this study was to identify a model to support the evaluation and selection of those actions of the redevelopment plan for Biancavilla's historic town centre which were capable of ensuring sustainable development. In this regard, the proposed model makes use of a theoretical approach and a more operational approach of reference which have been presented in Sect. 2.1 and 2.2. The final model has identified a ranking of the alternatives for the redevelopment plan of Biancavilla's historic town centre (Fig. 2). This ranking has highlighted some contradictions with the forecasting evaluation for the action plan. For example, the redevelopment of the Botanical Garden on the original project, which was initially seen as an opportunity to enhance has had a lower evaluation than the new project for the Botanical Garden; the project for the multi-storey car park that was considered a priority for the local administration was eventually recognized as dangerous because of the risk of pollution from fluo-roedenite. The proposed model takes account of the approach to the systems theory that has allowed to identify the values of the urban systems involved in the actions of the plan.

References

1. Della Spina, L., Scrivo, R., Ventura, C., Viglianisi, A.: Urban renewal: negotiation procedures and evaluation models. In: Gervasi, O., et al. (eds.) Computational Science and Its Applications, ICCSA 2015. LCNS, vol. 9157, pp. 88–103. Springer, London (2015)
2. Nesticò, A., Sica, F.: The sustainability of urban renewal projects: a model for economic multi-criteria analysis. J. Prop. Invest. Finance **35**(4), 397–409 (2017)
3. Giuffrida, S., Ventura, V., Trovato, M.R., Napoli, G.: Axiology of the historical city and the cap rate the case of the old town of Ragusa superiore. Valori e Valutazioni **18**, 41–55 (2017)

4. Trovato, M.R., Giuffrida, S.: The choice problem of the urban performances to support the Pachino's redevelopment plan. Int. J. Bus. Intell. Data Min. **9**(4), 330–355 (2014)
5. Rizzo, F.: Valore e valutazione. La scienza dell'economia o l'economia della scienza. FrancoAngeli, Milano (1999)
6. Luhmann, N.: Sistemi sociali: Fondamenti di una teoria generale. Il Mulino, Bologna (1990)
7. Maturana, H., Varela, F.: Autopiesi e cognizione, la realizzazione del vivente. Marsilio, Venezia (1985)
8. Maturana, H., Varela, F.: L'albero della conoscenza. Garzanti, Milano (1987)
9. Piaget, J.: L'epistemologia genetica. Laterza, Rome (2000)
10. Prigogine, I.: La complessità. Esplorazioni nei nuovi campi delle scienze. Einaudi, Torino (1991)
11. Rizzo, F.: Il capitale sociale della città. FrancoAngeli, Milano (2003)
12. Dyer, J.S.: MAUT. In: Figueira, J.R., et al. (eds.) Multiple Criteria Decision Analysis: State of the Art Surveys, pp. 265–295. Springer, Berlin (2005)
13. Ishizaka, A., Nemery, P.: Multi-Criteria Decision Analysis, pp. 81–104. Wiley, UK (2013)
14. Rizzo, F.: Economia del patrimonio architettonico-ambientale. FrancoAngeli, Milano (1989)
15. Bucolo, P.: Storia di Biancavilla. Edizioni Gutemberg, Adrano (1953)
16. Portal, S.: Catalogus Plantarum Horti Botanici Savatoris Portal Albaevillae in Sicilia. Tipografia Francesco Longo, Catania (1826)
17. Gabrielli, L., Giuffrida, S., Trovato, M.R.: Functions and perspectives of public real estate in the urban policies: the sustainable development plan of syracuse. In: Gervasi, O., Murgante, B., Misra, S., et al. (eds.) ICCSA 2016, IV. LNCS, vol. 9789, pp. 13–28. Springer, London (2016)
18. Gabrielli, L., Giuffrida, S., Trovato, M.R.: Gaps and overlaps of urban housing sub market: a fuzzy clustering approach. Green Energy and Technology, pp. 203–219. Springer (2017). Issue 9783319496757
19. Gabrielli, L., Giuffrida, S., Trovato, M.R.: From surface to core: a multi-layer approach for the real estate market analysis of a central area in Catania. In: Gervasi, O., Murgante, B., Misra, S., et al. (eds.) ICCSA 2015. LCNS, vol. 9157, pp. 284–300. Springer, London (2015)
20. Napoli, G., Giuffrida, S., Trovato, M.R., Valenti, A.: Cap rate as the interpretative variable of the urban real estate capital asset: a comparison of different sub-market definitions in palermo, Italy. Buildings **7**(3), 80 (2017)

Historical Cultural Heritage: Decision Making Process and Reuse Scenarios for the Enhancement of Historic Buildings

Lucia Della Spina$^{(\boxtimes)}$ (iD)

Mediterranea University, 89100 Reggio Calabria, Italy
lucia.dellaspina@unirc.it

Abstract. In the last twenty years, conservation policies of cultural heritage have become key policies of the European community. This is due to two main factors: the importance attributed to the use of heritage, and the need to support the role of cultural values. More specifically, heritage is seen as cultural capital and as a potential driver for tourism, while cultural values have a crucial role in shaping the identity of territories because of their intrinsic value and their value as an investment for the cultural, social and economic development. Choosing among different alternative scenarios of reuse, valorisation and conservation of unused cultural heritage is generally a complex decision-making process, given the multidimensional nature of the scenarios and the wide set of values they represent.

These decisions are not always consensual, also because the political choice on the potential use of the asset concerns the maintenance of its physical integrity, of its intangible values and of the economic development that can be achieved from its new functions.

In this context, the paper proposes the use of Social Multi-criteria Evaluation (SMCE), a multi-dimensional approach applied to a real case study that can support the choice of possible reuse scenarios.

Keywords: Cultural Heritage · Enhancement · Highest and Best Reuse
Stakeholders Analysis · Social Multicriteria Evaluation (SMCE)
NAIDE method

1 Introduction

Choosing among different alternative scenarios of reuse, valorisation and conservation of cultural heritage is generally a complex decision making process, given the multi-dimensional nature of the decisions and the wide set of values they represent. These decisions are not always consensual, also because the political choice on the potential use of the asset concerns the maintenance of its physical integrity, of its intangible values and of the economic development that can be achieved from its new functions. Since the 1990s the European Community has developed policies that focus more on the valorisation of heritage and its relations with communities and society. This has generated a long and intense debate on these issues, which sees cultural heritage as a crucial resource for the integration of the different dimensions of cultural, ecological,

© Springer International Publishing AG, part of Springer Nature 2019
F. Calabrò et al. (Eds.): ISHT 2018, SIST 101, pp. 442–453, 2019.
https://doi.org/10.1007/978-3-319-92102-0_47

economic, social and political development. Furthermore, in a context of increasing globalization, cultural heritage contributes to protect cultural diversity and sense of place, besides fostering dialogue, democratic debate and openness among cultures [1].

The idea of reusing and valorising cultural heritage [2] by making places accessible while respecting tangible and intangible values, seems to be an increasingly promising strategy to achieve balance between different needs such as: the preservation of existing buildings that require changes consistent with the requirements set by new uses; the conservation of the symbolic values of historical buildings; the achievement of sustainability principles; the community's commitment; the improvement of the territorial development processes. Therefore, the choice of intervening with Highest and Best Reuse should be supported by adequate analytical tools, which, during the evaluation process, can take into account feedbacks from a technical and social point of view. For that purpose, the Social Multi-criteria Evaluation (SMCE), developed by Munda [3–6], seems to be the most appropriate theoretical framework to support multi-value public policies in complex decision-making contexts, in which the legitimate interests at stake are different and often conflicting. This document proposes to apply SMCE to define the intervention of Higher and Best Reuse of historical cultural assets located in Calabria (Southern Italy).

This article is divided into 3 main sections: the first section illustrates the use of the methodological background used to solve the decision problem, with a focus on the use of SMCE and on the application of the discrete method *New Approach to Imprecise Assessment and Decision Environments* (NAIADE); the second section describes the application of the aforementioned methodology to the case study; the third section discusses the results and proposes future research lines.

2   Methodological Background

To address the complexity of the decision problem under review, an integrated assessment framework has been implemented in the present study. In particular, the methodological framework was structured according to a multi-methodological approach that has been organized in two main phases (Fig. 1):

The first phase, which is based on the Stakeholder Analysis (SA) application [7], was aimed at designing a series of alternative scenarios for the reuse of the asset [8]. The SA has allowed the identification of different categories of stakeholders: promoters, operators, and users. For each category the respective values, points of view and perceptions were identified. The analysis of points of view and perceptions has been structured by carrying out structured interviews with the stakeholder sample.

Furthermore, a panel of experts was organized to better define the decision problem. The group of experts also helped to determine the relevant attributes that needed to be considered in the evaluation model, which were used to structure the experimental design of alternative reuse scenarios. These scenarios will be presented in detail in Sect. 3.1 of this document.

The second phase is based on the development of a Social Multi-Criteria Evaluation [3] and is aimed at selecting the most effective alternative scenario, concentrating on the social actors involved in the decision making.

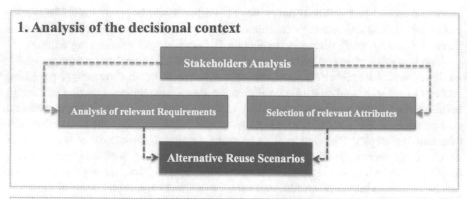

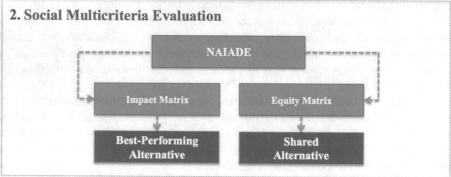

Fig. 1. The integrated evaluation framework

This study mainly dwells on the application of the SMCE. Consistently with the SMCE approach, and applying the NAIADE method (*New Approach to Imprecise Assessment and Decision Environments*) [3–6], it was possible to structure a multi-criteria and multi-group decision model in the fuzzy field, able to reduce the level of conflict and reach a certain degree of consensus.

2.1   Novel Approach to Imprecise Assessment and Decision Environments (NAIADE)

NAIADE (Novel Approach to Imprecise Assessment and Decision Environments) is a multicriteria evaluation discrete method developed by Munda [3, 9] that allows to use information that has been influenced by different types and degrees of uncertainty. The impact (or evaluation) matrix includes either crisp, stochastic or fuzzy measurements of the performance of an alternative with respect to an evaluation criterion, thus it is very flexible for real-world applications.

A specificity of NAIADE is the use of conflict analysis procedures to be integrated with the multi-criteria results. NAIADE can give the following information:

– ranking of the alternatives according to the set of evaluation criteria (i.e. technical compromise solution/s);

- indications of the distance of the positions of the various interest groups (i.e. possibilities of convergence of interests or coalition formations);
- ranking of the alternatives according to actors' impacts or preferences (social compromise solution).

The method involves two types of evaluation:

1. A **Multicriteria Analysis**, performed on the **Impact Matrix** (alternatives vs criteria), which is based on a comparison algorithm of the alternatives made up by the following steps:

   - completion of the criteria/alternatives (impact) matrix
   - pairwise comparison of alternatives using preference relations
   - aggregation of all criteria
   - ranking of alternatives.

   Following the NAIADE methodology [5], a semantic distance is used in order to compare the criteria values for the alternatives. It is possible to give a value using a qualitative evaluation expressed by pre-defined linguistic variables such as "Good", "Moderate", "Very Bad" and so on. The linguistic variables are treated as fuzzy sets defined in the 0–1 scale.

   The final result of the impact matrix is the ranking of alternatives based on the set of criteria preferences.

2. A **Social Impact Matrix,** which analyzes conflicts between different interest groups and, through an **Equity Matrix**, gives a linguistic indication of the interest group judgement for each of the alternatives. In this instance, semantic distance is also used to calculate the similarity indexes among interest groups. A similarity matrix is then computed starting from the equity matrix. The similarity matrix gives an index for each pair of interest groups i, j, of the similarity of judgement over the proposed alternatives. This index s_{ij} is calculated as $s_{ij} = 1/(1 + d_{ij})$ where d_{ij} is the Minkovsky distance between group i and group j which is calculated as follows:

$$d_{ij} = \sqrt[p]{\sum_{k=1}^{N} (S_k(i,j))^p} \tag{1}$$

where $S_k(i, j)$ is the semantic distance between group i and group j in the judgement of alternative k, N is the number of alternatives and p > 0 is the parameter of the Minkovsky distance [10, 11].

Lastly, through a sequence of mathematical reductions the dendrogram of coalition formation is built. It shows possible coalition formation for decreasing values of the similarity index and the degree of conflict among interest groups.

3 Case Study

In this study, the multi-methodological evaluation has been applied to Palazzo S. Anna, a public asset. Palazzo S. Anna is a prestigious historic building located in the village of Gerace (Italy), one of the most beautiful towns in southern Italy, renowned for its rich tangible and intangible cultural heritage. The building, located above the 'Bombarde' belvedere, used to be a monastic complex, whose origins seem to date back to the 14th century. Though the building underwent several modifications over time, the original building plan remained unvaried and is still visible.

As public financial resources keep being reduced, reviving unused building becomes an opportunity to foster economic development [12, 13], while improving the quality of life of local communities [14]. With regard to the analysis of the positional and intrinsic characteristics of the asset under evaluation, its potential reuse and alternative uses have been defined. The aim is to prevent its decadence and the abandonment of cultural heritage through the enhancement of economic and social resources and preservation of the architectural object from a historical and cultural point of view. Furthermore, the adoption of a conservative reuse approach requires the choice of reversible and compatible functions that could play a key role in strengthening local identity, traditions and local practices.

In order to support the design of reuse alternatives, the main categories of stakeholders with different levels of interest/power were analysed. Each stakeholder was analysed in respect to final users requirements, technical and functional aspects, and business and normative criteria, all of which to be fulfilled in the designing phase.

3.1 Alternative Scenarios

One of the main features of the evaluation framework for multidisciplinary criteria is that alternatives are designed by considering information from different sources [6] such as participatory process, technical interviews and so on.

In this application, alternative scenarios correspond to different hypotheses of reuse that have been elaborated by referring to requirements and preferences of the main stakeholder categories with different levels of interest/power: Superintendence (G1); Municipality of Gerace (G2); Tourism and trade sector (G3); Experts (G4); Local community (G5); Entrepreneurs (G6), according to the results of the Stakeholder Analysis (SA) [7].

More precisely, according to the SA, the most preferred attributes have been selected as fundamental elements of the design of the reuse project of Palazzo S. Anna [14].

With reference to Palazzo S. Anna, the alternative scenarios considered for reuse are the following:

Scenario 1. Exhibitions and Regional Food Tasting. This first scenario aims to enhance and promote the best of Calabrian culinary tradition. The building would be designed to showcase local culture and food and wine. The first floor of the building would house cooking workshops, as well as spaces for educational workshops or cooking classes. The panoramic terrace would be an outdoor space designed for tasting sessions and food stalls. The upper part of the building is conceived as a flexible space and would allow different uses (events, stalls for typical local products, festivals, or conference rooms).

Scenario 2. Accommodation Facility. This scenario involves the renewal of the current accommodation structure, which currently features 11 rooms, 25 beds, a restaurant with 110 seats, a pizzeria, and a meeting room maintaining the pre-existing internal functional distribution (see Fig. 2).

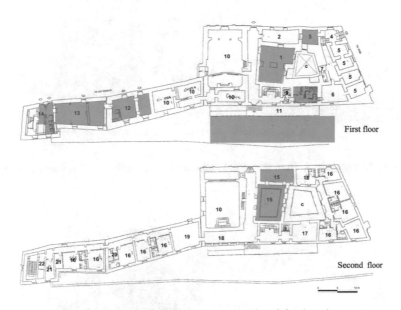

Fig. 2. Scenario 1. Exhibitions and regional food tasting

Scenario 3. Upgrade of the Accommodation Facility and Wellness Centre. The third scenario involves the upgrade the accommodation facility and the creation of a wellness centre. The number of beds will increase from 25 to 38, thanks to the construction of wooden mezzanines in some of the rooms. Outdoor gazebos with a capacity of 200 seats will be built on the panoramic terrace on the ground floor, in order to host banquets and events. The original restaurant area will be used exclusively as meeting room (see Figs. 3 and 4).

Scenario 4. Higher Education and Research Activities. The fourth scenario foresees the reuse of the palace for higher education and university research, included using the venue as headquarters of the "Unesco Centre on plans and models of cultural heritage management". The idea is to offer courses on valorisation and management of cultural heritage, courses on production techniques of local products, and on regional culinary culture. Additional spaces are dedicated to the exhibition and sale of typical local products (food stalls) and to laboratories and commercial kitchens. The reception spaces of the building on the upper floor are designed as spaces for conferences and public meetings.

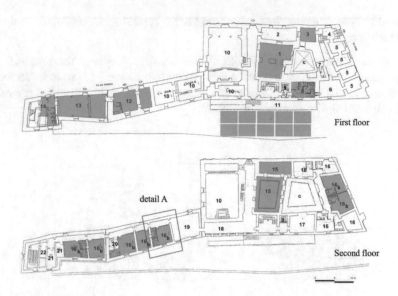

Fig. 3. Scenario 3. Upgrade of the accommodation facility and Wellness Centre

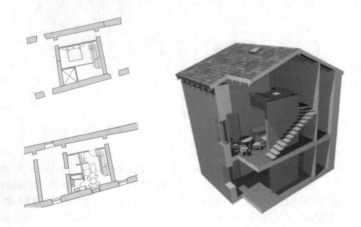

Fig. 4. Scenario 3. Detail A: wooden mezzanines

3.2 Multicriteria Analysis

Before applying the NAIADE method, the stakeholder analysis was studied to clarify values and preferences and to define the set of criteria for the comparison of the four reuse scenarios.

In a social multi-criteria domain, the criteria are not provided but should be defined in relation to the role and position of the stakeholder in the scenario in question. Therefore, a collaborative approach was assumed, and the stakeholders groups were involved in some brainstorming sessions to discuss and identify the main goals associated with conservation and enhancement of the asset [8].

With the help of a questionnaire followed by a collective brainstorming, a series of quantitative and qualitative criteria that represent the translation by the experts of stakeholder expectations have been defined [6, 8].

More specifically, the criteria are defined as follows: 1. Flexibility, the possibility of reusing the historic building for a different function according to changes in environmental and social systems over time; 2. Presence of Public spaces; 3. Events for residents of the village; 4. Invasivity, new installations and equipment required by the new function; 5. Cultural Promotion for the municipality of Gerace; 6. Investment costs; 7. Profitability during the management phase; 8. Target, population's needs met by the scenario.

Each of the 4 scenarios was evaluated on the basis of this set of 8 criteria, as shown by the Impact matrix (see Table 1) and a first technical classification was obtained by applying the NAIADE method.

Table 1. Multicriteria analysis. Impact matrix (Criteria/Scenario)

Criteria/Scenario	Scenario 1	Scenario 2	Scenario 3	Scenario 4
1. **Flexibility** reusing	More or Less Bad	Good	Good	More or Less Good
2. **Public spaces**	Bad	More or Less Good	Very Good	Moderate
3. **Events** for residents	More or Less Bad	Good	More or Less Good	More or Less Good
4. **Invasivity** new install	Very Good	Very Good	Good	Good
5. **Cultural Promotion**	More or Less Bad	Good	Good	More or Less Good
6. **Investment** costs	Very Good	More or Less Good	Moderate	More or Less Bad
7. **Profitability**	Very Bad	Very Good	More or Less Good	Moderate
8. **Target**	Bad	Good	More or Less Bad	More or Less Bad

From the evaluation of the scenarios it has emerged that the most preferred alternative from a technical point of view is Scenario 3: Upgrade of the accommodation facility and Wellness Center, followed by Scenario 1: Exhibitions and regional food tasting, Scenario 4: Higher education and research activities and lastly by Scenario 2: Accommodation facility (see Fig. 5).

Scenario 3	Scenario 1	Scenario 4	Scenario 2
(0,73)	(0,65)	(0,52)	(0,37)

Fig. 5. Multicriteria analysis results.

3.3   Social Impact Analysis

According to the NAIADE approach, a second matrix must be defined and is represented by the Social Impact Matrix, which is modelled on the evaluation expressed by
each stakeholder through the use of a questionnaire (see Table 2). Unlike the Impact
Matrix, which represents a technical translation independent of the stakeholders'
preferences, through the Equity Matrix social actors can evaluate each alternative using
linguistic variables. In general, linguistic variables are very useful to characterize
phenomena that are too complex to be described in quantitative terms and are a natural
representation of cognitive observations [5, 12, 13]. Fuzzy set theory provides a
framework for the development of approximate calculations of linguistic variables [15,
16]. In the case under examination, following the NAIADE methodology [5], a
semantic scale was used for the evaluation of alternatives, in which nine linguistic
judgments were considered: "Perfect", "Very Good", "Good", "More or Less Good",
"Moderate", "More or Less Bad", "Bad", "Very Bad" and "Extremely Bad".

Table 2. Social Impact Analysis. Equity matrix result.

Stakeholders/Scenario	Scenario 1	Scenario 2	Scenario 3	Scenario 4
G1 Superintendence	Good	Moderate	Very Good	More or Less Good
G2 Municipality of Gerace	More or Less Bad	Good	Very Good	Very Good
G3 Tourism & Trade sector	Moderate	More or Less Good	Very Good	Good
G4 Experts	Very Good	Moderate	Moderate	Good
G5 Local Community	Moderate	Moderate	Very Good	Good
G6 Entrepreneurs	Good	Good	More or Less Bad	More or Less Bad

Starting from the Social Impact Matrix, by using a distance function d_{ij} as conflict
indicator, for each pair of interest groups i and j, a coalition dendrogram can be
obtained (Fig. 6). The graph helps to visualize the proximity of the actors' objectives,
as in the case of the first coalition: the Tourism & Trade sector (G3) and Local
Community (G5), whose credibility is very high (0.8247). The interests of G3 are also
shared by the Municipality of Gerace (G2) and the Superintendence (G1), whose
preferences focus on the same alternatives (credibility index: 0.7564). On the other
hand, the Entrepreneurs (G6) and the Expert (G4) show a medium-high degree of
credibility but a significant distance. Both have the same goal: to improve local
economy and reduce costs, preferring scenario 1 and 2.

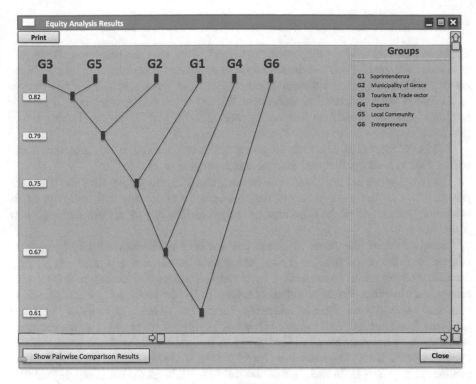

Fig. 6. Social impact analysis results. Dendrogram of coalitions.

4   Results and Discussion

In light of the results of the NAIADE application it is possible to state that from a technical point of view (see Table 1 and Fig. 5), the best performing solution is Scenario 3, followed by Scenarios 1 and 4 and lastly by Scenario 2. From the point of view of the analysis of social conflicts (see Table 2 and Fig. 6), it seems that Scenario 1 is the alternative that generates the maximum conflict as it is not much appreciated by the coalition of G3 actors (Tourism and Commerce sector), G5 (Local community) and G2 (Public Administration of Gerace) which has a high level of credibility (0. 7564). Furthermore, it is also possible to note that Scenario 2 is not appreciated by most social actors, while Scenarios 3 and 4 are classified in a medium-high position by many stakeholders.

In summary, it seems correct to state that Scenario 3 is the "most defensible" project from both a technical and a social point of view, while the other scenarios either maximize social conflict (Scenarios 1 and 2) or are not effective from a technical point of view (Scenario 4).

5 Conclusions

This research proposed an integrated framework to support the decision-making process related to the renewal and enhancement of an unused historical asset located in the historic village of Gerace, Calabria (southern Italy). In particular, in this study the use of the Social Multi-criteria Evaluation (SMCE) was developed by applying the discrete method New Approach to Imprecise Assessment and Decision Environments (NAIADE).

NAIADE was used to select the reuse scenario for Palazzo S. Anna that could reduce the level of conflict and reach a certain degree of consensus.

These decisions are not always consensual, also because the political choice on the potential use of the asset concerns the maintenance of its physical integrity, of its intangible values and of the economic development that can be achieved through new functions.

In this context, the paper proposes the use of a multi-dimensional approach to support the choice of possible reuse scenarios applied to a real case study, in a decision-making arena characterized by a multiple subjects with different legitimate values and objectives linked to cultural heritage [17]. The methodology used is able to take into consideration different technical criteria and values and goals of the actors involved. The choice of indicators, their definition of political priorities or the choice of benchmarks is not only a technical matter, it is mainly a socio-political issue. For this reason, a participatory approach is strongly recommended to ensure the quality of the evaluation process [3]. In fact, one of the main strengths of the proposed approach is the possibility of structuring the assessment as a social learning process where decision maker and the stakeholders learn the problems while solving them [18].

Moreover, the method creates a common pool of knowledge for decision makers [19], local communities and tourists, thus ensuring the strengthening of the social capital.

The main advantages of this procedure are:

- the information obtained are readily useful for political purposes
- avoidance of compensation between the different dimensions because the indicators are not aggregated
- transparency of the whole process.

Despite the congruence of the results, future research studies will consider the development of a specific sensitivity analysis on the technical ranking, as well as of an analysis on the degree of credibility of the coalitions in order to verify the model and the results obtained and provide the decision maker with sounder political advice.

References

1. Council of Europe: Framework Convention on the Value of Cultural Heritage for Society (2005). www.coe.int
2. Plevoets, B., Van Cleempoel, K.: Adaptive reuse as a strategy towards conservation of cultural heritage: a literature review. WIT Trans. Built Environ. **118**, 155–164 (2011)

3. Munda, G.: Social multi-criteria evaluation (SMCE): methodological foundations and operational consequences. Eur. J. Oper. Res. **158**(3), 662–677 (2004)
4. Munda, G.: Social Multi-Criteria Evaluation for a Sustainable Economy. Springer, Berlin (2008)
5. Munda, G.: A conflict analysis approach for illuminating distributional issues in sustainability policy. Eur. J. Oper. Res. **194**(2009), 307–322 (2009)
6. Gamboa, G., Munda, G.: The problem of windfarm location: a social multi-criteria evaluation framework. Energ. Policy **35**, 1564–1583 (2007)
7. Dente, B.: Understanding Policy Decisions. Springer, Berlin (2014)
8. Della Spina, L.: Integrated evaluation and multi-methodological approaches for the enhancement of the cultural landscape. In: International Conference on Computational Science and Its Applications, pp. 478–493. Springer, Cham (2017). https://doi.org/10.1007/978-3-319-62392-4_35
9. Munda, G.: Multicriteria Evaluation in a Fuzzy Environment: Theory and Applications in Ecological Economics. Physica-Verlag, Heidelberg (1995)
10. Joint Research Centre of the European Commission. NAIADE: Manual and Tutorial (1996)
11. Merigó, J.M., Gil-Lafuente, A.M.: Using the OWA operator in the Minkowski distance. Int. J. Comput. Sci. **3**(3), 149–157 (2008)
12. Nesticò, A., Sica, F.: The sustainability of urban renewal projects: a model for economic multi-criteria analysis. J. Prop. Invest. Finance **35**(4), 397–409 (2017). https://doi.org/10.1108/JPIF-01-2017-0003
13. Giuffrida, S., Napoli, G., Trovato, M.R., Ventura, V.: Axiology of the historical city and the cap rate. The case of the old town of Ragusa Superiore. Valori e valutazioni. Teorie ed esperienze **18**, 41–55 (2017)
14. Calabrò, F.: Local communities and management of cultural heritage of the inner areas. An application of break-even analysis. In: Gervasi, O., et al. (eds.) Computational Science and Its Applications - ICCSA 2017. Lecture Notes in Computer Science, vol. 10406. Springer, Cham (2017). https://doi.org/10.1007/978-3-319-62398-6_37
15. Morano, P., Locurcio, M., Tajani, F., Guarini, M.R.: Fuzzy logic and coherence control in multi-criteria evaluation of urban redevelopment projects. Int. J. Bus. Intell. Data Mining **10**(1), 73–93 (2015)
16. Zadeh, L.A.: The concept of a linguistic variable and its application to approximate reasoning. Inf. Sci. **8**(3), 199–249 (1975)
17. Calabrò, F., Cassalia, G.: Territorial cohesion: evaluating the urban-rural linkage through the lens of public investments. In: Bisello A., Vettorato D., Laconte P., Costa S., (eds.) Smart and Sustainable Planning for Cities and Regions, Results of SSPCR 2017. Green Energy and Technology, Springer (2018). https://doi.org/10.1007/978-3-319-75774-2_39
18. Della Spina, L.: Evaluation decision support models: highest and best use choice. Procedia Soc. Behav. Sci. **223**, 936–943 (2016). https://doi.org/10.1016/j.sbspro.2016.05.323
19. Morano, P., Tajani, F., Locurcio, M.: Multicriteria analysis and genetic algorithms for mass appraisals in the Italian property market. Int. J. Hous. Markets Anal. **11**(2) (2018)

Cities and Cultures in Movement

A Pragmatic and Place-Based Approach to Social Integration

Bakary Coulibaly[2], Maurice Herkrath[1(✉)], Silvia Serreli[2],
and Valeria Monno[3]

[1] Wageningen University & Research, 6708 PB Wageningen, Netherlands
`maurice.herkrath@wur.nl`
[2] Università degli studi di Sassari, 07100 Sassari, Italy
[3] Politecnico di Bari, 70126 Bari, Italy

Abstract. In Europe migration has become a prevailing topic in the 21st century and integration policies play an important role. Yet, the current outcome of social integration lags behind the ambitions such as stipulated in various policy documents, hinting to a theory-practice gap. In order to reverse the most negative social and spatial implications resulting from ineffective integration management, it is necessary to adopt an approach that can bridge the theory practice gap. The approach forwarded in this paper is one that is pragmatic, process-oriented and place-based. Evidenced by two examples, it is argued that practical interventions that take place on the local scale and involve multiple and diverse actors, have the potential to positively influence the migrants' integration. Additionally, an intercultural collaboration can foster capacity-building, thereby improving a cities capacity to deal with complex spatial problems.

Keywords: Migration · Social integration · Intercultural · Place-based
Collaborative

1 Introduction

In Europe, migration and the management of refugee streams has become a prevailing topic in the 21st century, up to a point that it is often labelled as refugee 'crisis' [1]. While this choice of words indicates the extent of the migrations events, the numbers regarding migrant arrivals in Europe draws an image of the situation. In 2015 over 1 million people arrived in Europe through the Mediterranean basin [2], while the trends for 2016 and 2017 look similar. Consequently, the constant and increasing flow of migrants from Africa and the Middle-East requires management on a national and European level. Coordination and cooperation on the political plan are not to be omitted, but require a coherent and structural approach from all member states [3]. While providing a brief overview on the situation on the macro-level, the focus in this paper lies on policies and interventions on the local level, and more specifically on the resulting outcomes that become visible and tangible, or not. First, integration policies and how they are defined, is analysed. Then the reality and current state of migrant

© Springer International Publishing AG, part of Springer Nature 2019
F. Calabrò et al. (Eds.): ISHT 2018, SIST 101, pp. 454–463, 2019.
https://doi.org/10.1007/978-3-319-92102-0_48

integration are assessed, pointing at a prevalent theory-practice gap. Subsequently, and based on two examples, the paper highlights a pragmatic, process-oriented, place-based and intercultural approach that can potentially reduce the theory-practice gap and improve the social and spatial outcome of integration policies.

2   Integration Policies

With the increasing influence of migrant streams on the daily events of a country, it is evident that integration ranks high on political agendas [6]. Besides the interventions that happen in a first place, being the rescue actions and emergency housing of refugees in concentrated spots and camps, there is need for an organised structure that deals with the inclusion and integration of the new arrivals. In the following paragraphs we describe how these integration policies have been defined over the years and throughout different documents and discuss why the idealistic wording permeating policy documents is seldom achieved in actual integration measures on the local level, thus creating a theory-practice gap [4].

2.1   Integration Policy in Theory

From the mid 20th century on, migration has been a growing topic on the European continent, and also affronted by organisations such as the UN. And although the reasons and types of migration have widely diverged over the years and decades, the broad definition of integration policies has more or less retained the same idealistic tone. In the KING report [3] on migration some good examples are provided, which show that files on integration often take equal chances and opportunities as well as the elimination of discrimination as core values. While *"the conceptualization of the integration as a two ways process is something that is commonly recognized and well accepted"*, recently there has been an added focus on the participation of the local population [3].

In other policy oriented documents such as 'The case for a common European refugee policy' [1], the need for a better coordinated long-term integration policy on the European level are also asserted and supported by a number of arguments. Bordignon and Moriconi call for better information exchange on migrants, a reduction of waiting times for insertion into integration programs, a good allocation process that prevents segregation, monitoring of the integration process and its success and the involvement of civil society. Also in the ISMU Foundations 'Country Report Italy' [5], as well as in their issue on 'Migration: a picture from Italy' [6], they provide an in-depth analysis of figures and conditions of the integration process and further provide concepts that are sound for improving the success of integration efforts. They assess the *"Italian Integration model"* [4] and provide an elaborate analysis on migration and integration models, especially in the case of Italy. They define 4 concepts of integration interpretation, being *'cultural distance'*, *'anticipated socialisation'*, *'transnationalism'* and *'segmented assimilation'* [6]. These 4 concepts stand in relation to the values and traditions that are held by migrants and the receiving society, thereby looking at how cultural differences persist or intertwine. Integration is further labelled as *"a multi-dimensional process aimed at pursuing peaceful coexistence, within a particular historical and social reality, among culturally and/or ethnically different persons and groups based on mutual respect of ethno-cultural differences, on condition it does not prejudice any fundamental right and does not damage democratic institutions"* [2].

In the more recent version from 2016 the ISMU Foundation added the important distinction between multiculturalism and interculturalism, defining that *"multiculturalism, in its multiple forms, puts the accent on cultural differences, and assimilationism tends towards a more or less forced homogenisation and thus towards monoculturalism, interculturalism fosters relationships between different cultures, founded on bidirectional, symmetrical and personal exchange"*, meaning that *"interculturalism puts the accent on the relationships between different cultures"* [2].

In 'World Migration report 2015' [7], 'Rebuilding after crisis statement' [8] or also the OECD's 'Indicators for Immigrant Integration 2015' [9] publications, the same holistic approach to integration policy is taken and presents reasons and arguments for a long-term, peaceful coexistence of multiple cultures in a geographically confined area, such as the city.

In terms of research, knowledge and policy composition, the most important aspects for effective integration measures have been identified and called for in a prescriptive manner. However, words are not always translated into actions, which is why it is necessary to have a closer look at the current situation around migration and integration policies.

2.2   The Theory-Practice Gap

Whilst policies can be considered a first step towards the recognition of integration oriented intervention, they do so foremost all on a (supra-)national and administrative level. This means that the reality on the ground, the real integration that we experience at the urban level for example, the actual effectiveness of the stipulated integration models, often lags behind the idealised propositions in the policy documents, hence constituting a theory-practice gap. The gap becomes visible when analysing some statistics provided by surveys on migrants and refugees, such as the 'Immigrant Citizens Survey' [10]. It is observable that integration policy has not yet managed to achieve its proclaimed goals and immigrants still perceive major hurdles regarding employments, civic and political participation, residence and citizenship during their integration process.

The gap between the theory outlined in policy documents and the practice such as applied in cities has a plurality of reasons. Firstly, it is often the case that a policy or theoretical text represents an ideal scenario that stands as a vision or goal. This is in line with a statement made in the 'Country Report for Italy' [4] saying that the limits of the integration policy do not exist in the policy text itself, but rather in its enforcement. Though integration is local, many policies are national and, increasingly, affected by European law and trends [10]. Thus efforts for implementation and action lag behind the policy's ambitions. The report takes it further stating that *"along with the well-known farrago and ineffectiveness of the Italian bureaucracy, several researchers have noticed in norm application an excessive administrative discretion, and, quite often, a considerable territorial diversification in the treatment reserved by the public administration to immigrants, further increased by operators' and managers' insufficient information and sensitization action"* [5].

Furthermore immigration and integration are seen as problematic and worrisome from a big part of the population. Consequently integration also became *"a long-lasting issue on the Italian political agenda"* [5] and causes the persistence of various positions between and within political parties. Therefore it is also observable that *"the periods in which immigration is most present among the acts approved coincide with the years of centre-right government, at the regional level we saw that the region that the issue the most times in its acts is of the centre-left and the region that approved the least is of the centre-right"* [6]. This means that the political agenda strongly influences the composition, as well as the implementation of integration policies. 'Diversity Management' is a concept that is also applied in the case of Italy, yet *"one of the major problems is the persistent gap between the espoused discourse on diversity,…, and the real consistency of the practices implemented to enhance and valorise diversity"* [5].

In addition to the elements that hinder effective integration policy, there is the prevailing attitude of large parts of the population towards immigrants. This attitude is influenced by the media and the aforementioned populist party politics that describe the presence of migrants as a problem or threatening situation [2]. Due to misinformation and prejudice people have a tendency to withdraw from, rather than engage in intercultural exchanges. This consequently fosters discrimination and segregation and leads to a lack of opportunities of encounter where the receiving society and migrants can exchange reciprocally [5]. Yet encounters create understanding of one another resulting in a situation of tolerance and acceptance at the least, and cooperation and trust at the most desirable level.

In sum the theory-practice gap can be ascribed to a missing implementation on the local level politics and the preconceived ideas from both parties, the migrants and the local population. In order to deal with the insufficient linkages between the social orientation and the urban ground and this specific theory-practice gap underlying the failure of several integration policies, it can be useful to focus on integration as a social construction resulting of the interaction among diverse ways of inhabiting also the physical space of a city.

2.3   Spatial Implications

A less mentioned aspect at the base of the theory-practice gap, is related to the role of the city and urban space in the integration process. It is often seen as a dead background or fabric on which the integration occurs. On the contrary, as many studies argues, society and space are mutually constitutive.

Such a divide between the social and the spatial dimension generates a situation in which migrants face difficulties to grasp hold in the local society and get their economic as well as social lives in harmony with their new surroundings. In absence of integration measures which cope with the urban dimension and the identification of appropriate measures on an administrative, formative, work-oriented or social plan, the new arrivals run risk of being clustered together, forming a sub-culture. In other words, segregation is a direct outcome of integration procedures and creates an isolated enclave within the organisation of a city [12, 13]. Examples in history show that if such a situation is reproduced perpetually, then it will create a segregated status where interaction between various ethnic groups become prejudice-ridden and discriminative. Spaces of suspension, as Petti [11] calls them, emerge and persist as a sign of disruption between the territory, the state and the population.

The outcome on the small scale, being that of a city or town, lags behind the aim of an intercultural and solidary society, but often creates a distorted landscape of isolated individuals or groups that remain strangers to their direct environment and the residing population [11]. Segregation arises on ethnic, but foremost all on socio-economic aspects. If migrants cannot profit from a structured integration process that allows them to take a foothold in society, they face barriers preventing them from mingling with that society. In other words, the spatial implications are that the city is dissected into multiple entities of different cultures, respectively ghettos, potentially creating conflicts within a divisive society.

3   A Pragmatic and Place-Based Approach

We think that the in order to deal with the insufficient linkages between the social orientation and the *urban ground,* underlying the failures of integration policies, it is necessary to overcome this specific theory-practice gap. In the following sections we describe two different practice of integration of migrants carried out in the Alghero Municipality: "ResPublica", a social centre managed by citizens which is a cultural meeting spot the city and thereby presents a space for intercultural encounters. "Fertilia in Movimento", a participatory regeneration project carried out through a workshop by researchers of DADU-Alghero. "Fertilia in movimento" represents a case of an occasional event that spurred collaboration between participants of various cultural backgrounds.

Pragmatism in planning proposes an approach that finds practical solutions to real and existing problems in a process-oriented mode, rather than a priori theorising [14]. With that philosophy it presents itself as an applicable concept to bridge the identified theory-practice gap. Since one of the main shortcoming of integration policy was lacking enforcement and insufficient implementation, it can be valuable to adopt a pragmatic and democratic process that involves all actors, whether politicians, citizens or migrants, in a collaborative and direct manner [2]. In that way the city functions as an educational environment [15] where dialogue emerges to improve reciprocal understanding, while producing fitting solutions.

Identified earlier, the misinformation and prejudice that inhabitants hold towards migrants can cause a situation of division between two, or more, entities that inhabit a city. One potential remedy to this situation is created by opportunities for a reciprocal and intercultural dialogue that helps all those involved to gain a better understanding of the situation and the culture of others [20]. If citizens don't know any other than the information they receive from media or the stories that are being told, the prejudice they hold will likely prevail. However, a single and first moment of encounter between locals and foreigners in a reassuring and outspoken environment can prompt curiosity and understanding for both sides [21]. This constitutes an important precondition for solidarity to arise. Whilst solidarity in itself can take a variety of forms and present itself with different facets, the Oxford Dictionary defines it as "*unity or agreement of feeling or action, especially among individuals with a common interest; mutual support within a group*".

Movement and change within the built environment influences the perception of ourselves, of the others and of the built [17]. Consequently, it is direct interventions that create movement and change in order to produce a liveable city [18]. Exploring the environment through movement and practical applications in the surroundings, puts importance on the involved subject and on the location thereby creating a link between perception and action [19]. A pragmatic approach to integration in the urban space signifies that interactive and experimental spaces and events create intercultural exchange [16]. This is exemplified by the two cases of small scale interventions that are rooted to the urban space and the society.

3.1   "ResPublica"

"ResPublica" was an informal place that was self-organised and run by dedicated citizens. When the municipality recognized its usefulness and helpfulness, they formalised it in order to assimilate its cultural services as complementary to the administrative ones offered by the municipality. A lot of organisations and associations that are active in the field of social support hold their grounds there, respectively offer courses, meeting and events taking place there. Henceforth there is a rather open-minded and tolerant society that tries to be inclusive of everyone and is identified as place where citizens as well as newly arrived migrants can find reciprocal and trust-based relationships. Relationships that potentially support them psychologically in their integration process. This welcoming, helpful and formative environment is an outcome of the relative small scale, the a-political and human-centred stance, as well as of a pragmatic approach that induces action and interaction.

With "AFRIKASIALGHERO" a group was created within the "ResPublica". Their purpose is to be solidary and make a positive contribution to the reception, integration and socialization of migrants in the local society. The principle was to organise regular meeting sessions with a network of diverse people with the idea to bring the migrants out of isolation from the reception centre into touch with the city and the society.

For the case of the "ResPublica" solidarity presents itself in the form of supportive measures that help people to expand their knowledge on the receiving society as well as their skills in multiple areas such as communication, language, arts, work. These can be considered to be directly linked to a smooth integration process that helps migrants to get a foothold in society. Additionally, the support arises from personal relationships that form during these opportunities for encounter. These personal and mutual relationships are crucial for providing the migrants with an unprejudiced, trustworthy and psychologically supportive friendship that allows them to gain self-confidence and autonomy.

3.2   "Fertilia in Movimento"

Fertilia in Movimento" is another type of example for a small scale practically oriented intervention. In essence it was a workshop that lasted from the 21st to the 27th of May 2017 and took place in Fertilia, a "frazione" of the municipality of Alghero. The workshop basically combined research and action, and had an interesting composition as it involved the University of Sassari, Emergency Architecture and Human Rights Copenhagen, numerous other associations concerned with social integration, arts or architecture as well as a reception centre for refugees placed in Fertilia. The location for the workshop was deliberately chosen to be accessible and close to the refugee centre in Fertilia, in order to re-establish a possibility for exchange between migrants and inhabitants, after that the refugee centre had been made inaccessible for outsider for security reasons. Consequently one central aim of the project was to create dialogue and cooperation on the ground with a variety of actors so as to produce a common project of urban regeneration. In a collaborative and intercultural fashion, the workshop fostered formal and informal encounters between people of diverse cultural backgrounds. Thus it offered an opportunity for the refugees, living in the temporary reception centre, to make connections, gain self-confidence and produce something characteristic while getting in touch with their direct urban surroundings.

"Fertilia in Movimento" presents a different form of solidarity, which might not be as deeply rooted as that unfolded in the "ResPublica", but just as powerful in creating a relations of support with migrants. The format of a workshop clearly represents a more intensive and condensed form of encounter, where during one week people of diverse cultures are in constant exchange. Compared to "ResPublica" where the interaction is mostly confined to locals, respectively Italians, and migrants, the workshop is based on communication between a larger variety of actors with diverse cultural backgrounds and motivations. Consequently the challenge to collaborate is bigger, but also the outcome of the workshop can constitute an important co-produced message.

Furthermore the workshop also contains an effect of solidarity, in the sense that it provides all participant with the experience that it is possible to collaborate and build capacity together in a reciprocal way. For the migrants it can additionally provide a feeling of empowerment to involve in the receiving society and thereby raise their self-confidence and willingness to engage. Lastly also personal relationships and friendships develop during such projects, that can create opportunities for employment in the future, and create a sense of belonging and acceptance for the new citizens.

When reconsidering the identified theory-practice gap and the shortcoming that integration policies remained within political realm of inaction, thus lacking effective implementation on the local level, the two examples of "ResPublica" and "Fertilia in Movimento" provide a remedy to this with their place-based and process-oriented approach. Both examples, one more permanent and receptive, the other ephemeral and collaborative, have in common that they are interlinked with the territory and occur on a relatively small scale which allows them to offer a more personal environment that spurs dialogue, understanding and capacity-building between a multitude of actors. They created an experimental sphere for a democratic and inclusive exchange with tangible, visible and perceptible results for the city and its inhabitants, whether newly arrivals or established ones.

3.3   Spatial Outcome

In Sect. 2.3. the consequences of a malfunctioning integration process have been elucidated. In which ways can a pragmatic, place-based approach to integration undermine or revert a potential division and segregation within the local population? The benefits of accessible spaces that focus on intercultural exchange has been demonstrated on hand of the two employed examples in regard of social integration. How this translates into the spatial configuration of a city is given by an explicit relationship. When more opportunities for intercultural exchange, co-creation, peer-research and capacity-building are provided [22, 23], then mutual understanding is enhanced and may transform into tolerance, acceptance and possibly solidarity and cooperation within the city [24].

Consequently, and opposed to the situation where a segregated landscape of divided entities emerges, the receiving society and the migrants will more likely engage in a dialogue that reduces the fear and thereby creates more opportunities for the new arrivals to fully integrate into the culture with less hurdles of prejudice and discrimination. The idea of interculturalism entails that migrants should not omit their own culture by being forced to fully adopt the receiving society's culture, but rather share it with those open for it, and interested in it. In workshops as well as permanent spaces, where co-creation of urban regeneration, cultural events, construction, arts, etc. is promoted, all the participants help to shape and re-shape their city and its identity. Reflecting the concept of 'cityness' [25] the citizens can appropriate and reinterpret the city together and rely on the most suitable interventions and solutions that each individual, and in the broader sense each culture, provides. 'Cityness' is further characterised by the interrelation between ephemeral interventions and stable relations, both creating fruition and inducing knowledge exchange. On the long-term, the global and the local intersect and interact so as to form new aspects of the city, new places of encounter, possibilities for rooting as well as for mobility and movement [20].

A revived interpretation of 'citizenship', proposed by Luigi Mazza, describes it as *"a social process and combination of practices, an experience and an activity of citizens who act to redesign rights, duties and a sense of belonging"* [26]. A pragmatic, adaptable and solidary society, involved into the spatial development of its city, is likely better prepared for an uncertain future with complex challenges lying ahead [20]. The recognition of multiculturalism and diversity as resource can thus improve the intellectual, economic and physical development of cities by stimulating innovation, creativity and competiveness [27].

4   Conclusion

With large amounts of refugees and migrants arriving in Europe every year, many of them considering long-term residence and the acquirement of the receiving country's citizenship [10], it is evident that social integration has to be high up on the political agenda. If the asserted theory-practice gap in integration policy persists, the societal and spatial results might negatively affect European cities in the short- and long-term.

However the proposed pragmatic approach is apt to reverse the most considerable implications by:

- Closing the gap between theory and practice in relation to social integration. With a stronger focus on the local level implementation of approved measures.
- Increasing the opportunities for intercultural exchange. Places and events of encounter constitute a platform for dialogue. Whether formal or informal, communication enhances reciprocal understanding, tolerance and solidarity, thereby reducing social discrimination and spatial segregation of migrants in the city.
- Adopting a process-oriented and place-based method. Co-creation in a collaborative way enhances capacity-building. The local population can share information and increases its knowledge, whether about cultural and social factors, or whether in relation to the urban space. Established and new citizens thereby constantly re-interpret and re-appropriate the city, entailing flexibility and adaptability of the city's spatial configuration and the interrelated social cohesion.

Acknowledgments. This research has been carried out within the existing research program called "Territori e Culture in Movimento" funded and led by the Laboratorio LEAP of the Department of Architecture, Design and Urbanism of the Università degli Studi di Sassari, coordinated by Silvia Serreli.

References

1. Bordignon, M., Moriconi, S.: The case for a common European refugee policy. Policy Contrib. **8**, 1–13 (2017)
2. Cesareo, V.: The Twenty-second Italian Report on Migrations 2016. Fondazione ISMU, McGraw-Hill Education, Milan (2017)
3. Gilardoni, G., D'Odorico, M., Carillo, D.: KING, Knowledge for Integration Governance. Fondazione ISMU, Milano (2015)
4. Shumaker, S.A., Brownell, A.: Toward a theory of social support: closing conceptual gaps. J. Soc. Sci. Issues **40**(4), 11–36 (1984)
5. Zanfrini, L., Monaci, M., Mungiardi, F., Sarli, A.: Country Report Italy. Fondazione ISMU, Milan (2015)
6. Cesareo, V.: Migration: a picture from Italy. Quaderni ISMU 2 (2013)
7. Editorial: World Migration Report. International Organisation for Migration, Geneva (2015)
8. Papademetriou, D.G., Benton, M., Banulescu-Bogdan, N.: Rebuilding After Crisis: Embedding Refugee Integration in Migration Management Systems. MPI, Washington DC (2017)
9. OECD/European Union: Indicators of Immigrant Integration 2015: Settling In. OECD Publishing, Paris (2015)
10. Huddleston, T., Tjaden, J.D.: Immigrant Citizens Survey. King Baudouin Foundation and Migration Policy Group, Brussels (2012)
11. Petti, A.: Arcipelaghi e enclave, Architettura dell'ordinamento spaziale contemporaneo. Bruno Mondadori, Milano (2007)
12. Sassen, S.: Cities in a World Economy, 4th edn. Pine Forge Press, California (2011)
13. Grassi, M., Giuffrè, M.: Vite (il)legali. SeidEditori (2013)
14. Allmendinger, P.: Planning Theory, 3rd edn. Palgrave and Macmillan, London (2017)

15. Clemente, F.: I contenuti formative della città ambientale. Pacini, Pisa (1974)
16. Chaoy, F.: Espacements: l'évolution de l'espace urbain en France. Skira, Milano (2003)
17. Proulx, M.J., Todorov, O.S., Aiken, A.T., De Sousa, A.A.: Where am I? Who am I? The relation between spatial cognition, social cognition and individual differences in the built environment. Front. Psychol. **7**(554), 64 (2016)
18. Tagliagambe, S.: Le due vie della percezione e l'epistemologia del progetto. FrancoAgneli, Milano (2005)
19. Berthoz, A.: Le sens du mouvement. Odile Jacob, Paris (1997)
20. Agostini, I., Attili, G., Decandia, L.: La città e l'accoglienza. Manifestolibri, Roma (2017)
21. Minerva, F.P.: Mediterraneo-Europa, dalla multiculturalità all'interculturalità. Quaderno IRRSAE Puglia (36), Bari (1997)
22. Giuffrè, M.: Genere e Migrazioni in Riccio B. Antropologia e Migrazioni. CISU, Roma (2014)
23. Nairn, K., Smith, A.: Young People as Researchers in Schools: The Possibilities of Peer Research. University of Otago, New Zealand (2003)
24. Colucci, M.: La città Solidale. Franco Agneli, Milano (2012)
25. Sassen, S.: Cityness. In: Ruby A, Urban Trans Formation. Ruby Press, Berlin (2008)
26. Mazza, L.: Spazio e cittadinanza. Donzelli, Roma (2015)
27. Khovanova-Rubicondo, K., Pinelli, D.: Evidence of the Economic and Social Advantages of Intercultural Cities Approach: a Meta-analytic assessment. Council of Europe (2012)

Unused Real Estate and Enhancement of Historic Centers: Legislative Instruments and Procedural Ideas

Serena Mallamace[(✉)], Francesco Calabrò[iD], Tiziana Meduri, and Carmela Tramontana[iD]

Mediterranea University of Reggio Calabria, 89100 Reggio Calabria, Italy
serena.mallamace@libero.it

Abstract. Private real estate in historic centers plays a key role in development dynamics as their efficient use is functional in the competitiveness of the territories they belong to. This estate is often neglected and with complicated ownership circumstances hardly compromising the implementation of evaluation plans. Generally, this paper aims to contribute to the debate on historic centers and has the specific goal to suggest those tools and preliminary procedures to help Administrations on the issue of private unused buildings within an integrated process of evaluation. We brought to light two approaches: one related to regulatory issues that define the available legislative instruments; the other one related to evaluative issues, which show the role of evaluation in finding out those preparatory instruments able to create integrated plans in terms of feasibility and sustainability of investment recovery. This should be realized with the help of private funds, so it is essential a benefit estimation.

Keywords: Unused real estate · Historic centers · Enhancement
Sustainability · Legislative instruments · Inner areas

1 Introduction

This paper aims to define the feasible preparatory procedures that local governments could adopt in order to plan enhancement actions on private real estate of town centers and in particular on disadvantaged inner urban areas that, as a shared opinion, play an important role in the dynamics of the sustainable development [1, 2]. Conforming to this widespread realization, the enhancement of historic centers appears to be like a remarkable opportunity thanks to their value and despite the considerable rearrangements suffered over time, before their definitive abandonment. The resulting depopulation due to economic difficulties gradually caused a status of physical degradation in which most of the ancient units pour - normally minor housing constructions -, this

---

This is the result of the joint work of the authors. Although scientific responsibility is equally attributable, the abstract and the Sect. 6 were written by all authors; the Sect. 1 was written by F. Calabrò, Sect. 2 by S. Mallamace, Sects. 2.1 and 4 by S. Mallamace and C. Tramontana, Sect. 3 by T. Meduri, Sect. 5 by C. Tramontana.

© Springer International Publishing AG, part of Springer Nature 2019
F. Calabrò et al. (Eds.): ISHT 2018, SIST 101, pp. 464–474, 2019.
https://doi.org/10.1007/978-3-319-92102-0_49

highlights a specific cultural identity, which now appears to be unused and doomed to disappear. The definition and the realization of the enhancement plan of these areas are something that local administrations must deal with but one of the greatest difficulties is the real fulfillment of regular recovery plans able to value the cultural landscape drawn on territories [3–6]. Often, disused buildings and plots of land are private belongings and, in most cases, these "privates" are abroad residents or people living in other Italian regions, so they probably have no more interests to invest on their properties. Not to mention those properties and lands with complicated ownership circumstances affecting the application of integrated enhancement plans based on endogenous resources whose historical building heritage is part. Generally, real estate is essential for its effective and functional use related to the competitiveness of territories they belong to, since it could be intended as the perfect place for business leading local enhancement and since it could host collective services able to improve social cohesion, raising the well-being of communities [7]. In this context, the present paper aims directly to contribute to the ongoing discussion on the rehabilitation of historic town centers and clearly pursues the goal by suggesting tools and preliminary procedures in support of Administrations, on the issue of unused private estate within an integrated process of revaluation of the territories. The discussion, although embryonic, reveals two types of complementary lines. The first one is closely related to regulatory issues that define the available legislative tools or those to be arranged; the other one related to evaluative issues, which show the role of evaluation in finding out those preparatory tools able to create integrated plans in terms of feasibility and sustainability of investment recovery. This should be realized with the help of private funds, so it is essential a benefit estimation. This paper focuses on a wide research perspective and leads to an acquisition approach through which Administrations, in the light of current Legislative Instruments, might act on unused buildings getting investments for tourist accommodation services and according to the different cropping up scenarios. To support Administrations in getting innovative evaluation programs for real estate of smaller historic centers, the proposed idea could also:

- Foster the depopulation drop in town centers;
- Achieve repercussions on lands in terms of accommodation activity;
- Start enhancement actions to promote historic real estate spread;
- Promote a better urban regeneration.

The methodology used for research purposes has a multidisciplinary and integrated approach that connects a set of problems posed by the evaluation of real estate heritage of town centers, either those under protection, and the concerning socioeconomic implications. In specific terms, this research follows an economic estimate approach leading to the analysis of management hypothesis and to the sustainability of enhancement actions.

2   Regulations

The management procedures of actions designed to save and estimate cultural assets, according to Legislative Decree No. 42/2004 [8] are quite well defined. The challenge point is the preservation and the accomplishment of those cultural areas, having a so-called council-owned lodging, landscaped assets and even those within the property complexes that draw the aesthetic and traditional value trait (Art. 136, par. c). In the first case, the legislator, in addition to outlining the protective measures of these assets and those who guarantee preservation, of both public and private belonging, regulates restoration and other private conservative works providing an administrative stream-lining activity to encourage owner initiative. This is essential to achieve public interest in the heritage protection and is, amongst other things, an aim shared with the Parliament in order to consider and regulate a compulsory form of conservative action whether private owners appears to be inactive. Therefore, the Art. 32 acknowledges that the Ministry has the right to work ex officio for cultural heritage preservation by granting the execution of the needed actions imposed on owner or by taking replacement measures. The case of council-owned lodgings, located in the town centers, is quite different. In relevant cultural and landscaped areas is almost easy to find disused properties or plots of land, where the inertia of private owners represented (and, unfortunately, still is) an obstacle hard to bypass. City Councils have attempted for so long to oppose to this widespread practice of neglecting with standards in building regulations aimed to persuade the owners to preserve buildings and lands or rather to respect the architectonic and decorative rules required from communal regulations. Unfortunately, these were sterile attempts and in addition to owner inactivity, the regulation was not suitable and did not bestow the required authority to Mayors. The modification of art. 54 of legislative decree 267 of 2000 (TUEL) introduced in Art. 6 of Decree Law 92 of 2008, passed into Law 125 of 2008 and called Security package [9], took a tiny step towards the fulfillment of the requirement. It's a legislative regulation intended to provide for an extension of the Mayor powers' in terms of security and public order letting him able to implement "even compelling" ordinances to avoid serious risks that threaten "public safety and urban security". This Ordinance is a form of measure, which can be aimed at a general information of subjects to specific persons: according to this second hypothesis, in case of non-compliance and in addition to the remedies under penal law, the Mayor can provide for ex officio, exerting a replacement power, at interested parties expenses. Among the most relevant features of the legislative amendment, there is the extension of ordinance power that can be adopted by Mayor including "urban safety". The Decree Law No. 92/2008 [10] explains clearly the idea: it is like a measure of public security in preventing those illegal events through the municipal areas that affect not only the city security and the orderly cohabitation, but also the environment and the local quality of life. The objectives pursued by this Decree are further explained by the Ministerial Decree of 5 August 2008 [11] fixing the implementation policy of new powers given to the Mayors. Indeed, Art. 2 states that the Mayor has to prevent and hinder all those situations of urban degradation and those in which "occurring behaviors, such as damage to public and private heritage, that prevent the usability or determine the deterioration of urban quality". Not to mention also the neglect, the illicit occupation of real estate and those situations affecting the city decor. With the help of these guidelines and with these powers, many Municipalities provided for their properties preservation using ordinances in order to oppose to degradation ensued from the neglect of buildings and lands. For example, the Mayor of Olbia [12] highlights the decay of urban quality derived from the neglect and affecting those areas and buildings exposed to any maintenance or custody. Whereas the Municipal Administration of Viareggio [13] shows how neglect and degradation of buildings and lands can affect residents' health. In both cases, it is clear that promoting a better quality of urban life depends on lands protection and renovation; it is furthermore obvious that these aims can be achieved only with a full and active cooperation among estate and property owners. If this lacks, it is vital for Administrations to have efficient tools and powers to overcome private inactivity even through a different but authentic management of unused assets. Another legislative work - D.L. No. 14 - introduces urgent provisions relating to protection of city security [14]. This regulation about urban security, seen as public assets connected to city decor and quality of life, intends, firstly, to achieve a cross-sectional model of governance also included among different tiers of government with the subscription of special agreements between State and Regions or with local authorities. Secondly, it works on the administrative penalty section in order to prevent events that affect negatively city security and decor, thus, it guarantees the accessibility of public areas. The leading aim is to improve and enhance the quality of life and lands, and to promote social inclusion and rehabilitation of involved socio-cultural areas.

2.1   The New Law on Town Centers

With the approval of Law No. 158/2017, "Measures for the support and development of small towns, as well as provisions for the renovation and recovery of historic town centers" [15], the Senate has marked a turning point for the enhancement of small municipal areas identifying the basic role of little villages. Small municipal areas are the real beneficiaries of this regulation, namely those centers with up to 5000 residents, but also the municipalities established by merger among centers, each with a population up to 5000 inhabitants. The purpose of the law is to foster and promote an economic, social, environmental, cultural sustainable development to endorse the National demographic balance encouraging the residence in small municipalities, which account for more than half of those present on the national territory. Besides, this law works on protection and enhancement of environmental, rural, historical, cultural and architectonic heritage and aspires to adopt measures for resident people and productive activities against depopulation and in favor of touristic growth. The law provides for an allocation of EUR 100 million over the 2017–2023 period; EUR 10 million for 2017 and EUR 15 million for each year from 2018 to 2023. These resources are bound to financing investments in environmental and cultural heritage protection, in the mitigation of the hydrogeological risk, in the urban preservation and renewal of historic centers, in the security regulation of road infrastructures and schools, in the economic and social promotion and growth and in the establishment of new production activities. However, this is a framework law and can reorganize the existing standard as well as enable additional funds to pursue the ultimate goal of the development of small towns.

The use of these resources requires an arrangement of a National Plan rehabilitating small towns and listing priority actions. Furthermore, there are other standards to consider since all municipalities should be located in those areas affected by hydro-geological instability, economic backwardness, terms of settlement disadvantage, everything based on specific parameters defined by old age index, the percentage of employed persons compared to the resident population and rural index, for example. There are different intervention measures addressed to small municipalities leading to an integrated development approach, such as:

- Broadband deployment;
- Promotion, use and marketing of agricultural products coming from short supply chain or local purchases;
- Film promotion as means of tourist and cultural enhancement.

The law considers also the chance to establish multifunctional centers, even in a partnership, to provide services also in the environmental, social, energetic, educational and postal fields. The main point of this law focuses on the recovery and rehabilitation of historic town centers, thus it works on identifying privileged areas, referring to cultural and architectonic assets, and their renewal with the help of public and private actions and respecting the types of original structures. Moreover, the regulation allows these municipalities to foster hotels, to acquire and rehabilitate buildings opposing to land abandonment, to avoid the risk of hydrogeological instability and to manage clearing operations, both on lands and buildings, to stipulate agreements in order to recover road-renting houses and railway stations that are no longer used. We are talking about recovery interventions for real estate: renewal, preservation, emergency maintenance, rehabilitation, static and anti-seismic consolidation and urban service enhancement. This law proves the attention of the Government towards small municipalities in need and with different situations compared with urban areas or metropolitan cities. Until yesterday, progress was portrayed as the movement of people from the countryside and from the mountains to cities, and the laws favored, or even promoted, this trend reversing the real meaning of the sustainable development as well as the preservation of the landscape and the balanced relationship between man and nature. Thus, small municipalities are the ideal areas of trial and processing with the help of this regulation that connects various institutional levels: indeed local administrators have a new set of indications and can take the chance (or should take it) to be creators and implementers of local development of their territories. As the H.M. Realacci claims, the first signer of the legal text, small municipalities "are not a legacy of the past but an extraordinary opportunity to defend our identity, our qualities and project them into the future. This is an ambitious view of Italy that passes also by the right promotion of territories, communities and talents".

3 The Role of Public-Private Partnership

In addition to the above-mentioned Legislative Instruments, the new Law for historic centers and municipalities' ordinances is not able to leave the importance of the public-private partnership out of consideration especially in the politics and dynamics of territorial development. The use of public-private partnership forms has been growing over the years and now represents an interesting perspective considering the low availability of public funds [16–18]. Therefore, a strong incentive to put public-private partnership back on track also comes from the cohesion policy 2014– 2020 that increases the concept of Community-led local development involving local representatives of the socio-economic interests [19]. The cooperation between public and private sectors aims to the pursuit of the public interest: it spreads positive repercussions on the community allowing an efficient public action with the help of private resources; therefore, it is important and essential to test the sustainability of the investment. Public-private partnership is not only a legal procedure but also a technical and financial process that can only work if there is a structured system of benefit between public and private subjects and the latter is adequately remunerated. In the context of the proposed research, the cooperation between the public sector and private operators carries out with the fulfillment of a direct activity aimed at pursuing public interests. These are focused on recovering the private historic pattern that Administrations cannot manage directly for the existing stuck or hindered situations. Among other interests connected to the rehabilitation of real estate, there are also consequences on territory in terms of accommodation activities considering the possible business activities with touristic aims as well as the depopulation reduction. According to the proposed model, public-private partnership is possible only through contractual links [20] with which Administrations and privates (property owners or future administrators of planned activities) regulate their relations exclusively on formal basis [21]. The main point is to direct a dynamic economic development focused on the bottom-up principle, based on expectations, ideas, projects and on the participation of local population [22–25]. At the same time, disused properties can be revaluated as well as their ordinary and extraordinary maintenance that would otherwise be treated with difficulty.

4 The Proposed Procedure

As explained before, the purpose is to be able to include private unused estate, frequently crumbling, in a system of recovering actions with the help of a systemic intervention of Local Governments. To achieve the aim some preparatory procedures will be promoted in order to see how Administrations deal with the recovering of private properties. Taking into account the legal framework we exposed, Municipalities should equip themselves with means, as the ordinances previously mentioned, in order to force owners to provide independently for their properties recovery or at least keep them safe to contain degradation. If Municipalities had already these means, the fate of many centers would be different. Besides, the recent Law No. 158/2017 opens up new perspectives of rebirth for these centers, something well bound to the proposed idea. The same Law connects historic town centers with hotels promotion: this key of accommodation supply - entirely Italian - combines aspects of preservation and protection with economic, social and environmental features, in accordance with the principles of a sustainable development. The value generated by the main actors-factors of tourism industry [real estate, enterprises, local institutions and consumers] presents itself in a well-balanced system, born from the interaction of the interests of all parties involved in the evaluation process. This approach relies on a proficient plan of assimilated evaluation that features complicated cross-sectorial actions, bound together, and leading to the common goal of local development. Authors provided a list of premises and possible construction phases of the procedural model included in the drafting of the potential implementing process. In order to have a recovery plan, as a part of integrated planning process, Administrations will:

1. establish a register of the assisted buildings, after making a historical town mapping and testing proprietary conditions;
2. publish the evaluation plan of the center and of the expression of interest aimed at owners of properties included in the register;
3. conclude a concession agreement [free loan] with property owners;
4. organize a recovery plan with suitable codes of practice and guidelines in order to suggest the best attitude on the historical building respecting its identity;

a. find out the different appropriate use classifications;
b. estimate investment costs, through synthetic procedures, considering the planning phase in which the estimation is carried out;
c. engage the suitable actors for the implementation: this step is essential to the plan since these private people will take part to the implementation managing the expected activities. Indeed, only those privates who can:

- manage only the planned activity in the restored estate with public funds,
- invest in estate and manage it,
- invest in estate's co-financing of the property and manage it, will be consulted.

Once the pledged grant is expired, the owner can demand for the rehabilitated building or it will remain to who managed it, but with new rules; when accommodating activities take a break for a period, the real owner can benefit from it.

5   Feasible Scenarios and Evaluation Role

According to the explained setting, evaluation culture serves as a wide source and is bound to the above-mentioned process. Pre-feasibility studies are the main key of supporting decisions: over the years, Administrations have dealt with lots of investments - even of considerable importance - that turned to be untenable from every point of view. Indeed, evaluation culture is an essential step to achieve the suitable level among all those factors fixing sustainability, economy, environment and society.

Besides, these studies take a deep glance on functional terms, economic and financial features such as the definition of intervention costs, the management ideas, the testing of public-private partnership contract, the profitability of investment, all to pursue the sustainability of the intervention [26, 27].

Turning back to the proposed idea, once had in granting the unused buildings for accommodations, we have supposed the essential part of the process in three potential and possible scenarios defined according to the fulfillment of investments and how to manage them.

Scenario 1:

– Investments with public funds
– Management activities of private property

Scenario 2:

– Investments with private funds
– Management activities of private property

Scenario 3:

– Investments with private and public funds
– Management activities of private property

Each outlined scenario raises questions, thus the following table explains briefly the aim of evaluation (See Table 1).

Table 1. Potential scenarios and connected aim of the evaluation (source: personal revision)

Scenario 1	Aim of the evaluation
– Investments with public funds – Management activities of private property	– Verify the sustainability of the investment including the cost of ordinary and extraordinary maintenance – Alternatively, estimate a considerable rent payable to Administrations in order to cover the cost of ordinary and extraordinary maintenance
Scenario 2	**Aim of the evaluation**
– Investments with private funds – Management activities of private property	– Verify the profitability of the investment – Identify the payback period as a function of a following rental fee to pay to Administrations, in order to cover the cost of ordinary and extraordinary maintenance
Scenario 3	**Aim of the evaluation**
– Investments with private and public funds – Management activities of private property	– Estimate the profitability of the rehabilitated real estate – Estimate the possible amount of the public contribution and consider the possible presence of privates who can afford the investment with a payback period not exceeding 5 years. – Estimate the rental fee to pay to Administrations once the investment is regained

To answer these questions from the profitability tier to be pursued depending on the scenario and according to the type of the managing subject, since, compared to it, those standards governing the right economical and financial sustainability could change [28], as reported on the following table (See Table 2).

Table 2. Potential scenarios and connected profitability tier, (source: personal revision)

Scenario 1	Profitability tier
– Investments with public funds – Management activities of private property	Medium profitability: revenues have to cover the management cost with an appropriate profit margin
Scenario 2	**Profitability tier**
– Investments with private funds – Management activities of private property	High profitability: revenues have to cover the management cost and remunerate fully and adequately the invested capital
Scenario 3	**Profitability tier**
– Investments with private and public funds – Management activities of private property	Medium-high profitability: revenues must cover the management cost and remunerate adequately the private portion of invested capital

6 Conclusions

The discussion of themes above shows that the case of historic centers is a current issue as well as the private unused real estate, especially those affected with complicated ownership circumstances. Furthermore, we identified lots of results achieved starting from the deserved acknowledgement for Inner Areas within policies supporting the sustainable development and from the role played by town centers in the development dynamics of territories with the correct use of endogenous resources. This went straight up to the "Measures for the support and development of small towns, as well as provisions for the renovation and recovery of historic town centers" law approval. We mentioned also all those measures easy to implement, such as the ordinances that Administrations could use to arrange their local area. In addition to this, thanks the public-private partnership forms, it is possible to create a virtuous growing trend. The output of this paper is a proposal according to the current regulations about the chances that Administrations have to acquire unused private estate from historic centers, in order to benefit from them and to contribute to the environmental development. There are also lots of points and questions referring to evaluation culture, in terms of testing the feasibility and the sustainability over the proposal and the identified scenarios. The suggested proposal takes advantage of an integrated approach combining resources, awareness and knowledge based on a multidisciplinary logic able to understand the difficulties of an area, its cultural, human and economic resources, and that includes inevitably the real estate of historic centers.

We therefore refer to future contributions for detailed studies in addition to the above-mentioned, as an assembly of the evaluation process that arises from the proposed measure, in order to suggest an adoptable preparatory practice for a real reborn of disadvantaged lands such as the small municipalities of the Inner Areas.

References

1. Mollica, E.: Le aree interne della Calabria. Rubbettino Editore, Soveria Mannelli (1997)
2. Strategia Nazionale per le Aree Interne: Definizione, Obiettivi, Strumenti e Governance, Materiali Uval (2014)
3. Della Spina, L.: Integrated evaluation and multi-methodological approaches for the enhancement of the cultural landscape. In: Gervasi, O., et al. (eds.) Computational Science and its Applications, ICCSA 2017, LNCS, vol. 10404, pp. 478–493. Springer, (2017). https://doi.org/10.1007/978-3-319-62392-4_35
4. De Mare, G., Nesticò, A., Tajani, F.: Building investments for the revitalization of the territory: a multisectoral model of economic analysis. In: Gervasi, O., et al. (eds.) Computational Science and its Applications, ICCSA 2013. LNCS, vol. 7973, pp. 493–508. Springer-Verlag, Berlin Heidelberg (2013). https://doi.org/10.1007/978-3-642-39646-5_36
5. Della Spina, L., Lorè, I., Scrivo, R., Viglianisi, A.: An integrated assessment approach as a decision support system for urban planning and urban regeneration policies. Buildings 7, 85 (2017). MDPI, Basel, Switzerland. https://doi.org/10.3390/buildings7040085
6. Della Spina, L.: The integrated evaluation as a driving tool for cultural-heritage. enhancement strategies. In: Bisello, A., et al. (eds.) Smart and Sustainable Planning for Cities and Regions. Green Energy and Technology. Springer, Cham (2018). https://doi.org/10.1007/978-3-319-75774-2_40
7. Stanghellini, S.: Modalità di acquisizione dei suoli e problematiche valutative nei progetti di trasformazione urbana. In: Atti del XXXIII Incontro di Studio del 24–25.10.2003, pp. 85–115, Cagliari (2003)
8. Decreto legislativo 22 Gennaio 2004 n. 42, Codice dei beni culturali e del paesaggio
9. Legge 24 Luglio 2008 n. 125, "iConversione in legge, con modificazioni, del decreto-legge 23 maggio 2008, n. 92, recante misure urgenti in materia di sicurezza pubblica
10. Decreto legge 23 Maggio 2008 n. 92, Misure urgenti in materia di sicurezza pubblica
11. Decreto ministeriale 5 Agosto 2008, Misure urgenti in materia di sicurezza pubblica
12. Ordinanza del Comune di Olbia n. 124 del, 4 Settembre 2009
13. Ordinanza del Comune di Viareggio n. 69 del, 24 Settembre 2010
14. D.L. 20 Febbraio 2017 n. 14, Disposizioni urgenti in materia di sicurezza delle città
15. Legge 06 Ottobre 158/2017: Misure per il sostegno e la valorizzazione dei piccoli comuni, nonché disposizioni per la riqualificazione e il recupero dei centri storici dei medesimi comuni
16. Morano, P., Tajani, F.: Decision support methods for public-private partnerships: an application to the territorial context of the apulia region (Italy). In: Stanghellini, S., et al. (eds.) Appraisal: From Theory to Practice: Results of SIEV 2015. Green Energy and Technology. Springer, Cham (2017). https://doi.org/10.1007/978-3-319-49676-4_24
17. Sagalyn, L.B.: Public/Private development. J. Am. Plan. Assoc. 73(1), 7–22 (2007)
18. Savas, E.S.: Privatization and Public-Private Partnerships. Chatham House Publishers, New York (2000)
19. Linee Guida Partenariato pubblico Privato: normativa, implementazione metodologica e buone prassi nel mercato italiano. EPAS (2013)
20. Libro Verde relativo ai Partenariati Pubblico-Privati ed al.: Diritto Comunitario degli Appalti Pubblici e delle Concessioni Bruxelles, p. 327 (2004)
21. Dipace, R.: Partenariato pubblico privato e contratti atipici. Giuffrè, Milano (2006)
22. Gastaldi, F.: Il ruolo del capitale sociale nella promozione dello sviluppo locale. In: Buratti, N., Ferrari, C. (eds.) La valorizzazione del patrimonio di prossimità tra fragilità e sviluppo locale, Franco Angeli, Milano (2011)

23. Mangialardo, A., Micelli, E.: Social capital and public policies for commons: bottom up processes in public real estate property valorization. Procedia Soc. Behav. Sci. **223**, 175–180 (2016). https://doi.org/10.1016/j.sbspro.2016.05.343
24. Barca, F.: Towards a place-based social agenda for the EU. Report Working Paper (2009)
25. Torre, C.M., Morano, P., Tajani, F.: Social balance and economic effectiveness in historic centers rehabilitation. Lecture Notes in Computer Science, vol. 9157(III), pp. 317–329 (2015). https://doi.org/10.1007/978-3-319-21470-2_22
26. Scrivo, R., Rugolo, A.: Sostenibilità e Fattibilità nella programmazione delle Opere Pubbliche. Metodologie e Strumenti per un centro regionale di controllo dei costi negli Appalti Pubblici, LaborEst, vol. 12, pp. 52–56 (2016). https://doi.org/10.19254/LaborEst.12.08
27. Dolores, L., Macchiaroli, M., De Mare, G.: Sponsorship for the sustainability of historical-architectural heritage: application of a model's original test finalized to maximize the profitability of private investors. Sustainability **9**(10), 1750 (2017). https://doi.org/10.3390/su9101750
28. Calabrò, F.: Local communities and management of cultural heritage of the inner areas. an application of break-even analysis. In: Gervasi, O., et al. (eds.) Computational Science and its Applications, ICCSA 2017, vol. 10406, pp. 516–531. Springer, Cham (2017). https://doi.org/10.1007/978-3-319-62398-6_37

A Cultural Route on the Trail of Greek Monasticism in Calabria

Daniele Campolo$^{(\boxtimes)}$, Francesco Calabrò[iD],
and Giuseppina Cassalia[iD]

Mediterranea University of Reggio Calabria, 89100 Reggio Calabria, Italy
daniele.campolo@unirc.it

Abstract. The discovery of the greek origins of Calabria is critical to under-stand the cultural and social evolution of this Italian region. In some historical periods, the cultural richness of Calabria became landmark in the whole Europe for the production of precious codex, and gave birth to great philosophers and theologians. Further, the Greek culture, which was widespread in the area, has influenced the history of many regions across Europe.

This project develops from a thorough analysis of an historical period that is not yet fully known and from the best practices of some European countries in the promotion of slow travel.

Slow Travel and slow tourism have all the potential to use the cultural resources already existing in the territory through a sustainable repurposing of currently abandoned infrastructures.

The creation of slow routes as greenways, velorails, etc., may result in the creation of a "Cultural Route" on the trail of Greek monasticism in Calabria.

Keywords: Cultural route · Byzantine monasticism · Slow travel

1 Introduction

The coastal territory of Calabria, with its peculiar geological characteristics spread across the Tyrrhenian and Ionian coasts, welcomed the settlement of monks coming from eastern countries that were escaping from the invasions of Islam, first, and later on from the iconoclastic persecutions. Such process, lead to the creation in this territory of rocky settlements in caves and hypogea.

Upon their arrival in Calabria, the Greek cultural background characterizing this area revealed itself as a good fit for Monks, since it shared some similarities with their own cultural origins. A prosperous and active civilization was built during these years, with the monks being fully aware of the cultural importance of their Greek origins, and

---

This is the result of the joint work of the authors. Although scientific responsibility is equally attributable, the abstract and Sects. 4 and 5 were written by Daniele Campolo; Sects. 1 and 3 were written by Giuseppina Cassalia; Sect. 2 was written by Francesco Calabrò; Sect. 6 is the result of the joint work of the authors.

© Springer International Publishing AG, part of Springer Nature 2019
F. Calabrò et al. (Eds.): ISHT 2018, SIST 101, pp. 475–483, 2019.
https://doi.org/10.1007/978-3-319-92102-0_50

bringing a linguistic unity (the "κοινή", the Greek language) and an oriental culture, that trace their roots back to two millennia.

Such a cultural heritage is of great historical, cultural, religious and artistic value, and may constitute a starting point for the enhancement of this territory.

The creation of an itinerary "on the trail of Greek monasticism of Calabria" could be a keystone in the valorization policies of the Calabrian cultural heritage. This route could set off a virtuous process of sustainable development, due to its value as a testimony of culture and history, easily readable and recognizable, and spread throughout the region of Calabria.

Caves, hermitages, ruins of monasteries, byzantine churches, and rock settlements are evidence of the presence of the Greek Monks in the territory, which lasted for over seven centuries in this region.

2 On the Trail of "Greek of Calabria" Monasticism

Byzantine monasticism, often inappropriately termed as "Basilian", initially developed from the Hellenic East and was subsequently increased by the emigration of monks. Ever since the 7th century, Calabria had been a refuge for the Monks fleeing from the eastern provinces of the Empire to escape iconoclastic persecution.

Instead of being of Greeks' origin - a common misconception - among the first monks that moved from the East to the West there were the so-called *Melkite* monks, a group with Christian origins following the Byzantine rite from the patriarchs of Alexandria, Antioch and of Jerusalem (Syria, Palestine, Greece, Egypt).

In the 8th century, byzantine Iconoclasm was fueled by a ban on religious images by Emperor Leo III and continued under his successors. It was accompanied by widespread destruction of images and persecution of supporters of the veneration of images.

Because of these controversies in 732–733 the ecclesiastical patrimony of Calabria and of Sicily passed to the Byzantine fiscal control with an edict of Leo III Isaurico, and, contextually, these regions underwent the control of the patriarchate of Constantinople's jurisdiction.

The Iconoclast period (which officially ended with the Council of Nicea in 786–787, but that in practice lasted until the Synod of Constantinople in 843) was a critical factor that favored the emigration of many monks. Upon their arrival in Southern Italy, and particularly in Calabria, they were not only tolerated, but even protected due to the existing Greek cultural substratum.

It is rather difficult to characterize the access routes traversed by the different migratory flows that followed each other in the inland of the Ionian and Tyrrhenian sides of the Aspromonte, but the settlement systems, caves and cenobi are evident, although many of them are yet to be thoroughly studied.

The migratory flow of these monks that was caused by the Arabs' occupation, initially streamed from the Mesopotamia to Sicily, passing through Syria, Palestine, Egypt, Libya and ended in Calabria only after the Arab's conquest of Sicily.

The coastal ionic routes constituted inward and outward arteries from the coasts to the inland areas, which were rich of geological conformations and hypogeal settlements dating back to the Neolithic period, easily used as dwellings.

A second migratory phase of Greek monks took place from the ninth century A.D., when the Arab invasion of southern Sicily (as well as other lands in the East incorporated by Islam) caused further migratory flows.

The 10th century (third migration) was a century of great migration for Greek monks who crossed the Strait of Messina from Sicily to Calabria due to the fall of the Eastern Roman Empire. In 902 the defeat of Taormina and, later in 965, the defeat of Rometta (last city of the Eastern Roman Empire) amplified the migration phenomenon.

These monks climbed Calabria: some of them stopped in already existing monastic places; others moved to the monastic eparchy of the Mercurion, located on Mount Pollino and known throughout the Byzantine East; others still began to expand their culture in the lands of West. This third phase lasted until the year 1000 and caused in Calabria a process cultural, religious and linguistic renewal that greatly influenced also the local population that pertained to a Greek-speaking ethnic minorities. In this period, the Greek language was used at every level: from the legal to the religious, from the toponymical, to the agricultural and pastoral levels.

In this period, thanks to byzantine monks that were promoters of Eastern culture, Calabria became a cultural center, known throughout the West for the presence of Greek-origin coenobics, along with those of Calabrian origin. Among others, Saint Nilo from Rossano, S. Fantino, San Nicodemo, San Bartolomeo from Rossano and San Luca from Isola Capo Rizzuto are renowned monks that started new byzantine monasteries. In this period, we can find a creation and diffusion of amanuensis centers that produced precious codes with their calligraphic skills, which are nowadays sought for. The transcription of codes for Italian-Greek monasticism became a real art that gave rise to numerous libraries.

Around 1100 Reggio Calabria and Rossano were the most active centers of Greek culture, and in this period Calabria became not only a cultural place, but also a important production and trade center thanks to the knowledge of these monks: in this period, we can see an increase of the production of wheat and oil - also to be exported to Constantinople - and of wine and the livestock breeding.

Great development of this territory also came from the introduction of silkworm breeding in Calabria by the Byzantines. This breeding lasted until the mid twentieth century, with the production of considerable quantities of a excellent quality silk.

The "damask" fabric, made in Calabria, takes its name from Syria (Damascus) and it was requested throughout Europe.

The Greek culture also flourished thanks to Byzantine rite that was maintained across the territory of Calabria in spite of the Norman domination and that lasted until 1300 in the episcopate of Crotone and in Oppido, until 1350 in Catanzaro; until 1460 in Rossano; until 1497 in Locri; until 1570 in Bova, one of the last stronghold of the Byzantine rite and of the Calabrian Greek culture. Nowadays, Bova remains the center of the Greek linguistic minority of Calabria with the last surviving Greek-speaking community [1].

One of the most important characteristics of this Greek monasticism of Calabria resides in the fact that these monks were not only import vectors of culture from the

East but also influence for the Western countries, and their knowledge permeated throughout Italy and Europe.

For example, S. Nilo first founded S. Adriano near S. Demetrio Corone, then near Montecassino (Gaeta) and finally Grottaferrata, which is considered the most important Greek center of all the West, and yet a territorial abbacy of the Italo-Greek Catholic Church.

The monks' influence spread also outside the territory of Calabria, since a great number of them returned to their own countries of origin after their training in the ascetic schools of Calabria. Among others, congregations influenced by the monasticism of Calabria include St. Stephen of Muret, founder o fit order of Grandmont; the famous monastery of Orval, in Belgium, likely created by a group of monks coming from the Valle del Crati).

3 Cultural Production and Manuscript Tradition on the Territory

Following the first monastic phase, which was fundamentally dedicated to the solitary asceticism in caves and had no influence on the local populations, Greek monasticism became a point of reference for the economic, social and cultural development of the local populations.

After the initial period of hermitage and isolation, the monks helped the local populations to plow lands, taught them advanced agronomic techniques, and contributed to education.

The Greek monks were also amanuensis calligraphers and illuminator, and were able to transmit and to spread the Greek culture radically rooted in these territories (see Fig. 1).

Examples of the cultural relevance of some of the Calabrian monks for both the Italian and the European culture are Barlaam of Seminara (1290–1348), master of Greek of Petrarca in Avignon; Leonzio Pilato, a pupil of Barlaam, who translated into Latin the Iliad and the Odyssey for Petrarca and Boccaccio;

The manuscript tradition as well as the calligraphic and miniaturistic art are a prestigious wealth, somehow yet to be discovered. The main monasteries had their own "scriptorium", a space dedicated to the transcription of the Greek manuscripts. Calabria was very active for the production of Greek codes, a patrimony that was well sought after in the 15th and 16th century, although many manuscripts were dispersed or are still to be identified. Among other, a mayor event that contributed to the dispersion of both greek codes and manuscripts was the earthquake of 1783 which destroyed most of the Calabrian-Greek monasteries.

4 Best Practices in Europe

Recently, Tourism has changed its shape with the introduction of a new type of traveling: named "slow travel".

Fig. 1. Codex Purpureus Rossanensis. (by Michele Abastante - Opera propria, CC BY-SA 4.0, https://commons.wikimedia.org/w/index.php?curid=46115570).

Slow travel is not only about moving from one place to another, but is also about immersing oneself in a destination. It often consists in residencies that are more prolonged, allowing to develop a deeper connection with the local territory; to discover the beauty of the historic centers, the typical local food, and visiting local places. Spending time with locals and discovering their habits and customs can turn a regular trip into a slow travel experience. The key is to take one's time and to let oneself be carried along [2, 3].

This type of tourism supports sustainable development and focuses tourists' attention towards inland areas, developing on "greenways" or on old disused railways that can be used as slow mobility routes [4].

These routes of slow mobility generally cross inland areas of low population density and are a driving force for tourism, hospitality, craftsmanship, for the dissemination of the historical heritage and the growth of small historic villages, nature and parks: they are also an opportunity to avoid the abandonment of these territories and contrast with the hydrogeological instability of the slopes.

In 2011, France has decided to lighten the special regulations concerning the rules of operation for the use of tourist railways. Such change allowed the experimentation of the "Velorail" to ride on the tracks. These abandoned railway lines have been redeveloped with a project that takes different names (velorail, railbike, railway draisinie) and consists of pedal vehicles adapted to move on rails (see Fig. 2).

Fig. 2. Vélo-rail de Médréac (France) (by Daniele Campolo).

Although this method is widespread in most of the Nordic countries, in France the velorail is mainly proposed as a tourist solution: 38 circuits are already active and a National Federation of Velorail was established to take care of the technical and organizational details of the individual tracks, and to promote different events.

5 Enhancement Hypothesis

Currently, over 1600 km of railway lines have been dismissed by the Italian Ministry of infrastructure and transportation, and these have been abandoned for a long time. These railways have the potential to become "greenways", traffic-free routes for walkers, cyclists and horse-riders for living a slow travel in contact with local communities. One of the main objectives of the Italian Strategic Plan of Tourism (PST) is to invest in new paths through the recovery and the sustainable reuse of state property (Objectives A3.2 and A3.1 of the PST): a concrete tool to identify sustainable tourism as one of the most important policies for Italian economic and social development; to promote a new way of tourist use of Italian tangible and intangible heritage.

In August 2017, the Ministry of Infrastructures and Transportation, the MiBact, and the Italian Regions have agreed on realizing new bicycle lanes by using dismissed railway lines, service buildings and roadman's houses of F.S. and ANAS. The new bicycle lanes are part of the national network of Italian tourist cycle tracks and are added to the "Ciclovia del Sole", "Ven-To" (Venice-Turin), the "Apulian Aqueduct" and the "Ciclovia Grab" (Source Mibact - Direzione Generale del Turismo). according to this protocol, more than 5 thousand km of bicycle paths are planned to be built by 2024 throughout the Italian territory, reaching 20 thousand km by 2030. For the realization of this national system of cycleways, € 89 million have been allocated for the period between 2016–2018. Additional resources of € 283 million will finance the construction of the cycle routes that will be identified by MIT in the period 2017/2024.

As part of the plan, the "Magna Grecia bycycle Route", which will develop in the Calabrian territory, will run for approximately 1000 km, passing by the territories of the Basilicata, Calabria and Sicily. It will mainly take place on service roads, starting from Metaponto up to Reggio Calabria, then going up again on the Tirrenica coast and arriving to Maratea in Basilicata.

The town of Reggio Calabria will be the hub that connects to the "line1" of the "Bicitalia network", the "line11" of the "Ciclovia degli Appennini" network and the Sicilian line "Eurovelo7" that connects Messina to Catania, Syracuse and Pachino.

The rock settlements heritage of the Greek monasticism in Calabria would benefit from a new way accessing those places, a goal that could be achieved by exploiting the potential of the slow tourism. In fact, the sites involved in the research project are located in the immediate vicinity of an abandoned railway path that could become a possible greenways or cycle.

Further, a variety infrastructures currently not used, such as level-crossing keeper's, stations, roadblocks or service buildings, are present along these abandoned tracks, and could be repurposed for hospitality activities or tourism promotion, tourist and cultural purposes or for service supply.

The sections of the abandoned railways concern the Southern Tyrrhenian Railway and the Ferrovie della Calabria (see Fig. 3).

In the sixties of the last century, a fast increase in passenger traffic in the "Costa Viola" and the consequent saturation of the line, were counteracted by the creation of a second track in the Southern Tyrrhenian Railway. However, the original track still exists, and develops among terraces overlooking the Mediterranean sea with a panoramic views from the Costa Viola to Capo Vaticano, and with a breath-talking view of the Aeolian Islands (see Fig. 4).

Within the same territory, the Calabro-Lucane Railways (Ferrovie della Calabria) in the Gioia Tauro-Sinopoli line (Km 26) were built to allow connections with the inland areas. This section, reduced in size in 1994 and definitively closed in 2011, crosses an area with a natural and cultural heritage of particular interest, closely linked to the local economy.

The idea of the railways enhancement projects was born with the intention to promote not only the Rock Settlements of greek monasticism but also to re-power a rail system that has entered in the identity of local communities and that holds a special beauty in its contrast between the engineering solutions of steel bridges and tunnels in stones and bricks; these structures were made in the early twentieth century in an area with uncontaminated nature and in a territory dedicated to agriculture and the use of local resources [5].

Fig. 3. Abandoned railway lines of the Ferrovie della Calabria, Municipality of Melicuccà (RC) (by D. Campolo).

Fig. 4. The Costa Viola: view of the Aeolian Islands (RC) (by D. Campolo).

6 Conclusion

It is well known that tourism, with its connected goods and services, is one of the major instruments for the economic development of a territory, a trend that has also been shown within the European Union, where tourism has been the only growth industry sector despite the economical crisis [6].

Studies on tourism trends have been shown both a rapid increase in production and fruition of cultural attractions, as well as a trend toward touristic activities devoted to elder, and more educated, citizens requiring forms of ecotourism, of cultural travel and relaxation. Therefore tourism will expand with a trend towards forms of "slow travel", with art, culture and environment at the center of interest.

Starting from this data, the explained case study highlights how the project for the recovery and enhancement of greek monastic settlements could be able to create a competitive and innovative interactions and synergistic connection among the area's resources and constitutes a great opportunity to create a solid network for the enhancement and promotion of activities in the territory through the idea of an "eco-museum" system, a museum focused on the identity of the place [7–9].

References

1. Campolo, D.: The cultural landscape of the "Grecanic Area" and the recovery of the genius loci of its historical centres. In: Advanced Engineering Forum, vol. 11, pp. 464–469. Trans Tech Publications, Zürich (2014)
2. Barcelona Slow Travel. https://www.barcelonaslowtravel.com/slow-travel/. Accessed 15 June 2017
3. Della Spina, L., Lorè, I., Scrivo, R., Viglianisi, A.: An integrated assessment approach as a decision support system for urban planning and urban regeneration policies. Buildings 7, 85 (2017). https://doi.org/10.3390/buildings7040085
4. Della Spina, L., Scrivo, R., Ventura, C., Viglianisi, A.: Urban renewal: negotiation procedures and evaluation models. In: Gervasi, O., et al. (eds.) Computational Science and Its Applications - ICCSA 2015. Lecture Notes in Computer Science, vol. 9157. Springer, Cham (2015). https://doi.org/10.1007/978-3-319-21470-2_7
5. Campolo, D., Bombino, G., Meduri, T.: Cultural landscape and cultural routes: infrastructure role and indigenous knowledge for a sustainable development of inland areas. In: 2nd International Symposium New Metropolitan Perspectives, Reggio Calabria (2016). Procedia - Social and Behavioral Sciences, vol. 223, pp. 576–582 (2016)
6. Della Spina, L.: Integrated evaluation and multi-methodological approaches for the enhancement of the cultural landscape. In: Gervasi, O., et al. (eds.) Computational Science and Its Applications - ICCSA 2017. Lecture Notes in Computer Science, vol. 10404. Springer, Cham (2017). https://doi.org/10.1007/978-3-319-62392-4_35
7. Cozzupoli, F., et al.: Greenway della Costa Viola: ipotesi di valorizzazione delle Ferrovie della Calabria, LaborEst, vol. 10, pp. 23–28 (2015)
8. Della Spina, L.: The integrated evaluation as a driving tool for cultural-heritage enhancement strategies. In: Bisello, A., Vettorato, D., Laconte, P., Costa, S., (eds.) Smart and Sustainable Planning for Cities and Regions. Results of SSPCR 2017. Green Energy and Technology. Springer (2018). https://doi.org/10.1007/978-3-319-75774-2_40. ISSN 1865-3537
9. Della Spina, L., Ventura, C., Viglianisi, A.: A multicriteria assessment model for selecting strategic projects in urban areas. In: Gervasi, O., et al. (eds.) Computational Science and Its Applications, ICCSA 2016. Lecture Notes in Computer Science, vol. 9788. Springer, Cham (2016). https://doi.org/10.1007/978-3-319-42111-7_32

A Model for Defining Sponsorship Fees in Public-Private Bargaining for the Rehabilitation of Historical-Architectural Heritage

Luigi Dolores[✉], Maria Macchiaroli, and Gianluigi De Mare

University of Salerno, 84084 Fisciano, Italy
ldolores@unisa.it

Abstract. This paper proposes a model of support for public administrations aimed at determining the rates for the exploitation of image rights by those companies that intend to sponsor rehabilitation or restoration projects in order to enhance the historical and architectural heritage present in Italy. This model has been applied to the city of Salerno (Italy). Indeed, it was assumed that the municipal administration is looking for sponsors intent on financing the restoration works of four city monuments. Through the model it was possible to determine the amount of funding, to be paid by the sponsors, including the amounts necessary for restoration works and advertising costs. The advertising fees for the city of Salerno have been determined starting from those established by Naples Municipal and applied to similar cases of cultural sponsorship. The parameter used for the comparison is the average monthly number of attendances that characterizes each location (direct audience). The costs of restoration work, for each monument of Salerno, have been determined through expeditious bills of quantities.

Finally, the total cost of sponsorship is equal to the sum of the restoration works costs and advertising costs.

Keywords: Sponsorship · Advertising rates
Redevelopment of public assets

1 Introduction and Objectives

The urban landscape policy makes use of a large number of instruments whose objective is the sustainable management of the urban environment aimed at improving the quality of life. Of primary importance, with regard to this topic, is the conservation of the aesthetic, cultural and emotional values of the man-made environments. In fact, in order to transmit the cultural performances of urban landscapes to future generations, it is necessary to act in a provident manner. Within the latter concept, this work can be contextualized. In fact, the paper focuses attention on the growing interest of the private world on the theme of the redevelopment of the historical-architectural heritage.

The contribution of private companies in the field of cultural heritage - above all in the face of the lack of public resources for the sector - is becoming increasingly

© Springer International Publishing AG, part of Springer Nature 2019
F. Calabrò et al. (Eds.): ISHT 2018, SIST 101, pp. 484–492, 2019.
https://doi.org/10.1007/978-3-319-92102-0_51

essential to the rehabilitation and use of public heritage. This happens above all in Italy, where the need for interventions appears infinite [1].

Private intervention in the cultural heritage sector takes place in various forms and has involved various types of subjects that are also very different from each other. In particular, in this work we focused on cultural sponsorship, whose relevance within the national scene is evident. In fact, there are numerous examples of private sponsorships for the recovery of world-famous monuments. The best known case is the restoration of the Coliseum financed by Diego della Valle's Tod's.

In Italy has increased in recent years the search for private sponsors by municipalities and state in order to finance restoration or rehabilitation projects impossible to support for the impoverished public coffers [2, 3].

However, companies are interested in investing in sponsorship only if the investment is able to generate a certain profit margin and an acceptable return in terms of image and reputation [4].

The point of view of the companies has already been analyzed by the authors in a previous study [5]. In the present work we will focus instead on the point of view of public administrations. In particular, a model will be proposed below, to support institutions, whose objective is the determination of the tariffs for the exploitation of image rights by those companies that intend to sponsor projects for the rehabilitation or restoration of the historical-architectural heritage.

The model in question was validated through the following case study: the restoration of four monuments of the city of Salerno financed by private companies through sponsorship. The objective is to determine the amount of financial resources deemed sufficient by the public administrations to proceed with the signing of the sponsorship contract. In this regard, the total cost of the sponsorship will include not only the amount necessary for the implementation of the interventions, but also a further amount that depends on the level of attractiveness of the locations; in fact, the latter factor is the one that most influences the economic return due to the company. By virtue of this, it is necessary to determine, for each location, adequate tariffs for the exploitation of image rights.

2 The General Logic of the Model

Figure 1 shows the scheme that summarizes the logic adopted in the construction of the model aimed at determining the tariffs representing advertising costs for those companies that invest in cultural sponsorship.

The total cost of the sponsorship can be considered equal to the sum of the cost to be incurred for the realization of the rehabilitation/restoration interventions plus an additional rate representative of the surcharge generated by the free sponsorship market [6]. This surcharge is directly proportional to the estimate of the image return and the economic return due to the company. In turn, the returns of the company are a function of different parameters, among which the attractiveness of the locations hosting the monuments plays a fundamental role. The latter depends mainly on the direct audience, that is, on the number of those present who recognize the message conveyed by sponsorship [7]. Generally speaking, it is possible, estimated the direct audience for a

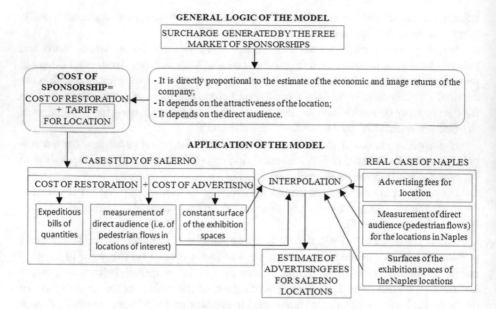

Fig. 1. Flow chart of the logical steps carried out for the construction of the model.

given location, to establish a tariff that represents the advertising cost. Therefore, the total cost of the sponsorship will be equal to the sum of the cost of the restoration works and the cost for the acquisition of advertising space on the scaffoldings of the construction sites.

3   The Application of the Model to the Case Study of Salerno

The proposed model was applied, through a case study, to the city of Salerno. It has been assumed that the municipal administration is intent on seeking some sponsors to finance the restoration of the following monuments:

- monument 1: Don Tullio's Fountain (location: Villa Comunale);
- monument 2: *Fishs' Fountain* (location: *Piazza Largo Campo*);
- monument 3: *Porta Nova* (location: *Piazza Flavio Gioia*);
- monument 4: *Dolphins' Fountain* (location: *Piazza Flavio Gioia*).

3.1   Monitoring Measures for Estimating Direct Audience

In order to establish the amount of publicity costs borne by the sponsors, measurements were taken to estimate the direct audience [8–12], that is to quantify the pedestrian flows at the locations of the city of Salerno where the 4 monuments, selected for the case study, are located.

Following the same procedures, monitoring campaigns were also conducted in the city of Naples. In fact, in this city the restoration work of some city monuments was sponsored through the Monumentando initiative promoted by the municipal administration [13].

The monuments chosen for Naples, and the relative locations, are the following:

- monument 1: Via Chiaia Bridge (location: Via Chiaia);
- monument 2: *Benches in Piperno stone* (location: *Villa Comunale*);
- monument 3: *Virgil's Temple* (location: *Villa Comunale*);

Each survey took place in a specific location where a monument is located. Exceptions are Piazza Flavio Gioia in Salerno and Villa Comunale in Naples, as they both host two monuments.

The measurements of direct audience carried out for each location have been grouped into 5 monitoring campaigns, each of which corresponds to a different day of the week. Two days chosen to make the measurements are midweek (Monday and Tuesday) instead the remaining three days coincide with the weekend (Friday, Saturday and Sunday).

For the locations of both cities, each monitoring campaign consists of a certain number of hourly measurements distributed in two time slots: one morning and the other afternoon-evening. Each measurements of pedestrian flows is one hour long.

Both for Salerno and for Naples have been chosen only those locations that are limited traffic areas. In this way it was possible to exclude the measurement of vehicular traffic.

Two techniques of measuring pedestrian flows were used: the first is the technique of *visual measurement*, with which we proceeded to the direct count of passers-by, while the second is that of *video measurement*, in which the counting of pedestrian flows is occurred after the realization of video recordings.

Finally, the result of each visual measurement was mediated with the corresponding result of the video measurement, obtaining a single reference value.

3.2 The Results of the Monitoring Campaigns Carried Out in Salerno and Naples

From the data obtained from the hourly measurements it was possible to obtain the average number of visitors present for each time slot (morning and afternoon-evening) and for each pre-established weekly day.

The average values obtained were in turn mediated and merged into three macro-groups: "number of visitors from Monday to Thursday", "number of visitors on Friday" and "number of visitors from Saturday to Sunday". This was possible because there was a homogeneity of the number of visitors on certain weekly days. For example, the number of visitors detected in a certain time slot on Monday is roughly the same as that recorded, in the same time slot, on Tuesday. The data was extended until Thursday, as it was assumed that in the weekdays the number of presences in the different time slots is maintained more or less constant. On the other hand, a significant increase in the number of presences was recorded on Friday: this day, in fact, it was not merged with anyone else because no homogeneity was found, in terms of number of presences, with other weekdays. On Saturday and Sunday there was a further increase in the number of visitors, which nevertheless remains almost constant for these two days.

Furthermore, through simple mathematical steps, it was possible to obtain the average monthly number of visitors for each location studied.

The data relating to the two areas of the Villa Comunale of Naples, initially thought of as two different locations due to the considerable extension of the large park, have been added together and merged as even for them there a homogeneity of pedestrian flows was recorded.

Below are the summary tables of the average number of visitors found for both Naples and Salerno in the two time slots (morning and afternoon-evening) and throughout the day for each of the three macro-groups (number of visitors from Monday to Thursday, number of visitors on Friday, number of visitors from Saturday to Sunday). The last column shows the average monthly number of visitors for each location of interest (Tables 1 and 2).

Table 1. Average values of visitors for the locations of Naples.

Location	Morning (10:30–14:30)			Afternoon–evening (15:00–19:00)			Average of daily visitors			Average monthly visitors
	Mon–Thu	Fri	Sat–Sun	Mon–Thu	Fri	Sat–Sun	Mon–Thu	Fri	Sat–Sun	
Via Chiaia	654	10860	16510	4905	8145	12383	11445	19005	28893	490280
Villa Comunale	395	650	1915	296	488	1436	691	1138	3351	42420

Table 2. Average values of visitors for the locations of Salerno.

Location	Morning (10:00–14:00)			Afternoon-evening (17:00–22:00)			Average of daily visitors			Average monthly visitors
	Mon–Thu	Fri	Sat–Sun	Mon–Thu	Fri	Sat–Sun	Mon–Thu	Fri	Sat–Sun	
Villa Comunale	798	2690	5375	3990	13450	26875	4788	16140	32250	399168
Piazza Largo Campo	1425	4370	8325	7128	21850	41625	8553	26220	49950	641328
Piazza Flavio Gioia	1128	3250	5675	5640	16250	28375	6768	19800	34050	438688

3.3   The Determination of the Advertising Fees of Salerno

The advertising fees for the locations of Salerno were determined using as data of the problem those established by the Municipality of Naples and assuming as independent variables the results obtained from the monitoring campaigns carried out in both cities.

The advertising fees of Naples have been extrapolated from the appropriate technical sheets drawn up by the municipal technical office for the *Monumentando* project [14].

The following table contain all the most significant information referring to the three locations of Naples, including the square meters of the exhibition spaces, the monthly unit costs of advertising and the monthly unit prices for the resale of advertising space (Table 3).

Table 3. Data relating to the sponsorship for the restoration of monuments located in the three locations in Naples where the monitoring campaigns were carried out.

Location	Via Chiaia	Villa Comunale	Villa Comunale
Monument	Via Chiaia Bridge	Virgil's Temple	Benches in Piperno stone
Exhibition area [m^2]	150	190	500
Exposure period [months]	8	4	1,5
Cost of sponsorship [€]	260.000	80.000	68.000
Cost of restoration [€]	210.000	63.000	52.000
Cost of advertising [€]	50.000	17.000	16.000
Unit cost of advertising [€/(m^2 x month)]	42	22	21
Resale price of advertising [€/month]	120.000	80.000	100.000
Unit resale prize of advertising [€/(m^2 x month)]	800	421	200

In the case of Naples, the *Monumentando* project was given the opportunity for the sole sponsor to be able to resell advertising space to third parties who wish to promote their image [15]. The same principle can also be applied to the case study of Salerno: for this reason, in addition to advertising costs, the resale prices of the exhibition spaces were also determined.

Linear interpolation was used to determine the tariffs and resale prices of the Salerno exhibition spaces, shown in Table 4.

Table 4. Advertising costs and resale prices of advertising spaces determined for monuments located in the Salerno locations.

Location	Exposure period [months]	Exhibition area [m^2]	Cost of advertising [€/(m^2 x month)]	Cost of advertising [€]	resale prize [€/ (m^2 x month)]	Resale price [€]
Villa Comunale	3	100	34	10.258	651	195.399
Piazza Largo Campo	3	100	55	16.482	1.046	313.940
Piazza Flavio Gioia	3	100	39	11.788	748	224.535

The results obtained have been approximated as shown in Table 5.

Table 5. Advertising costs and resale prices of advertising space for each of the locations in Salerno.

Location	Cost of advertising [€]	Resale price [€]
Villa Comunale	10.500	200.000
Piazza Largo Campo	16.500	320.000
Piazza Flavio Gioia	12.000	225.000

To obtain the total cost of the sponsorship, the respective costs for the restoration work must be added to the advertising costs (see Table 6); the latter have been determined through the processing of expeditious bills of quantities.

Table 6. Costs of advertising, costs of restoration work and total costs of sponsorship for monuments located in Salerno locations.

Monument	Cost of advertising [€]	Cost of restoration [€]	Total cost of sponsorship [€]
Dolphins' Fountain	12.000	27.000	39.000
Porta Nova	12.000	205.000	217.000
Don Tullio's Fountain	10.500	55.000	65.500
Fishs' Fountain	16.500	43.000	59.500

4 Conclusions

The public administrations, in resorting to the sponsorship contract, are interested in the identification of criteria and parameters to which to refer to maximize the public return in the application of the instrument [16–25].

The model proposed in this paper aims to identify these criteria and parameters, in full compliance with the relevant legislation. The amount of the funding must not only cover the costs for the works, services and supplies necessary for the recovery/restoration of cultural assets, but must also include an additional value representative of the cost for advertising, in turn linked the attractiveness of the locations and, for this reason, function of direct audience.

The estimate of the direct audience, despite being a very used technique both by companies and by the public administrations for the evaluation of the effects of a sponsorship campaign, presents a limitation. In fact, the knowledge of the number of individuals potentially able to recognize a specific advertising message should not be sufficient to evaluate the effectiveness of the sponsorship. This is because this measurement does not take into account the emotional reaction of target audience. The estimation of direct audience should be accompanied by additional methods of investigation through which, thanks to the help of surveys, questionnaires, focus groups and interviews, it is possible to know the opinion of each individual on the sponsorship.

Finally, the indirect audience should also be monitored, i.e. the public who has become aware of sponsorship through the mass media (media coverage).

References

1. Nesticò, A., Macchiaroli, M., Pipolo, O.: Costs and benefits in the recovery of historic buildings: the application of an economic model. Sustainability 7(11), 14661–14676 (2015)
2. Fidone, G.: Il ruolo dei privati nella valorizzazione dei beni culturali: dalle sponsorizzazioni alle forme di gestione, Aedon, pp. 1–2 (2012)
3. Mollica, A.: L'Italia si aggrappa agli sponsor per salvare i suoi monumenti. In: il Giornalettismo (2004)
4. Siano, A., Siglioccolo, M., Vollero, A.: Corporate communication management: Accrescere la reputazione per attrarre risorse. G Giappichelli Editore, Turin (2015)
5. Dolores, L., Macchiaroli, M., De Mare, G.: Sponsorship for the sustainability of historical-architectural heritage: application of a model's original test finalized to maximize the profitability of private investors. Sustainability 9(10), 1750 (2017)
6. Legislative Decree of December 19, 2012, Approval of technical standards and guidelines regarding sponsorship of cultural assets and similar or related cases
7. Cornwell, T.B., Maignan, I.: An international review of sponsorship research. J. Advert. 27(1), 1–21 (1998)
8. Pham, M.T.: The evaluation of sponsorship effectiveness: a model and some methodological considerations. In: Gestion 2000, pp. 47–65 (1991)
9. Ensor, R.J.: The corporate view of sports sponsorship. Athl. Bus. 9, 40–43 (1987)
10. Hulks, B.: Should the effectiveness of sponsorship be assessed, and how? AdMap 12, 623–627 (1980)
11. McDonald, C.: Sponsorship and the image of the sponsor. Eur. J. Mark. 25(11), 31–38 (1991)
12. Sparks, R.E.C.: Rethinking media evaluation: tobacco sponsorship messages and narrative conventions in motorship telecasts. In: Grant, K., Walker, I. (eds.) World Marketing Congress Proceedings, pp. 111–115. Academy of Marketing Science, Melbourne (1995)
13. Monumentando Napoli (2015–2017). https://monumentandonapoli.com/. Accessed 02 Dec 2017
14. Website of the Unooutdoor s.r.l. (2015–2017). http://www.unooutdoor.it/?pagina=monumentando. Accessed 02 Dec 2017
15. Website of the Municipality of Naples (2015–2017). http://www.comune.napoli.it/flex/cm/pages/ServeBLOB.php/L/IT/IDPagina/23775. Accessed 02 Dec 2017
16. Nelli, R.P., Bensi, P.: La sponsorizzazione e la sua pianificazione strategica. Modelli di funzionamento e processi di selezione. Vita e Pensiero, Milan (2005)
17. De Mare, G., Nesticò, A., Macchiaroli, M.: Significant appraisal issues in value estimate of quarries for the public expropriation. Valori e Valutazioni 18, 17–23 (2017)
18. De Mare, G., Granata, M.F., Nesticò, A.: Weak and strong compensation for the prioritization of public investments: multidimensional analysis for pools. Sustainability 7(12), 16022–16038 (2015)
19. Bencardino, M., Nesticò, A.: Demographic changes and real estate values. A quantitative model for analyzing the urban-rural linkages. Sustainability 9(4), 536 (2017)

20. Nesticò, A., Galante, M.: An estimate model for the equalisation of real estate tax: a case study. Int. J. Bus. Intell. Data Min. **10**(1), 19–32 (2015)
21. Nesticò, A., Sica, F.: The sustainability of urban renewal projects: a model for economic multi-criteria analysis. J. Property Invest. Fin. **35**(4), 397–409 (2017)
22. D'Alpaos, C.: Methodological approaches to the valuation of investments in biogas production plants: Incentives vs. Market prices in Italy. Valori e Valutazioni **19**, 53–64 (2017)
23. D'Alpaos, C., Marella, G.: Urban planning and option values. Appl. Math. Sci. **8**(157–160), 7845–7864 (2014)
24. Canesi, R., Antoniucci, V., Marella, G.: Impact of socio-economic variables on property construction cost: evidence from Italy. Int. J. Appl. Bus. Econ. Res. **14**(13), 9407–9420 (2016)
25. Antoniucci, V., Marella, G.: Is social polarization related to urban density? Evidence from the Italian housing market. Lands. Urban Plann. (2017, in Press)

Risk Management, Environment, Energy

Institutional Relations in the Small-Scale Fisheries Sector and Impact of Regulation in an Area of Southern Italy

Monica Palladino[1]([⊠]), Carlo Cafiero[2], and Claudio Marcianò[1]

[1] Mediterranea University of Reggio Calabria, 89100 Reggio Calabria, Italy
monicapalladino@hotmail.com
[2] Statistics Division, Food and Agriculture Organization of the United Nations, 00100 Rome, Italy

Abstract. This article is part of a broader study of the fishery system in a coastal area of Southern Italy. It presents a description of the main features of the fishing sectors in the five ports of the southernmost part of Tyrrhenian coast of the Reggio Calabria province, as derived from personal interviews to key informants. Such description contributes to an understanding of the key elements of the economic, social and political environment surrounding the small-scale fishery sector and its performance, both from a narrow economic and a broader development perspective. Further data, obtained by structured interviews to a sample of fishermen, provide information on the relations that fishermen entertain with cooperatives and other professional organizations. The main conclusion is that the system of such relations is still weak, heavily conditioned by some of the problems that resulted from the imposition, long time ago, of environmentally-motivated restrictions to the fishing activity in the area, which have all but been resolved. Lack of transparent, reliable information on the real environmental impact and sustainability of traditional practices contributes to create an environment that does not encourage active participation.

Keywords: Small-scale fishery · Environmental regulation in the EU fishery Cooperatives and other institutions

1 Introduction

This article elaborates on information gathered through qualitative interviews to key informants, conducted during the preparatory phase of a broader study aimed at analysing in the fishery system of the southernmost part of the Thyrrenian Coast of the province of Reggio Calabria, in Southern Italy, and from the data collected through structured interviews with fishermen in the area in the context of the same broader study, part of whose results have been already reported elsewhere [1–3]. The original study - commissioned by the *Stretto* Fishery Local Action Group (FLAG) - had the specific objective to provide information useful in developing new programming tools for the local governance of fishery in Calabria.

The objective of this article is to further elaborate on some of the conclusions of the broader study, reflecting on what emerges as one of the main obstacles to the possibility

© Springer International Publishing AG, part of Springer Nature 2019
F. Calabrò et al. (Eds.): ISHT 2018, SIST 101, pp. 495–504, 2019.
https://doi.org/10.1007/978-3-319-92102-0_52

to enact sustainable development strategies, namely, the difficulty in involving fishermen in a truly participated, bottom-up, policy making process.

The article begins by summarizing methodological aspects of the broader research that are relevant for this specific article. It then continues with a description of the study areas, highlighting the main features of the fishery value chain, as revealed through the eyes of the key witnesses, whose in-depth knowledge of the industry and willingness to share it, permitted us to form an understanding of the social and economic fabric on which the fishery sector is woven. Special focus is put on the cooperatives, given the traditional presence of this form of organization in the fishery sector in Italy.

Following the descriptive analysis, attention is given to the impact on the local fishing economy of norms and regulations issued to respond to sustainability concerns mostly raised by "others" and on which some of the local actors feel that their voices have not been sufficiently heard. The presence of different opinions and uncertainty on how serious the environmental threat that justifies the restrictions is, on which instrument could be most effective in address it, and on how effective are current monitoring and enforcing activities, contributes to fuel the belief that these regulations might be unfair, unduly penalizing the fishermen. By taking as examples the regulations concerning red tuna and swordfish catches, we identify critical aspects that, if properly addressed, could contribute to better integrated and more effective marine resource management and economic development of the local community.

2 Methodology

The research was conducted in the fishing ports of Gioia Tauro, Palmi, Bagnara Calabra, Scilla and Cannitello of Villa San Giovanni along the southernmost trait of the Tyrrhenian coast of the province of Reggio Calabria in Southern Italy (Fig. 1).

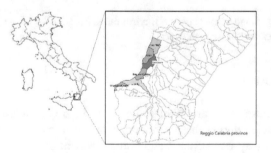

Fig. 1. The study area

In each of the area ports, a series of meetings and conversation with key witnesses allowed gathering information on the organization of the fishery value chain and the structure of existing relationship between the institutional actors. A series of in person, unstructured interviews to key informants on the small-scale fisheries sector were conducted by one of the authors between April and August 2015. In total,

15 professionals were interviewed (5 presidents or vice presidents of cooperatives and 10 professional fishermen). Each interview lasted for about one hour, and was based on open-ended questions. The interviews were aimed at (1) identifying the various actors and their role in the value chain in each of the port; (2) understanding the type of link between fishermen and the cooperatives, and (3) identifying the major institutional counterparts for both fishermen and their cooperatives when in need of information, technical advice and political support.

Additional information on the relations between fishermen and institutions in the area is provided by the data collected by detailed, structured interviews conducted on a convenience sample of 26 fishermen identified based on their willingness to participate in the research, covering the five ports (10 from Bagnara, 7 from Palmi, 4 from Scilla, 4 from Gioia Tauro and 1 from Villa San Giovanni)[1].

3 Key Actors and Their Role in the Fishery Value Chain in the Study Area

Of the five ports surveyed, *Bagnara Calabra* is the largest in terms of infrastructures, capital and number of actors involved. The entire economy of Bagnara is linked to fishing, and, in particular, to swordfish (*Xiphias gladius*). The harbour is almost entirely occupied by 66 fishing vessels, despite the presence of a recreational fleet linked to an active tourism sector. There is no wholesale fish market; fishermen sell their catch directly to retail fish sellers, with no agents or other intermediaries. Apparently, fishermen and fish seller do not sign formal contracts. The quantities traded depend on the season and the volume of catches. Retailers, in turn, sell fishes to restaurants and to smaller retailers in the main city of Reggio Calabria and across the province. While most of the fish traded by these retailers is indeed sourced locally, they integrate their offer with fish bought in fish markets from other regions. Part of the reason is that several local species quite common in the past, have now practically disappeared due to the excessive pressure of the fishing activity. Some describe the way in which fishing is operated in Bagnara as "wild"; bottom trawlers operate close to the shore, possibly in violation of existing regulations[2]. In the winter, when there are very limited local catches, even traditional oily fishes like the *alici* (*Engraulis encrasicolus*) come from abroad. When available, though, local fish does obtain a premium and, in the informal agreements with the retailer, the price is set by the fisherman. In some cases, as for example for fishermen who use the "palangaro" – a longline that needs to be armed with baits that the fishermen procure from the fish seller, the fish retailer has some advantages in the trade. This happens when catches are limited, and the bait ends up costing more than the fish caught, so that, in the end, fishermen have to pay a balance to the fish seller.

---

[1] See [2] for a detailed description of the quantitative study.

[2] REG. CE 1967/2006.

The second port, in terms of fleet size, is *Palmi*, where 33 vessels are listed in the local fishery license archive. Fishermen from *Palmi* operate out of the *"Tonnara"*, a small harbour village where all fishermen live, located north of the main town centre.

Most vessels in Palmi are engaged in what in Italy is defined *piccola pesca*[3] and operate with the *"lampara"* or *"cianciolo di notte"*, a nocturnal fishing practice based on surrounding nets operated from small boats; also, commonly used are the *"tremaglio"* and a version of longlines called *"conzo"*. Catches are sold directly to the only local fish seller and to other small traders that come from close-by towns. During the summer, when tourists abound in the area, temporary stands are set up along the coast by fishermen and their family members.

Third in terms of number of registered vessels is the port of *Gioia Tauro*, hosting 22 vessels, even though only about 10 of them, licensed for trawling are effectively active. Most of the fishermen in *Gioia Tauro* are originally from San Ferdinando. As opposed to the case of *Bagnara* and *Palmi*, in *Gioia Tauro* a wholesale fish market exists where fishermen sell half of their catches, while the other half is sold directly to final consumers.

With only 17 registered boats, the fishery of *Scilla* is renowned especially for the traditional swordfish "hunting" with the "passerella", a typical boat driven by a steersman who operates from the top of the "antenna", a 15–25 m tall tower where also a spotter stands. When the spotter and the steersman see a swordfish swimming at a distance, they operate the boat to bring the harpooner, who stands at the end of a 30-meter-long bridge mounted at the bow of the boat, close enough to strike his shot. Those who do it, lament that swordfish hunting is no longer as profitable as it once was, even though it is still part of a strong social and cultural local identity. Swordfish "hunting" used to be the centre of an activity to which entire families in *Scilla* will devote their time during the season from May to September. Today, the only 2 remaining *passerellas* (out of the many more that existed up to 10 years ago) still operate, despite the high operating costs, mostly out of a passion that they say they have in their "blood", handed down from generations, and somehow integrate their return by engaging in fish-tourism activities [4]. The other fishing boats operating out of *Scilla* use longlines and gillnets operating from small boats. As in *Bagnara* and in Palmi, fish is sold through informal agreements to local fish sellers, with no agents.

Last in term of size of the ports in the area, *Cannitello* (a fraction of *Villa San Giovanni*) lists only 5 vessels, only 2 of which appear to be still active, and thus insufficient to supply even just the local market, which mostly comes from the neighbouring fleets of Scilla and Bagnara.

4 Links Between Fishermen, Cooperatives and Other Institutions

One very common institution in the fishing sector throughout Italy is the cooperative. Pushed by a legislation created in the 1950's, which granted several fiscal benefits to cooperatives that were not available to single entrepreneurs, they have become a very

---

[3] A form of coastal fishing system regulated by the Ministerial Decree 14 September 1999 and its modifications.

popular form of association among fishermen. However, cooperatives seem to be little more than specialized service providers, rather than being the expression of a true cooperation spirit among fishermen who remain fundamentally individualistic [5].

Bagnara hosts the larger number of cooperatives in the study area, and 55 of the 66 vessels listed in the local fishing license archive are identified as being members of a cooperative. However, only three of them have proper sizes. Records show that the other 5 coops have only 1 or 2 active members, something that calls into question the very concept of a cooperative. The few vessels which are not members of cooperatives are defined as "autonomous" and are assisted by other firms or single professionals for the legal advice and accounting services that cooperatives provide. Nevertheless. the cooperatives remain the main counterpart for fishermen when technical, administrative or managerial functions are needed, but play no role in the marketing of fish.[4] Nevertheless, each fisherman individually maintains direct and frequent contacts with the cooperatives, which therefore still play a role in networking. Contacts by fishermen become instead very occasional, and only indirect, with the other organizations that revolve around fishing in the area, such as UNCI pesca, Lega Pesca, Federpesca, Federco-pesca Calabria and the National Observatory for Fisheries (Osservatorio Nazionale della Pesca). The relationship between the cooperatives and these other organization is direct, as cooperatives are members of these higher-level organizations, with which they are obviously in contact, but respect to which the key informants declare a rather low level of satisfaction. Except for direct acquaintance, for most fishermen one opportunity to meet each other is created by the participation to congresses or other events of regional or national scope.

Most fishermen in *Gioia Tauro* are not members of a cooperative. When in need of assistance, they use the services of the only professional that operates in the area, which are, reportedly, very expensive. Also, it is reported that relevant information (such as the publication of a call for application to local development plans that might provide financial support to the fishing activity) may not circulate timely and effectively. According to the interviewed informant, the circulation of this kind of information is left to the goodwill of the few who may find out, perhaps by browsing the Internet, and inform the others.

The situation is not very different in the other ports. There is only one cooperative in *Palmi*, named "Scoglio dell'Ulivo", with 12 members. The cooperative provides its members with technical support, while for administrative assistance fishermen make recourse to the services sold by a local professional accountant. As in *Bagnara*, also in *Palmi* the local cooperative is itself a member of a higher-level organization, with which, other fishermen, not members of the coop, have occasional relationships, particularly when in seek of information. Information is also sought independently by fishermen, with the help of someone who can help them browsing the Internet. The only cooperative in *Scilla*, the "Cooperativa Pontillo", lists many, but not all the local fishermen as their members. Some of them are members of the cooperatives based in *Palmi* and *Bagnara*. There are no cooperatives in *Cannitello*.

---

[4] Technical support includes, for example, the assistance needed for vessels inspections, or guidance on the use of specific equipment, such as GPS, etc.

To further explore the actual relevance of cooperatives and other institutions in the sector, we can refer to the results of the structured interviews that were conducted with fishermen, asking whether they know the institutions and cooperatives identified during the preliminary phase of the research, and whether they keep any relation with them. Figure 2 shows the results.

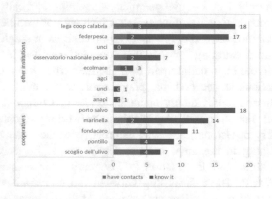

Fig. 2. Knowledge of and relationships with cooperatives and other institutions

The listed cooperatives are generally known by the interviewed fishermen. Three cooperatives based in Bagnara ("Porto Salvo", "Marinella" and "Fondacaro") have been reported to be known also by fishermen who operates in other ports. Several of the interviewed declared not only to know them, but also to have contacts with them.

For those who declared to have contacts, we also inquired on the degree of satisfaction, but many of the interviewed declined. Direct interaction of fishermen with higher level organizations, however, is quite limited, confirming the conclusions drawn from the interviews to key witness described above.

One institution in particular which should play a key role in support of the local fishing community is the Fishery Local Action Group (FLAG) [6], a reason why it was important to inquire on the opinion that fishermen had on its usefulness. Only 12 of the 26 fishermen we asked declared to be aware of the FLAG, and only 6 of them responded to the question concerning its usefulness and effectiveness in promoting the development of the coastal area, giving an overwhelming negative opinion (Table 1).

Table 1. Value judgement on the usefulness of the services provided by the "Stretto" FLAG

	For his or her activity	For the development of the fishery sector
Very useful	0	1
Rather useful	1	0
Not useful at all	5	3

5 EU Fishery Regulations and Their Impact on the Study Area

During the preliminary survey of the area, when we were looking for fishermen to interview, we faced the rather disappointing outcome of being able to find only few who were willing to participate, particularly from *Bagnara*. When approaching them, one could sense an aura of distrust. We soon realized that the difficulties fishermen had in releasing interviews might be linked to an episode dating back to 2008, when reporters working on an investigative TV documentary, used hidden cameras to record some of the interviews. Ever since, anyone asking for an interview in *Bagnara* would be seen with suspicion.

The documentary[5] denounced how some of the fishermen in *Bagnara* and other ports of Calabria still used the *spadara*, a drift gillnet traditionally used to catch swordfish. *Spadaras* had been officially banned by the European Commission back in 1998, with the ban formally adopted in Italy in 2002, even though its full implementation has proven rather difficult. Despite dating back to more than 15 years ago, the ban is still a highly debated issue in *Bagnara*. A fundamental problem seems to be that the fishermen keep thinking that, with the ban, they have been "unfairly" deprived of an important source of income - swordfish catching - now legally possible only with longlines or harpoons, which present much higher operation costs than drift nets.[6]

Without questioning whether the protection of marine mammals and other species accidentally caught by driftnets in the Southern Tyrrhenian sea would justify the restriction,[7] and irrespective of whether the *spadara* ban has been enforced properly, one interesting aspect that emerged from talking to the fishermen in the area is that there might have been unexpected consequences of the ban that do not seem to be captured in the prevailing legal and public debate on the issue.

One possibility is that the banning of the *spadara* nets might have had a negative impact on the sustainability of other fish stocks in the area. This is an opinion reported by fishermen interviewed in the other ports, who confirmed how fish was more abundant in the waters close to the shore at the time when the *spadaras* where legally in use. The explanation they provide is that when *spadaras* could be used freely, fishermen from *Bagnara* did not need to target other species. As a result, during the

---

[5] "Report" MARE NOSTRUM episode. Available at: http://www.report.rai.it/dl/Report/puntata/ContentItem-1c6411c7-2f60-490d-bd5a-2829c1d233ff.html.

[6] It is a common opinion among fishermen that there was no valid reason to ban the *spadara* and, even more so, the *ferrettara*, (a drift-net with smaller mesh size than the *spadara*, included in the ban a few years later) considered to be among the most highly selective fishing gears. In supporting their argument, local fishermen usually refer to a study according to which: "the number of species caught, on average, by a drift gillnet is generally lower than the number of species caught by any other type on fishing net" [7]. The study, however, qualifies the finding by stating that it does not consider the "problem of the accidental catches of cetaceous" which is one of the main reasons why the drift net ban was issued in the first place, following a 1995 UN General Assembly resolution calling for a moratorium of all large-scale drift nets.

[7] For discussions on the extent of by-catches of protected species by driftnets in the Mediterranean, see [8, 9].

swordfish fishing season (from April to October) many fewer boats would operate close to the shore in Costa Viola, thus allowing for local fish stocks to replenish. Since the ban, many of the fishermen who used the *spadara* or the *ferrettara* are now reported to be looking for alternatives, thus creating what is perhaps an excessive pressure on whitefish such as cod or dolphinfish.

Another relevant aspect of fishery regulation is the one linked to the quota system that regulates red tuna (*Thunnus thynnus*) catches. Currently, red tuna can be legally fished only by a small number of vessels historically registered in an explicit list, to which a quota is assigned every year. The allowed gears include surrounding nets, longlines (*palangari*) and traps (*tonnara fissa*) but not the harpoons.

6 Main Results and Discussion

Several results obtained from the analysis of the information gathered with the interviews in the area are worth being summarized. First, there is some evidence of possible overfishing lead to disappearance of traditional catches. Second, swordfish hunting is reported to be no longer as profitable as it once was. Third, fishermen have been reluctant to report an explicit opinion on the degree of satisfaction with the services provided by the cooperatives and other institutions they are in contact with.

While it is difficult to determine the reason for not giving an answer, the perception of the author who conducted the field interviews is that about a half of the interviewed fishermen do feel being supported by the activity of the cooperatives, but not from the higher-level organizations. This is telling, as especially fishermen who are not members of a cooperative would need to interact with these higher-level organizations. The other half informally shared the view that cooperatives might not always respond to the fishermen needs, even though they did not want their opinion to be kept "on record". Even the FLAG, an institution created specifically to allow participation of local actors, does not appear as a very participated initiative. While a thorough analysis would require more extensive interviews, the information reported here is sufficient to raise a few serious concerns regarding the success of the FLAG which has been unable to involve the fishermen who should be the key actors in a development plan designed around the fishing activity. As highlighted by Marcianò and Romeo [6], one of the reasons could be linked to the delays that occurred in the implementation phases of the FLAGs in Calabria, due to bureaucratic complexities arisen at the national and regional level.

Considering the need to address issues that involve the entire sector, rather than single fishermen (such as the need to address the issues generated by the *"spadara"* ban), lack of an active involvement by fishermen with the institutions that should be representing them, is a problem. There is a clear risk that relevant, specific information possessed by them and that could be crucial to design effective policies, gets neglected. One example is the unforeseen impact of the *spadara* ban on the depletion of other species - noted by those engaged in the small-scale, traditional fishing operations - that does not seem to be captured in the prevailing legal and public debate on the issue. Another example is given by the apparent contradiction between a system intended to prevent overexploitation of tuna stocks - the "quota system" - and the ban imposed to catches with the harpoon. It is undoubted that this is the most selective gear one could think of, as the harpooner must see the prey before catching it, and therefore can spare juvenile forms and avoid any by-catch. From the environmental point of view, this type of fishing should be promoted, but the high cost of operation, associated with the limited catches of swordfish they can make, make it unsustainable form an economic point of view. Why then, for example, do not give them also the possibility to catch red tuna? Even a small share of the current red tuna quota assigned to them might be sufficient to make the activity profitable again.

Local cooperatives have advanced proposals to the Regional Government and to the Italian Ministry of Agricultural Policies regarding the possible authorization to use in Calabria of alternative gear, such as smaller driftnets, as specifically foreseen by the EU regulation that allows for the regionalization of the fishing regulation. Similarly, the possibility to create a specific Producer Organization (PO) has been considered, but for that, participation of at least 70% of the fleet is required and, thus far, it has been impossible to reach an agreement among a sufficient number of fishermen. Proposals have also been made to revisit the system of attribution of red tuna quotas, by bringing it from the national to the regional government responsibility and abandoning the criterion based on the historic distribution, which excludes those who have not received a quota in the past.

At the time we write, these proposals have yet to materialize in concrete acts, and the prospects for a real change in the Calabrian fishery are still dim.

7 Conclusions

This article presented an overview of the fishing economy in an area of Calabria, in Southern Italy and its main problems, based on field interviews conducted in 2015. It highlights some of the key issues that condition the economic, social and environmental sustainability of an historically important sector in the area. One central theme has been the debate on regulations that impose restrictions on the fishing activity motivated by environmental sustainability concerns. These regulations come from policy making at the EU level, which is perceived as distant by most of the people we interviewed, imposed by people who are not aware of the local context, and who may be dominated by international interest.

One of the conclusions that can be drawn from the analysis presented above is that the lack of sufficiently detailed, reliable data on the performance of the fishing industry in the area, or on the actual selectivity of different fishing gears, makes it difficult to distinguish "facts" from "opinions". Information is disseminated mostly by environmentalists or by representatives of fishermen's interests, with the resulting suspicion that it may be conditioned by their competing special interests. On one hand, this contributes to exacerbate conflicts and make cooperation particularly difficult, in an environment where many of the possible sustainable development strategies would depend on stronger cooperation. On the other, it represents an area where initiatives like the FLAG and other forms of public/private partnerships for local development could contribute.

Acknowledgments. This study has been supported by GASTROCERT, a Project on Gastronomy and Creative Entrepreneurship in Rural Tourism, funded by JPI Cultural Heritage and Global Change - Heritage Plus, ERA-NET Plus action "Development of new methodologies, technologies and products for the assessment, protection and management of historical and modern artefacts, buildings and sites", 2015–2018. It provides further elaboration of the data collected and the results obtained in a project supported by the "Stretto" Coast FLAG, Calabria Region - European Fisheries Fund (2007–2013), Priority Axis 4: Sustainable Development of Fisheries Areas - Measure 4.1. Strengthening the Competitiveness of Fisheries Areas. Special thanks are due to Mr Vincenzo Bagnato and Mr Mimmo Zagari for their availability.

References

1. Palladino, M.: Social Network Analysis del sistema pesca in Calabria: un'analisi sul capitale relazionale nell'area del "GAC dello Stretto". Unpublished Report. Dipartimento di Agraria. Università degli Studi Mediterranea di Reggio Calabria, Italy (2015)
2. Palladino, M., Cafiero, C., Marcianò, C.: Relational capital in fishing communities: the case of the "Stretto" Coast FLAG area in Southern Italy. Procedia Soc. Behav. Sci. **223**, 193–200 (2016)
3. Palladino, M., Cafiero, C., Marcianò, C.: The role of social relations in promoting effective policies to support diversification within a fishing community in Southern Italy (2018, Forthcoming)
4. Nicolosi, A., Sapone, N., Cortese, L., Marcianò, C.: Fisheries-related Tourism in Southern Tyrrhenian Coastline. Procedia Soc. Behav. Sci. **223**, 416–421 (2016)
5. Buonfiglio, G., Coccia, M., Ianì, E.: L'associazionismo cooperativo nella pesca. Chapter 3.2. In: Cataudella, S., Spagnolo, M. (eds.) Lo stato della pesca e dell'acquacoltura nei mari italiani. Ministero delle politiche agricole alimentari e forestali, Rome (2011)
6. Romeo, G., Marcianò, C.: Integrated local development in Coastal Areas: the Case of the "Stretto" Coast FLAG in Southern Italy. Procedia Soc. Behav. Sci. **223**, 379–385 (2016)
7. Ferretti, M., Tarulli, E., Palladino, S. (eds.): Classificazione e descrizione degli attrezzi da pesca in uso nelle marinerie italiane con particolare riferimento al loro impatto ambientale. Istituto centrale per la ricerca scientifica e tecnologica applicata al mare - ICRAM (2002). http://www.isprambiente.gov.it/contentfiles/00010100/10119-icram-vol3.pdf. Accessed 10 June 2017
8. di Natale, A., Mangano, A., Maurizi, A., Montaldo, L., Navarra, E., Pinca, S., Schimmenti, G., Torchia, G., Valastro, M.: A review of driftnet catches by the Italian fleet: species composition, observers data and distribution along the net. Col. Vol. Sci. Pap. ICCAT **44**(1), 226–235 (1995)
9. di Natale, A.: Driftnets impact on protected species: observers data from the Italian fleet and proposal for a model to assess the number of cetaceans in the by-catch. Col. Vol. Sci. Pap. ICCAT **44**(1), 255–263 (1995)

The Value of Water: an Opportunity for the Eco-Social Regeneration of Mediterranean Metropolitan Areas

Alessandro Sgobbo$^{(\boxtimes)}$ ⓘ

University of Naples Federico II, 80138 Naples, Italy
alessandro.sgobbo@unina.it

Abstract. The quantitative limitation policies, adopted for environmental, ecological and socio-economic aims, in the Mediterranean metropolitan areas, do not seem to have had the desired effect. In fact, the coercive restraining approach of the metropolitan public Authorities which have tried to impose ecological and cultural objectives collide with the expectations and concrete needs of the citizen. In this way, there is among the stakeholders the distorted perception for which the public and the private interests seem naturally contrasting elements such that the protection of one must necessarily result in the mortification of the other.

On the contrary, the Mediterranean Water Sensitive Urban Planning research Project has assessed the hypothesis that the key to the effectiveness of the strategies aimed at ecology, safety and environment, both because of the evident multiscalarity and the potential multi-functionality of the issues and solutions, is in the ability to make the necessary actions consistent with the legitimate aspirations of the various stakeholders, especially when the low available financial resources impose stringent limits on public spending.

Keywords: Water Sensitive Urban Planning · Sustainability · Resilience

1 Climate Change: Conflicts and Opportunities

The emergency of climate change and its direct correlation with human activity and particularly energy production, has offered the opportunity for new forms of investment and economic development generally summarized in the so-called Green Economy. Unlike other fields and previous experiences, the very important aspect of this approach is that green economy is not focused on repairing the effects, developing mainly in adaptation, in order to make compatible the built environment to the new condition in the medium-term, and in mitigation and containment of the climate phenomenon in order to reverse the processes at the origin of climate change in the long run [1–6].

The analysis of the conditions surrounding the issues of resilience and ecological footprint in the metropolitan areas has been an opportunity to question the effectiveness of quantitative limitation policies that have been adopted with growing severity, in some regions, with respect to those environmental, ecological and socio-economic goals that are rightly pursued [7–10]. Or if it is also necessary to note that these policies did not achieve the desired aims and that in many areas the environmental and social

© Springer International Publishing AG, part of Springer Nature 2019
F. Calabrò et al. (Eds.): ISHT 2018, SIST 101, pp. 505–512, 2019.
https://doi.org/10.1007/978-3-319-92102-0_53

impairment level is now such high to require deep regeneration processes. These, under strict public budgets, find economic sustainability just in presence of appropriate forms of public-private partnerships which, in urban planning, become incentives for eco-oriented interventions supported by the granting of development rights [11–14].

The individuation of the transformations intensity that can be predicted in municipal planning is today the subject of significant disputes that cross both urbanistic and sociological, demographic and geographical aspects [15]. In urban studies, the aspect of major contemporary debate is in the definition of a balanced condition between new housing supply and environmental impact. These themes are still intimately interwoven with those of the residence and the renewed housing tensions in the Mediterranean metropolitan areas also highlight its social relevance [16–18].

The reasons that push, especially the small Municipalities, to develop excessive settlement forecasts are quite evident. In fact, although strategies aimed at containing land take and ecological footprint of urban settlements seem to have general agreement in public opinion, this availability to the public well-being immediately stops when this results in the lack of individual economic benefit [19].

In metropolitan public Authorities, farther from the citizen and more influenced by an evident fear of granting than the flaunted public interest [20], prevails the idea that the most effective way to follow the aims of environmental sustainability is first to limit new buildings construction [21]. However, the urban history of the Mediterranean metropolitan areas has taught that policies based only on prohibitions and constraints have limited hope of being able to achieve the pre-established objectives, resolving themselves in effects often more devastating than those which wanted to fight [22–24]. Starting from this observation, in the Mediterranean Water Sensitive Urban Planning research Project, the effectiveness of an alternative approach, aimed at transforming sustainability into a shared development, has been verified.

2   The Methodological Approach

The scientific literature and the review of the European best practices of urban renewal highlight the achievement of a high maturation of the products offer aimed at regeneration in terms of social, ecological and environmental quality as well as at increasing resilience. However, in Mediterranean metropolitan areas, although some institutions have tried to repeat these virtuous experiences, the conflictual dimension that leads such innovative pushes to self-depletion generally prevails [25]. In particular, the coercive restraining approach with which metropolitan public Authorities try to impose flaunted ecological and cultural aims collide with the expectations and concrete needs of the citizen. In this way, there is among the stakeholders the distorted perception for which the public and the private interests seem naturally contrasting elements such that the protection of one must necessarily result in the mortification of the other [26, 27]. On the contrary, the thesis of the Mediterranean Water Sensitive Urban Planning research Project is that the key to the effectiveness of the strategies aimed at ecology, safety and environment, both because of the evident multiscalarity and the potential multi-functionality of the issues and solutions, is in the ability to make the necessary actions consistent with the legitimate aspirations of the various stakeholders [28].

The Project is funded by the Department of Architecture of the University of Naples Federico II and by the CeNSU[1], as an intermediate product of the European research Project INTENSSS PA (funded by Horizon 2020). Further funding sources, as well as experimentation opportunities, came from the agreements for study and scientific research signed with various municipalities in the Metropolitan City of Naples.

The first step of the research Project involved the construction of an abacus of urban renewal best practices that respond to the prefixed objectives of resilience and ecological sustainability, selected because of various criteria. Social criteria, above all, where the measurement of the impact generated by each solution was implemented according to the theory of the capabilities approach, as re-elaborated by Nussbaum [29]: not in terms of synthetic well-being indicators but in terms of the number, quality and reach of the opportunities actually available to citizens for integration, inclusion and the development of personal freedoms. For the purposes of financial evaluation, the study employs the multi-criteria methodology based on MC Greal's research [30]; for the environmental and ecological indicators, the study refers mainly to the Monitor Urban Renewal [31].

Compatibility of every solution with local contexts was verified through their insertion in urban interventions that had common social and ecological objectives and which would be concretely realized. The evaluation was carried out by testing a pool of experts and decision-makers who were entrusted with the institutional responsibility of evaluating the feasibility of the intervention. The results were obtained with the help of scientific mediation through the comparison of alternative transformation strategies, while applying the multi-criteria and multi-group ANP evaluation method [32].

About the third phase, focused on the comparative assessment of the social and urban quality effects achieved in alternative intervention scenarios, an example is given in the following paragraph. Having set the results to be achieved in the following five-year period, three scenarios have particularly been constructed: S1 as the situation that would be caused without external interventions; S2 provides a new plan that achieves the aims based on the "traditional" strategies emerging from the debate between political decision makers and stakeholders. S3, the result of the WSUP approach, is developed through the synergistic implementation of the best practices that are suitable for the context according to the ANP assessment.

The selected scenarios were studied comparatively to evaluate the ecological and social effects. Furthermore, citizen inclusion, participation and integration effects were tested through repeated tests of the pool of non-professional end-users by way of in-depth interviews using the CATWOE approach [33].

3   A New Neighbourhood at the Edge of the Metropolis

Volla is a suburban municipality of the Metropolitan City of Naples whose settlement history is easily traced to three phases: the first, up to the 50s of the last century (about 4,000 inhabitants) is conditioned by the marshy nature of the sites and the settlement

---

[1] National Center for Urban Studies.

develops along the existing road network, focusing around the farms; the urbanism age, up to 1980 (about 8,000 inhab.), when the growth continued by planned subdivisions contiguous to the original core area; the sprawl phase, from the Irpinia earthquake up to about 2010 (23,000 inhabitants), essentially conditioned by the illegal building activity induced by the policies that accompanied the progressive expulsion of the middle and lower middle classes from the city of Naples [34, 35] (Fig. 1).

Fig. 1. The test scenarios S2 and S3.

Despite the demographic data indicate a stable population trend, the family dynamics and the migratory flow trends determine a housing need of over 2,200 new homes, of which about 1,850 for conditions of previous disadvantage. The current services provision is not very effective, although it is nominally acceptable[2]. In fact: the surfaces actually transformed for services purpose amount to about 7 sqm/ab, mostly dedicated to education; the areas of municipal property that have never been transformed amount to 4 sqm/ab; most of the areas used for green, sports and leisure are in a situation of neglect and urban decay and so they cannot be utilized by citizens.

The analysis of needs has determined that an area of about 276,000 m^2 has to be allocated to new services of which at least 195,000 for green, sports and leisure. 92,000 m^2 must be realized on areas already owned by the municipality and 184,000 on areas that have to be acquired. From the infrastructural point of view, there are serious conditions of hydraulic stress. Because of the marshy nature of the site that, although reclaimed, presents an average piezometric altitude at about −2 m and because of the territory high waterproofing, the pluvial flooding is an event which occurs at least every month. The financial resources amount to about € 12 million for the sewage infrastructure and € 2 million for ecological services related to urban waste management.

---

[2] About 11 m^2 per inhabitant, although much lower than the minimum legal requirement, it is more than two times the average endowment of the Metropolitan City of Naples municipalities.

The building rights availability is equal to 810,000 cubic meters for residential use (2,500 houses) and 180,000 m^2 of tertiary surfaces.

The scenarios S1 and S2 are very similar. Both envisage the construction of new, almost monothematic residential neighbourhoods on currently agricultural areas and the saturation of vacant lots in areas which are probably illegally subdivided. They also deal with the precarious hydraulic condition by increasing the capacity of the sewage network while the service deficit is entrusted to a hypothetical access to European structural funds and to the urbanization charges collected because of the new buildings. The S3 scenario plans to use the entire urban availability in the central area of the city, partly conditioned by the presence of the cemetery and mostly characterized by residual spaces waiting for a transformation.

Glossing over the description of the urban project[3], an element of particular interest of the solution is the role played by green and blue infrastructures. These form a multi-scale and multi-functional structure that meets the hydraulic requirements, the needs of green spaces, sports and leisure and those of ecosystem services. A network of raingarden and bioswales form the draining network of the rainwater, which is partly infiltrated and partly conveyed to the retention basin. This consists of uncovered canals and swimming ponds which: guarantee the necessary accumulation function; enrich the landscape quality of the urban environment; form highly inclusive social spaces thanks to the management model that involves both citizens' committees and local micro-enterprises. The adjacent green areas, ordinarily used for sports and leisure, are able to manage the water surplus in case of very intense rains (with a return time of 30 years - T30). In the presence of even more extreme events (100-year return time - T100) even part of the paved area is flooded but the efficiency of urban functions is still guaranteed (Fig. 2).

Fig. 2. WSUP approach implemented in the urban renewal case study: the drainage network in ordinary conditions, in heavy rain conditions – T30 and in extreme rain conditions – T100. *Source: own work with Carbone, Corrado, De Nicola and Faiella.*

4 Test Results and Conclusions

The prototype test, although with some critical issues[4], confirmed the thesis. S2 and S3 are able to achieve the minimum targets set while S1 was not able to guarantee the required service supply or to meet the financial limits. From the ecological point of

---

[3] For the detailed description, please consult the complete research report [39].

[4] Mainly due to the local property market mistrust.

view, the greening indicators have provided the results shown in Table 1 (where: Ga expresses the ratio between green and uncovered surface, Gs the ratio between green and non-waterproofed surface, Gb the ratio between permeable green surface and habitable/commercial; Gh represents the green surface dotation per person). Furthermore, for the S3 scenario: the RIE index [36] is doubled compared to S2; TEP consumption is reduced by 58.4% thanks to the implementation of an innovative Indirect Low Temperature District Heating[5] (I-LTDH). The TEP reduction is due to the innovative plant, which offers energy to the inhabitants of the new district with decreasing costs[6], and to the widespread growth of green roofs, supported by development rights incentives. The microclimatic conditions benefit from the significant evapotranspiration phenomenon guaranteed by the presence of green and permeable surfaces with average night temperatures that, following the project sunshine, are about 2° lower than those recorded in S2. To conclude, the volume of undifferentiated waste was 65% lower thanks to the incentive on the distributed heating that is a consequence of the organic fraction transfer by citizens to the biogasifier.

Table 1. Greening test results.

Indicator	S2	S3
Ga (the ratio between green and uncovered surface)	31%	48%
Gs (the ratio between green and non-waterproofed surface)	4,5%	87%
Gb (the ratio between green and built surfaces)	29%	84%
Gh (the green surface dotation per person)	8,15 m^2	19,50 m^2

As concerns the social aspects, the relation between the resiliency and improvement of the urban landscape gets on average 53% approval. This rises to 91% when the improvement is accompanied by the self-management of the spaces for social purposes. The in-depth interviews show the reasons for this: civil management of public space increases its perception as a common asset; the collaboration of citizens in the administration of services for increased socializing amplifies the collective effort toward inclusiveness objectives.

Participation in management gives rise to identity suggestions that go beyond the aesthetic-symbolic quality of the product, and shared responsibility and profit emphasizes the inclusive efficacy of each project. Although it has been checked at every scale (from co-housing to self-build, from co-working to social cooperative), the socio-ecological efficiency of the model seems to peak at the urban dimension when its effects reach beyond the boundaries of the project, including significant sectors of the

---

[5] A description of the plant, especially designed for the purposes of this research Project, can be found in the articles by Moccia and Sgobbo [37, 38].

[6] The amount of savings is dependent on the quantity and quality of organic waste delivered to the anaerobic biogas plant by this group of residents. For the others, according to a classical scheme already implemented in other cities, the profit boils down to a decrease in local taxes however very limited because of legislation, that depend on various parameters not directly controllable by the citizens.

city. An innovative role of social housing interventions emerges which, built with the help of scientific mediation through bottom-up processes that are intensive in participation, become a development driver for the urban community and, going beyond a simple response to the housing crisis, generate common assets with widespread inclusivity, safety and environmental quality effects.

From the hydraulic point of view, the system abolishes the use of sewage network for rainwater that, instead, will be disposed of partially by deep infiltration and partially by taking advantage of the reclaimed Bourbon canals. The use of ponds for bathing purposes regenerates the relationship of the inhabitants with the water that, from a waste to be rid of, returns to be a characterizing element of the territory identity.

References

1. Cianciullo, A., Silvestrini, G.: La corsa della green economy: come la rivoluzione verde sta cambiando il mondo. Edizioni Ambiente, Milano (2010)
2. Beatley, T.: Green Cities of Europe: Global Lessons on Green Urbanism. Island Press, Washington DC (2012)
3. Moccia, F.D., Sgobbo, A.: Flood hazard: planning approach to risk mitigation. WIT Trans. Built Environ. **134**, 89–99 (2013)
4. Frey, M.: La green economy come nuovo modello di sviluppo. Impresa Progett. Electron. J. Manag. **3**, 1–18 (2013)
5. Leigh, N.G., Blakely, E.J.: Planning Local Economic Development: Theory and Practice. Sage Publications, Thousand Oaks (2016)
6. Tira, M., Giannouli, I., Sgobbo, A., Brescia, C., Cervigni, C., Carollo, L., Tourkolia, C.: INTENSSS PA: a systematic approach for INspiring Training ENergy-Spatial socioeconomic sustainability to public authorities. UPLanD – J. Urban Plan. Landsc. Environ. Des. **2** (2), 65–84 (2017)
7. Moccia, F.D.: L'urbanistica nella fase dei cambiamenti climatici, Urbanistica 140 (2009)
8. Moccia, F.D., Sgobbo, A.: La polarizzazione metropolitana. L'evoluzione della rete della grande distribuzione verso un sistema policentrico sostenibile. Liguori, Napoli (2013)
9. Ceccarelli, T., Bajocco, S., Perini, L., Salvati, L.: Urbanisation and land take of high quality agricultural soils-exploring long-term land use changes and land capability in Northern Italy. Int. J. Environ. Res. **8**(1), 181–192 (2014)
10. Sgobbo, A.: Mixed results in the early experience of a place-based european union former program implemented in campania. Procedia-Soc. Behav. Sci. **223**, 225–230 (2016)
11. Della Spina, L., Calabrò, F., Calavita, N., Meduri, T.: Trasferimento di diritti edificatori come incentivi per la rigenerazione degli insediamenti abusivi. LaborEst **9**, 71–75 (2014)
12. Sgobbo, A.: Risk Economy: the effectiveness of urban supportive policies for the safety and resilience in town centres. UPLanD-J. Urban Plan. Landsc. Environ. Des. **1**(1), 77–119 (2016)
13. Gisotti, M.R.: La riqualificazione delle periferie romane nel nuovo prg e nelle politiche comunali. Strumenti e realizzazioni. Macramè **3**(2), 61–68 (2009)
14. Sgobbo, A.: La città che si sgretola: nelle politiche urbane ed economiche le risorse per un'efficace manutenzione. BDC Bollettino Del Centro Calza Bini **16**(1), 155–175 (2016)
15. Olagnero, M.: La questione abitativa ei suoi dilemmi. Meridiana **62**, 21–35 (2008)
16. Tosi, A.: Le case dei poveri: ricominciare ad annodare i fili. La vita nuda, pp. 151–162 (2008)

17. Indovina, F.: Appunti sulla questione abitativa oggi. Archivio di studi urbani e regionali **82**, 15–49 (2005)
18. Sgobbo, A.: Eco-social innovation for efficient urban metabolisms. TECHNE J. Technol. Archit. Environ. **14**, 335–342 (2017)
19. Mannarini, T.: Comunità e partecipazione. Franco Angeli, Milano (2004)
20. Cappiello, V.: Terraced Landscapes. The "Penisola Sorrentino - Amalfitana" Case. UPLanD – J. Urban Plann. Lands. Environ. Des. **2**(2), 299–318 (2017)
21. Bencardino, M.: Consumo di suolo e sprawl urbano. Drivers e politiche di contrasto. Boll. Soc. Geogr. Ital **13**(8), 217–237 (2015)
22. Di Lorenzo, A.: L'anticittà della camorra: la condizione disurbana della provincia di Napoli. Meridiana **73–74**, 173–190 (2012)
23. Moccia, F.D., Sgobbo, A.: Città Metropolitana di Napoli. In: De Luca, G, Moccia, F.D. (eds.) Pianificare le città metropolitane in Italia, Interpretazioni, approcci, prospettive, pp. 289–326. INU Edizioni, Roma (2017)
24. Couch, C., Petschel-Held, G., Leontidou, L. (eds.): Urban Sprawl in Europe: Landscapes, Land-use Change and Policy. Wiley, Hoboken (2007)
25. Sgobbo, A., Moccia, F.D.: Synergetic temporary use for the enhancement of historic centers: the Pilot project for the Naples waterfront. TECHNE J. Technol. Archit. Environ. **12**, 253–260 (2016)
26. Campbell, S.: Green cities, growing cities, just cities? urban planning and the contradictions of sustainable development. J. Am. Plan. Assoc. **62**(3), 296–312 (1996)
27. Pacione, M.: Private profit, public interest and land use planning - A conflict interpretation of residential development pressure in Glasgow's rural-urban fringe. Land Use Policy **32**, 61–77 (2013)
28. Hamdouch, A., Depret, M.H.: Policy integration strategy and the development of the 'green economy': foundations and implementation patterns. J. Environ. Plan. Manage. **53**(4), 473–490 (2010)
29. Nussbaum, M.C.: Women and Human Development: The Capabilities Approach. Cambridge University Press, Cambridge (2001)
30. Adair, A., Berry, J., McGreal, S., Deddis, B., Hirst, S.: Evaluation of investor behaviour in urban regeneration. Urban Stud. **36**(12), 2031–2045 (1999)
31. Häkkinen, T.: Assessment of indicators for sustainable urban construction. Civil Eng. Environ. Syst. **24**(4), 247–259 (2007)
32. Saaty, T.L., Vargas, L.G.: Decision Making with the Analytic Network Process. Springer Science, New York (2006)
33. Rosenhead, J., Mingers, J.: Rational Analysis for a Problematic World Revisited: Problem Structuring Methods for Complexity, Uncertainty and Conflict. Wiley, Chichester (2001)
34. Mangoni, F., Sgobbo, A.: Pianificare per lo sviluppo. Un nuovo insediamento ai margini della metropoli. Edizioni Scientifiche Italiane, Napoli (2013)
35. Sgobbo, A.: Pianificare alla frontiera della metropoli: tra identità e centralità. In: Coppola, E. (ed.) La pianificazione comunale nel Mezzogiorno, pp. 329–344. INU Edizioni, Roma (2015)
36. Comune di Bolzano. http://www.comune.bolzano.it/. Accessed 30 Nov 2017
37. Sgobbo, A.: Recycling, waste management and urban vegetable gardens. WIT Trans. Ecol. Environ. **202**, 61–72 (2016)
38. Moccia, F.D., Sgobbo, A.: Pertnership pubblico-privato, infrastrutture ed ecologia. Planum. J. Urban. **25**(2), 1–7 (2012)
39. Sgobbo, A.: Water Sensitive Urban Planning: approach and opportunities in Mediterranean metropolitan areas. INU Edizioni, Roma (2018)

Post Carbon City: Building Valuation and Energy Performance Simulation Programs

Alessandro Malerba[1], Domenico Enrico Massimo[1(✉)] (iD),
Mariangela Musolino[1] (iD), Francesco Nicoletti[1] (iD),
and Pierfrancesco De Paola[2] (iD)

[1] Mediterranea University of Reggio Calabria, 89100 Reggio Calabria, Italy
demassimo@gmail.com
[2] Federico II University of Naples, 80125 Naples, Italy

Abstract. Today, world carbon energy consumption has increased dramatically. The survival of the Earth is endangered by pollution, produced by excessive oil and carbon over use. The sector that consumes this 40% of total energy is construction which needs innovative models for integrated ecological-energy - economic forecast. The research set up an integrated model for the overall assessment of a building having alternative characteristics: sustainable, vs. unsustainable or Common or Business As Usual BAS. The research takes into consideration the energy consumption for the thermal management and climate metabolism of the buildings and the consequent impacts in terms of CO_2 emissions. The results obtained validate the adoption of ecological cork panels for passivation and insulation in sustainable building vs. common. The research scientifically and accurately quantifying <u>two</u> alternative (sustainable vs. BAS) prototype buildings, comparatively testing three Energy Performance Simulation Programs (Energy Plus; Termus; Blumatica Energy) ascertaining the coherence and convergence of all their output and results.

Keywords: Post Carbon City · Green buildings · Passivation using cork
Energy consumption · Energy Performance Simulation Programs
Energy Plus

1 Introduction

Today, energy consumption has increased dramatically, especially in urban areas. One cause of this global problem is the high percentage of migration of the rural population to the mega-cities, and the consequent deforestation and urbanization of all the available rural and wood areas surrounding the existing towns. Buildings are among the largest consumers of energy [1]. Recent research has shown how energy consumption in the world has increased also because air conditioning in urban buildings has raised increased significantly in order to reach comfortable living temperature [2]. Energy

A. Malerba authored Sects. 7–9. D. E. Massimo authored Sects. 3–5. M. Musolino authored Sects. 1 and 2. F. Nicoletti authored Sect. 6. P. De Paola authored Sect. 10.

© Springer International Publishing AG, part of Springer Nature 2019
F. Calabrò et al. (Eds.): ISHT 2018, SIST 101, pp. 513–521, 2019.
https://doi.org/10.1007/978-3-319-92102-0_54

policies must be based on the more efficient use of energy, especially in construction, planning to regenerate existing settlements and not to continue to develop an the territories bordering urban areas. Consequently, the two coordinated following results would be achieved at the same time. First, the physical rehabilitation of the buildings through restoration interventions. Second, the implementation of an ecological process of building energy structural permanent saving through ecological retrofitting [2, 3]. It'is important to assess the energy consumption of buildings with high precision instruments. In the research that follows a prototype building has been analyzed valuated in two different construction scenarios. *I.e.* two alternatives using opposite techniques and materials: sustainable and ecological (oil free; chemical free) *versus* commonly used in current building process. Three Energy Performance Simulation Programs (EPSP) have been used comparatively to verify if sustainable and ecological buildings can achieve significant savings in heating and cooling costs [2].

2   Research Problem

The planet's environmental crisis is caused, among others, by energy over consumption after people metropolitan concentration: more than 50% of the population live in urbanized areas and this will rise to 66% in 2050. Consequence is unequivocal Climate Change in direction of Global Warming and its negative spillovers: increasing of global air and ocean temperatures, rising of global average sea level, reductions of glacier, ice and snow surfaces. The forecast for the end of the century is: temperature increases of 3 to 6 °C, extreme weather phenomena intensification, reduced levels of regular rain in various areas. The sectors that consumes this 40% of total yearly used energy are the civilian and the building industry (see Fig. 1). Contemporary buildings (80% of the existing total) are mostly to blame because they have an excessive consume of energy. It consumes (per unit) much more energy than historical building.

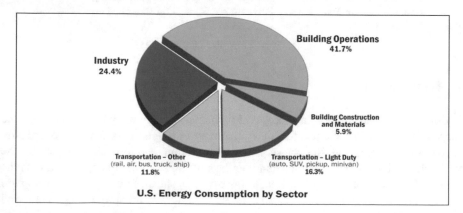

Fig. 1. The sectors that consumes 40% of the total energy are both the civilian and the building industry. Source: U.S. Energy Information Administration (2011) Energy annual review.

Contemporary buildings are mostly to blame. Modern construction (80% of the existing total) has an excessively consume of energy. It consumes (per unit) much more energy than historical building. The metabolism of modern building is different that of historical building, which used larger thicker walls and natural materials. Consequently, it is important to decrease energy consumption in modern buildings.

3 Aim of Research: Prototypes of Alternative Scenarios

The scope of this research is focused on integrated ecological - energy - economic valuation of buildings. The research includes the innovative comparative test of different buildings. In fact, testing will not be carried out on a single case study, but on two prototypes buildings which are equal in size but alternative in the materials used and then in the respective thermal behaviors. The two "alternative" prototypes used are very small, simply built units where the building energy performances can be easily checked. These are the minimum units of meters ($5 \times 5 \times 4$) that can be valued for multiple residential, tertiary and productive uses (including agricultural, zoo-technical and forestry). The prototypes are simplified architectures with extreme characteristics, like single - storey cubes. The use of cork panels is in the underlying conditions of certain energy assessment instruments (**EPSP**). **Use of cork** shall constitute a general advantage for: society, economy, ecology, energy balance, **all** actors of the process. The verification of a positive and significant impact of the insulation and specifically of adoption of cork panels in the prototype case makes it highly that these positive effects of cork panels will be amplified (and therefore more favorable) in the majority of real cases. These prototypes, previously designed, are described in Fig. 2.

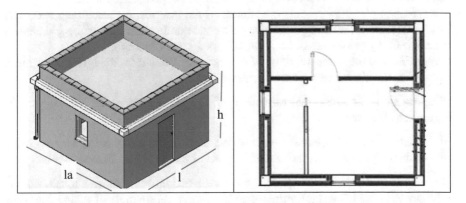

Fig. 2. BAS Prototype. The conceptual model in axonometric view.

3.1 First Scenario. "Common" (Business as Usual = BAS)

The "Business As Usual, BAS" building units have been constructed using the commonly–used and the materials procedures of southern Italy. It consists of a common point structure in reinforced concrete (base beam, pillars, flat roof slab) and the usual buffering in common bricks (flanked, not confined, non-cooperative).

Commonly - used (external - internal) plasters is in cement-based mortar, or in industrial hydrated lime plus cement - based mortar. This plaster is made up of three or four layers (bridge of adhesion, plaster = rustic, civil = shaving = finish; eventual putty or smooth finishing with American metallic spatula) plus the generally synthetic color (Table 1).

Table 1. Comparison between the BAS prototype and BAS + Cork prototype dimensions.

Prototype	l (m)	w (m)	h (m)	S (m^2)	V (m^3)
01.SuberType BAS	5,05	5,05	3,19	25,50	81,35
02.Subertype BAS + Cork	5,12	5,12	3,19	26,21	83,62

3.2 Second Scenario. Sustainable "BAS + Cork"

It is the same of BAS without plaster, unless a flat surface for panels. There is therefore an addition of cork panel components with a thickness of 6 cm, both horizontal (above the floor or attic, under crawl space) than vertical (external walls) either: or during the new construction or as an ecological retrofitting of the same BAS unit.

4 Research Methodology

The experiment uses two different alternatives, which are the adoption of the components in cork versus the non - use of cork.

The research activity carried out is structured in four main steps shown below.

- The design of the different architectures (buildings; units; details) in order to simulate: adoption of the components in cork; against their failure to adopt.
- The creation of a first module containing all the information machining necessary for construction, including a relative micro-economic Analyses of the Elementary Factors (EFA) employed and the market price repertoires of the factors.
- The estimating the monetary costs of the resources needed for the initial investment, for the realization of two alternative buildings.

The research involves the formation of the databases which will allow the researchers to forecast:

- the different thermal behaviors of materials used in the buildings to be compared;
- the ecological footprint in terms of emissions of the different alternative prototypes.

In order to obtain scientific quantitative results this information was translated into purely monetary and financial terms of induced impacts (or not) of taking into consideration the adoption and implementation of cork panels.

There are several aspects to the impacts that the use of cork can make, the most important and the most immediate of which are physical and financial:

- The increased energy efficiency, *i.e.* lower energy consumption, expressed in kWh, and then perpetually smaller energy bill;

The consequent increased health and sustainability of the everyday life into ecological shelter, and the more effective climatic management of buildings.

5 Research Expected Results

This research aims to verify, in advance, the possible positive impact of insulation using cork panels in terms of energy efficiency in common buildings and to estimate both initial investment costs and the permanent saving in energy management.

Simulation is performed on two alternative "prototypes" *i.e.* simple buildings which create two alternative scenarios.

- The first scenario: buildings built along traditional lines which are been always followed in the region and macro region. It can be called a "Business As Usual (BAS)" or "common" scenario. BAS is the building technique that a technical designer, a builder, a contractor would spontaneously use in the majority of cases.
- The second scenario is designed with the aim of creating a major improvement in thermal behavior by means of external total insulation using cork panels.

6 Compare Energy Performance Simulation Programs, BEPSP

Research performs the valuation of energy consumption in kWh and CO_2 emission in kilos in two different scenarios. This will be performed by means of three very different Building Energy Performance Simulation Programs, below described.

Energy Plus. (Version 8.3.0) together with Design Builder (Version 4.5.0.178) is one of the best known energy simulation software tools [4–7]. It is a software for thermal simulation and energy diagnosis in dynamic building arrangements. There are external graphical interfaces that facilitate the creation of the thermal model of the building and the inclusion of its characteristics, like Design Builder and others.

TerMus. (Version 30.001) is an Italian software used for the thermal engineering and energy performance of buildings. The energy certification (APE-AQE), calculation of transmittance and drafting Protocol Ithaca are some of the outputs of this software. It is regarded as standard software in Italy.

Blumatica Energy. (Version 6.1) is a software that allows the planner to design the thermal insulation of buildings and the management of their energy certification.

7 Estimate of Energy Consumption and CO_2 Emissions

Estimate were carried out on two prototypes (BAS and BAS + Cork), using the three EPSP above cited, each having its own characteristics. Output is below (Table 2).

The output of the three Energy Performance Simulation Programs are convergent (Table 3).

Passivation using cork paneling improves the overall thermal performance of buildings and research in the field has shown that healthier buildings and with better thermal performance have are worth more and their selling prices is consequently

Table 2. Comparison of output from three various simulation software tools.

Scenarios	Termus		Blumatica energy		Energy Plus	
	EPgl kW/m² y	CO₂ Kg/m² y	EPgl kW/m² y	CO₂ Kg/m² y	EPgl kW/m² y	CO₂ Kg/m² y
01.BAS	114	24	116	11	129	15
02.BAS + cork	69	15	71	8	73	9
Δ	−45		−45		−56	

Table 3. D (kWh/m²year consumption, Kg/m²year emissions) between BAS and Eco Scenarios.

	Termus Δ	Blumatica Energy Δ	Energy Plus Δ
EPgl kWh/m²y	−40%	−39%	−44%
CO₂ Kg/m²y	−36%	−26%	−43%

higher in the real estate market [8–19]. Valuation compares energy consumption (kWh) and CO_2 emissions (Kg) assesses and compare them with the monetary costs of construction and insulating materials. The software provided the following output regarding:

- Global Primary Energy (EPgl) which demonstrates the efficiency of the building and the system used for the heating and hot water.
- CO_2 that the building and the systems release in the environment (Table 4).

Table 4. Total energy consumption (kWh) and CO_2 emissions (Kg) in 01 and 02 Scenarios (Energy Plus tool) in the year.

Scenarios	A m²	EPgl kW/m² y	EPgl kWh	CO₂ Kg/m² y	CO₂ Kg
01.BAS	25,50	129	3.313	15	382
02.BAS + cork	26,21	73	1.913	9	235

8 Cost of Cork in the Construction

Based on analytical and detailed estimate, this research has forecasted the financial costs involved in the construction of the two alternative scenarios (Table 5).

Table 5. Comparison of the building costs of the 2 prototypes. Light Δ cost of sustainability.

Prototipo	01.BAS	02.BAS + cork	Δ	%
Tot €	37.156	40.378	+3.221	08,66
Tot €\m²	1.456	1.540	+83	
Tot €\m³	364	385	+20	

9   Results

The research carried out here shows in quantitative terms that the adoption of cork panels creates a measurable difference in the energy and ecological management of building. Even in though there is an initial higher technical cost in construction. In this way the planner can succeed in achieving multiple perpetual benefits, regarding energy saving and consequent CO_2 emission reduction that can be quantified:

$$From\ type\ 01\ (BAS)\ to\ type\ 02\ (BAS + Cork = Eco) : \qquad (1)$$

- Energy consumption, [kWh\m²\Year] in theory can be reduced from 129 to 73, the difference in saving being −44%, with conventional class improvement from F to C;
- CO_2 emissions can be reduced from 15 to 9 kg\m²\Year, i.e. −6 minus which is −43%. In conclusion, research shows that just the only one improvement of the cork panels adoption have a significant positive ecological, economic, financial impact (Fig. 3).

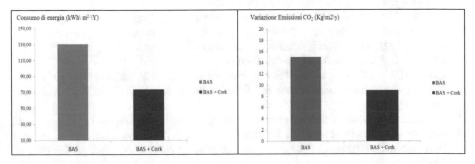

Fig. 3. Differentials (Δ) annual Energy consumption (–44%) and CO_2 emissions (–40%) in the two scenarios.

Passivation using bio cork paneling have multiple impact. Researches detected that bio eco green buildings have selling price higher (if compared to common building, no green) in the real estate market [7–12, 19, 22, 24]. Green building strategy extended at urban level allow to start the Post Carbon Historic Centers, University and City built up [14–17, 19, 23, 25, 26]. Using Cork enlarge area devoted to Cork Oak forest and enhances landscape [13, 18, 20, 21].

10   Conclusion

The comparison of the 2 prototypes buildings allows for a valuation in quantitative terms of their different energy consumption and CO_2 emissions due to peculiar thermal behaviors. At the same time it provides checks of the two buildings and proof of the effectiveness of Cork in its application in construction as a thermo-insulating material.

Energy efficiency is the final goal of the Strategy. From the energy and economic valuation the high positive impact of the use of cork panelling in buildings is evident when compared to its non-adoption. The two most immediate and visible results are the consequent: lower ecological emissions; lower energy consumption. Future developments in research will valuate in multi-dimensional terms the additional "fundamental benefits" i.e. healthier indoor and outdoor environment due to mitigated pollution and independence from the cycle of oil due to radical saving above demonstrated. All the above possess arelevant economic and ecological value. Also this research ascertained the very important coherence, convergence and similar outcomes of three different "Building Energy Performance Software Program", or BEPSP, namely EnergyPlus, Termus, Blumatica.

References

1. Yilmaz, Z.M.: Evaluation of energy efficient design strategies for different climatic zones. Energy Build. **39**, 306–316 (2007)
2. Stoakes, P.J.: Simulation of airflow and heat transfer in buildings. M.Sc. thesis, Virginia Polytechnic Institute and State University (2009)
3. Crawley, D.B., Hand, J.W., Kummert, M., Griffith, B.T.: Contrasting the capabilities of building energy performance simulation programs. Build. Environ. **43**(4), 661–673 (2008)
4. Crawley, D.B., et al.: Energy Plus: new capabilities in a whole-building energy simulation program. Build. Simul. **33**(4), 51–58 (2001)
5. Rallapalli, H.S.: A comparison of EnergyPlus and eQuest whole building energy simulation results for a medium sized office building. Technical report, Arizona State University. Mimeo (2010)
6. Sousa, J.: Energy simulation software for buildings: review and comparison. Energy (2012)
7. Del Giudice, V., De Paola, P., Manganelli, B., Forte, F.: The monetary valuation of environmental externalities through the analysis of real estate prices. Sustainability **9**(2), 229 (2017)
8. Salvo, F., De Ruggiero, M., Forestiero, G., Manganelli, B.: Buildings energy performance in a market comparison approach. Buildings (2017)
9. Tajani, F., Morano, P., Locurcio, M., Torre, C.: Data-driven techniques for mass appraisals. Applications to the Residential Market of the City of Bari (Italy). Int. J. Bus. Intell. Data Min. **11**(2), 109–129 (2016)
10. Massimo, D.E., Del Giudice, V., De Paola, P., Forte, F., Musolino, M., Malerba, A.: Geographically weighted regression for the post carbon city and real estate market analysis: a case study. In: Calabrò, F., Della Spina, L., Bevilacqua, C. (eds.) Local Knowledge and Innovation Dynamics Towards Territory Attractiveness Through the Implementation of Horizon/E2020/Agenda2030. Springer, Berlin (2018). ISBN: 978-3-319-92098-6
11. Del Giudice, V., Massimo, D.E., De Paola, P., Forte, F., Musolino, M., Malerba, A.: Post carbon city and real estate market: testing the dataset of reggio calabria market using spline smoothing semiparametric method. In: Calabrò, F., Della Spina, L., Bevilacqua, C. (eds.) Local Knowledge and Innovation Dynamics Towards Territory Attractiveness Through the Implementation of Horizon/E2020/Agenda2030. Springer, Berlin (2018). ISBN: 978-3-319-92098-6

12. De Paola, M.P., Del Giudice, V., Massimo, D.E., Forte, F., Musolino, M., Malerba, A.: Isovalore maps for the spatial analysis of real estate market: a case study for a central urban area of Reggio Calabria, Italy. In: Calabrò, F., Della Spina, L., Bevilacqua, C. (eds.) Local Knowledge and Innovation Dynamics Towards Territory Attractiveness Through the Implementation of Horizon/E2020/Agenda2030. Springer, Berlin (2018). ISBN: 978-3-319-92098-6
13. Spampinato, G., Massimo, D.E., Musarella, C., De Paola, M.P., Malerba, A., Musolino, M.: Carbon sequestration by cork oak forests and raw material to built up post carbon city. In: Calabrò, F., Della Spina, L., Bevilacqua, C. (eds.) Local Knowledge and Innovation Dynamics Towards Territory Attractiveness Through the Implementation of Horizon/E2020/Agenda2030. Springer, Berlin (2018). ISBN: 978-3-319-92098-6
14. Massimo, D.E., Malerba, A., Musolino, M.: Green district to save the planet. In: Oppio, A., et al. (eds.) Integrated Evaluation for the Management of Contemporary Cities. Springer, Berlin (2017)
15. Massimo, D.E., Malerba, A., Musolino, M.: Valuating historic centers to save planet soil. In: Oppio, A., et al. (eds.) Integrated Evaluation for the Management of Contemporary Cities. Springer, Berlin (2017)
16. Massimo, D.E., Fragomeni, C., Malerba, A., Musolino, M.: Valuation supports green university: case action at Mediterranea campus in Reggio Calabria. Soc. Behav. Sci. **223**, 17–24 (2016)
17. Massimo, D.E.: Green building: characteristics, energy implications and environmental impacts. case study in Reggio Calabria, Italy. In: Coleman-Sanders, M., (ed.) Green Building and Phase Change Materials: Characteristics, Energy Implications and Environmental Impacts, vol. 01, pp. 71–101. Nova Science Publishers (2015)
18. Massimo, D.E., Musolino, M., Barbalace, A., Fragomeni, C.: Landscape quality valuation for its preservation and treasuring. In: Advanced Engineering Forum, pp. 625–633. TTP Publications, USA (2014)
19. Massimo, D.E., et al. : Sustainability valuation for urban regeneration. The Geomatic Valuation University Lab research. In: Advanced Engineering Forum, pp. 594–599. TTP Publications, USA (2014)
20. Massimo, D.E., Fragomeni, C., Musolino, M., Barbalace, A.: Landscape and settlements. Historic center qualitative and quantitative valuation. In: Agribusiness, Paesaggio & Ambiente, vol. 17, pp. 309–322 (2014). Special Edition 1
21. Massimo, D.E., Musolino, M., Barbalace, A., Fragomeni, C.: Landscape and comparative valuation of its elements. In: Agribusiness, Paesaggio & Ambiente, vol. 17, pp. 53–60 (2014). Special Edition 1
22. Massimo, D.E.: Emerging Issues in Real Estate Appraisal: Market Premium for Building Sustainability, Aestimum, pp. 653–673 (2013)
23. Massimo, D.E., Musolino, M., Barbalace, A., Fragomeni, C.: Multi dimensional valuation of monuments. In landscape context. In: Sabiedriba, Integracija, Izglitiba, vol. 3. pp. 89–100 (2013)
24. Massimo, D.E.: Stima del green premium per la sostenibilità architettonica mediante Market Comparison Approach, Valori e Valutazioni (2010)
25. Massimo, D.E.: Valuation of urban sustainability and building energy efficiency. A case study. Int. J. Sustain. Dev. **12**(2-3-4), 223–247 (2009)
26. Musolino, M., Massimo, D.E.: Mediterranean Urban Landscape. Integrated Strategies for Sustainable Retrofitting of Consolidated City, Sabiedriba, integracija, izglitiba III, pp. 49–60 (2013)

Ecological Resilience and Care of the Common House to Build the Landscape of Contemporaneity and Future Scenarios of Territories and Cities

Stefano Aragona[✉]

Mediterranea University of Reggio Calabria, 89100 Reggio Calabria, Italy
stefano.aragona@gmail.com

Abstract. The paper has its main base in the ecological vision of territory and city. It is linked to the holistic philosophy of social and spatial transformations. Transformations that involve attention to the short term but which must be connected to the medium and long term. This requires actions to act immediately, included in broad strategies. All this considering the territory and the city as "Common Good", citing the 2015 Encyclical *Laudato Sii for the Care of the Common House* [1]. Thus going far beyond the idea of cities as "public space" and emphasizing the care of it, i.e. of its management. In this debate a very important contribution comes from scientific works that also require from a legal point of view the modification of the relationship between man and nature as done by F. Capraand U. Mattei U. *Ecology of Law. Science, Politics, Common Goods* [2] (Ecologia del diritto. Scienza, politica, beni comuni). The paper continues the research studies about the anthropization way, started in 1987, considering the so call "crisis" as opportunity for a radical change of it. It indicates a path in line with the landscape indications of the homonymous Florence Convention, a document particularly relevant for our country.

Keywords: Ecological approach · Integrated town planning · Scenarios

1 Territory and Development: New Paths

"We destroy the beauty of the landscape because the splendours of nature, freely available, have no economic value. We would be able to extinguish the sun and the stars because they do not pay a dividend" (John Maynard Keynes) [3].

These words light in a clear way the question faced by the paper. The outcomes of about 350 years of industrialization are evidencing many unsustainable aspects. The spatial effects and, at the same time, origins of industrial society, consist in the towns and territories where we live. For the most of towns there are problems of pollution, traffic congestion, above all peripheral areas characterized by a very low quality. At the same times a lot of territories are devastated because the use of many natural resources and, destruction of their cultural heritage. In 1972 the book *The Limits of Growth*[1] described many of these rising questions and it foresaw the coming crisis.

---

[1] Report made by the Meadows & Meadows research group of MIT, commissioned by Aurelio Peccei, President of Club of Rome.

© Springer International Publishing AG, part of Springer Nature 2019
F. Calabrò et al. (Eds.): ISHT 2018, SIST 101, pp. 522–533, 2019.
https://doi.org/10.1007/978-3-319-92102-0_55

The anthropization processes may radically change, thus this so-called crisis, the Greek ancient word κρίσις, would recover the meaning of substantial transformation of path. This paper continues the research – started in 1987 and extending until the ERSA Congress 2016 - about these structural transformations[2]. It is not said that they go in the sense of a more equitable distribution of wealth and therefore of a more just or more "social and inclusive" city[3]. The new ways of anthropization could also be a further exploitation of natural resources according to the line of thought that thinks of them "only" as a base on which to build the world[4].

First of all it is necessary to overturn the logic of action: that is to have the local conditions as a design suggestion and not vice versaIdea recalled by Settis in 2014 [7], during his Lectio Magistralis L'etica dell'architetto e il restauro del paesaggio (The architect's ethics and landscape restoration), held for his "ad Honorem" bachelor in Architecture. He reminded us of Vitruvius' approach that today we call multidisciplinary with the fundamental role played by context. This is one of the characteristics of what I call "ecological approach". It means overcoming the industrial paradigm - recalling metaphorically the word that Khun [8] used for scientific revolutions - evolving over 350 years to build another developmental path.

It is essential to have a vision that starts from the territory: the EU itself, in the 2007 Charter of Leipzig, calls for "... integrated strategies between rural, urban, small, medium, large and metropolitan areas". Indication also useful to support "closing the cycle" of the materials produced and their reuse: important elements in the construction of the ecological scenario. In line with all this, the event Metropolis, nature, agriculture, development: for an ecology of the territory (see Fig. 1) was organized for the First Metropolitan Cities Festival, held in Reggio Calabria in 2015.

---

[2] At the annual AISRe scientific conferences, since 1987, with S. Macchi we started to face and publish about on the issue Telematics and Territory. The following year there was participation in the MPI 40% INTRA research project Technological Innovation and Territorial Transformations, DipPiST, Faculty of Engineering, Naples. In 1989 this was followed by the programs Technological innovation, territorial transformations and protection of the natural and anthropic environment and Technological innovation, territorial transformations, Dep. TECA, Fac. of Engineering, Rome La Sapienza. Subsequent researches are then exposed in La città virtuale: Trasformazioni urbane e nuove tecnologie della informazione (The Virtual City: Urban Transformations and New Information Technologies, [4]) and Ambiente urbano e innovazione. La città globale tra identità locale e sostenibilità (Urban Environment and Innovation. The Global City between Local Identity and Sustainability, [5]) as well as in the various essays presented, many published, at the annual AISRe Conferences, at the INU and SIU Congresses. To give continuity to these studies, since 2011 there is an "ad hoc" session, during the Annual Scientific Conferences AISRe, entitled Integrated planning and design for ecological territories and cities." Every year there is a specific declination, that of 2016 was "between transformations and risks". Topics addressed at international level also in the Metropolis, nature and anthropization: between the earth's resources and those of culture Session of the 2nd International Symposium NEW METROPOLITAN PERSPECTIVES - Strategic planning, spatial planning, economic programs and decision support tools, through the implementation of Horizon/Europe2020. ISTH2020, Reggio Calabria, May 18–20, 2016.

[3] These are the goals of Smart City.

[4] This philosophy is the opposite of that which requires Del Nord [6] when it calls for a "cultured technology".

2 Ecological Territories and Human Ecology

The need for ecological territories is closely linked to the goal of modern urban planning. That is to improve the living conditions of the inhabitants, of those that increasingly must become cum-cives, that is, sharers of civitas and its spatial representations [9]. This taking into account that space is a finite resource and that, therefore, we need to think about transformation and/or protection of the existing and not to have new expansions. Here is another basic difference with the industrial city built in the West until the mid-70s. Considering the increasingly high awareness of the environmental sustainability and of the need for risks reduction.

As previously mentioned, all this requires rethinking the modalities of anthropization from the bases. This work, begun some decades earlier, has among its various references a text of 1990 *Innovazione Tecnologica e Nuovo Ordine Urbano* (Technological Innovation and New Urban Order) edited by Gasparini and Guidicini [10]. Among the various essays there is that of Appold and Kasarda in which the authors pointed out the necessity of an "human ecology". The same word used in many pages (pp. 5, 115, 118, 119, 120) of Encyclical *Laudato Sii* (Laudato Be) by Pope Francis edited in 2015. This document has been elaborated by a group of 40 scientists coming from diverse e different disciplines. The existential condition is placed as the center of human action and of the construction of the world. It talks of integral ecology regarding the economic field and social and cultural questions. Continuing along this philosophy, a chapter is titled "Educate to Alliance between Humanity and Environment" (pp. 209 – 215). This part is, in an impressive way, very similar to man - nature alliance required by Scandurra in the text *L'ambiente dell'uomo* (The Environment of Man) [11] where, rightly, he speaks of the need to develop projects for the sustainable city[5].

Italy, and especially Calabria, has an ancient history with respect to these topics, this philosophy of the world. *De rerum natura juxta propria principia* (The nature of things according to their own principles) was written, from the mid 16th century, in more times, by Bernardino Telesio, born in Cosenza [12]. Some years later, in 1602, his disciple Tommaso Campanella, of Stilo (RC), published *La città del sole* (The city of the sun) [14]. The author, with relevant influences by philosophers such as Thomas More with his Utopia of 1517, went back to Platone (V century AC). While the II edition, *Civitas Solis idea republicae philosophica* of 1623 was written in vulgar Florentine and published in Freiburg, just one year before of F. Bacon's *New Atlantis*, even if it was edited in 1627.

It is interesting to note that over the centuries there has been a fruitful interweaving of religious presence, initially of a spiritual nature and protection of territories which will have a different destiny, much more material. So at the end of the 11th century, on August 15, 1094, at the presence of Ruggiero I of Calabria and Sicily, there was the solemn consecration of the church of Santa Maria of Turri or of the Wood: the former

---

[5] This paper has among the references the *anthropocosmos model* by Doxiadis [13] based on the relations between οίκος, environment, λόγος, analysis, and behaviour. Studies he did before the cooperation with the fascist, dictatorial, Junta of Colonels in Greece.

Fig. 1. The program of the event *Metropolis, nature, agriculture, development: for an ecology of the territory*, First Metropolitan Cities Festival, Reggio Calabria, 2015.

Hermitage founded by Bruno of Cologne in 1091, in the Calabria Ulterior, at the present the central - south part of the Region (VV). The king, to highlight this event, expanded the donation of land to Bruno and gave him other areas of Stilo and the farmhouses of Bivongi and Arunco. Centuries later, these areas became places of the iron industry which will be mentioned later. Later the church become the Certosa of Serra of Saints Stefano and Bruno, despite many alternate and destructive events, becomes well known in Europe. Often it was almost abandoned and many reconstructions had to be undertaken, especially due to earthquakes that struck Calabria. In the same period in Paola (CS), the future Saint Francis, created an Order practically vegetarian. It is therefore important to underline how a "transversal" spirit goes through the various components of the world, also tenaciously linked to the places and their use.

And the complete title of the encyclical Laudato Sii, that is "for the Cure of the Common House", highlights the relationship that must exist between the space of life, the societas, therefore the city, and its care, i.e. its management. Argument has become increasingly relevant in the issues related to town planning. And there is the term "Common House", that goes far beyond the concept of "public good". There seems to be a strong assonance between the feelings, the wishes of the citizens, and the direction indicated by the Encyclical. Which, it is repeated, is a document of religious orientation but based on the studies of a large number of scholars belonging to different and varied disciplines. The assonance referred to is clearly reflected in the 2011 referendum when there was a strong majority in maintaining public water management. The sense of that popular pronouncement was that water, scarce resource, had to be governed by politics. So the economic choices that concern it must be traced back to this logic. On the contrary, the judgment of the Council of State n.2481/2017 concerning the tariffs has instead reiterated the supremacy of economic aspects in the definition of them, as Marotta writes, professor of Economics at the Suor Orsola Benincasa University [15].

The theme is so relevant and now clearly interdisciplinary that the physicist F. Capra[6] and the jurist U. Mattei[7] in their recent book *Ecologia del diritto. Scienza, politica, beni comuni* (Ecology of law. Science, Politics, Common goods) [2] require a new approach in interpreting the world. This is seen as a system of a vast network of fluid communities of which to study the dynamic interactions to which to associate a profound revision of the law of the conception of ownership of parts, so far considered separate and often private: it is a kind of Copernican revolution.

Issues highlighted by the emerging theme of the so-called "ecosystem services" [16]. Many of the features of ecosystem services refer to what were "civic uses". They referred to the essential elements that were necessary for the life of inhabitants of a territory. In fact, the beneficiary of these uses was one who belonged to the local community, therefore not a private or public subject but a subject who could enjoy, in the measure of his needs of a common good. With the evolution of modalities of anthropization and the pressing of radical innovations, necessities of life changed and the requisites too. The ecosystem services become basic since the use of finite and therefore non-renewable resources is increasingly evident, as already anticipated over 40 years ago the cited *I limiti dello sviluppo* (The Limits to Growth)[8] [17]. In addition, environmentally harmful uses are increasingly emerging. Harmfulness that for years environmental organizations, such as Legambiente, are denouncing [19]. Also paying great attention to the relationship with the issues of legality as shown in the annual Ecomafie Report published since 2013. This not only or mainly for ethical reasons - certainly relevant - but above all for reasons of public utility and safety: just remember the phrase of a telephone interception in which it was said *"... okay, but what does it matter if the water table is polluted, we drink bottled water"* [20].

It should also be remembered that for some years now also National Bodies like ISTAT together with CNEL [21], in assessing the conditions of life and the state of a nation, go well beyond the GDP or per capita income. So in 2013 considering that these questions are related at quantity but, with growing importance, involve the quality, they elaborated the 134 indicators of the Fair and Equitable Wellbeing (BES). It has become one of the elements of the Budget Law of 2017. If there had not been a bold researcher like Ezechieli with his research on the measurement of happiness in 2003, perhaps today we would still be very far from these considerations [22].

---

[6] Author and co-author of many books including *The Tao of Physics* and *Life and Nature*, PhD, director and founder of the Center for Ecolfabetization of Berkeley, California, fellow of the Schumacher College in Gb and member of the International Earth Charter Council.

[7] Active in the European movement of common goods and author of essays and academic publications. He has the teacher's desk Alfred and Hanna Fromm of International and Comparative Law, Hastings College of Law at the University of California and is a professor of Civil Law at the University of Turin.

[8] See Aragona [18] *Servizi Ecosistemivi e Contesto Locale* (Ecosystem Services and Local Context).

3 Land Resources and Ecological Development

It is necessary to highlight how international surveys on the quality of life give medium-sized city the tops of rankings (see Fig. 2). Vienna, that has "only" 1,840,000 inhabitants. for the Mercer Consultant, is as first[9] [23]. It is also second, after Melbourne, in the Economist's annual ranking [24].

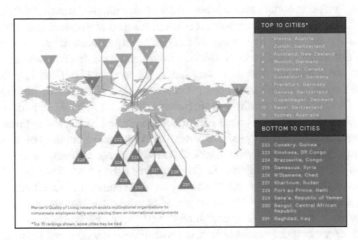

Fig. 2. World rankings for quality of life 2017. (Source: [23])

These rankings, thus, belie the usefulness of inter-national and national policies such as those of the Urban Agenda. Policies that are based on the competitiveness of cities and not on the quality of life. And they highlight how and how bad it was to implement policies that gradually have resulted in the destructuring the most of the, small and often internal, territorial realities that make up the precious fabric of the Italian landscape.

These decisions, being of a structural nature, have effects in the medium to long term. A relevant example is the so-called, wrong, "cut of the dry branches of the railways" started with Amato in 1992 [25]. And it was also wrong to make all the choices - between privatization and liberalization - which led to disappear historical territorial offices such as post offices, barracks, judicial offices, hospitals, schools, stations, etc. They increasingly disarticulate the local communities and push the abandonment of the land not only increasing the hydrogeological risk, etc. of them (and therefore "downstream" or where there are often major urban centers) but also increasing pollution and pressure on the few attractors centers.

For Italy there is a further consideration linked to its particular geographic con-figuration and to the ancient and diversified history. It is no coincidence that the country is also called the country of the "100 Bell Towers". This means that the

---

[9] Each year Mercer Consultant, an American consulting firm, publishes the Mercer Quality of Living Survey of 221 cities in the world [23].

landscape that has been designed over the millennia has resulted in a multitude of small and very small local communities. Result, more in general, of a socio-cultural process as affirmed in the European Landscape Convention 2000 [26]. This is a great cultural and social wealth and represents a source of attraction. We as planners have the great opportunity and challenge of enhancing local conditions to make them opportunities for ecological development.

It is no coincidence that foreign newspapers often mention the small urban realities of central Italy and in some cases of areas such as Salento as the most livable cities and territories in the world. At this regard, it is interesting to note, years ago Todi has been indicated as the best place by a study of the University of Kentucky, and retaken from the New York Times. And in 2010 the city organized a conference entitled "From the viability to sustainability, the ideal city 20 years later" [27]. Thus highlighting the further development of the concept of liveability that is enriched, highlights, more explicitly, thanks to the multiple components linked to the theme of sustainable development. In this way, the above mentioned integrated ecological approach is being built. A great opportunity to propose a new territory based on the before mentioned man - nature alliance: considering the local conditions, not as design constrains, but as suggestions for plans and projects: i.e. starting from the place, from the geomorpho-logic elements, the historical events, etc., with the responsibility of all actors involved in the organization and physical structure of territory and towns. All that is a big opportunity to move from the unsustainable industrial, mass town towards an eco-logical path of anthropization. In coherence with the philosophy of "Smart city" that is to construct local inclusive communities that are sustainable, both materially and socially[10].

This philosophy is in line with what Barca suggests for the development of many areas of Southern Italy when he claims that it should be "Place based" [28]. To realize this scenario it is essential to inform and involve the local population of the essential importance of the environmental sustainability and therefore of the possibility/need for a different development from that existing for over 40 years. Positive signals on the growing sensitivity regarding these topics come from the *Ecomuseo delle ferriere e fonderie di Calabria, Parco archeologico, monumentale, ambientale delle comunità e delle testimonianze della prima industrializzazione Meridionale* (Ecomuseum of the ironworks and foundries of Calabria - Archaeological, monumental, environmental park of the communities and of the testimonies of the first Southern industrialization) [29]. The idea of this Museum started in 1984 is useful to remember that even in the South, in the past, this opportunity has been exploited [29] until this has been allowed. The iron present in the area of the Serre (VV) for about 150 years was used for a flourishing metallurgical activity in Mongiana and Feridandea. This until 1861 when the Kingdom of the Two Sicilies was conquered by the State of Savoy. Then in about ten years the plants stopped producing and in central northern Italy, unified, this type of industry grew significantly. In addition to using the locally present mine resource, the woods were protected by "ad hoc" laws, as the wood was used for industrial activities. In this way, which today we would call naturalistic engineering, the hydrogeological

---

[10] This is our Wisdom as the name of the first University in Rome born about the XVI century.

risk diminished because these trees have a deep and widespread root system, being of tall trees. These elements then created horizontal linkages related to the natural components present and acted as "local multipliers" in the triggering of more work activities[11]. It is still necessary to highlight how the landscape and the structure of the territory have always been dependent on political choices. In fact, the Bourbons, in order to reduce transport costs, built an "ad hoc" road, connecting the industrial plants to Pizzo (VV), where an industrial port was built[12].

The ecological approach is consistent with the concept of Landscape of the homonymous Florence Convention (2000). A concept that is well conjugates with the territorial specialization that Dematteis [31] has been talking about for years and which calls for the presence of primary urbanization, an adequate local financial, a good administrative/policy level, the local knowledge, the absence of organized crime. But many, often all, these conditions, are lacking in the South. In this regard, Calabria is even more penalized because the geomorphological conditions have meant that the settlements are very small, widespread but isolated: the largest city, Reggio Calabria has only 183,000 inhabitants[13]. Attempts to build networks between different urban centers did not achieve great results in terms of synergic strengthening[14].

An important element in this reasoning is the competitiveness of the territory. It is linked to creativity, and this depends on three T, Talent, Technology and Tolerance, i.e. openness to different, new, innovation. Unfortunately the South of Italy, is in the last places regarding this last element[15].

The ecological approach that transforms local conditions, constraints, into opportunities has a wide field of work and landscape experimentation in the seismic and hydro-geological risk. Studying and experimenting with solutions to these risks would be an opportunity to create poles of excellence related to local universities[16]. Through these we could build a strategy of information/education for people, technicians, and politicians to explain the effects between localizations and risks in the short, medium, long period. That is, rethink territory and cities using indicators of environmental

---

[11] The workers were about 4000, at that period many more then the entire Sabaudo Kingdom.

[12] But all this should not be surprising as The Borbons were at the forefront of productive and cultural innovation at San Leucio very innovative textile production site, for that time a sort of Olivetti at Ivrea, and the magnificent Royal Palace of Caserta that nothing had less of Versailles is a very relevant testimony of all that ([30], cap. 1).

[13] The only big neighbouring city is Messina, over the Strait.

[14] On a regional scale, the 1999 POR (Regional Operational Plans) [30], with the "Network of Small Municipalities" Action, has already tried to trigger virtuous processes of collaboration/competition between small cities, that is the most of its cities. The result was not satisfactory, as for other instruments, e.g. the Integrated Territorial Projects.

[15] These results derive from a study developed by Tinagli [33] between 2004 and 2005 on the (then) 103 Provinces of Italy using the method developed in 2003 by prof. Florida of Carnegie Mellon University of Pittsburgh [34].

[16] An example was the creation of the useful "Seismic Risk Laboratory" of prof. Fera with the collaboration of architect De Paoli (Department of Environmental and Territorial Sciences, University of Reggio Calabria), but has been abandoned since years.

sustainability and to construct or reconstruct landscapes, in a wider scenario of integrated ecological town planning[17].

But, regarding all that, in Italy the great difficulty is that the territory is considered not as a public good but first of all as a private good, this especially in the South[18].

4   A Note in Closing: The Ecological Territories Need Political Guidance

The anthropization processes determine the vocation of a territory. And they are driven by the political choices[19]. Because they can result in unsustainable or became unsustainable it is necessary to be aware of the consequences and sustainability must always be considered with respect to the specific context and degree of scientific knowledge and technological level. In any case crucial is the cultural element. Alongside the infrastructures needed for realizing networks, the population, the local and regional authorities must be persuaded of the unsustainability to continue in the cementification of the territory. That, first of all, for reasons connected to seismic and geomorphological risk and, also, to prevent that the magnificent views are destroyed.

The first mission of us as territorial and urban planners consists in better motivate the reasons of the "ecological approach" for having a sustainable space. That means to act for the implementation of the many proposals and to reinforce the education of local urban actors[20].

But it is very important to highlight that the best dimension for the well-being of the inhabitants is not the big metropolis. They have large peripheral areas, mostly lacking in urban quality. The main motivation of their creation is the hypothesis that increases the competitiveness. But this is for the benefit of the few while the citizens are in a vast but little liveable territory. This evolution of the modern city, with its focus on the large urban areas, metropolitan areas, is not effective because the well-being of local residents and communities is not growing. Therefore it is necessary to propose to "different" scenario really useful to citizens rather than the globalized economy without control, that is to start the "smart globalization" required by Rodrik in 2011 [39].

It is necessary to overturn the point of view, that is, when it is raining it can been seen as good weather: it is a basical change for going towards an ecological vision of the territorial and town planning. This is an example of how a "problem", in this case water can, must become a resource to be managed. River contracts are a great

---

[17] E.g. Faenza with its urban sustainable regeneration strategy as illustrated in 2016, at the Congress of the Italian Society of Urbanists (SIU), from E. Nanni, head of the city's Town Planning Department [35].

[18] According to the Associate Professor of Ethics of Law, Cananzi [36], individualism would be an "anthropological" peculiarity of many inhabitants of Reggio Calabria, although I think it would be interesting to know how it was before the unification of the nation.

[19] Already in 1988, Valeria Erba highlighted the central function of the political choices in addressing the transformations of the territories [37].

[20] This to participate to the "city pedagogy" described by Gennari in 1995 [38]. Relevant example is mobility, now in Italy is based on the use of private car while in other places of the world as Copenhagen or Norway for the most is sustainable using bicycle or walking street.

opportunity in this regard. From the regulatory point of view and the launching of procedures, the institutions have been active. In fact, Regions such as Calabria were among the first to incorporate the law on River Contracts (Contratti di fiume) and start meetings to implement this tool. Now they have to start, and in the meantime, the devastating actions such as the covering of watercourses and the cementing of the coasts must interrupted. And plans and projects for renaturalization start as soon as possible since this is the only way to try to stem, at local level, the effects of the larger phenomenon of the global warming.

We must be aware that we propose ideal scenarios, but that are not utopistic, unreal. And we must propose actions in the short, medium term, so that we can contribute to realization of the scenarios mentioned above. That is to approach the meeting point to the infinity of these two parallel lines, i.e. the scenarios and the le actions. The first smartness is to help citizens to become "ecologicus" people. At the Faculty of Architecture of Reggio Calabria some time ago there has been a three-year initiative, called *Archisostenibile* (Sustainable Architecture), to in/form awareness about these issues. Topics on which the University of Reggio Calabria is at the forefront and is in the European Universities Network for Energy[21]. Equally innovative is NOW's REWECH experimental project, *Laboratory for Electric Power Conversion of the sea waves*[22], one of the priorities in EU energy policy.

References

1. Encyclical Letter Laudato Be of the Holy Father Francis on the Care of the Common House. Tipografia Vaticana, Città del Vaticano (2015)
2. Capra, F., Mattei, U.: Ecologia del diritto. Scienza, politica, beni comuni. Aboca Edizioni, Sansepolcro (AR) (2017)
3. Keynes, J.M.: Collected Writings, London, 1971–1989, vol. XXI, pp. 242, Columbia University Press, (2012)
4. Aragona, S.: La città virtuale. Trasformazioni urbane e nuove tecnologie della informazione. Gangemi Editore, Roma-Reggio Calabria (1993)
5. Aragona, S.: Ambiente urbano e innovazione. La città globale tra identità locale e sostenibilità, Gangemi Editore, Roma-Reggio Calabria (2000)
6. Del Nord, R.: Presentazione. In: Mucci, E., Rizzoli, P. (eds.) L'immaginario tecnologico metropolitano. Franco Angeli, Milano (1991)
7. Settis, S.: L'etica del architetto e il restauro del paesaggio, Lectio Magistralis for Honorary Degree in Architecture, University Mediterranea of Reggio Calabria (2014)
8. Khun, T.S.: The Structure of Scientific Revolutions, Chicago University Press. 1962, 1970, It. tr. of II ed., La struttura delle rivoluzioni scientifiche. Einaudi, Torino (1979)
9. Cacciari, M.: Aut civitas, aut polis. In: Mucci, P., Rizzoli, P. (eds.) L'immaginario tecnologico metropolitano. Franco Angeli, Milano (1991)

---

[21] Another useful contribution to the formation of the new ecological mentality was the participation in Trondheim (No), in 2016, of the Prorector C. Morabito at the meeting "Human resources and new knowledge to build the future energy system".

[22] It is one of spin-off activities of the University of Reggio Calabria, whose director is prof. F. Arena and founder is Prof. P. Boccotti, with researchers and young scientist, all from this University.

10. Appold, S.J., Kasarda, J.D.: Concetti fondamentali per la reinterpretazione dei modelli e dei processi urbani. In: Gasparini, A., Guidicini, P. (eds.) Innovazione tecnologica e nuovo ordine urbano. Franco Angeli, Milano (1990)
11. Scandurra, E.: L'ambiente dell'uomo. Verso il progetto della città sostenibile. Etas Libri, Milano (1995)
12. Telesio, B.: De rerum natura iuxta propria principia, libri IX (1565, 1570, 1586) (anast. rep..) (curator Giglioni, G.), Carocci editor, Roma, Collana Telesiana (2013)
13. Doxiadis, C.: Ekisticks: An Introduction to the Science of Human Settlements. Oxford University Press, New York (1968)
14. Campanella, T.: La città del sole (1602, 1623) (Curators: Ernst, G., Salvetti Firpo, L.). Laterza, Bari. IX Edizione (2006)
15. Marotta, S.: Acqua pubblica tra referendum e mercato in economia e politica Rivista online di critica della politica economica, anno 9 n. 14 sem. 2. Accessed 14 Sept 2017. http://www. economiaepolitica.it/industria-e-mercati/mercati-competizione-e-monopoli/acqua-pubblica-tra-referendum-e-mercato/
16. ISPRA: I Servizi ecosistemici. Accessed 14 Oct 2017. http://www.isprambiente.gov.it/it/temi/biodiversita/argomenti/benefici/servizi-ecosistemici
17. Meadows, D.H: I limiti dello sviluppo. Club di Roma. Milano: Mondadori. Meadows, D.L. (et al.) (1972). The Limits to Growth. Universe Books, New York (1972)
18. Aragona, S.: Servizi Ecosistemici e Contesto Locale, in Special Session Challenges, resistances and opportunities for the inclusion of ecosystem services in urban and regional planning. In: Moccia, F.D., Sepe, M. (eds.) 10° INU STUDY DAY Crisis and rebirth of Cities, s.i. Urbanistica Informazioni n. 272 (2017)
19. Legambiente: Rapporto Ecomafie. Accessed 15 Oct 2017. https://www.legambiente.it/contenuti/dossier/rapporto-ecomafia
20. Ciolla, P.: CAMPANIA VIOLATA/4, Rifiuti, quarta fonte di reddito criminale. Accessed 15 Mar 2018. https://www.avvenire.it/attualita/pagine/rifiuti–inchiesta-4
21. ISTAT-CNEL: Bes 2013. Il Benessere Equo e Sostenibile in Italia. Tipolitografia CSR, Roma (2013)
22. Ezechieli, E.: Beyond Sustainable Development: Education for Gross National Happiness in Bhutan, Stanford University (2003)
23. Mercer: Quality of Living City Rankings (2017). Accessed 27 Jul 2017. https://mobilityexchange.mercer.com/Insights/quality-of-living-rankings
24. The Economist Intelligence Unit, A Summary of the Liveability Ranking and Overview. Accessed 03 May 2017. http://www.eiu.com/public/thankyou_download.aspx?activity=download&campaignid=Livabilty2016
25. Aragona, S.: Infrastrutture di comunicazione, trasformazioni urbane e pianificazione: opzioni di modelli territoriali o scelte di microeconomia? In: Proceedings of the XIV Conference of the Italian Association of Regional Sciences, vol. 2, Bologna (1993b)
26. Council of Europe: European Landscape Convention, Florence (2000)
27. Redazione tuttoggi info: Convegno a Todi sulla "città più vivibile del mondo". Accessed 09 Jun 2017. http://tuttoggi.info/convegno-a-todi-sulla-citta-piu-vivibile-del-mondo/85797
28. Barca, F.: Report An Agenda for a Reformed Cohesion Policy, chps. I, IV (2010)
29. Franco, D.: Ecomuseo delle ferriere e fonderie di Calabria, Parco archeologico, monumentale, ambientale delle comunità e delle testimonianze della prima industrializzazione Meridionale. Accessed 27 Sept 2017. http://web.tiscali.it/ecomuseocalabria/
30. Aragona, S.: Costruire un senso del territorio Spunti, riflessioni, indicazioni di pianificazione e progettazione. Gangemi Editore, Roma - Reggio Calabria, chp.1 (2012)
31. Dematteis, G.: Modelli Urbani a Rete: Considerazioni Preliminari". In: Curti, F., Diappi, L. (eds.) Gerarchie e Reti di Città: Tendenze e Politiche. Franco Angeli, Milano (1990)

32. UE: Regional Operative Plan (POR) Calabria, Structural Funds, Axis City, Measure 5.1, Action 5.1.c-Networks of small municipalities, 1999–2006 (2006)
33. Tinagli, I.: Creatività ed Innovazione: Le nuove sfide del sistema economico globale. Convegno APQ_Firenze. 13 May. Accessed 05 Jun 2017. http://online.cisl.it/qattualita/I04724173.6/Convegno%20APQ_Firenze%202006.doc
34. Florida, R.: L'ascesa della nuova classe creativa. Stile di vita, valori e professioni. Mondadori, Milano (2003)
35. Nanni, E.: Il caso di Faenza, II Sessione Plenaria Le città nel cambiamento, buone pratiche ed esperienze in corso. XIX National Conference of Società italiana degli urbanisti, Cambiamenti. Responsabilità e strumenti per l'urbanistica al servizio del Paese, 16–19 June, Catania (2016)
36. Cananzi, D: Speech at REGGIO 1946–REGGIO 2016. Dalla ricostruzione della città di 70 anni fa alla costruzione della città Metropolitana. Un percorso comune di riflessione tra ricordi del dopoguerra e la prospettiva del futuro Round Table, Officine Miramare, 12 February 2016 Reggio Calabria, Centro Internazionale Scrittori della Calabria (2016)
37. Erba, V.: L'Efficacia dello strumento 'piano regolatore' letta attraverso la produzione di modelli e di generazioni di piano. In: Gibelli, M.C., Magnani, I. (eds.) Pianificazione Urbanistica come Strumento di Politica Economica. Coll. Di Scienze Regionali Franco Angeli, Milano (1988)
38. Gennari, G.: Semiologia della città. Marsilio, Padova (1995)
39. Rodrik, D.: La globalizzazione intelligente. Laterza, Bari, The Globalization Paradox. Democracy and the Future of the World Economy Oxford University Press, WW. Norton & Company (2011)

Prioritization of Energy Retrofit Strategies in Public Housing: An AHP Model

Chiara D'Alpaos and Paolo Bragolusi[✉]

University of Padova, 35131 Padova, Italy
paolo.bragolusi@dicea.unipd.it

Abstract. The design and implementation of buildings energy retrofit strategies is a complex process involving a great number of decision variables and actors, especially when public housing is concerned. This complexity is exacerbated by stringent public budget constraints and lack of financial resources that make public housing energy retrofit currently a critical issue in Italy. In this context, multiple objectives related to energy saving, thermal comfort and conservation compatibility need to be pursued and multiple criteria approaches provide a proper theoretical and methodological framework to address economic, technical, social and environmental issues that characterize investments in energy saving and retrofit strategies.

In this paper, we analyze different energy efficiency measures to be implemented in public housing and we propose an AHP (relative) model for multi-criteria prioritization of energy-retrofit strategies on public-housing existing stock.

Keywords: Buildings energy retrofit · Public housing · AHP

1 Introduction

The built environment accounts for approximately 30% of global energy consumption and generate about 20% of all energy-related greenhouse gas (GHG) emissions [1, 2].

As cities are host to the majority of the world's population and building stock, they are responsible for approximately 60% of global energy use and over 75% of energy related GHG emissions [2, 3]. It is widely recognized that cities represents a primary area to climate change mitigation and the built environment provides low-cost and short-term opportunities to reduce emissions by improving energy performance of buildings: the implementation of energy efficiency measures can reduce the global cost of limiting warming to 2 °C by up to $2.8 trillion by 2030 [4]. Although nowadays new buildings should be designed as zero or nearly zero-energy buildings, the greatest challenge worldwide is the refurbishment of the existing stock [5–8].

In compliance with the EU 2030 Climate and Energy Framework, the Italian "National Energy Strategy" (SEN) and the Italian "Action Plan for Energy Efficiency" (PAEE, 2014) identify the building sector as a key element for achieving the 2030 objectives set by the Country.

The Italian housing stock is one of the least energy-efficient in the EU-27. The residential sector accounts for 36% of primary energy use in Italy: nearly 76% of Italian

© Springer International Publishing AG, part of Springer Nature 2019
F. Calabrò et al. (Eds.): ISHT 2018, SIST 101, pp. 534–541, 2019.
https://doi.org/10.1007/978-3-319-92102-0_56

dwellings were built before 1981 and almost 90% of the Italian building stock exhibits an excessive energy demand [9]. This condition widely affects public properties and specifically public housing.

The design and implementation of buildings energy retrofit strategies is a complex process involving a great number of decision variables and actors, especially when public housing is concerned [10–12]. This complexity is exacerbated by stringent public budget constraints and lack of financial resources, which make public housing energy retrofit currently a critical issue in Italy [13–16]. Significant European experiences have proved the importance of integrated approaches to the renovation of public housing, aiming at leveraging on environmental sustainability, creating urban identity, reducing social disadvantage but offering at the same time high quality housing standards [3, 17]. In this context, in which multiple objectives related to energy saving, thermal comfort and conservation compatibility need to be pursued, multiple criteria approaches provide a proper theoretical and methodological framework to address the complexity of economic, technical, social and environmental factors that characterize public investments in energy saving and retrofit strategies.

In this paper, we analyze different energy efficiency measures to be implemented in public housing and we propose an AHP (relative) model for multi-criteria prioritization of energy-retrofit strategies on public-housing existing stock.

The remainder of the paper is organized as follows. Section 2 presents the hierarchical approach to support the decision maker in the prioritization of energy retrofit strategies; Sect. 3 provides the decision model and discusses the results; Sect. 4, concludes.

2 Method

Energy retrofit of existing buildings plays a primary role in mitigating climate change. The performance of existing buildings can be improved by using different retrofit options, which varies from energy-consumption reduction measures to the adoption of low carbon technologies. Their selection can be very challenging [18].

Decisions related to the implementation of energy retrofit strategies in public-housing ought to be addressed as decision-making problems where multiple criteria, often conflicting, must be taken into account. Multi-criteria Decision Making (MCDM) methods have been extensively proposed in the literature to select green technologies and support design decisions for low carbon buildings [19–21].

Among the different MCDM methods, the Analytic Hierarchy Process (AHP), presented by Saaty in the Eighties [22], proved to be a well-established technique to address complex decisions, which requires different and multidisciplinary know-hows in several research areas [23–26].

The AHP allows for measurement of tangible and/or intangible criteria and factors and assumes that the decision-maker is always able to express a preference and judge the relative importance of (or preference for) the evaluation parameters. The AHP deconstructs the initial problem into several levels, developing a hierarchy, with unidirectional hierarchical relationships between levels. The top of the hierarchy is represented by the main goal of the decision problem, whereas criteria and sub-criteria which contribute to the goal are placed at lower levels and alternatives to be evaluated are at the bottom level [27]. Once criteria are agreed upon, and supporting data are collected for each alternative, it is then possible to rank a finite number of alternatives by evaluating them with respect to a finite number of attributes (criteria, sub-criteria, etc.). Attributes relative importance is determined through pairwise comparisons expressed in semantic judgments, which are converted into numerical values according to Saaty's fundamental scale [22]. By pairwise comparisons between criteria, experts provide their subjective preference (relative importance) on the dominance of one criterion over another with respect to the goal.

The pairwise-comparison procedure results in square matrices of preferences where the dominance coefficient a_{ij} represents the relative importance of the component on row i over the component on column j [22, 28, 29]. The weights/priorities are determined according to the eigenvalue approach to pairwise comparisons and pairwise comparisons of the elements in each level are conducted with respect to their relative importance towards their control/parent criterion [23, 28]. Unlike other MCDM methods, AHP tolerates some inconsistency in experts judgments: the consistency of pairwise comparison matrices is verified by determining the inconsistency index IC:

$$IC = \frac{\lambda_{max} - n}{n - 1} \tag{1}$$

where λ_{max} is the maximum eigenvalue and n is the rank of the pairwise comparison matrix. IC < 0.10 is usually considered as acceptable [22].

The global ranking of alternatives is then obtained via a weighted-sum, bottom-up, aggregation procedure throughout hierarchical levels [27]. In other words, by local priorities of criteria in a node are multiplied by local priorities of a corresponding parent criterion [22]. Finally sensitivity analysis is performed to validate the solution and test for rank reversal.

3 Model and Results

Italian public-housing building stock is generally characterized by inadequate envelopes and low performance HVAC and water heating systems [3, 30]. To prioritize energy efficiency strategies on existing buildings, we conducted an extensive literature review and selected a pool of seven experts to identify key factors and define the hierarchy [29, 31–34]. We organized focus groups to develop the set of criteria and sub-criteria and validate the hierarchy by dynamic discussion. The panel of experts structured the decision problem and disaggregated it into sub-problems by identifying three hierarchical levels (goal, criteria and sub-criteria) and four decision nodes (see Fig. 1). Three criteria (Economic, Technical and Socio-Environmental) and ten sub-criteria (Indirect costs[1], LCC 30 years, Pay back, Compatibility, Efficiency,

---

[1] Costs related e.g. to shutting of parts of the building/residential unit, decanting residents to other location, annoyance to the general public and *occupants* of neighboring property.

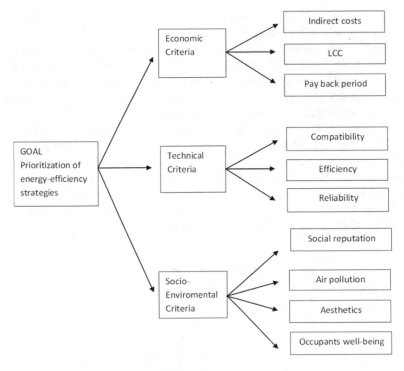

Fig. 1. Hierarchy

Table 1. Descriptions of criteria and sub-criteria.

Criteria	Sub-criteria	Description
Economic	Indirect costs	Costs related to inconveniences to occupants
	LCC	Life cycle cost over a 30-year period
	Pay back period	Time to recover investment costs
Technical	Compatibility	Compatibility of new with older
	Efficiency	Improvement in technical performance
	Reliability	Frequency of failures and system safety
Socio-Environmental	Social reputation	Reputation capital increase
	Air pollution	CO_2 emissions reduction
	Aesthetics	Facade attractiveness improvement
	Occupants' well-being	Comfort improvement

Reliability, Social reputation, CO_2 reduction; Aesthetics; Occupants well-being) were selected. Table 1 summarizes description of criteria and sub-criteria.

Alternative energy-saving measures were discussed by the focus groups. The panel of experts identified 7 alternatives consisting of basic measures or combinations of them: (a) installment of condensing boilers (alternative 1); installment of double-glazed windows (alternative2); application of insulating layers on the external walls, roofs and ceilings (alternative 3); installment of condensing boilers and double-glazed windows (alternative 4); installment of condensing boilers and application of insulating layers on the external walls, roofs and ceilings (alternative 5); installment of double-glazed windows and application of insulating layers on the external walls, roofs and ceilings (alternative 6); installment of condensing boilers and double-glazed windows and application of insulating layers on the external walls, roofs and ceilings (alternative 7). We asked each expert to compile the fourteen pairwise-comparison matrices and calculated ICs, which proved to be within the acceptability threshold. We then aggregated individual judgments (see Table 2) by calculating the geometrical mean, which allows for synthesizing individual judgments, given in response to a single pairwise comparison, as the representative judgment for the entire group [35, 36]. In the final phase of the group decision process, we obtained the final priority vector and the prioritization of the alternatives with respect to the goal (see Table 3).

Table 2. Criteria and sub-criteria priority vectors.

Criteria	Priority vector	Sub-criteria	Priority vector
Economic	0.637	Indirect costs	0.105
		LCC	0.636
		Pay back period	0.259
Technical	0.105	Compatibility	0.143
		Efficiency	0.715
		Reliability	0.142
Socio-Environmental	0.258	Social reputation	0.260
		Air pollution	0.381
		Aesthetics	0.232
		Occupants' well-being	0.127

Table 3. Ranking of alternatives and final priority vectors (ideals and normal).

Alternatives	Normal	Ideals
1	0.100	0.370
2	0.162	0.600
3	0.270	1.000
4	0.057	0.212
5	0.105	0.389
6	0.188	0.694
7	0.118	0.435

According to priority vectors displayed in Table 2, Economic Criteria play a major role in the achievement of the goal because their relative importance is the greatest, as well as Socio-Environmental Criteria which are ranked as second in terms of relative importance. Retrofit strategies that consist in the application of insulating layers on the external walls, roofs and ceilings are ranked as first and strategies that provide for installment of double-glazed windows and application of insulating layers on the external walls, roofs and ceilings are ranked as second. In addition, results show that installment of condensing boiler are not considered to be as much preferable as application of insulating layers and installment of double-glazed windows as they become cost-effective only near the end of their useful life.

4 Conclusions

Improving energy efficiency and comfort conditions of existing buildings in a cost-effective manner requires careful consideration of many issues. Evaluation of energy-retrofit strategies is a multidimensional and complex problem that incorporates multiple, often conflicting, criteria and objectives. In this paper, we provide an AHP model for multiple-criteria prioritization of energy retrofit strategies in public housing. The model results may have interesting effects in terms of policy implications. According to experts' judgments, priority should be given to thermal insulation, minimization of additional capital and management costs, delivering of cost and comfort benefits throughout the useful life of the building. High insulation values as well as tightening building envelopes can reduce energy demand significantly. If efficiently designed, retrofit measures represent a driver in achieving the 2020 (and 2030) targets on energy efficiency and emissions reductions. In addition, proper energy-retrofit may play a key role in reducing fuel poverty, which is currently a priority within national and European energy policies. Fuel poverty affects a wide range of individuals and families and it is caused by a convergence of factors: low-income household, high fuel prices and poor energy efficiency of homes due to low levels of insulation and old or inefficient heating systems. Fuel poverty is consequently a problem, which can be effectively tackled jointly to gas emissions reduction by retrofitting existing buildings.

References

1. Lucon O., Ürge-Vorsatz, D., Zain Ahmed, A., Akbari, H., Bertoldi, P., Cabeza, L. F., Eyre, N., Gadgil, A., Harvey, L. D.D., Jiang, Y., Liphoto, E., Mirasgedis, S., Murakami, S., Parikh, J., Pyke, C., Vilariño, M.V.: Buildings. In: Edenhofer, O., Pichs-Madruga, R., Sokona, Y., Farahani, E., Kadner, S., Seyboth, K., Adler, A., Baum, I., Brunner, S., Eickemeier, P., Kriemann, B., Savolainen, J., Schlömer, S., von Stechow, C., Zwickel, T., Minx, J.C. (eds.): Climate Change 2014: Mitigation of Climate Change. Contribution of Working Group III to the Fifth Assessment Report of the Intergovernmental Panel on Climate Change, Cambridge University Press, Cambridge, United Kingdom (2014)
2. UNEP: Why Buildings? Buildings Day at COP21, 3 December 2015, Paris, France (2015). Accessed 30 Nov 2017. http://web.unep.org/climatechange/cop21?page=7
3. Beccali, M., Ciulla, G., Lo Brano, V., Galatioto, A., Bonomolo, M.: Artificial neural network decision support tool for assessment of the energy performance and the refurbishment actions for the non-residential building stock in Southern Italy. Energy **137**, 1201–1218 (2017)

4. Fraunhofer ISI, How energy efficiency cuts costs for a 2-degree future, Fraunhofer Institute for Systems and Innovation Research ISI, Karlsruhe, Germany (2015). Accessed 30 Nov 2017. http://www.isi.fraunhofer.de/isi-en/service/presseinfos/2015/press-release-34-2015-energy-efficiency-two-degree-target.php
5. Machairas, V., Tsangrassoulis, A., Axarli, K.: Algorithms for optimization of building design: a review. Renew. Sustain. Energy Rev. **31**, 101–112 (2014)
6. Si, J., Marjanovic-Halburda, L., Nasirib, F., Bell, S.: Assessment of building-integrated green technologies: a review and case study on applications of Multi-Criteria Decision Making (MCDM) method. Sustainable Cities and Society **27**, 106–115 (2016)
7. Visscher, H., Sartori, I., Dascalaki, E.: Towards an energy efficient European housing stock: monitoring, mapping and modelling retrofitting processes. Energy Build. **79**, 1–3 (2016)
8. Becchio, C., Corgnati, S.P., Delmastro, C., Fabi, V., Lombradi, P.: The role of nearly-zero energy buildings in the transition towards post-carbon cities. Sustain. Cities Soc. **27**, 324–337 (2016)
9. ISTAT. 15° Censimento della popolazione e delle abitazioni. Accessed 12 Jan 2011. http://www.istat.it/it/censimento-popolazione/censimento-popolazione-2011
10. Diakaki, C., Grigoroudis, E., Kolokotsa, D.: Performance study of a multi-objective mathematical programming modelling approach for energy decision-making in buildings. Energy **59**, 534–542 (2013)
11. Lizana, J., Barrios-Padura, A., Molina-Huelvab, M., Chacartegui, R.: Multi-criteria assessment for the effective decision management in residential energy retrofitting. Energy Build. **129**, 284–307 (2016)
12. Marinakis, V., Doukas, H., Xidonas, P., Zopounidis, C.: Multicriteria decision support in local energy planning: An evaluation of alternative scenarios for the sustainable energy action plan. Omega **69**, 1–16 (2017)
13. Ma, Z., Cooper, P., Daly, D., Ledo, L.: Existing building retrofits: methodology and state-of-the-art. Energy Build. **55**, 889–902 (2012)
14. Harvey, L.D.: Recent advances in sustainable buildings: review of the energy and cost performance of the state-of-the-art best practices from around the world. Annu. Rev. Environ. Resour. **38**, 281–309 (2013)
15. Corrado, V., Ballarini, I., Paduos, S.: Assessment of cost-optimal energy performance requirements for the Italian residential building stock. Energy Procedia **45**, 443–452 (2014)
16. Tan, B., Yavuz, Y., Otay, E.N., Çamlıbel, E.: Optimal selection of energy efficiency measures for energy sustainability of existing buildings. Comput. Oper. Res. **66**, 258–271 (2016)
17. Gianfrate, V., Piccardo, C., Longo, D., Giachetta, A.: Rethinking social housing: Behavioural patterns and technological innovations. Sustain. Cities Soc. **33**, 102–112 (2017)
18. Antoniucci, V., D'Alpaos, C., Marella, G.: Energy saving in tall buildings: from urban planning regulation to smart grid building solutions. Int. J. Hous. Sci. Appl. **39**(2), 101–110 (2015)
19. Dawood, S., Crosbie, T., Dawood, N., Lord, R.: Designing low carbon buildings: a framework to reduce energy consumption and embed the use of renewables. Sustain. Cities Soc. **8**, 63–71 (2013)
20. Re Cecconi, F., Tagliabue, L.C., Maltese, S., Zuccaro, M.: A multi-criteria framework for decision process in retrofit optioneering through interactive data flow. Procedia Eng. **180**, 859–869 (2017)
21. D'Alpaos, C.: Methodological Approaches to the Valuation of Investments in Biogas Production Plants: Incentives vs. Market Prices in Italy. Valori e Valutazioni **19**, 53–64 (2017)

22. Saaty, T.: The analytic hierarchy process: planning, priority setting, resource allocation. McGraw-Hill, New York (1980)
23. De Felice, F., Petrillo, A.: Absolute measurement with analytic hierarchy process: a case study for Italian racecourse. Int. J. Appl. Decis. Sci. **6**(3), 209–227 (2013)
24. Ferreira, F.A., Santos, S.P., Dias, V.M.: An AHP-based approach to credit risk evaluation of mortgage loans. Int. J. Strateg. Prop. Manag. **18**(1), 38–55 (2014)
25. Grafakos, S., Flamos, A., Enseñado, E.M.: Preferences matter: A constructive approach to incorporating local stakeholders' preferences in the sustainability evaluation of energy technologies. Sustainability **7**(8), 10922–10960 (2015)
26. Garbuzova-Schliftern, M., Madlener, R.: AHP-based risk analysis of energy performance contracting projects in Russia. Energy Policy **97**, 559–581 (2016)
27. Banzato, D., Canesi, R., D'Alpaos, C.: Biogas and biomethane technologies: an AHP model to support the policy maker in incentive design in Italy. In: Bisello, A., Vettorato, D., Laconte, P., Costa, S. (eds.) Smart and Sustainable Planning for Cities and Regions. SSPCR 2017. Green Energy and Technology, pp. 319–331 (2018). https://doi.org/10.1007/978-3-319-75774-2_22. ISBN 9783319757735
28. Saaty, T.L.: Fundamentals of decision making and priority theory with the analytic hierarchy process. RWS Publications, Pittsburgh (2000)
29. Saaty, T.L., Peniwati, K.: Group Decision Making Drawing Out and Reconciling Differences. RWS Publications, Pittsburgh (2012)
30. Ciulla, G., Lo Brano, V., D'Amico, A.: Modelling relationship among energy demand, climate and office building features: a cluster analysis at European level. Appl. Energy **183**, 1021–1034 (2016)
31. Peniwati, K.: Criteria for evaluating group decision-making methods. Int. Ser. Oper. Res. Manag. Sci. **95**, 251–273 (2006)
32. Senge, P.M.: The Fifth Discipline: The Art & Practice of the Learning Organization. Currency Doubleday New York (2006)
33. D'Alpaos, C.: The value of flexibility to switch between water supply sources. Appl. Math. Sci. **6**(125–128), 6381–6401 (2012)
34. Bertolini, M., D'Alpaos, C., Moretto, M.: Do smart grids boost investments in domestic PV plants? evidence from the Italian electricity market. Energy **149**, 890–902 (2018)
35. Xu, Z.: On consistency of the weighted geometric mean complex judgement matrix in AHP. Eur. J. Oper. Res. **126**(3), 683–687 (2000)
36. Grošelj, P., Zadnik Stirn, L.: Acceptable consistency of aggregated comparison matrices in analytic hierarchy process. Eur. J. Oper. Res. **223**(2), 417–4201 (2012)

The Ex-ante Evaluation of Flood Damages for a Sustainable Risk Management

Francesca Torrieri[1(✉)] and Alessandra Oppio[2]

[1] University of Naples "Federico II", 80121 Naples, Italy
frtorrie@unina.it
[2] Politecnico of Milan, 20133 Milan, Italy

Abstract. The present paper presents an integrated model for the ex-ante evaluation of flood damages based on "stage damages curve" as a decision support tool for public and private authorities in the sustainable management of risk areas.

The proposed model aims to estimate the damage both at a micro-scale and on an extended area. In particular, the dynamics of flooding in flood areas and their effects on the degree of vulnerability of the different types of buildings are considered. The proposed model, already experimented in different territorial contexts, has been applied to the case of the Senigallia flood, which took place in 2014. The first results show that the approach based on damage curves applied to different types of exposed goods at a micro-scale can reliably estimate the damages to buildings. In contrast, the ex-ante estimation of indirect damages is still uncertain due to the difficulty of evaluating, according to reliable criteria, unique assets such as intangibles and cultural and environmental assets.

Keywords: Ex-ante evaluation · Flood damages · Integrated model
Stage damages curves

1 Introduction

The damages caused by the floods represent in Italy and in the world a problem of significant social and economic impact. In Italy about 17 million people are exposed to risk, as well as 1.642.000 local units of companies, and most of the infrastructures and cultural heritage are located in areas with hydraulic risk [1]. In a study developed by the Free University of Amsterdam [2], it is assumed that within the next 40 years, exceptional floods such as those that have affected Europe in the recent months could double in frequency, every 10 years (and not once every 16 as it happens now), and that the annual costs for repairing hydrogeological damages are expected to increase fivefold, reaching 23 billion euros in 2050, in respect to 4.9 billions a year spent from 2000 to 2012.

The recurrence of the flood events and the huge costs to manage the post-disaster emergency have led, in various international contexts, to adopt a "Risk management" policy, going from a basic approach focused only on the reconstruction damages to an approach more addressed to forecasting and prevention, based on a correct

© Springer International Publishing AG, part of Springer Nature 2019
F. Calabrò et al. (Eds.): ISHT 2018, SIST 101, pp. 542–550, 2019.
https://doi.org/10.1007/978-3-319-92102-0_57

identification of the risk conditions and on the adoption of interventions aimed at minimizing the impacts of the events [3].

In Italy the Legislative Decree 49/2010, by implementing the European Directive 2007/60/EU, has in fact introduced the obligation for areas at flooding risk he preparation of the so called "Risk Management Plans" (PGR) in order to reduce the consequences of flooding on human health, environment, cultural heritage and economic activities. The PGR, according to logic of prevention, aims to identify structural and non-structural measures for the reduction of flood risk and consequently to define an intervention plan on the territory based on an economic analysis of the policies to be adopted. Benefits - Costs analysis is referred as an ex-ante evaluation tool to support the choice of the interventions to be adopted, so as to prefer those options that have a higher economic value, in addition to being able to minimize the expected residual damage. Under this perspective, it is clear the importance of reliably estimating the expected damages in order to support investment decisions aimed at securing the territory, not only from an economic and financial point of view, but also consistently with the instance of environmental and social resources' protection. The experiences carried out so far have shown the limits of the evaluation models based only on statistical analysis of historical data. Limits, above all, linked to the difficulty of reliably estimating the economic damage expected at the territorial scale in high complex contexts. The purpose of this contribution is to propose an integrated model for assessing the expected damages, by the approach of "stage damage curves" at the building scale. The model is combined with Geographic Information Systems (GIS), and is able to estimate the damage both on a single building and on a large area. In particular, the dynamics of flooding and their effects on the degree of vulnerability of the different types of buildings in the flood areas are considered. The proposed model, already experimented in different territorial contexts, has been applied to the case of the Senigallia flood, which took place in 2014. The first results have shown that the damage curves approach applied to the micro-scale for different types of exposed goods, can reliably estimate the damage to buildings. In contrast, the ex-ante estimate of indirect damages is still uncertain due to the difficulty of evaluating, according to reliable criteria, unique assets such as intangibles as well as cultural and environmental assets. The contribution is therefore structured as follows: the first part will describe the integrated model of ex-ante assessment of damages, with specific reference to the construction of "stage damage curves"; the second part will show the main results obtained by applying the model to the case study; then preliminary conclusions and future research perspectives are pointed out.

2 The Integrated Evaluation Model for Ex-ante Damages Assessment

The ex-ante estimation model for flood damage is shown in Fig. 1.

The risk is estimated in accordance with the European legislation (Directive 2007/60) and Italian law (Decree Law 49/2010) as a function of: (i) the frequency of occurrence of the potential flooding (H); (ii) the value of exposed assets at risk

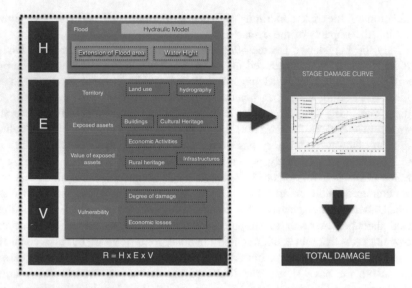

Fig. 1. Integrated model of ex-ante evaluation of flood damages.

(buildings and economic activities) present in the overflowing area (E); (iii) by their degree of vulnerability (V), according to the equation:

$$R = H * E * V \tag{1}$$

Where:

H = hazard, or the probability of occurrence of the flood event and the related extension of the floodable areas;
E = value of the exposed assets localized in the flood area;
V = Vulnerability of the exposed assets, i.e. the percentage of loss of the elements at risk;

so the damage:

$$D = E * V \tag{2}$$

And so

$$R = H * D \tag{3}$$

From this formulation it is clear that the damage caused by a flood is a function of the intensity of the event, but above all it depends on the value of the exposed goods and of their vulnerability. The proposed model is implemented thanks to the use of GIS, in order to combine data related to different hazard scenarios with data on the economic value of damages for each of the elements located in the flood area under investigation.

In particular, as shown in Fig. 1, the model is structured in several modules:

1. The first module concerns the simulation of hydraulic hazard scenarios based on a hydrodynamic simulation model [4].
2. The second module concerns the identification and classification of the different elements exposed in the risk area according to the typologies and land uses provided by the municipal 3D cartography, and the estimation of the economic value by the criterion of the Depreciated Reproduction Cost on the basis of the assets' age [5, 6];
3. Finally, the third module is aimed to estimate the vulnerability on the basis of damage functions [7].

The total expected damage function would therefore be provided by the sum of the estimated damages for the each single element. The phases of the model and the main results will be described below with specific reference to phase 2 and phase 3 and to the definition of expected damage functions.

3 The Case Study

3.1 The Territory

The Italian town of Senigallia, is a town of 44.796 inhabitants in the province of Ancona (Marche Region) and is one of the main tourist destinations in the region. The city is located on the mid-Adriatic coast at the outlet of the Misa river and is 28 km far from Ancona (North) and 35 km from Pesaro (South. The territory (with an area of 115.7 Km^2) is predominantly flat even if surrounded by hills that slope towards the sea. On 3 May 2014, the city was hit by a flood that occurred following the break of a stretch of bank (right bank) along 50 m of the Misa river.

The Fig. 2 shows the area affected by the flood event. As can be seen from the image, the breakage of the embankment occurred just above Borgo Bicchia and the areas subsequently flooded were: Borgo Molino, via Capanna, the area of the former Regulatory Plan, via Rovereto area, the waterfront, the areas close to the Rotonda and Ponte Rosso, as well as via della Chiusa and the Cannella area.

The quantity of water that has invested the territory has been calculated to be equal to 14 million m^3 and has caused the death of 3 people, has destroyed movable and immovable properties, has brought serious damages to public and private structures, damaged thousands of houses emptying furniture, furnishings, memories, affections and irreparably compromising many economic activities. The residents involved in the flood were 9.707, the families that have lost all are 1.250,542 and the affected companies divided into the following categories are: 280 commercial buildings, 95 service activities, 85 craft activities, 5 agricultural activities and 77 other economic activities. Individuals have reported damages of 133.000.000 euros (39.000.000 euros to real estate properties and 94.000.000 euros to movable property), while economic activities damaged 46.000.000 euros (16.000.000 euros to assets) properties and 30.000.000 euros for movable assets). The costs incurred by the Municipality for the restoration of

Fig. 2. Extension of the flood area.

public assets amounted to 8.325.000, while those for emergencies and emergency services amounted to 4.200.000.

3.2   Phase 2: The Identification and Classification of the Elements Exposed in the Flooded Area and the Estimate of Reconstruction Costs

The first phase in the definition of the model has concerned the classification of the elements located in the territory. The division in categories has been developed by the use of the Territorial Information System defined for the territorial context under analysis and based on the 3D maps provided by the Municipality as well as by information on individual buildings taken from Google Maps. Therefore, different layers have been defined for each category affected by the flood event. More in deep, six categories have been identified:

1. Residential buildings
2. Commercial buildings
3. Industrial buildings
4. Storehouses
5. Services
6. Public assets.

Therefore a perimeter has been assigned to the buildings, and the delimiting polylines have been converted into polygons in order to have, for each building, the data relating to the area and, moreover, to the height. The ArcMap screen is presented below in which it is possible to distinguish non-residential buildings (delimited by a green section) from residential buildings (identified by a purple section). For each non-residential building, a further classification has been carried out with respect to the type of activity carried out. The analysis, developed through Google Maps, has shown that most of the activities belong to the commercial or service sector and, in addition, there are also several public buildings and some industrial warehouses (Fig. 3).

Fig. 3. Building in the flood area (green and purple section).

For each category of buildings, an ordinary average cost of new reconstruction has been estimated, and then compared to the level of damage by each building on the basis of the damage curves. The cost of new reconstruction for each type of building has then been estimated by the use of a parametric procedure based on functional elements [6, 8]. The historical data about reconstruction cost for functional elements have been taken from the Lombardy Region Typological Price List (DEI 2014). In the next section the damage curves used will therefore be described.

3.3   Phase 3: Evaluation of Vulnerability: Stage Damage Curves

Damage curves represent one of the most used methods in the literature for the ex-ante estimation of flood damage [3, 9, 10]. They represent functions that relate the potential damages to water height, with reference to a flood event. The studies examined [7] show that the damage curves refer to different categories of buildings (residential, non-residential) and within each category, to different typologies. In most of the cases, the only direct damage to buildings is considered, while few examples exist in the literature that re-examine the damage to the activities [11] and indirect damages to activities [12], as well as to cultural and environmental assets [13].

In this case, different damage curves have been used for each category of the classified buildings. In addition, the residential buildings have been divided into different categories (multi-story building, villa). The identified damage curves and their average functions are shown in Figs. 4, 5, 6, 7, 8, 9, 10, 11 and 12. As can be seen from the figures, the average damage function has been considered, as well as its variance. The average functions have been derived from a comparative analysis of the existing literature with specific reference to homogeneous case studies for territorial context and conditions to the case study under analysis, in which micro-scale damage functions have been defined for each type of building [7]. The results obtained are described in the next paragraph.

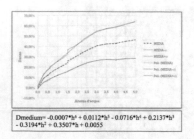

Dmedium= -0.0007*h⁶ + 0.0112*h⁵ - 0.0716*h⁴ + 0.2137*h³ - 0.3194*h² + 0.3507*h + 0.0055

Fig. 4. Commercial building

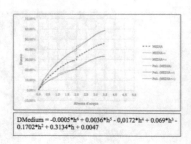

DMedium = -0.0005*h⁶ + 0.0036*h⁵ - 0,0172*h⁴ + 0.069*h³ - 0.1702*h² + 0.3134*h + 0.0047

Fig. 5. Industrial building

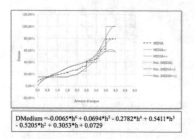

DMedium =-0.0065*h⁶ + 0.0694*h⁵ - 0.2782*h⁴ + 0.5411*h³ - 0.5205*h² + 0.3053*h + 0.0729

Fig. 6. Storage

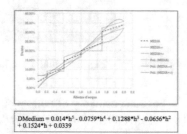

DMedium = 0.014*h⁵ - 0.0759*h⁴ + 0.1288*h³ - 0.0656*h² + 0.1524*h + 0.0339

Fig. 7. Services

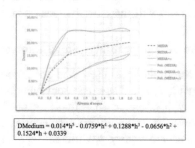

DMedium = 0.014*h⁵ - 0.0759*h⁴ + 0.1288*h³ - 0.0656*h² + 0.1524*h + 0.0339

Fig. 8. Public building

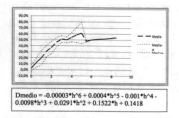

Dmedio = -0.00003*h^6 + 0.0004*h^5 - 0.001*h^4 - 0.0098*h^3 + 0.0291*h^2 + 0.1522*h + 0.1418

Fig. 9. Residential building 1 floor with-out basement

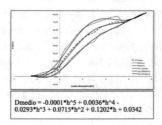

Dmedio = -0.0001*h^5 + 0.0036*h^4 - 0.0293*h^3 + 0.0715*h^2 + 0.1202*h + 0.0342

Fig. 10. Residential building 1 floor with basement

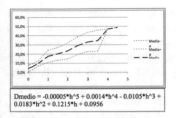

Dmedio = -0.00005*h^5 + 0.0014*h^4 - 0.0105*h^3 + 0.0183*h^2 + 0.1215*h + 0.0956

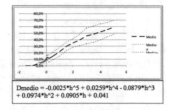

Dmedio = -0.0025*h^5 + 0.0259*h^4 - 0.0879*h^3 + 0.0974*h^2 + 0.0905*h + 0.041

Fig. 11. Residential building multiple floor with-out

Fig. 12. Residential building multiple floor with-out basement

4 Conclusions and Future Research Perspectives

The application of the damage functions (Figs. 4, 5, 6, 7, 8, 9, 10, 11 and 12) to the Senigallia case study has led to estimate an average damage to non-residential buildings in the flooded area of € 19.270.484 and an average damage to residential buildings equal to € 38.188.977. On the other hand, by using the curves called "media $-\sigma$" and "media $+\sigma$" respectively, damage values of € 13.702.300 and € 26.023.930 for the non-residential segment are respectively obtained, and values equal to € 29.343.732 and € 47.162.240 for the residential one. The previously estimated damages seem to confirm the goodness of the ex-ante estimation provided by the damage curves.

It can be noticed that the values obtained are very similar to those reported to the municipality of Senigallia by private individuals and local firms. On the other hand, it should be pointed out that the proposed model could be integrated by extending the field of investigation to the estimation of indirect damages to economic activities, as well as to the estimation of damages to cultural and environmental assets and activities. Under this perspective, further applications based on multicriteria spatial evaluation models are considered as promising [14–19], in order to include the social and environmental values in addition to the mere economic values of the resources at risk in the territorial context under investigation.

References

1. ISPRA, Istituto Superiore per la Protezione e la Ricerca Ambientale: Dissesto idrogeologico in Italia: pericolosità e indicatori di rischio, Rapporto 233 (2015)
2. Jongman, B., Kreibich, H., Apel, H., Barredo, J.I., Bates, P.D., Feyen, L., Gericke, A., Neal, J., Aerts, J.C.J.H., Ward, P.J.: Comparative flood damage model assessment: towards a European approach. Nat. Hazards Earth Syst. Sci. **12**, 3733–3752 (2012)
3. Merz, B., Kreibich, H., Thieken, A., Schmidtke, R.: Estimation uncertainty of direct monetary flood damage to buildings. Nat. Hazards Earth Syst. Sci. **4**, 153–163 (2004)
4. Montaldo, N., Ravazzani, G., Mancini, M.: On the prediction of the Toce alpine basin floods with distributed hydrologic models. Hydrol. Process. **21**, 608–621 (2007)
5. Del Giudice, V.: Estimo e valutazione economica dei progetti: profili metodologici ed applicazioni al settore immobiliare. Loffredo Editore (2010)
6. Manganelli, B.: Il deprezzamento degli immobili urbani. FrancoAngeli Editore (2011)

550 F. Torrieri and A. Oppio

7. Mancini, M., Lombardi, G., Mattia, S., Oppio, A., Torrieri, F.: An integrated model for ex-ante evaluation of flood damage to residential building. In: Appraisal: From Theory to Practice. Part of a Series on Green Energy and Technology, pp. 157–170. Springer (2016)
8. Del Giudice, V., Torrieri F., De Paola P.F.: The assessment of damages to scientific building: the case of the "Science Centre" Museum in Naples, Italy. In: Advanced Materials Research Vols. 1030–1032, pp. 889–895 (2014)
9. Thieken, A.H., Ackermann, V., Elmer, F., Kreibich, H., Kuhlmann, B., Kunert, U., Maiwald, H., Merz, B., Muller, M., Piroth, K., Schwarz, J., Schwarze, R., Seifert, I., Seifert, J.: Methods for the evaluation of direct and indirect flood losses. In: 4th International Symposium on Flood Defense: Managing Flood Risk, Reliability and Vulnerability, Toronto, Ontario, Canada, Maggio 6–8 (2008)
10. Molinari, D., Ballio, F., Menoni, S., Handmer, J.: On the modeling of significance for flood damage assessment. Int. J. Disaster Risk Reduct. 10, 381–391 (2014)
11. Arrighi, C., Brugioni, M., Castelli, F., Franceschini, S., Mazzanti, B.: Urban micro-scale flood risk estimation with parsimonious hydraulic modelling and census data. Nat. Hazards Earth Syst. Sci. 13, 1375–1391 (2013)
12. Bubeck, P., Kreibich, H.: Natural Hazards: direct costs and losses due to the disruption of production processes. CONHAZ Report 2011. http://conhaz.org/CONHAZ%20REPORT% 20WP012. Accessed 12 Sept 2017
13. Markantonis, V., Meyer, V., Schwarze, R.: Valuating the intangible effects of natural hazards - review and analysis of the costing methods. Nat. Hazards Earth Syst. Sci. 12, 1633–1640 (2012)
14. Comino, E., Bottero, M., Pomarico, S., Rosso, M.: The combined use of Spatial Multicriteria Evaluation and stakeholders analysis for supporting the ecological planning of river basin. Land Use Policy 58, 183–195 (2016)
15. Kubal, C., Haase, D., Meyer, V., Scheuer, S.: Integrated urban flood risk assessment – adapting a multicriteria approach to a city. Nat. Hazards Earth Syst. Sci. 9, 1881–1895 (2009)
16. Meyer, V., Scheuer, S., Haase, D.: A multi-criteria approach for flood risk mapping exemplified at the Mulde River, Germany. Nat. Hazards 48, 17–39 (2009)
17. Rosso, M., Bottero, M., Pomarico, S., La Ferlita, S., Comino, E.: Integrating multicriteria evaluation and stakeholders analysis for assessing hydropower projects. Energy Policy 67, 870–881 (2014)
18. Torrieri, F., Batà, A.: Spatial multi-criteria decision support system and strategic environmental assessment: a case study. Buildings 7(4), 96 (2017)
19. Torrieri, F., Oppio, A.: The sustainable management of flood risk areas: criticisms and future research perspectives. Green Energy and Technology. Springer, under press

The Energy Performance in the Construction Sector: An Architectural Tool as Adaptation to the Climate Challenge

Najoua Loudyi$^{(\boxtimes)}$ and Khalid El Harrouni

Ecole Nationale d'Architecture, Rabat Instituts, 6372 Rabat, Morocco
njsbai@gmail.com

Abstract. The building sector constitutes one of the most "energivore" sectors, which take part in the increase of the emission of the GES. Various energy policies carried out in several countries in the world show that it is possible to make important energy saving in this sector either by: seeking alternative energies (hydraulics, wind, solar) or by implementing the energy efficiency measures and laws (DPE, thermal regulation, taxes…), or by modifying the daily behaviors. The acts of heating, air conditioning, producing hot water or using devices household appliances constitute the first sections of energy consumption in a building. Designing a performant building is precisely controlling its environmental impact and optimizing the use of passive means with the purpose to reduce its Carbone Print. To decrease energy consumptions in the construction sector constitutes a true challenge to take up by the effective mobilization of all the involved actors, introduction and application of regulation frameworks, financial incentive, but also the support with the research actions and development.

Keywords: Performant building · Energy policies · Climate change

1 The Building and Energy Saving Deal

The energy is considered today as a question in the center of the concern of the governments, experts and public opinions. It systematically challenges the question of the environment and the climate change by pushing the reflection to find solutions to take up the climate challenge effects.

A political scenario, which doesn't put the energy problem at the heart of its concerns, is not reliable, of the least bearable, because not suiting to the objectives of the sustainable development.

The energy issue in the construction sector is not to be any more demonstrated because the building constitutes one of the most energy-consuming sectors, which participates in the increase of the emission of the Greenhouse Gas. In Morocco, the part of the building sector exceeds 25% of the energy balance of the country [1] (Fig. 1).

Morocco, following the example of other countries, which were mobilized to fight against the climate warming and to promote renewable energy, is gradually adopting a national energy strategy. The objective is to implement a national

© Springer International Publishing AG, part of Springer Nature 2019
F. Calabrò et al. (Eds.): ISHT 2018, SIST 101, pp. 551–556, 2019.
https://doi.org/10.1007/978-3-319-92102-0_58

Indicators	Weighting	Score	Rank
Emissions Level			
Primary Energy Supply per Capita	7.5%	100.00	4
CO_2 Emissions per Capita	7.5%	99.56	5
Target-Performance Comparison	10%	97.45	6
Emissions from Deforestation per Capita	5%	32.83	42
Development of Emissions			
CO_2 Emissions from Electricity and Heat Production	10%	32.18	47
CO_2 Emissions from Manufacturing and Industry	8%	50.54	52
CO_2 Emissions from Residential Use and Buildings	4%	31.72	52
CO_2 Emissions from Residential Use and Buildings	4%	9.11	58
CO_2 Emissions from Aviation	4%	41.07	37
Renewable Energies			
Share of Renewable Energy in Total Primary Energy Supply	2%	19.57	35
Development of Energy Supply from Renewable Energy Sources	8%	28.35	43
Efficiency			
Efficiency Level	5%	51.78	38
Efficiency Trend	5%	65.07	56
Climate Policy			
International Climate Policy	10%	78.90	10
National Climate Policy	10%	100.00	4

Fig. 1. Country scorecard morocco CCPI 2017 Source: «The Climate Change Performance Index 2017: Results»

plan of energy efficiency, which aims the reduction of 15% of the energy consumption by 2030 and which targets three sectors:

- The Tertiary sector,
- The residential one (through the use of the energy efficiency building code, especially the Thermal Regulation),
- Industry and transport.

"However, national experts value the country's ambitious targets and its solid policy framework for their implementation, which has resulted in a leading position in the national policy ranking and a top-10 placement in international policy." [2].

In order to carry out a policy of Energy efficiency in the building sector, Morocco joined in an energy prospect through several actions of which mainly:

- Introduction of an institutional framework with the adoption in 2014 of the Thermal Regulation of Construction in Morocco (RTCM). It is about a number of standards relating to the orientation of the buildings, thermal insulation, glazing specifications and the principles of sustainable design generally. According to the experts estimations, the application of the regulation will contribute, by 2035, to the reduction of approximately 2.97 million Teq CO2 [1].
- The use of the new energy strategy was also reinforced by:
  - Institutional and legislative tools: (Law 13-09 relating to renewable energy, law 16-09 relating to the creation of the ADEREE: National Agency for the Development of Renewable, Energy and Energy Efficiency which becomes AMEE the Moroccan Agency for Energy Efficiency, law 47-09 relating to energy efficiency),
  - Financial tools: (Creation of the Energy Development Fund FDE).

- The initiation by ADEREE (in collaboration with the UNDP) of the program of energy efficiency building that aims to introduce the minimal requirements of energy performance into the construction and rehabilitation projects. Through the transfer of knowledge and good practices, the program wants to be a way of promotion of the political dialogue between government, industrial, institutional, and professional organizations (architects, promoters, companies…) and of stimulation of public/private partnerships.
- The launching of the program "*Shemsi*" for the solar-fired heaters development, which was   conceived to   support the heating of   domestic   water by solar energy instead of electrical gas or fuel one. Being conscious of the extent of the request, the government placed renewable energy and particularly solar one at the row of the priorities by setting up this project, which aims the saving in emission of 3.7 million tons of $CO2$ annually [1]. It is a means of optimization of the environmental impact that target the development of 1.7 million m^2 of surface equipped with these solar heaters by 2020 against 350 000 m^2 available today. This program is based on the reinforcement of the quality of the heaters by the certification of the products and the approval of the fitters according to rigorous criteria in conformity with the international standards (tests of control, performance, warranty period…).
- At the territorial level, and following the example of some European cities, the AMEE (the Moroccan Agency for Energy efficiency) has launched a territorial strategy program named Jiha Tinou (my region in amazigh), which is a kind of variation of the national energy strategy on the level of the country's communities. This program is based on the methodological framework MENA Energy Award (MEA) of European inspiration that focuses on:
  - Facilitating integrated territorial energy planning,
  - Developing governance models,
  - Achieving projects.

Three urban communes in three pilot Moroccan cities (Agadir, Chefchaouen and Oujda) have been selected for this program. Thus, the accompaniment throughout the cycle of planning and the implication of the citizens and the elected officials in the dynamics of change constitute the first asset of this inclusive strategy.

In order to adapt to the impacts of climate change and to transit towards a green worldwide economy, Morocco has also started in 2016 the program Cleantech Morocco that aims the promotion of the innovation as regards clean technologies (the green entrepreneurial, innovations in energy efficiency, renewable energies, and green buildings …).

The various energy policies carried out in several countries of the world have shown that it is possible to make important energy saving in this sector without compromising the comfort of the users. Strong energy consumption does not go hand in hand inevitably with a high comfort. The orientation of the building and its compacity, the size and position of the openings, the height of ceiling, the daylight, the distribution of volumes, the thermal insulation of the walls and the roof, constitute passive architectural measures, which make it possible to reach the goal naturally with little contribution of energy.

2 Inciting Policies

Regarding the national energy strategy, the use of renewable energy or the adoption of new laws, the bioclimatic approach remains however a reliable tool for the construction of performing buildings. Indeed, the various agreements and strategies in this field aiming the reduction of gas emissions encourage the development of a "climate responsive architecture". The concept of energy performance in the building constitutes of this fact a true "construction site" to open on several faces namely:

- The encouragement of the bioclimatic design whose main goal is to privilege the use of supplied renewable energy to external oncs and to the mechanized techniques. The following table gives an idea on the realized economies on bill of electricity generated by a passive conception (Fig. 2).

Consommations conventionnelles	Maison classique RT 2000	Maison passive	Economie	%
Chauffage	849 €	90 €	-759 €	-89%
Rafraichissement estival	0 €	24 €	24 €	240%
Eau chaude sanitaire	194 €	58 €	-136 €	-70%
Auxiliares chauffage ventilation	37 €	46 €	9 €	24%
Autres usages (éclairages, electromenager...)	422 €	254 €	-168 €	-40%
Abonnements	189 €	105 €	-84 €	-44%
TOTAL	1 691 €	577 €	-1 114 €	-66%

Fig. 2. Forecast of electricity bills at the time of the conception Source: «*Practices of energy consumption in the performant buildings: theoretical consumptions and real consumptions*» [3].

- The development of the technologies, which allow increasing the energy efficiency through the adoption of certain measures (lowering the price for LED lightings, encouraging the use of technological innovations in the daily consumption of household electrical appliances, heating, settings in stand-by mode). By mobilizing massively the technologies development and research, great advances can be achieved to meet the energy challenge. In the same register, we cannot ignore the role of the NIT that allowed technological breakthroughs offering new spatiotemporal flexibilities. We speak today about telecommuting, which became in the Western companies an economic and social phenomenon of mass. We even think that this new kind of economy creates a new urban ecology and a new quality of life since it would generate:
  - Less displacement and mobility, therefore less use of personal vehicles and thus less CO_2 emissions.
  - Less traffic jam thus a decrease of the stress attributable to the daily displacements what is reflected on the wellness and the productivity of employees.
  - Real profits due to the replacement of the real offices by virtual offices.
- To request the public authorities to support the recourse to the climatic building techniques by introducing the concept of sustainability into the conception of cities

policies. Knowing that the eleventh objective among seventeen Sustainable Development Goals (SDG) fixed by United Nations is of *"Do so that cities and human establishments are opened to all, safe, resilient and sustainable»* , the concept of resilience and durability of the cities joins perfectly that of performing building and climatic approach in architecture. The conception of the new towns must de facto take the energy and environmental considerations through the optimization of transport between the habitat zones of and the economic activities ones. The evaluation of the risks of catastrophe must also be taken into account in the standards relating to the establishment, the conception and construction of the buildings and the infrastructures. *"The application of the codes of the construction and the mechanisms of planning and follow-up during the use of the city grounds, constitute an invaluable means to reduce the vulnerability to the catastrophes and the risks related to the extreme events, such as the earthquakes, the floods, the fires, the emission of dangerous matters or other similar phenomena"* [4].

- To reinforce and value the institutional and statutory aspects considering that any proposal or strategy do not have any value unless they are effectively applied. Moreover, in other countries, mentalities were hustled by imposed spending patterns like carbon taxes, or the dioxide tax aiming the limitation and regulation of climate warming. Concerning the residential sector, France, as an example, set the goal of a 38% decrease of the energy consumptions of the existing buildings park in the horizon 2020 by encouraging heavy renovations (at the rate of 400 000 important renovations a year) [5].

In order to encourage citizens to complete works of energy improvement of their houses, the French state set up, since 2009, devices of incentive: the credit tax for sustainable development (CIDD). This tax allows financing part of the related spending to the acquisition of materials or equipment judged necessary to reduce energy consumption (thermal insulation materials, powerful boilers, heat pumps…).

The posting of the label of the energy diagnosis of performance (DPE) is an obligation also in France. The DPE is a document, which gives an outline of the energy performance of housing by an estimate of its energy consumption, and its rate of gas emission (compared to the heating, the production of sanitary warm water and cooling, brought back to the surface of the building).

Except in special cases, the DPE is valid for 10 years and includes recommendations allowing the purchaser, the owner, the lessor or the tenant, to know the most effective measures to save energy. Since January 2008, the DPE must be posted in the lobby of some public buildings so that the occupants can know the energy performance of the building and its impact on the greenhouse effect (Fig. 3).

- To educate the public to the challenges of sustainable development in the construction sector and the utility of recourse to the energy performance. Sharing and exchanging knowledge and experiments remain the best way to show the benefits of these practices.
- To invest more in the scientific research and the promotion of new technologies and the renewable energy.

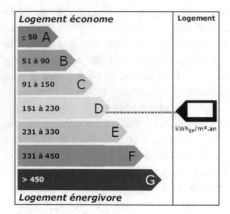
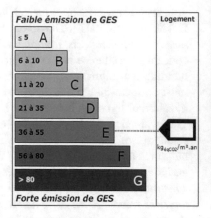

Fig. 3. (a) The energy label to know the primary energy consumption (b) climate label to know the amount of greenhouse. Source: ADEME "Energy savings in the building, New Buildings - Existing Buildings: The set of devices to improve the energy performance of buildings".

As most of the countries, the first concern of public authorities is to place the safety of the supply in the foreground of the objectives of their energy policies. However, the economic balance in Morocco is widely impacted by the value of the purchases of oil which accounts for 20% of the total imports [6]. To control energy is to save energy and the sector of the habitat precisely constitutes an opportunity for the promotion of new behaviors toward the energy. By setting minimum energy performance requirements for new buildings (espepcially related to heating and cooling systems, roofs, walls and so on), architectural tools can really pretend to be an effective way for implementing energy efficiency measures in the construction sector.

References

1. PNUD   MAROC.   http://www.ma.undp.org/content/morocco/fr/home/operations/projects/environment_and_energy/CEEB.html. Accessed 24 Oct 2017
2. Jan, B., Franziska, M., Christophe, B.: The Climate Change Performance Index 2017: results, German watch, Bonn, pp. 26–27 (2016)
3. Gaëtan, B.: Pratiques de consommation d'énergie dans les bâtiments performants: consommations théoriques et consommations réelles, p. 10. ADEME. LeroyMerlinSource, Paris (2013)
4. United Nations: Rendre les villes plus résilientes: Manuel à l'usage des dirigeants des gouvernements locaux, Une contribution à la Campagne mondiale 2010–2015, United Nation Office for Disaster Risk Reduction UNISDR, Geneva, pp. 41–44 (2012)
5. Comité Rio +20 CAHIER D'ACTEUR: Les bonnes pratiques à mutualiser en vue de Rio +20, Des énergies vertes pour une économie verte et équitable. Ministère de l'écologie, du développement durable, des transports et du logement, Paris, pp. 1–4 (2011)
6. Haut-Commissariat au Plan Maroc: Prospective Maroc 2030: actes du séminaire, prospective énergétique du Maroc: enjeux et défis. CP, Casablanca, pp. 96–113 (2006)

The Evaluation of the Economic Impact of University Campuses Energy Initiatives: The UPC Case Study

Lorenzo Uri, Patrizia Lombardi, Luigi Buzzacchi, and Giulia Sonetti

Politecnico di Torino, 10125 Torino, Italy
giulia.sonetti@polito.it

Abstract. Whilst the academia has always contributed to the social and cultural development of the places in which they are located through a sense of civic responsibility, the emerging climate change challenge require a "glocal" engagement to foster the so called "third mission" also in the sustainability management of the campuses. The impacts of a sustainable campus policy by University are essential for remaining competitive and productive internationally while meeting the UN SDGs. Yet, they are notoriously difficult to quantify and often materialise over a number of years. Therefore, this study tries to find out a possible economic quantification of energy activities performed by the universities and to test the proposed over the Polytechnic University of Catalunia (UPC) via a cost-effectiveness analysis (CEA) and an impact assessment analysis. This study provided valuable information about the economic impact of such activities on the local economy, enhancing the role of the university as source of social and economic growth. Indeed, despite the relative low percentage of investments with respect to the total annual budget, the repercussion in terms of employment, output and GDP have been important in the case of UPC. Moreover, it can be said that the induced effects of such initiatives are much more impacting than the investments themselves. Possible future researches can be focused on a guideline for a homogeneous data collection, trying to extend the number of universities considered and therefore have thresholds and influencing factors for each case study.

Keywords: Economic impact · University · Energy

1 Introduction

Nowadays, one of the most discussed challenge we have to face is surely the climate change [1]. The main cause of this global environmental problem are related human activities: we are indeed in a new epoch, the anthropocene, in which humans and our societies have become a global geophysical force [2]. After the industrial revolution and the exploitation of combustions and chemicals for human wellbeing and reproduction, global growth both in the economy and in the population index increased so quickly that the quantities of carbon dioxide, methane and nitrous oxide in the atmosphere have been following an exponential path toward never reached levels.

© Springer International Publishing AG, part of Springer Nature 2019
F. Calabrò et al. (Eds.): ISHT 2018, SIST 101, pp. 557–570, 2019.
https://doi.org/10.1007/978-3-319-92102-0_59

Specifically, according to the Intergovernmental Panel on Climate Change (IPCC), from the 1750 to the 2011 have been emitted 2040 ± 340 GtCO$_2$, of which about the 40% was held by the atmosphere, the 30% absorbed by the oceans, and notably half of the total emissions between the two aforementioned years was produced in the last just four decades [3]. This increasing amount of anthropogenic emissions implies several issues, such as the global warming and the ocean acidification [4]. Furthermore, overpopulation is the source of several other consequences [5]. If one thinks that in less than a century the world population is more than doubled, it should be quite easy to have an impression of the efforts and of the resources needed to feed all these people. The results are a fast natural materials depletion, which clearly are not infinite (e.g. deforestations), as well as an enormous waste disposal issue and a style of wicked consumption proper of a city life [6].

All these environmental issues are strictly correlated one to another because they produce effects and substances that interact in the global dimension [7]. In particular, since the forests act as strong carbon holders, the 15% of all the GHG emissions depends on the deforestations. It emerges that some of the environmental issues are dramatically affecting our current and, specifically, future wellbeing and resource availability [8]. In the past, economy told us that the trade was an efficient method to overcome some possible lack of natural resources [9]. Now, the problem is different: the global reserves are decreasing exponentially. The world is finite. The link to the economy it is quite straightforward: the market rule adjusts the price in relation to the quantity available in the market itself. So, in the scarcity condition we are entering, the prices of the natural resources are going to be extremely high. In addition, in a catastrophic scenario, if these resources will terminate, the loss of the jobs associated to the industries related to the exploitation and the re-work of these raw materials will worsen the economic situation even further. But what is the role of universities to this respect? Whilst the academia has always contributed to the social and cultural development of the places in which they are located through a sense of civic responsibility, the emerging climate change challenge require a "glocal" engagement to foster the so called "third mission" [10]. The requirements for global engagement therefore embraces many facets of the "responsive university" which are being generated by evolving priorities within the higher education system and the SDGs agenda [11]. These priorities include meeting the needs of a more diverse stakeholders set alongside the physical campus sustainability management; a lifelong learning created by a changed skill demands (interdisciplinarity and soft skills, sustainability education) [12]; more engagement with the end users of the academic research and first of all the city in which the University live in [13]. On the economic side of the discourse, when universities pursue the previous goals they are meanwhile supporting the Country's economic growth in many ways, since they [14]:

- generate and translate world-class research into new products, services, processes, and ways of working [15]
- drive innovation with industry, government and the third sector [16]
- help improve policy-making and the delivery of public services develop the existing workforce and new talent – graduates, apprentices, entrepreneurs – that our economy needs to grow and become more productive [17]

- act as civic and local growth leaders for their local areas, attracting high-quality investment and talent, investing into the regeneration of their towns and cities, supporting local businesses to grow and export, and improving the well-being of local communities [18].

These impacts of a University are essential for a country to remain competitive and productive internationally while meeting the SDGs [19]. Yet, they are notoriously difficult to quantify and often materialise over a number of years [20]. Also, the often limited budget of public bodies such as universities involves more and more effort in the evaluation of projects related to campus operations, the rigid selection of investments and a strategic allocation of funds that are not always aware of the opportunities given by a sustainable choice over another [21]. Therefore, the main challenges of this study are:

- To find out a possible economic quantification of the impact of energy-related initiatives;
- To develop an assessment of these and new initiatives from an economic point of view;
- To test the proposed assessment comparing the performance over the Polytechnic University of Catalunia (UPC).

The methodology proposed to meet the objectives of this study, is articulated in the following two approaches: a cost-effectiveness analysis (CEA) and an impact assessment analysis. These methodologies assessed the different aspects implied in the words "economic impact", such as the internal efficiency of budget expenditure, the external economy and the related employment rate are illustrated in section number two. It is important to explicit that the focus of this study are the initiatives and the projects related to the energy consumption and production in Universities, taking the UPC as cases study. Section 3 presents the test of the two methodologies with input output data collected in the university. Section 4 discusses the results and provides final conclusions.

2 Methodologies

Ambargis et al. [22] proposed a method to estimate the economic impact of a university, focused on a bill of goods approach. Particularly, this approach uses detailed data on the purchases of the locally produced inputs (including local labour). The impacts of these purchases are then added to the initial change in final demand to arrive to the total impact [23].

Another interesting approach was presented by Johnson [24]. He elaborated a methodology for the assessment of the economic impact of a university upon a region where it is not located (i.e. a non-local economy). Despite the first part of the method is quite standard (i.e. he considered the four economic sources (i.e. university, students, visitors and staff) which affect the employment and the business volume through a direct, indirect and induce impact), the formulae used to estimate the non-local impact are quite original. Starting from the Caffrey and Isaacs model [25], he modified the equations to account for changes in spending patterns that result when an expenditure occurs in a non-local instead of a local economy. Clearly, the estimation of these economic leakages from the local to the selected non-local economy is essential to determine the indirect impacts and the multiplier effects.

According to Ohme [26], another methodology could be used to estimate the economic impact of a university. Such method is called "ACE" and it is based on the expenditures of the university, the students, the employees and the visitors to local vendors. Then two regional multipliers are applied to compute the economic impact and the one on the local employment.

Valero et al. [27] shifted the attention towards the totality of the universities, analysing their economic impact in terms of GDP per capita. Particularly, they associated the number of universities present in a selected region throughout the history, clearly limited to the data available, to the economic growth. The study comprehends also some feedbacks about the development of the society closed to the universities in terms of culture, innovations and politics.

Therefore, summarizing the literature described above, it can be said that the solution to the selected problem and the relative application could reasonably be a mix of the two macro-arguments proposed so far. Indeed, some of the techniques presented will be taken into consideration in the development of the methodology, which will include also some existing methods seldom applied in the past studies.

2.1   The Cost-Effectiveness Analysis

When assessing the efficiency and the impact on the allocated budget of the energy-related activities, cost effectiveness analysis (CEA) represents a functional and feasible tool. The reasons are linked with the main features of such method:

- Its primary goal is to find out how the selected activity is reaching the objectives which is supposed to satisfy and its resulting economic implications. To this purpose, the method permits also the use of non-monetary values. This represents a key characteristic since the outcomes of the energy-related policies may be hardly valuable in a financial way.
- It is not affected by the energy prices changes. This feature permits to avoid considering the information, deeply simplifying the application.
- It can be used for ex-post evaluations, which is the current case.

CEA is usually composed by four steps:

1. Select the objectives and the physical measures (i.e. indicator) that describe the effectiveness criteria we want to assess.
2. Collect the cost incurred in the operations. Generally, only direct monetary resources are included, although also the eventual indirect and induced costs can be included. In this second scenario, the amount of uncertainty and variables will be reasonably higher, and the measured outcomes cannot be included so as not to count them twice.
3. Measure the impacts, namely the physical quantities of the desired outcomes.
4. Compute the final values which allow to evaluate the considered activities.

This last point usually consists of the cost effectiveness ratio vs the cost incurred and the effects achieved, quantified trough an indicator. Namely, the resulting value represent the amount of money spent per unit of outcome. If there are several activities which have comparable effects, they can be ranked by the common indicator following an increasing order. It also suggested to make a sensitivity analysis, simply repeating the calculations varying some key factors, so that making the results more robust. This methodology permits to assess the effectiveness of the activities and it will reasonably help the decision makers about the future allocations of the internal budget.

2.2   The Impact Assessment Analysis

As seen in the literature reviewed in Sect. 2, the actions undertaken by a university in the field of energy can be also related to a regional economic dimension.

In order to assess them under an economic impact point of view, this study performed a methodology called "Regional Input-Output Modelling System" (RIMS II) as explained in the RIMS II user guide edited by the Bureau of Economic Analysis of the U.S. Department of Commerce.

The goal is to find out the multipliers that are fundamental to compute the impact of the activity in the regional economy, in terms of output (sales), value added (Gross Domestic Product), employment and earnings (households' income).

Before starting to explain how the RIMS II works, it is important to clarify what is the logic upon which it is based. Namely, an initial change in one economic activity results in other rounds of spending. Indeed, the output of an industry can be the input of another one and obviously it can have consequences on their output and on the employment. To this purpose, the economic relations among the industries in terms of products and services exchanged, are expressed by the Input-Output (I-O) tables which almost every national statistical office provides periodically. Simply, the RIMS II adjusts these relations in order to adapt them for the selected region.

In order to formalize the principle of the I-O tables, in the following will be described the way with which the said tables are generated.

Producers are grouped into n industries, where businesses in an industry are assumed to use the same production process. Each industry i produces gross output, X_i, which is measured in dollars. This output is sold to industries j as intermediate inputs, z_{ij}, or to final users, Y_i,

$$X_i = z_{i1} + z_{i2} + z_{i3} + \ldots + z_{in} + Y_i \tag{1}$$

$$a_{ij} = z_{ij}/X_j \tag{2}$$

Each coefficient shows how much of industry i's output is needed to produce a dollar of output in industry j. These coefficients show how I-O models assume that industries always use the same proportions of inputs to produce output.

This framework implies the following assumptions before applying the RIMS II methodology:

- Backward linkages: I-O models can measure the impact of an industry's production on other industries in two ways. In a backward-linkage model, an increase in demand for output results in an increase in the demand for inputs. In a forward linkage model, an increase in the supply of inputs results in an increase in the supply of output. RIMS II is a backward-linkage model.
- Fixed purchase patterns: I-O models assume that industries do not change the relative mix of inputs used to produce output. They also assume that industries must double their inputs to double their output.
- No supply constraints: I-O models are often referred to as "fixed price" models because they assume no price adjustment in response to supply constraints. In other words, businesses can use as many inputs as needed without facing higher prices.
- Local supply conditions: RIMS II is based on national I-O relationships that are adjusted to account for local supply conditions. These adjustments account for the fact that local industries often do not supply all of the intermediate inputs needed to produce the region's output. Industries must purchase some intermediate inputs from suppliers outside the region. These purchases are often called leakages because they represent money that no longer circulates in the local economy. RIMS II accounts for these leakages by considering each industry's concentration in the region relative to its concentration in the nation. This method does not explicitly account for what is often called cross-hauling. Cross-hauling is when a good or service is both an import and an export of a region.
- No regional feedback: RIMS II is a single region I-O model. It ignores any feedback that may exist among regions.
- No time dimensions: The length of time that it takes for the total impact of an initial change in an economic activity to be completely realized is unclear because time is not explicitly included in I-O models. The actual adjustment period varies and is dependent on the initial change in an economic activity and the industry structure that is unique to each region.

The RIMS II model starts from the I-O tables which can be of three kinds: the make, the use and the import tables.

Particularly, the transactions reported in these documents are measured in respect to a precise year and are in the so-called "producer values", which means that the trade and the transportation margins are excluded.

The regional total requirements table provides information on the first and subsequent rounds of intermediate inputs required to produce another dollar of output. These rounds include goods and services produced in the region that are used by industries and purchased by households in the region. This table is also often called the "output multiplier" table. The sum of the entries in a column equals an industry's final-demand output multiplier. These final-demand multipliers measure the total change in output across all local industries per dollar of change in final demand. The output multiplier table is created by taking the Leontief inverse of the regional direct requirements table. This inversion is named after Wassily Leontief, who won a Nobel Memorial Prize for his finding: a final-demand change could be used to predict how an economy would react as measured by a change in total output. This result can be seen by substituting the set of equations describing the "technical coefficients" into the set of equations describing the calculation of the gross output produced by each industry, and expressing the result in linear algebra form:

$$X = AX + Y \tag{3}$$

where A is the regional direct requirements matrix, X is the gross output matrix and Y corresponds to the final users' matrix. The first equation can be rearranged as

$$X = I - A - 1Y \tag{4}$$

where I is the identity matrix and $(I - A)^{-1}$ is the Leontief inverse matrix. The predictive form of this last equation is

$$\Delta X = (I - A)^{-1} \Delta Y \tag{5}$$

which shows how a final-demand change can be multiplied by the coefficients in the total requirements table to predict total changes in output. The intuition behind this calculation is that a change in demand for an industry's output will result in a change in demand for the output of industries that supply intermediate inputs. This will result in a subsequent change in demand for the output of industries that supply intermediate inputs, which creates further change in a diminishing manner.

In addition to the final-demand output multipliers, three other types of final-demand multipliers are available earnings, employment, and value added.

The earnings multipliers measure the total change in local household earnings per dollar of final-demand change. Earnings consist of wages and salaries of proprietors' income, which is the net earnings of sole-proprietors and partnerships. These multipliers are calculated by multiplying each entry in the final-demand output multiplier table by the household-row entry in the regional direct requirements table.

The employment multipliers measure the total change in the number of local jobs per dollar of final-demand change. Employment consists of full and part-time jobs. These multipliers are created by multiplying each entry in the final-demand earnings multiplier table by the region-level employment-to-earnings.

The value-added multipliers measure the total change in local value added per dollar of final-demand change. Value added is comparable to regional measures of GDP. These multipliers are calculated by multiplying each entry in the final-demand output multiplier table by the value-added-to-output ratio in the national use table.

3   Application of the Methods to the Case Study

In order to test the efficacy of the methodologies proposed to calculate the cost effectiveness of the initiatives and the projects related to the energy consumption and production in Universities, the Polytechnic University of Catalunya (i.e. UPC) has been selected. In the following subsections, a brief context will be presented, as well as the related data collection for the applications of the aforementioned methodologies.

3.1 The UPC Context and Its Energy Policy

The UPC was founded in 1971 in Barcelona, Spain. Nowadays, it is the largest technical university of the Catalunya and one of the most important in Europe, thanks also to the supercomputer that they host since 2007, which is the biggest of the continent. Particularly, due to the high number of students (i.e. 30864), both international and local, and to the significant budget they can rely on (i.e. 283 million euros), the UPC represents a good example for the purpose of the study. The economic crisis started in 2008 forced the governments to cut, among the others, the education funds. For that reason, the UPC was forced to reduce the expenses trying to provide the best possible services anyway. One of the fields which was affected by the decreasing policies was the energy one. Indeed, the energy expense was one of the most significant and, at the same time, due to the high quantity of wastes, it was one of the most suitable for the cost reduction activities. Furthermore, the continuous increasing attention in sustainable policies contributed to the implementation of the related activities.

In this kind of environment, there was the necessity to formulate a plan of optimization and efficiency aimed to reduce the wastes and therefore the expenses, as well as installing and using new energy production facilities and innovative procurement ways. The main cause of the implementation of this plan was, in 2011, the forecasted increase of the energy prices which would have reasonably created many problems to the already critical economic conditions of the university. Especially, an estimated increase of the 10% would have caused an energy bill of about 7 million euros. The key actions of the energy-saving plan were:

- Teamworking and POE (Energy Optimization Projects).
- Institutional decisions and investments in energy efficiency.
- 600 k€ invested in efficiency improvements;
- 80 k€ is the cost of the monitoring system;
- 134 k€ are associated to the coordination of the plan.

  which gave the following results:

- Reduction by the 27% of the consumption of energy (from 2010 to 2014): from 53.8 GWh to 38.5 GWh. Especially, the reduction is divided in the 20% less concerning electricity and the 42% less relating to gas. Totally, in these 4 years the amount of GWh saved is 41.26, equal to more than 1 ycar of consumption.
- From 2011 to 2014, these measures permit to save 3932697 €.
- In the 2014 the 95% of the total area was covered by the POE measures.
- 932 training hours per person and 166 participants from 2011 to 2014.
- 4000 employees educated.
- 10587 visits to Slideshare for 25 presentations.
- 337 followers on Twitter.
- 200 is the number of professors and university staff people which have been educated concerning the importance of behave sustainable, during the period 2010–2014.
- 20 students actively involved in these projects.

The UPC purchases around the 98% of the total energy consumed, with costs running around the 0,25% of the estimated volume expenditure for the year. An important difference, with respect to the traditional paradigm, concerns the way through the UPC buys the energy. Indeed, in 2013, the UPC contributed to the birth of a consortium, called CSUC (i.e. Consorci de Serveis Universitaris de Catalunya), which offers the possibility to purchase the energy on behalf of the universities and the associated entities. Whether clients decide to start exploiting this opportunity they must pay a fee at the beginning of the year depending on the volume of energy required, to cover the costs associated to it.

The advantages of this joint procurement are several and they explain why an increasing number of entities is joining the consortium.

- Savings: both the volume and the professionalization of the purchase make the group push for lower prices.
- Simplification of procedures: thanks to the collaborative work, tendering and processing tasks are centralized in CSUC, reducing the workload of the participating entities.
- Best practices and sharing experiences as the collaborative process promotes knowledge transfer between participating institutions.
- Strength in dialogue in front of providers.
- Benchmark with other universities' strategies and procedures (basic for the sustainable development).

From the 2014 to the 2017, the cumulated energy saving is estimated to be 13.584.964 €. Considering the entities which joined the consortium, the single average savings is around the 15/20% of the annual expenditure. The total amount which is bought grew from 125 GWh in 2013 to 443 GWh in 2017.

3.2 The Economic Impact Evaluation of UPC's Energy Activities

This section deals with the application of the methodology described before. The first part concerns the cost-effectiveness analysis. In order to understand the different impacts of the single activities, assuming comparable lifecycles of the investments, it has been chosen to keep the division already presented in the description of the initiatives. Results are displayed in the table below (Fig. 1):

As already mentioned before, the impacts have been divided per macro-area, in order to estimate the multitude of the differentiated consequences. The cost considered when analysing the optimization/efficiency activities, is simply the sum of the investments reported above, in the presentation of the initiatives. Instead, while the social outputs have been already quantified before, the kilograms of CO_2 saved have been calculated by considering the Spanish energy mix which leads to 0,308 kg/kWh.

The economic results of the study seem to be quite satisfactory, indeed, the savings obtained are much more than the initial expense as it is represented by the ratio. Furthermore, under the social point of view, the budget invested per employee educated or student involved seems to be very high, obviously because the numbers at the denominator are quite low. Finally, the investment with respect to the kilograms of emission saved is very low, which means that the results are very impressive and, at the

UPC	Economical	Social	Environmental
Period 2011-2014			
OPT/EFF	Cost/savings	Cost/# employees educated	Cost/Kg CO2 saved
	€ 0,21	€ 203,50	€ 0,06
		Cost/training hour per person	Cost/Certified area (m^2)
		€ 873,39	€ 1,90
		Cost/students involved	
		€ 4.070,00	
		Cost/slideshare visits	
		€ 76,89	
Period 2013-2016			
Purchase	Cost/savings	Cost/# entities joined	
	€ 0,02	€ 2.028,44	
Production	Cost/annual savings	Cost/# students involved	Cost/annual Kg CO2 saved
	€ 6,68	€ 24,60	€ 3,62
4 years			
Production	Cost/savings	Cost/# students involved	Cost/Kg CO2 saved
	€ 1,67	€ 6,15	€ 0,91

Fig. 1. The CEA analysis for UPC.

same time, also the budget corresponding to the squared meters certified, results to be quite low, which imply a very good level of efficiency.

Switching to the jointly purchase of the energy, it is evident the blank space corresponding to the environmental impacts. The reason is that the company chosen in the contractual phase, can change several times even in the same year, which means that the energy mix could not be necessarily the same. Moreover, giving the average same value of consumption, even considering the Spanish mix, the number of renewable sources does not change, leading to the same kilograms of CO_2 emitted. Therefore, this purchase solution has primarily economic benefits rather than environmental. It can be said that the consortium has the power to foster the renewable investments by the companies, due to the high volume purchased, but so far, more emphasis was put in the economic impacts. It is confirmed by the ratio corresponding to the financial consequences, which highlights the huge amount of savings with respect to the initial allocated budget. Looking at the social implications, the result is difficult to be interpreted. Indeed, the number of entities which joined the consortium is very important, but at the same time, it is due to the economic results which the consortium reached, not only the UPC. So, the ratio is very high because the number of entities is relatively low with respect to the initial investment, even if considering just the UPC expense for the service.

Finally, considering the production for the auto-consumption, the UPC is moving its first steps. However, the results, considering the limited investments, are quite satisfactory. Indeed, they are comparable, in the 4 years' estimation, with the results of the optimization/efficiency activities even if the production deals lower amounts and quantities. Conversely with respect to the outputs corresponding to the social impacts of the first mentioned activities, in this case, the number of persons reached, in particular students, are very high considering the investment made. average of 3400 students attending the library.

The second part of the methodology deals with the impacts of the activities on the regional economy. To do so, the investments of such actions are multiplied by the coefficient shown in the table below, already computed by the Catalan national office.

Assuming that the multipliers are the same of the 2011, the results showing the impacts on the economy are reported in Fig. 2. In this case, the value added can be considered as the GDP because of the exclusion of the intermediate inputs.

	Output	Employment	Value Added (GDP)
Investments	€ 1.116.998,93	2,86	€ 843.076,54
Savings	€ 6.638.275,08	16,99	€ 5.010.366,50
Tot	€ 7.755.274,01	19,85	€ 5.853.443,04
% of the Barcelona's in 2011			0,008%

Fig. 2. Impact of the UPC energy activities on the Catalan economy.

In Fig. 3 are represented the multipliers with respect to the Catalan economy in the 2011, in fact it is the most recent update that the statistical office emitted. Particularly, the yellow line corresponds to the post-diploma education and the multipliers computed are related to the total output of the industry, to the employment and to the value added brought by the activities and especially, by the investments made. When computing the impacts, it is important to consider the investments as well as the savings which the said actions entail. This is because it is supposed that the savings deriving from the implementation of the plan are reinvested in the regular operations of the university, increasing the level of the services provided and, generally, the total output. So, when computing the impacts, are necessary the total investment made by the UPC (i.e. 935.500 euros) as well as the amount saved by the increase in efficiency, the reduction of the consumption and the different procurement activities (5.559.635 euros).

Multiplicadors de producció, ocupació i valor afegit
Marc Input-Output de Catalunya 2011. Taula simètrica

			Producció	Ocupació (per milió d'eur)	Valor afegit
71	854	Serveis educació superior	1,19	3,06	0,90
72	855-856	Altres serveis educació	1,42	19,13	0,91
73	861	Serveis hospitalaris	1,32	14,75	0,76
74	862	Serveis mèdics i odontològics	1,30	10,79	0,90
75	871-879	Serveis socials amb allotjament	1,32	33,44	0,89
76	90-91	Serveis creació, artístics, espectacles, biblioteques i museus	1,66	12,06	0,78
77	92	Serveis relacionats amb jocs atzar i apostes	1,48	6,34	0,82
78	93	Serveis esportius, recreatius i entreteniment	1,47	11,59	0,84
79	94	Serveis proporcionats per associacions	1,50	15,35	0,83
80	95	Serveis reparació ordinadors, efectes personals i domèstics	1,51	20,20	0,82
81	96	Altres serveis personals	1,29	9,47	0,90
82	97-98	Serveis de les llars	1,00	80,09	1,00

Fig. 3. Catalunya multipliers. Source: Idescat.

4 Discussion and Conclusion

The cost-effectiveness methodology has shown the differences among the activities with respect to the different impact field. The initiatives at UPC which permits to generate more savings with respect to the cost associated, are the jointly purchase of energy and the optimization/efficiency measure.

The input-output analysis presented the use of I-O tables accounting for local transactions. Clearly, the assumptions involving the linearity of these transactions, the aggregation of the single businesses, into a unique industry characterised by the same processes, and the fact that the multipliers are calculated on the base of past data, can be quite misleading, not reflecting the exact current relations. However, such considerations have been made to simplify the study, also due to the presence of limited and not precise data.

This study provided valuable information about the economic impact of such activities on the local economy, enhancing the role of the university as source of social and economic growth. Indeed, despite the relative low percentage of investments with respect to the total annual budget, the repercussion in terms of employment, output and GDP have been important in the case of UPC. Moreover, it can be said that the induced effects (i.e. savings) of such initiatives are much more impacting than the investments themselves.

Possible future researches can be focused on a guideline for a homogeneous data collection, trying to extend the number of universities considered and therefore have thresholds and influencing factors for each case study.

Finally, this study could be a useful starting point for the economic assessment of the sustainable initiatives implemented in the universities, hoping to contribute to foster the commitments towards these kind of initiatives, to spread the information and the awareness about the impacts of such initiatives on the economy, the society and the environment, and to improve the communication and the collaboration for sustainability leadership among all higher education institutes.

References

1. Klein, N.: This Changes Everything: Capitalism vs. the Climate. Simon and Schuster, New York (2015)
2. Steffen, W., Persson, Å., Deutsch, L., Zalasiewicz, J., Williams, M., Richardson, K., Crumley, C., Crutzen, P., Folke, C., Gordon, L., Molina, M., Ramanathan, V., Rockström, J., Scheffer, M., Schellnhuber, H., Svedin, U.: The anthropocene: from global change to planetary stewardship. Ambio **40**, 739–761 (2011)
3. Revi, A., Satterthwaite, D., Aragón-Durand, F., Corfee-Morlot, J., Kiunsi, R.B.R., Pelling, M., Roberts, D., Solecki, W., Gajjar, S.P., Sverdlik, A.: Towards transformative adaptation in cities: the IPCC's fifth assessment. Environ. Urban. **26**, 11–28 (2014)
4. Scheffer, M., Carpenter, S., Foley, J.A., Folke, C., Walker, B.: Catastrophic shifts in ecosystems. Nature **413**, 591–596 (2001)
5. United Nations Population Division, D. of E. and S. A. World Urbanization Prospects, the 2011 Revision. New York (2012)

6. Buhaug, H., Urdal, H.: An urbanization bomb? Population growth and social disorder in cities. Glob. Environ. Change **23**, 1–10 (2013)
7. Schuetze, T., Chelleri, L.: Urban sustainability versus green-washing-fallacy and reality of urban regeneration in downtown Seoul. Sustainability **8**, 33 (2015)
8. Chelleri, L., Kua, H., Rodríguez Sánchez, J., Nahiduzzaman, K., Thondhlana, G.: Are people responsive to a more sustainable, decentralized, and user-driven management of urban metabolism? Sustainability **8**, 275 (2016)
9. Holling, C.: Understanding the complexity of economic, ecological, and social systems. Ecosystems **4**, 390–405 (2001)
10. Ferrer-Balas, D., Lozano, R., Huisingh, D., Buckland, H., Ysern, P., Zilahy, G.: Going beyond the rhetoric: system-wide changes in universities for sustainable societies. J. Clean. Prod. **18**, 607–610 (2010)
11. Lozano, R., Ceulemans, K., Alonso-Almeida, M., Huisingh, D., Lozano, F.J., Waas, T., Lambrechts, W., Lukman, R., Hugé, J.: A review of commitment and implementation of sustainable development in higher education: results from a worldwide survey. J. Clean. Prod. **108**, 1–18 (2014)
12. Holden, M., Elverum, D., Nesbit, S., Robinson, J., Yen, D., Moore, J.: Learning teaching in the sustainability classroom. Ecol. Econ. **64**, 521–533 (2008)
13. Ceulemans, K., Molderez, I., Van Liedekerke, L.: Sustainability reporting in higher education: a comprehensive review of the recent literature and paths for further research. J. Clean. Prod. **106**, 127–143 (2014)
14. Martin, R.: Regional economic resilience, hysteresis and recessionary shocks. J. Econ. Geogr. **12**, 1–32 (2012)
15. Lozano, R., Ciliz, N., Ramos, T.B., Blok, V., Caeiro, S., van Hoof, B., Huisingh, D.: Bridges for a more sustainable future: joining Environmental Management for Sustainable Universities (EMSU) and the European Roundtable for Sustainable Consumption and Production (ERSCP) conferences. J. Clean. Prod. **106**, 1–2 (2015)
16. Velazquez, L., Munguia, N., Platt, A., Taddei, J.: Sustainable university: what can be the matter? J. Clean. Prod. **14**, 810–819 (2006)
17. Cortese, A.D.: The critical role of higher education in creating a sustainable future. Plann. High. Educ. **31**, 15–22 (2003)
18. Venetoulis, J.: Assessing the ecological impact of a university: the ecological footprint for the University of Redlands. Int. J. Sustain. High. Educ. **2**, 180–197 (2001)
19. Lombardi, P., Sonetti, G.: News from the Front of Sustainable University Campuses. Edizioni Nuova Cultura (2017)
20. Sonetti, G., Lombardi, P., Chelleri, L.: True green and sustainable university campuses? Toward a clusters approach. Sustainability **8**, 83 (2016)
21. Baboulet, O., Lenzen, M.: Evaluating the environmental performance of a university. J. Clean. Prod. **18**, 1134–1141 (2010)
22. Ambargis, Z.O., McComb, T., Robbins, C.A.: Estimating the local economic impacts of university activity using a bill of goods approach. In: The 19th International Input-Output Conference, pp. 13–19. International Input-Output Association, Vienna (2011)
23. Bess, R., Ambargis, Z.O.: Input-output models for impact analysis: suggestions for practitioners using RIMS II multipliers. In: 50th Southern Regional Science Association Conference, New Orleans, Louisiana, pp. 23–27 (2011)
24. Johnson, M.H., Bennett, J.T.: Regional environmental and economic impact evaluation: an input-output approach. Reg. Sci. Urban Econ. **11**, 215–230 (1981)
25. Caffrey, J., Isaacs, H.H.: Estimating the Impact of a College or University on the Local Economy (1971)

26. Ohme, A.: The economic impact of a university on its community and state: examining trends four years later. Univ. Delaware (2003)
27. Valero, A., Van Reenen, J.: The economic impact of universities: evidence from across the globe. National Bureau of Economic Research (2016)

A Contribution to Regional Planning Finalized for Fire Resilience

Alessandra Casu[✉] [ID] and Marco Loi

University of Sassari, 07041 Alghero, Italy
casual@uniss.it

Abstract. The study deals with a plan hypothesis for a sustainable development in the inner region of Barigadu (Sardinia), based on the prevention of fires.

Common aspects of all fires are the abandonment of the territory and the lack of properly managed agro-pastoral practices, which favor the formation of highly flammable plant material which, together with an increase in the High Temperature Day (HTD) and the decrease in annual averages of rain, leads to increasing the probability of triggering fires and the areas they can affect.

The work provides a forecast model (burn probabilities) in the current conditions, a set of practices based on land use, seen both in terms of fire prevention and in terms of improving the socio-economic conditions, and a simulation of the burn probabilities if the proposed plan was implemented.

Keywords: Fire prevention · Fuel model · Rural development

1 Introduction

Fires are of particular importance in the countries of the Mediterranean basin, considered as a hot spot, with a probable evolution towards a warm and dry climate, with a significantly greater risk of episodes with intense heat waves, as well as an increased risk of fire [1–3]. From the 70s to today, fires show an increase in the trend regarding number and areas. Flammability of the plant fuel is related to spatial distribution of vegetation, and its interaction with the climatic variability [4]. In addition, studies focused on this area have shown that rainfall, in terms of annual sum and number of rainy days, are gradually decreasing, while days of extreme heat are increasing: this could cause an increase in summer drought risk, which in turn, depending on the type of vegetation, could cause an increase in fire risk [5].

Among Mediterranean regions, Sardinia is one of the most affected by these phenomena. Forest fires are concentrated from June to September, with peaks in July. If the average of forest burned in the first forty years of XX century was between 1,000 and 1,500 ha, in the forty years between 1971 and 2014 the hectares burned on average in Sardinia are 7,294 [6]. In 1995–2009, from June to September, in Sardinia the incidence was of about 2,500 fires/year, with an average of about 17,000 hectares/year of affected areas [7]. The increase in the number of exceptional events increases the risks: heat peaks during the summer season increase the fire risk, peaks of rain increase the hydro-geological one. Sardinia also has a high density of urban-rural interface areas (WUI), that are increasingly threatened by serious fires.

© Springer International Publishing AG, part of Springer Nature 2019
F. Calabrò et al. (Eds.): ISHT 2018, SIST 101, pp. 571–578, 2019.
https://doi.org/10.1007/978-3-319-92102-0_60

In the study area - the historical inner region of Barigadu - there is a strong forest presence, which is not linked to a strong economy. In addition to agro-pastoral and forest activities with different densities, also a Regional Park (the Oasis of wildlife protection of Assai) makes the incidence of so-called flammable areas very high.

The phenomenon of fires in this area is similar to the whole island. The abandonment of customs and habits has generated the propagation of incendiary phenomena, alien to the local populations, found to use a tool (the fire) that culturally was not so present and, for these reasons, difficult to manage. The most affected areas are those of the commons and largely wooded that, damaged by fire or by a no longer wise cut, also affect landslides.

2   The Study Area

The Barigadu shows a landscape determined by low mountains and by the important presence of woods. In this region the agricultural practice is exerted on a limited portion, and from an economic viewpoint it is a depressed area, also because no projects have ever been developed to exploit its resources. The river Tirso and its tributaries have been a fundamental resource for the entire region, influencing all the activities: in particular the agricultural ones, due to the presence of water resources, but now they have partially given way to vast, completely abandoned forests, limiting agro-pastoral practices to the plains. The geology and the morphology strongly condition the present plant species, incising in different way also on fire danger. The composition of soils and sub-climates of the region have led to the development of four main different species of vegetation, which include herbaceous species and oaks.

As described by Angius [8] and Le Lannou [9], between the mid-1800s and the mid-1900s the area presented a territory covered by forests, which were decreasing as agriculture was developing, especially vine and olive trees. These fruitful crops, combined with reckless policies of exploitation of woodland resources, allowed a slight decrease in the fire phenomena, as pastoral practices decreased.

These uses are currently forgotten or forbidden, for reasons of conservation and protection which, in some cases, increased the fire risk: within some now protected areas, there are no longer those activities that also had the purpose of cleaning the undergrowth. Furthermore, the loss of experience in fire use meant that fires hit large forest areas, and also became urban interface fires, with a considerable danger for settlements and inhabitants.

3   Methods to Map the Fire Risk

The simulation and use of models to analyze spatial variations of forest and interface fires has gone through the use of GIS software (*ArcMap®*), *FlamMap* (used to map the incendiary danger), and *WindNinja* (for the simulation of the wind behavior, based on the orography and the dominant vegetation). They are among the most used in scientific literature related to fire prevention, mainly because they have outputs that can be used in several other softwares and a variable scale of detail [10].

The study of orography, based on the GIS elaboration of the DEM reworked with a resolution of 50 m for the relief map (elevation), is useful to the construction of the display maps of the slopes (aspect) and reliefs expressed in degrees (slope).

The next phase is a Fuel Model map, representing land uses and their aspects concerning the combustion and its variable values, according to the season. The model was drawn up by an interpolation between the CORINE Land Cover Map updated to 2012 and the official land use map developed in 2008, reclassified according to their characters in 14 categories, each with 5 values (which constitute the Fuel Moisture Content, hereafter abbreviated as FMC) relative to 1-hr, 10-hr, 100-hr (time-lag values in which the moisture is reduced at 63.2% of the previous or initial value, based on the corresponding thickness of 0–0.635 cm, 0.635–2.54 cm, 2.54–7.62 cm), Live Herbaceous and Live Woods, present in each category of the Fuel Model.

The hazard map due to the winds is made with *WindNinja*. After loading the elevation model with a resolution of 50 m and the dominant vegetation as input, the values related to prevailing winds were inserted, i.e. the mistral (NW), the libeccio (SW) and the sirocco (SE), which characterize the highest number of fires throughout Sardinia. In particular the mistral (although, unlike libeccio and sirocco, it is not a hot wind) has a higher average speed (over 25 km/h), which then leads to a rapid advancement of fires, so that it is the most dangerous wind. Moreover, it rarely transports humidity, so fuel is drier and increases the chances of a primer.

Outputs realized with *WindNinja* (wind direction and speed), along with Elevation, Slope, Aspect, Fuel Model and FMC are the inputs that are useful to process the hazard map using *FlamMap*. Fires are simulated using the Minimum Travel Time (MTT) of Finney diffusion algorithm [11]. This is a two-dimensional model of fire growth and calculates growth and development from the search for a series of paths, in which the fire spreads in the shortest possible time from linear or polygonal ignitions [11]. The results are identical to the expansion wave used in other software (e.g.: FARSITE), but climatic and humidity conditions remain constant. In this sense, *FlamMap* shows an extreme danger: prevention actions can be envisaged, reloading the inputs into the software and verifying that the changes have effectively served to reduce the danger, always at a fixed and not variable criticality. The MTT can be used to calculate the probability of burning for a specific number of randomly positioned ignition points for a constant duration, as in the case of this work, in which 20,000 random ignition points are expected for a duration of 500 min (about 8 h). The larger the area and the more ignition points are necessary to have a homogeneous coverage, and a more realistic simulation of the phenomenon: several tests were performed to estimate the number of ignitions (500-1,000-5,000-10,000-15,000-20,000). This feature produces a single map containing the fraction of the number of fires encountered per each node (0.0 = null or minimum; 1.0 = maximum). The map of burn probabilities thus provides an almost real situation of the response to fire by the vegetable fuel present in the territory, or the development of fire according to the different land uses. This already provides a clear example of what most limits the fire: not only the land uses (as the example of the vineyards), but also the cleaning of the undergrowth, or the presence of natural or anthropogenic barriers: streets are a peculiar case, as they represent areas where ignition points are most present, but at the same time they act as barriers.

4 A Planning Model

The proposed model does not impose individual uses, but offers the possibility of using resources with actions aimed primarily at limiting their degradation, restoring the ecosystem equilibrium, and developing the region. The proposal, therefore, has not a regulatory form, but objectives and actions, consistent with the inherent two-way community-site relationships, that are currently being lost. We will try to establish measures for rural development, divided in sectors according to the landscape units (forest, agricultural, peri-urban areas, urban) and to the limitation of the fire risk. The planning measures will take into consideration the role of local communities, their practices and economic activities. Actually, other ways to encourage the implementation of the measures concern again the function of agro-pastoral activities, directed towards the protection and safety from degradation, landslides and fires.

The forest area in Barigadu plays a primary role and is the main feature of the region, and agro-pastoral activities are almost an outline of this important resource. In the now rooted collective imaginary, the presence of extensive woods is synonymous with high environmental and landscape quality, and rules that aim to protect and preserve this heritage meet a wide margin of favor. The reduction of forests, therefore, is not proposed as deforestation that had affected Sardinia in the second half of the nineteenth century for industrial purposes, but becomes a consequence of the resumption of activities where they had historically taken place.

The forest uses will be proposed on the basis of three general objectives: Control of the wooded area expansion; Forest maintenance; Protection from fires.

Controlling forest expansion is aimed in particular at whom carries out activities within the eco-tonal areas, adjacent to the forests, such as shepherds and farmers. Part of these activities will go hand in hand with a thinning of forest areas, to decrease the probability of incendiary expansion.

The second general objective deals with forest maintenance, and needs the best agreement between public and private. About 20,000 ha are privately managed, with a strong vegetative presence with more incendiary danger. These woodlands cannot be fully acquired by public bodies but probably, in some parts, the management method of Assai Oasis could be applied. Among the proposed collaboration between public and private bodies there is the example of Suni, in which the sheep pastures are burned with the Forestry Corps, and the use of fire for agro-pastoral purposes is taught to the owners. The Corps can help to selectively cut the wood, indicate which plants are better for the collection of cork or other practices for improving the quality of the environment, which would contribute to an economic return. Another important measure is the gradual replacement of some non-endemic tree species, such as eucalyptus, which for its rapid development and strong soil hold-up was used on some slopes with a very high landslide hazard. It can be implemented with naturalistic engineering practices applied to the settling of the slopes.

The proposed measures try to make the work of shepherds and farmers economically advantageous. The two actions proposed concern the promotion of a protocol for managing some protection areas (especially firebreaks) defined by the workers themselves. The definition of the firebreak has a very strong impact on the morphology and the landscape, but with the direct care of the shepherd can bring benefits both for what concerns the decrease of the fire danger, and for the improvement of environmental conditions of the forest, thanks to the control of biomass. Secondly, agreements are promoted for processing and trading the products of these activities, also finding a diversification with respect to the rest of the pastoralism and aiming, through niche markets, to other market segments outside Sardinia.

The recovery of abandoned crops and a better management of those already present could offer the Barigadu an exit from the current situation and an insertion into economic circuits, that could increase general well-being. The proposal separates the objectives and actions between agriculture and pastoralism, without interrupting the relations between them. The work concentrated in a limited space brings minimal benefits to the agro-pastoral world, concentrated in the short term, while if we aim at extensive methods we would have a fruition of resources for a much longer period.

The first general objective is the reduction of hydrogeological and fire risk. Although they may seem separate, the occurrence of one affects the other and, therefore, contrasting one also means acting against the other. Among the proposed actions, the recovery or construction of terraces is included: compared to the hydrogeological risk, they strongly limit the solid transport process, typical of landslides and floods, but also limit the advance of the fire upwards, because trees have a longer distance, that also allows a better care of the plants. Terracing also allows the cultivation in slopes with significant acclivities, with species such as olive trees, which suffered a significant drop in this area. Shepherds would have the opportunity to use extensive pastures in the *saltus*, parts of the territory where pastures were present until the 1960s.

The second general objective must necessarily be based on a productive but sustainable farming culture, on associative forms of producers and on the possibility of placing products with DoO brands on the market. The opportunity of entry into niche markets could mean better profits for producers, compared to the current ones, also because management costs would be strongly reduced, especially for pastoralism, considering the reintroduction of three-year rotating the cultivated land.

In order to reduce fire risk, the integrated fuels management is essential in densely populated areas located near forests, and also in this case the work identifies general and specific objectives and actions in order to minimize the danger. The contrast to the advancement of fire is the first specific objective, reachable through various actions, including the establishment of fireproof curtains, i.e. forest areas adjacent to urban centers which tree density is lower than the forest, in order to leave only tree species of contrast to fire, as for example the cork oak.

The high density of vegetable gardens surrounding the villages has historical, cultural (they still are an important source of sustenance), and pedological (they usually are the most humid land in the area) reasons, which have given the population a set of unwritten guidelines, in order to continue these activities protecting against degradation processes such as fires, floods or landslides. Many vegetable gardens located near Barigadu villages (especially from the border between Busachi and Ula Tirso proceeding to the north) develop on terraces built in antiquity that allow a decrease of acclivities, so an easier cultivation and a much lower erosion in case of heavy rains and a much lower fire propagation. In any case, even in the absence of acclivities and therefore of terraces, the gardens have the same effect as the firefighting bands, so their recovery - even in flat areas - would have the same effect.

The second specific objective, related to the protection of inhabited areas, is to favor a rapid rescue, aiming at better accessibility to roads, to water resources or other practices of extinguishing. Accessibility to roads presents difficulties with regard to secondary roads, as the dense network of paths structured by ancient uses has given way to nature and dense forests, so this action must necessarily pass from their reopening and from the creation of new ones to facilitate access to vehicles.

5   Assessing the Proposal: Simulation at a Detailed Scale

The focus includes the territory of Ula Tirso and part of Ardauli, Neoneli, Busachi and Ortueri. The area shows a very high forest density, with a consistent presence of Mediterranean vegetation, with high probability of fire ignition. The slopes facing north and those facing Lake Omodeo are the only higher humid areas and the sole contrast to the fire spread. Agricultural activities practiced on small plateaus don't help to combat fires for two main reasons: the tall vegetation is sparse so, in case of high temperatures, the humidity of lower vegetation decreases considerably, and many crops are non-irrigated arable lands, which increase the ignition probability. Arboreal cultivations are not diffused, and natural grazing is even less: human lives, agricultural activities and the environment are exposed to high risk, especially in the case of Busachi and Ula Tirso, located at the edge of the plateau and in direct contact with the uncontrolled vegetation grown on the slopes.

The output of *FlamMap* has returned a situation that can be found in the current state with 10,000 ignition points. The mapped event (Fig. 1) is characterized by mistral wind (NW) at 9 m/s and provides a simulation of fire behavior. The areas with the highest BP (Burn Probability) are concentrated in the western part of the plateau and south of the urban center of Busachi, where the land is flatter, the forest density is very low and forage crops are not irrigated, and where woodland density is about the same as on Lake Omodeo slopes, but the humidity is lower, and vegetation dry.

The reliability of this simulation is given by the comparison with the fires that occurred in the same area, such as those in 2013 and in 2011, in which there is no diffusion of the vine as a protection zone. The values tending to zero (lighter colors in the map) are exclusively in areas where there is strong presence of tree crops, such as olive groves or vineyards, but also in areas where family gardens are located. This shows how the garrison affects the development and spread of fires (Fig. 2).

Fig. 1. Current burn probabilities in the focus area.

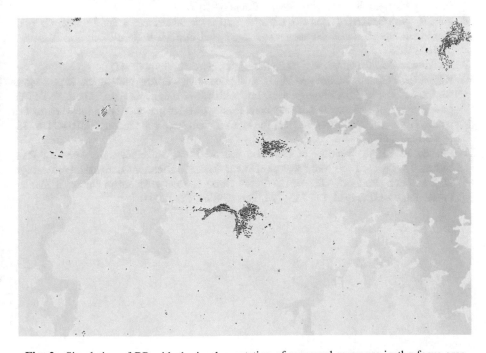

Fig. 2. Simulation of BP with the implementation of proposed measures in the focus area

References

1. Giannakopoulos, C., Le Sager, P., Bindi, M., Moriondo, M., Kostopoulou, E., Goodess, C.: Climatic changes and associated impacts in the mediterranean resulting from a 2 °C global warming. Glob. Planet. Change **68**(3), 209–224 (2009)
2. Dimitrakopoulos, A., Vlahou, M., Anagnostopoulou, C., Mitsopoulos, I.: Impact of drought on wildland fires in Greece; implications of climatic change? Clim. Change **109**(3–4), 331–347 (2011)
3. Koutsias, N., Xanthopoulos, G., Founda, D., Xystrakis, F., Nioti, F., Pleniou, M., Mallinis, G., Arianoutsou, M.: On the relationships between forest fires and weather conditions in Greece from long-term national observations (1894–2010). Int. J. Wildland Fire **22**(4), 493–507 (2013)
4. Westerling, A.L.: Climate change impacts on wildfire. In: Schneider, S.H., Rosencranz, A., Mastrandrea, M.D., Kunz-Duriseti, K. (eds.) Climate Change Science and Policy. Island Press, Washington (2010)
5. Arca, B., Pellizzaro, G., Duce, P., Salis, M., Bacciu, V., Spano, D., Ager, A., Finney, M.A., Scoccimarro, E.: Potential changes in fire probability and severity under climate change scenarios in Mediterranean areas. In: Spano, D., Bacciu, V., Salis, M., Sirca, C. (eds.) Modelling Fire Behaviour and Risk. PROTERINA-C Project, EU Italia-Francia Marittimo 2007–2013 Programme, Sassari (2012)
6. Salis, M., Ager, A.A., Arca, B., Finney, M.A., Bacciu, V., Duce, P., Spano, D.: Assessing exposure of human and ecological values to wildfire in Sardinia, Italy. Int. J. Wildland Fire **22**(4), 549–565 (2012)
7. Regione Autonoma della Sardegna: Piano regionale di previsione, prevenzione e lotta attiva contro gli incendi boschivi 2014–2016, Cagliari (2016)
8. Angius, V.: Geografia, storia e statistica dell'Isola di Sardegna. In: Casalis, G. (ed.) Dizionario storico-statistico-commerciale degli stati di S.M. il Re di Sardegna, Maspero, Torino (1853)
9. Le Lannou, M.: Pâtres et Paysans de Sardaigne. Arrault, Tours (1941)
10. Keane, R., Frescino, T., Reeves, M., Long, J.: Mapping wildland fuels across large regions for the LANDFIRE prototype project. In: Rollins, M., Frame, C. (eds.) The LANDFIRE prototype project: nationally consistent and locally relevant geospatial data for wildland fire management. USDA, Forest Service, Rocky Mountain Research Station, RMRS-GTR-175, Ogden, UT (2006)
11. Finney, M.A.: An overview of FlamMap fire modeling capabilities. In: Andrews, P.L., Butler, B.W. (eds.) Fuels Management - How to Measure Success: Conference Proceedings, 28–30 March, Portland, OR. USDA Forest Service, Rocky Mountain Research Station, Proceedings RMRS-P-41, pp. 213–220, Fort Collins, CO (2006)

Urban Spaces and a Culture of Safety

Antonio Taccone[✉]

Mediterranea University of Reggio Calabria, 89124 Reggio Calabria, Italy
ataccone@unirc.it

Abstract. Cities have to be more inclusive and put citizens at the base of development projects, interpret their needs by relating their civic sense and their perception of urban quality and train and educate to safety with precise information actions but above all with the design of quality public spaces. The recent planning experiences of some European cities show us that it is possible to introduce quality in the connective space thanks to a different cultural approach and under the pressure of innovative territorial tools, attentive to the quality of urban facilities in the constant search for urban safety and security. The spaces designed for mobility are the main component of the public space, where transport and pedestrians perform their functions of walking and socializing. They are urban places with multiple uses that have to favor social relationship contributing to create a sense of safety and inclusion within the urban context. Moving, walking, socializing are activities that must be carried out serenely and in conditions of comfort because these are above all the indicators of the civilization of a society that make one perceive the feeling of self confidence.

Keywords: Public spaces · Safety and security · Mobility

1 Self Confidence in Urban Settlement

The awareness of the close link between safety and ways of designing and understanding the urban spaces of the city occurred in Italy only in the nineties, when, thanks to the influence of the policies of some nations such as Great Britain, France and United States, the theme has been linked to quality in urban areas. We have therefore became from a simple response of emergencies to prevention policies consisting of design practices capable of raising the quality of life and improving functionality, social relations and in particular the livability (Fig. 1) of the urban environment.

On the one hand there has been a push towards new attributions to local authorities, to which some safety responsibilities previously belonged to the activities of the State are nowadays, on the other hand the planning and design of the city and the urban passages of a more active dimension capable of affecting the quality of life of citizens by acting on the functionality, efficiency and timing of services.

From the definitions of the studies of Jane Jacobs (Death and Life of Great American Cities, 1961) on spontaneous control linked to the vitality of the city and the early guidelines for urban planning by Oscar Newman (Defensible Space, crime prevention through urban design, 1972), we have only recently reached a veritable validation of the role of planning and maintenance of urban spaces as practices that "can

© Springer International Publishing AG, part of Springer Nature 2019
F. Calabrò et al. (Eds.): ISHT 2018, SIST 101, pp. 579–584, 2019.
https://doi.org/10.1007/978-3-319-92102-0_61

Fig. 1. Livability urban spaces in Copenhagen. National Library and Superkilen Park.

contribute significantly to the safety of cities, objective and perceived"[1] (Technical Report TC 14383-2, CEN, European Committee for Standardization).

Annex D of the Technical Report was the reference for the "urban safety guidelines manual, Urban Design Planning, Safety Space Management", drawn up as a result of the Agis - Action SAFEPOLIS project and conducted under the guidance of the Laboratory for the urban quality and safety of the Politecnico di Milano[2], Diap, which perhaps represents the most advanced study to provide technical support to professionals and clients with the aim of making our cities safer [1].

Contrary to all the manuals present in Europe and America aimed at providing criteria and guidelines to the building scale or, at the limit, to that of public space, the volume "Planning urban design management of space for safety" deals with design urban planning to propose a management model.

In recent years, in our cities in constant transformation, the concept has expanded towards the construction of social justice [2] and the specializations of the quality of places, which are read as spaces in which different degrees of inequality must be recognized due to to the functioning capacity of urban life [3].

2 Plurality and Urban Spaces

Today the cities present a particular system where the high resilience constituted through the continuous exchanges and the sum of important diversities that have constituted the wealth is now undergoing aggressions and actions that in some connective areas are approaching the limit point. In particular, in areas with a low degree of

---

[1] Technical Report TC 14383-2. Formed as a summary of the work of the working group on "Prevention of Crime by Urban Planning" which ended in 2006 with the issuance of the Report, which was finally adopted by CEN in 2007. The document is intended as support for good practices and not as a standard with a binding scope.

[2] Urban design planning, safety space management, urban safety guidelines manual published by Politecnico di Milano - DiAP, IAU Ile-de-France, Emilia-Romagna Region, as partners of the project Agis - Action SAFEPOLIS 2006–2007, co-funded by the European Commission. Director of the Diap, Prof. Alessandro Balducci, Scientific Coordinator of the project, Prof. Clara Cardia.

industrialization, the exodus from rural areas determines the phenomenon of abandoned centers with the consequent lack of even a simple defense of the territory causing serious hydro geological disruptions. In cities, urban policies have failed to prevent the consequent formation of new peripheral areas without services and infrastructures and often of low quality. All this contributes to giving the common spaces less livability with the appearance of increasingly recurrent problems of safety, degradation and abandonment.

Today, in the strategy of Europe 2020, in addition to focusing on smart and sustainable companies, cities will have to be inclusive and put citizens at the base of development projects, interpret their needs by relating their civic sense and their perception of quality. urban and to train and educate to safety with precise information actions but above all with the design of quality public spaces. For this reason, promoting actions to rehabilitate the physical space, supporting the vitality of the neighborhoods, mobilizing citizens to encourage spontaneous territorial control mechanisms, thinking about safe and efficient mobility systems, means "taking action on the urban environment, in particular on large cities or parts of them, to prevent them from becoming insecure places, acting both on prevention and on the reassurance of the populations" [4].

In recent years there has been a growing safety culture: the recent Decree Law of 20/02/2017 nr. 14[3], an articulated package of measures, whose objective is to enhance intervention and planning in the fight against the degradation of urban areas, with an approach that privileges the coordination and planning of integrated interventions. In the text of the Decree urban safety is defined as "the public good related to the livability and decorum of cities, to be pursued also through the joint contribution of local authorities". The implementation of the measures, however, is left to the local authorities only through: the redevelopment and recovery of the most degraded areas or sites; the elimination of the factors of marginality and social exclusion; prevention of crime, in particular of predatory nature and the promotion of respect for legality in order to achieve higher levels of social cohesion and civil coexistence.

The intent is to stimulate processes based on opportunities aimed at improving urban quality in our increasingly multiethnic society. Historically, especially in the Mediterranean area, the city has been articulated in urban parts strongly connoted from the ethnic point of view, result of the long duration of the coexistence between Islam and Christianity and that has materialized with the formation of peculiar urban and territorial fabrics. The different cultures have in fact left a strong imprint throughout the territory of the various dominations, spreading an urban civilization in which life revolves around the city, understood as an urban ensemble composed of clans and villages, closed in themselves, but which find in some urban poles the key to undermining the isolation. Moreover, globalizing policies are the source of a loss of the economic role of most cities, not only with economic marginalization, but also through the global homologation of consumption patterns that have additional destructive

---

[3] Decree Law 14/2017 on "Urgent provisions on the safety of cities", published in the Official Gazette no. 93 of 21 April 2017 together with the conversion law n. 48/2017.

effects as they contribute to depleting the fabric of cities, favoring the progressive sense of insecurity perceived by its inhabitants.

One of the strong themes is represented by plurality, which can be considered as a richness of our urban places even if it often creates situations of conflict because of the different models of life. Such situations must be answered in the policies of inclusion and design of quality spaces, especially those intended for exchange, trade and in general for the meeting of new city desires. This condenses the will to open up to any type of relationship, commercial, social and cultural in a path where collective spaces and emergencies participate in a unique urban design, differentiated by functions. These are the cities that have always constituted the hinge between the coasts and the hinterland and assumed political functions that went beyond the government of only economic and commercial activities. Especially the port cities, which in the collective imaginary are considered the least safe, have always represented the history and the way of communication of their civilizations and, in every corner of the ancient part, represent a historical mixture that constitutes a great laboratory to read and interpret the identity.

In modern times the city has slowly begun to lose these characteristics of the city-market, leaving space for service and exchange activities in place of residential and commercial ones that have caused the loss of vitality. For this reason a correct policy of valorization of the characters of the city must take into account "integration" rather than the exclusive "conservation" to be able to respond to the new needs of a multicultural and multiethnic society that requires different approaches and great openings. cultural heritage and take into account customs, customs, ways of understanding the territory and the city, which can be quite peculiar.

To contribute to the formation of the culture of safety, we must relate the civic sense of the inhabitants, their identity and their perception of urban quality and activate design practices capable of raising the quality of life, social cohesion and the pleasantness of the places.

3   Connective, Mobility and Urban Security

Recognizing the value of uniqueness and diversity is one of the objectives of the EU that is working towards inclusive cities through development projects that underpin the recognition of cultural identity to form integrated companies, excluding the risks of social exclusion. A policy of these areas capable of developing superior communication skills, where the integration and development functions add to the cultural exchange in the broadest sense of the word, should lead the urban connective to the role of showcase or better "limelight" of great mass phenomena and social towards a renewed vitality. The search for a revitalization of these areas to instill a higher perception of safety should be linked to a careful redesign of public and collective spaces as well as a redistribution of the connective, without losing the identity of the various neighborhoods. Basically every corner of the city will have to be turned over of itself.

All this requires much more careful policies aimed at allowing the peculiarities, the characters of the cities, not only to not be suffocated, but to be placed at the foundation of a new vitality to reverse the perception of safety and build an original path of development. If the characters express themselves in places, these are to be understood both physically and mentally. We must allow ourselves to "think" the places we live, to know them and recognize them as an indigenous source of value, to capture the outside as an opportunity and not as a model to imitate and guarantee vitality to promote spontaneous surveillance and therefore greater safety in the cities.

In these "places", urban spaces, especially those designed for mobility. Therefore, it is not just a network made to house transport vehicles and traffic flows, but is to be understood as the main component of public space where citizens perform their functions of walking and socializing. They are urban places with multiple uses that must favor social bonds contributing to create a sense of safety and inclusion within the urban context. It becomes urgent to act on the planning of these areas (which do not necessarily presuppose a nucleus, but can be disseminated to the network in the territory), understood as "emerging areas", the places that are an expression of interest and that can constitute, intervening on systems of soft mobility, also in the form of aggregation, new centralities, and whose targeted planning could lead to overcoming the sense of insecurity. Moving, walking, socializing are activities that must be carried out serenely and in conditions of comfort because these are above all the indicators of the civilization of a society that make one perceive the feeling of safety.

We must therefore promote this form of wellbeing by focusing on new soft mobility systems and redesigning some spaces, freeing them if improperly occupied by parking areas, regenerating public space to foster the vitality of exchanges and rediscovering the cultural and natural historical attractions of the urban environment., all strategies to improve the living conditions and health of citizens [5]. Also the activity of education and sensitization of the inhabitants towards this "Safety culture" can prove to be a winning strategy, as experienced in some European cities, where good urban regeneration and socio-economic practices have succeeded in triggering new forms of participation, integration and entrepreneurial initiative that allowed risk reduction. In fact, the city of Malmö (Fig. 2) has successfully adopted integration policies based on the involvement of the community in public decision-making processes by activating "living labs" and "urban farming" in the neighborhoods most at risk of social segregation of the city [6].

Fig. 2. Urban spaces in Malmö.

These design practices and the activity of urban laboratories will create alternatives capable of stimulating awareness of the most appropriate choices in moving around the city and the spread of a real culture of safety, reducing risks and promoting a better quality of life even in terms of social and cultural relations.

References

1. Cardia, C.: La sicurezza nella progettazione urbanistica ed edilizia in Manifesto delle città. In: La Sicurezza nella Progettazione Urbana, Atti del Convegno, Bologna, 13 novembre 2000, Quaderno n° 2 supplemento al n° 17 di Metronomie, a. VII (2000)
2. Secchi, B.: La città dei ricchi e la città dei poveri. Laterza, Bari (2013)
3. Belli, A., (ed.): Oltre la città Pensare la periferia. Cronopio, Napoli (2006)
4. Cardia, C., Bottigelli, C.: Progettare la città sicura. Pianificazione, disegno urbano, gestione degli spazi pubblici. Hoepli, Milano (2011)
5. Fallanca, C.: Gli dèi della città. Progettare un nuovo umanesimo. FrancoAngeli, Milano (2016)
6. Boeri, A., Testoni, C.: Rigenerazione urbana e società multietnica: Torino e Malmö a confronto". In: Smartinnovation. http://smartinnovation.forumpa.it/story/110089/rigenerazion-urbana-e-societa-multietnica-torino-e-malmo-confronto. Accessed 8 Dec 2017

ECOSITING: A Sit Platform for Planning the Integrated Cycle of Urban Waste

The Case of Study of the City of Rome

Roberto Panei[1]([✉]), Giovanni Petrucciani[1], Dario Bonanni[1],
and Patrizia Trovalusci[2]

[1] AMA S.p.A., 00142 Rome, Italy
roberto.panei@amaroma.it
[2] Sapienza University of Rome, 00142 Rome, Italy

Abstract. Urban planning has long been introduced into the territorial classification elements as belonging to integrated waste cycle management. Within such framework, types of urban hygiene are defined and described. In particular, the General Regulatory Plan of the City of Rome has established that areas and facilities for separate collection of waste belong to the secondary urbanization works to be identified by executive planning, as well as for temporary collection, compacting and conveying inert and bulky waste.

The aim of this research is to find a shared method of selection and siting of areas compatible with the various "objects" of the integrated cycle: decentralized territorial offices, municipal collection centers, reuse centers, eco-plots, urban waste valorization plants, defined as nodes, spots, targets for a GIS based planning method. This study proposes the translation into an algorithm of operational research the needs of localization of areas to be used for infrastructure and urban hygiene, taking into account the structural and authorization factors, but also the anthropic factors. Once a sufficiently populated territorial database has been set up, thematic cartographies are developed as a decision support in the integrated cycle planning, moving away from subjective and improvised methods.

From the General Regulatory Plan, the Solid Waste Urban Planning is completed so that its infrastructure can be integrated as much as possible with the urban and living needs of the citizens, reporting the operations and activities related to waste cycle within the daily metabolism of the city organism.

Keywords: Recycle · Waste · Siting · GIS · Planning

1 Introduction

The municipal territory of Rome extends for about 1,300 km^2. Within its perimeter there are different kinds of "city": the Ancient Town, densely built and inhabited, the nineteenth-century expansion, more regular and urbanized, the Second World War suburbs, including the spontaneous agglomerations and the townships, to reach the wide cultivated countryside that stretches to the sea, where Ostia's neighborhood is located. Several cities gathered in one municipality, all dependent on the same

© Springer International Publishing AG, part of Springer Nature 2019
F. Calabrò et al. (Eds.): ISHT 2018, SIST 101, pp. 585–592, 2019.
https://doi.org/10.1007/978-3-319-92102-0_62

municipal government. The road infrastructure partially redraw the consular routes of ancient Rome according to a stellar scheme that runs from the center to the suburbs and to the sea. In this extremely varied environment, AMA operates the Urban Property Company, which is a municipal property, which carries out garbage collection and garbage disposal. AMA is present in the country with a large number of coordinated infrastructure to ensure an efficient service. In particular, the real-estate assets consists of 5 garages (4 in the "cardinal directions", and the fifth in the neighborhood of Ostia), about 80 decentralized territorial offices and sites, from which the solid waste collection system starts and spreads, 14 collection centers to be completed, where the bulky and non-organic waste destined for separate collection is carried directly by citizens, and 6 transshipment centers where material transfers between small vehicles into larger ones.

The treatment plant component is so far made up of two Mechanical Biological Treatment plants (MBT) located at the "Salario" and "Rocca Cencia" locations, respectively in the North and East of the city, and the Composting plant at Maccarese (Municipality of Fiumicino). The waste-to-energy plant located at the "Ponte Malnome" location, West of the city, is built for hospital sanitary waste, and is currently out of order for revamping. The vast majority of business real-estate assets, with the exception of "Salario" and "Tor Pagnotta" sites, North and South of the city respectively, have been inherited by the Old Municipality of Rome, and only a few are the result of new developments. Although the organization and co-ordination between the sites, the garages, the collection centers and the facilities can be deepened and refined, the situation remains a manifest shortage in the proper distribution on the municipal territory. With this contribution, we want to set up a tool to locate the municipal property areas and properties needed to balance the business infrastructure in strict relation to the inhabitants and the different urban situations in the city [1–3].

2   The Ideal City

2.1   The Architectural Types

The General Regulatory Plan of Rome, approved in 2008, defines for the first time the urban reference areas in the planning of integrated waste cycle. This work stems from the need to transform into urban planning as defined by the PRG. Taking inspiration from the theoretical needs of the urban hygiene service, we want to come up with a scheme that, once established service standards respecting all constraints and all the priorities present in the territory, can be an ideal city model for Rome. First of all, the types of buildings that are the core of the urban hygiene service are set up, including all the possible declinations of the service itself and representing, in a dimensional fashion, different functional needs.

Collection Center. Area, accessible to citizens, used for temporary storage of waste, not organic, organized in a differentiated manner for subsequent recycling, such as furniture, household appliances, computers, mobile phones, demolition and building waste, iron, green waste and other materials.

Service Hub. A building complex including the main activities related to the urban hygiene service performed by the company, which are necessary and sufficient to ensure the proper management of the territory of a city district with a population of about 50,000. In these places is possible to optimize the various activities in sufficiently limited spaces, covering a total area of approximately 10,000 m^2. The activities of Service Hubs concern in special waste collection, logistics, small maintenance and vehicles wash as a service support. Various features are provided, some for citizenship, such as the Collection Center (also known as "ecological islands"), and the Creative Reuse Center (CRIC); other for business operators such as waste transshipment, vehicle refueling and regular maintenance, and also dressing rooms for workers for about 150/200 employees, and operating offices as well as warehouse and archives.

Composting Plant. Composting is a natural biological process, also known as litter box formation ("humus"), which is conducted in an industrial way, in order to reduce its duration of organic matter stabilization from multi-season to about a month. In an area with a surface area of 60,000 m^2, it is possible to create a composting plant for processing capacity of 60,000 tons per year, destined for a basin of about 500,000 inhabitants.

2.2 Planning Parameters

The need to operate according to objective and preordained approaches, setting rules on one part of "usability", on the other of "well-being" in terms of using the system by citizenship, has addressed the methodology towards an operational "best oriented" research system. The so-called "optimum conditions", theoretically defined, translate into practice in jointly "usability" and "well-being" conditions, and occur in the assumption or starting variables and parameters tendency, a set of values or standard ranges, which are defined on the basis of the different types of settlements identified.

   This reconstructive method of "ideal city" is summarized in a rules generation, conditions and criteria for identifying the needed areas:

   Capacitance: Identification of invariant, structural internal features that deal with maximum or minimum dimensional criteria (eg minimum operating space, minimum transit space and stop, etc.) that affect the "shape" of the investigated area;
   Conformity: City-specific, consistent use destinations compatible with the insertion of artifacts;
   Adequacy: extension of the area related to the number of inhabitants served, supposedly internal to a "radius of influence", predefined;
   Accessibility: distance from the adequate classroom through which you can reach the site;
   Proximity: nearness to waste production facilities (in the case of urban waste: proximity to dwellings), understood as a positive factor related to the best usability. This criterion is linked to the design choice to prefer waste management networks with smaller and distributed sites. This choice implies a significantly smaller fluctuation range than that of large plants and as a consequence reduces the time and costs of transport;

Harmoniousness: The "influence radius" of similar sites must overlap to a minimum and determined measure, and the "cloud" of sites in the survey territory should not leave uncovered areas.

In response to the definition of the "ideal management system" criteria, the "guiding parameters" for the implementation of the operational research method have been identified. These parameters within the research method have the potential to be integrated in response to different localization needs. Some of the limit values directly used in the search are dependent on the "population" (POP) parameter. Attached to this, based on territorial needs, identify the others listed below.

AREA (A): The size, as a minimum, of the site location area to be built, in compliance with the dimensions established by reference norms and on the basis of the already obtained information.

The calculation equation, depending on the population served, is as follows:

$$A = f_A * \left(\sum\nolimits_{i=1}^{n} \frac{PPC_i}{FCONV_i} * POP \right) \tag{1}$$

On the basis of the pro-capite production PPC of the n waste classes in place, the conversion factor (FCONV) between weight and site occupancy at site, and a factor (fA) that takes into account the presence of other areas (maneuver, discharge, accumulation, etc.).

Coverage/Influence Radius (RIN): The coverage or influence radius represents the territorial reference extension for the site, and of course depends on housing density. It represents the radius of the equivalent circumference (same area) which contains the served population.

$$R = \sqrt[v]{\frac{\frac{POP}{DENS}}{\pi}} \tag{2}$$

The sphere of influence allows an additional advantage in research, called "territorial hierarchy", that is, to favor areas not served than those already served.

Distance of Protected Goods (DBT): the distance from protected property, in particular from water bodies and archaeological areas, is that laid down by the reference regulations (Hydrogeological Plan, Regional Territorial Plan, etc.)

Distance Sensitive Sites (DSS): distance from sensitive sites should be evaluated according to site typology and criticality related to the proximity of this and the various categories of urbanized textiles considered "sensitive" (schools, cultures, hospitals).

Diagnosis Distance (DVIA): the location of the installations must also be established in relation to the proximity to the urban road network, as required by the norm and by agreement, it is possible to establish a maximum distance from the main road axes within which to select the areas for new locations.

Use Destination (DUS): Current site uses of site compatible with the site are obviously related to its specific features, both from a functional point of view and from an urban point of view (compliance).

The combined disposition of settlement types linked to the waste cycle and the values attributed to the various parameters of composition of related conditions and constraints determines a table of the Parametric Standards of the Urban Project (SPPU).

This table, exemplified herein and not exhaustive, summarizes the conditions to "intersect" to obtain regions that are suitable for fruition. Both the settlements typology and the parameters identified should be understood as the true strength of the model proposed here, as dynamic and "fluid", subjected to a continuous verification and evolution in response to evolution of both territorial needs and knowledge of the basin of investigation. Therefore, the numerical values given here are to be considered (see Table 1).

Table 1. Example of parametric standards of the urban project for the typologies coded by AMA: the collection center, the service hub, the composting plant.

	POP	A	DBT	DSS	DVIA	DUS
CRAC	50.000		cfr PAI, PTPR		100	cfr PRG
CSER	50.000	10.000	cfr PAI, PTPR		100	cfr PRG
ICOM	500.000	50.000	cfr PAI, PTPR	500	1000	cfr PRG

2.3 From Ideal City to Real City

In model exemplification, an ideal pattern based on identified standards is set up for each urban area, where hygiene infrastructures are inserted into nodes, regardless of local conditions and contingent constraints. The result is an abstract ideogram, dropped in the urban reality of Rome city, which defines the minimum number of "nodes" needed to manage the territory [5]. This scheme will be compared with constraints and environmental, housing and infrastructure situations, to reach a project synthesis that can identify the integrations needed to optimize the urban hygiene service [4].

3 The Real City

3.1 Scheduling Rules

The rules underlying the "usability" of the best tied areas can not act independently of pre-established classification and territorial planning rules. Landscape is part of territory including all the goods into local community identity from historical, cultural and natural point of view. Regional Territorial Plans represent the planning tool through which the actions are governed, aimed at protecting and enhancing sites specific peculiarities. They are superordinate to municipal planning tools, so it is necessary to take into account the requirements set forth within the NTAs of the individual Plans before they can act on the territory and the same is what is happening in waste management.

3.2   The Involved Themes

Among the themes involved in the construction of the Royal City we find: 1. Town Planning Destination; 2. Use of soil; 3. Roads; 4. housing Density.

Such themes, which can be expanded at will, are the "filter" of selection to narrow the initial ideal horizon to a set, generally a subset of areas compatible with the purposes.

3.3   GIS Automation

The adoption of computer land-based information systems (Global Information Systems - GIS), with adequate digitized mapping information and associated information databases, provides a high degree of rapid resolution of complex localization problems. In fact, GIS allows to link different data as long as georeferenced (associated with a geographic reference system) so as to generate new information content.

Spatial analysis consists in the transformation and elaboration of pure geographic elements (areas, paths, points) with derivative elements ("numerical attributes" of geographic locations) according to different algorithms:

1. overlay: overlapping (union, intersection, difference) between topological elements of two themes to create a new result thematic;
2. spatial query: querying the database with spatial criteria imposed on the values of the themes (distance, code or text values);
3. buffering: given a geometric theme a "respect" or "buffer" geometry at a fixed or variable distance depending on the element's characteristics;
4. segmentation: search for long polygonal points based on numeric criteria;
5. network analysis: identifying minimum paths in a network of linear elements;
6. spatial analysis: spatial analysis on raster data of various typologies, eg: color gradation;
7. geostatistic analyzes: spatial correlation analysis of georiferous variables.

3.4   Shapefile

A shapefile is a vector storage format for associated objects and relational databases.

The shapefiles register "base" forms: Points, Lines, Polygons, Texts. In a database record table are then stored the properties and attributes for each "shape". The "shape" and its attributes can determine infinite geographic themes, which will influence the power and accuracy of the geospatial analysis of GISs.

4   Case Study of Composting Plants

To briefly illustrate the method power, it is a simulation of applied research to a composting plant, type and size as indicated above. The simulation is still provisional, as related to provisional data, which is still very widespread in the horizons of the map data and available themes. Below are the progressive representations of relative compatibility that are highlighted by the calculation model to today's introduction of the Parametric Standards of the Urban Project (SPPU), in the background of the territory of the City of Rome.

The sequential order by which the themes are filtered results from the intention to effectively reduce, from the beginning, the redundant or not significant information content from cartographic progression. For this reason, it begins to filter the territory with the aforementioned use intended as apt to localization, thus resulting in overlapping and immutable contents to features subject to rediscussion, even in the absence of the final result with respect to the particular sequence selected.

The first thematic passage investigates the land use map, which immediately identifies those "qualitative" locations that are well adapted to the realization of industrial composting, thus identifying the uses compatible with the settlement in question. Discrimination of the smallest dimension, since it takes into account the final operation of the site, leads to the detection of a first set of 2252 areas among the compatible ones (see Fig. 1).

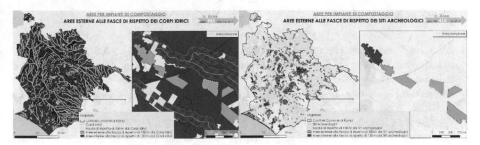

Fig. 1. ECOSITING Composting plant -Respect for water bodies (1) and archeology (2) (source: Lanni, Delogu 2017 graduation thesis) - Detected areas = 415/2282 (initial)

5   Conclusions

The search for new operating venues has been concentrated within the Municipality to offer solutions to the company's goals: to be able to cover as much as possible the entire municipal territory in order to be able to serve all municipalities equally. Resolution 129/14 of the Capitoline Assembly states, in fact, at point 10: "… to create an adequate plant network at the disposal of the differentiated collection, starting from the Collecting Centers, with the aim of reaching at least one for the City Hall, also providing for the treatment of the differentiated organic fraction (FORSU) through anaerobic digestion and/or aerobic composting plants".

The method developed in this work provides high potential for objective identification for each type of settlement chosen once established, and appropriately georiferized, the features necessary for the functional exploitation of sites located in the territory.

This operational potential, which becomes a precise method of investigation, is deeply correlated with the quantity, detail, distribution and quality of data available on the net, or by specialized bodies and institutions; data that must already allow a return that is compatible with the major GIS software. For this reason, in addition to the flexibility and progressiveness in the development of the survey method, which is reiterated, the necessary progressivity in the research and validation of geo-rigid map data is highlighted, trusting in the progressive structuring and continuous verification of the material published in Open Data Systems of Public Administrations.

The comparison with the existing, which is constantly evolving beyond the computer artifacts, ultimately determines the ultimate value of such methodologies as the formidable operational accelerators, which can not, however, be left out of the final human evaluation, being among other issues sustainability in general and the well-being of citizens in particular [6].

Aknowledgements. Thanks to the graduates Gabriele Delogu and Luca Lanni who attended the Lab Recycling Laboratory at the Faculty of Architecture of Sapienza in Rome, for the decisive contribution provided to the processing of the starting data and for graphic support.

References

1. Arici, F.: Pianificazione urbanistica e sostenibilità urbana: il ciclo di gestione dei rifiuti solidi urbani. Folio, Rivista del Dottorato di Pianificazione Urbana e territoriale dell'Università di Palermo **25**, 21–24 (2011)
2. Brunner, P.H., Fellner, J.: Setting priorities for waste management strategies in developing countries. Waste Manag. Res. **25**(3), 234–240 (2007)
3. Geddes, P.: Cities in Evolution. Williams & Borgate, London (1915)
4. Hostovsky, C.: Integrating planning theory and waste management - an annotated bibliography. J. Plan. Lit. **15**(2), 305–332 (2000)
5. Lynch, K.: Wasting Away. Sierra Club Books, San Francisco (1990)
6. Scandurra, E.: L'ambiente dell'uomo. Verso il progetto della città sostenibile. EtasLibri, Milano (2003)

Boosting Investments in Buildings Energy Retrofit: The Role of Incentives

Marta Bottero[1], Chiara D'Alpaos[2(✉)], and Federico Dell'Anna[1]

[1] Politecnico di Torino, 10125 Turin, Italy
[2] University of Padova, 35131 Padua, Italy
chiara.dalpaos@unipd.it

Abstract. More than 40% of the EU building stock was built before 1960 and 90% before 1990. It is common wisdom that older buildings typically exhibit greater energy demand than new ones. The renovation of existing buildings is therefore a cornerstone in the reduction of energy consumption and relative CO_2 emissions under the post-carbon city paradigm.

In the present work, we analyze various energy retrofit strategies, evaluate their impact on buildings energy performance and determine their relative cost-benefit tradeoffs to address the multiple benefits of renovations and the financial barriers to their implementation and taking up.

Aim of the paper is to identify cost-effective energy retrofit strategies which match technological advancements and knowledge in energy retrofitting with environmental needs and end-user's behavior. To determine how far (and how much) it is optimal to push on retrofitting of existing buildings, we investigate the role of incentives and their impacts on private investment decisions.

Keywords: Energy retrofit · Nearly zero energy building · Fiscal incentives

1 Introduction

With about 40% of global energy consumption and 36% of CO_2 emissions in the European Union (EU), the building sector plays a key role in the mitigation of climate change and energy and environmental issues [1]. In this context, the potential of residential buildings should be exploited to take an active role in the transition to the low-carbon concept and to meet the new requirements set by the EU for 2050 [2].

In fact, the focus is shifting more and more towards new concepts of post-carbon buildings (i.e. zero or nearly zero energy buildings) and post-carbon cities [3]. More than 40% of the EU building stock was built before 1960 and 90% before 1990. It is common wisdom that older buildings typically exhibit greater energy demand than new ones. The rate at which new real estate assets replace this old stock is about 1% per year, therefore the renovation of existing buildings is a cornerstone in the reduction of energy consumption and relative CO_2 emissions.

The original version of this chapter was revised: Belated correction has been updated. The erratum to this chapter is available at https://doi.org/10.1007/978-3-319-92102-0_76

© Springer International Publishing AG, part of Springer Nature 2019
F. Calabrò et al. (Eds.): ISHT 2018, SIST 101, pp. 593–600, 2019.
https://doi.org/10.1007/978-3-319-92102-0_63

In the present work, we discuss the adoption of green technologies to reduce energy and environmental impacts of private residential properties. We analyze various energy retrofit strategies, evaluate their impact on buildings energy performance and determine their relative cost-benefit tradeoffs to address the multiple benefits of renovations and the financial barriers (e.g. renovation costs, relatively low energy prices) to their implementation.

Aim of the paper is to identify cost-effective energy retrofit strategies which match technological advancements and knowledge in energy retrofitting with environmental needs and end-user's behavior. To determine how far it is optimal to push on retrofitting of existing buildings, we investigate the role of incentives and their impacts on private investment decisions.

In detail, we examine deep energy-retrofit interventions on a residential single-family building, located in the Piedmont Region (Northern Italy): we identify and analyze 15 retrofit strategies to increase energy efficiency which involve both the building envelope and HVAC (Heating, Ventilation and Air Conditioning) systems [4]. The different strategies are compared to a baseline scenario in which the reference building (RB) satisfies minimum standards required by national legislation on energy consumption reduction.

The remainder of the paper is organized as follows. Section 2 presents materials and methods by introducing the case study and providing the theoretical and methodological framework for the assessment of improvements in building energy performance and the evaluation of their economic outcomes. It also provides a review of incentive policies to encourage investments in energy retrofit. Section 3 illustrates and discuss results and Sect. 4 concludes.

2   Materials and Methods

2.1   The Project for a New NZEB in Northern Italy

In 2010 the European Union introduced the EBPD (European Building Performance Directive) recast, which defined two concepts for assessing the energy performance of buildings [5]. In particular, the nearly zero energy building (NZEB) is defined at legislative level as a building with very high energy performance with a very low energy consumption that is largely covered by renewable energy sources aiming at energy self-sufficiency. The EPBD recast obliges Member States to ensure that all new buildings are NZEBs by 31 December 2020, and that new public buildings by 31 December 2018. Furthermore, the Cost-optimal methodology for calculating the optimal level of energy performance in terms of costs is established.

The case study analyzed is located in Northern Italy, in a rural context of the Piedmont Region. The current state is a typical old barn currently not inhabited and represents a basis to create a house in accordance with the new energy requirements to achieve the NZEB goal. The retrofit actions hypothesized in this study refer to a deep re-functionalization of the construction that involves the conversion of the intended use into a residential building and the respect of the binding energy limits. With this objective in mind, 16 combinations of Energy Efficiency Measures (EEMs) related to the building envelope and HVAC system features are compared. The retrofit projects envisage construction strategies that limits thermal losses, including four high energy performance building envelope levels, and four efficient system configurations, which reduces energy consumption and running costs ensuring good indoor thermal conditions in both winter and summer. The building envelope levels refer to different energy performance requirements stated by national and European regulations [6]. On the other hand, the HVAC systems used in this study are the condensing boiler and the water-to-water heat pump, combined with natural ventilation or Controlled Mechanical Ventilation (CMV). All configurations use the radiant panels as emission system for heating, while different systems were set for cooling (radiant panels or split). As stated by NZEB definition, a reduction of energy consumption by optimizing RES is configured through solar collectors and photovoltaic panels installation.

2.2 Energy and Economic Evaluation

The first step for the application consists in evaluating the energy performance of each energy configuration. The aim of energy evaluation was to determine the annual energy consumption in primary energy terms (kWh/m^2y).

The energy performance was evaluated through a whole building dynamic simulation carried out by the EnergyPlus software, considering delivered and self-produced energy. The RB scenario represents the baseline scenario, characterized by traditional technologies, with a low level of insulation of the envelope. As shown in Table 1, the 2D, 3C, 3D, 4C e 4D packages reach the primary energy consumptions close to zero, respecting the NZEB limits. In particular, 4D scenario can be defined as "positive energy" configuration, constructing a building which produces more energy than it consumes.

Economic evaluation considered the initial costs, the running costs and the final value of components at the end of the calculation period. The investment costs include the construction cost of the structural part of the building (e.g. stairwell, excavations, structural frame) and the part linked to the building elements referred to opaque and transparent envelope components and to the HVAC systems. The estimate of the energy costs requires the quantification of the required energy to cover the thermal and electrical needs. The consumption of thermal energy concerns the heating of indoor spaces, while the electrical energy is calculated for the provided electrical equipment (e.g. lighting, CMV and water-to-water heat pump). The maintenance costs were calculated as a percentage of the initial investment cost of every building component, according to EN 15459:2007.

2.3 Incentives Policy to Energy Efficiency Improvements

To reduce GHG emissions associated with consumption in the home, in recent years, many countries have implemented incentive policies to encourage residential energy-efficiency upgrades and retrofit, which typically include specific home renovations (e.g. insulation, new windows, etc.) and equipment (e.g. high-efficiency heating and cooling systems, appliances, etc.) [7, 8]. The improvement of energy-efficiency in residential buildings is a crucial issue in Italy where 55% of the building stock is older than 40 years and about 89% of the existing buildings were built prior to 1991, i.e.

Table 1. Energy and economic performances.

Scenario	Primary energy kWh/m^{2}y	Investment costs €/m^{2}y	Energy costs €/m^{2}y	Maintenance costs €/m^{2}y	Residual value €/m^{2}y	Replacement costs €/m^{2}y
RB	113.61	1,578	37	2	265	76
1B	87.08	1,637	28	4	290	106
1C	40.7	1,722	11	4	335	145
1D	32.06	1,751	9	6	344	160
2A	79.06	1,615	21	2	268	76
2B	55.77	1,674	15	4	293	106
2C	12.19	1,759	7	4	338	145
2D	5.57	1,787	6	6	344	160
3A	71.59	1,705	17	2	270	76
3B	49.19	1,764	11	4	295	106
3C	7.31	1,849	7	4	341	145
3D	0.03	1,878	6	6	341	160
4A	65.16	1,733	14	2	274	76
4B	45.64	1,792	8	4	298	106
4C	2.61	1,878	6	4	344	145
4D	–4.69	1,906	5	6	352	160

before the entry into force of national Law n.10/1991[1] on the implementation of the national energy plan that set the milestones for efficient energy use, energy savings and renewable energy sources development. Real estate assets are highly energy-consuming: data on energy consumption show that residential and tertiary buildings are responsible for 33% of primary energy consumption. Consequently, the Italian residential sector offers considerable potential for reducing energy use GHG emissions, particularly through energy-efficient renovations [9–11].

Except for direct financial investments, Governments can introduce a wide range of policy instruments to encourage households in undertaking energy-efficient renovations: regulatory instruments, economic and market-based instruments, supports information and voluntary actions [8, 12, 13]. Since the 1970s, the most common financial instruments that have been introduced in Europe include grants and subsidies, loans, and tax incentives [14, 15].

Since 2006, the Italian Government has introduced fiscal incentive programs to enhance energy efficiency in residential buildings. Effective February 19, 2007, a national law allowed homeowners to deduct from their income taxes up to 55%[2] of the expenses incurred to implement some types of energy efficiency renovations or source of renewable energy in existing buildings. More recently, national Law n. 232/2016

---

[1] Law n.10/1991 is the first specific and comprehensive national regulations for the reduction of energy consumption in buildings.

[2] Caps of €30,000, €60,000 and €100,000 per residential unit were applied, depending on the type of renovation.

increased the tax deduction to 65% with a maximum cap of €90,000[3] per residential unit. This fiscal incentive proved to be successful in boosting capital investments in energy retrofitting, and energy efficiency became a main driver of the economic recovery of the building sector, although most renovations did not involve entire buildings but specific systems or components of single residential units [16]. According to the National Energy Agency [16], during the period 2014–2017 total capital investments were about 9.5 billion Euros and about 1 million renovations were undertaken, 50% of which involved the installment of new windows, whereas 25% the application of insulating layers on the external walls and roofs and 20% the replacement of HVAC systems. The projection value for tax deduction over the next 10 years amounts to 2.1 billion Euros and estimated energy savings attributable to fiscal incentives during the period 2014–2020 are about 1.38 Mtoe/year.

The economic impact of fiscal incentives during the period 1998–2016 is significant: 237 billion Euro investments that generated a cost for the Government amounting to 108.7 billion Euros due to fiscal incentives (i.e. tax deductions) and a tax revenue amounting to 89.8 billion Euros. The final balance is negative: 18.9 billion Euros (nearly 1 billion Euros per year). This confirm the general perception that incentives are excessively costly and not cost-effective.

Nonetheless it can be considered that as the Government obtains tax revenues on the renovation works by homebuilders and installers, projections show a net capital gain for the Government of about 0.3 million Euros.

3 Results and Discussion

Firstly, we carried out a feasibility analysis of the retrofit options under investigation. We compared each scenario (i.e., the 15 best technical alternative configurations) to the RB scenario and estimated relative potential incremental costs and benefits to determine the profitability of investing in buildings which exhibit improved energy performances compared to ones that satisfy minimum requirements set by national laws (i.e., RBs).

In detail, we determined the incremental life cycle costs of each alternative scenario with respect to the baseline scenario and compared their present value with the present value of incremental benefits deriving from increased energy efficiency. In our analysis the benefits coincide with the avoided costs obtained by reduction in energy consumption (i.e. energy saving)[4]. To determine the profitability of investments, we developed a discounted cash flow (DCF) analysis over a period of 30 years and determined the additional the Net Present Value (NPV) of each EEMs with respect to RB by estimating a 2% discount rate in accordance to the current rate of return on savings in Italy. Secondly, to determine how far it is optimal to push on deep

---

[3] Specific deductions and caps applies to residential units in condominiums.

[4] Although in our paper we focus on avoided costs due to energy savings, energy retrofit of existing buildings generates a wide range of (direct, indirect, tangible, untangible) benefits, among which it is worth mentioning that, usually, statistically-significant increases in property market values arise. In this respect, see e.g. [17–20].

retrofitting of existing buildings, we investigated the role of incentives and their impacts on private investment decisions by including in our DCF analysis the present value of Government incentives (i.e., fiscal incentives) to EEMs implemented on the existing building. The results of the two DCF analyses we performed are shown in Table 2.

Table 2. DCF analyses results.

Scenario	Without incentives	With incentives
	NPV (€)	NPV (€)
1B	14,435	21,305
1C	70,616	87,465
1D	65,458	85,620
2A	59,278	63,594
2B	53,671	64,857
2C	82,161	103,325
2D	70,761	95,239
3A	58,531	73,359
3B	66,693	88,391
3C	66,124	97,801
3D	54,618	89,608
4A	68,078	86,224
4B	70,325	95,341
4C	64,983	99,977
4D	53,311	91,619

Our results show that the proposed interventions for the improvement in energy efficiency with respect to the RB scenario are economically sustainable. It is always profitable to invest in deep retrofit strategies which guarantee a significant reduction in energy consumption, by exceeding the minimum standards set by national laws and regulations. In other words, the present value of incremental benefits deriving from increased energy efficiency exceeds the present value of incremental life cycle costs for each alternative scenario both when incentives are not accounted for and when they are. Incentives obviously increase NPVs. In addition, it worth noting that according to our findings, fiscal incentives are not key drivers in encouraging investments in NZEBs, as these buildings are per se positive NPV investment projects. Savings (in present-value terms) in household energy expenditures generated by NZEBs more than offsets their higher investments costs and make households better off. Our results apparently contradict data on the relatively low rate of deep energy retrofitting of buildings in Italy: 85% of the market is in fact minor renovations. Finally, it is worthy of note that incentives favor those interventions that are more innovative from the point of view of the envelope and HVAC systems. Indeed, these scenarios show higher NPVs compared to the situation without incentives.

4 Conclusions

The renovation of existing buildings is a cornerstone in the reduction of energy consumption and relative CO_2 emissions under the post-carbon city paradigm. In this paper we analyzed various energy retrofit strategies, evaluated their impact on buildings energy performance and determined their relative cost-benefit tradeoffs to address the multiple benefits of renovations and potential financial barriers to their taking up. In details we compared energy savings against the cost of alternative EEMs over a thirty-year period. We then investigated the role of fiscal incentives in encouraging investments in deep energy retrofitting of existing buildings. Our findings show that it is always profitable to invest in energy renovations which exceed the minimum standards set by national laws and regulations independently from fiscal incentives. In addition, according to the results of our investigation, NZEBs are per se positive NPV investment projects and consequently incentives are not key drivers in encouraging investments. By contrast, they may attract homeowners that would have undertaken the investment without incentives (i.e., free-riders) and represent a superfluous cost for the Government due to income-tax deductions.

References

1. BPIE: The BPIE data hub for the energy performance of buildings. In: BPIE Data Hub (2015). https://www.buildingsdata.eu/. Accessed 28 Nov 2017
2. EC: Energy Roadmap, 2050 (2013). https://ec.europa.eu/energy/en/topics/energy-strategy-and-energy-union/2050-energy-strategy. Accessed 28 Nov 2017
3. Becchio, C., Corgnati, S.P., Delmastro, C., Lombardi, P.: The role of nearly-zero energy buildings in the definition of post-carbon cities. Energy Procedia 78, 687–692 (2015)
4. Barthelmes, V.M., Becchio, C., Bottero, M., Corgnati, S.P.: Cost-optimal analysis for the definition of energy design strategies: the case of a nearly-Zero energy building. Valori e Valutazioni 16, 61–76 (2016)
5. EU: Directive 2010/31/EU of the European Parliament and of the Council of 19 May 2010 on the energy performance of buildings (recast). In: Official Journal European Union, pp. 13–35 (2015)
6. Barthelmes, V.M., Becchio, C., Corgnati, S.P., Guala, C.: Design and construction of an nZEB in Piedmont Region, North Italy. Energy Procedia 78, 1925–1930 (2015)
7. Alberini, A., Bigano, A.: How effective are energy-efficiency incentive programs? Evidence from Italian homeowners. Energy Econ. 52, 576–585 (2015)
8. Conticelli, E., Proli, S., Tondelli, S.: Integrating energy efficiency and urban densification policies: two Italian case studies. Energy and Build. 155, 308–323 (2017)
9. CRESME, CNAP, ANCE: Riuso 2012. Città, mercato e rigenerazione. Analisi del contesto per una nuova politica urbana, Roma (2012). www.old.awn.it. Accessed 11 Dec 2017
10. ISTAT: 15° Censimento della popolazione e delle abitazioni (2011). http://www.istat.it/it/censimenti-permanenti/censimenti-precedenti/popolazione-e-abitazioni/popolazione-2011. Accessed 30 Nov 2017
11. Evola, G., Margani, G.: Renovation of apartment blocks with BIPV: energy and economic evaluation in temperate climate. Energy Build. 130, 794–810 (2016)
12. Dresner, S., Ekins, P.: Economic instruments to improve UK Home energy efficiency without negative social impacts. Fisc. Stud. 27(1), 47–74 (2006)

13. Dongyan, L.: Fiscal and tax policy support for energy efficiency retrofit for existing residential buildings in China's northern heating region. Energy Policy **37**, 2113–2118 (2009)
14. Atanasiu, B., Maio, J., Staniaszek, D., Kouloumpi, I., Kenkmann, T.: Overview of the EU-27 building policies and programmes. WP5 ENTRANZE (2014). https://www.entranze.eu. Accessed 17 Feb 2017
15. D'Alpaos, C.: Methodological approaches to the valuation of investments in biogas production plants: incentives vs. market prices in Italy. Valori e Valutazioni **19**, 53–64 (2017)
16. ENEA: Rapporto annuale 2017 - Le detrazioni fiscali del 65% per la riqualificazione energetica del patrimonio edilizio esistente, Roma, p. 108. www.efficienzaenergetica.enea.it. Accessed 30 Nov 2017
17. Cerin, P., Hassel, L., Semenova, N.: Energy performance and housing prices. Sustain. Dev. **22**, 404–419 (2014)
18. De Ruggiero, M., Forestiero, G., Manganelli, B., Salvo, F.: Buildings energy performance in a market comparison approach. Buildings **7**(1), 16 (2017)
19. Canesi, R., D'Alpaos, C., Marella, M.: Foreclosed homes market in Italy: bases of value. Int. J. Hous. Sci. Appl. **40**(3), 201–209 (2016)
20. Canesi, R., D'Alpaos, C., Marella, M.: Forced sale values vs. Market values in Italy. J. Real Estate Lit. **24**(2), 377–401 (2016)

The PrioritEE Approach to Reinforce the Capacities of Local Administrations in the Energy Management of Public Buildings

Monica Salvia[1(✉)], Sofia Simoes[2], Norberto Fueyo[3],
Carmelina Cosmi[1], Kiki Papadopoulou[4], João Pedro Gouveia[2],
Antonio Gómez[3], Elena Taxeri[4], Filomena Pietrapertosa[1],
Karlo Rajić[5], Adam Babić[5], and Monica Proto[1]

[1] Institute of Methodologies for Environmental Analysis - National Research Council of Italy, 85050 Tito Scalo, Italy
monica.salvia@imaa.cnr.it
[2] CENSE - Center for Environmental and Sustainability Research, Universidade Nova de Lisboa, 1099-085 Lisboa, Portugal
[3] University of Zaragoza, 50018 Zaragoza, Spain
[4] Centre for Renewable Energy Sources and Saving, 19009 Pikermi, Greece
[5] North-West Croatia Regional Energy Agency, 10000 Zagreb, Croatia

Abstract. In the Mediterranean area most of the public authorities need to enhance their institutional capacity in the field of Energy Efficiency (EE) and use of Renewable Energy Sources (RES) in order to contribute to the Energy Performance of Buildings and the Energy Efficiency Directives, developing solutions suited to various regional contexts. The PrioritEE project, funded by the Interreg MED programme, aims at reinforcing the capacities of public administrations in selecting and implementing eco-friendly and cost-effective energy planning measures. This paper aims to describe the main efforts carried out by local public authorities and professional institutions from five MED countries (Italy, Portugal, Spain, Greece and Croatia) in order to reduce energy consumption and prioritize EE investments in Municipal Public Buildings (MPBs). In particular, it focuses on the methodological framework describing the main components of the proposed toolbox, the main objectives and expected outcomes but also the current achievements and the way forward.

Keywords: Energy efficiency · Innovation capacity and Awareness-Raising Municipal Public Buildings

1 Introduction

Energy dependence and an increasing concern about climate change are currently major challenges faced by European Union (EU) countries. Energy efficiency (EE) is a privileged driver to reduce EU energy and climate vulnerability. The 2012 Energy Efficiency Directive (2012/27/EU), addressing the need to reduce energy consumption at all stages of the energy chain, establishes a set of binding measures to achieve the 20% EE target by 2020. Buildings are a main focus due to their high contribution to

© Springer International Publishing AG, part of Springer Nature 2019
F. Calabrò et al. (Eds.): ISHT 2018, SIST 101, pp. 601–608, 2019.
https://doi.org/10.1007/978-3-319-92102-0_64

final energy consumption and GHG emissions (respectively about 40% and 36% [1]) and their high untapped energy saving potential. However, the implementation of the 2012/27/EU Directive is hampered by an insufficient knowledge of owners, managers and public authorities about buildings features [2–6] and their consumption as well as on the available options to improve energy management and efficiency [5–7]. Moreover, in the public sector the refurbishment and the implementation of new technologies is heavily influenced by the lack of financial resources supporting investment [5, 6, 8]. Ruparathna et al. [9] group the opportunities to improve energy performance in existing commercial and institutional buildings into three categories: incorporating technical measures for EE and use of renewable energy (technological change), improving the building energy management (organizational/managerial change), promoting awareness programmes among building users (behavioral change). Deciding on EE improvements of building stocks requires usage of building typologies [10], a detailed characterization of the buildings (e.g. area, conservation status, implemented EE/RES measures, equipment features) and their users (e.g. number of users, hours of usage) [11, 12] as well as decision-support tools [6]. The potential for energy savings due to measures targeting behavior is supported by an emerging body of literature, which highlights the crucial role of the behavior of building operators and occupants to ensure the success of implemented measures [13–15]. Nonetheless, it should be mentioned that most of current work has covered the residential sector, while "the tertiary sector is relatively unexplored" [15].

The project PrioritEE "Prioritise energy efficiency measures in public buildings: a decision support tool for regional and local public authorities" aims at contributing to these needs and priorities common across all Europe and particularly in the Mediterranean countries. PrioritEE aims to reinforce the capacities of public administrations in selecting and implementing eco-friendly and cost-effective energy consumption reduction measures and increase the use of renewable energy sources (RES) in Municipal Public Buildings (MPBs).

This paper aims to describe the main challenges the project seeks to address, as well as to present the proposed methodological approach. Section 2 focuses on the development and testing of customized tools to prioritize EE investments, and the initiatives aimed at enhancing capacity building. In Sect. 3, the methodological progresses and current achievements are presented and discussed.

2 Materials and Methods

The overall aim of PrioritEE is to strengthen the policy making and strategic planning competences of public authorities in the energy management of MPBs in five Mediterranean countries (Italy, Portugal, Spain, Greece and Croatia). This aim will be pursued working on a three-fold set of activities: Improving the knowledge on the local energy systems; Monitoring energy consumption and assessing different strategies and policies through a comprehensive set of indicators; Promoting collaborative learning schemes, transfer of knowledge and virtuous energy consumption behaviors.

2.1   The PrioritEE Toolbox: Development and Test

PrioritEE activities are structured along the development and testing of a toolbox which has been designed to support local authorities in managing and monitoring the energy consumption of MPBs, assessing the cost-effectiveness of a predefined set of EE and RES measures, and prioritizing investments. The main benefit for public authorities using such a tool is to get comprehensive strategic guidance on the development of EE plans for municipal public buildings (MPBs). To this end, the toolbox will include five main components:

- A spreadsheet-based *analytic database of EE measures* collected through literature review, expert interviews and national energy certification systems;
- A spreadsheet-based *Decision Support Tool (DST)* for comparing a portfolio of EE interventions on the overall set of MPBs of a given local authority delivering a transparent and objective evaluation of the investment opportunities in terms of costs, energy savings and CO_2 emissions avoided;
- Guidelines for *strategic actions to* enhance sustainable energy awareness and *foster behavioral changes*;
- A total of eight *How-to briefs* compiling in a transparent and easy to use manner the best practices on Stakeholder engagement, Sustainable Energy Action Plans, Innovative financing, Roof-top uses, Building envelope, Behavioral changes, Centralized energy management and ICT;
- An *open data & knowledge access infrastructure* on sustainable EE strategies and measures for MPBs.

2.2   The Pilot Case-Studies

The PrioritEE toolbox will be implemented on 5 pilot case-studies (one per involved MED country) and its adequacy tested to refine the proposed planning tool based on concrete experience. The pilots were set and are being developed within the project, based on an in-depth analysis of local priorities and covering different key EE issues (Table 1). Three rounds of Local Living Labs (LLLs) are foreseen in the pilot regions to promote an active stakeholder engagement emphasizing their role in the entire decision process and providing them with additional competences and knowledge through targeted activities.

2.3   Transferring Knowledge

The main aim of this project component is to provide an effective transfer of all the generated knowledge to the local public authorities engaged, through: Development of training material and training courses; Capacity building of each local public authority; Collaborative and transnational learning; and Building a general protocol for the use of the PrioritEE toolbox.

In this framework, three rounds of workshops are planned during the project in each of the five MED regions to collect experiences and suggestions from local public authorities and energy technicians, and to valorize their inputs in the training phase: *Workshop 1*, where the status quo of pilots and requirements are collected, *Workshop 2*,

Table 1. Main expectation about the project in the local pilots.

Pilot	Main objectives
Pilot 1 – Karlovac (HR)	Better understanding and control of the energy bill. Large open source DB with updated information on the cost-effectiveness of potential investments and energy audits. Assessment of the effectiveness of technical and soft measures and investments. Training sessions for MPBs managers.
Pilot 2 – Potenza, (IT)	Better management and understanding of energy costs. DST with a full DB on the energy behavior of the MPBs. Effects of selected EE strategies and prioritization of investments. Support decision-makers and energy planners in coordination with SEAPs. Cost-effective policies and interventions
Pilot 3 – Teruel (ES)	Additional knowledge and skills on EE in MPBs. Focus on municipalities with low population densities and budgetary restrictions. Training on DST for public authorities of 3 representative counties. Local plan to improve EE in MPBs. Outputs adapted to be used by end-users without any specific technical training, of any local government in Aragon.
Pilot 4 – CIM LT (PT)	Focus on 11 municipalities - rural communities with small municipal bureaus and 220 MPBs with diverse function and construction type. Robust DST to prioritize EE investments in the MPBs portfolio and ensure cost-effective budget allocation. Results capitalized to other neighboring municipalities.
Pilot 5 – Western Macedonia (EL)	Collect information on current energy consumption and analyze improvement via EE, with special focus on heat demand in MPBs. Full training on the DST. Development of a regional plan for improving EE in MPBs. Ensure effective transfer of know-how to local authorities.

where the toolbox responses to the requirements specified in Workshop 1 are assessed; and *Workshop 3,* where public authorities will test the general protocol to support the PrioritEE toolbox usage, including technical tools, good practices, methods or financing options. The protocol will guide the user to choose the most appropriate tool/set of tools to deal with a specific situation (e.g., improvement of EE in schools) and to elaborate a local plan for improving EE and RES in MPBs. The PrioritEE protocol will be elaborated by the technical partners, while local public authorities will test it by developing local plans.

3   First Results and the Way Ahead

The work on developing the toolbox started with a clarification of the operating objectives to match requests and expectations of the local authorities with the availability and/or necessity of suited methodological tools. The main discussion items were: Different tools per pilot or only one common but that can be tailored? How many typologies of MPBs? What will be the criteria for cost-effectiveness decision? What time frame? What actions to enhance awareness and foster behavioral changes of building occupants? This discussion allowed identifying common goals and functionalities of the toolbox. Meanwhile, a review of the status-quo of the pilots' MPBs was performed regarding their energy consumption and potential for EE/RES measures, including: status of the policies and plans, success stories and best practices, weaknesses and barriers, data needs and availability for decision-making. A template table was distributed to collect baseline information on MPBs typologies based on parameters considered relevant in existing literature on EE improvements for both public and residential buildings [6, 10–15] (Table 2).

Table 2. Collecting information on MPBs in pilots.

Building characterization	Equipment characterization
• Building identification and typology • Total useful and implementation area • Construction date and conservation status • Major renovations or EE improvements • Energy certificate and class & Identified measures (if there is a certificate) • Implemented smart meters • No. workers and of daily public users and working hours • Solar PV and annual average generation • Annual energy consumption *(per fuel)* & allocation of energy use per end-use *(by fuel)*	• Building identification (name) – Heating and cooling equipment (type of equipment, power, no. of units, age) • Equipment for water heating (Type of equipment, power, no. of units, age) • Types and characteristics of the lighting (type of equipment, power, no. of units, age)

Complementarily, an extensive desktop research was conducted to identify previous and current projects, initiatives, platforms as well as existing tools and methods that could provide synergies with the project. This will help to refine the selected and/or ad-hoc implemented tools and support the elaboration of a strategic plan on how to improve EE of MPBs in the partner regions. The review of the status-quo of the pilots' MPBs regarding their energy consumption and potential for EE and RES measures is ongoing. For each of the five pilots, the status of the EE policies and plans, success stories, weaknesses and barriers, data needs and availability for decision-making were characterized. The development of the PrioritEE toolbox is also ongoing, especially the draft *Decision Support Tool (DST)*, which will allow ranking EE/RES interventions in about 100 MPBs in the pilots. The DST is spreadsheet based, considers the different MPBs typologies (schools, sport centers, swimming pools, offices, culture buildings, health centers) and its complexity can be adapted to the data availability and objectives in each pilot. A structured *Analytical database* of EE/RES technical measures and the first *How-to brief* on building envelope and sustainable thermal are also under finalization.

Several activities have started to provide an effective stakeholders engagement and prepare the ground for the transfer of all the knowledge generated during the project to the local public authorities. The first round of local workshops (*Workshops 1*) was organized in the partner regions between May and July 2017 to characterize the status quo of pilots and gather requirements from local authorities. The Workshops 1 experience (Table 3) showed some common aspects, e.g. regarding motivation and barriers to implement EE measures, but also requirements for the toolbox (user-friendly tool and with some degree of customization to the local needs). On the other hand, a diversity of outcomes was found regarding the pilots' end-users, the target buildings and the final use of the toolbox. This observed diversity is considered a pre-condition for activating synergies within the project and enhancing the transferability opportunities in other MED territories.

Table 3. Main outcomes of the first round of workshops (WS1) in the pilots.

	Pilot 1 (HR)	Pilot 2 (IT)	Pilot 3 (ES)	Pilot 4 (PT)	Pilot 5 (EL)
Attendees	104	17	49	13	7
Energy efficiency status quo					
Staff with EE experts	Yes	Yes	No	No	Yes
Who implements EE measures	Heads of departments managing the MPBs	Legal representatives and technical officers	Mayor of municipalities (mainly)	Mayor of municipalities	Tech. departments of municipalities
Main EE motivation	Energy cost saving	Energy bill awareness, maintenance	Annual energy cost saving	Energy cost savings, financial	Financial benefits
Prior experiences EU EE projects	Yes (lots) Lack of funding sources	Yes Lack of data & training, low skills	No Lack of budget & knowledge	No Lack of funding & knowledge	Yes Difficult access to funding
Technical expectations for the PrioritEE toolbox					
Buildings to apply the toolbox	MPB stock (different types)	Schools, public offices	MPB stock (different types), street lighting	MPB stock (varied portfolio)	Schools, sports, offices, cultural buildings
Main requirements for the toolbox	Easy to use, yet sophisticated Customized to local needs	Easy to use, open source, free and cross-platform	Easy to use Designed for non-technical users	Easy to use, not time consuming	Simplified approaches Ad-hoc design for the pilots
Final use	Investment planning and asset managing	Technical support on public funding Energy monitoring Training	Prioritize small options. Quantify larger options for funding proposals	Technical support to implement & prioritize EE measures	Enhance local governments Local plans to prioritize investments

The first round of *Local Living Labs (LLL1)* was also organized in the five pilots between October and December 2017 establishing a synergic cooperation between the project's actuators and the main stakeholders, and promoting energy awareness and capacity building.

In March–May 2018 *Workshops 2* will be organized to prioritize EE investments in local pilots using the PrioritEE's DST while the *LLL2* will allow awareness raising activities with selected target groups. *Workshops 3* are planned for the end of 2018 to transfer the general protocol to be elaborated to support a wider set of local/regional public authorities in the use of the PrioritEE toolbox. Finally LLL3 will be held in May 2019 to maximize the knowledge transfer and evaluate the toolbox.

4   Conclusions

There is an urgent need to improve the effectiveness of local sustainable energy policies, enhancing the skills of public authorities and stakeholders in the assessment, definition, adoption and implementation of EE practices. The PrioritEE project aims to contribute to these issues creating a collaborative environment among local public authorities, energy agencies and research institutions of five Mediterranean countries on the development of decision-making support tools, capacity building and knowledge transfer. The overall impact of the proposed activities in each pilot is two-fold: an increased knowledge on energy management, monitoring and planning, as well as an enhanced capacity of the involved public authorities to implement EE measures in municipal public buildings. In this context, transnational cooperation fosters an active participation of public authorities and stakeholders contributing significantly to augment their perception for the multiple benefits of EE and local energy production and for the implementation of acquired good practices. The main outcomes of the proposed activities will converge in the PrioritEE toolbox that will include both 'hard' (analytic database, decision support tool, open data & knowledge access infrastructure) and 'soft' components (guidelines for changing behavior, how-to briefs) in order to provide strategic guidance to all local administrations struggling with the development of energy consumption management plans for municipal public buildings (MPBs). Although the project is still at a comparatively early stage, this paper outlines the importance of combining cost-effective technical actions with actions to enhance awareness and foster EE behavior in order to achieve a substantial and lasting reduction of energy consumption of MPBs.

Acknowledgements. This research was carried out in the framework of the project PrioritEE "Prioritise energy efficiency (EE) measures in public buildings: a decision support tool for regional and local public authorities" (Project Number: 1MED15_2.1_M2_205, Duration: 01/02/2017-31/07/2019). PrioritEE was funded under The Interreg MED Programme 2014-2020; Priority Axis: 2. Fostering low-carbon strategies and energy efficiency in specific MED territories: cities, islands and remote areas. We would like to thank the five partner regions on which this work is based: Karlovac County (Croatia), Municipality of Potenza (Italy), Aragón region (Spain), Lezíria do Tejo Intermunicipal Community (Portugal), and Region of Western Macedonia (Greece).

References

1. BPIE - Buildings Performance Institute Europe: Europe's buildings under the microscope. A country-by-country review of the energy performance of buildings. Brussels (2011)
2. Borgstein, E.H., Lamberts, R., Hensen, J.L.M.: Mapping failures in energy and environmental performance of buildings. Energy Build. **158**, 476–485 (2018)
3. Christen, M., Adey, B.T., Wallbaum, H.: On the usefulness of a cost-performance indicator curve at the strategic level for consideration of energy efficiency measures for building portfolios. Energy Build. **119**, 267–282 (2016)
4. Ozawa-Meida, L., Wilson, C., Fleming, P., Stuart, G., Holland, C.: Institutional, social and individual behavioural effects of energy feedback in public buildings across eleven European cities. Energy Policy **110**, 222–233 (2017)
5. Webb, A.L.: Energy retrofits in historic and traditional buildings: a review of problems and methods. Renew. Sustain. Energy Rev. **77**, 748–759 (2017)
6. Bertone, E., Sahin, O., Stewart, R.A., Zou, P., Alam, M., Blair, E.: State-of-the-art review revealing a roadmap for public building water and energy efficiency retrofit projects. Int. J. Sustain. Built Environ. **5**(3), 526–548 (2016)
7. European Commission, Directorate-General for Energy, Unit C.3 Energy Efficiency: Public Consultation for the Review of Directive 2012/27/EU on Energy Efficiency, Final Synthesis Report, Brussels (2016)
8. EPEC - European PPP Expertise Centre: Guidance on Energy Efficiency in Public Buildings. European PPP Expertise Centre, Luxembourg (2012)
9. Ruparathna, R., Hewage, K., Sadiq, K.: Improving the energy efficiency of the existing building stock: a critical review of commercial and institutional buildings. Renew. Sustain. Energy Rev. **53**, 1032–1045 (2016)
10. Ballarini, I., Corgnati, S.P., Corrado, V.: Use of reference buildings to assess the energy saving potentials of the residential building stock: the experience of TABULA project. Energy Policy **68**, 273–284 (2014)
11. Gouveia, J.P., Seixas, J., Long, G.: Mining households' energy data to disclose fuel poverty: lessons for Southern Europe. J. Clean. Prod. **178**, 534–550 (2018)
12. Lee, S.H., Hong, T., Piette, M.A., Taylor-Lange, S.C.: Energy retrofit analysis toolkits for commercial buildings: a review. Energy **89**, 1087–1100 (2015)
13. Barbu, A.D., Griffiths, N., Morton, G.: Achieving energy efficiency through behaviour change: what does it take?, EEA Technical report, no. 5/2013, Copenhagen (2013)
14. Wolfe, A.K., Malone, E.L., Heerwagen, J., Dion, J.: Behavioral Change and Building Performance: Strategies for Significant, Persistent, and Measurable Institutional Change, Washington (2014)
15. Ucci, M., Domenech, T., Ball, A., Whitley, T., Wright, C., Mason, D., Westaway, A.: Behaviour change potential for energy saving in non-domestic buildings: development and pilot-testing of a benchmarking tool. Build. Serv. Eng. Res. Technol. **35**(1), 36–52 (2014)

Planning for Climate Change: Adaptation Actions and Future Challenges in the Italian Cities

Grazia Brunetta[(⊠)] and Ombretta Caldarice

Politecnico di Torino, Responsible Risk Resilience Centre (R3C),
10125 Torino, Italy
grazia.brunetta@polito.it

Abstract. Climate change is a prominent concern of the 21th-century daily life so much that cities worldwide have been widely engaged in contrasting it. In this scenario, urban resilience is becoming one of the top priorities of development agendas and a guiding principle of the policy governance of contemporary cities. This paper discusses the spread of the urban resilience paradigm within the field of Italian spatial planning focusing in particular on the national approach to adaptation and the recent - and unique - experiences of Bologna and Ancona local plans. The cases are discussed about their approach to adaptation within the academic debate on climate change and to their contents and procedures. The paper puts light on one side on the fundamental role of cities to be pivotal to contrast the climate change and on the other side on the implications of adaptation in the local policy-making processes advancing future challenges for Italian spatial planning.

Keywords: Spatial planning · Climate change · Adaptation plans

1 Introduction. Cities Dealing with Climate Change

In Europe and worldwide, cities are facing significant challenges because they are involved in an unprecedented process of dynamism and a fast rate of change. The emerging futures for cities will be related to their growing complexity, the ageing of the population, the emergence of environmental issues, and the unexpected course of development. This rapid and often unplanned expansion of cities is exposing a higher number of people living in urban areas to natural and anthropic risks and, consequently, to climate change [1].

City authorities worldwide are increasingly placing resilience at the heart of their policy-making activity to provide a broad answer to these urban uncertainties. In a nutshell, resilience has become a significant component of the climate adaptation, environmental management, regional economic development, and strategic planning [2]. In successful adaptation policies, therefore, urban resilience is one of the top priorities for risk-related issues, and it is considered the primary guiding principle of policy governance and a key political category of the contemporary time [3].

The main global agreements on this topic are based on the EU strategy on adaptation to climate change (2013) that aims to make Europe more climate-resilience

© Springer International Publishing AG, part of Springer Nature 2019
F. Calabrò et al. (Eds.): ISHT 2018, SIST 101, pp. 609–613, 2019.
https://doi.org/10.1007/978-3-319-92102-0_65

enhancing the preparedness and the capacity of all governance levels to respond to the impacts of climate change. The EU strategy is expected to be further implemented directly by the Member States in specific National Adaptation Strategies (NASs) and related National Adaptation Plans (NAPs) following the Paris Agreement (UNFCCC, 2015)[1]. Also, the UN 2030 European Urban Agenda fosters for cities and human settlements environmentally sustainable and resilient, socially inclusive, safe and violence-free, economically productive, and better connected to and contributing towards sustained rural transformation. Lastly, the 2015 UN Sendai Framework for Disaster Risk Reduction 2015–2030 (SFDRR) (UNISDR, 2015) acknowledges climate change as one of the drivers of disaster risk and requires countries to take risk pre-vention and reduction measures. In this institutional framework, urban areas are con-sidered the places for innovation where policies and actions oriented to climate change are mainly mobilised[2]. But, at the same time, there is little empirical evidence to prove that current urban adaptation plans are useful as they focus more on a broad vision rather than on specific actions [5]. In general terms, this shows as the traditional spatial planning tools are often insufficient to tackle these problems and challenges and are even unfit to govern these kinds of transformation processes [6]. In a nutshell, although the principal global cities are engaged in climate planning actions, the struggle to mitigate and adapt to climate change in urban areas will arguably be the most impactful issue of the 21th-century [7].

In light of this, the goal of this paper is to critically review the Italian approach to climate change planning both at national and local level. From this analysis, the paper discusses the role of the current planning theory and practice about its capacity to adequately address climate change issues. The paper supports the idea that fostering resilience needs to involve spatial planning not only for recovery from shocks but also to encourage preparedness and seek potential transformative opportunities that emerge from changes.

2   Actions. The Italian Climate Adaptation Planning

Italy is the European country most disturbed by risk dynamics especially related to the soil vulnerability and the high frequency of natural hazards. As reported in the Inter-national Disaster Database (www.emdat.be), risks affected in Italy 4 million people from 1900 to today, leading to the death of more than 138 thousand of people and causing estimated economic damages for about one billion USD (Table 1 and Fig. 1).

Nevertheless the issue's salience, the Italian climate adaptation planning is in its infancy. Following the EU strategy on adaptation to climate change (2013), the Italian Adaptation Strategy was elaborated in 2015 by the Italian Environment Ministry,

---

[1] As reported by the European Environment Agency, today 25 EU Member States have adopted NASs while 15 EU Member States have developed NAPs. Over the last five years, there has been a steady increase in the number of NASs and NAPs being adopted by countries. Over the same period, several countries that adopted their NAS some years ago reviewed and adopted a revised NAS.

[2] As outlined by [4], 75% of the 350 cities that are members of ICLEI - Local Government for Sustainability - are defining a national climate adaptation strategy.

Table 1. Natural disasters in Italy from 1990 to 2017.

Disaster typology	Occurrences	Total deaths	Affected	Total damage ('000 US$)
Drought	3	0	0	1.990.000
Earthquake	36	115.971	1.084.705	54.684.852
Extreme temperature	8	20.169	0	4.532.601
Flood	46	1.101	2.879.573	25.346.600
Landslide	21	280	6.524	4.498.900
Storm	5	735	21.024	3.100
Volcanic activity	7	21	320	1.700.000
Wildfire	21	280	6.524	4.498.900
Total	**147**	**138.557**	**3.998.670**	**97.254.953**

Fig. 1. The Italian positioning in the European natural disaster scenario.

which will follow - probably at the middle of 2018 - by the National Action Plan that will report measures and interventions with a defined governance for the operational management of the climate change issue. At the regional level, the Lombardy Region approved in 2014 an adaptation strategy and in 2016 the action plan, while the strategies of Emilia Romagna and Abruzzo Regions are being drafted. Other regional actions are being studied in Piedmont, Veneto and Marche Regions. Finally, at the local level, even in the absence of a regulatory framework that foresees them, the Municipalities of Bologna and Ancona have drafted their own Climate Adaptation Plan, while Rome and Milan have joined the 100 Resilient Cities experience funded by the Rockefeller Foundation. Alongside these initiatives expressly aimed at resilience, the Covenant of Majors energy and climate plans - PAES 2.0 - also start to have a high number of memberships among the Italian municipalities. To sum up, Italian cities focus more on mitigation than adaptation. Only two large cities (Bologna and Ancona) have adaptation plans and address climate change impacts on extreme events, while other cities have civil protection plans that focus on these issues as well, but they do not have an adaptation plan that integrates all impacts of climate change.

In particular, the Bologna Adaptation Plan focus on making the city less vulnerable and able to react to floods, droughts and other consequences of climate change by providing some real measures and actions. The Plan, approved by the City Council in October 2015, is the outcome of the European project LIFE + BLUE AP - Bologna

Local Urban Environment Adaptation Plan for a Resilient City - and its primary goal is the definition of policies for land management oriented to limiting the climate change effects. Bologna has activated a participatory process that has involved the main stakeholders through two open conferences and some thematic meetings to discuss the received proposals and to evaluate their implementation and their inclusion in the Adaptation Plan. The Plan focuses on specific actions concerning the three main vulnerabilities of the city: (i) drought and water scarcity; (ii) extreme weather events and hydrogeological risk; and (iii) heat waves in urban areas. For each of these areas, the Plan identifies objectives for 2025 and describes the actions to achieve them by defining a series of pilot projects. Lastly, the implementation of the Plan is accompanied by a monitoring phase.

In the same way, the Adaptation Plan of Ancona is the outcome of the LIFE ACT project - Adaptation to Climate change in Time. The project is oriented to the construction of a planning tool that takes into account the environmental, social and economic impacts of climate change to increase the resilience of the city to climate change. The Plan is based on four city visions: (i) Ancona polycentric, habitable and accessible city; (ii) Ancona interconnected and competitive city; (iii) Ancona ecological city; and (iv) Ancona landscape and beautiful city. The main goal of the Plan is to promote the response of the city by safeguarding its natural and cultural heritage and managing development changes. The Plan focuses on specific actions related to the potential vulnerabilities of Ancona, identifying four priority areas: (i) landslides and hydrogeological structures; (ii) coastal erosion; (iii) infrastructure; and (iv) historical and cultural heritage. The core of the Plan is the construction of stable governance - through the establishment of the Local Adaptation Board composed of institutional representatives and members of the Universities and civil society - and the monitoring phase.

3   Challenges. How Can Cities Plan for Climate Change?

Today, contemporary cities are involved in profound changes in urban processes and in social dynamics that require a reformulation of the rational dimension of spatial planning [8]. A new planning mindset is requested, and it is called to be to able to control the organisation of human processes drastically reducing impacts on the ecosystem. In this scenario, the growth control, the reduction of soil sealing, and the adaptive regeneration become the focus of planning for resilience. In this view, spatial planning plays a fundamental role in addressing the causes and impacts of climate change reducing the sensitivity and vulnerability of urban areas to extreme events [9].

However, a careful analysis of the adaptation practices highlights that integration between climate protection and spatial planning seems to have occurred mainly at the theoretical level, lacking the practical declination of resilience [10]. Even if urban planning is theoretically able to respond to adaptation, the perception of experiences seems to suggest that it is not able to fully understand the relationship between change and adaptation. In particular, the Italian approach to spatial planning - characterised by the predominance of the territorial dimension, by quantitative parameters and by the central role of rights [11] - is today mainly inadequate to respond to the challenges posed by climate change. While focusing on environmental and ecological issues, Bologna and Ancona Adaptation Plans are not correctly successful and effective adaptation policies, as they do not show demonstrable results.

In light of this, this papers sustains that adaptation should be incorporated into the culture of spatial planning in a comprehensive and integrated way. The adaptation should, therefore, be systematised into the planning policies so that they need to shift from specific responses to specific exposures to a resilient perspective that integrates adaptive strategies, dynamic processes and urban development [12]. This idea implies that adaptation becomes a fundamental principal of local policies to activate the transition of urban ecosystems. In this perspective, adaptation should not be planned as a single objective linked to a specific action but should trigger a process of combined and synergistic actions in response to system vulnerabilities. This would help to interpret the resilience in spatial planning, declining it as the capacity of each urban system to develop its co-evolutionary perspective. In summary, spatial planning alone can not address the challenges of climate change adaptation, but it can be part of and drive the adaptation process for resilience-oriented cities.

References

1. Jones, S.: Cities Responding to Climate Change. Copenhagen, Stockholm and Tokyo. Springer, Dordrecht (2018)
2. Davoudi, S., Brooks, E., Mehmood, A.: Evolutionary resilience and strategies for climate adaptation. Plan. Pract. Res. **28**(3), 307–322 (2013)
3. Chandler, D.: Beyond neoliberalism: resilience, the new art of governing complexity. Resilience **2**(1), 47–63 (2014)
4. Aylett, A.: Progress and Challenges in the Urban Governance of Climate Change: Results of a Global Survey. MIT, Cambridge (2014)
5. Siders, A.R.: A role for strategies in urban climate change adaptation planning: lessons from London. Reg. Environ. Change **17**, 1801–1810 (2017)
6. Albrechts, L.: Some ontological and epistemological challenges. In: Albrechts, L., Balducci, A., Hillier, J. (eds.) Situated Practices of Strategic Planning. An International Perspective, pp. 1–11. Routledge, London (2017)
7. Merrow, S., Mitchell, C.L.: Weathering the storm: the politics on urban climate change adaptation planning. Environ. Plan. A **49**(11), 2619–2627 (2017)
8. Rauws, W.S.: Embracing uncertainty without abandoning planning exploring an adaptive planning approach for guiding urban transformations. disP – Plan. Rev. **53**(1), 32–45 (2017)
9. Campbell, H.: Is the issue of climate change too big for spatial planning? Plan. Theor. Pract. **7**(2), 201–230 (2006)
10. Carmin, J., Nadkarni, N., Rhie, C.: Progress and Challenges in Urban Climate Adaptation Planning: Results of a Global Survey. MIT, Cambridge (2012)
11. Servillo, L., Lingua, V.: The innovation of the Italian planning system: actors, path dependencies, cultural contradictions and a missing epilogue. Eur. Plan. Stud. **22**(2), 400–417 (2014)
12. Brunetta, G., Caldarice, O.: Putting resilience into practice. The spatial planning response to urban risks. In: Brunetta, G., Caldarice, O., Tollin, N., Rosas-Casals, M., Morató, J. (eds.) Urban Resilience for Risk and Adaptation Governance. Theory and Practice. Springer, Dordrecht (forthcoming)

Contemporaneity of Floods and Storms. A Case Study of Metropolitan Area of Reggio Calabria in Southern Italy

Giuseppe Barbaro[1(✉)], Olga Petrucci[2], Caterina Canale[1], Giandomenico Foti[1], Pierluigi Mancuso[3], and Pierfabrizio Puntorieri[1]

[1] Mediterranea University of Reggio Calabria, 89100 Reggio Calabria, Italy
giuseppe.barbaro@unirc.it
[2] National Research Council of Italy, Institute for Geo-Hydrological Protection, CNR-IRPI, 87036 Rende, Italy
[3] Department of Infrastructure Public Works and Mobility, Sector 8-Defense Interventions, Calabria Region, Regional Citizen, 88100 Catanzaro, Italy

Abstract. The environmental balance is being increasingly altered by mankind's direct and indirect actions. Waterproofing of territory and continuous climate change are amongst the main factors of hydrogeological risk. In the presence of complex orographies and particular geographical exposure, meteorological phenomena can have devastating consequences. Calabria, located in the southern part of Italy, stands at the confluence of the Tyrrhenian and Ionian Seas and is particularly exposed to such phenomena. Its unique geomorphological formation makes it subject to flooding and sea storms that have revealed the fragility of its territory. When such phenomena occur concurrently, the effects can be devastating, both in terms of infrastructure damage and inconvenience to the local population, as downstream flooding interacts with wave run-up. We will analyze the history of contemporary flooding and storms in the Metropolitan Area of Reggio Calabria, which is located in the south of Calabria and in the middle of the Mediterranean Sea. Two case studies will be illustrated, one relating to the Ionian coast and the other to the Tyrrhenian coast.

Keywords: Floods · Sea storms · Meteorological phenomena
Torrents · Wave run-up

1 Introduction

Anthropisation and waterproofing are amongst the main causes of the irreversible process of land consumption. The situation is exacerbated by climate change which affects both the rainfall regime, and wave action on the coasts, causing extreme events such as storms and flooding [1]. These represent worrying phenomena if they occur singularly, but devastating when occurring together. Storms and flooding frequently occur across the Italian territory, but are particularly violent in Calabria. Calabria's particular orography, geomorphology and exposure to the winds of the Scirocco in the South and the Mistral in the North influence the response of the territory to

© Springer International Publishing AG, part of Springer Nature 2019
F. Calabrò et al. (Eds.): ISHT 2018, SIST 101, pp. 614–620, 2019.
https://doi.org/10.1007/978-3-319-92102-0_66

precipitation and to wave motion, with consequent river flooding in urban areas, and inundation in coastal areas.

These two events are almost always studied separately in existing literature. As far as flooding is concerned, many studies deal with the urban flood risk [2–4] and the dangers of such flooding [5]. From the coastal point of view we mainly study shoreline changes and erosion processes, important concepts for planning and management of areas near the sea [6–8]. It is important to analyse the main phenomena affecting coastal dynamics [9–13], in particular wave action [14–17] and the interaction between coastal and river transport [18–24]. Calabria, due to remarkable coastal development, has more than 700 km of coast, much of which is subject to erosion [25], and in particular many areas are prone to the risk of sea inundation with disastrous consequences for coastal urban areas. The separate analysis of such events can lead to an underestimation of the severity of the outcome when such phenomena occur simultaneously. It follows that events with low return times can significantly affect the territory even when such events are concomitant. In this paper the history of storms and floods in Reggio Calabria has been analysed. This analysis begins with the historical data of alluvial events, provided by the CNR-IRPI of Cosenza, and of wave data provided by ABRC MaCRO, and by identifying the storms by means of Boccotti's theory. In the first analysis, it was possible to evaluate the simultaneous manifestation of events that strongly affected the vulnerability of the territory. The concomitance is recorded both on the Ionian coast, struck several times by the Scirocco wind, and on the Tyrrhenian coast, where wave agitation is mainly due to the action of the Mistral currents. Case studies will present chronologically distant events in the two geographically distant towns of Scilla (Tyrrhenian coast) and Monasterace (Ionian coast). The results obtained allow us the possibility of extending the study to territories with the same climatic, orographic and morphological characteristics as those of the Metropolitan Area of Reggio Calabria.

2 Description of the Territory

The Metropolitan Area of Reggio Calabria is located in the south of Italy, in the centre of the Mediterranean, as shown in Fig. 1:

The territory is characterised by the presence of "fiumare", typical rivers of southern Italy with torrential regime. In them, the high slope of the riverbeds, the continuous erosion of the banks and the dragging action of the water favour the transport of solid material [26], while the alternation of dry periods and periods of flooding help to reactivate landslides inside the riverbeds. Most of the river basins are very small, so the response to precipitation is rapid. For the reasons outlined above, flooding is frequent especially due to the presence of short and intense rainfalls [27]. From a coastal point of view, the most violent storms are due to the exposure of the Scirocco and Libeccio winds in the south and the Mistral in the north [28]. The southern perturbations generate storm waves and intense rainfall affecting most of the Mediterranean. These phenomena are of great violence on the Ionic front, where the pluviometric regime is highly influenced by the humid currents coming from Africa, and the clash with the mountain ranges pushes these currents repeatedly on the sea.

Fig. 1. Geographic position of the metropolitan area of Reggio Calabria.

On the Tyrrhenian coast, winds from the north can generate small Mediterranean hurricanes, meteorological phenomena with an extraordinary nature. Precipitation is very frequent but not very intense, in contrast to that of the Ionian coast.

3 Methodology

The contemporaneity of flooding and storms was studied by comparing two databases. The first of these was the CNR-IRPI in Cosenza and is related to alluvial events occurring in the coastal areas of the Metropolitan Area of Reggio Calabria. Storms and flooding which posed a serious danger to the population and to the territory itself were selected. For each event it was possible to identify the date of the event, its effect on the territory and the characteristics of the rivers involved. The amount of rain data was not evaluated, only the occurrence of the event itself. The second database is related to the wave data processed by the ABRC MaCRO software, starting with the anemometric data provided by the Met Office, referring to a depth of − 200 m. From this data, time series were extrapolated for the period in which the alluvial events analysed occurred. For the study of wave motion, Boccotti's theory of equivalent triangular storms (MTE) was used. By obtaining the critical height as being 1.5 times the mean of the significant heights recorded for each time series, it was possible to specify the events during which this threshold was exceeded. Where a sea storm was identified, any concurrence with flooding was also verified.

In this phase the entity of the events were not evaluated, but only the combined events. Future studies will focus on the analysis of values of rain data, wave data and the return period of the same events.

4 Case Studies

The territory of Reggio Calabria has numerous flooding events of particularly exceptional nature. These phenomena have affected the hinterland and the coastal areas, causing devastation and irreversible damage to the territory and to its inhabitants. In certain climatic conditions such effects are amplified by the combined action of flooding and storm surges. This section describes two cases, one for each of the Ionian and Tyrrhenian coasts.

As far as the Ionian coast is concerned, the city of Monasterace was considered. Located on the border with the province of Catanzaro, it has been repeatedly hit by strong Scirocco storms that have affected and damaged the Kaulon Archaeological Park. As described above, the Ionian area is very well suited to an analysis of contemporary events. A flood took place on 5th October 1996 which involved the Fiumarella di Guardavalle torrent on the border between Monasterace and the town of Guardavalle. This river is characterized by a basin area of 28.58 km^2 and a main stream length of 18.26 km. The sea storm occurred from 4th to 6th October 1996. The highest H_s was recorded on October 5th 1996, and coincided with the alluvial event, and took on the value of $H_{Smax} = 3.41$ m. The critical wave height, evaluated with the expression obtained from Boccotti, was a $h_{crit} = 0.93$ m. From the trend of significant heights as a function of time, it wass observed that the height threshold was exceeded for more than 40 h, as shown in Fig. 2:

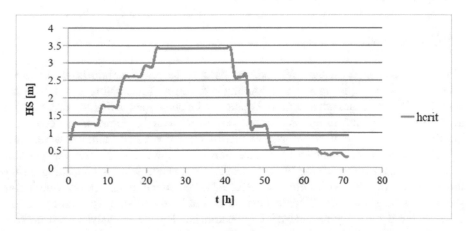

Fig. 2. Trend of the storm that hit the town of Monasterace from 4th to 6th October 1996.

On the Tyrrhenian coast the city studied was Scilla, located at the entrance to the Strait of Messina. It has just under 5,000 inhabitants and is one of Calabria's tourist attractions. On 18th December 2003 heavy rains caused the flooding of the Vallone Oliveto. This river has a basin area of reduced extension, approximately equal to 1.5 km^2, and the length of the main stream is less than 1.5 km. From the coastal point of view a critical height equal to $h_{crit} = 0.82$ m was obtained.

The sea storm that struck Scilla from 16th to 18th December 2003, at the same time as flooding, had a maximum significant height of just over 2 m and a duration of about 48 h (Fig. 3).

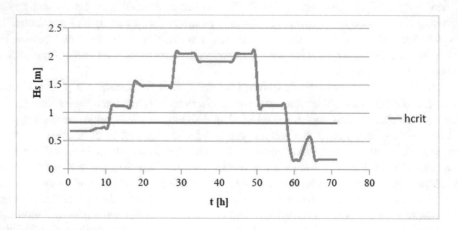

Fig. 3. Progress of the storm that hit the municipality of Scilla from December 16 to 18, 2003.

The cases analysed are typical examples of a condition common to the entire Metropolitan Area of Reggio Calabria.

5   Conclusion

The study conducted on the Metropolitan Area of Reggio Calabria is the basis for future research. From the analysis of flooding and storms, it is possible to conclude that simultaneous events are not a territorial feature but a phenomenon that is closely related to meteorological and orographic conditions which can be found in several other locations. Specifically, it was observed that in the presence of Scirocco winds, precipitation is often accompanied by intense storms. On the Ionian coast it was possible to identify a great coincidence of such events. Contrary to what is happening on the Ionian coasts, the Tyrrhenian areas are affected by precipitation and storms only in the presence of particular meteorological events caused by the Mistral wind. These results were achieved by a simple analysis of phenomena. The cases analysed can be considered as pilot studies for further exploration in the field of risk from coastal inundation and flooding. Starting from this point it will be possible to more concretely analyse the effects of such event in terms of return period, precipitation and wave height. By correlating the geomorphologic features of the river basins with the intensity of precipitation, it is also possible to identify those areas most at risk from flooding.

By correlating coastal morphology to wave run-ups, it is possible to detect coastal areas that are susceptible to flooding. Such information can be a useful tool for local and state administrations in order to improve the planning of mitigation actions for hydrological and coastal risks.

References

1. Breil, M., Catenacci M., Travisi C.: Impatti del cambiamento climatico sulle zone costiere: Quantificazione economica di impatti e di misure di adattamento – sintesi di risultati e indicazioni metodologiche per la ricerca future. Fondazione Eni Enrico Mattei (FEEM), Centro Euro-Mediterraneo per i Cambiamenti Climatici (CMCC) (2007)
2. Mascarenhas, F.C.B., Miguez, M.G.: Urban flood control through a mathematical cell model. Water Int. 27(2), 208–218 (2002)
3. Prestininzi, P., Fiori, A.: A two-dimensional parabolic model for flood assessment. Ital. J. Eng. Geol. Environ. 1, 5–18 (2006)
4. Kim, Y., Han, M.: Rainfall-Storage-Drain (RSD) model for Runoff Control Rainwater Tank System Design in Building Rooftop. Rainwater Research Center in Seoul National University (2007)
5. Barbaro, G., Scionti, F., Foti, G., Tripodi, G.: La pianificazione degli interventi di minimizzazione del rischio: analisi idrologico-idraulica e proposta innovativa e sostenibile per le aree inondabili del torrente Forio di Cittanova. III Convegno Italiano sulla riqualificazione fluviale, Reggio Calabria (2015)
6. Phillips, M.R., Jones, A.L.: Erosion and tourism infrastructure in the coastal zone: problems, consequences and management. Tour. Manag. 27(3), 517–524 (2006)
7. Marin, V., Palmisani, F., Ivaldi, R., Dursi, R., Fabiano, M.: Users' perception analysis for sustainable beach management in Italy. Ocean Coast. Manag. 52(5), 268–277 (2009)
8. Addo, K.A.: Shoreline morphological changes and the human factor: case study of Accra Ghana. J. Coast. Conserv. 17(1), 85–91 (2013)
9. Komar, P.D.: Coastal erosion-underlying factors and human impacts. Shore & Beach 68(1), 3–16 (2000)
10. Maiti, S., Bhattacharya, A.K.: Shoreline change analysis and its application to prediction: a remote sensing and statistics based approach. Mar. Geol. 257(1–4), 11–23 (2009)
11. Arena, F., Barbaro, G., Romolo, A.: Return period of a sea storm with at least two waves higher than a fixed threshold. Math. Probl. Eng. 2013, 1–6 (2013)
12. Barbaro, G., Foti, G., Sicilia, C.L.: Coastal erosion in the South of Italy. Disaster Adv. 7, 37–42 (2014)
13. Barbaro, G., Fiamma, V., Barrile, V., Foti, G., Ielo, G.: Analysis of the shoreline changes of Reggio Calabria (Italy). Int. J. Civil Eng. Technol. 8(10), 1777–1791 (2017)
14. Barbaro, G.: A new expression for the direct calculation of the maximum wave force on vertical cylinders. Ocean Eng. 34, 1706–1710 (2007)
15. Barbaro, G., Foti, G., Malara, G.: Set-up due to random waves: influence of the directional spectrum. In: Proceedings 30th International Conference on Ocean, Offshore and Artic Engineering (OMAE), Rotterdam, The Netherlands (2011)
16. Barbaro, G., Foti, G., Malara, G.: Set-up due to random waves: influence of the directional spectrum. Int. J. Marit. Eng. 155, A105–A115 (2013)
17. Barbaro, G., Foti, G.: Shoreline behind a breakwater: comparison between theoretical models and field measurements for the Reggio Calabria sea. J. Coastal Res. 29, 216–224 (2013)

18. Boccotti, P., Arena, F., Fiamma, V., Romolo, A., Barbaro, G.: Estimation of mean spectral directions in random seas. Ocean Eng. **38**, 509–518 (2011)
19. Tomasicchio, G.R., D'Alessandro, F., Barbaro, G.: Composite modelling for large-scale experiments on wave-dune interaction. J. Hydraul. Res. **49**, 15–19 (2011)
20. Sicilia, C.L., Foti, G., Campolo, A.: Protection and management of the Annunziata river mouth area (Italy). J. Air Soil Water Res. **6**, 107–113 (2013)
21. Barbaro, G., Foti, G., Sicilia, C.L., Malara, G.: A formula for the calculation of the longshore sediment transport including spectral effects. J. Coastal Res. **30**, 961–966 (2014)
22. Boccotti, P.: Wave Mechanics and Wave Loads on Marine Structures. Elsevier BH, Oxford (2015)
23. Tomasicchio, G.R., D'Alessandro, F., Barbaro, G., Musci, E., De Giosa, T.M.: Longshore transport at shingle beaches: an independent verification of the general model. Coast. Eng. **104**, 69–75 (2015)
24. Borrello, M.M., Foti G., Puntorieri P.: Shoreline evolution near the mouth of the Petrace River (Reggio Calabria, Italy). In: Proceedings 9th International Conference on River Basin Management, Prague, Czech Republic (2017)
25. Barbaro, G.: Master Plan of solutions to mitigate the risk of coastal erosion in Calabria (Italy), a case study. Ocean Coast. Manag. **132**, 24–35 (2016)
26. Sorriso- Valvo, M., Terranova, O.: The Calabrian fiumara streams. Zeitschrift für Geomorphologie **143**, 109–125 (2006)
27. Petrucci, O., Pasqua, A.A., Polemio, M.: Flash flood occurrences since the 17th century in steep drainage basins in Southern Italy. Environ. Manage. **50**, 807–818 (2012)
28. Terranova, O.: Regional analysis of superficial slope instability risk in Calabria (Italy) through a pluviometrical approach. Risk Anal. IV **77**, 257–266 (2004)

Environmental Assessment of a Solar Tower Using the Life Cycle Assessment (LCA)

Fausto Cavallaro[1]([envelope]) [iD], Domenico Marino[2] [iD],
and Dalia Streimikiene[3] [iD]

[1] University of Molise, Campobasso 86100, Italy
cavallaro@unimol.it
[2] Mediterranea University of Reggio Calabria, Reggio Calabria 89100, Italy
dmarino@unirc.t
[3] Lithuanian Institute of Agricultural Economics,
V. Kudirkos g. 18, Vilnius 01113, Lithuania
dalia.streimikiene@knf.vu.lt

Abstract. That renewable energy technologies, particularly in the production phase, are currently those that generate a lower environmental impact compared to traditional fossil fuel systems is now well-established. Despite this, many studies fail to include an evaluation of the impacts generated by systems designed and built for energy production over their entire life cycle. The aim of this paper is to provide, with the aid of LCA, a preliminary environmental assessment of a solar power tower.

Keywords: Concentrated solar thermal · Life Cycle Assessment (LCA)
Sustainability assessment

1 Introduction

The principle of conversion of thermal solar power has been known for more than one century. However, the first industrial-scale solar power installations were built only in the mid-1980s. While the Solar power tower technology is commercially less mature than linear parabolic collectors, a number of experimental systems have been tested on the field in various sites worldwide, and their design feasibility has been demonstrated. Today, development of high-temperature thermal solar projects is on the agenda of many government's policies and plans.

One important, key aspect to be assessed in this technology is its potential impact on the environment. The aim of this paper is to make a preliminary analysis, with the Life Cycle Assessment (LCA) approach, of the main environmental modifications caused by the electricity generation processes of a solar tower power installation based on an air volumetric receiver.

Similar studies have already been developed. Some of the most significant are: [1–16].

© Springer International Publishing AG, part of Springer Nature 2019
F. Calabrò et al. (Eds.): ISHT 2018, SIST 101, pp. 621–628, 2019.
https://doi.org/10.1007/978-3-319-92102-0_67

2 Solar Tower Technologies

Thermal solar power can only exploit direct beam radiation or direct normal irradiation (DNI), i.e. the fraction of light that hits directly the surface of the earth directly without any deviations caused by clouds, moisture and/or other atmospheric agents [17]. Generally, these types of technologies are best used in sites with high direct solar radiation, with a minimum of 1,800 kWh/m^2/year and, especially, in regions where climate conditions and vegetation do not generate high levels of humidity, dust and atmospheric events able to interfere with direct solar radiation [17].

The installation comprises a system of mirrors, called heliostats, which follow the motion of the sun by means of electronic-control electromechanical actuators and which reflect solar energy on a receptor (boiler) located on the top of a tower located at the centre of the field of mirrors (see Fig. 1).

A heat transfer fluid (water, air or a mixture of molten salts with liquid sodium) flows inside the receiver; it absorbs the captured heat and conveys it to a steam generator. The steam generated at high temperature and high pressure then powers a turbine according to a classic thermodynamic cycle [18]. The system uses hundreds of heliostats and is suitable for large-scale applications. As a rule, each mirror covers a surface area of about 30–50 m^2; however, the latest systems can cover up to 100–150 m^2. The mirrors are usually formed by a concave metal plate on which a layer of glass is placed.

The support structure consists of a grid-type framework, supporting a few dozen mirrors.

Since the purpose of the receiver is to transfer solar heat to the heat transfer fluid, it must be designed appropriately taking into due account resistance to high temperature and heat loss due to convection and irradiation [19].

In pilot installations in Europe, the project designers have preferred to use air as heat transfer fluid instead of molten salts or synthetic oil.

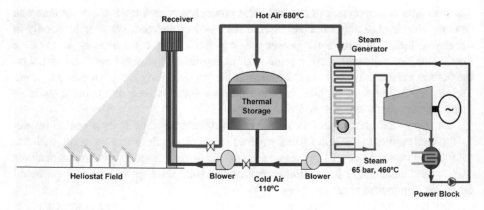

Fig. 1. Diagram of solar tower power plant based on a volumetric air receiver. Source [19]

3 Structure of the LCA Study

The installation chosen for this study is a solar tower power plant of the PHOEBUS type (plant with atmospheric air as heat transfer fluid) [19, 20] with output of 30 MW, mirror surface area of about 160,428 m^2 and net efficiency of 14.2%. The functional unit (f.u.) chosen is 1 kWh of electricity. It has been assumed that the components (solar mirrors, tower, heat storage system and power block) would be manufactured in Germany and then assembled, installed and commissioned with all associated services, including servicing during operation and dismantling at the end of the plant's useful life in an industrial area in southern Sicily, planned as the power plant site.

Data quality is an essential requirement for performing LCA analysis. Therefore, correct study procedure starts from data collection. This stage is an iterative process, meaning that as data are collected and the system's structure is developed, it is possible to identify the degree of relevance of elements and hence collect additional data or correct the data set. The data selected in the inventory will be the basis for assessing the environmental impact of the life cycle of the product or service; therefore the data must be inputted in accordance with a well-defined and transparent plan.

The primary data come mainly from the paper of Weinrebe et al. (1998); however, some uncertainties persist regarding the exact quantities of some of the materials used. The secondary data, in particular the data on the extraction of raw materials, on transport and recycling have been obtained from the libraries available in computing code SIMAPRO and in the Ecoinvent 2.0 database.

3.1 System Description and Boundaries

It is assumed that the solar thermal power plant addressed by this study would be installed in a site in southern Sicily, benefiting from optimal sun irradiation. The study includes the steps concerning manufacture of the various components of the power plant, the extraction and supply of consumables, transport and plant decommissioning and recycling.

The diagram in Fig. 2 reproduces the system boundaries. It makes it possible to identify the main process units considered for analysis.

In particular, the following activities are included:

1. *Manufacture of power plant components:* The power plant comprises the following components: solar field (consisting mainly of the heliostats), tower, energy storage system, power block and buildings. The main materials employed in construction of the power plant are: steel, copper, zinc, plastic, concrete, glass, aluminium, lubricant oil for mechanical parts, fuel (diesel oil) and electricity;
2. *Construction:* this step includes: (1) transport from the factory, by means of a large transoceanic ship, of the power plant's various components; (2) transport of the various materials necessary for construction of the power plant (e.g. concrete and steel for foundations) by means of trucks. Besides transport, the operation of earth moving machinery has been taken into consideration;

3. *Use and maintenance:* The power plant operation phase involves demineralised water, which is used in the power bloc and turned into steam, and natural gas supplementing solar energy when radiation is at low levels. For servicing/ maintenance activities, replacements for parts subject to wear and tear have been considered;

4. *Demolition and dismantling:* The reasonable assumption has been made that at the end of the power plant's life (that is after about 40 years) 50% of the material will be recovered and reused to build another power plant, while the remaining 50% will be sent to landfill. In this stage, the use of machinery and vehicles for dismantling and transport to recycling centres and to landfill has been factored in.

Among the different impact assessment methods available in the literature, we have selected Eco-indicator-99. This is a method developed by Pré (Product Ecology Consultants) for the Ministry of the Environment of the Netherlands and it is a powerful tool, useful for aggregating the outputs of an LCA. The methodology groups environmental impacts in three macro-categories of damage: impact on human health, on the quality of ecosystems and on the consumption of natural resources. Figure 4 captures the environmental impact of the solar power plant on the basis of these macro-categories.

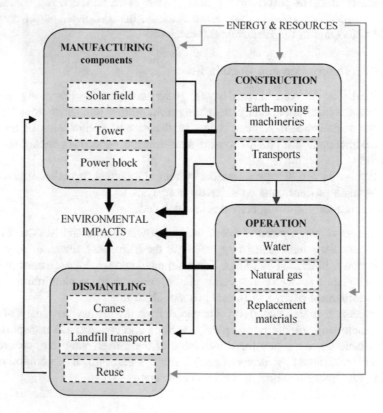

Fig. 2. System boundaries.

4 Preliminary Results

Analysis of results (Fig. 3) shows that the most critical phase in terms of environmental impact is that concerning the assembly of the different power plant components; next in terms of importance is the dismantling process. For end-of-life dismantling, it has been assumed that 50% of the materials of the various components can be recovered and reused to build another power plant, while the remaining 50% will be sent to landfill. The environmental impact of the construction process relative to the plant's entire life cycle is negligible, as it will only involve the use of earth-moving machinery and transport vehicles. On the other hand, the power plant operation and maintenance stages will have a modest environmental impact. In particular, operation of the plant involves the consumption of a certain amount of water to feed the steam cycle of the power unit and of natural gas when the solar component is unable to supply sufficient power. Figure 4 shows the results expressed in terms of damage; the main component involved is that of resources. The calculated $CO_{2-eq.}$ emissions are about 44.7 g/kWh. In the literature, the figure falls within a range of 11 to 50 g/kWh, except for Uchiyama, whose analysis yielded 213 g/kWh. With regard to potential acidification calculated in terms of $SO_{2eq.}$, the value obtained is 0.8 g/kWh, while the figure in the literature is in a range of 0.098 to 0.107 g/kWh. Lastly, Fig. 5 presents the result in the form of single individual scores.

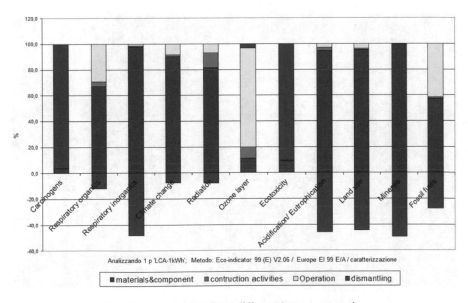

Analizzando 1 p 'LCA-1kWh'; Metodo: Eco-indicator 99 (E) V2.06 / Europe EI 99 E/A / caratterizzazione

■ materials&component   ■ contruction activities   □ Operation   ■ dismantling

Fig. 3. Assessment of the different impact categories

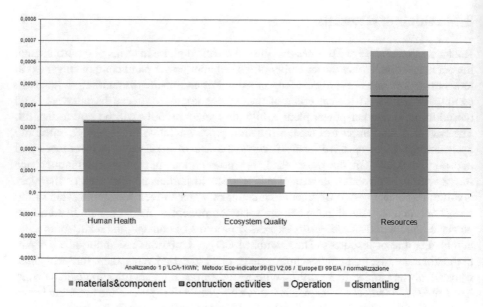

Fig. 4. Damage assessment

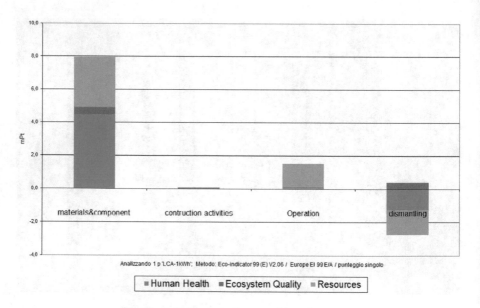

Fig. 5. Individual single score for each process

5   Conclusion

As shown by the data presented above, the overall environmental impact produced by the entire life cycle of a thermal solar power plant using solar tower technology, is very low and largely insignificant compared with the impact of traditional fossil fuel power plants. For each kWh of electricity generated, emissions of $CO_{2\ eq.}$ are about 44 g, while the acidification potential calculated in terms of $SO_{2eq.}$ is 0.8 g. The results obtained are quite encouraging and deserve further study; in particular, comparison with other thermodynamic solar technologies is planned.

References

1. Weinrebe, G., Böhnke, M., Trieb, F.: Life cycle assessment of an 80 MW SEGS plant and a 30 MW PHOEBUS power tower. In: Proceedings ASME International Solar Energy Conference Solar Engineering, Albuquerque, USA, pp. 417–424 (1998)
2. Burkhardt III, J.J., Heath, G., Turchi, C.: Life cycle assessment of a model parabolic trough concentrating solar power plant with thermal energy storage. In: Proceedings ASME 4th International Conference on Energy Sustainability, Phoenix, Arizona, USA, pp 599–608 (2010)
3. Piemonte, V., Falco, M.D., Tarquini, P., Giaconia, A.: Life Cycle Assessment of a high temperature molten salt concentrated solar power plant. Sol. Energy **85**(5), 1101–1108 (2011)
4. Kuenlin, A., Augsburger, G., Gerber, L., Maréchal, F.; Life cycle assessment and environomic optimization of concentrating solar thermal power plants. In: Proceedings of the 26th International Conference on Efficiency, Cost, Optimization, Simulation and Environmental Impact of Energy Systems, ECOS 2013, Guilin, China (2013)
5. Whitaker, M.B., Heath, G.A., Burkhardt, J.J., Turchi, C.S.: Life cycle assessment of a power tower concentrating solar plant and the impacts of key design alternatives. Environ. Sci. Technol. **47**(11), 5896–5903 (2013)
6. Klein, S.J.W., Rubin, E.S.: Life cycle assessment of greenhouse gas emissions, water and land use for concentrated solar power plants with different energy backup systems. Energy Policy **63**, 935–950 (2013)
7. Corona, B., San Miguel, G., Cerrajero, E.: Life cycle assessment of concentrated solar power (CSP) and the influence of hybridising with natural gas. Int. J. Life Cycle Assess. **19**(6), 1264–1275 (2014)
8. San Miguel, G., Corona, B.: Hybridizing concentrated solar power (CSP) with biogas and biomethane as an alternative to natural gas: Analysis of environmental performance using LCA. Renew. Energy **66**, 580–587 (2014)
9. Asdrubali, F., Baldinelli, G., Scrucca, F.: Comparative life cycle assessment of an innovative CSP air-cooled system and conventional condensers. Int. J. Life Cycle Assess. **20**(8), 1076–1088 (2015)
10. Corona, B., Ruiz, D., San Miguel, G.: Environmental assessment of a HYSOL CSP plant compared to a conventional tower CSP plant. Procedia Comput. Sci. **83**, 1110–1117 (2016)
11. Lalau, Y., Py, X., Meffre, A., Olives, R.: Comparative LCA between current and alternative waste-based TES for CSP. Waste Biomass Valoriz. **7**, 1509–1519 (2016)

12. Ehtiwesh, I.A.S., Coelho, M.C., Sousa, A.C.M.: Exergetic and environmental life cycle assessment analysis of concentrated solar power plants. Renew. Sustain. Energy Rev. **56**, 145–155 (2016)
13. Telsnig, T., Weinrebe, G., Finkbeiner, J., Eltrop, L.: Life cycle assessment of a future central receiver solar power plant and autonomous operated heliostat concepts. Sol. Energy **157**, 187–200 (2017)
14. Lamnatou, C., Chemisana, D.: Concentrating solar systems: Life Cycle Assessment (LCA) and environmental issues. Renew. Sustain. Energy Rev. **78**, 916–932 (2017)
15. Batuecas, E., Mayo, C., Díaz, R., Pérez, F.J.: Life Cycle Assessment of heat transfer fluids in parabolic trough concentrating solar power technology. Sol. Energy Mater. Sol. Cells **171**, 91–97 (2017)
16. Cavallaro F., Ciraolo L.: A life cycle assessment (LCA) of a paraboloidal-dish solar thermal power generation system. In: Proceedings 1st International Symposium on Environment Identities and Mediterranean Area, ISEIM, Corte-Ajaccio, France, pp. 260–265 (2006)
17. Concentrated solar thermal power - Now! - ESTIA SolarPaces, Settembre 2005
18. Cavallaro, F.: La tecnologia solare a concentrazione: tra innovazione e mercato (in italian). Energia **4**, 48–55 (2014)
19. Pitz-Paal, R., Dersch, J., Milow, B.: ECOSTAR: European Concentrated Solar Thermal Road-Mapping; Roadmap Document (WP 3 Deliverable No. 7) (2005)
20. Ávila-Marín, A.L.: Volumetric receivers in solar thermal power plants with central receiver system technology: a review. Sol. Energy **85**(5), 891–910 (2011)

Off-site Retrofit to Regenerate Multi-family Homes: Evidence from Some European Experiences

Alessia Mangialardo[1](✉) and Ezio Micelli[2]

[1] University of Padua, 35122 Padua, Italy
alessia.mangialardo@dicea.unipd.it
[2] University IUAV of Venice, 30135 Venice, Italy

Abstract. The research of new ways to make cities more sustainable has become a central theme in the international agenda. The focus is on retaining, refurbishing and recycling the existing components of the city, converting the current linear economy model into a circular one in the urban sphere. Many economic sectors are moving straightforward in this direction, but construction industry does not seem to be focused on this goal. Representing one of the main responsible of waste production and energy consumption, the construction industry has to undertake a radical change of perspective along with a new social, cultural, and economic interest to regenerate the existing city in a sustainable way. The aim of the research is to highlight how to industrialize the construction industry in a sustainable way. This can be performed through off-site retrofit interventions, where all components are realized in the factory and assembled in situ, guaranteeing at the same time high performance, long lasting interventions and a big aesthetic variety. The paper critically analyses some European case studies through which it is possible to consider and to evaluate the relevance in terms of building components and the business model.

Keywords: Deep-retrofit · Construction industry · Off-site interventions
Industrialization construction industry

1 Introduction

The recovery of the existing asset, the reuse of brownfields, the containment of land use and the energy performance improvement of buildings are just some of the objectives that actual public policies aim to achieve to make sustainable territories. If the new agenda seems widely shared at national and international level, less obvious is how the construction industry works to ensure the reuse of existing cities that are sustainable and long lasting [1, 2].

The great complexity and variety of interventions for the reuse of the existing building heritage make the technologies and therefore the costs related to each operation highly changeable. In response to a demand that drastically changed its availability of expenditure as a result of the economic crisis, retrofits and recovery of the existing building usually have the same costs as the new constructions. For these

© Springer International Publishing AG, part of Springer Nature 2019
F. Calabrò et al. (Eds.): ISHT 2018, SIST 101, pp. 629–636, 2019.
https://doi.org/10.1007/978-3-319-92102-0_68

reasons, the themes related to the innovation of retrofit technologies play a fundamental role in renewing the construction sector [3, 4].

Itard et al. [5] identified the two main barriers to successful sustainable retrofitting of residential buildings. The first one is represented by the lack of knowledge: the current knowledge of technical solutions is now obsolete and it is necessary that the construction industry innovate the retrofit technologies by improving the level of efficiency and productivity. The second one is represented by the economic feasibility of these interventions. The current retrofit solutions are not cost effective, there are funding problems or the investors do not profit from the investments [6].

These two elements concerning the construction industry are strictly linked to the economic efficiency of processes. To achieve this purpose is necessary a competitive jump only possible thanks to the new 4.0 manufacturing. The advent of 4.0 technologies opened new perspectives for the construction industry. Off-site industrialization increases the energy efficiency of buildings, generating better returns in terms of economic and environmental sustainability, becoming also an enormous opportunity in the construction value chain.

The aim of the paper is to analyze the economic feasibility conditions of off-site retrofit operations, identifying and measuring their value drivers. The increasing demand for these processes show that they are a great opportunity to develop the global energy efficiency also improving the construction industry, demonstrating these processes are able to create new economic and environmental value on obsolete buildings.

To examine in detail how these retrofit operations take place, authors describes some of the most original and famous project in Europe, which are not only a prototype but also, they have been effectively carried out. These case study for their peculiar features, can be considered representative of a large faction of the existing obsolete real-estate stock in Italy and Europe.

The paper is divided into four parts. The first one highlights the keywords about the off-site constructions. The second part analyse four case studies taken by the literature.

The thirth one interprets the results of the analysis. Finally, the last parts summarize the results and reveals the future research.

2   On-site vs. Off-site Constructions: Towards a Circular Approach

The process of modular construction is quite different than on-site construction. Traditional construction is based essentially into two phases: the design and assemblage of components fixed-in-place. Traditional buildings are produced on-site: for this reason, each intervention is unique and original. The construction phase employs provisional teams and this is completely different from other production industries. The productivity in the construction industry can be implemented removing the site, not make the building exclusive, and retain workforce intact from project to project. Eliminating the site is obviously not possible, but removing as much as possible the phases on site transferring many construction processes to the factory positively increase the efficiency [7, 8].

Off-site retrofit interventions embrace modular and componentized elements that include the service of structural, enclosure and internal partition systems. Modules may be constructed from many different materials including: wood framing, cold-formed steel framing, hot rolled steel, concrete, or a combination of material assemblages. The construction sector changes into an industrial area, with many positive returns. First of all, the component systems are made with advanced technologies able to make the building energetically self-sustainable, converting the economic value of the energy bill into resources for the stock renovation [9].

Off-site interventions guarantee a great degree of predictability of the costs of interventions. The use of digital technologies - such as BIM - and the careful preliminary design of the intervention constitute the main off-site design phase and make it possible to predict with extreme precision all the phases of the retrofit operation. The off-site construction helps to reduce the costs associated with expensive unforeseen during construction and to reduce the waste of materials that are pre-assembled at the factory and no more in the pipeline.

Through the use of off-site interventions, the on-site construction phase is drastically reduced, which contributes to limiting the on-site disturbance and CO2 emissions due to the transport of materials in situ. Clearly, material waste is reduced because fall off is recaptured in the manufacturing stream.

However, off-site construction drastically contribute to the reduction in transportation energy and carbon because of workers traveling to a factory instead of the job site that change once the construction is finished [10].

Another important aspect related to off-site interventions is the necessity to make more sustainable the construction industry. Today almost the 80% of the obsolete building components at the end of their life became a waste. The negative externalities and the collective costs related to this issue are evident. In the future, the availability and price of commodities, resources and material will be increasingly scarce. To this, add the considerable costs of disposing of waste and the environmental and economic damages deriving from the high value of the materials that are squandered not being able to disassemble the elements. The linear economy model – that is the current approach by the construction industry – has becoming always more unsustainable [11–13]. The construction industry has to overturn its field of action assuming a more circular approach.

The circular model can be pursued in different ways [13]. The first one is represented by a new concept to consider the entire lifespan of a building, starting by consideration of wastes as opportunities to be transformed into a new resource. Repairing elements - and all the operations linked to this like refitting and refurbishment - are preferable to interventions such as demolition and reconstruction.

The use of reclaimed or remanufactured elements has to be encouraged to limit demand for raw resources. The second element is to design the retrofit operation differentiating the various elements that constitute a building and their different lifespans. In order to recycle and replace them, it is important that each component presents different rules that are independent each others. In this way, layers make it possible to easily identify and replace something damaged, without changing the adjacent layers. For the construction industry, this means to produce a new market for salvaged products and materials by using them for other operations.

3   Off-site Retrofit: Some European Experiences

To understand how it is possible to conceive a retrofit intervention through off-site technologies this paragraph analyses four European experiences derived by the literature [14]. These case studies are multi-family buildings built in the second half of the twentieth century that now present many issues related to an insufficient insulation, many thermal bridges and, therefore, inadequate thermal comfort. These experiences can be considered representative of many Italian building stock, expect for the property: they have just one owner that rents apartments and there are two cases of social housing. Furthermore, these cases are the state of the art in this field, representing the tip of the research and their characteristics cling perfectly to Italian building stock.

In particular, for these case studies have been researched two aspects. The first one is represented by the project strategy, that means what kind of technologies are used and what components of the buildings have been manufactured off-site. Table 1 shows the solution that each case study adopted to improve the thermal comfort and to refurbish the buildings. The facades have been distinguished into two typologies: those with integrated heating systems (the most employed) and the others having modular elements of thermal insulation included ventilation ducts. Only three case studies have rebuilt the roof through modular components and solar thermal system. Finally, into two cases were built new pre-fab balconies.

Table 1. Project strategy for each case study analyzed. Source: elaboration by authors.

	Pre-fab facade with modules for new insulation and new windows	Pre-fab facade with modules for new insulation and new windows and integrate thermal systems	Modular roof	Balconies
Zug (SW)	x		x	x
Zurich (SW)		x	x	x
Roosendaal (NL)	x		x	
Dieselweg 3- 19, Graz (A)		x		

More in detail, Table 2 shows the typology of the installed plants. The totality of the case studies presents a mechanical ventilation system (installed in facade) and a solar thermal system (installed in the roof or in the external façade). On the contrary, the thermal system is in majority geo-thermal (installed in situ) and just the case of Roosendal presents a condensing boiler.

All the interventions have provided for the use of natural materials and components: in some cases, existing materials have been reused, in others natural and completely recyclable elements have been employed, pursuing circular optics. All the façades have been designed in the factory (this is the longer phase of the project) and once they are

Table 2. Technologies employed to innovate the buildings. Source: elaboration by authors

Low enthalpy geothermal plant with heat pump		Geothermal system with water-water heat pump	Condensing boiler	Mechanical ventilation system	Photovoltaic system	Solar thermal with accumulation	Rainwater recovery
Zug (SW)	x			x	x	x	x
Zurich (SW)	x			x	x	x	
Roosendaal (NL)			x	x		x	x
Dieselweg 3-19, Graz (A)	x			x		x	x

definitively completed the components arrive in situ to be assembled. In environmental terms, all the interventions contributed to drastically reduce the energy bills.

The second element that was been analysed concern the costs and the business model adopted to financially support retrofit interventions. Table 3 examine the total costs of the retrofit operations and the business model for each case study. The business model was distinguished into was calculated considering the total costs for the refurbishment and the new construction compared to the increased rents and the non-cost of the energy bill - thanks to NZEB interventions are almost 0 - that continue to pay by tenants as in the pre-intervention phase.

In the first experience, Zug, to cover the costs of the refurbishment was necessary to increase the existing volume with 300 sqm, to add the current energy bill to the rent and to increase the rent more than the 30%. In the second case study, in Zurich, there has been a minor increase in volume but a considerable increase in the rent and the energy bill that was included in the rent. The sustainability of the interventions depends on the type of technologies employed. If in Rosendaal - that represents a low-cost intervention - the cost of retrofit was equal to 25.000 euros due to the lack of increase in volume and installation of the elevator and thanks to the use of the condensation boiler which is much cheaper than geothermal solutions, other projects have realized techno- logical components more complex and expensive, but at the same time with a greater respect for the environment. On the contrary, in Rosendaal, there wasn't an increased volumetric adding and all the costs were recovered through the rent – that was increased of 0,3% - and the costs of the energy bill. Finally, in the last case study were used many performative prefab technologies (like the heating pipes that were inserted in XPS board that are mounted on the existing walls). In this case the retrofit costs were about 812 Euro/sqm and amount to a total of 8.8 milion of Euro and, being a social housing, they were totally financed by public funds. Despite the consumption of bills has decreased by 93%.

Table 3. Business model for the retrofit operations of the case studies. Source: elaboration by authors

	Total costs of the refurbishment	Business model
Zug (SW)	2,5 milion of Euro	+3 apartments + 30% rent
Zurich (SW)	1.285.000 Euro	+1 apartment +40% rent
Roosendaal (NL)	25.000 Euro/apartments	Energy bill included in the rent +0,3% rent
Dieselweg 3-19, Graz (A)	8,8 milion of Euro	+22 apartments Public funds

4   The Great Variety of Off-site Interventions: An Interpretation

The are many benefits by retrofit experiences through off-site interventions that were analysed in the previous paragraph. In environmental terms, more than the 90% of the energy consumption were reduced. This means a drastic drop in CO2 production and greater environmental comfort for tenants. From an aesthetic point of view, the buildings have been completely renovated with the most disparate external facade. The off-site interventions, in fact, while maintaining the principles of prefabrication, are able to design prototypes that can be modified easily according to the needs of the client, guaranteeing variety and originality to each individual project [8–14].

Off-site retrofit interventions represent a valid alternative to on-site constructions because, thanks to the use of modularity, the construction in several leyers and the use of recyclable materials respond better than on-site technologies to the principles of circularity. Nevertheless, the great variety of technologies employed in these case studies, if on the one hand it is a strong element because it avoids the homologation of the prefabrication which has made these practices unsuccessful in the 60s and 70s, on the other it shows that the industrialization of the construction industry is still at an early stage and it needs further refinement to become a consolidated practice.

The analyzed case studies highlight, from a technological point of view, that the ability to retrofit existing buildings through the off-site interventions are numerous and vary according to the characteristics of the buildings and the availability of investor spending. As the case of Rosendaal it is possible to produce low-cost technologies and ready to use. These components, although less environmentally friendly than others that are more sophisticated, they are cheaper and do not require to provide for planivolumetric rewarding to get to pay-back reasonable timeframe. These technologies are in line with the current needs of the middle-class market which often does not have sufficient financial resources to enable the traditional retrofit processes. In these cases, the hybridization with the energy business is a very interesting perspective [9].

On the contrary, the most sophisticated and pioneering technological experiences, even if they are more respectful of the environment, require increasing the volume and the intervention of external financing - for the most part of public type - to reach reasonable pay-back times. In this case, the Government totally funded the retrofit interventions, without asking a higher rent to the users. In this case the total cost in sqm of the renovation, including the new volume, is about 800 Euro/sqm.

Although they are more economical than on-site retrofit processes, the possibilities of retrofitting buildings using off-site technologies are different and for this reason there is great variety of costs. Off-site interventions require that the construction industry will be able to find technologies with a great capacity for adaptation to the great heterogeneity of existing residential building types. In order to minimize costs and increase at the same time the level of productivity is necessary that constructive components are produced in large-scale series. The experiences analyzed show that the off-site interventions are often not yet enough competitive: the technological paradigm upon which the off-site construction is still at a preliminary stage. It is necessary to implement and generate easily replicable prototype for the great diversity of existing buildings to be retrofitted in order to generate economies of scale and learning that reduce costs and consequently the payback period, guaranteeing, at the same time, performative intervention.

5 Conclusions

Off-site interventions represent the new frontier for innovating the construction sector, both from the point of view of new constructions and from the recovery of the existing building. When understood and organized by all the stakeholders through early design phase, it is a well-matched solution to control project schedules and budgets. At the same time, it is possible to increase the quality and reduce the environmental issues related to interventions.

Nevertheless, there are still problems that limit the use of off-site construction. The prefabricated façade panels arrive from the factory to the building site in their real size: often the structural bulk of the components is considerable and requires transport restrictions, limiting the module and the dimensions of the panel.

Another important aspect is to find solutions to manage the collective costs that characterize the majority if Italian residential buildings, that make retrofit operations more difficult to afford.

References

1. Johnson, M., Hollander, J., Hallulli, A.: Maintain demolish, re-purpose: policy design for vacant land management using decision models. Cities **40**, 151–162 (2014)
2. Power, A.: Does demolition or refurbishment of old and inefficient homes help to increase our environmental, social and economic viability? Energy Policy **36**, 4487–4501 (2008)
3. Addis, W., Schouten, J.: Design for Deconstruction: Principles of Design to Facilitate Reuse and Recycling, 1st edn. CIRIA, London (2004)
4. Kohler, N., Yang, W.: Long-term management of building stocks. Build. Res. Inf. **34**(3), 287–294 (2007)
5. Itard, L., Meijer, F., Vrins, E., Hoiting, H.: Building renovation and modernisation in Europe: state of the art review. OTB, TU Delft, Delft (2008). http://www.erabuild.net. Accessed 11 Dec 2017

6. Micelli, E., Mangialardo, A.: Recycling the city - new perspective on the realestate market and construction industry. In: Bisello, A., Vettorato, D., Stephens, R., Elisei, P. (eds.) Smart and Sustainable Planning for Cities and Regions, Springer International Publishing AG, Cham (2017)
7. Canesi, R., Antoniucci, V., Marella, G.: Impact of socio-economic variables on property construction cost: evidence from Italy. Int. J. Appl. Bus. Econ. Res. **14**(13), 9407–9420 (2016)
8. Smith, R.E., Rice, T.: Permanent modular construction. University of Utah, Salt Lake City (2015). http://www.modular.org/Html-Page.aspx?name = foundation_order_report. Accessed 11 Dec 2017
9. Munkhof, J., Erck, R.: A house makeover paid for your energy bill. Responsabilitè Environnement **78**, 85–88 (2015)
10. Quale, J., Eckelman, M.J., Williams, K.W., Sloditskie, G., Zimmerman, J.B.: Construction matters. J. Ind. Ecol. **16**(2), 243–253 (2012)
11. Andersen, M.S.: An introductory note on the environmental economics of the circular economy. Sustain. Sci. **2**(1), 133–140 (2007)
12. Braungart, M., McDonough, W.: Cradle to Cradle: Patterns of the Planet. Vintage books, London (2009)
13. Cheshire, D.: Building Revolutions - Applying the Circular Economy to the Built Environment, 1st edn. Riba Publishing, London (2017)
14. IEA. Prefab Systems for Low Energy/High Comfort Building Renewal, Stuttgart: Fraunhofer IRB Verlag, Annex 51, IEA: Prefab Systems for Low Energy/High Comfort Building Renewal, Stuttgart: Fraunhofer IRB Verlag, Annex 51 (2013)

Economic Value Assessment of Forest Carbon Sequestration and Atmospheric Temperature Mitigation in the Metropolitan City of Reggio Calabria (South Italy)

Fortunato Alfredo Ascioti, Vincenzo Crea, Giuliano Menguzzato, and Claudio Marcianò[✉]

Mediterranea University of Reggio Calabria, 89100 Reggio Calabria, Italy
claudio.marciano@unirc.it

Abstract. Forest CO_2 sequestration and effective atmospheric temperature mitigation, two important aspects of the global warming mitigation strategy, are evaluated for the forested area of the Metropolitan City of Reggio Calabria (South Italy). Carbon Stock in different living (above- and below-ground biomass) and non-living (litter, necromass, and soil) pools, along with Carbon Flux, is estimated using a ground-based method that relies on National Forest Inventory data. The results are compared with those analogously estimated in a Northern Italian area (Trento Province) and both the similarities and the differences found in the way carbon gets into the diverse C-pools are discussed. C-Flux estimated values are also reported and discussed. An analysis of the qualitative composition of our Southern forest stands indicates that they are dominated by broadleaf trees, which are capable of effective atmospheric temperature mitigation. The contribution to the mitigation of global warming *via* CO_2 sequestration, although relatively low, is nevertheless noteworthy (15.000 Gg of C). Moreover, the economic values of this environmental asset are remarkably high, even when a low C-emission-trading market price is considered. Therefore, there is an urgent need to protect this forested area from negative human impacts (pollution, illegal cutting, and fires), while developing sustainable ways of exploiting it (e.g. by FSC-certification and Carbon Credits).

Keywords: Climate change · Ecosystem service
Carbon stock and flux estimates · Metropolitan cities · Environmental asset

1 Introduction

The ongoing climate change and the consequent global warming are the effects of a phenomenon known as Dangerous Anthropogenic Interference (DAI) with the Earth's climatic system [1]. It is likely that we have even skipped an expected glacial era just because of our Green House Gasses (GHGs) emissions during the past and the present century [2, 3]. The release in the atmosphere of these GHGs, primarily CO_2 (as clearly shown in [4]) and various aerosols, if kept at the present rate, may lead to a 1,000 years of climate change. This change would be irreversible if the emission of these GHGs does not cease within a relatively short time horizon (ca. three decades) [5].

© Springer International Publishing AG, part of Springer Nature 2019
F. Calabrò et al. (Eds.): ISHT 2018, SIST 101, pp. 637–644, 2019.
https://doi.org/10.1007/978-3-319-92102-0_69

To avoid this dangerous drift of our planetary climate system (but irreversible does not necessarily mean unavoidable [6]), and in order to keep the global temperature increase, due to this climate "mess up", well below 2 °C, a three-levers strategy has been proposed [7]: (i) the carbon neutral lever to achieve zero net emissions of CO_2; (ii) the SP (Super Pollutant) lever to mitigate short-lived climate pollutants; and (iii) the third lever, i.e. CO_2 extraction from the source (e.g. a coal power plant) and/or its sequestration from the air to reduce the CO_2 atmospheric blanket. Forests (along with oceanic phytoplankton [8]), thanks to their photosynthetic activity, are efficiently sequestering CO_2 from the atmosphere [9, 10]. Hence, they contribute to the (third lever) mitigation of temperature increase, specifically if their "qualitative" composition is also appropriate to this end, i.e. they have more broadleaf tress than conifers (as stressed in the work of Naudts et al. [11]).

In this paper, we consider the extensive forest area of the Metropolitan City of Reggio Calabria (in South Italy, i.e. the Metropolitan Area of Reggio Calabria, henceforth abbreviated as MARC) and the Ecosystem Service that this green-area provides: CO_2-sequestration, along with an effective atmospheric temperature mitigation. Therefore, we have estimated the carbon stock and carbon flux that these forest ecosystems produce within the perimeter of the MARC, and show how the composition of the trees in these same forests is well suited for effective atmospheric temperature mitigation [11]. Finally, we computed the economic value of this MARC "environmental asset" by considering both the present and the expected (in the case of worsening global climate change) average carbon prices.

2 Methods

In order to estimate C-Stock and C-Flux in the MARC forests, we used a *ground-based method* whose computation relies on National Forest Inventory (NFI) data [12, 13]. This approach is by no means free from inaccuracies and uncertainties [14], but none of the available methods for C-Stock and C-Flux estimates (e.g. remote sensing via MODIS satellites or Eddy covariance estimates) are 100% reliable [15, 16].

The C stocked in the living biomass pool is estimated according to the method reported in [17]. That is to say, the Above-Ground Biomass (AGB, woody tree dry matter, d.m., in Mg of d.m.) for each forest type is computed using the relation:

$$\text{AGB (d.m.)} = \text{GS} \times \text{BEF} \times \text{WBD} \times \text{A} \qquad (1)$$

where

- GS = the volume of growing stock in m^3/ha (from NFI 2005 [13]),
- BEF = the Biomass Expansion Factor which transforms GS into above-ground woody biomass (from Table 6.4 of [17] using the values for Stands, since there is only a negligible difference among Stand, Coppice, and Plantation values for the different forest types),
- WBD = Wood Basic Density, which converts fresh volume to dry weight, in Mg/m^3(we get the values from the same Table 6.4 of [17], as for the BEF),

– A = the area of forest occupied by a specific forest type, in ha (NFI, 2005 [13]).

The Below-Ground Biomass (BGB, in Mg d.m.) is computed as follows:

$$BGB \ (d.m.) \ = \ GS \ \times \ BEF \ \times \ WBD \ \times \ R \ \times \ A \tag{2}$$

where:

– R = Root/Shoot ratio which converts growing stock biomass into BGB, for each forest type, while the other variables have the same meaning as above (in Eq. 1).

The R values were obtained from Table 6.5 of [17] (again using the values for Stands, since there is only a negligible difference among Stand Coppice, and Plantation values for the different forest types).

We assume a default conversion factor of Mg d.m. biomass to Mg of carbon content (Mg C) equal to 0.47, according to [17].

Carbon in non-living pools, i.e. litter, necromass, and Organic Carbon (OC) in organic and mineral soil horizons, is estimated as follows:

$$Litter \ C \ (Mg \ C) \ = \ litter - C \ content \ (Mg \ C/ha) \ \times \ A \tag{3}$$

where:

– the litter-C content was derived for each forest types from NFI data (2005 [13]); and
– A = the area of forest occupied by a specific forest type (in ha).

$$Total \ Necromass \ (Mg) \ = \ LN \ + \ FFN \tag{4}$$

$$LN \ (Mg) \ = \ Large \ Necromass \ Mg/ha \ \times \ A \ (ha) \tag{5}$$

$$FFN \ (Mg) \ = \ Fine \ Fraction \ Necromass \ Mg/ha \ \times \ A(ha) \tag{6}$$

where the values of LN and FFN, in Mg/ha, as well as the area values A (in ha), are derived from the NFI per each forest type [13]. Total C in the necromass (for each forest type) is, thus, the Total Necromass Mg $\times$ 0.47.

As for the soil OC, the Mg C/ha data available in the NFI (2005 [13]) for both organic and mineral soil horizons (for each forest type) times A (the area of forest occupied by a specific forest type, in ha) gives the Soil OC content in Mg C for each soil horizon. The sum of these two quantities (i.e. OC in the organic + OC in the mineral soil horizons) gives the Total SOC (Soil OC in Mg C) in our forests.

We also computed the average values of Mg C/ha for each C-pool, and for the total area of forest in the MARC (Table 1: see the Results section).

Once we had estimated the Carbon Stock in biophysical units (those given in Table 1: see the Results section), we obtained the Economic Values (EV, in €) of the carbon stored in the forests of the MARC in the following way:

$$EV_{CS} \ = \ CS \ \times \ P \tag{7}$$

where:

- CS = Carbon Stock, in Mg C (from Table 1: see the Results section);
- P = unitary prices, in €/Mg C.

We chose two unitary prices, P: (i) a minimum of € 4.0/Mg C; and (ii) a maximum of € 53.0/Mg C. The former minimum value is derived from the recent average price for carbon quoted on the C-emission-trading market [18], while the latter maximum price could likely be achieved, according to [19], in the case of a future worsening of climate change (i.e. it gives a "future"-projected possible economic asset).

Carbon Flux (CF, in Mg C/year) can be estimated as follows [20]:

$$CF = CI \times BEF \times WBD \times A \times 0.47 \tag{8}$$

where:

- CI = the Above-Ground Current Increase in m^3/ha/year for each forest type (from NFI 2005 [13]),
- A, BEF, and WBD have the same meaning (and dimensions) as in Eqs. (1) and (2), for each forest type (source NFI 2005 [13]).

Considering the same carbon unitary prices as above (i.e. min € 4.0/Mg C, Max € 53.0/Mg C), and applying a similar equation as (7) we get the Economic Values (EV, in €/year) of C-Flux in the forests of the MARC. That is to say:

$$EV_{CF} = CF \times P \tag{9}$$

where:

- CF = Carbon Flux, in Mg C/year (for each forest type);
- P = unitary prices, in €/Mg C.

3   Results

The "Metropolitan City" of Reggio Calabria (MARC) has an area of 3.210 km^2 and includes 97 municipalities. Its territory is very complex with extremely heterogeneous areas, including natural zones of high biodiversity and even a National Park (NP) (i.e. the Aspromonte National Park). The presence of an NP within the perimeter of a Metropolitan Area (MA) represents a unique case in Europe and probably worldwide (Fig. 1). The forests of this MA cover a surface of 108,493 ha, and this makes the MARC rank 5th, among the 14 Italian MAs, in terms of forested area, and 1st for the extent (96.8%) of non-urbanized area.

The areas of forest considered here (those available in the Italian National Forests Inventory, NFI 2005 [13]) cover 69,773 ha, i.e. 64% of the entire forest area (108.493 ha) of the MARC. 82% of these forest trees are broadleaves while conifers represent the remaining 18% (Fig. 1). Carbon Stock and Carbon Flux biophysical quantities (in Mg of C) for the 64% of the MARC forests (computed according to the previously described methods) are reported in Table 1 (below). The Total living

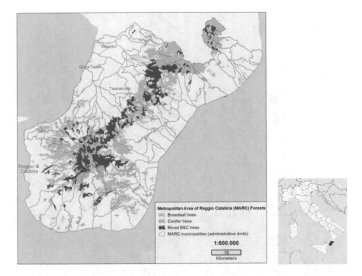

Fig. 1. Forests in the metropolitan area of Reggio Calabria

biomass C (Above-Ground Biomass C + Below-Ground Biomass C) in Mg C/ha is comparable to that found in a Northern area of Italy (i.e. Trento Province) [21].

Organic Carbon (OC) stored in non-living biomass shows lower values (Mg C/ha) for litter and necromass (as large and fine fractions), and higher values for soil (both organic and mineral horizons) when compared to the above-mentioned Norther Italian region (Trento Province) [21]. Consequentely, the total Mg of C stocked per ha is approximately equal in both regions, and this gives us more confidence in our C-Stock estimates.

There is a simple possible explanation for the differences found in the way carbon is allocated in the diverse C-pools when the Northern Trento area is compared with the Southern MARC, i.e. the differences in the rate of litter (and also necromass) decomposition [22]. Both the predominance of broadleaf trees in, and the relatively higher temperatures of our Southern area (the MARC) may well cause a higher decomposition rate of dead organic matter [22], and therefore the lower litter and necromass Mg C/ha values and higher soil Mg C/ha content that we found in our area (MARC) compared with the values of the Northern area (the Province of Trento [21]).

The C-Flux estimates (in Mg C/ha/year) obtained using the current annual incre-ment in Eq. (8) should represent the stand-level (volume) increase in the absence of natural disturbances and management practice [23]. However, the low value per ha we obtain here (i.e. 2.16 Mg/ha/year, Table 1) indicates either that this kind of compu-tation gives an underestimate of Net Primary Production (NPP) or that it rather mea-sures the Net Biome Production (NBP): that is to say, an estimate of NPP minus the heterotrophic respiration and the net losses due to harvest and natural disturbances (e.g. fires) [23]. This latter C-Flux result needs further investigation.

Transforming the biophysical C quantities into economic values (considering the C-market prices discussed above), we get the results shown in Table 2. It can be observed that the values found for the MARC in terms of both C-Stock and C-Flux are

Table 1. C-Stock and Flux in 64% of the forests in the MARC

	Beeches	Oaks	Chestnut	Holms	Med. Pines	Larch Pines	White Spruces	Totals	Total per ha
Surface ha	22,014.00	7,835.00	11,194.00	16,044.00	373.00	10,074.00	2,239.00	**69,773.00**	
Carbon Stock (all values in Mg C)									Mg C/ha
AGB - C	2,626,557.97	357,904.31	604,012.93	939,185.66	22,134.14	955,098.73	140,713.27	5,645,607.01	80.91
BGB - C	525,311.59	71,580.86	169,123.62	939,185.66	7,304.27	343,835.54	39,399.72	2,095,741.26	30.04
Total living Biomass C	**3,151,869.56**	**429,485.17**	**773,136.55**	**1,878,371.32**	**29,438.40**	**1,298,934.28**	**180,112.99**	**7,741,348.27**	110.95
Litter C	96,861.60	11,752.50	20,149.20	83,428.80	484.90	40,296.00	8,508.20	261,481.20	3.75
Necromass (total) C	33,109.06	4,050.70	48,402.86	14,327.29	753.83	23,200.42	3,051.76	126,895.91	1.82
Soil (organic + mineral) OC	3,145,800.60	459,131.00	689,550.40	1,397,432.40	36,292.90	992,289.00	312,340.50	7,032,836.80	100.80
Total OM pools C	**3,275,771.26**	**474,934.20**	**758,102.46**	**1,495,188.49**	**37,531.63**	**1,055,785.42**	**323,900.46**	**7,421,213.91**	106.36
Total C stock	**6,427,640.82**	**904,419.37**	**1,531,239.00**	**3,373,559.81**	**66,970.04**	**2,354,719.70**	**504,013.45**	**15,162,562.18**	217.31
Carbon Flux (Mg C/year)									Mg C/ha/y
Carbon Flux	54,934.55	7,817.47	28,199.40	29,128.14	668.15	25,453.51	4,608.28	**150,809.50**	2.16

Source: our elaboration

Table 2. Economic values of C-Stock and Flux in 64% of the forests in the MARC

Economic values of C/Mg	EV × Total C-Stock (€)	EV × C-Flux (€/y)
min 4.00 €/Mg	60,650,249	603,238
Max 53.00 €/Mg	803,615,796	7,992,903

remarkably high, even when the min. unitary carbon price (€ 4.0/Mg) is used in Eqs. (7) and (9).

Moreover, these values account for 64% of the total forest area of the MARC. Therefore, even higher economic values are reasonably to be expected for the whole forested area (ca. 36% more, i.e. approximately € 22 million of C-Stock and € 217,000/year of C-Flux to be added to the above Table 2 minimum price values).

4 Conclusions

Forest carbon sequestration and effective atmospheric temperature mitigation can be considered important parts of the strategy aimed at keeping the global temperature increase, due to our Dangerous Anthropogenic Interference with the planetary climate system [1], "well below 2 °C" [7]. The contribution that the MARC forests make to this strategic global plan, although relatively small, is nevertheless noteworthy. 64% of the forests of the MARC contribute about 15,000 Gg of stored carbon, and gives a C-flux of 2.16 Mg/ha/year over a surface of ca. 70,000 ha. This suggests that the forests of the MARC would provide ca. 8% more stored carbon than that provided by the Province of Trento (i.e. 45,000 Gg of C over 346,000 ha [21]). Moreover, the predominance of broadleaf trees in the MARC forests should contribute more effectively to the mitigation of climate warming than the trees in the Northern forests which are dominated by conifers, as stressed in the previously quoted work [11]. The economic value of these MARC forest ecosystem services, emerging from our analysis, stresses the urgent need to protect this profitable "environmental asset" from negative human impacts (pollution, illegal cutting, and fires), while exploiting it in sustainable ways, e.g. with FSC-certification and carbon credit acquisition by public and private owners of MARC forest stands. The possible ways in which these latter goals may be achieved will be explored in future applied research.

Aknowledgments. This study has been supported by GASTROCERT, a Project on Gastronomy and Creative Entrepreneurship in Rural Tourism, funded by JPI Cultural Heritage and Global Change - Heritage Plus, ERA-NET Plus 2015-2018. It provides a new application of the methodology developed in the PAN LIFE project, LIFE13NAT/IT/00107.

References

1. Smith, J.B., et al.: Assessing dangerous climate change through an update of the Intergovernmental Panel on Climate Change (IPCC) "reasons for concern". PNAS **106**(11), 4133–4137 (2009)
2. Crucifix, M.: Earth's narrow escape from a big freeze. Nature **529**, 162–163 (2016)

3. Ganopolski, A., Winkelmann, R., Schellnhuber, H.J.: Critical insolation–CO_2 relation for diagnosing past and future glacial inception. Nature **529**, 200–203 (2016)
4. Feldman, D.R., et al.: Observational determination of surface radiative forcing by CO_2 from 2000 to 2010. Nature **519**, 339–343 (2015)
5. Solomon, S., et al.: Irreversible climate change due to carbon dioxide emissions. PNAS **106** (6), 1704–1709 (2009)
6. Matthews, H.D., Solomon, S.: Irreversible does not mean unavoidable. Science **340**, 438–439 (2013)
7. Xu, Y., Ramanathan, V.: Well below 2 °C: Mitigation strategies for avoiding dangerous to catastrophic climate changes, PNAS early edition 1–9 (2017)
8. Sabine, C.L., et al.: The oceanic sink for anthropogenic CO_2. Science **305**, 367–371 (2004)
9. Pan, Y., et al.: A large and persistent carbon sink in the world's forests. Science **333**, 988–993 (2011)
10. Fernández-Martínez, M., et al.: Atmospheric deposition, CO_2, and change in the land carbon sink. Nature Sci. Rep. **7**(9632), 1–12 (2017)
11. Naudts, K., et al.: Europe's forest management did not mitigate climate warming. Science **351**, 597–600 (2016)
12. Gasparini, P., Tabacchi, G. (eds): L'Inventario Nazionale delle Foreste e dei serbatoi forestali di Carbonio INFC 2005. Edagricole-Il Sole 24 ore, Bologna (2011)
13. SIAN. http://www.sian.it/inventarioforestale/jsp/pubbl_scient.jsp. Accessed 5 Sept 2017
14. Guo, Z., et al.: Inventory-based estimates of forest biomass carbon stocks in China: a comparison of three methods. For. Ecol. Manage. **259**, 1225–1231 (2010)
15. Nicolini, G., et al.: Impact of CO_2 storage flux sampling uncertainty on net ecosystem exchange measured by eddy covariance. Agric. Forest Meteor. **248**, 228–239 (2018)
16. Petrokofsky, G., et al.: Comparison of methods for measuring and assessing carbon stocks and carbon stock changes in terrestrial carbon pools. How do the accuracy and precision of current methods compare? A Syst. Rev. Protoc. Env. Evid. **1**(6), 1–21 (2012)
17. Romano, D., et al.: Italian Greenhouse Gas Inventory 1990-2015. National Inventory Report 2017, pp. 224–239 (Vitullo, M. and Fioravanti, G.). ISPRA Rapporti 261 (2017)
18. Economist. www.economist.com/blogs/freeexchange/2015/12/schr-dinger-semisionstraing-system. Accessed 5 Sept 2017
19. Ding, H., Nunes, P.A.L.D., Teelucksingh, S.: European Forests and Carbon Sequestration Services: An Economic Assessment of Climate Change Impacts. ESE Working Paper Series. UNEP-DEPI paper no. 9 (2011)
20. Schirpke, U., Scolozzi, R., De Marco, C.: Modello dimostrativo di valutazione qualitativa e quantitativa dei servizi ecosistemici nei siti pilota. Parte1: Metodi di valutazione. Report Progetto LIFE-Making Good Natura. EURAC-research, Bolzano (2014)
21. Pilli, R., Grassi, G., Cescatti, A.: Historical analysis and modeling of the forest carbon dynamics using the carbon budget model: an example for the Trento Province (NE, Italy). Forest@ **11**(1), 13–28 (2014)
22. Zhang, D., et al.: Rates of litter decomposition in terrestrial ecosystems: global patterns and controlling factors. J. Plant Ecol. **1**(2), 93–95 (2008)
23. Pilli, R., et al.: The European forest sector: past and future carbon budget and fluxes under different management scenarios. Biogeosc. **14**, 2387–2405 (2017)

Montalto di Castro - Sustainable Tourism as an Opportunity for Urban and Environmental Regeneration

Maria Rita Schirru[✉]

La Sapienza University of Rome, 00185 Rome, Italy
mritaschirru@gmail.com

Abstract. The goal of this paper is to carry out an in-depth analysis of the tourism sector in Montalto di Castro, aimed primarily at identifying issues that have led to the current critical situation in tourist services and facilities, with a view to devising a project of sustainable tourism development and enhancement that highlights both its existing strengths and defines new strategic assets, through both tradition and innovation. Therefore, after analysing the area of Montalto di Castro, potential enhancement strategies will be prepared, to implement through projects in line with established objectives. In particular, the enhancement project will focus on strengthening and upgrading the whole environmental system made up of the River Fiora and the surrounding area. This will take place also through the systemisation of historical artefacts found throughout the territory and its wooded areas, by developing a set of outdoor trails and cultural routes for sports and recreational activities. The project will improve access and usability of the River Fiora for both tourists and local residents through the introduction of environmental recovery and upgrading processes and the creation of walking and cycling paths in line with a vision of sustainable tourism.

Keywords: Sustainable tourism · Outdoor/cultural routes
Urban and environmental regeneration

1 Sustainable Tourism and Themed Routes

In the late Eighties, the concept of sustainable tourism evolved following the introduction of a more general concept of "sustainable development" put forward in the Burtland Report in 1987, with a view to conserving environmental heritage for future generations [1]. It is fundamentally a type of "low-impact" tourism, which views and uses the environment as a precious resource, relying on core values to attract tourists and highlighting the need to preserve it by accepting necessary limitations [2].

Sustainable tourism involves parks, protected natural areas and particularly important environmental areas, where a necessary balance between protection and use must be found by limiting environmental impact to a minimum and promoting pragmatic but sensible socio-economic development.

"Walking tourism" would appear to be the best solution for the most valuable environmental areas. This is a type of tourism not necessarily "for all" as it is aimed at

© Springer International Publishing AG, part of Springer Nature 2019
F. Calabrò et al. (Eds.): ISHT 2018, SIST 101, pp. 645–653, 2019.
https://doi.org/10.1007/978-3-319-92102-0_70

people with a significant awareness of environmental issues who visit mainly from cities of a certain size [2]. The benefits of walking tourism are clear to most and include: physical activity in the open air, close contact with the environment, and access to archaeological, natural, and ethnographic sites and activities.

Walking tourism develops mainly along walking routes or trails, supported by hoteliers and the like who offer assistance to the traveller from their departure station to the next stop, through the management of baggage and local transfer services.

Within the context of walking tourism, the rediscovery of the "path", which some might claim tourism itself is based on, takes centre stage: it can easily be turned into a "themed" itinerary, thus becoming an opportunity to rediscover both one's cultural heritage and local landscapes.

An example of a success story of a network of walking routes is the "Path of Love" (Strada dell'Amore) initiative in the Cinque Terre area, where natural-cultural itineraries have actively contributed to the steady increase of tourist numbers over a wider territory, stimulating both the local economy (farmhouse hotels, restaurant services, trade, arts and crafts, agricultural production, etc.) and an interest in local history.

The growing demand for walking tourism has led many European countries to promote policies at various levels aimed at protecting and marketing natural-cultural itineraries: France exemplifies perfectly the enhancement of this type of walking route, in the mountains and the flatlands, as well as along its coasts. The coast, in fact, represents a meeting point between the marine and land environment and constitutes a unique landscape containing a rich and fascinating ecosystem.

In Italy, we can also point to significant examples in a similar vein: in the Belluno Prealps, approximately 250 km of walking, mountain biking and horse trails have been created with clearly registered details, featuring well-equipped rest areas (also for horses), restaurants, hotels, farmhouse hotels, shelters, local hamlets, architecture and art.

On the Island of Elba, an important initiative has seen the setting up of 80 itineraries for excursions on foot or by bicycle, managed by local organisations and cooperatives that also offer guided tours. In marketing terms, these natural-cultural itineraries have often been split into themes ("Napoleon's Path", "The Path of the Mines", "The Wine Trail", etc.) where each trail narrates the present and reconstructs history from a specific viewpoint.

Thus, both in Italy and abroad, various kinds of so-called "themed itineraries" have been developed - some surprisingly imaginative - and have subsequently become a key motivation for visitors and an effective marketing tool for the growth of tourism. Some notable examples include: botanical itineraries with signs that give details about featured species (common in France and Switzerland); historical-cultural routes (like the "The Wine Route" near Friborg); astronomy routes (such as "The Planet Walk" in Santa Cristina in Val Gardena); and paths of natural discovery (such as the trail on Lake D'Allos and the Port Cros submarine trail, both in France).

In 2017, as part of its management role in the enhancement of natural and cultural paths and trails, the Italian Ministry for Cultural and Environmental Heritage (MiBACT) defined "cammini" or trails as «cultural itineraries of particular national and/or European importance, which can be followed on foot or with other forms of so-called "soft mobility" and also represent an effective tool for conveying their natural and cultural heritage, as well as being an opportunity to promote related natural and cultural attractions» [3]. The Ministry has also begun promoting the "Digital Atlas of the Trails of Italy", a collection of routes and trails devised and created following the guidelines and criteria set out by the ministerial directive issued by the Trails Committee, made up of MiBACT, regional and provincial representatives.

Also the Lazio Region, through its approval of law no. 2/2017 called "Provisions for the construction, maintenance, management, promotion and enhancement of the trail network in Lazio" has set itself the goal of enhancing and promoting a network of major European cultural itineraries (such as the Via Francigena) along routes which, due to their historical, religious, cultural and territorial importance, are deemed worthy of protection in law [4]. This includes historically important roads, as well as hiking trails in protected natural areas and age-old pilgrimage routes (where a basilica is included).

In this context, the Lazio town of Montalto di Castro, in the Province of Viterbo, with its wealth of cultural resources and landscape features (such as the archaeological site of Vulci, the Litorale NW at the mouth of the River Fiora and the Pian dei Cancani), has undeniable potential to develop and support this kind of sustainable, low-impact tourism through trails and walking routes, as we will discuss below.

2 A Territorial Analysis of Montalto di Castro

History

Montalto di Castro was most probably founded in the fifth century by the inhabitants of a coastal town who moved inland to this small hill in order to defend themselves from the raids of the Saracens who poured into the coasts from the eighth to the twelfth century.

Montalto was first recorded in the history books as "Montis Alti" in 853 in a papal bull issued by Pope Leo IV to Virobono, the bishop of Tuscania. In the papal document, it says that the "castrum Montis Alti" belonged to the Tuscania Diocese, for which it also served as a port and a border towards both the sea (crawling with Saracens) and towards Tuscany, occupied by the Longobards.

For centuries, the territory of Montalto was devastated by the struggle against the Popes and their allies by the powerful Vico barons, who had been feudal lords of Montalto since the mid-twelfth century. This long battle between the Popes and the Vico barons led to the almost total destruction of Montalto Castle and the surrounding countryside, as well as the abandonment of houses, displacement and consequent depopulation.

After the defeat of the Vico lords in 1359, the Castle of Montalto started to change hands: from the Orsini family, to Angelo Ventura (also known as Tartaglia) and to other lords and Popes.

With the establishment of the Duchy of Castro, in 1537, created by Pope Paolo III Farnese on behalf of his son Pier Luigi, Montalto (and the other areas included in the duchy) was able to benefit from a period of relative peace until its destruction in 1649 at the hands of Pope Innocent X. The Duchy of Castro was repossessed by the church and joined other assets controlled by the apostolic chamber and the unity of the territory was ended. It was instead granted, on a leasing basis, for a short period to several different lords, thus preventing the introduction of any significant architectural intervention. In 1870, the power of the popes definitively ended and Montalto became part of unified Italy.

Demography

In the 1950s, the local authority of Montalto di Castro registered considerable demographic growth (+44.13% from 1951 to 1961) following the recovery by the Ente Maremma which resulted in the settlement of families belonging to economically and culturally diverse groups, coming from neighbouring towns as well as other areas in Italy. Another significant consequence was the development of seaside tourism, coupled with a growth in second home ownership. In the following decade, from 1961 to 1971, the population only grew by 1.15%. Since the Seventies and Eighties, however, the population has grown considerably, mainly because of the construction of a thermoelectric power station which brought new residents to the area (see Table 1).

With the completion of the power station in 1989, many of the people who had been employed in its construction lost their jobs and moved elsewhere but, despite this, the local population continued to rise even after the power station was up and running (see Table 1).

Despite the power station's production being downsized between 2004 and 2006 (due to the impact of renewable energy and a general reduction in demand caused by the economic crisis), the population grew by 12.74% between 2001 and 2011 (see Table 1) and, to a lesser degree, also in the following period (2011–2016), by 2.39%.

The local authority area currently consists of 8,985 inhabitants located in the three main populated areas: the capital Montalto, Pescia Romana and the tourist town of Marina di Montalto.

It is worthy of note that in the last twenty-five years this growth trend (21.39%) has been significantly higher than that found in the Province of Viterbo (12.69%), in Lazio (12.85%) and also in Italy as a whole (6.29%).

Montalto di Castro stretches over an area of 189.4 km^2 and has a population density of 47.44 inhabitants per square kilometre, again significantly lower than the provincial (88.25), regional (342.86) and national (201.11) figures.

Foreigners officially resident in Montalto di Castro in 2016 represented 13.21% of the general population and have increased in the last twenty-five years by 97.21%, a slightly higher increase when compared to the Province (94.99%), the Region (90.82%) and to Italy as a whole (92.89%). The vast majority of foreigners in these figures are Romanian (63.3% of the whole foreign population), followed by Albanians (10.3%) and then Tunisians (4.5%).

Looking at the breakdown of the population by age group in Montalto, in the period from 1991 to 2016, children up to the age of 14 represented 12.68% of the general population but have declined in the last twenty-five years by almost four percentage points, in line with the provincial (down by 3%), regional (down almost 2%) and national decreases (down more than 2%). The number of people aged between 15 and 64 also fell by almost 6%, standing at only 64.43% of the wider population from 70.42% in 1991, again in line with the decreases registered provincially (almost 3%), regionally (more than 5%) and nationally (almost 5%). At the same time, the number of people over 65 increased by almost 10% points, representing 22.89% of the wider population and up from 13.31% in 1991, in line with provincial (up almost 6%), regional (up about 7%) and national changes (up almost 6%).

The old age index, which compares the number of over-65s per 100 individuals to the 0–14 age group, has gone from 81.80% to 180.60% in the last twenty-five years, highlighting an underlying upward trend in the general age of the population (more than double in the period in question) and higher than equivalent figures for the province (75.27%), the region (63.53%) and the country (68.76%) [5]. This means that in Montalto in 1991 there were 82 people over 65 for every 100 young people (aged 0–14), while in 2016 there were 181.

The structural dependence index, which measures individuals in non-active age for every 100 individuals of active age, has in the last twenty-five years gone from 42% to 55.21%, showing an increase of 23.92%, higher again than the provincial (11.30%), regional (21.53%) and national increase (18.83%) [6]. In other words, in Montalto there is a tendency towards the growth in the number of children and elderly that the "active" population must look after: in 1991, there were 42 children and elderly for every 100 active individuals, while in 2016 the number of non-active individuals per 100 active is 55.

The elderly percentage, which measures the number of elderly people compared to the total population, has increased over the last twenty-five years from 13.31% to 22.89%, an increase of 41.87% which is higher than its provincial (24. 69%), regional (33.23%) and national equivalent (31.39%), demonstrating that Montalto is ageing faster than its context of reference [7].

Table 1. Population of Montalto di Castro from 1871 to present

Total figures								
1871	**1951**	**1961**	**1971**	**1981**	**1991**	**2001**	**2011**	**2016**
700	3.411	6.105	6.176	6.604	7.063	7.653	8.770	8.985
Percentage increases								
	1871-51	**1951-61**	**1961-71**	**1971-81**	**1981-91**	**1991-01**	**2001-11**	**2011-16**
	79,48	44,13	1,15	6,48	6,50	7,71	12,74	2,39

The economy and employment

From a comparison between the active population and how many of them are actively employed in the various sectors of the economy, the dominance of the tertiary sector emerges (56.90%); followed by industry (23.88%) and agriculture (19.30%) (see Table 2). The agriculture and tertiary sector, in the period from 1991 to 2011, increased its collective number of employees, while industry lost manpower.

The industrial sector, based above all on service companies dependent on the power station, is currently in a severe state of crisis which has led local government agencies and departments to research and devise proposals for the decommissioning and sustainable conversion of the power station and area for the benefit of sectors like tourism (including accommodation), entertainment and culture.

Furthermore, since 2009, the site adjacent to the plants has been home to one of the largest Italian solar panel plants (built by SunPower).

The building work for the power plant, during the 1980s and early 1990s, absorbed up to 20% of the active population, as well as providing business for many small and medium-sized companies spread out throughout the Viterbo area. While the work was still ongoing, the local authority signed an agreement with ENEL (pursuant to the ministerial decree from 1988) which provided for the allocation of "exceptional and unrepeatable" funds for "environmental and territorial rebalancing". However, at the end of the building work, employment levels fell critically which led to the local authority of Montalto di Castro being declared a crisis area in 1997 (pursuant to Law 662/96 and the CIPE Resolution on 21 March 1997). In 2000, an "Area Contract" for Montalto di Castro and Tarquinia was signed, which was a negotiating tool designed to promote new production initiatives able to take on the previously employed but now available workforce, as well as starting new socio-economic development in the area.

The employment rate between 1991 and 2011 increased by more than 4% points and was higher than the equivalent provincial, regional and national levels – this becomes an almost 7% increase when we focus just on youth employment, again higher than the provincial, regional and national figures.

The unemployment rate fell by almost 8% points over the 1991–2011 period and was higher than the equivalent rate for province, region and country while youth unemployment fell by almost 25% points over the same period and was below provincial, regional and national levels.

Local numbers increased in all the periods considered, although always less than the previous period; while employee numbers went up across all periods, except from 1961 to 1971 (−0.36%) and 1991–2001 (−80.09%) (see Table 3).

Table 2. % incidence of employment in various sectors (agriculture, industry, other)

	1991			2001			2011		
	Agr.	Ind.	Oth.	Agr.	Ind.	Oth.	Agr.	Ind.	Oth.
Montalto di Castro	28,10	25,10	46,80	21,00	25,80	53,20	19,30	23,80	56,90
Province of Viterbo	26,20	31,50	42,30	9,60	24,70	65,70	8,20	20,10	71,70
Lazio	3,95	22,20	73,85	3,40	21,99	74,61	2,96	16,58	80,46
Italy	7,64	35,65	56,71	5,50	33,48	61,02	5,55	27,07	67,38

Table 3. Local numbers and employee numbers of Montalto di Castro

Total figures						
Years	**1961**	**1971**	**1981**	**1991**	**2001**	**2011**
LN	286	309	432	527	661	748
EN	827	824	1.404	3.618	2.009	2.304
Percentage increases						
Periods	**Var. %** **1951-61**	**Var. %** **1961-71**	**Var. %** **1971-81**	**Var. %** **1981-91**	**Var. %** **1991-01**	**Var. %** **2001-11**
LN	46,85	7,44	28,47	18,03	20,27	11,63
EN	45,10	−0,36	41,31	61,19	−80,09	12,80

Tourism
Montalto is fundamentally a seaside destination for tourism and over the years has seen a consistent increase in numbers, showing a marked seasonal trend with peaks in numbers during the summer months. Recently, cultural tourism has begun to develop thanks to the potential the area offers in terms of natural beauty, landscapes and archaeological heritage. One of the main attractions is in fact the archaeological area of Vulci, followed by the old hamlet of Montalto di Castro and other emerging historical-artistic sites, especially inland.

In 2016, there were 40,171 visitors in Montalto (14.17% of whom were foreigners) for a total of 191,306 nights stayed (14.30% due to foreign bookings), ranking second only to Fiumicino, both in terms of single visitors and total nights stayed, in a comparison with other local authorities (Santa Marinella, Ladispoli and Fiumicino). The average residing time totalled 5 days, ranking first compared to these other Lazio local authorities.

Residential tourism facilities available in Montalto - 8 hotels and 32 non-hotel accommodation types in 2016 - are insufficient to meet tourist demand in the area. The one 4-star hotel, five 3-star hotels and two 2-star hotels (there is no 5 star hotel) offer a total of 551 beds, while the non-hotel facilities offer 2,456 beds. In addition, there are five campsites and twenty-two farmhouse hotels (agriturismi) which together offer the highest number of available beds.

The area features a high number of second homes which satisfy most of the seaside demand. Since the 1980s, empty homes exceed those that are occupied and currently total 57% of total homes.

Overview of the SWOT Analysis
The strength of Montalto lies undoubtedly in its culture and rich landscapes. This is also demonstrated by the fact that two local sites (namely the Litorale NW at the mouth of the River Fiora and the Pian dei Cancani), considered important for the community at large, were put forward for inclusion in the 2000 Nature Network. However, adequate infrastructural access to these sites and facilities for those who wish to visit do not currently exist to satisfy emerging tourist demand.

Given its cultural wealth and the beauty of its landscapes, there exists a real opportunity to develop sustainable nature tourism in Montalto di Castro, linked to a system of routes and trails centred around the River Fiora and dedicated to sports and recreational activities, which can then be added to the MiBACT Digital Atlas "Cammini d'Italia" (Trails of Italy) and the Lazio Region's network of paths in order to be able to take advantage of the available funding.

A threat to consider is the potential competition with neighbouring towns with similar features, which actively compete for the attention of tourists, as well as the emergence of coastal tourism which has, however, a significant environmental impact. Other considerable threats include the rise in the old age index, the elderly percentage and the structural dependence index to which we may add an unknown factor related to the future of the power station and the success of its conversion.

3   A Proposal for Sustainable Tourism for Montalto di Castro: The River Fiora as a Natural "Knowledge Infrastructure"

Montalto di Castro sits within a group of tourist destinations that after periods of development and consolidation are now going through a so-called "mature phase" (or mid-life crisis), a critical moment that sees the destination lose some of its appeal. From this stagnation comes the need to implement a process of transition from undifferentiated tourism to more diversified forms of supply where a qualitative emphasis is placed on archaeological and landscape features. This has become necessary because, on the one hand, tourism demand is increasingly unpredictable, varied, competitive, flexible and sector-specific and, on the other, because its tourism services and facilities have not been able to adapt to ongoing changes. The strategy for tourism in Montalto has, in fact, placed its main emphasis on coastal development, supported by its two marinas (Marina di Montalto and Marina di Pescia Romana) and by the growth in second homes, without attempting to adapt or to diversify (this is however happening to an extent in sectors like culture). Other resources in the area (the archaeological area of Vulci, the old hamlet of Montalto di Castro and other emerging areas of historical-artistic interest, especially away from the coast) have not as yet been the focus of a strategic project aimed at so-called "4E tourism" (i.e. Environment and clean nature, Educational tourism, Event and mega events, Entertainment and fun).

Finding itself at this "mature", stagnant stage, Montalto faces two possible scenarios: decline or renewal. Decline would occur if stagnation continued over time, if it was no longer able to attract new visitors, if investment fell, if the feeling of distrust or resentment towards tourists grew and if other, new destinations took away custom. A period of renewal, however, would aim instead to improve accommodation facilities and infrastructure, to launch a new marketing campaign and to better identify target markets.

On the back of an analysis of Montalto di Castro and its surrounding territory, the aim of this proposal is to identify possible local enhancement strategies which can be implemented through schemes designed on the basis of clearly defined positioning objectives.

A development strategy for the Montalto tourism system must seek to include all the key local elements which will most likely become qualifying features of a fresh tourist plan.

To sum up within the general frame of reference, it can be said that the development in and around Montalto should base itself on the relationship between local archaeology, its natural features and landscapes, and the maritime and coastal area.

In this context, the River Fiora represents a primary support infrastructure, a link between the inland area and the coast and a path which includes all of the above features: landscape (the banks and its flora), nature (the woods and areas which have been recognised for their importance within the community) and archaeology (specific, contained sites like Vulci and extended sites with heritage of notable value).

The place where the River Fiora flows into the sea can and must represent a focal point to effectively reposition and re-evaluate its worth and role in the redevelopment of the whole river mouth/coastal area, with the goal of finding a balance between coastal and other environmental features.

A strategic and legislative opportunity arises in the context of the imminent reform in the regulation of beach licences which, by incorporating the relevant European directive, will allow the reorganisation of the current system in favour of a more balanced one.

In Montalto, this new focus can allow the inclusion of new additions to the Plan for Sandy Beach Use (PUA) that enhance and redefine poorly used parts of the coastline as areas with an alternative vocation to swimming or traditional beach activities. This can be implemented emphasising a strong, close relationship with the River Fiora which will then become a source of knowledge about the inland area and a sort of "knowledge highway" focusing on the whole area up to the mouth of the river.

The parts of the coastline not designed for swimming or sunbathing could be identified by future PUAs as arrival points for themed routes or trails where, in addition to general tourist facilities, information would be provided on further exploration of the area via a network of walking or cycle paths (to be included also, as previously mentioned, in the MiBACT Digital Atlas of Italian Trails and the Lazio Region's Lazio trail or route network in order to take advantage of dedicated funds).

The project will improve the usability of the River Fiora for both tourists and local residents through the setting up of both environmental recovery and redevelopment initiatives, as well as the creation of walking and cycle paths (or trails) in the context of sustainable tourism.

References

1. Word Commission on Environment and Development: Report of the World Commission on Environment and Development: Our Common Future, (1987). http://www.un-documents.net/our-common-future.pdf. Accessed 2 Dec 2017
2. Di Meo, A.: Il marketing dell'ambiente e della cultura. Lupetti, Milano (2002)
3. Ministero dei Beni e delle Attività Culturali e del Turismo: Direttiva del Ministro dei Beni e delle Attività Culturali e del Turismo "2016-Anno dei cammini d'Italia", Roma, (2016). http://www.camminilazio-focalpoint.org/cammini/direttiva-del-ministro-dei-beni-e-delle-attivita-culturali-e-del-turismo/. Accessed 12 Dec 2017
4. Legge Regionale Lazio 10 marzo 2017 n. 2, "Disposizioni per la realizzazione, manu-tenzione, gestione, promozione e valorizzazione della rete dei cammini del Lazio"
5. The old age index represents the percentage ratio between over-65 s and the 0–14 population
6. The structural dependency ratio represents the percentage ratio of the non-active age popula-tion (0–14 and 15–24) and those in active age (15–64)
7. The elderly percentage shows the percentage ratio between over-65s and the general population

Evaluation of Benefits for Integrated Seismic and Energy Retrofitting for the Existing Buildings

Benedetto Manganelli, Monica Mastroberti, and Marco Vona$^{(\boxtimes)}$

University of Basilicata, 85100 Potenza, Italy
marco.vona@unibas.it

Abstract. Currently, a significant part of the residential building stock has been built without modern provisions for earthquake resistance and energy efficiency, resulting in high seismic vulnerability and very expensive management and environmental cost due to low energy performance buildings. An integrated approach to seismic and energy retrofitting is need to reduce the effects seismic events and energy consumption. In this work, the economic effects of the seismic events are evaluated for existing Reinforced Concrete buildings are considered. The energy performances of buildings are also evaluated. Then based on an integrated approach, the seismic and energetic Expected Annual Loss is evaluated. It is obtained combining both the cost of the seismic risk consequence and energy cost. Finally, the economic feasibility is based on the global Economic Feasibility Index.

Keywords: Existing RC buildings · Integrated retrofitting approach
Feasibility analysis

1 Introduction

Recent earthquakes in European seismic prone countries have highlighted the fragility of existing buildings while their big losses in energy and consequent environmental impact are always more evident. Moreover, both energy and seismic topics are responsible of high economic losses resulting from high operational energy consumption and from post-earthquake reconstruction interventions [1].

The building stock typically consists of multi-story residential buildings. In terms of volume and inhabitants, the Reinforced Concrete (RC) frame structures are the main type. They are often characterized by poor architectural features, obsolete equipment and finishing, and on the other hand by non seismic system and details [2].

Based on raised drastically energy prices for building (since 1990), in last decades the intervention strategies for existing buildings has been usually focused on the reduction of the energy consumption. The economic impact and convenience for owners are linked to the energy performances, as detailed by the European Directive [3]. It is to be highlighted that only energy retrofitting is unable to provide extension of structural service life of buildings. Moreover, the structural seismic safety level is not considered. On the contrary, if only energy retrofitting is carried out it could be an

© Springer International Publishing AG, part of Springer Nature 2019
F. Calabrò et al. (Eds.): ISHT 2018, SIST 101, pp. 654–662, 2019.
https://doi.org/10.1007/978-3-319-92102-0_71

increase of expected seismic losses due to seismic damage. In this study, an integrated approach to assess the seismic and energy performance of RC existing buildings is proposed.

2 Proposed Methodology to Evaluate the Economic Feasibility

The trend of depreciated cost of an existing building during its life is reported in Fig. 1. If at give year an integrated intervention is applied, there will be a recovered in depreciated cost and increase of the residual economic life of building (Fig. 1a). On the contrary, after seismic event there will be a significant depreciated cost due to the seismic damage (Fig. 1b). This cost could be the building over its life economic life. In this study, only the benefit on depreciated cost of seismic intervention has studied. As preliminary step, the economic return of an integrated seismic plus energy retrofitting have evaluated only based on a financial approach, considering the payback time.

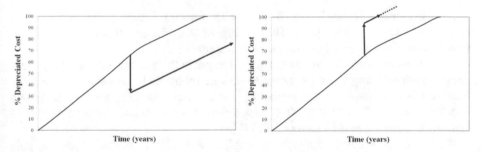

Fig. 1. Trend of depreciated cost of existing building with retrofitting (a, on the left) or after seismic event (b, on the right).

The methodology is based on the Pay Back Time concept. It is the time at which the return of initial investment will be achieved or in other terms the number of years after which the cost of the intervention is fully amortized by its benefit [4]. In the case of integrated seismic and energy retrofitting the benefit are the total prevented seismic losses after future seismic event and the total prevented energy consumption during the residual life of building after the retrofitting. The Pay Back Time of an integrated approach (PBT_{S+E}) is evaluated based on the correspondence between the total benefit and the cost of the integrated solution, as reported in Fig. 2.

The red curve describes the cumulative benefits provided by the integrated approach considering the residual building life and the retrofitting intervention. It is obtained summing the seismic and energy retrofitting benefit (green and yellow curves respectively).

The economic feasibility depends to the relation between the payback time and the maximum acceptable return time for monetary investment (IT). In this study, it is setting to 25 years. PBT is defined based on the retrofitting cost or, in dual way, in

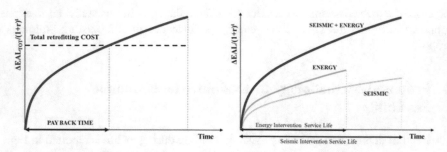

Fig. 2. Evaluation of benefit using the proposed integrated approach.

terms of seismic damage repair cost and energy cost. Thus, an Economic Feasibility Index (EFI) to measure the economic feasibility has been defined as:

$$EFI = 1 - \frac{PBT_{S+E}}{IT_{Tot}} \tag{1}$$

The EFI index is ranging in 0–1, moving forward 1 the economic feasibility improve, while if the Pay Back Time is higher than the investment one it is not economic feasible. In this case the index is negative.

The economic scenarios are useful and widespread tools in seismic risk mitigation. They allow forecasting about the expected seismic economic losses on a wide range of possible seismic intensity. From economic point of view, the integration of economic losses scenarios with hazard seismic ones results in quantification of seismic risk in terms of Expected Annual Losses (EAL) [5].

In this study, a novel approach has applied. It is able to evaluate the economic effectiveness of combined seismic and energy losses. It is the first step to evaluated the benefits for integrated seismic and energy retrofitting for the existing buildings. In Fig. 3, the flowchart summarizes the approach.

The Expected Annual Losses due to seismic damage is a synthetic parameter to evaluate the seismic risk level. The reduction in EAL can evaluated as: $\Delta EAL_S = EAL_{S,as\,built} - EAL_{S,retrofitted}$. It measures the economic return for owners based on the reduction due to retrofitting strategy.

The energy interventions can be evaluated in terms of energy expected annual losses: $\Delta EAL_E = EAL_{E,as\,built} - EAL_{E,retrofitted}$. Consequently, the integrated seismic and energy retrofitting can be evaluated as the total reduction in expected annual losses: $\Delta EAL_S + \Delta EAL_E$. In this work, only the $EAL_{as\,built}$ is evaluated.

The procedure to evaluate the seismic Expected Annual Losses as built ($EAL_{as\,built}$) is based on Fragility Curves (FCs) and Repair Cost Functions (RCs). The definition of FCs and repair cost functions could be solved according to the analytically or experimental approach. In this study, the analytical approach has been employed to defined the FCs and the repair cost functions in existing condition. Analytical FCs and repair cost functions convolution lead to evaluate the seismic direct economic losses.

In order to use the seismic direct economic losses in terms of EAL, the economic scenario needs to be replaced with annual probability of seismic intensities. It is possible

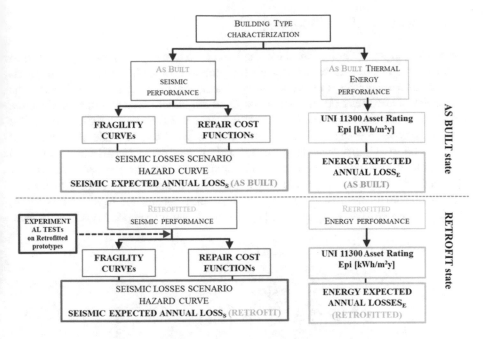

Fig. 3. Simple flowchart of the proposed procedure.

based on the relationship between seismic intensity parameter and corresponding annual frequency.

About the energy performance, the heating has been considered as the main energy use in buildings. The annual heating requirement has evaluated based on the required heating, according to the thermal balance. Geometry building and thermo-physical properties are the methodology input data. The evaluation has performed based on some hypothesis: the windows total surface, exposure of windows surface, the thermal bridge and so on. Based on this simple procedure, the energy performance index is evaluated in Kilowatt for hour/square meter for year. The building as it built is considered. Thus, the annual expected cost has evaluated for integrated approach.

The expected annual losses due to seismic damage in a given time period (in general the residual life cycle of existing building after retrofitting intervention) is a useful synthetic parameter to communicate the seismic risk level with stakeholders, insurance companies, governmental agencies etc.

To evaluate the economic benefits due to integrated approach, the current value of energy and seismic losses avoided in service life play a fundamental role. In general, a modern energy intervention has a service life about of 25 years, after that the system must be replaced. On the other hand, according to seismic code, a seismic rehabilitation should guarantee an increment of building life of 50 years. However, an increment in residual building life equal to 25 years is more reasonable.

3   Application of Methodology

The residential RC buildings are the considered type to carry out a first application of methodology. In order to better explain the proposed procedure and to highlight its accuracy and reliability, the proposed approach has been applied considering the assessment of economic direct seismic losses of the widely used Reinforced Concrete with Moment Resistant Frames (RC - MRF) Italian and European building types. The considered types are the well know and studied Pre Seismic Code (PC) and Old Seismic Code (OC) types. Several dimensional characteristics (in plan and elevation) are considered and several infill panel distributions are also considered. The studied types and seismic performances are widely described the previous studies [2, 6]. More explanations about numerical analyzes of types are outside the topic of this work, developing the procedure reported in [7]. In Fig. 4, the investigated types are shown.

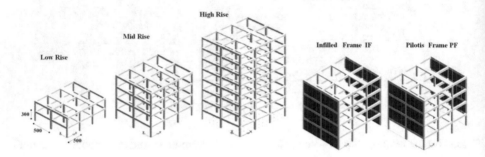

Fig. 4. Characteristics of considered types.

Based on obtained numerical results, fragility curves [8], repair cost functions, seismic direct losses scenarios, and consequent EAL values are evaluated for each types in their as built condition. The fragility curves for damage states coherent with performance states provided by actual Italian and European Seismic Code have been obtained. Thus, damage criteria directly related to ductility ratio of more critical structural elements have been employed. Fragility curves and seismic direct losses scenarios condition in Housner intensity have been built. The Housner intensity, unlike others seismic intensity parameters, is capable of appropriately representing the potential damage of real seismic events, and it is strictly correlated with engineering response damage parameters [9]. Moreover, the Housner intensity is linked with macro-seismic EMS-98 intensity scale, commonly used in seismic risk analysis. Macro-seismic scales are commonly used for scenarios on wide territorial scale.

To repair cost functions definition for each damage state (and range of seismic intensity of achievement) the numerical damage distributions have been interpreted on the basis Damage Consequence Model (DCM). The DCM for column, beam, and infill masonry panel are employed. Based on convolution operation, seismic direct economic losses scenarios have been defined. The expected annual losses values have been evaluated as the area identified by the seismic direct losses scenario but reporting in as cites the mean annual frequency of seismic intensities. Based on seismic hazard of

Italy, relationships Pga - mean annual frequency (considering a reference life of 50 years) for four seismic Italian region have been built. They result from the Italian hazard maps for the probabilities of exceedance in 50 years of 99%, 81%, 63%, 50%, 30%, 10%, 5%, 2%. The lognormal parameters of Fragility Curves for investigated types have evaluated. Finally, EAL values for each type have obtained. The EAL values highlight the influence of the seismic demands in the buildings expected annual loss. The same building type is characterized by EAL values progressively increasing with the hazard. Moreover, the EAL values are influenced by the number of storeys and the design code respectively. In Fig. 5, the procedure for seismic losses evaluation is reported.

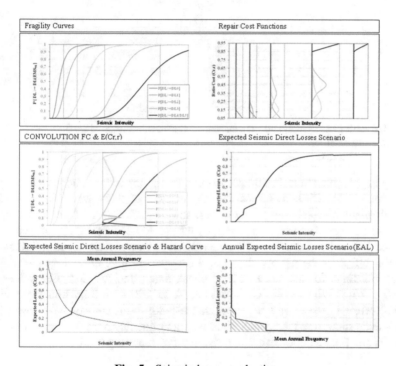

Fig. 5. Seismic losses evaluation.

The energy assessment of the building types has been performed according to the two construction details and consequent periods, identified as Pre Code and Old Code. The TABULA project [10] database has been considered to choose the technological system of the building and to assign the U-values of transmittance according to climatic zones and building and architecture tradition. Values for two periods of constructions and for different climatic zones are reported in the Fig. 6.

Days have identified according the different climatic zones, by choosing an average value that represents each city that fall into both each climatic category and a specific seismic region. The assessment of the energy performance has been developed for considered types of buildings and for the six Italian climatic zones (from A to F),

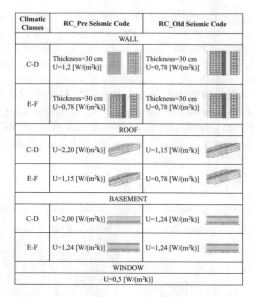

Climatic Classes	RC_Pre Seismic Code	RC_Old Seismic Code
WALL		
C-D	Thickness=30 cm U=1,2 [W/(m²k)]	Thickness=30 cm U=0,78 [W/(m²k)]
E-F	Thickness=30 cm U=0,78 [W/(m²k)]	Thickness=30 cm U=0,78 [W/(m²k)]
ROOF		
C-D	U=2,20 [W/(m²k)]	U=1,15 [W/(m²k)]
E-F	U=1,15 [W/(m²k)]	U=0,78 [W/(m²k)]
BASEMENT		
C-D	U=2,00 [W/(m²k)]	U=1,24 [W/(m²k)]
E-F	U=1,24 [W/(m²k)]	U=1,24 [W/(m²k)]
WINDOW		
U=0,5 [W/(m²k)]		

Fig. 6. Technological system and transmittance values according to climatic zones and building and architecture tradition.

resulting in 24 case study buildings. Results of the procedure show that the Energy Performance Index (EPI) increases with the number of days, which implies a higher energy demand in colder areas. However, the characteristics of buildings elements show a wide range of U-values that already takes into account the climatic differences. Moreover, it is to be highlighted the dimensional characteristics of the buildings play a significant effect in the energy demand: 4-storey buildings have a smaller EPI than 2-storey buildings for the same year of construction and location. EPI values range from $60 \div 150$ kWh/m²y for climatic zone A to $280 \div 450$ kWh/m²y for climatic zone F. Results in terms of EAL show that recently built edifices have lower losses, which is even more true if they belong to warmer climatic zones. In Fig. 7, as example the results for Bare Frame with 2 and 4storey types are reported.

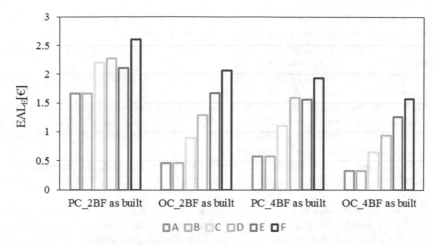

Fig. 7. Seismic and energy losses evaluation; results for: Pre seismic Code Bare Frame with 2 and 4 storey (PC_2BF, PC_4BF) and Old seismic Code (OC) Bare Frame with 2 and 4 storey (OC_2BF, OC_4BF) types.

4 Conclusion

Then, the economic advantages due to integrated solutions are evaluated and com-pared. To provide an economic assessment, risk – based seismic performance and energy performance methodologies are crucial. A new procedure for the assessment of energy and seismic economic performance of "existing" and "retrofitted" buildings has been proposed.

Based on this novel approach, the economic effectiveness of different integrated (energy+seismic) solutions have been studied on selected building types. Seismic interventions are able to provide unitary index for performance level. For energy levels, the reduction of the energy consumption of 30% and 60%, have been considered. The economic feasibility has been also studied according to the geographical location characterized by specific seismic and climatic parameters.

The most economic feasible integrated solutions have been identified. Among these, the financial benefits offered by a recent Italian law (December 11, 2016, No. 232) are further investigated.

References

1. Dolce, M., Di Bucci, D.: Comparing recent Italian earthquakes. Bull. Earthq. Eng. **15**(2), 497–533 (2017)
2. Masi, A., Vona, M.: Vulnerability assessment of gravity-load designed RC buildings, evaluation of seismic capacity through nonlinear dynamic analyses. Eng. Struct. **45**, 257–269 (2012)
3. DIRECTIVE 2010/31/EU of the European Parliament and of the Council, 19 May 2010 on the energy performance of buildings. Official Journal of the European Union L 153/13

4. Manganelli, B.: Real Estate Investing: Market Analysis, Valuation Techniques, and Risk Management. Springer International Publishing, Switzerland (2015)
5. Porter, A.K., Beck, J.L., Shaikhutdinov, R.: Simplified estimation of economic seismic risk for buildings. Earthq. Spectra **20**(4), 1239–1263 (2004)
6. Vona, M.: Fragility curves of existing RC buildings based on specific structural performance levels. Open J. Civil Eng. **4**, 120–134 (2014)
7. Mastroberti, M., Vona, M., Manganelli, B.: Novel models and tools to evaluate the economic feasibility of retrofitting intervention. In: COMPDYN 2017 6th ECCOMAS Thematic Conference on Computational Methods in Structural Dynamics and Earthquake Engineering Rhodes Island, Greece, 15–17 June (2017)
8. Masi, A., Vona, M., Mucciarelli, M.: Selection of natural and synthetic accelerograms for seismic vulnerability studies on reinforced concrete frames. J. Struct. Eng. **137**(3), 367–378 (2011)
9. Chiauzzi, L., Masi, A., Mucciarelli, M., Vona, M., Pacor, F., Cultrera, G., Gallovič, F., Emolo, A.: Building damage scenarios based on exploitation of Housner intensity derived from finite faults ground motion simulations. Bull. Earthq. Eng. **10**(2), 517–545 (2012)
10. Corrado, V., Ballarini, I., Corgnati, S.: D6.2 National scientific report on the TABULA activities in Italy (2012)

Carbon Sequestration by Cork Oak Forests and Raw Material to Built up Post Carbon City

Giovanni Spampinato[1] , Domenico Enrico Massimo[1(✉)] ,
Carmelo Maria Musarella[1,3] , Pierfrancesco De Paola[2] ,
Alessandro Malerba[1], and Mariangela Musolino[1]

[1] Mediterranea University of Reggio Calabria, 89100 Reggio Calabria, Italy
demassimo@gmail.com
[2] Federico II University of Naples, 80125 Naples, Italy
[3] University of Jaén, 23071 Jaén, Spain

Abstract. Over the last few decades, there has been widespread awareness that global warming is linked to the introduction of CO_2 into the atmosphere from the use of fossil fuels. Urban areas play a very important role in CO_2 emissions. Cork, a natural and renewable material (which in itself is the result of a storage of C) can effectively contribute to improving the quality and the insulation of buildings, reducing energy waste, preserving environment, saving landscape [20, 21], the design of the post-carbon city [12–17, 25]. The increase in the area occupied by the cork oak forests would increase the storage of carbon in a permanent way, as the use of cork does not compromise the forest resource and does not involve the introduction into the atmosphere of CO2: indeed its use in thermal insulation of buildings reduces CO_2 emissions for domestic heating and cooling. Cork oak forests take on a multi-functionality that includes economic, environmental and landscape values. Their protection requires the adoption of a territorial governance that takes into account the commitment to lower down climate change.

Keywords: Carbon sequestration · Cork for passivation · Cork oak forests
Green building · Post Carbon City

1 Introduction

Most of the scientific community is unanimous in considering that the increase of the concentration of CO2 in the atmosphere, is responsible for global climate changes observed over the last decades and in particular for increase in the Earth's temperature [1]. Responsible for this increase of CO2 in the atmosphere are the human activities related to the burning of fossil fuels and deforestation. Global changes, induced by the rapid growth of the human population and the consequent high consumption of resources, are phenomena that alter the structure and functions of the ecosystems [2]. The Intergovernmental Panel on Climate Change (IPCC) estimates that the average global temperature has increased by about 0,6 °C since 1860. On the basis of current

---

Attributions: G. Spampinato auth. Sects. 1, 2, 4., D.E. Massimo auth. Sects. 6, 7., C.M. Musarella auth. Sect. 3., P. De Paola auth. Sect. 9., A. Malerba auth. Sect. 5., M. Musolino auth. Sect. 8.

© Springer International Publishing AG, part of Springer Nature 2019
F. Calabrò et al. (Eds.): ISHT 2018, SIST 101, pp. 663–671, 2019.
https://doi.org/10.1007/978-3-319-92102-0_72

trends of emission of greenhouse gases, it is estimated a further increase of the Earth's temperature between 1,4 and 5,8 °C in the period 1990–2100. The emissions limitation and the sequestration of CO_2 excess has become a worldwide concern. The taking of the political consciousness has allowed the adoption and then the ratification of the Framework Convention on Climate Change of the United Nations in 1994. The signatory states are committed to carry out national inventories, which quantify the greenhouse gas emissions and the potential of "tanks" for the storage of carbon (C). In 1997, during the third Conference held in Kyoto, the signatories of the Protocol took the commitment to reduce greenhouse gas emissions between 2008 and 2012 by at least 5% compared to 1990 levels also by increasing the capacity of absorption of C. In the global level, C is circulating among the four major forms of tanks: hydrosphere, lithosphere, biosphere and atmosphere. Each tank (pool) becomes a source or well of CO_2 in relation to the direction that the exchange flows take with the atmosphere. A source will increase the amount of C in the atmosphere (CO_2 emissions), while a well will capture in the long term the C of the atmospheric tank (CO_2 sequestration). About half of the CO_2 emitted into the atmosphere, as a consequence of human activities, is absorbed in marine and terrestrial ecosystems: the remaining part contributes to increasing the concentration of CO_2 in the atmosphere (see Fig. 1A).

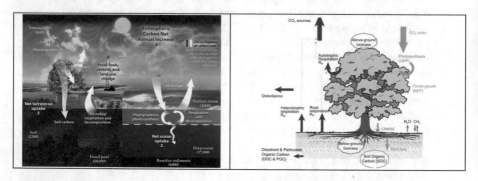

Fig. 1. A. Global Carbon cycle (Modified from http://genomicscience.energy.gov/carboncycle/index.html). B. Main Carbon stock and flux in a forest (from https://www.forestry.gov.uk).

Thanks to the photosynthetic activity (see Fig. 1B), plants use CO_2 to increase total biomass, while a small amount of C is also lost through respiration and decomposition. Soils are the main terrestrial tanks C [3]. About 75% of the total terrestrial C is stored in the soil [4] and, among these, the forest soils retain about 40% of the total C.

2   Cork Oak Forests in the Global Carbon Cycle

Understanding and quantifying the carbon and greenhouse gas balance in forests is a key part of forest management programs. In quantifying the carbon balance of the forest ecosystem and understanding how this can be influenced by climate change, all the main components (soil, trees, undergrowth vegetation) should be taken into account.

Significant influence has the type of management, the structure and functionality of the forest as well as the type of forest. For this purpose various bio-geochemical models have been prepared, such as BIOME-BGC. It is a model developed at the University of Montana [5] that is able to estimate deposits and flows of water, carbon and nitrogen in many terrestrial ecosystems considered homogeneous. The original model, although it does not consider the specific species composition, takes into account the main types of biomes: evergreen and deciduous forest, evergreen forest and deciduous broadleaf. Appropriately modified, this model can also be applied to Mediterranean forest ecosystems. According to the FAO [6], carbon stocks in Mediterranean forest ecosystems (between biomass and soil) are on average $64,9$ Mg ha^{-1}. Many forest ecosystems in the Mediterranean basin are characterized by low density. Among the most particular forest formations of the Mediterranean are the *Quercus suber* forests, with a great environmental and economic value. Like other forest ecosystems, as well as providing goods and services directly valued in the market, also generate environmental services that are essential for human survival.

3   The Cork

The cork is the result of the activity of the subero-fellodermal change, a secondary meristem present in a peripheral position in the stem and in the roots, capable, in different forest species, of giving rise to thick layers (See Fig. 2A) with protective purposes. In *Quercus suber* cork has a considerable thickness and lends itself to being picked up by skilled workers without damage to the tree. The bark removal is carried out in the late spring or early summer every 8–10 years.

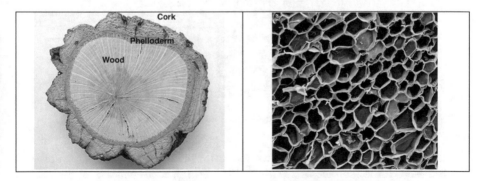

Fig. 2. A. Cross section of cork oak trunk. B Colored SEM of the cork.

The peculiar properties of cork depend on the suberification process to which peridermal cells are subjected, with the formation of a thick layer of suberin (a polyester of organic acids) which is deposited in lamellae inside the cell wall (see Fig. 2B). The suberification of the cell wall involves the death of the cell and the replacement of the protoplasm with air [7]. Agglomerates of cork are an ideal material for creating lightweight and highly insulating structures and have even been used in aerospace applications [8]. For this purpose the bottle caps for oenology could be re-used. Almost 80% of the total value of world cork production is destined for oenology for the production of corks, but increasingly large percentages are destined to the production of insulating panels for the building industry.

4   Case Study: The Cork Oak Forest of Angitola (Calabria, Italy)

The forests dominated by cork oaks in Calabria [9, 10] occupy an altimetric range variable between 50 and 700 m s.l.m. Structurally they can be monoplane, or more rarely, biplane forests. Other species such as Quercus ilex, Quercus virgiliana, Fraxinus are associated in the tree layer. The shrub layer is usually well represented by species of the Mediterranean maquis. From a phytosociological point of view, the cork oak forests present in Calabria belong to the Helleboro-Quercetum suberis association [11]. Three different types of cork oak forests have been distinguished, each of which can present different facies in relation to the intensity of the anthropic impact and the silvicultural interventions carried out: thermo-xerophile cork oak forests, meso-thermophile cork oak forests and mesophilic cork oak forests.

The cork oak forest considered as a Case Study (Fig. 3), located in the Angitola valley, on the Tyrrhenian side of the Calabria region, in the district Divisa (Municipality of Pizzo Calabro, Maierato), belong to the last type above mentioned.

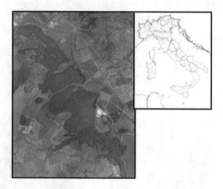

Fig. 3. Angitola cork oak forest. Aerial photo.

For the definition of the dendrometric and structural characteristics of the cork oak, transects were made (surface of 1,000 m^2, 50 * 20 m), representative of the average conditions of the populations. In transect, all the plants present starting from a minimum diameter of 2.5 cm were traced. Of each plant were found: the polar coordinates with respect to a vertex of the transect, the total height and the height of insertion of the foliage and the projection on the ground of the same in the four cardinal directions. The dendrometric analyses have shown an heterogeneity of the population (Table 1).

The amount of carbon fixed in the forest ecosystems, and the equivalent carbon dioxide (CO_2) subtracted from the atmosphere are quantified by the biomass measurement, expressed in terms of dry weight. The carbon fixed in the plant tissues (wood, leaves, etc.) constitutes in fact about 50% of the total biomass, while the equivalent in carbon dioxide is obtained by multiplying the carbon content of the biomass by the ratio between the molecular weights of the anhydride carbon and elemental carbon (44/12 = 3,67).

Table 1. Main dendrometric parameters.

Valle Angitola cork oak forest	Forest density (N. trees/ha)	Cork oak density (N. trees/ha)	Average diameter (cm)	Average height (m)	Basimetric area (m^2/ ha)	Volume (m^3/ha)	Age (years)
Cork oak forest	896	590	38	13,50	56	422	44

For forest ecosystems, estimates of absorption and emissions are made for each of the tanks provided by the Good Practice Guidance report: living biomass (aboveground biomass + belowground biomass), dead organic matter (deadwood + litter soils), soil organic matter:

$$\Delta CFF = \Delta CLB + \Delta CDOM + \Delta CSoils \tag{1}$$

ΔC_{FF} = annual variation of the carbon stock from Forest Land remaining Forest Land [t C y−1]

ΔC_{LB} = annual variation of the carbon stock in Living biomass (above and belowground) in the forest land remaining forest land [t C y−1]

ΔC_{DOM} = annual variation of the carbon stock in dead organic matter (dead wood and litter) in the forest land remaining forest land [t C y−1]

ΔC_{Soils} = annual variation of the carbon stock in soils in forest land remaining forest land [t C y−1] Forest land.

Applying this equation to the present case it has been obtained that on average the cork in question absorbs 5 t/ha/year of CO2 (50,000 kg). On the whole, the cork oak in question, covering about 150 ha, absorbs 7,500 tons of CO2 every year. If the cork oak was registered in the "National Register of Agro-Forestry Carbon Tanks", then it would receive an allowance of 20 €/tonne/year [in 2005 the cost of a tonne of CO2 fluctuated between 8 and 32 €/tonne/year] and therefore € 150/ha/year. The interesting aspect of this analysis is that the carbon stored from a cork forest will remain immobilized in the future, as the use of the forest does not involve the clearing\demolition of the trees but only the extraction or removal of the cork barks. Indeed this raw material, grinded, granulated and transformed into panels will contribute to isolated buildings and further reduce the release of CO2, due to building heating and summer air conditioning, into the atmosphere. Experiences conducted in various Italian cork oaks have shown that cork production is about 4–5 t/ha/every 10 years of dry cork (10 years are the period between a bark removal and the next). In order to evaluate cork production, 10 bark removal trials of cork plants have been carried out with a diameter equal to the average diameter.

The tests carried out have shown that a cork plant with an average diameter of 38 cm produces 11 kg of cork every 10 years. It is therefore possible to estimate a production of about 6,590 kg per hectare in 10 years (659 Mg/ha/10 years). This figure agrees with those found for other Italian regions. In Sicily, for the territory of the Hyblaean Mountains, 12 kg/plant production and 6,000 kg/ha production per hectare have been highlighted, while for the Nebrodi Mountains a production of 10 kg/plant and 5,000 kg/ha has been evaluated.

5 Cork Products for Green Buildings: KWh and CO_2 Reduction

It is important to estimate roughly the quantity of raw material which can be extracted after carrying out doing a careful bark removal every ten years. The parameters for an preliminary estimate are as follows:

- an average tree possessing an average diameter, average height and average age;
- the observed average quantity of raw material which can be bark removed from each average tree over a ten-year period;
- the average tree density per hectare.
- the size expressed in hectares of the wood which can be deducted from base maps.

Experimental samples have been collected, with the cooperation of the owners and cork related expert workers, in a cork forest which can be compared to that of the Case Study.

In this way, the amount of bark removed cork (in kilos) obtained from each plant was established. The average quantity produced by each plant was 11 kilos the lowest amount being 4 kilos with the highest being 37 kilos.

6 Oil Free Products

Medium to low quality bark removed cork is defined as being "grindable" and it is not suitable for wine stopper and the food industry. Therefore, by manufacture processing this medium to low quality cork bark, it can then be used in the bio architecture industry.

The first product in this processing is melted granulates or "granulated cork". It is both a final product (which can be immediately used) and, alternatively, an intermediate product or input. As an innovative final product, it can be used to partially (and eventually totally) substitute sand in the mix for conglomerate, concrete and mortar.

This results in a significant improvement in thermo acoustic insulation which, of course, makes desiderably lighter all buildings. When used as an intermediate input, the granulated cork can be further transformed into high density rolls for acoustic insulation. It can also be turned into one to ten centimeter thick cork panels. Their function is to prevent mould, act as igro regulators, provide insulation from the cold, and, of utmost importance, act as a retardant buffer (phase displacer) slowing down the overheating of the construction by summer outdoor heat while improving the overall thermal performances of buildings [15, 16, 18, 19, 24]. Additionally, on the field researches has shown that "healthier" buildings with better thermal performances were sold at higher selling prices in the real estate market [12–14, 22, 23]. Additionally, cork oak forests enhance landscape.

7 Green Building Prototype Experiment

This research aims to initiate an investigation into the relationship between:

- the reduction in energy consumption together with the abatement of CO_2 emissions;
- the quantity of cork to be used in order to bring about such improvements.

In order to formulate a valuation as complex as this, a simple and basic building was simulated carefully in minute details. This building is about $5 \times 5 \times 3$ m. Such small dimensions make for a construction which is particularly prone to thermal dis- persion [15, 19]. So, it constitutes an ideal testing ground for the objective of the research. Several identical copies of the prototype were simulated. The different Scenarios are:

- A "Business As Usual (BAS)" type building. This was simulated following the tried and trusted procedures commonly used in the building industry of the region.
- A sustainable building. In this case, an entire layer of cork insulation is applied on the 6 sides of the building. The first appraisal goal is to design the building based on a previous climate valuation of its thermal behavior and metabolism. Well before drawing up all final building designs, the ideal thickness of the cork panels must be calculated and simulated. Climatic valuation is carried out using 3 tools or Energy Performance Simulation Programs, which are different in origin, structure and kind. At the same time, in both Scenarios, the detailed costs of both buildings are esti- mated using the analytical approach and method. All this is performed in order to evaluate in Euro and in a percentage the greater initial investment cost of ecological passivation that is *i.e.* the price of cork added to the BAS building costs.

8 Results

It was found that a mere 6 cm of cork reduced annual energy consumption by 44%, from 3,111 to 1,782 kWh, i.e. 1,329 kWh less. Furthermore, there was a consequent permanent reduction of CO2 emissions of 37%, from 359 kilos per year to 213 kilos per year i.e. 146 kilos per year less. If 6 cm thick cork insulation is used, the difference in initial investment cost amounts to only an increase of 9% in Euro i.e. 40,378 as opposed to 37,174. The quantity of cork panels used to insulate Prototype building was estimated at about 5,83 m^3, coming to a total weight of 9,31 quintals, i.e. 931 kilos. In the simulated Prototype, every kilos of cork used reduces the building consumption of energy (on a permanent basis) by 1,42 kWh per year forever. Also it mitigate the quantity of CO2 emissions on permanent basis by 0,15 kilo per year forever.

9 Conclusion

There are many measures that governments (and even individual consumers, families, companies, construction companies or contractors, real estate promoters and owners) can take to promote post-carbon cities. The thermal insulation is definitely one of the most important investment which are necessary to achieve these future goals. Cork, being a natural and renewable material, which in itself is the result of a CO2 absorption and sequestration, and also a storage of C, can effectively contribute to the design of the post-carbon city by reducing energy waste, improving the quality and the insulation of buildings. The increase in the area occupied by the cork oak forests would increase permanently the absorption and sequestration of carbon.

Forest policy guideline must aim at the restoration of the existing cork oak forests, simplified into the structure and impoverished in their biological diversity, as well as the restoration of cork oak forests destroyed by man. This guideline involves adoption of management systems capable of combining the production needs with the maintenance of the resource. This will increase and strengthen the contribution made by forest resources to the construction of the Post Carbon City. Some benefits of such policy are as follows: (a) the sequestration of CO2 eliminated from the atmosphere (creating carbon credits), because it is used for the growth of trees and the creation of cork bark; (b) the availability of cork planks which are the raw material for bio-building; (c) the positive consequences of using cork panels and granules in Bio Green Buildings, such as energy saving for heating and cooling and the consequent reduction of CO2 emissions.

References

1. IPCC: Climate change 2013. Contribution of Working Group I to the Fifth Assessment Report of the Intergovernmental Panel on Climate Change. Cambridge University Press, Cambridge and New York, NY, USA (2013). (Editors: Stocker, T.F., Qin, D., Plattner, G.K., Tignor, M., Allen, S.K., Boschung, J., Nauels, A., Xia, Y., Bex, V., Midgley, P.M.)
2. Vessella, F., Schirone, B.: Predicting potential distribution of Quercus suber in Italy based on ecological niche models. For. Ecol. Manag. **304**, 150–161 (2013)
3. Batjes, N.H., Sombroek, W.G.: Possibilities for carbon sequestration in tropical and subtropical soils. Glob. Change Biol. **3**, 161–173 (1997)
4. Henderson, G.S.: Soil organic matter: a link between forest management and productivity. In: McFee, W.W., Kelly, J.M. (eds.) Carbon Forms and Functions in Forest Soils, pp. 419–435. Soils Science Society of America Inc., Madison (1995)
5. Chiesi, M., et al.: Application of BIOME-BGC to simulate Mediterranean forest processes. Ecol. Model. **206**(1–2), 179–190 (2007)
6. FAO: Global Forest Resources Assessment. Progress towards sustainable forest management. FAO Forestry Paper 147, FAO, Rome, Italy. http://www.fao.org/forestry/fra/fra2005/fr. Accessed 06 Dec 2017
7. Palma, J.H.N., Paulo, J.A., Tomé, M.: Carbon sequestration of modern Quercus suber L. silvo arable agroforestry systems in Portugal: a YieldSAFE-based estimation. Agrofor. Syst. **88**(5), 791–801 (2014)
8. Gil, L.: Cork composites: a review. Materials **2**, 776 (2009)
9. Spampinato, G., Crisarà, R., Cameriere, P.G., Musarella, C.M.: Phytotoponyms a tool to forest landscape analysis and its transformations. The case study of Calabria (Southern Italy). In: Chiatante, D., et al. (eds.) Sustainable Restoration of Mediterranean Forests-Fl. Medit, vol. 27, pp. 5–76 (2017)
10. Spampinato, G., Crisarà, R., Cannavò, S., Musarella, C.M.: I fitotoponimi della Calabria meridionale: uno strumento per l'analisi del paesaggio e delle sue trasformazioni. Atti Soc. Tosc. Sci. Nat., Mem., Series B **124** (2017)

11. Mercurio, R., Spampinato, G.: Primo contributo alla definizione tipologica delle sugherete della Calabria. S.I.S.E.F., Atti 3, pp. 483–490 (2003)
12. Massimo, D.E., Del Giudice, V., De Paola, P., Forte, F., Musolino, M., Malerba, A.: Geographically weighted regression for the post carbon city and real estate market analysis: a case study. In: Calabrò, F., Della Spina, L., Bevilacqua, C. (eds.) Local Knowledge and Innovation Dynamics Towards Territory Attractiveness Through the Implementation of Horizon/E2020/Agenda2030. Springer, Berlin (2018)
13. Del Giudice, V., Massimo, D.E., De Paola, P., Forte, F., Musolino, M., Malerba, A.: Post carbon city and real estate market: testing the dataset of reggio calabria market using spline smoothing semiparametric method. In: Calabrò, F., Della Spina, L., Bevilacqua, C. (eds.) Local Knowledge and Innovation Dynamics Towards Territory Attractiveness Through the Implementation of Horizon/E2020/Agenda2030. Springer, Berlin (2018)
14. De Paola, P., Del Giudice, V., Massimo, D.E., Forte, F., Musolino, M., Malerba, A.: Isovalore maps for the spatial analysis of real estate market: a case study for a central urban area of reggio calabria, Italy. In: Calabrò, F., Della Spina, L., Bevilacqua, C. (eds.) Local Knowledge and Innovation Dynamics Towards Territory Attractiveness Through the Implementation of Horizon/E2020/Agenda2030. Springer, Berlin (2018)
15. Malerba, A., Massimo, D.E., Musolino, M., De Paola, P., Nicoletti, F.: Post carbon city: building valuation and energy performance simulation programs. In: Calabrò, F., Della Spina, L., Bevilacqua, C. (eds.) Local Knowledge and Innovation Dynamics Towards Territory Attractiveness Through the Implementation of Horizon/E2020/Agenda2030. Springer, Berlin (2018)
16. Massimo, D.E., Malerba, A., Musolino, M.: Green district to save the planet. In: Oppio, A., et al. (eds.) Integrated Evaluation for the Management of Contemporary Cities. Springer, Berlin (2017)
17. Massimo, D.E., Malerba, A., Musolino, M.: Valuating historic centers to save planet soil. Oppio, A., et al. (eds.) Integrated Evaluation for the Management of Contemporary Cities. Springer, Berlin (2017)
18. Massimo, D.E., Fragomeni, C., Malerba, A., Musolino, M.: Valuation supports green university: case action at Mediterranea campus in Reggio Calabria. Soc. Behav. Sci. **223**, 17–24 (2016)
19. Massimo, D.E.: Green building: characteristics, energy implications and environmental impacts. Case study in Reggio Calabria, Italy. In: Coleman-Sanders, Mildred, Green Building and Phase Change Materials: Characteristics, Energy Implications and Environmental Impacts, pp. 71–101. Nova Science Publishers (2015)
20. Massimo, D.E., Musolino, M., Barbalace, A., Fragomeni, C.: Landscape and comparative valuation of its elements. In: Agribusiness, Paesaggio & Ambiente, vol. 17. Special Edition 1, pp. 53–60 (2014)
21. Massimo, D.E., Musolino, M., Barbalace, A., Fragomeni, C.: Landscape quality valuation for its preservation and treasuring. In: Advanced Engineering Forum. TTP Publications, USA (2014)
22. Massimo, D.E.: Emerging issues in real estate appraisal: market premium for building sustainability. Aestimum, 653–673 (2013)
23. Massimo, D.E.: Stima del green premium per la sostenibilità architettonica mediante Market Comparison Approach. Valori e Valutazioni (2011)
24. Massimo, D.E.: Valuation of urban sustainability and building energy efficiency. A case study. Int. J. Sustain. Dev. **12**(2–4), 223–247 (2009)
25. Musolino, M., Massimo, D.E.: Mediterranean urban landscape. integrated strategies for sustainable retrofitting of consolidated city. In: Sabiedriba, integracija, izglitiba, vol. 3, pp. 49–60 (2013)

Ecological Networks in Urban Planning: Between Theoretical Approaches and Operational Measures

Angioletta Voghera(ID) and Luigi La Riccia(✉)(ID)

Politecnico di Torino, 10125 Torino, Italy
luigi.lariccia@polito.it

Abstract. The ecological network can be considered in different ways: as a strictly interrelated system of habitats, as parks and protected areas network, as a multi-purpose ecosystemic scenario, as a sequence of natural, rural and open landscapes. Nevertheless, all the interpretations of natural landscapes not always have been considered in the lexicon of urban and regional planning, relegating natural and rural areas to an "inessential" role (and generically defining them as "in state of pre-urbanisation").

This contribution reflects about the ecological meaning of landscape, and therefore about its primary ecosystemic role, introducing a review proposal of the current programs and planning paradigms, highlighting its importance in the economic, entrepreneurial and policy debates in Europe. The main objective is promote new clear and specific local planning regulations, direct to the project of new ecological corridors with a more and useful consideration of the binominal value "landscape-biodiversity", and in general of the "natural-rural-urban" correlation, as an essential condition for defining a new vision of sustainable urban and regional development.

Keywords: Ecological networks · Regional and urban planning
Guidelines · Sustainability

1 Ecological Networks, Protected Areas and New Urbanizations

Despite the Protected Areas and Natura 2000 sites are now considered the "backbone" of the European policy for biodiversity, at the local level they are included with a clear difficulty within the urban policies and plans. The policies for the improvement of ecological networks are in fact necessary to overcome the fragmentation of the habitats and natural areas, which is the main cause of biodiversity loss in Europe. From this point of view, in fact, the Natura 2000 network, now implemented in 28 Member States and considered, at Community level, such as the exclusive policy for the conservation of biodiversity values, covering a total of 18.36% of the surface of the member states and It includes a set of sites of Community interest for about 60 million hectares. Then, there is a considerable overlap of these with the surface of Protected Areas that instead corresponds to approximately 22% of the surface of the Member States [1].

© Springer International Publishing AG, part of Springer Nature 2019
F. Calabrò et al. (Eds.): ISHT 2018, SIST 101, pp. 672–680, 2019.
https://doi.org/10.1007/978-3-319-92102-0_73

In view of these data, in recent years we have seen an exponential growth of urban land use towards more natural spaces: external urban areas (uncultivated land, cultivated land abandoned, the burnt areas, degraded forests) are often been confined to a "inessential" position and sometimes simply considered as "waiting for a new urbanisation". Too often, this is due to poor operability of local plans to lead an urban development coherent with the preservation of natural areas and ecological connectivity. We can identify the consequences of these processes in 6 significant phenomena [2]:

1. the substantial loss of natural areas: urban development has led in recent years, a reduction of natural areas (in the world, in the years 2000–2010, the rate of decline amounted to about 16 million hectares lost each year);
2. the fragmentation of natural areas: a process that determines a breakdown of structural areas of ecological networks into smaller patches, and consequently more isolated from the point of view of connectivity;
3. the degradation of wetlands, which have always been an important ecological function for the control of water flows, for the ability to block the sediments, for the support of plant and animal species (stepping stones function) and for the ability to provide nutrients for the ecosystems;
4. the inability to ecosystems to respond to change and find a new ecological balance: that is to say a significantly reduced resilience;
5. the loss of ecosystem services: natural systems have important "services", such as the control of water, the filter functions for pollutants, the preservation of the climatic risks;
6. the increased costs for public services, due to the response to natural disasters as a result of the ecological footprint by man.

Nature conservation in the city is one of the biggest challenges for sustainable urban development, as a result of a social and ecological coevolution. The value of nature in the city, however, goes far beyond its influence on the inhabitants' quality of life or rather an intrinsic value: urban areas are surprisingly rich in biodiversity. The conservation and management of nature and biodiversity in urban areas is often vary complex [3, 4]: there are more people, stronger development pressures, less space, a multiplicity of actors involved, etc. Often, the analyses reveal that the urban natural reserves are few but large and have a high density.

Large natural reserves can be especially important in urban landscapes, as the difference between the urban and natural environment can be high [5]. It should be noted, however, that the strategies of urban planning and those of nature are in Italy generally separated. One possible reason is that the protection of nature has favoured a purely "conservative" vision towards nature outside the city and has made trivial and distorted the vision of urban nature conservation.

However, the identification of urban nature is also part of a broader change in perspective within the conservation policies and remains as a necessary point of reference for a sustainable urban development. In many cities, this change of perspective was manifested through the institution of urban areas for nature conservation, supported by a general concept of "urban landscape". In the urban context, the establishment of these areas has been started during the twentieth century as a reaction to the rapid degradation of the urban environment due to industrialization and the consequent urban growth. It was therefore seen as a necessary step to keep nature and landscape away from private exploitations. Today, instead, the public interest is more oriented to the preservation of social values, biodiversity of nature and landscape. In the recent decades, in effect, the nature conservation and landscape policies have changed: today, a possible alliance between nature and landscape [6, 7] is assumed to be an essential condition for sustainable development and lays itself at different scales [8–11] and lays itself at different scales [12, 13].

Until the 1970s, in Italian urban planning, we could not speak about a real ecological paradigm, but of "urban greening", the distribution of which was generally expected in new districts as well as in historical centres. The creation of urban parks also became one of the focal points of the urban plans. Keeping them indicated a fundamental aspect of environmental continuity in urban space. The consideration of nature and landscape in the Italian urban planning tradition has privileged the aesthetic approach, oriented to the historical and cultural heritage of excellence. During those years, when in Italy the debate was focused on the general "crisis of planning" [14], at the international level an important shift on focus could be observed towards the "landscape planning" [15], a new way of understanding the landscape in the plan, closer to the urgency of reducing ecological problems and supported by an emerging environmentalist currency in the cultural and political scenes. On the one hand, there was a growing need to put an end to environmental disasters; on the other hand, the issue of landscape merged forcefully in different disciplinary contexts.

Town planning, rewritten through a new ecological paradigm [16], does not appear to be capable of solving the identification, convenient, of landscape with the natural environment, still promoting an approach, especially design, which remains "promotional". Sustainable development requires more than designed landscapes that are created using sustainable technologies. Design is a cultural act, a product of culture made with the materials of nature, and embedded within and inflected by a particular social formation; it often employs principles of ecology, but it does more than that, enabling social routines and spatial practices, from daily promenades to commuting to work.

In the Italian experience, indeed, green areas acted as a common element for re-joining city and countryside, that is to say, for the redevelopment of the modern districts in order to reduce the pressures on both historic centre and new districts. The ecological paradigm is therefore a different vision and has guided the practice of urbanism towards a new direction [16]. The environmental provisions now seem to articulate the new practices, coordinating behaviours and reconfiguring the spaces of the city: this means defining new and more ecological functionalist provisions.

2 The Construction of the Ecological Network at the Local Level and the Rules for the Urban Planning

Attributing ecological significance and therefore an ecosystem role, not necessarily secondary, to territory means reflecting on a general renovation of the urban planning paradigms, considering the importance of productive, business and policy interests. Therefore, a clear need to define the objectives, which avoid simple "territorial schemes" of new ecological corridors, maybe excellent in aesthetic terms, but lacking of all meanings from the point of view of biodiversity. For this reason it is important not to stop to analyse only the state of naturalness and diversity at different scales, but it is necessary go further to give priority to the pursuit of ecological coherence of the whole territory: that is to say to link the network with the impacts deriving from human activities and, more generally, to define a framework for urban planning operability.

In this context, several interesting experiences about this issue have been launched in the Piedmont region (Italy) with the aim to improve the overall ecological quality of the natural and landscape areas and specifically indicate the operational procedures to avoid the ecological fragmentation. Between 2014 and 2016 the research "Guidelines for the Green System of PTC2" (convention between Metropolitan City of Turin, ENEA and Polytechnic of Turin) and the "Operational proposals for the ecological network of Chieri" (Polytechnic of Turin and Comune di Chieri, Turin) were conducted with the objective of defining a proposal for the implementation of the ecological network at the local level in two municipalities of Turin (Ivrea and Chieri).

In these experiences, the approach proposed by ENEA was reconsidered to guide governments with specific measures to limit anthropogenic land use and, where possible, orient and qualify the conservation of ecosystem services. Habitats, natural areas and landscape have not been interpreted only by exclusively ecological point of view (a mosaic of ecosystems) but also considering a broader perspective that embraces cultural, social and economic aspects of the Ivrea area. The proposed methodology identifies the ecological character of the territory and defines the criteria for the evaluation of different types of land use: in the Ivrea area 97 types of use, according to Corine Land Cover database, were identified. Subsequently, we applied five key indicators for assessing the ecological status (see Fig. 1):

- *Naturalness:* the types of land use are classified into 5 levels of naturalness, considering the closeness to the formations that would be present in the absence of disturbance (climax). So, the natural levels ranging from the 1st which includes all natural formations up to at maximum the 4th considering the types of land use at total anthropic determinism but not artificial (like almost all cropland) and the 5th level which includes the types of land use corresponding to artificial areas.
- *Relevance for conservation:* the types of land use are classified on four levels of relevance based on the relevance/suitability of land use for biodiversity conservation at the same time considering the importance for habitats and species. It introduces the concept of interest habitats for species of the Natura 2000 network including not only the habitats of Community interest but also the complex habitats whose conservation is necessary for the protection of species of the Natura 2000 Network.
- *Fragility:* the types of land use are classified in terms of intrinsic fragility due pressures such as pollution, ingression of exotic and invasive species, human disturbance in general. The 1st level includes types of land use that define both natural environments with very low resilience as rock fields or glaciers is semi-natural areas and significant anthropic determinism but easily fragile for both types of land use and poor resilience such as artificial water reservoirs or areas with sparse vegetation.
- *Extroversion:* the types of land use are classified on the basis of the potential "capacity" to put pressure compared to the neighbouring patches. We have

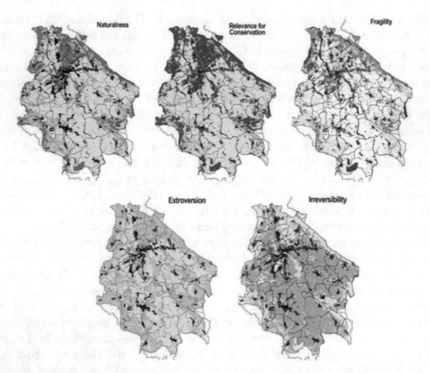

Fig. 1. Maps of Ivrea territory according to the considered five indicators (Processing ENEA 2015).

considered the pressures in an integrated way that goes from pollution of productions to the spread of invasive alien species. It ranges from level 1st, which includes types of land use that coincide with the areas with the highest human settlement and able to exert pressure, to the 5th level, containing natural types of land use types of use of the natural ground.

- *Irreversibility:* the types of land use are classified on the basis of the potential possibilities of change in the intended use. The 1st level includes all artificial types of land use totally characterized by the irreversible intended use (for example: urban, commercial industrial zones).

From the integration of the results of different indicators the so-called "Structural map of the ecological network" has been obtained (see Fig. 2). This map shows the elements of the Local Ecological Network system, chosen on the basis of the levels of naturalness, ecological functionality, geographical continuity, and consists of three main elements:

- *Structural elements of the network* (primary ecological network), namely the areas of high and moderate ecological functions, as well as areas that hosting the specific conservationist emergencies, i.e. of natural and significant importance for the conservation of biodiversity.

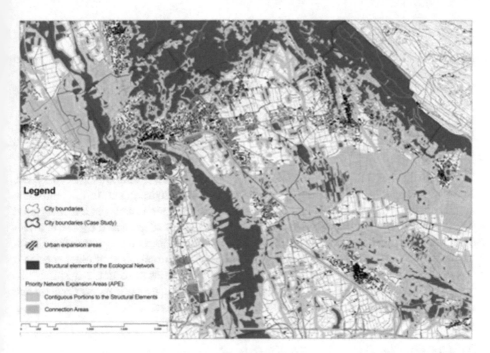

Fig. 2. Map of the ecological structurality of Ivrea territory. The picture shows the three components of ecological structurality and the relationship with the urban expansion areas (processing: Politecnico di Torino 2015 on ENEA data).

- *Priority Network Expansion areas*, namely the at residual ecological function areas where priority action to increase the functionality of the primary ecological network and for which the implementation of protection measures for the maintenance of primary ecological network. These areas are further divided into: *Connection areas* and *Contiguous portions to the structural elements.*
- *Possible expansion of the network areas*, i.e. areas at residual ecological functionality, but on which it is possible implement new interventions aimed at increasing naturalness useful to protect the habitat and species of interest for the conservation of biodiversity.

In the considered case studies, from an analytical process (framing of the territorial ecological system and public consultation through negotiating tables) it has come to drafting of rules, directly integrated with the urban plans, which include provisions for the implementation tools, such as spatial equalization measures, compensation and mitigation of impacts and provisions for the urban green management.

These implementation mechanisms are designed to intervene where projects and actions included in the urban plan could lead to changes of the level of the local ecological functionality. The procedure for the definition of the compensatory measures for impacts not mitigated includes an analytical phase, an assessment phase, a phase of planning/design, an implementation phase and a phase of management and monitoring:

1. recognition and evaluation of the ecological relevance of the compensatory areas, through the evaluation of urban-environmental state;
2. definition of possible measures for improvement and protection of the ecological and landscape value, for each area identified for compensation;
3. setting priorities for action, to increase biodiversity and the sustainable use of the territory;
4. choice of the compensatory measures;
5. design of compensatory measures, based on the characteristics of each lot chosen;
6. updating the natural value of the areas subject to compensation.

Some rules are introduced for the urban green: the idea is that urban green spaces can contribute, with the green development of the rural environment, to the landscape quality of the territory. The defined parameters for the green management integrated (i.e. in the case of the City of Ivrea) the list of plant species adapted to the general urban conditions (climate and soil), as well as the conditions imposed by the urban environment, such as the resistance to pollution and pests. In the selection of plant species it is indicated to have to consider: at least 50% of native species or particularly suitable to the urban environment and less than 25% of non-native species or naturalized (hence excluding the weeds or plant with relevant on-going diseases).

3 Conclusions

The urban nature conservation requires also new conditions: ecosystems, such as landscape, transcend the scales, beyond just the urban area. We need to understand, within the rules and projects, that green is no longer just a mere architecture of context but contributes, primarily, to create a complex system, unitary consistent with historical heritage and environmental dynamics. We can identify five key passages through which to build this system [17]:

1. Transposing the ecological network elements at regional level and verifying the implementation and the possible expansion at local level (the network project must become an integral part of the territorial vision).
2. Defining the appropriate modalities for intervention favouring the natural use for the areas included in the network.
3. Making the local ecological network also through the institution of urban and territorial equalization models giving priority to the protection of rivers areas and public lands.
4. Ensuring the correct inclusion of allowed building work and the prohibition of definitive elimination of trees and shrub formations, including rows, hedgerows, etc.
5. Defining compensations and mitigation measures of impacts deriving from urban transformations, consistent with the goals of enhancing the local ecological network and the landscape quality.

Urban planning is therefore called to consider this aspect, going beyond the mere response to environmental and ecological issues and enabling to understand and appreciate the values of cultural processes underlying the urban and natural landscape, as well as the qualitative effects of choices considered in some way "environmentally sustainable". Ecological networks describe the structure of existing real ecosystems: in a recent work we have proposed a OWL [18] ontology for the representation of ecological networks, through an informatics language that we called GeCoLan [19] for expressing specifications about ecological planning considering land-use restrictions. GeCoLan can be automatically translated to GeoSPARQL queries for implementing the validation checks in an efficient way. Moreover, the language can support not only the verification of constraints in a geographical area, but also other reasoning tasks, such as making constructive suggestions, possibly optimizing some desired measure. Whereas, at the current stage, we implemented the validation as a stand-alone proto-type, the main motivation and application of our work lies in its possible integration within Participatory Geographical Information Systems (PGIS), in order to support online interaction with stakeholders in inclusive processes aimed at collecting feedback and project proposals from stakeholders. This would be a novel feature of PGIS (e.g., see Ushahidi (ushahidi.com), PlanYourPlace (planyourplace.ca) and OnToMap (ontomap.ontomap.eu)), which only support feedback collection, and fail to provide validation functions to check the feasibility of the proposed actions.

The conservation of nature in the city is therefore not possible without a broader consideration of the concept of the urban landscape, where the areas for nature con-servation may play a central role for the new image and the ecological rehabilitation of the city.

References

1. European Environment Agency: An introduction to Europe's Protected Areas. https://www.eea.europa.eu/themes/biodiversity/europe-protected-areas. Accessed 08 Dec 2017
2. Benedict, M.A., McMahon, E.T.: Green Infrastructure: Smart Conservation for the 21st Century. Sprawl Watch Clearinghouse Monograph Series, Washington DC (2002)
3. Antrop, M.: The language of landscape ecologists and planners - a comparative content analysis of concepts used in landscape ecology. Landsc. Urban Plann. **55**, 163–173 (2001)
4. Antrop, M.: Landscape change and the urbanization process in Europe. Landsc. Urban Plann. **67**, 9–76 (2004)
5. Powell, J., Selman, P., Wragg, A.: Protected areas: reinforcing the virtuous circle. Plann. Pract. Res. **17**, 279–295 (2002)
6. Gambino, R., Peano, A. (eds.): Nature Policies and Landscape Policies. Towards an Alliance. Springer, Dordrecht (2015)
7. La Riccia, L.: Nature conservation in the urban landscape planning. In: Gambino, R., Peano, A. (eds.) Nature Policies and Landscape Policies. Towards an Alliance, pp. 157–164. Springer, Dordrecht (2015)
8. United Nations Environment Programme: Convention on Biological Diversity. UNEP, Rio De Janeiro (1992)
9. European Council of Town Planners (ECTP): The New Charter of Athens (2003)
10. Hooper, D.U., et al.: Effects of biodiversity on ecosystem functioning: a consensus of current knowledge. Ecol. Monogr. **75**, 3–35 (2005)

11. International Union for Conservation of Nature, Nature+. IUCN World Conservation Congress, Jeju, 6–15 Sept 2012
12. Potschin, M.B., Haines-Young, R.H.: Landscapes and sustainability. Landsc. Urban Plann. **75**, 155–161 (2006)
13. Selman, P.: Planning at the Landscape Scale. Routledge, New York (2006)
14. Gabrielli, B.: Contro i piani di settore. In: Muscarà, C. (ed.) Piani, parchi, paesaggi, pp. 281–287. Laterza, Rome (1995)
15. Turner, T.: Landscape planning: a linguistic and historical analysis of the term's use. Landsc. Plann. **9**, 179–192 (1983)
16. La Riccia, L.: Landscape Planning at the Local Level. Springer, Cham (2017)
17. Voghera, A., La Riccia, L.: Landscape and ecological networks: towards a new vision of sustainable urban and regional development. LaborEst **12**, 89–93 (2016)
18. W3C: Web ontology language (OWL) (2012). https://www.w3.org/TR/owl2-overview/. Accessed 31 Jan 2018
19. Torta, G., Ardissono, L., La Riccia, L., Savoca, A., Voghera, A.: Representing ecological network specifications with semantic web techniques. In: KEOD 2017-Conference Proceedings (2017). http://www.keod.ic3k.org/. Accessed 31 Jan 2018

The Urban Question in Seismic Risk Prevention. Priorities, Strategies, Lines of Action

Concetta Fallanca[(✉)] [iD]

Mediterranea University of Reggio Calabria, 89100 Reggio Calabria, Italy
cfallanca@unirc.it

Abstract. The widespread inability to correctly perceive risk favors a discontinuity of interventions for securing urban centers and territories and does not facilitate a correct social demand for seismic prevention, even in the most dangerous areas. However, in the face of a weak culture of prevention, seismic risk can be considered an opportunity to increase urban quality in an integrated creative effort and the systemic logics that the security requires can have interesting repercussions on the quality of common spaces and the overall connective. The urban safety project is also an opportunity to create an urban network of safe places in areas with a high seismic and environmental risk, both natural and cultural. Designing urban security, in fact, also means enhancing the cultural level of the community to create awareness and commitment on the concept of the common good. Moreover, the topics of territorial prevention and defense are closely linked to the search for a fine balance between the protection of the historical and identity heritage and the process of renewal and regeneration of the city. The Metropolitan City of Reggio Calabria could represent, in this sense, the engine of effective programming and strategic planning tools, and could host a "prevention for safety" laboratory designed as a place to open up a more focused debate on the relationship between coexistence with risk and quality of urban planning. Lastly, the invitation is to support the institutional nature set of seismic adaptation measures and the accountability of the world of technical professions for a vast and coordinated set of social actions, to activate urban laboratories, active civic networks, also at the district level, for the affirmation of a culture of prevention.

Keywords: Urban safety · Safe places · Heritage · Soft mobility

1 Three Preliminary Considerations

The widespread inability to correctly perceive risk and to maintain a collective memory of events creates serious damages due to the discontinuity of initiatives in favor of interventions for securing urban centers and territories. In this field the continuity and the ability to work with incremental effect are fundamental both for the extent of the phenomenon, which requires coordinated and strictly finalized investments, and for the complexity of the interacting factors.

© Springer International Publishing AG, part of Springer Nature 2019
F. Calabrò et al. (Eds.): ISHT 2018, SIST 101, pp. 681–690, 2019.
https://doi.org/10.1007/978-3-319-92102-0_74

This difficulty can not be attributed exclusively to the known "policy of direct benefits". If the political world does not care what it should, in fact, it is also because there is no correct social demand for seismic prevention, and this happens, incomprehensibly, even in the most dangerous areas. Furthermore, interventions aimed at safety and environmental protection offer long-term benefits and are therefore not very useful for acquiring electoral support [1].

It is a widespread cultural problem that makes our Country unprepared to face the problems that have always been part of the intrinsic characteristics of cities and territories [2].

We know that there is still no a culture of prevention, even if some examples, such as Norcia, which reported contained damages in the latest earthquake in central Italy precisely because rebuilt with anti-seismic logics and according to interventions aimed at prevention, following the earthquakes of the late 90s, start to create a public awareness. Attention is drawn to the safety of one's home, to the main buildings that perform a public function - schools, hospitals, civic centers, sports centers - but it is difficult to establish a collective willingness and capacity to contribute to the processes of intelligent cohabitation with the risk conditions with the activation of prevention and safety measures, with a stubborn, progressive process, attentive to the priority scales.

The systemic logics that the security requires concern the urban organism and can have interesting repercussions on the quality of common spaces and the overall connective. Some studies and intuitions have led to an increasingly refined definition of the *Emergency Boundary Conditions* and of a *Minimum Urban Structure* [3], to include in this, as the Umbria region does, "the historical cultural heritage" because identity is essential for the life of urban centers and therefore must be included among the vital aspects as the collective services.

The research initiated at first in the academic sphere has become over time a matter of cultural interest in the administrative and political sector; the Higher Council of Public Works prepares the "Preliminary Study for the development of orientation tools for the application of seismic regulations to historical settlements" and cities such as Ferrara transform a serious problem into a new opportunity. Following the 2012 earthquake, participatory workshops for the prevention of earthquake damage are established "to elaborate together strategies, intervention procedures, daily practices and habits for the ordinary management of seismic risk, with a view to prevention rather than emergency". Following this trace the "Casa Italia Project" with its "shipyards" that will test the measures for securing a "vital" urban fabric, with dwellings used during all phases of the intervention. The purpose is to derive useful guidelines for the safety of the entire national territory, to involve "a common wisdom" and to train specialized and at the same time "expeditious" skills on the topic of seismic prevention and safety.

Three brief preliminary considerations can be summarized.

The seismic risk, but also the risk factors of various kinds, can be considered an opportunity to increase the urban quality in a creative commitment integrated with the interpretive corpus inherent to the design thinking on urban and territorial ecological networks, the enhancement of the "common good", the reflection towards the project of an increasingly effective material and immaterial connectivity.

It is time to re-evaluate the role of technical/design professions knowledge that can offer a responsible direction for the transformation interventions for the seismic risk mitigation, already recognized for the seismic risk assessment of buildings, both with the conventional method and with the exemplified one, to be tested in the urban areas, neighborhoods, in the entire urban body.

It is correct that a "expeditious" operation be promoted and that it can be expressed with the current state of knowledge with the awareness that experimentation can follow a path parallel to scientific and technical research that is increasingly profound (micro-zonations and multidisciplinary specialists) that must not however limit the planning, the interventions, the observations of the outcomes and the possible feedbacks.

2 The Invitation to Design Urban Security as an Opportunity for a New Quality of Common Spaces

The discussion session invites us to reflect on urban security as a precondition of development that requires continuous interventions to achieve and maintain satisfactory and ever-increasing levels of quality. In order to improve urban security, the most innovative technologies and methodological processes becoming more sophisticated, allowing the comparison of the effects in the different realities can be used.

The urban safety project can be understood as an opportunity to increase the urban quality of the common spaces experienced in everyday life and in areas with a high seismic and environmental risk (natural and cultural) must tend to create an urban network of safe areas. It is important to transform "safe islands", too often places "shelved" for the contingency, in areas of sociality that raise liveable and urban quality. The topic goes beyond the security of the building that must respond to anti-seismic measures and is aimed at its inclusion in the urban area that must ensure adequate places that can be reached in calamity, because it makes no sense to create safe "islands" if they cannot be "safely" reached.

Flooding, landslide, and run-off phenomena accentuate the overall risk and must be considered by the seasonal and annual maintenance plans that are an integral part of safety-related projects [4].

Designing urban security also means enhancing the cultural level of the community to create awareness and commitment on the concept of the common good and respect for the heritage dimension in terms of urban civilization and of living together. Educating for safety means transmitting the perception that places are easy to reach and enjoyable to live. It also means networking and facilitating the accessibility of areas and places where cultural activities can be carried out to facilitate individual and collective growth and the exchange of experiences. An investment on human heritage and on the sense of citizenship requires a redesign of public mobility towards *soft mobility* and a rethinking of the times and ways of access and enjoyment of the fundamental places of the city. Working in a network logic means, finally, connecting the significant places for urban identity and safety and seeking a fine balance between the protection of historical-architectural values and the renewal and regeneration process that is essential to continuous creative evolution of the city and its parts.

The topics of territorial prevention and defense are closely linked to the effectiveness of local public transport policies on a metropolitan scale; territories well connected to metropolitan level services, in fact, remain densely inhabited because they do not live the difficulties of a stict marginality - as happens, for example, for the internal centers of the Metropolitan City of Reggio Calabria - and, at the same time, the distance from the main center offers settlement advantages which offset a comfortable commuting.

The new Authority will have to coordinate and plan local public transport policies on a metropolitan scale, focusing in particular on the road network and the connections between the inland and coastal territories and between the centers that animate the main territorial systems that can be thought of as homogeneous zones by which to co-govern the metropolitan dimension.

In Calabria, a project for the revival of railway transport and the recovery of stations in smaller urban centers has been proposed for a long time, as a sign of attention to the autonomy of travel not strictly conditioned by the use of a personal vehicle (Fig. 1).

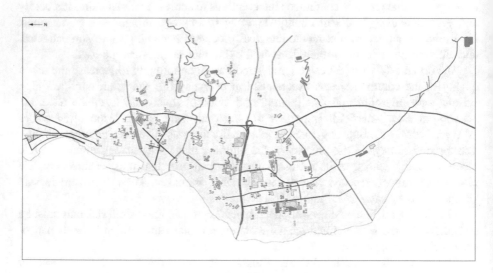

Fig. 1. Reggio Calabria and its safe places.

3   The Experimentation Laboratory for Urban and Territorial Safety of the Metropolitan City of Reggio Calabria

The territorial area of the Strait, with the cities of Reggio, Messina, Villa San Giovan-ni, with a potential of 500.000 inhabitants, and 1.200.000 inhabitants of the two metropolitan cities, has homogeneous characteristics due to its geomorphological and environmental structure, with remarkable common elements concerning the historical and settlement development and the chronological and territorial contiguity of the phenomena that took place there.

These territories are joined by the culture of coexistence with the seismic risk and with homogeneous economic and social characteristics, such as to allow the identification of territorial needs and requirements, and related common and shared objectives, to be achieved concretely through the development of an integrated urban-territorial planning.

The territorial context of the Metropolitan City of Reggio Calabria presents a framework of the emergencies attributable to environmental risks, to the settlement system, to the relational, mobility and production system, with socio-economic problems due to the difficulties of tracing a common path for the affirmation of a culture of legality and transparency to support the fight against crime.

The Metropolitan City of Reggio Calabria can be the engine of effective programming and strategic planning tools, as has been the case in the best European experiences of the last decade: Barcelona, Valencia, Lyon, Paris, Munich, Amsterdam, Stockholm. These cities have been able to identify resources, timeframes, subjects and implementation procedures, enhancing the local planning and outlining a shared vision of the vocations and prospects for the development of the territories. Experience indicates that advanced mobility services are the outcome of multilevel *governance* procedures, which widely involve the territories of large areas, guaranteeing them the efficiency of local services and a direct access to metropolitan level services, from innovative, comfortable, quick and frequent connections.

It is now essential to know how to outline an idea of a metropolitan city and of a metropolitan area for the elaboration of the Metropolitan Strategic Plan and its close link with the Development and Cohesion Funds. A first critical assessment may be directed to what seems to be an imbalance of interest much in favor of the Strategic Plan which does not take into account the shared purpose of development of the territory and urban fabric in the principles of subsidiarity, differentiation, adequacy that unites it to the less considered Metropolitan General Territorial Plan.

In the process of reorganizing the functions assigned and to be assigned to the metropolitana city - own functions, provincial functions, functions delegated by the State and the Region, relations with the Municipalities and the Unions of municipalities - the choice towards simplification to build a concrete capacity for governing issues of "proximity" would be desirable in response to the new demand for quality of the urban space that comes from a multi-ethnic society that expresses increasingly complex needs in terms of social and cultural *mixité*.

The topics on which to focus are the environment, parks, economic development, joint services, public mobility, reducing land use and the restraint of settlement dispersion for the purpose of defending the territory to mitigate the hydrogeological instability and reduce the vulnerability of urban and territorial heritage; the metropolitan city, in fact, assumes all the functions relating to the environment and soil protection, including those of the management of the civil protection system on a metropolitan scale.

A look back shows how in the field of seismic hazard, of exposure to these events, in the relationship between natural disasters and social life, it is possible to glimpse, between the two metropolitan cities, traits of an ancient and, at the same time, very current culture: the culture of cohabitation with the earthquake, which once was and could, should, return to be a culture of prevention and not of the removal of risk, of the unexpected, of the chaotic.

The City of the Strait should aspire to be a privileged place of observation on the topic of seismic hazard on the wider dimension of the Mediterranean basin, a whole basin of "moving lands", where Reggio as Messina, but as Istanbul, Athens, Ankara, Lisbon, share both high levels of urban vulnerability and high levels of unpreparedness in terms of intervention procedures, daily practices and habits for the ordinary management of seismic risk, in keeping with the spirit of prevention rather than emergency.

In agreement with the Universities through the promotion of innovation and research and investing also in the field of training and professional growth *life learning*, a "prevention for safety" laboratory could be thought of as a place to start more focused reflections on the relationship between coexistence with risk and quality of urban planning; where to promote programs and policies that can be put in place to protect the environment and the urban habitat from the consequences of a natural disaster; where to experiment the ability to use the continence of the calamitous danger to improve planning techniques and qualities.

Finally, in a laboratory conceived in this way, monitoring networks, surveys, information and training campaigns aimed at the population could be activated to provide the necessary knowledge, theoretical and practical, to reduce vulnerability to seismic risk.

In the context of the reflection on the enhancement of Internal Areas and in the direction of risk reduction comes the law "Salva borghi", the 158/2017, laying down *Measures for the support and enhancement of small municipalities, as well as provisions for the requalification and recovery of the historical centers of the same municipalities*. The law promotes the National Plan for the redevelopment of small municipalities and ensures the priority of the qualification and maintenance of the territory, through recovery and redevelopment of existing buildings and brownfields, as well as interventions aimed at reducing hydrogeological risk and making safe and redevelopment of road infrastructures and public buildings, with particular reference to schools and buildings destined to services of early childhood, to public structures with social-welfare functions and to most frequently used structures. The law recognizes that "the villages constitute a factor in preserving the land, especially for the activities against hydrogeological instability and for the activities of small and widespread maintenance and protection of common goods" (art.1).

As part of the safety measures and the increase in the resilience of the territories, Reggio Calabria is one of the chosen cities for the experimentation of Casa Italia Project, the anti-seismic prevention program born with the aim of making safe public residential buildings all over the Country. We do not appreciate yet the extent of the project at the urban level, that from the first public documents seems aimed at a building dimension and/or at urban level, with the main objective of experimenting, even consciously accepting the presumed increased costs, a method for the adjustment of buildings to seismic risk while maintaining the residential function and the location of the residents.

4 Need for Rapid Methods, Concrete Interventions for Damage Prevention Activities

The world of technical professions is called to an accountability, useful for modifying the course of the complex history for the affirmation of a prevention culture. Architects and engineers (but also surveyors and industrial experts following protests for exclusion) are qualified to classify seismic risk of buildings, to design interventions aimed at reducing seismic risk and attesting the effectiveness of interventions carried out. This procedure could be borrowed on a larger scale, involving the professionals who animate the municipalities' Plan Offices in the preparation of programs and projects to reduce the seismic risk in a similar task at the urban level.

The guidelines approved by the Ministry of Infrastructures and Transport Decree (28/02/2017) introduce a new way of understanding prevention measures. The projects of the seismic risk reduction interventions involve the qualified professionals and every member of the community that chooses to pursue the "safe house" path and that considers it useful to invest in this direction obtaining tax benefits. An intelligent way of involving, empowering and sensitizing the competent professionals of the sector and the whole society that, informed, chooses to improve the security of their homes. In this reflection the aspect most interesting is joining the conventional method for the attribution of the risk classes (which requires for the evaluation of the PAM class, the calculation of the Limit status, from the beginning of damage to the extreme one of total reconstruction), a simplified method to be applied limited to masonry types and which allows the passage of only one class of risk. The simplified method was introduced to evaluate and operate with those building aggregates in which the identification of the structural unit is complex, sometimes indistinguishable from the whole. Two parallel lines are then activated: on the one hand, the search for increasingly reliable and rigorous methods, also with the aim of determining priority scales of intervention, and on the other, putting in place all the simplifications that lead to the factuality that it has been lacking in the last two centuries in terms of seismic risk prevention. In the face of increasingly sophisticated theoretical and technical acquisitions, operational indications and above all effective measures to improve settlement conditions have been lacking [5].

It is important to always acquire new scientific knowledge and technological innovation, but it is equally important to activate all the energies and resources to create progresses, even small but continuous and progressive, to mitigate the seismic risk conditions and, in general, the risk of the settlements.

The technical professionalism is prepared and formed to a synthesis approach that tends to extract the intelligible meaning of increasingly sophisticated data, while other skills finalize the study to an increasingly analytical and certifiable knowledge, at a stage of deepening that sometimes may not be necessary for the purpose of territorial transformation. There are categories of common sense intervention that must be carried out regardless of the microzonization class and the seismic risk level, which also surpass the concept of priority of intervention because they appear to be basilar for the safety of the communities and for the prevention of risks in general. In this case it is correct that scientific research invests more and more in knowledge but the project must be able to use optimally what is already in place to mitigate the widespread attitude of postponing solutions in expectation of increasingly refined cognitive assets, while the calamities continue in succession, with unsustainable social costs.

At the *territorial level* it is possible to underline two macro aspects debated for decades but still unrealized.

The first concerns route alternatives, the opportunity to offer "junction" possibilities to vulnerable connection elements and which, in case of disasters, even a trivial landslide, can create disruption of population centers. Even now, the collapse of a masonry bridge, the flooding of a river, creates interruptions in the connections not only by road but also by rail. In the territory of the Metropolitan City there are numerous critical elements that can be solved with serious planning that, over a five-year period, could invest in overall accessibility, defined as "redundant" but in reality essential because it offers "plan B" in all conditions, a valid route alternative. Reports of the Plan offices, the observation of the effects of floods and landslides, a serious press release of the last decades, the design thinking aimed at thinking of a coordination at the level of vast territory, in overcoming the perceptions of cutting emergencies localistic, would like to create an overall and shared project of the metropolitan city community suitable to involve and bring together the necessary funding channels.

The second aspect is related to network infrastructures and the opportunity that they are directed in the direction of active and capable faults[1]. We continue to design and build pipeline networks, electrical, hydraulic, sewage, road, highway and railways grids "transversely" to particularly dynamic faults.

In drawing up the Structural Plans, the geological studies identify a band of respect for the projection on the surface of the fault lines considered active with the requirement to prohibit or strongly discourage new buildings, even if often the belt crosses urban centers and hamlets that should be affected from widespread, and therefore less and less probable, "safety measures". This consideration underlines that universally recognized useful design criteria are sometimes disregarded and that an ever-increasing geological knowledge does not always offer information that benefits the design and security of settlements. Consider also all the coastal settlements that occupy territories that in the past have been hit by the waves (which have reached the thirteen meters) of the tsunami. These are areas of territory that the PAI indicates with a severe risk factor. What to do in these cases? Common sense interventions could favor the exodus routes towards the first hinterland, investing and trusting in instruments of detection and immediate diffusion of alert.

---

[1] A fault is considered capable if has been activated at least once in the last 40,000 years and if is capable of breaking the topographic surface producing a fracture/dislocation of the ground as defined in "Linee Guida per la gestione del territorio in aree interessate da faglie attive e capaci", in Messina P., D.D.L. 2734 (Cartografia geologica d'Italia e della microzonazione sismica) Audizione presso la XIII Commissione territorio, ambiente, beni ambientali del Senato della Repubblica, CNR (2017).
(http://www.protezionecivile.gov.it/resources/cms/documents/Linee_Guida_Faglie_Attive_Capaci_2016.pdf).

At the urban level, the civil protection security plans had promptly faced the identification of the elements of endowment of the minimum structure, including the system of waiting areas, but also in cities with a high seismic risk such as Reggio Calabria, after a first draft of the 2008 plan that identified areas sometimes fenced and accessibility limited to some hours of the day, such as the school grounds, there is no intervention to improve the suitability of waiting areas.

A fundamental issue concerns the incorrect spatial organization of functions within the urban system that creates concentrations at different times of day and with service central points, sometimes imprudently close together, such as to make the possible interruption of flows not only extremely risky of exchange due to disastrous events, but sometimes simply limits the daily mobility of people, vehicles, goods, especially at peak times. The importance of the organization of the urban system is also inherent to changes in the concentration of people, services and goods between the day and night hours that determine changes in the intensity of use of the buildings, with fluctuations in the "exposed value" but also with the increased danger due to slowing of the flow through the escape routes [6].

The strategy of equitable distribution of "central points" in the urban organism would avoid the concentration of functional systems in a few works in favor of a widespread and flexible system for the distribution of the collective equipment of the school, social-health, entertainment, cult systems. In the case of an excessively hier-archized organization of the territorial and functional system, due to the effect of the "deferred vulnerability in space" [7], in fact, the seismic damage has repercussions on a much wider area than that directly damaged.

All we said is amplified in the *urban fabrics of ancient formation* that characterize the villages of small towns, historic centers and art cities [8]. The irregular settlement configurations, the stratifications, the lack of maintenance create conditions that are difficult to assess from the point of view of vulnerability[2], so as to require "synthesis" skills capable of interpreting the overall operating system. The road system, with narrow and irregular streets, the presence of shops and historical markets, makes the exodus and access to relief also unfavorable due to the vulnerability induced by buildings to roads aggravated by the possible collapse of "critical" elements much slender as towers and bell towers [9].

The invitation is therefore to support the complex of seismic adaptation measures of an institutional nature and a vast and coordinated set of social actions, to activate urban laboratories, active civic networks also at the district level to extend observations and interventions aimed at remove obvious impedance factors.

This in support of a position that would lead to empowerment and return a sig-nificant role to professionals of design-technical knowledge [10], essential (but inex-plicably under-utilised) also in order to create a sensitivity to the issues of prevention, as recently proposed by the Professional Associations for schools and training of professionals.

---

[2] Indicator of settlements' resilience to earthquakes, Law 77/2009, art.11.

References

1. Pileggi, M.: Terremoti: Gli incubi notturni del capo della Protezione Civile, i sonni tranquilli delle classi dirigenti meridionali, l'urgente necessità di agire per mettere in sicurezza milioni di vite umane e il "Piano Casa sicura" del Governo, Il Quotidiano del Sud 8 September, (2016)
2. Agenzia per la Coesione Territoriale: PON Governance e Capacità Istituzionale 2014-2020, Rischio Sismico e Vulcanico (2017)
3. Bramerini, F., Cavinato, G.P., Fabietti, V. (a cura di): Strategie di mitigazione del rischio sismico e pianificazione. Cle: Condizione limite per l'emergenza. Urbanistica dossier, INU edizioni 130 (2013)
4. Regione Emilia Romagna: Linee guida regionali per la riqualificazione integrata dei corsi d'acqua naturali dell'Emilia Romagna - Riqualificazione morfologica per la mitigazione del rischio di alluvione e il miglioramento dello stato ecologico (2015)
5. Benetti, D., Cara, P.: Linee guida per la compilazione della scheda WEB Centri Storici e Rischio Sismico - CSRS, Presidenza del Consiglio dei Ministri, Dipartimento della Protezione Civile, Ufficio Valutazione e Prevenzione del Rischio Sismico, Servizio valutazione della vulnerabilità, normativa tecnica e interventi post-emergenza (2008)
6. Caldaretti, S. (a cura di): Politiche insediative e mitigazione del rischio sismico. Un'esperienza su Rosarno e Melicco. Rubettino, Soveria Mannelli (2002)
7. Fabietti, V.: Centri storici a rischi. Tra leggi inadeguate e pericoli naturali. Palermo (2016)
8. Consiglio superiore dei lavori pubblici: Studio propedeutico all'elaborazione di strumenti di indirizzo per l'applicazione della normativa sismica agli insediamenti storici (2012)
9. Lagomarsino, S.: Ugolini, P.: Rischio sismico, territorio e centri storici. Atti del Convegno Nazionale Sanremo (IM), 2–3 luglio 2004
10. Rete Professioni Tecniche, Ispra, Enea: Proposta per la definizione di un piano di prevenzione del rischio sismico (2016)

The Ecological Challenge as an Opportunity and Input for Innovative Strategies of Integrated Planning

Gabriella Pultrone$^{(\boxtimes)}$ (ID)

Mediterranea University of Reggio Calabria, 89100 Reggio Calabria, Italy
gabriella.pultrone@unirc.it

Abstract. The current complex challenges brought about by development, the need to make it increasingly sustainable and the human origin of the environmental crisis require an integrated ecological approach that may combine seemingly different aspects, which, actually, are closely linked to each other. Since the concepts of sustainability, resilience and inclusiveness, as qualities of cities and human settlements, are among the goals of the UN 2030 Agenda for Sustainable Development (*Goal 11*), the real ambitious challenge is to start synergic collaborations between ecology, economy, legislation, spatial planning and territorial governance and society, trying to pave the way to circular economy and to a widespread culture of sustainability. In this context, the *green economy* is a must for spatial and urban planning in order to turn challenges into extraordinary opportunities of regeneration and safeguard of cities and territories, above all of those which are more at risk and environmentally deteriorated. Many European cities have started policies, programmes, and innovative actions that include projects combining regeneration and adaptation, invest in the natural capital, promote green infrastructures and more sustainable mobility, and improve the effective use of resources. This paper deals with the above-mentioned issues and examines the possible developments of an integrated ecological approach in ordinary spatial planning.

Keywords: Ecological approach · Integrated planning · Local identity

1 Complexity of Challenges and Need for an Urgent Breakthrough

Climate changes and the awareness of the limits of the current models of development are a global problem entailing serious environmental, social, economic, political risks of access to, distribution and use of resources. In this scenario, the simple reduction of negative impacts is not enough; on the contrary, a brave ecological breakthrough is required, a real cultural revolution that contemplates the principles of prevention and responsibility. Therefore, it is more and more crucial to combine often contrasting issues and face these systematic crosscutting challenges with an integrated approach [1]. At a time when the speed of anthropic actions clashes with the natural slowness of biological evolution, a new *governance* system for sustainable development, which is based on the circularity of ecology, economy and ethics, and in which participation and

© Springer International Publishing AG, part of Springer Nature 2019
F. Calabrò et al. (Eds.): ISHT 2018, SIST 101, pp. 691–698, 2019.
https://doi.org/10.1007/978-3-319-92102-0_75

joint responsibility in decision-making processes play a fundamental role, is needed. As a matter of fact, they are not separate crises and issues. It is rather a socio-environmental and cultural crisis which requires an integral ecology able to put together the various dimensions and issues in a cohesive and unitary framework with a view to achieving common good following the fundamental principles of respect and protection of nature, justice, equity and social commitment [2].

In this wide and complex context of multiple and interrelated challenges, where mankind lives in a "metropolitan planet", this paper is meant as a reflection on an ongoing research activity, which is still in its early stage. It highlights the crucial role urban planning can and must play in making human settlements and cities sustainable, safe, and resilient as well as in promoting a better quality of life for all, in line with *Goal 11* of the UN 2030 Agenda for Sustainable Development and with the "UN-Habitat Wheel of Prosperity", which identifies government institutions, legislative tools and town planning as drivers of development [3]. Actually, thanks to forms of synergic collaboration between ecology, economy, legislation, planning, territorial governance and society, it is possible to outline a concrete pathway leading to a widespread culture of sustainability through the elaboration of new paradigms and the experimentation of planning practices at different scales, from the territory, to the city and the neighbour-hood. A specific document on cities, climate change and disaster risk management, published in preparation for the United Nations Habitat III Conference, underlines that the ideas of urban ecology and resilience are intrinsically intertwined and fundamental to wellbeing and transformative change and that the concept of resilience has originated from ecology and from the principle that cities are unique and complex systems [4]. As a consequence, stresses and shocks should be evaluated in a holistic way, through a system approach, in order to understand what the major long-term issues for the health of cities and of their inhabitants are – *e.g.* climate change, energy demand, social cohesion, economic stability, governance, access to resources (particularly to water) and population growth –, and to elaborate suitable policies and strategies to face them. If they are well planned and governed, cities can become catalysers of environmental sustainability and positively contribute to resilience with impacts going well beyond their administrative borders and being closely linked to the regional and macroregional level. Furthermore, in spite of the huge diversity existing between cities, and between the local conditions within the same cities, certain results concerning urban ecology and urban environmental sustainability are widely shared and can benefit from common tools, such as solutions based on the nature and the evaluation of disaster risk. Sustainable urban development and its management are fundamental to the quality of life, just as it is the collaboration with local authorities and communities to renew and plan cities and human settlements that may facilitate cohesion and personal security and boost innovation and employment [5]. That is the reason why it is crucial to reduce the negative impacts of urban activities and of the chemicals that are hazardous to human health and to the environment as well as to reduce and recycle waste and use water and energy more efficiently [4].

With the purpose of finding new global equilibriums for sustainability, the UN General Assembly has decided to start a virtual dialogue, the so-called *Harmony with Nature* [6], which involves world's experts in the field of natural and social sciences to lead citizens and society to reconsider their interaction with the natural world in order to achieve sustainable development goals in harmony with nature. It is essential to put an end to the rational imperative of *homo oeconomicus,* focussed on maximizing profit, controlling the market, consuming and accumulating material goods and originating the short-sighted objective of short-term gains for the few. Such a virtual dialogue is centred on how to reform the governance systems following an *Earth-centred,* rather than a *human-centred,* perspective, so that everybody can live as a responsible member of the Community of the Earth. *Report* A/71/266 of 2016 gives a brief overview of the possible contribution of the various disciplines and, among them, of "arts, media, design and architecture", which are dealt with in *section G.* For instance, great emphasis is given to the implementation of public policies to decentralize the training of professionals in the field of architecture and planning, going beyond cities and incorporating local aspects, or to promote the analysis of local case studies and of other projects and their following dissemination and sharing of knowledge and dialogue.

Among the most significant European initiatives, worth mentioning is the launching, in 2013, of the seventh Environment Action Programme (EAP) by the European Commission. It defines a strategic programme to create environmental policies with nine priority objectives to be achieved within 2020. It also establishes a common understanding of the main environmental challenges Europe must face and what should be done to tackle them effectively. The programme underpins the *European Green Capital Award* (EGCA), which is related to sustainable urban planning and design policies. The protection and enhancement of the natural capital, greater efficiency of resources and a faster transition to a low-carbon economy are key elements of the programme, which also tries to face new emerging environmental risks, to contribute to protecting the citizens' health and well-being, to stimulating sustainable growth and creating new jobs [7]. However, the effectiveness of its results depends on the overcoming of the traditional approaches of linear planning, which do not always take into account the complex interactions existing in an urban system, and on the consequent need for a local *bottom-up* approach that considers urban areas as part of their surrounding territory, also in terms of flows of resources, people, water and energy. Therefore, resources, flows, interdependences of urban, peri-urban and rural areas, as well as the relation between a city and its natural environment should be paid due attention. This is also the wide context of the concept of *decoupling* – meant as "misalignment" between economic growth and increase in the emissions causing pollution and global warming, or between GDP growth rate and rate of environmental pressure – which is deemed as a necessary condition for sustainability and as a strategic approach to advance a global green economy entailing an improvement of human well-being and of social heritage, and reducing, at the same time, environmental risks and ecological scarcities (International Energy Agency-IEA).

The need to promote *green economy* measures and to limit the excessive use of natural resources should be met through the introduction of suitable legislative tools that should involve, first of all, public administrations. This was the case of the Italian law 221/2015 concerning environmental measures (GU n.13 of 18-1-2016), which significantly modified the current Code for public contracts introducing green procurement and the application of minimum environmental criteria in public procurements. Moreover, its attachment on the environment provided for measures to encourage the adoption of product certifications (*e.g.,* Ecolabel, PEFC, Plastica

Seconda Vita) or system certifications (e.g., ISO14001/EMAS). It is a significant start, yet it should be borne in mind that single sectorial measures, though undoubtedly necessary, are not enough if they are not accompanied by a cultural change involving all the actors of territorial transformations in a wider integrated scenario. This vision is outlined in the document *The City of the Future: A Green Economy Manifesto for Architecture and Urban Planning* (2017), elaborated by the Working Group of the States General of the Green Economy 2016 [8]. It highlights the role of architecture and urban planning in identifying objectives and a path to follow to effectively face challenges. Furthermore, it proposes the *green economy* as a fundamental, unavoidable and necessary choice to transform such challenges into extraordinary opportunities of relaunch and regeneration of cities. With particular reference to ongoing climate change and to the ecological issue, which cause considerable effects on the quality of life in a broad sense, cities should be made more resilient by planning and implementing adaptation policies and measures and, at the same time, they should play a role of strategic hubs in the implementation of effective climate mitigation policies. On a world level, cities are moving towards this direction and, sharing the same goals, they are looking for the most suitable solutions to achieve and adapt them to the local context.

2 The "Ecological Conversion" of the European Cities: Work in Progress

Many European cities are facing the challenges mentioned above by starting policies, programmes, and innovative actions aimed at combining regeneration and adaptation through innovative projects; investing in the natural capital; promoting green infrastructures; increasing the effective use of resources; developing more sustainable mobility. A few of them are among the most ecological cities in the world because they pursue the goal to become zero emission cities between 2020 and 2050, thanks to their environmental policies [9]. For instance, Reykjavik, the capital of Iceland, focuses on renewable energies, while Malmö, in Sweden, boasts a record in energy saving in the building sector, the third biggest wind farm in the world, and a port completely fuelled by renewable energy sources [9–12]. The Danish capital, Copenhagen, like many other cities in northern Europe, is making important investments on cycling, while, Stockholm, the capital of Sweden, was elected 2010 Green Capital of Europe, owing to its green areas, which account for 40% of its surface area, and because it is one of the first cities in the world that have introduced provisions concerning waste recycling and have optimized electric power generation through renewable sources. Thus, since 1990, it has managed to reduce CO2 emissions by 25% and aims at becoming a free fuel city within 2050. Hamburg, 2011 Green Capital of Europe, has been considered for many years the German most ecological city thanks to its so-called *greening* practice, which consists in expanding urban green areas as much as possible by planting new trees and building parks, above all by reconverting old industrial areas. Moreover, its public transport is fuelled by hydrogen and, since 1990, CO2 emissions have been cut by 15%, which is expected to reach 40% in 2020 and 80% in 2050 [7].

Cities are equipping themselves also to face the consequences of climate change, such as the increase in extreme weather events like floods, storms and heat waves, which may have serious effects on infrastructure, *e.g.,* transport systems, sewage networks and even food and non-food distribution systems. Moreover, they may entail severe damage on the economy and human health, as it is more and more common not only in non-European countries but also in Italy, which is already well-known for being characterized by a high hydrogeological risk. Therefore, climate change exerts pressure both on "heavy" infrastructures, such as roads, houses and sewage systems, and on "light" ones, such as the health care system [13]. London, Bratislava, Almada (Portugal), Rotterdam, Gand and Bologna are cities with advanced adaptation plans. In particular, the municipal administrations of Rotterdam and Gand have created partnerships with research centres to identify the hottest places in the city during heat waves, thanks to the installation of fixed and mobile thermometers in various areas and on trams. This allowed implementing mitigation actions to reduce the "urban heat island" effect, *e.g.,* planting trees. The city of Bologna, which is subject to the triple risk of flooding of the Po River, of heavy rainfalls and of heat waves, has developed an app ("Blue AP") for mobile phones, which is part of the adaptation plan and has been funded by the EU. This app enables citizens to detect the damage caused by heavy rains or heat waves, report it to the authorities and suggest them how to prepare for future weather events [13].

Soil quality is also of the utmost importance for the ecological challenge and for the reduction of the negative effects of climate change, given that permeable soils protect us from heat waves by retaining huge amounts of water and keeping temperatures low. Madrid is among the cities which are trying to use this function of the soil. It has refurbished the Gomeznarro Park so that it now includes new permeable surfaces, vegetation and underground water retention areas. This solution has been adopted also in other areas of the city and of Spain [14].

Worth mentioning is also the case of the Belgian village of Velm, near Sint-Truiden, whose inhabitants – after their houses had been flooded by mud for five times in 2002 – put pressure on the local authorities to take action, since the problem was recurring and water was washing out uncultivated fields dragging sediments away. Local authorities decided to use soil to protect houses through actions that contributed to restoring natural systems and proved to be successful in preventing following floods, in spite of the heavy rainfalls. The specific measures adopted included starter crops in winter, when the bare soil of fields favours the risk of flooding and the residues of crops are left to reduce soil erosion [15].

One of the most common and complex problems cities must face is the organization of the response to ecological and climate challenges by coordinating the different sectors and levels of the administration, *e.g.,* in the case of river management. The issue is even more complicated for those rivers, such as the Rhine and the Danube, which cross several countries, since the prevention of the damage caused by the flooding of these rivers requires the experimentation of new types of *governance* involving cities and states. In the case of the river Rhine, Switzerland, France, Germany and the Netherlands jointly worked to construct containment areas for the overflowed water [15].

Though this varied overview of significant case studies is not exhaustive, it is clear that each city has a different starting point and its response depends on the existing infrastructure, and on the political priorities and objectives. However, those which have been mentioned above are ahead of the others thanks to their farsighted vision. Change proceeds in small steps and public authorities play a fundamental role in facilitating it, not only by providing a framework, but also by leading by example and resorting to state-of-the-art technological innovation. The implementation of certain projects, such as separate waste collection and waste recycling, may take 5-10 years, yet public perception may take a generation to change. Other cases, *e.g.,* the transformation of existing buildings, may take even longer; thus, the greatest challenge is to plan in the medium and long term, preparing for a future that goes beyond our 5- to 10-year action plans and being wishful and flexible thinkers [16].

Therefore, the path towards ecological cities demands a global vision, the capacity to better manage climate change, the design of integrated systems, above all in the sector of mobility, with transport networks that guarantee to all citizens the possibility to move easily around the city on foot and by bike, not only in their residence neighbourhood but also in areas at a distance of 5–10 km. The presence of green public areas is crucial as well, since they favours meetings and social relations, the sense of belonging to a community but also of freedom and space outside one's house. Each neighbourhood should comprise accessible public spaces that, through their varied nature, can help meet social and individual needs – especially if they can be easily reached by public transport – *i.e.,* children's playgrounds, local parks and quiet areas, which facilitate the contact with nature and contribute to individual well-being. Besides these kinds of public space, squares are also relational places of choice for cultural or commercial activities.

However, it should be taken into account that, in this system, if, on the one hand, infrastructural networks connect urban areas, rural areas and people, on the other, they create barriers fragmenting the landscape and interrupting the connection between the different habitats and their vulnerability. As a consequence, all actions should be planned on a much wider scale than that of a single infrastructural project and they should involve the various stakeholders. Furthermore, special emphasis should be given to the creation of green infrastructures, conceived as a network of high-quality green spaces, strategically planned according to a wider vision of all green spaces (in remote, rural and urban areas as well as beyond the national borders) in order to establish interconnections and facilitate the movement of the species, as stated by the relevant strategy adopted by the European Union.

3   Final Considerations

In conclusion, this paper, as well as *The City of the Future: A Green Economy Manifesto for Architecture and Urban Planning*, clearly show that the change in the relation between man and nature, and the models of economic development and urban sprawl typical of the last century are inadequate to face the new global challenges [8]. Indeed, going further in this direction may even worsen the quality of life and of the urban environment, the negative effects of climate change, the damage to the natural heritage and the crisis of local development. Thus, a transition towards forms of "green economy" should be started, which should imply structural changes in key economic sectors, such as energy and transport, as well as long-term investments to increase the local attractiveness and the quality of life in its broadest sense.

Since challenges mostly concern the urban context, settlement development should follow the rules of the environmental system, of a landscape that man built in the past observing those rules and pursuing the unavoidable goal of sustainability, which means taking into account the capacity of regeneration of each ecosystem in its different sectors, within an integrating vision and an integral ecology comprising various environmental, economic, social and cultural aspects in a dynamic and participatory approach [2]. Certainly, implementation policies can vary according to the different territorial contexts because they should be "taylor-made", *i.e.*, based on the specific need to enhance local resources in terms of human, social, relational and territorial capital. Therefore, progress in sustainable development requires vision, good governance, strong leadership and the participation of citizens, as shown by the experience of the cities that are trying to find the right balance between good practices and regulations [15]. It is equally essential to exchange experiences through networks of different cities and regions with a view to disseminating knowledge and mutual learning in the path towards common goals of sustainability.

Thus, themes, such as the relation between citizenship and use of space, balance of rights and duties, well-being and quality of life, fulfilment of civil rights, equality and democracy are the ground for working to obtain greener, fairer and more resilient cities, which use their resources to reduce poverty through best practices of sustainable development, of development of the green economy, of attention to the landscape, not only as a merely aesthetic factor, but also as a key functional element for urban prosperity [17].

These challenges should be faced above all by urban planning, since it can contribute to pursuing real sustainability by combining responsibility and intergenerational and intragenerational solidarity. Architects and planners are required not only technical and scientific skills, but also high ethical values, since they have the great responsibility to play a role in the fulfilment of civil rights, of the right to the city, to nature and to culture. It is a very interesting field of study which offers further possibilities of research because it gives a glimpse of hope for a better future.

References

1. World Economic Forum: The Global Risks Report 2016, 11th edn. World Economic Forum, Geneva. https://www.weforum.org/reports/the-global-risks-report-2016. Accessed 12 Nov 2017
2. Papa Francesco (Jorge Mario Bergoglio): Laudato si. Lettera enciclica sulla cura della casa comune. Published editions of various publishing houses (2015)
3. United Nations: Transforming our world: the 2030 Agenda for Sustainable Development. Resolution adopted by General Assembly on 25 September 2015 (2015). https://sustainabledevelopment.un.org/post2015/transformingourworld
4. HABITAT III: Issue Paper 17. Cities and Climate Change and Disaster Risk Management. New York, 31 May 2015. https://unhabitat.org/wp-content/uploads/2015/11/Habitat-III-Issue-Paper-Cities-Climate-Change-and-DRR.pdf. Accessed 12 Nov 2017

5. Van den Berg, L., Van der Meer, J., Carvalho, L.: Cities as Engines of Sustainable Competitiveness: European Urban Policy in Practice. Ashgate Publishing, Burlington (2014)
6. United Nations: Harmony with Nature, Resolution 71/232 adopted by the General Assembly on 21 December 2016. http://www.un.org/ga/search/view_doc.asp?symbol=A/RES/71/232. Accessed 21 Nov 2016
7. European Commission, European Green Capital Homepage. http://ec.europa.eu/environment/europeangreencapital. Accessed 16 Nov 2017
8. La città futura. Manifesto della green economy per l'architettura e l'urbanistica. Elaborated within a Working Group of the General States of the Green Economy 2016 (2017). http://www.statigenerali.org/cms/wp-content/uploads/2017/04/Manifesto-Citt%C3%A0-Futura-IT.pdf. Accessed 16 Nov 2017
9. Beatley, T.: Green Cities of Europe: Global Lessons on Green Urbanism. Island Press, Washington, D.C. (2012)
10. City of Malmö: Environmental Programme for the City of Malmö 2009–2020. Environment Department Malmö (2009a). http://malmo.se/download/18.6301369612700a2db9180006215/1491304408540/Environmental+Programme+for+the+City+of+Malm%C3%B6+2009-2020.pdf. Accessed 16 Nov 2017
11. City of Malmö: Improving Malmö's Traffic Environment. Malmö: Environment Department and Streets and Parks Department (2009b). http://malmo.se/download/18.58f28d93121ca033d5e800077. Accessed 21 Nov 2017
12. Anderberg, S., Clark, E.: Green sustainable Oresund region: ecobranding Copenhagen and Malmö. In: Vojnovic, I. (ed.) Urban Sustainability: A Global Perspective, pp. 591–610. Michigan State University Press, East Lansing (2013)
13. Agenzia Europea dell'Ambiente: EEA: Segnali 2016. Verso una mobilità pulita e intelligente. I trasporti e l'ambiente in Europa. AEA, Copenaghen (2016)
14. European Clime Adaptation Platform Homepage. http://climate-adapt.eea.europa.eu/metadata/case-studies/the-refurbishment-of-gomeznarro-park-in-madrid-focused-onstorm-water-retention. Accessed 16 Nov 2017
15. Agenzia Europea dell'Ambiente: Vivere ai tempi del cambiamento climatico. AEA, Copenaghen (2015)
16. Agenzia Europea dell'Ambiente, EEA: Segnali 2014. Benessere e ambiente. Creare in Europa un'economia circolare ed efficiente nell'impiego delle risorse. AEA, Copenaghen (2014). https://www.eea.europa.eu/www/it/publications/aea-segnali-2014-benessere-e-ambiente. Accessed 16 Nov 2017
17. Settis, S.: Architettura e democrazia. Paesaggio, città, diritti civili, Giulio Einaudi Editore, Torino (2017)

Erratum to: Boosting Investments in Buildings Energy Retrofit: The Role of Incentives

Marta Bottero, Chiara D'Alpaos, and Federico Dell'Anna

Erratum to:
Chapter "Boosting Investments in Buildings Energy Retrofit: The Role of Incentives" in: F. Calabrò et al. (Eds.):
New Metropolitan Perspectives, SIST 101,
https://doi.org/10.1007/978-3-319-92102-0_63

In the original version of the book, belated correction from the chapter corresponding author to include the additional reference [20] in chapter "Boosting Investments in Buildings Energy Retrofit: The Role of Incentives" has to be incorporated. The erratum chapter and the book have been updated with the change.

The updated online version of this chapter can be found at
https://doi.org/10.1007/978-3-319-92102-0_63

© Springer International Publishing AG, part of Springer Nature 2019
F. Calabrò et al. (Eds.): ISHT 2018, SIST 101, p. E1, 2019.
https://doi.org/10.1007/978-3-319-92102-0_76

Author Index

© Springer International Publishing AG, part of Springer Nature 2019
F. Calabrò et al. (Eds.): ISHT 2018, SIST 101, pp. 699–701, 2019.
https://doi.org/10.1007/978-3-319-92102-0

Printed in the United States
By Bookmasters